AF761800

Human Colonization of the Arctic: The Interaction Between Early Migration and the Paleoenvironment

Human Colonization of the Arctic: The Interaction Between Early Migration and the Paleoenvironment

Edited by

V.M. Kotlyakov, A.A. Velichko and S.A. Vasil'ev

ELSEVIER

ACADEMIC PRESS
An imprint of Elsevier

Academic Press is an imprint of Elsevier
125 London Wall, London EC2Y 5AS, United Kingdom
525 B Street, Suite 1800, San Diego, CA 92101-4495, United States
50 Hampshire Street, 5th Floor, Cambridge, MA 02139, United States
The Boulevard, Langford Lane, Kidlington, Oxford OX5 1GB, United Kingdom

Library of Congress Cataloging-in-Publication Data
A catalog record for this book is available from the Library of Congress

British Library Cataloguing-in-Publication Data
A catalogue record for this book is available from the British Library

ISBN: 978-0-12-813532-7

For information on all Academic Press publications visit our website at https://www.elsevier.com/books-and-journals

Publisher: Candice Janco
Acquisition Editor: Anneka Hess
Editorial Project Manager: Emily Thomson
Production Project Manager: Debasish Ghosh
Designer: Victoria Pearson Esser

Typeset by TNQ Books and Journals

This edition has been dedicated to the memory of our friend and colleague, the late Professor Andrey Velichko (1931–2015), the leading Russian specialist in Pleistocene paleogeography and human paleoecology, who had initiated this work within the framework of the International Polar Year. His sudden death was a great loss for all Quaternary scientists in the world.

Contents

List of contributors

Patricia M. Anderson Washington State University, Seattle, WA, United States

Oleg P. Balanovsky Vavilov Institute of General Genetics, Russian Academy of Sciences, Moscow, Russia

Evgeniy M. Besprozvanny AV-Kom Nasledie Ltd., Yekaterinburg, Russia

Hans Peter Blankholm University of Tromsø-Arctic University of Norway, Tromsø, Norway

Alexandra P. Buzhilova Lomonosov Moscow State University, Moscow, Russia

Mariya V. Dobrovolskaya Institute of Archaeology, Russian Academy of Sciences, Moscow, Russia

Paul M. Dolukhanov Newcastle University, Newcastle upon Tyne, Great Britain

Svend Funder University of Copenhagen, Copenhagen, Denmark

Margarita M. Gerasimova Institute of Ethnology and Anthropology, Russian Academy of Sciences, Moscow, Russia

Olga Y. Glushkova Northeastern Interdisciplinary Research Institute, the Far Eastern Division, Russian Academy of Sciences, Magadan, Russia

Ted Goebel Texas A&M University, College Station, TX, United States

Kelly E. Graf Texas A&M University, College Station, TX, United States

Yuriy N. Gribchenko Institute of Geography, Russian Academy of Sciences, Moscow, Russia

Bryan Hood University of Tromsø-Arctic University of Norway, Tromsø, Norway

Jens F. Jensen Arctic Center of Ethnographic Collections, the National Museum of Denmark, Copenhagen, Denmark

Jan Ingolf Kleppe University of Tromsø-Arctic University of Norway, Tromsø, Norway

Elena A. Kosheleva Herzen Russian Pedagogical University, St. Petersburg, Russia

Pavel A. Kosintsev Institute of Floral and Faunal Ecology, the Urals Division, Russian Academy of Sciences, Yekaterinburg, Russia

Elena I. Kurenkova Institute of Geography, Russian Academy of Sciences, Moscow, Russia

Yaroslav Y. Kuzmin Institute of Geology and Mineralogy, the Siberian Division, Russian Academy of Sciences, Novosibirsk, Russia

Sergey N. Lisitsyn Institute for the Material Culture History, Russian Academy of Sciences, St. Petersburg, Russia

Anatoly V. Lozhkin Northeastern Interdisciplinary Research Institute, the Far Eastern Division, Russian Academy of Sciences, Magadan, Russia

Olga A. Markelova Moscow, Russia

Sergey V. Markin Institute of Archaeology and Ethnography, the Siberian Division, Russian Academy of Sciences, Novosibirsk, Russia

Ulla Odgaard Arctic Center of Ethnographic Collections, the National Museum of Denmark, Copenhagen, Denmark

Liybov A. Orlova Institute of Geology and Mineralogy, the Siberian Division, Russian Academy of Sciences, Novosibirsk, Russia

Pavel Yu. Pavlov Institute of Linguistics, Literature and History, the Komi Centre, the Urals Division, Russian Academy of Sciences, Syktyvkar, Russia

Patrick Plumet Université de Québec, Montréal, Canada

Andrey A. Pogodin The Urals Federal University, Yekaterinburg, Russia

Julia M. Ross Nunavut Government, Iqaluit, NU, Canada

Tatyana V. Sapelko Institute for Lake Studies, Russian Academy of Sciences, St. Petersburg, Russia

Vladimir Y. Shumkin Institute for the Material Culture History, Russian Academy of Sciences, St. Petersburg, Russia

Sergey B. Slobodin Northeastern Interdisciplinary Research Institute, the Far Eastern Division, Russian Academy of Sciences, Magadan, Russia

Vadim F. Starkov Institute of Archaeology, Russian Academy of Sciences, Moscow, Russia

Dmitriy A. Subetto Herzen Russian Pedagogical University, St. Petersburg, Russia

Sergey A. Vasil'ev Institute for the Material Culture History, Russian Academy of Sciences, St. Petersburg, Russia

Andrey A. Velichko Institute of Geography, Russian Academy of Sciences, Moscow, Russia

Alexander V. Volokitin Institute of Linguistics, Literature and History, the Komi Centre, the Urals Division, Russian Academy of Sciences, Syktyvkar, Russia

Valentina S. Zykina Institute of Geology and Mineralogy, the Siberian Division, Russian Academy of Sciences, Novosibirsk, Russia

Vladimir S. Zykin Institute of Geology and Mineralogy, the Siberian Division, Russian Academy of Sciences, Novosibirsk, Russia

Foreword

One of the most important events in world science of the early 21st century was the worldwide research program "The International Polar Year" (IPY) 2007–09 that integrated efforts of scientists from various countries working in various domains of science. The researches performed as a part of the IPY included problems of geology, geography, geophysics, oceanology, glaciology, climatology, and biology of the Arctic and the Antarctic. Prominent among other lines of investigation were studies of the anthropological constituent, a connecting link between natural and social sciences. That included a thorough analysis of the adaptation of aboriginal inhabitants of the North to environmental changes, and the measures needed for the preservation of the unique language and cultural heritage of those peoples.

An important part in the investigations was assigned to the IPY Project "Initial Human Colonization of Arctic in Changing Paleoenvironments" aimed at the problems related to the initial penetration of early man to the high latitudes and human adaptation to the extreme landscapes and climate. Besides the general worldwide interest in polar region studies, the significance of the project was instigated by many states being keenly interested in studies of the Arctic. The general picture of the environmental dynamics is based on paleo-reconstructions that show also the evolution of human society lifestyle (from nomadic to sedentary) against the background of the global climate fluctuations; besides, it makes possible tracing the trends of the global climate fluctuations as they show themselves in local natural context (Fig. 1).

The studies of those problems called for coordinated efforts of scientists: specialists in various areas of knowledge from various countries. Russian scientists took the initiative in organizing the works on the project. That was quite reasonable, as it is in the high latitudes of Eastern Europe and Asia that most prehistoric sites have been found, including the earliest of them (transitional from the Mousterian to the Upper Paleolithic). The idea of summing up the project results in the form of an atlas-monograph was actively supported by the Russian Geographical Society and the Russian Academy of Sciences; the Institute of Geography RAS would have acted as the leading organization in the project implementation, in cooperation with the Institute for the Material Culture History RAS.

The works performed as a part of the IPY Project provided the basis for this publication; that is the first scientific work where the process of the initial colonization of the high latitudes has been considered at a scale of the entire Northern Hemisphere by the teams of scientists from many countries (the territories studied by the project participants are shown in the scheme). Such a global approach (with all the existing data being taken into account) gave us a better insight into the entire picture of human

International Project No 276 "Initial Human Colonization of Arctic in Changing Paleoenvironments"

Figure 1 Polar regions initially colonized by the man.

colonization of the northern ecumene under conditions of environmental and climatic changes widely varying in range.

The first meeting of the project participants took place at the end of 2006 in St. Petersburg, where the intended targets were discussed and dates for presenting results of regional studies were agreed on. Some intermediate results of the work were later reported at conferences "Russia in the IPY" in 2006–12 and discussed at a series of workshops.

A representative International Conference "Way to North: paleoenvironments and earliest inhabitants of the Arctic and Subarctic" was held in Moscow in December 2007. That meeting was attended by many specialists in Quaternary sciences (archaeologists, anthropologists, geologists, paleogeographers, palynologists, and paleoclimatologists) from many countries, including Russia, Ukraine, Lithuania, Norway, Great Britain, and the United States. It demonstrated achievements of the recent years, as well as some noticeable gaps in our knowledge. The results of the conference were published (Velichko and Vasil'ev, 2008).

The present atlas-monograph consists of two parts.

The first part includes a series of regional articles. First, the data are summarized on the oldest archaeological localities of Fennoscandia (comprising the present-day Leningrad Region, Karelia, and the Kola Peninsula, as well as Finland, Sweden, and Norway). The next sections outline the colonization of the north and northeast of the East European Plain, and provide some information on the recent settlement of Spitsbergen and Novaya Zemlya archipelagos.

The next section presents the data on the colonization of the Urals, the North of West Siberia, on Central and Eastern Siberia, including the Central Siberian Plateau, the lower reaches of the Yenisei River, the Taymyr Peninsula, and the Lena and Aldan valleys. The following, more sizeable, section includes materials on the so-called Great Beringia subdivided into Western Beringia (Indigirka and Kolyma drainage basins, Kamchatka and Chukotka peninsulas) and Eastern Beringia, including Alaska Peninsula and the Yukon Territory. This region is of particular importance in the studies of man's dispersal over the planet, as it provides information on the penetration of the earliest Americans from Eurasia. The next section considers the Northern Canada colonization by the ancestors of the Eskimo people. Part I of the monograph—regional overview of the Arctic colonization—is completed with essays on the early inhabitants of Greenland and Iceland.

The correlation of different cultural layers and individual archaeological sites is based on various archaeological, paleogeographical, chronostratigraphic, and physical methods of dating. The main problem with the radiocarbon dating of the Upper Paleolithic sites consists in that as more and more dates are obtained, their chronological range keeps increasing. It is quite understandable and unavoidable in principle, as the absolute majority of Paleolithic sites have been actually inhabited more than once, and the episodes of a site functioning could differ notably in time. Such discrepancies in the radiocarbon dating results have been recorded on many Paleolithic sites in Europe and West Siberia (Dolni Vestonice, Pavlov, Willendorf, Sungir', Zaozerye, Byzovaya, Lugovskoe, etc.).

The usage of advanced dating technique (^{14}C scintillating and AMS technique, as well as TL and OSL dating) provided means for accumulating a large number of dates. Besides, there have been developed various procedures for calibrating radiocarbon dates to adjust them to the reconstructed changes in ^{14}C content in the atmosphere (Mellars, 2006; Pettitt, 2000). The usage of such corrected (calibrated) dates brings them closer to the results of TL (OSL and IRSL) dating, though it does not solve completely the problem of discrepancies between the dates (Housley et al., 2006).

Considering that radiocarbon dates for most of the archaeological sites of Eurasia given in earlier publications are mostly uncalibrated, in this monograph, we undertook the work of the date conversion to calibrated variant. In calibrating ^{14}C dates, we used the last version of the widely used calibration curve IntCal 13 (Reimer et al., 2013). The upper part of the scale—to 13,900 yr BP—is based on tree ring age according to the continuous Central European dendrochronological scale. The scale has been continued to ~50,000 yr BP on the basis of integrated and independently dated series on corals, foraminifers, speleothems, and plant macro-remains found in varved clays. Calibrated dates are given in the text immediately after the radiocarbon ones. They are given in italics and in slashes 11,840±50 ^{14}C yr BP (GIN-8400)/*13,600±60 cal. yr BP*/ where 13,600 is mathematical expectation (that is, average value of the calibrated date distribution), and ±60 is the standard deviation. Such a procedure permits, on one hand, to keep intact the initial author's positioning of cultural layers in the stratigraphic sequence, and on the other to compare the stages of the initial people dispersal with the main natural events of the Late Pleistocene as they are reflected in various versions of the chronostratigraphic schemes.

The texts of all the listed sections are based on the results of multidisciplinary studies of the early man sites and on the data on the environment and climate dynamics in the regions of the initial colonization obtained using methods of paleogeography, geology, and archaeology. We consider the colonization of Arctic regions by early man as a many-stage long-term process, in the course of which migrations of the ancient hunters northward alternated with their retreat to southern regions. The process continued for many thousands of years and was controlled by the Late Quaternary changes in environments. The results are presented as integrated characteristics of the early man sites and of natural events in relation to the stages of the human dispersal (see Table 1).

The chapters included into Part II of the volume present a thorough analysis of various aspects of the initial human dispersal. There are given characteristics of the adaptations, the physical appearance and genetics of the population, nutrition and subsistence systems, as well as the history of the modern ethnic group development in the North.

Table 1 The main climatic events in glacial and periglacial regions of the Northern Hemisphere (Velichko and Faustova, 2009)

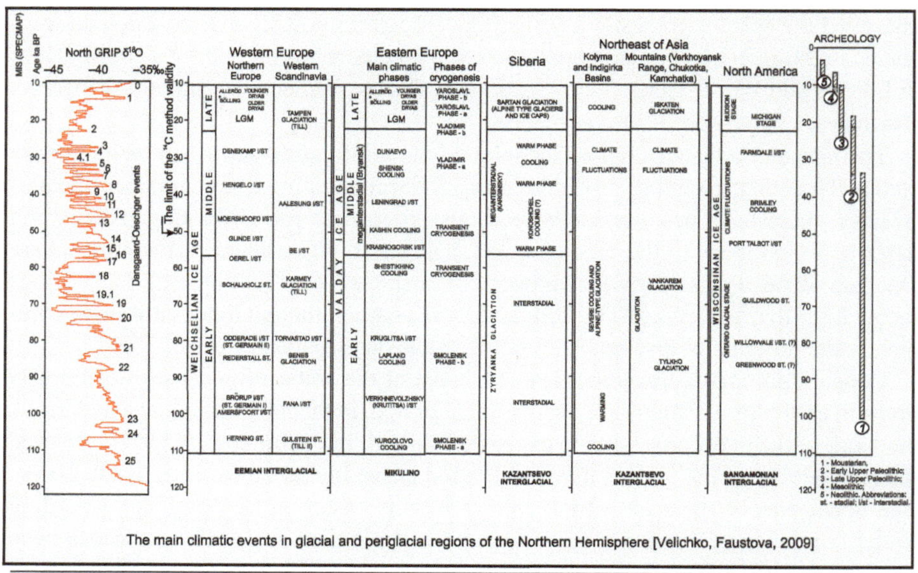

The main climatic events in glacial and periglacial regions of the Northern Hemisphere [Velichko, Faustova, 2009]

The synthesis of the data obtained is given in the final chapter.

Many difficulties were to be overcome in the project implementation. Most of them were due to the fact that not so many archaeologists work in the North, and the northern territories are still insufficiently studied. We hope, however, that the information available at the moment is presented here to the full extent and may give an insight into the process of the initial colonization of the high latitudes.

The offered atlas-monograph is a result of the combined efforts of specialists from many countries (Russia, Norway, Denmark, the United States, and Canada). The authors hope it will be of interest and practical use for a wide circle of paleogeographers, archaeologists, anthropologists, ethnographers, geologists, biologists, and everybody who cares for the remote past of life in the Arctic.

Acknowledgments

During a few years of the monograph preparation the work was being supported by a series of state research programs and institutions. Among them were the Russian Geographical Society (grant no. 04/2014-H3), Russian Foundation for Basic Research (grants nos. 07-06-00087 and 14-05-07020), the Presidium RAS programs "People and culture adaptation to the changes of environments and to social and technogenic transformations," and "Historical and cultural heritage and spiritual values of Russia." The authors want to express their deep gratitude to S. N. Timireva for her invaluable aid in preparation of the illustrations, T.A. Pivovarova and T.V. Rusina for preparing the maps, and I.I. Spasskaya for bibliographic and editorial work. We are deeply grateful to the Director of the Institute of Geography, Professor Olga N. Solomina for her help in organizing the English edition of the book.

V.M. Kotlyakov
A.A. Velichko
S.A. Vasil'ev

Part One

Regions of initial colonization

(photo by S.N. Timireva)

Section One

European North

(photo by S.N. Timireva)

Fennoscandia

Postglacial human colonization of Fennoscandia

1.1

Paul M. Dolukhanov
Newcastle University, Newcastle upon Tyne, Great Britain

The early man communities were settling Fennoscandia and the entire basin of the Baltic Sea throughout the Late Glacial time. As has been proven by numerous investigators, the Paleolithic population over a greater part of extraglacial Europe was dramatically reduced in density during the Last Glacial Maximum (Housley et al., 1997; Tarberger and Street, 2002). There existed, however, two refugia where the population density was steadily growing through that interval. One of the refugia developed in the southwest of Europe (Franco-Cantabrian region), and another developed in the east, including the drainage basin of the Vistula and the upper reaches of the Dniester, Dnieper, and Don rivers (Dolukhanov, 1986; Dolukhanov et al., 2002; Gamble et al., 2006). In the course of the Late Glacial, under conditions of increasing heat and moisture supply, along with growing efficiency of the hunting activities, the resource base of the Paleolithic groups became drastically diminished. That stimulated a gradual outflow of people from the refugia into other regions of Northern Europe and Fennoscandia noted for a greater food potential (Fig. 1.1.1).

Having analyzed the frequency distribution of the radiocarbon dates, Gamble et al. (2006) distinguished several stages in the demographic dynamics of the dispersal process from the western refugium as follows:

- initial: 29.5–16 ka BP;
- principal: 16–14 ka BP;
- static: 14–13 ka BP;
- recessionary: 13–11.5 ka BP.

It has been supposed (Dolukhanov, 1986; Nuñez, 1987; Bjerck, 2008; Zvelebil, 2008) that the European North was colonized by people arriving from both refugia (Fig. 1.1.1). The tool typology suggests that the west of Scandinavia was settled by newcomers from the area of the "Ahrensburgian technocomplex"; the new settlers arrived by way of the so-called "Northern Corridor," that is, part of the North Atlantic shelf between the Scandinavian ice sheet lobe and the coastline having retreated westward at that time. The further movement of the people northward and the appearance of the Komsa culture sites on the coasts of Norway and the Kola Peninsula resulted from a high-speed migration, probably with the usage of floating crafts, about 10,200–8900 yr BP (9500–8900 cal. yr BC). In the process of this migration the hunting activities were somewhat reoriented: instead of hunting reindeer exclusively, they began to use various marine resources, marine mammals in particular.

Human Colonization of the Arctic: The Interaction Between Early Migration and the Paleoenvironment
http://dx.doi.org/10.1016/B978-0-12-813532-7.00001-2

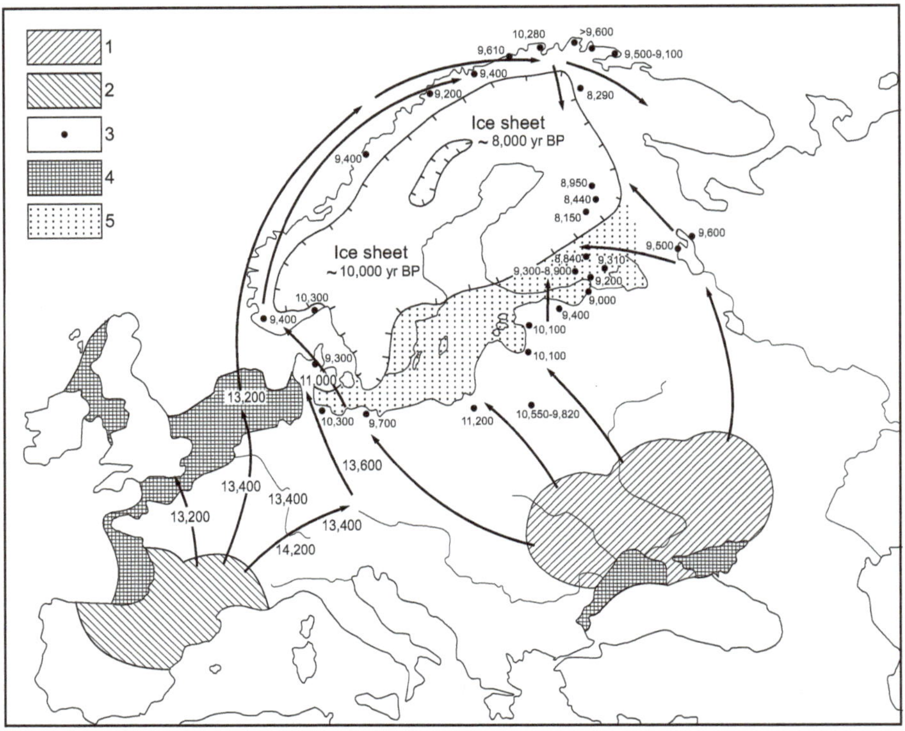

Figure 1.1.1 Postglacial colonization of Northern Europe (Zvelebil, 2008). 1: recolonization from Eastern Europe; 2: recolonization from Southwestern Europe; 3: first regional settlements and their age; 4: Late Glacial coastlines and emerged sea floor; 5: the Yoldia sea.

The earliest sites in the Scandinavian North (Finmarken) are traditionally attributed to the Komsa culture, while those in the south of western Norway to Fosna culture. The earlier phase (1) is dated to the Preboreal period, its age being established by correlation with coastlines as 10–9 ka BP (9500–8250 cal. yr BC) (Olsen, 1994; Woodman, 1993). Typological correlations with complexes of Ahrensburgian type in Germany and Holland confirm the notion of the initial peopling from south to north. At earlier stages the economy was based of the reindeer hunt; later, it was replaced by red deer and seals (*Phoca vitulina*, *Halichoerus grypus*) as the main objects of hunting.

The earliest site in the inner part of Finnish Lapland is Sujala on Vetsijarvi Lake dated at 8600–7800 cal. yr BC. Judging from the presence of Post-Swiderian elements in the stone inventory, one may suppose that the inhabitants of the site could come here from the drainage basin of the Upper Volga River (Rankama and Kankäänpaa, 2007).

The earlier stages in colonization of the northeastern Baltic region are marked by *sites of Mesolithic type*. The earliest dates were obtained for the Antrea Korpilahti locality northeast of Vyborg. Some implements of the Mesolithic type, including a birchbark fishing net and some floats, were dated at 9200–8250 yr BC (Matiskainen,

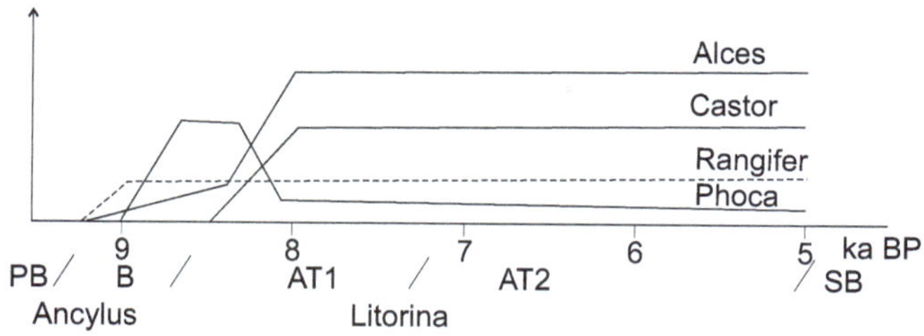

Figure 1.1.2 Changes in proportions of animals: objects of hunt in the Mesolithic of Finland (Matiskainen, 1989).

1989). Judging from the geological context, the Mesolithic type artifacts were initially deposited in a lagoon or a pool belonging to the Ladoga-Baltic strait system.

A similar date was obtained for Saarenoja locality, Joutseno region (9300±75 yr BC). In the course of time, there appeared a network of Mesolithic sites attributed to Suomusjärvi culture.

According to classification developed by Zvelebil (2008), the Mesolithic sites of the European North belong to the type of "foraging economy with some elements of long-term preservation of products (delayed consumption)." Having analyzed the available data on fauna obtained from the Mesolithic sites of southern Finland, Matiskainen (1989) identified a certain regularity in the hunting, fishing, and gathering activities and their specific distribution over seasons (Figs. 1.1.2 and 1.1.3). Since 9500 yr BP the seal hunting had acquired a considerable importance in the Mesolithic economy of coastal regions (Ukkonen, 2002), particularly so after the beginning of pottery production; the faunal materials recovered from the Early Ceramic sites include remains of two species of seals: *Phoca hispida* and *Phoca groenlandica.*

Judging from the data available, the economy of Mesolithic settlements in Karelia and in northeastern Russia was oriented to hunting forest animals (reindeer, red deer, boar, beaver) and waterfowl, with fishing being also of great importance (Oshibkina, 1996).

The Mesolithic sites were concentrated on the coasts of the Gulf of Finland, inner lakes, and streams connecting them (Fig. 1.1.4). Inland sites occurred usually on terraces of large rivers (such as the Severnaya Dvina, Pechora, Sukhona, Vychegda, and their tributaries). Large settlements have been discovered on low terraces of lakes Kubenskoye, Vozhe, and Lacha. At the Nizhnee Veretye 1 settlement excavated over an area of ~1500 m^2, there have been found remains of surface dwellings with post-holes and a hearth, indicative of a sedentary lifestyle.

Large Mesolithic burials are found in the north of Russia: Oleni Ostrov (Gurina, 1956) and Popovo (Oshibkina, 1996). The materials retrieved from the largest in Europe Oleni Ostrov burial ground on Onega Lake led specialists (O'Shea and Zvelebil, 1984) to the conclusion that the site belonged to "a large sedentary community with a distinct social stratification." The pieces of art recovered from the burial suggest sophisticated esthetic concepts typical of the communities of that type.

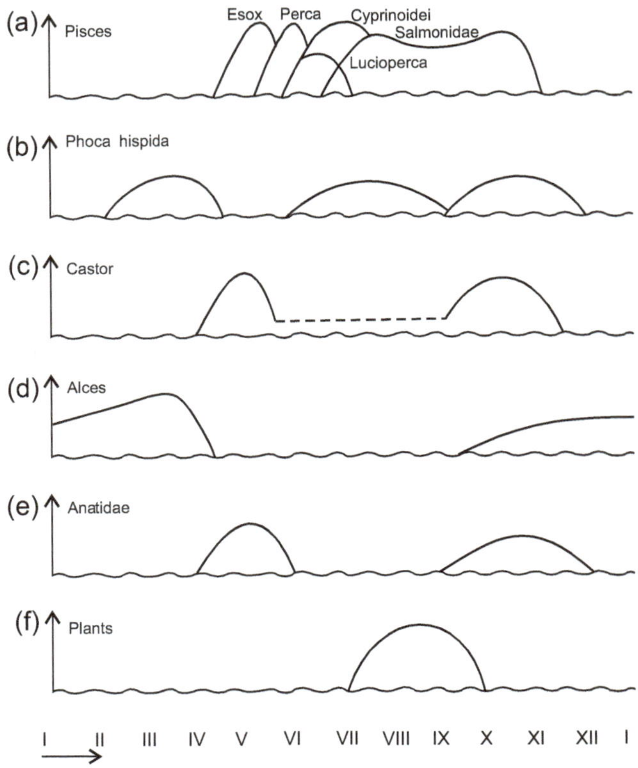

Figure 1.1.3 Seasonal cycles of hunting, fishing, and gathering in the Mesolithic of Finland (Matiskainen, 1989).

A large series of radiocarbon dates obtained for Mesolithic sites of the Russian North fall in the range of 9600–6000 yr BP.

The *Early Ceramic* sites (of ka I:1 or Sperrings 1 type) were first identified by Äyräpää (Europaeus) (1925). Hyyppä (1937) correlated the sites of this type with coastlines of the second Litorina transgression of the Baltic Sea. In the opinion of Siiriäinen (1970, 1982), the earliest sites of Sperrings type were related to the coastline of Päijänne Lake (southern Finland) at its maximum transgression (5200 yr BC). In common with the earlier settlements, the Sperrings sites were mostly located on the sea and lake coasts or on river banks, mainly in Southern Finland and Karelia. The type of economy was also much the same. Judging from the analysis of stable isotopes in the bones recovered from early ceramics sites in Sweden, the people's diet consisted mostly of seal meat and fish at that time (Fornander, 2006).

There are several types of Sperrings ceramics (German, 2002) (Fig. 1.1.5):

- large vessels with the vessel mouth diameter being 30–50 cm and an average wall thickness of 0.8–1.3 cm; the bottom is rounded or tapered;
- slightly smaller vessels, with mouth diameter of 20–30 cm, wall thickness of 0.4–0.7 cm, with a slanted or tapered bottom;

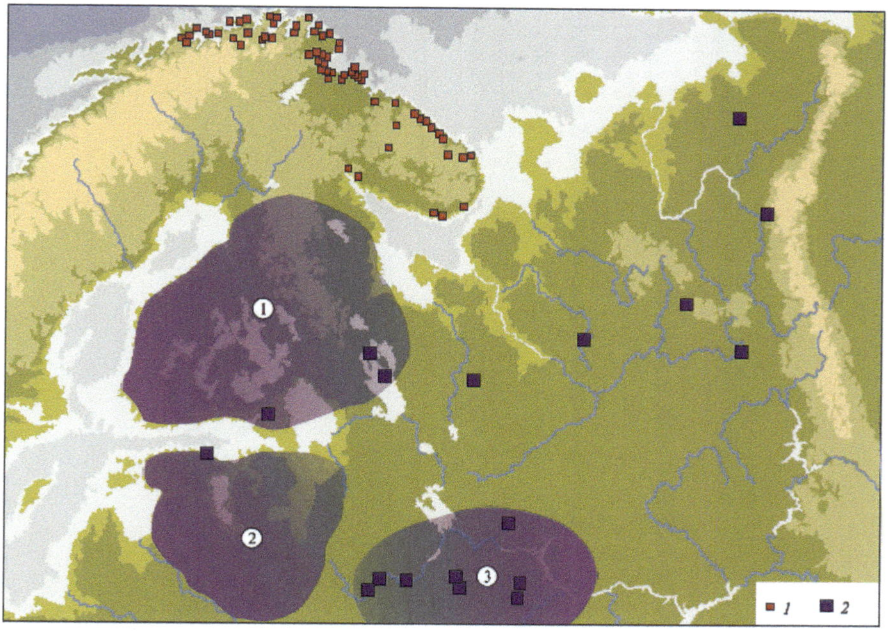

Figure 1.1.4 Mesolithic and Early Neolithic sites in Fennoscandia and in northern Russia. *1*: Mesolithic sites; *2*: Neolithic sites; Neolithic cultures of Eastern Europe marked by numbers: 1, Sperrings; 2, Narva-Neman; 3, Upper Volga.

- smaller vessels with round bottom, mouth diameter about 20 cm and wall thickness of 0.5–0.7 cm;
- small vessel-saucers with the mouth diameter of 12 cm and wall thickness of 0.5–0.7 cm.

Some impurities in the clay are found only in the ceramic material of larger vessels; the impurities are most often sand, as well as quartz grains and ground granite. An organic matter admixture is found in western regions. All the potteries were molded out of clay by hand and then subjected to firing in open kilns. The firing temperature was rather low, as follows from the dark color of pieces (from dark gray to grayish-brown) and a fragility of pieces.

The ornament consists mainly of fish vertebra impressions. The second most often used ornament is drawn lines. Many pots were ornamented with imprints of comb or of "twined cord." Pits were usually imprinted above the earlier made ornament.

The sites of the Säräisniemi type (Sär 1) are found in northern and northeastern Fennoscandia, including Karelia. Ceramics of that type have usually much in common with the Sperrings ones and are often considered to be a northern variety of the latter. Principal distinctions are in impurities: the Säräisniemi ceramics contain organic matter and micas. In most cases an ornament covers the entire surface of vessels (Fig. 1.1.6).

Many radiocarbon dates have been obtained for the Early Ceramics of Finland and Northern Russia. As is seen from the frequency distribution of the dates, they fall

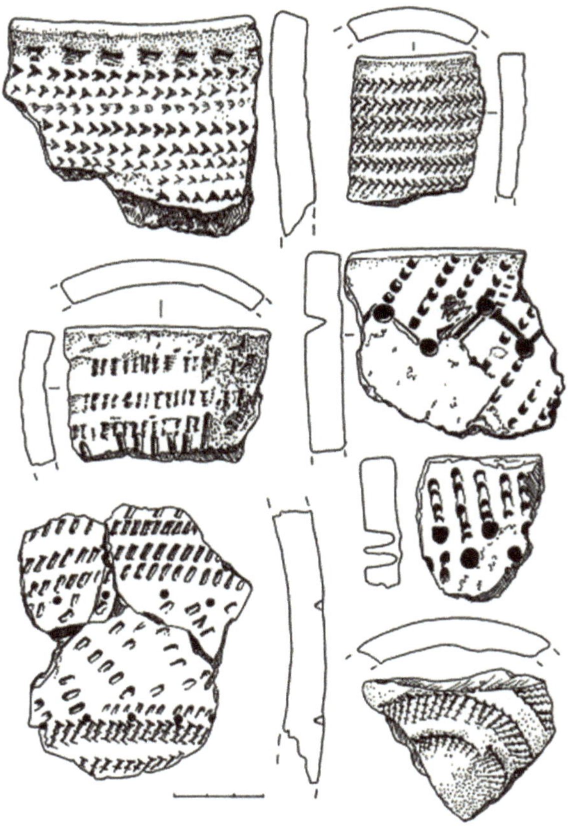

Figure 1.1.5 Ceramics of Sperrings type (German, 2006).

within a broad range of 7000–3500 yr BC, the greater frequencies corresponding to the interval of 6000–3500 yr BC (Fig. 1.1.7).

The dates of the Säräisniemi sites calculated from the shoreline gradients give the age of approximately 6100–5500 yr BC (Torvinen, 1997, 1998, 2000). Sites of the same type in the north of Norway correspond to the "Stone Age I" of the regional scale and are dated at 5800–5000 yr BC (Bjerck, 2008).

In common with the neighboring regions of Norway, the Mesolithic and Early Neolithic (Ceramic) sites on the Kola Peninsula are usually found close to the ancient marine coastlines in fiords or on small islands, as well as along river valleys and on lacustrine terraces (Gurina, 1997; Shumkin, 1996). Organic remains seldom occur at the sites. By analogy with the Early Neolithic sites, it is assumed that the economy was mostly based on sea mammal hunting.

The early ceramics described on the Kola Peninsula bear some features characteristic of the Sperrings and Säräisniemi type. Gurina (1997), however, regarded it as an individual Neolithic culture and described it under the name of "the Kola Culture."

The largest concentration of the sites is found at the Drozdovka Bay (fiord). The earliest dates have been obtained for sites Ust'-Drozdovka (5510 ± 100 yr

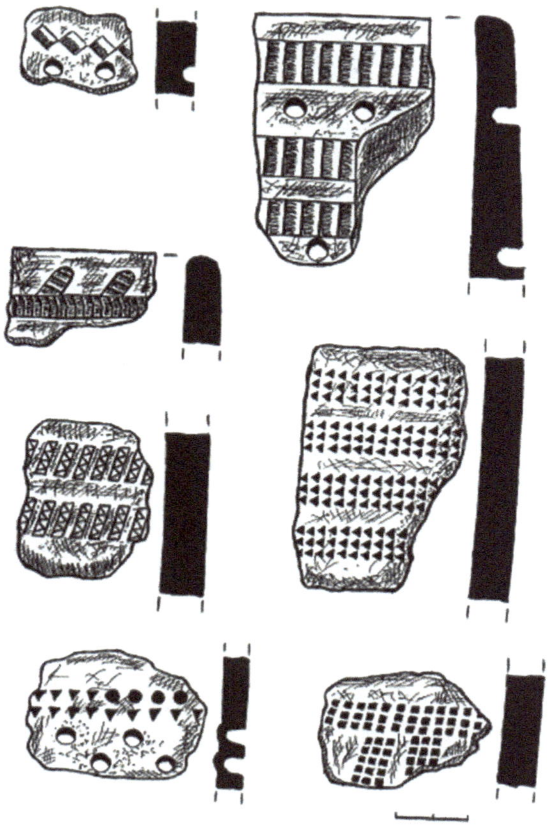

Figure 1.1.6 Ceramics of Säräisniemi type (German, 2006).

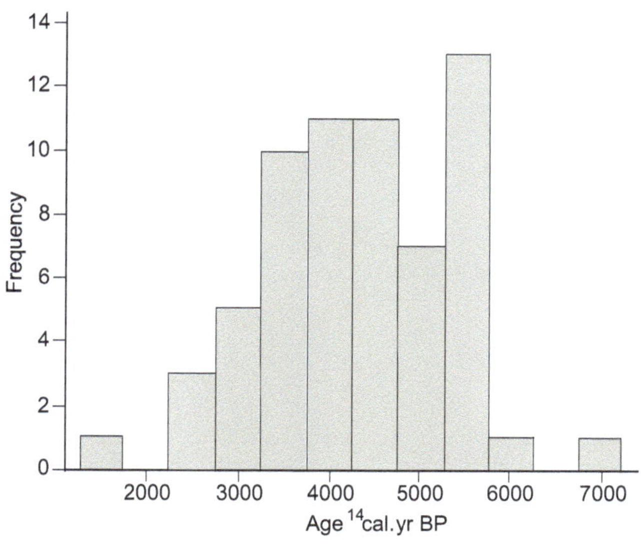

Figure 1.1.7 Frequency distribution of the Neolithic radiocarbon dates in Northern Russia.

BP/6300 ± 110 cal. yr BP) and Nerpichya (4630 ± 100 yr BP/5320 ± 160 cal. yr BP) on the northern coast and Chavan'ga site (5560 ± 80 yr BP/6360 ± 80 cal. yr BP) on the southern coast. As to the central part of the peninsula, there is a series of dates at the Tsaga site (Lake Lovozero), the oldest being 5760 ± 160 yr BP/6590 ± 180 cal. yr BP.

Northern Scandinavia: synthesis

1.2

Hans Peter Blankholm, Bryan Hood, Jan Ingolf Kleppe
University of Tromsø-Arctic University of Norway, Tromsø, Norway

Current radiocarbon dates suggest northern Norway was first colonized by groups moving northward along the Norwegian coast during the early—mid Preboreal period (10,000–9500 ucal. yr BP). Traditionally, these south Norwegian progenitors have been termed the "Fosna Culture" (Bjerck, 2008), but more recently, they have been linked with the Ahrensburgian technocomplex of the north European plain (Fuglestvedt, 2005). The earliest assemblages are characterized by tanged points, (occasional) blades, and flake axes, and by the use of a variety of raw materials including several types of cherts, quartzites, and vein quartz. Small pit-houses are known from some late Preboreal sites. The Sujala site in neighboring north Finland indicates the inland was colonized by 9200–8900 ucal. yr BP; this assemblage has been associated with a colonizing population derived from the northwest Russian Post-Swiderian cultures (Rankama and Kankaanpää, 2008). No directly comparable material has yet been found on the Norwegian coast. Currently, the earliest date for an inland site in north Norway is 8300 ucal. yr BP. The coastal colonists moved through regions characterized by birch forest in the inner fjords and arctic heaths on the outer islands, an environment similar to contemporary southern Greenland (Vorren et al., 2009). Sea surface temperatures warmed 3.5–4°C during the late Preboreal (Anderson et al., 2004). The early colonization of interior north Finland was synchronous with the establishment of birch forest in the inland. During the Boreal and early Atlantic periods (9000–7500 ucal. yr BP) the coastal technology was marked by micro-blades, and the settlement pattern included sites with one to five pit-houses that may represent seasonal base camps. The two dated interior assemblages from Finnmark consist of bipolar-reduced quartz, and they point to elk and reindeer hunting in a birch forest environment.

Pine was present on the northernmost coast of Norway by 8000 ucal. yr BP. It spread inland through northernmost Finland and Norway between 8000–7500 ucal. yr BP, reaching its maximum Holocene extent by 6000 BP (Seppä, 1996; Hicks and Hyvärinen, 1997; Allen et al., 2007). The currently earliest dated site in the interior Troms region of Norway (7000 ucal. yr BP) falls within the phase of pine expansion, but the site does not necessarily mark the first human colonization of the area.

Key sites 10,000 – 7000 BP (Fig. 1.2.1)

Finnmark county

Map ID: 1
Site name: Smellroren
Site setting: Coastal site on raised beach c.30–40 m a.s.l., northwest of Vardø town center
Age: Early Mesolithic (typological determination)
Environmental data: None
Function/subsistence: Coastal base camp and hunting/fishing camp
Literature: Bøe and Nummedal (1936), Odner (1966), Woodman (1993)

Human Colonization of the Arctic: The Interaction Between Early Migration and the Paleoenvironment
http://dx.doi.org/10.1016/B978-0-12-813532-7.00002-4

Figure 1.2.1 Key sites in Northern Norway dated to 10–7 ka BP. Numbers in the map correspond to site indexes given in the text (in brackets).

Map ID: 2
Site name: Seilmerket I and II
Site setting: Two coastal sites on raised beach, in fjord c.60–70 m a.s.l., southeast of Kirkenes city center
Age: Early Mesolithic (typological determination)
Environmental data: None
Function/subsistence: Coastal base camp and hunting/fishing camp(s)
Literature: Bøe and Nummedal (1936), Odner (1966), Woodman (1993)

Map ID: 3
Site name: Jernbanestasjonen
Site setting: Coastal site in fiord. 60–68 m a.s.l., south of Kirkenes city center
Age: Preboreal; Early Mesolithic (typological determination)
Environmental data: None
Function/subsistence: Coastal base camp or hunting/fishing camp in fiord
Literature: Bøe and Nummedal (1936), Woodman (1993)

Map ID: 4
Site name: Nii'beræppen-3
Site setting: Coastal site 53 m a.s.l.

Age: 9550 ± 55 BP
Environmental data: None
Function/subsistence: Base camp with pit-house
Literature: Grydeland (2006)

Map ID: 5
Site name: Mortensnes
Site setting: Coastal site on raised beach 45 m a.s.l.
Age: 8520 ± 130 BP
Environmental data: None
Function/subsistence: Coastal base camp or hunting/fishing camp
Literature: Schanche (1988), Blankholm (2008)

Map ID: 6
Site name: *Nesseby I*
Site setting: Coastal site, inner Varanger fjord
Age: Preboreal; Early Mesolithic (typological determination)
Environmental data: None
Function/subsistence: Base camp or hunting/fishing camp.
Literature: Bøe and Nummedal (1936), Odner (1966), Woodman (1993)

Map ID: 7
Site name: Čåkki-1
Site setting: Coastal site, 64 m a.s.l.
Age: 9782 ± 95 BP
Environmental data: None
Function/subsistence: Hunting/fishing camp
Literature: Grydeland (2006)

Map ID: 8
Site name: Stuorrasiida-2
Site setting: Coastal site, 37 m a.s.l.
Age: 8365 ± 50 BP.
Environmental data: None
Function/subsistence: Base camp with pit-house
Literature: Grydeland (2006)

Map ID: 9
Site name: Ovenfor Gropbakkeengen
Site setting: Coastal site, 48 m a.s.l.
Age: Preboreal/Boreal; Early/Middle Mesolithic (typological determination)
Environmental data: None
Function/subsistence: Hunting/fishing camp
Literature: Simonsen (1961)

Map ID: 10
Site name: Sæleneshøgda
Site setting: Coastal site, 53 m a.s.l.
Age: Boreal, Middle Mesolithic (typological determination)
Environmental data: None
Function/subsistence: Base camp with three pit-houses
Literature: Simonsen (1961), Woodman (1993)

Map ID: 11
Site name: Starehnjunni
Site setting: Coastal site, 43 m a.s.l.
Age: 7710 ± 480 BP
Environmental data: None
Function/subsistence: Base camp with several pit-houses
Literature: Engelstad (1989)

Map ID: 12
Site name: Karlebotn Sydvest
Site setting: Coastal site, 52–57 m a.s.l.
Age: Preboreal/Boreal (typological determination)
Environmental data: None
Function/subsistence: Hunting/fishing camp
Literature: Simonsen (1961)

Map ID: 13
Site name: Vest for Berlevåg
Site setting: Coastal site on raised beach, c.20 m a.s.l., west of Berlevåg town center
Age: Early Mesolithic (typological determination)
Environmental data: None
Function/subsistence: Coastal base camp and hunting/fishing camp
Literature: Bøe and Nummedal (1936)

Map ID: 14
Site name: Lagesidbakti
Site setting: Coastal site on raised beach, 71 m a.s.l.
Age: 9940 ± 101BP
Environmental data: None
Function/subsistence: Coastal base camp or hunting/fishing camp
Literature: Bøe and Nummedal (1936), Engelstad (1989), Woodman (1993), Grydeland (2000, 2005), Blankholm (2008)

Map ID: 15
Site name: Zoar
Site setting: Coastal site on raised beach, in fjord c.20 m a.s.l., west of Berlevåg town center
Age: Early Mesolithic (typological determination)
Environmental data: None
Function/subsistence: Coastal base camp or hunting/fishing camp.
Literature: Bøe and Nummedal (1936)

Map ID: 16
Site name: Skonsvika
Site setting: Coastal site on raised beach, in fjord c.20 m a.s.l., west of Berlevåg airport
Age: Early Mesolithic (typological determination)
Environmental data: None
Function/subsistence: Coastal base camp or hunting/fishing camp.
Literature: Bøe and Nummedal (1936), Woodman (1993)

Map ID: 17
Site name: Sarnes B4
Site setting: Coastal site on raised beach.

Age: 8120±75 BP
Environmental data: None
Function/subsistence: Base camp or hunting/fishing camp
Literature: Thommesen (1996), Blankholm (2004)

Map ID: 18
Site name: Sarnes B5
Site setting: Coastal site on raised beach
Age: 9800±60
Environmental data: None
Function/subsistence: Base camp or hunting/fishing camp
Literature: Thommesen (1996), Blankholm (2004)

Map ID: 19
Site name: Vedbotneidet
Site setting: Coastal site on raised beach, in fjord c.20–30 m a.s.l., Nordkapp county
Age: Preboreal; Early Mesolithic (typological determination)
Environmental data: None
Function/subsistence: Coastal base camp and hunting/fishing camp.
Literature: Bøe and Nummedal (1936), Woodman (1993)

Map ID: 20
Site name: Melkøya
Site setting: Coastal site on raised beach
Age: 7652±59 BP
Environmental data: Pollen sampling
Function/subsistence: Base camp or hunting/fishing camp
Literature: Hesjedal and Niemi (2003), Ramstad et al. (2005)

Map ID: 21
Site name: Stenseng
Site setting: Coastal site with two localities on raised beaches 52–62 m a.s.l.
Age: Early Mesolithic, Preboreal/Boreal (typological determination)
Environmental data: None
Function/subsistence: Base camp or hunting/fishing camp, small pit-houses
Literature: Nummedal (1929), Bøe and Nummedal (1936), Gjessing (1945), Woodman (1993), Bjerck (2008)

Map ID: 22
Site name: Tollevik
Site setting: Coastal site with several localities on raised beaches 55–74 m a.s.l.
Age: Early Mesolithic, Preboreal/Boreal (typological determination)
Environmental data: None
Function/subsistence: Base camp or hunting/fishing camp, small pit-houses
Literature: Nummedal (1929), Bøe and Nummedal (1936), Gjessing (1945), Woodman (1993), Bjerck (2008)

Map ID: 23
Site name: Storbukta
Site setting: Coastal site on raised beach, in fjord c.42 m a.s.l., Porsanger county
Age: Preboreal; Early Mesolithic (typological determination)
Environmental data: None

Function/subsistence: Coastal base camp or hunting/fishing camp
Literature: Bøe and Nummedal (1936), Woodman (1993)

Map ID: 24
Site name: Slettnes
Site setting: Coastal sites on raised beach
Age: 9610±80 BP
Environmental data: None
Function/subsistence: Base camp or hunting/fishing camp
Literature: Hesjedal et al. (1996), Blankholm (2008)

Map ID: 25
Site name: Virdnejavri 113
Site setting: Inland site on lake
Age: 8295±35 BP
Environmental data: None
Function/subsistence: Hunting camp for elk and reindeer
Literature: Simonsen (1986), Skandfer and Hood (personal communication)

Map ID: 26
Site name: Virdnejavri 101
Site setting: Inland site on lake
Age: 7880±35 BP
Environmental data: None
Function/subsistence: Hunting camp for elk and reindeer
Literature: Simonsen (2001), Skandfer and Hood (forthcoming)
Troms county

Map ID: 27
Site name: Tønsnes
Site setting: coastal site, 26–29 m a.s.l., overlooking junction of three sounds
Age: 19 radiocarbon dates averaging 8000 BP
Environmental data:
Function/subsistence: Base camp with four pit-houses
Literature: Skandfer et al. (2009), Finstad and Grydeland (2009)

Map ID: 28
Site name: Simavik
Site setting: Coastal site on raised beach, c.40 m a.s.l., facing narrow strait between two major islands
Age: 9200±200 BP
Environmental data: None
Function/subsistence: Base camp or hunting/fishing camp
Literature: Sandmo (1986), Blankholm (2008)

Map ID: 29
Site name: Målsnes 1
Site setting: Coastal site, 51 m a.s.l., on isthmus of former peninsula that at the time of occupation formed the southernmost tip of Målsnes Island
Age: 9500 (sea level determination)
Environmental data: None from site, but pollen diagram from nearby bog

Function/subsistence: Temporary (summer), broadly oriented exploitation camp primarily geared toward fishing and hunting marine mammals
Literature: Blankholm (2008)

Map ID: 30
Site name: Almenningen 1
Site setting: Inland site on narrow peninsula near outlet of lake Little Rostavatnet in the Målselv river system
Age: 7260 ± 95 BP
Environmental data: Charcoal
Function/subsistence: Fishing and hunting camp
Literature: Blankholm (2008)
Nordland County

Map ID: 31
Site name: Fure
Site setting: On the tip of a former island between two straits
Age: 9400 BP (sea level determination)
Environmental data: None
Function/subsistence: Hunting/fishing camp.
Literature: Hauglid (1993), Bjerck (2008)

Map ID: 32
Site name: Evjen 3
Site setting: On a c.20 m broad raised beach terrace surrounded by fell, access to Sundan fjord through narrow ravine
Age: 9580 ± 90 BP
Environmental data: None
Function/subsistence: Coastal base camp; hunting, fishing, gathering
Literature: Hauglid (1993)

Map ID: 33
Site name: Åsgarden 1
Site setting: On level ground of coarse glacial deposits surrounded by steep slopes in a wildly undulating landscape
Age: 8330 ± 90 BP
Environmental data: None
Function/subsistence: Base camp with dwellings
Literature: Bjerck (1989)

Map ID: 34
Site name: Mohalsen
Site setting: On a low promontory
Age: 9350 ± 270 BP; 8440 ± 190 BP
Environmental data: None
Function/subsistence: Coastal base camp.
Literature: Bjerck (1989, 1990), Hauglid (1993)

Map ID: 35
Site name: Middagskarheia 2
Site setting: On water-eroded fell in former skerry landscape

Age: 8490 ± 130 BP
Environmental data: None
Function/subsistence: Coastal hunting site with dwellings
Cultural affiliation: Fosna
Literature: Bjerck (1989, 1990)

Northern Scandinavia: paleogeography of the Kola Peninsula

1.3

Tatyana V. Sapelko
Institute for Lake Studies, Russian Academy of Sciences, St. Petersburg, Russia

Chapter Outline

1.3.1 Introduction

A radical change of environments at the Late Pleistocene–Holocene boundary (10,300–9300 BP, or 11,500–9500 cal. yr BP) was particularly distinct on the Kola Peninsula. In the White Sea the glacial-marine sedimentation gave place gradually to deposition of purely marine sediments at that time. In the Barents Sea the Preboreal–Boreal period was marked by a rapid contraction of the ice sheet. Several postglacial transgressions were recorded both in the Barents and White seas.

During and after deglaciation the evolution of environments was controlled by the topography of sea coasts, climatic factors, and tectonic movements. As the ice sheets shrank and decreased in thickness, the Earth's crust rebounded, the land surface rose, and the sea receded gradually from the coasts. As a result, a series of the late- and postglacial coastlines may be seen on the northern and southern coasts of the Kola Peninsula. The maximum altitudes of the uppermost level of the Late Glacial basin in the north of the peninsula are 125–130 m and descend to 55 m east of the Kolskiy Bay. On the southern coast the uppermost marine terrace occurs at a still lower elevation: no more than 35–40 m (Gurina et al., 1974). Accordingly, the oldest archaeological localities are found on the uppermost terraces. After the final deglaciation, several transgressions took place on the Barents and White seas.

Long-term paleogeographic investigations performed on the Kola Peninsula yielded a great many data resulting from the studies of lacustrine and marine deposits, coastal sequences, wetlands, etc. Next, we present most of the results of the recent paleolimnological investigations (Fig. 1.3.1).

Human Colonization of the Arctic: The Interaction Between Early Migration and the Paleoenvironment
http://dx.doi.org/10.1016/B978-0-12-813532-7.00003-6

Figure 1.3.1 Main Early Holocene cross-sections on the Kola Peninsula. 1, main Holocene cross-sections.

1.3.2 Northwestern part of the Kola Peninsula

That area became ice-free in the Allerød–Younger Dryas. The Barents Sea level changed repeatedly at that time, the coastlines being displaced alternatively landward and seaward. The compensatory isostatic rebound of the land resulted in that locally the uppermost parts of marine bays were separated from the sea and continued to exist as lagoonal lakes, quite commonly found on the coasts of the Barents and White seas. The bottom deposits of such lakes contain a record of the gradual transition from glaciolacustrine to marine sediments, and finally to lacustrine deposits.

The Barents Sea level fluctuations in the Holocene period have been reconstructed on the basis of data on the lakes at the northern coast of the peninsula situated at different altitudes (a.s.l.) (Snyder et al., 1997; Corner et al., 1999; Kremenetski et al., 2004).

At the Kolskiy Bay, near the Polyarny settlement, nine small lakes at various altitudes—from 12 to 57 m a.s.l.—were studied (Corner et al., 2001). The oldest lakes dated to the Preboreal are found at the maximum elevations (41–57 m a.s.l.), and the youngest ones are the lowest. The Preboreal age of the Lake 1 (69°12′ N, 33°20′ E, 57 m a.s.l.) is indicated by the presence of tree pollen (as is known from studies of other sections on the Kola Peninsula, the tree vegetation has expanded across the peninsula since the beginning of the Preboreal period). This age is confirmed by other results of integrated studies of the lakes.

At the Younger Dryas the landscapes in the north of the peninsula were dominated by tundra communities with dwarf birch and willow. Such landscapes have been reconstructed for Lake KP3 (69°04′19″ N; 36°00′40″ E, 75 m a.s.l.). The colder intervals are noted for a decrease in *Salix* percentage and increase in the *Artemisia* (Kremenetski et al., 2004) indicative of unfavorable conditions for the shrub growth and lowered mean air temperatures.

The transition from the Late Pleistocene to Holocene was recorded in sediments of other lakes on the Kola Peninsula, such as Poteryanny Zub (Lost Tooth) (68°48′91″ N, 35°19′32″ E, 131 m a.s.l.) (Gervais et al., 2002) and Yarnyshnoye Lake (69°04′ N; 36°04′ E, 54 m a.s.l.) (Snyder et al., 2000). The water level was at its maximum in all the lakes at that time. At the end of Younger Dryas and in the early Preboreal (11,500 cal. yr BP) the climate warming is indicated by the peaks of dwarf birch, *Ericaceae*, and *Poaceae* in the pollen diagram. The warming led seemingly to a more diversified shrub and herb tundra vegetation.

Further evolution of landscapes in the Preboreal was marked by expansion of birch (Fig. 1.3.2). Comparison of the pollen data from three lakes (Poteryanny Zub, KP3, Yarnyshnoye) shows the maximum expansion of birch around Lake Poteryanny Zub, the lake most distant from the Barents Sea coast. The pine expansion there occurred somewhat later than in the south of the peninsula, about 9500–8000 cal. yr BP as recorded in the pollen diagrams of all the three lakes.

Another lake thoroughly studied is Lake Soldatskoye (68°16′ N, 38°28′ E, 15.4 m a.s.l.) in Drozdovskaya inlet. Here the history of the lake has been traced since the Preboreal. The lake was dated by radiocarbon at ~10,000 cal. yr BP (Grönlund and Kauppila, 2002). Birch appeared around the lake before 9500 BP and pine about 8500 BP (Vuorela and Saarnisto, 2002), which is in reasonable agreement with the pollen data from northern and central parts of the peninsula (Snyder et al., 2000).

The studies of the northernmost areas of the Kola Peninsula—Rybachiy and Sredniy peninsulas—were aimed at reconstructing the deglaciation process in the region (Koshechkin et al., 1973; Evzerov et al., 2007; Møller et al., 2002). As follows from the results obtained, the region became completely ice-free at the Allerød. The ice sheet readvanced at the Younger Dryas did not affect the peninsula; the active ice boundary was much farther south at the time. That has been confirmed by materials from a peat bog on the Rybachiy Peninsula (69°38′ N, 32°22′ E, 60 m a.s.l.). The peat bog was not subjected to marine transgressions due to its elevated position (Kremenetski et al., 1997). The peat, 110 cm thick, began to form in the Boreal, at the same time as the peat accumulation began in the central and southern parts of the Kola Peninsula.

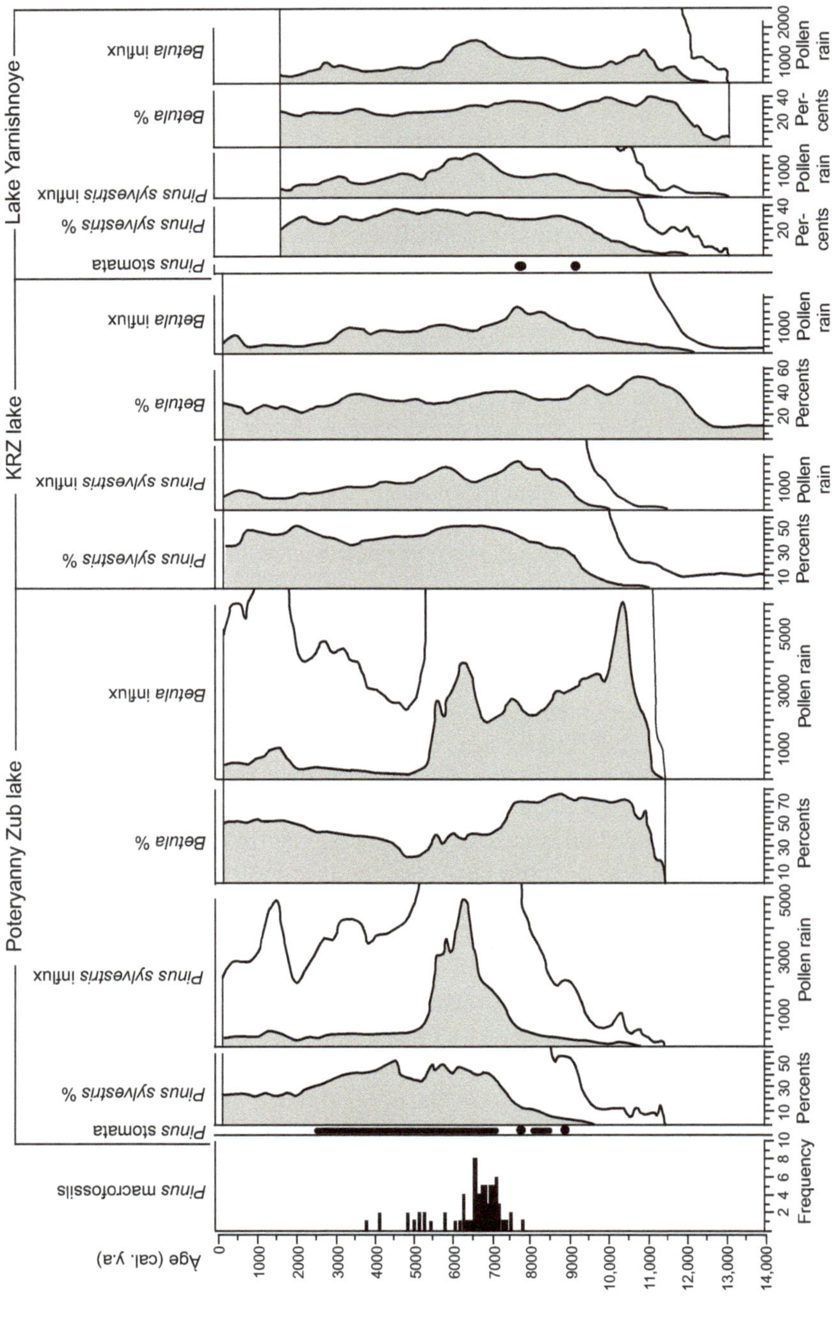

Figure 1.3.2 Correlation of pollen data from lake sediments in the north of the Kola Peninsula (Kremenetski et al., 2004).

1.3.2.1 Central part of the Kola Peninsula

The region abounds in large depressions with lakes within. Most of the lake basins, such as Imandra, Umbozero, Lovozero, and many others, are tectonic in origin (Kagan et al., 1992).

The main sections studied in the central part of the Kola Peninsula are mostly located in the Khibiny Mountains or in their foothills. The latter present hill-and-basin topography built of glacial, fluvioglacial, lacustrine, and boggy deposits. The region belongs to the Atlantic-Arctic area of the temperate zone, at the boundary of the northern taiga and tundra. The vegetation is dominated by spruce forests, and some lesser space is occupied by pine forests and grass and sphagnum swamps. In the Khibiny Mountains, there is a distinct altitudinal zonality with coniferous forests, elfin wood of birch, and mountain tundra replacing each other from foothills to the top. Paleoenvironmental reconstructions are mostly based on the results of biostratigraphic analyses of lake and bog deposits (Elina and Lebedeva, 1982; Elina et al., 2000; Kagan et al., 1992; Davydova and Servant-Vildary, 1996).

At the end of the Younger Dryas–Preboreal period of the Holocene, as soon as the ice melted and glaciofluvial landforms developed, a series of small, dammed lakes were formed in the hollows of the morainic topography. As follows from pollen data (Olyunina et al., 2005, 2008), the region was dominated by forest-tundra landscapes: arctic and subarctic tundra plant communities (dwarf willow, sedge, and green moss) with groves of *Alnaster* and of shrub and dwarf birch. Woodlands of birch, spruce, and pine occurred more rarely. The climate was noticeably colder than at present. The permafrost was characteristic of the region.

During the Boreal period, most small lakes were drained as the Umbozero water level dropped and the second terrace of the lake formed (Olyunina et al., 2005; Solovieva and Jones, 2002). The surface was gradually overgrown with forests, and the tundra coenoses decreased in proportion. The forests were dominated by pine, and dark coniferous and small-leaved species were present in negligible quantities. The climate became warmer and dryer. It is that time interval in which most of dates obtained for the center of the peninsula fall. It was marked by active peat accumulation. At that time, peat bogs formed in the Malaya Belaya River valley (Lutnermayok peat bog, 67°41′ N, 33°17′ E) on the old delta of the Kunyok peat bog (67°50′ N, 33°40′ E) (Kremenetski et al., 1997) and some others. Most lakes, such as Lovozero (Elina et al., 2000), Chuna (Solovieva and Jones, 2002; Solovieva et al., 2005), and some others also came into being in the Boreal period.

1.3.2.2 Southern part of the Kola Peninsula

This region has been thoroughly studied, with a special attention paid to the coastal formations. The first works aimed at dating and measuring marine terraces on the southern coast were performed by W. Ramsay (1898). Later on, numerous attempts were repeatedly made at specification of the terrace age and the dynamics of the White Sea level. M.A. Lavrova (1960, 1969) carried out the survey of the Terskiy and Kandalaksha coasts on a broad scale.

Investigations on the southern coast of the Kola Peninsula (Evzerov et al., 1976) have shown that a series of sand and silt deposits with some varve-like lamination was accumulating in the ice-dammed basin from the Allerød to the early Preboreal. Thickness of individual layers decreases upward, along with increasing proportion of sand at the expense of silt and clay, which suggests a gradual shallowing of the basin. The apparent thickness of the sediments is more than 14 m.

After the ice dam had been removed from the White Sea mouth, the sea water salinity grew since the second half of the Preboreal and a normal marine regime became established by the Boreal. The sediments dated to the second half of the Preboreal are sandy silts with varve-like lamination changing into fine sands up the sequence. The top of the sands was eroded by the subsequent marine transgression. The apparent thickness of the deposits does not exceed a few meters.

The data permitted a conclusion on some chronological discrepancy in the postglacial sedimentation in the near-coast zones of the Barents and White seas (Evzerov, 1979).

At the Late Glacial time, even under conditions of highly dynamic coastline, rather thick sedimentary members were deposited over considerable areas due to an input of large masses of clastic material by the glacier meltwater. A certain deficit of clastic material began to show itself only at a final stage, that is, at the beginning of the Holocene. As a result, sandy deposits of that time (during the regressive phase of the basin) formed only in places of the meltwater inflow into the bays. The close relationship of the sedimentation in the coastal regions and the deglaciation processes is a typical feature of the Late Glacial period.

Well-pronounced landforms in the zones of marginal formations are found also at the Ponoy Ice Sheet periphery. On the west of it, there existed a wide zone (3–8 km) of end moraine ridges and kame mounds. The zone extends east of Lake Lovozero and farther south. There are somewhat smaller marginal zones also at other margins of the Ponoy Sheet.

Glaciolacustrine deposits have been studied in the southwest of the Kola Peninsula. The deposits belonged to a paleolake known as Ust'-Pyalka (66°20′ N, 39°40′ E) and located at the periphery of the shrinking Belomorsk lobe of the Scandinavian Ice Sheet. As is shown by the studies, the age of the marginal formations Keiva I and Keiva II are older than was thought before, the Keiva I ridge being formed later than Keiva II (Bakhmutov, 2006). The paleolake was ~60 m deep and was fed by meltwater of the thawing ice sheet, the latter having formed the dam and at the same time the southern coast of the lake. The paleolake deposits compose two terraces 150–145 m and 142.5–140 m a.s.l., with surfaces gently sloping toward the Ust'-Pyalka River.

At the Late Glacial and Postglacial periods the inner regions of the Kola Peninsula were rising faster than its periphery. Judging from the isobath pattern, the glaciolacustrine deposits in the northern part of the water body are raised by about 10 m compared with the southern coast; so the lake bottom was tilted northward at the time of the varve deposition. Clastic material was supplied to the lake from the south.

The recent studies of the Keiva I and Keiva II ridges seemingly indicate their age to be older than the Younger Dryas (Hattestrand et al., 2007), though it still cannot be established with certainty.

At present, several small lakes located at different altitudes above the sea level are actively investigated. In this way, it is possible to gain an understanding of the White Sea coastline dynamics after the ice sheet recession. The lake sediments are also of some interest in paleoclimatic reconstructions. In particular, mean summer temperatures reconstructed from chironomids in the Lake Berkut sediments (Ilyashuk et al., 2005) for the Holocene beginning (10,100–9200 cal. yr BP) appeared to be rather high (12.3–12.5°C). Similar temperatures were also reconstructed from pollen data in the central part of the Kola Peninsula (Vashchanova and Klimanov, 1987), as well as in southern Sweden (Hammarlund et al., 2002) and in Finland (Seppä and Birks, 2001, 2002).

In the Early Holocene the climate in the northern and central Kola Peninsula was more continental than in its southern part, with warm and dry summers and cold and dry winters (MacDonald et al., 2000). In the south Kola Peninsula the climate was wetter, having acquired marine characteristics after the ice sheet recession. In the southeast of the peninsula, birch forests were spread most widely.

Since 10,000 cal. yr BP, pine forests have become common (Kremenetski and Patyk-Kara, 1997; Sapelko et al., 2009), which suggests warm and humid enough conditions favorable for the afforestation of the region.

The growing importance of forest landscapes, both in the north and in the south of the peninsula, is seen in the pollen zones recovered from the lakes sediments (Fig. 1.3.3). In particular, data on the pine and birch distribution have been obtained from sediments of three lakes: Babozero (66°22′30″ N, 37°31′ E), Krugloye (66°22′ N,

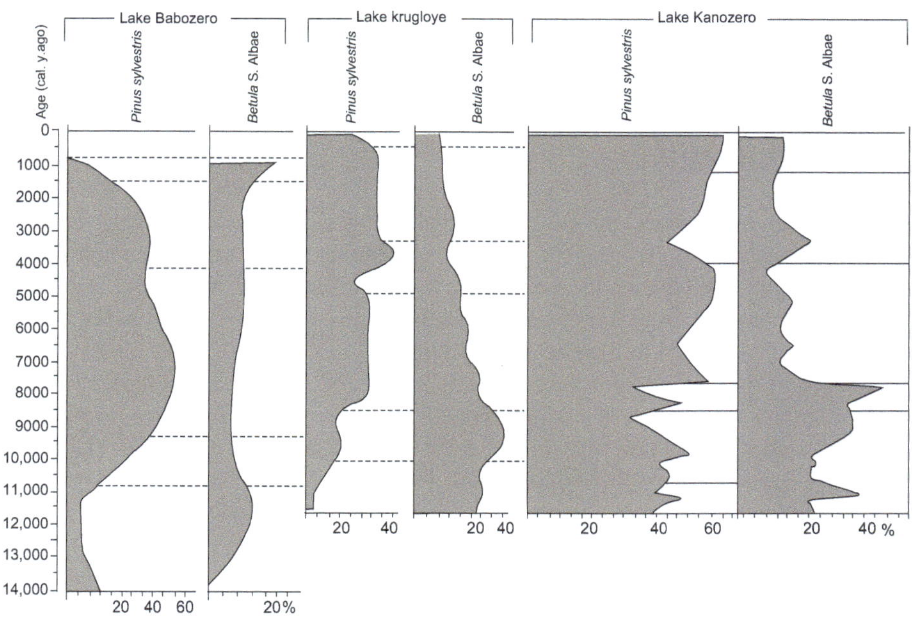

Figure 1.3.3 Correlation of pollen data from lake sediments in the south of the Kola Peninsula.

37°35′ E), and Kanozero (67°03′ N, 34°06′ E), the catchment basin of the latter being most densely forested. In the Preboreal, there were herb and dwarf shrub communities in the lake catchment basin, with occasional groves of birch (Sapelko et al., 2009). Pine was also common here at that time, that is, earlier than in the Lake Krugloye basin (Kremenetski and Patyk-Kara, 1997). As to Lake Babozero, areas of pine forests were also growing in importance at that time, while the birch percentage is reduced.

Occasional findings of macrophytes in combination with willow suggest the presence of shallow water locally.

Relatively warm and humid conditions favorable for the forest advance northward existed around Lake Berkut (66°20′69″ N, 36°39′77″ E) until 7000 cal. yr BP (MacDonald et al., 2000). Similar climatic conditions are recorded in the northern and central regions of the Kola Peninsula about 8500–7000 cal. yr BP. At the same time, relatively high lake levels and the beginning of the peat accumulation are recorded in the Khibines (Davydova and Servant-Vildary, 1996; Kremenetski et al., 1997).

It follows from the preceding that the beginning of the Holocene was an important stage in the environmental evolution on the Kola Peninsula.

At that time, lacustrine sedimentation became the dominant process in the north and in the south of the peninsula. After that time, with growing moisture supply and rising mean temperatures, some tree species—birch, and later, from the Boreal period on, pine—began expanding. In the central part of the peninsula the lake formation began somewhat later. The peat accumulation took place all over the peninsula since the Boreal period. The process of peat accumulation is slow—0.14 mm/year in the north (the Rybachiy Peninsula) (Kremenetski et al., 1997), 0.15 mm/year in the central part (the Khibines), or 0.1 mm/year at the Lake Lovozero (Elina et al., 2000)—and the resulting thickness of peat is small. Numerous earthquakes occurred all over the Kola Peninsula at the Pleistocene–Holocene transition.

1.3.3 Conclusion

The beginning of the Kola Peninsula peopling depended directly on the environmental changes at the beginning of the Holocene. A specific feature of this region consists in that the sea level fluctuations were of primary importance in man's appearance and dispersal. At present, there are more than 70 Mesolithic sites known in the region, 14 of them being attributed to the Early Mesolithic.

The oldest Mesolithic sites here appeared about 8 ka BC in the north of the Kola Peninsula, at the Barents Sea coast. Those are sites on the Rybachiy Peninsula, Rastiniemi, and two more sites in the northwest of the Kola Peninsula (Dolgaya Shchel Bay), and a site on the Kildin Island. The sites are 38–60 m above the sea level on Rybachiy and 45 m on Kildin up to 65 and 91 m in the northwest.

In the north of Norway the earliest settlements are dated to the Preboreal, but it was only by the end of Preboreal that small groups migrated to the Kola Peninsula north. By that time the regional environments were relatively stable. New lakes appear at 41–57 m a.s.l. The Late Pleistocene–Holocene transition is recorded in sediments of the lakes located at higher altitudes: 75–131 m a.s.l.

The environments, however, were extremely inhospitable to humans at that time. The conditions became more stable at the beginning of the Boreal period, as seen from the peat bog on the Rybachiy Peninsula at 60 m a.s.l. The peat accumulation began there in the Boreal and continued practically through the entire Holocene.

In the Late Boreal, people migrated eastward along the northern coast. Besides the places settled in the Early Mesolithic, the Middle Mesolithic sites are found farther east (the Lumbovka R. valley). Those sites occur on marine terraces 33–45 m a.s.l. and are dated at 7–6 ka BC.

It was only in 6–5 millennia BC that the Mesolithic people began to colonize the inner territories of the peninsula. Late Mesolithic settlements are found not only in the north, but also in the central and southern parts. On the sea coasts the sites of that age occur mostly at 20–40 m a.s.l. Besides, the people began to settle in river valleys and on lake coasts, presumably due to a necessity to find better protected places in connection with the beginning of a new transgressive phase of the Barents Sea.

The absence of early sites in the south of the Kola Peninsula might be related to difficulties of the human penetration into the region; as for the central part, the environments became more hospitable there much later than in other parts of the region: about 6–5 ka BC.

The Early Holocene (Mesolithic) sites on the Kola Peninsula

1.4

Vladimir Y. Shumkin
Institute for the Material Culture History, Russian Academy of Sciences, St. Petersburg, Russia

Chapter Outline

The sites attributed to the oldest stage of the Northern Scandinavia colonization are known in Norway and the Kola Peninsula. The postglacial Stone Age in the Scandinavian arctic corresponds to the Holocene and is traditionally subdivided into Mesolithic and Neolithic.

The beginning of the Mesolithic, that is, an appearance of the first settlements in Western Norway, corresponds to the Early Holocene of the chronostratigraphic scale; its final stage falls on the time of the Tapes transgression (~6 kyr BP). So the Mesolithic in Western Lapland lasted for about 4000 years and spanned the Preboreal, Boreal, and the first half of the Atlantic.

About 200 Mesolithic sites are known in northern Fennoscandia (Fig. 1.4.1). Not all of them are equally informative and well preserved, and many sites call for further investigations. Due to a very slow process of soil formation in the polar regions, most sites have thin cultural layers, where artifacts of organic materials are seldom preserved, so only stone tools and debitage have been discovered. Many collections consist only of surface finds, as the cultural layer either had been destroyed long ago, or the artifacts had not been buried under later deposits. Usually, there is no quantitative or qualitative difference between excavated materials and surface findings.

Recently, archaeological field studies were performed using more advanced techniques in Northern Norway (sites of Mortensnes, Karlebotn, and Devdis). It appeared that on excavated Late Mesolithic sites, there were dug-out dwellings (pit-houses), while at the open surface, only hearth remains are discovered.

In the studies of archaeological sites in Northern Fennoscandia, geomorphological materials, and data on marine terraces in particular, are of prime importance. There are well-preserved traces of various marine regressions and transgressions (Goretskiy, 1937; Zemlyakov, 1940; Holtedal, 1960). A possibility of repeated transgressions should be also taken into account (that happened in the western sector more than once). Besides,

Human Colonization of the Arctic: The Interaction Between Early Migration and the Paleoenvironment
http://dx.doi.org/10.1016/B978-0-12-813532-7.00004-8

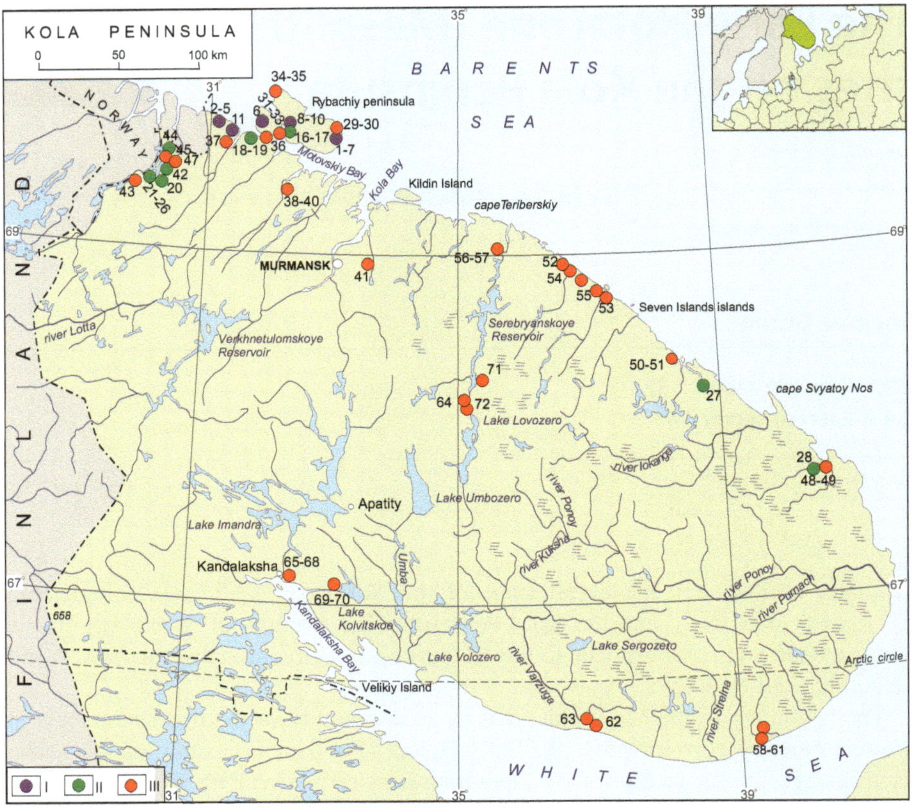

Figure 1.4.1 Mesolithic sites of the Kola Peninsula (numbers on the map). *1: Early Mesolithic*: 1–7: Tsip-Navolok; 8–10: Bolshaya Motka; 11: Rastimieni; 12–15: Dolgaya Shchel. *2: Middle Mesolithic*: 16, 17: Bolshaya Motka; 18, 19: Ambarnaya; 20: Kolasyoki; 21–26: Shuoniyoki; 27: Chernaya; 28: Lumbovka. *3: Late Mesolithic*: 29, 30: Tsip-Navolok; 31–33: Bolshaya Motka; 34, 35: Vayda; 36: Malaya Volokovaya; 37: Pechenga; 38–40: Zapadnaya Litsa; 41: Kamennoye Ozero (Lake); 42: Kolasyoki; 43: Patsoyoki; 44: Nyalyavr; 45–47: Kuetsyavr; 48, 49: Lumbovka; 50, 51: Morskaya; 52: Krasnaya I; 53: Kharlovka; 54: Rynda; 55: Zolotaya; 56, 57: Serebryanskaya, 58–61: Pyalitsa; 62: Kuzomen'; 63: Golye Gory; 64: Nivka; 65–68: Niva; 69, 70: Kolvitsa; 71: Uyma; 72: Nivka (burial).

sites of the same age may be positioned within certain altitudinal limits, and not strictly at one altitude (in that case the lower level would provide a more exacting time limit).

To take but a few examples, the oldest sites in the eastern Finnmarken and western regions of the Kola Peninsula lie now at ~90 m a.s.l., while sites of the same age on the Lofoten Islands are below sea level. There are also considerable variations in marine terrace elevations in other parts of Scandinavian coasts. An influence of the Tapes transgression should be also taken into account: the transgression dated to the end of the Mesolithic (6 kyr BP) destroyed some archaeological sites (Engelstad, 1991).

Northern Fennoscandia is deficient in silicified rocks; therefore the prehistoric people had to make their tools mostly of quartz, found in abundance, both in solid rocks and as pebbles and debris. Less common was the usage of other rocks (chert, quartzite, dolomite, sandstone, siliceous shale). As to the bone and wood materials, which were undoubtedly used, they are exceedingly rare (Figs. 1.4.2–1.4.4).

At the same time, the quartz inventory classification presents certain difficulties and may be successfully developed only on the basis of long-term experience in the work with that raw material and an acquaintance with specific features of its knapping and treatment. Such knowledge may be acquired from handling large collections of various tools and debris.

1.4.1 Western sector

All the Mesolithic sites in Northern Norway are traditionally attributed to the Komsa culture. In 1925, A. Nummedal, an archaeologist known for his discovery of the Fosna culture on the western coast of Norway, surveyed Finnmarken in search of similar sites (Fig. 1.4.5). When studying high marine terraces in the north, between Langfjord in the west and the Russian–Norwegian boundary (Varanger Fjord) in the east, he found sites with tools and flakes made mostly of quartz, quartzite, and other nonflinty rocks. Nummedal noted a similarity between all the objects discovered and described the sites as belonging to the Komsa culture, named after the Komsa Mountain in the community of Alta, Finnmark, where he found the first such site (Nummedal, 1929; Simonsen, 1975, 1988).

Archaeological assemblages attributed to the Komsa culture are highly variable in space and time, in the eastern regions in particular. New and extremely informative sites with intact (or buried under the late deposits) cultural layers have been discovered in the Tromsø county (Sandmo, 1989), near Bodo (Norland) and even in the well-studied region of the typical Komsa sites: at the Varanger Fjord (Schanche, 1988).

It should be kept in mind that some of the Late Mesolithic sites were destroyed during the Tapes transgression. But it is not inconceivable that new localities may be found on the large area still not studied.

Besides often mentioned “core-shaped forms” and “Levallois points,” all the studied sites have in common an extremely limited amount of retouched tools (no more than 50 in most localities). It is possible, however, that some flakes were used without secondary treatment (retouch) (Schanche, 1988).

Traditionally, 10 morphological and functional groups have been distinguished in Mesolithic stone tool assemblages recovered from sites of Northern Fennoscandia: points, burins, end scrapers, knives, retouched or utilized flakes, microliths, drills/borers, cores, blades, and axes/adzes (Hood, 1991). The Mesolithic people preferred to use silicified raw materials, but in case of necessity, they could even make points from any material available, such as quartz or quartzite. Large blades and flakes found in many sites were obtained from coarse-grained quartzite using hard hammer.

The size of the raw material available was likely more important than its quality. The *cores* are variable in types and raw material used. There are not many single- or double-platform cores with a distinct parallel reduction. More often, there are found crude, globular (amorphous) cores not suitable for obtaining blades, though blades are commonly found in the lithic inventories (e.g., Smellroren).

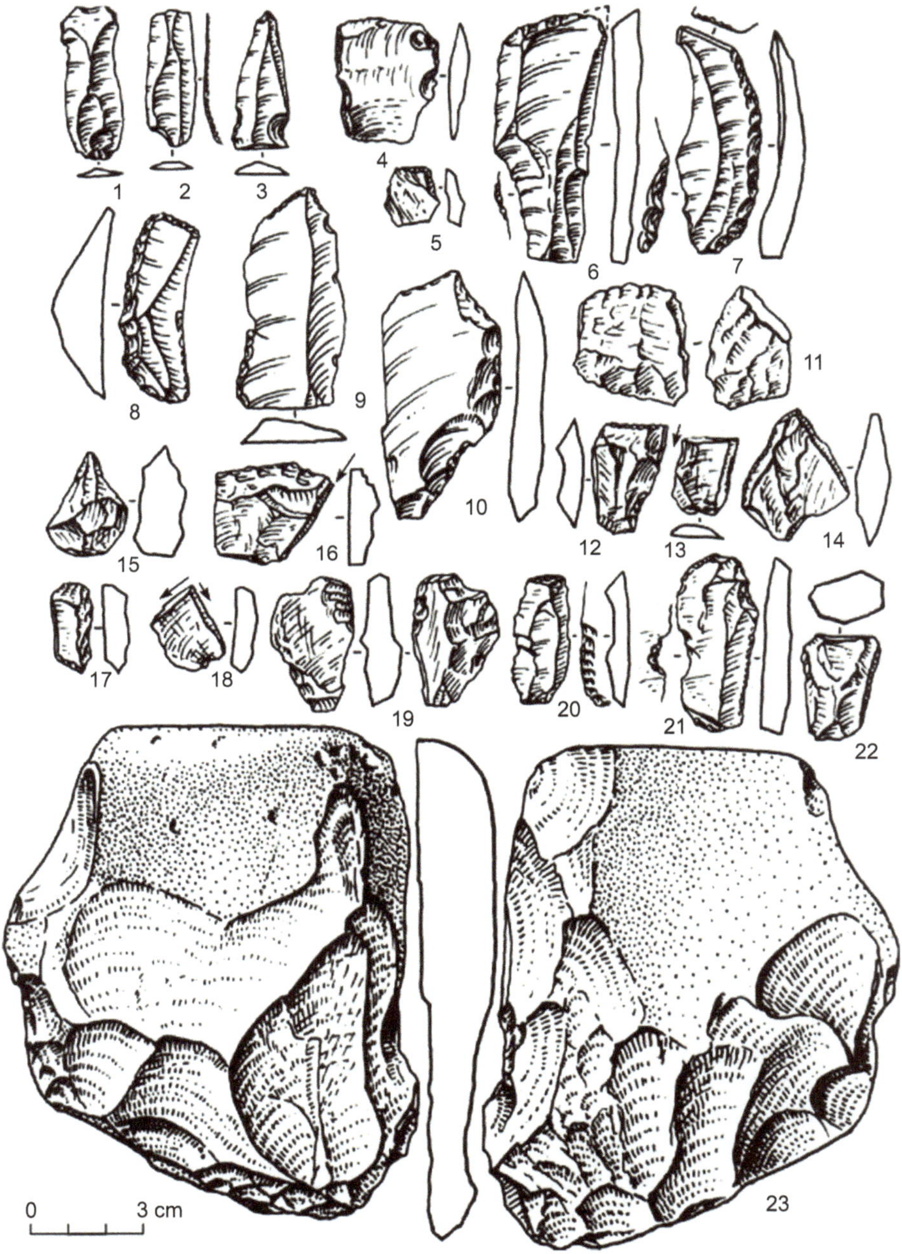

Figure 1.4.2 Stone tools of the Early Mesolithic. Sites: 1–10: Tsip-Navolok IV; 11–13, 15, 16: Tsip-Navolok VI; 14: Tsip-Navolok Xa; 17–20: Tsip-Navolok XIII; 21–23: Tsip-Navolok XIV. Stone tools: 1, 2, 4: flint; 3, 6–10, 23: siliceous schist; 5, 11–22: quartz.

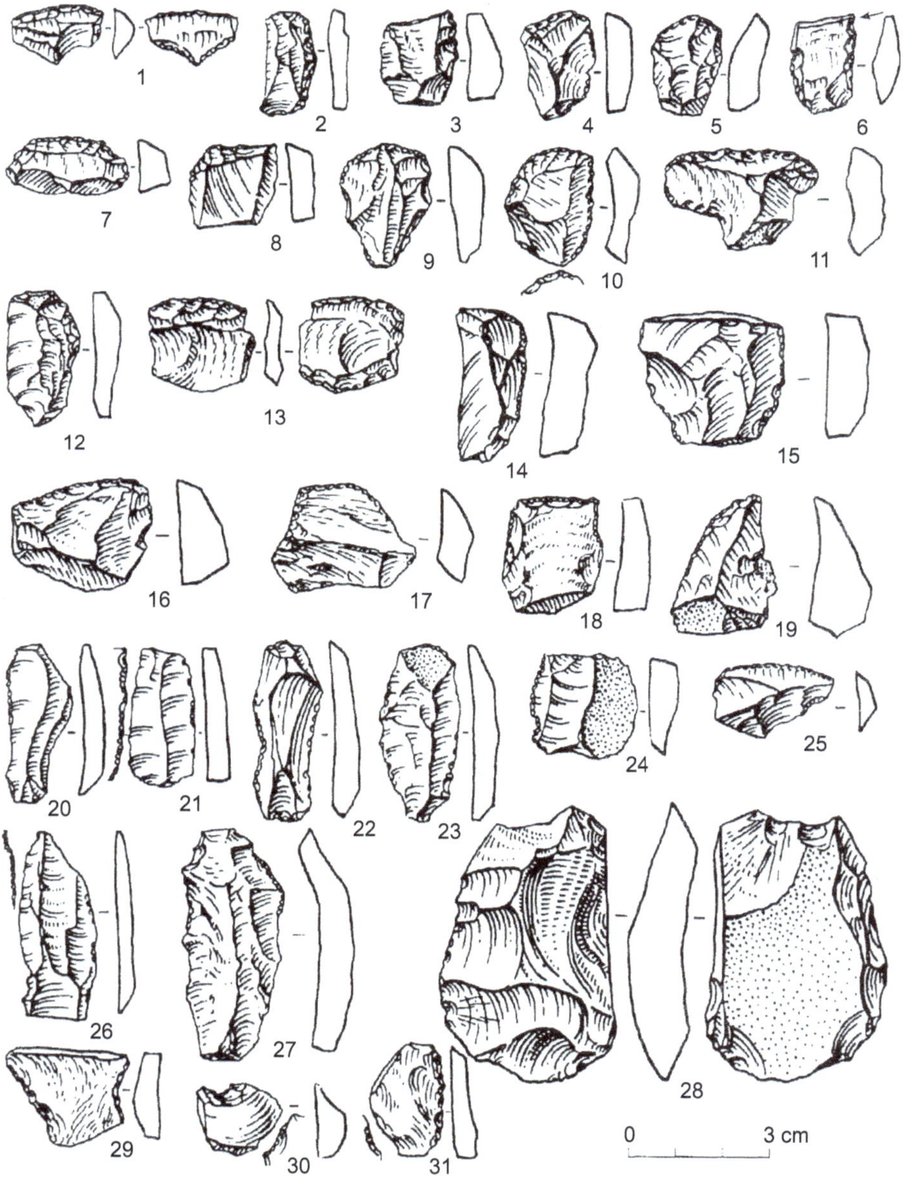

Figure 1.4.3 Stone tools of the Middle Mesolithic (from the site of Chernaya I). Stone tools: 2, 8, 12, 14, 19–24, 26, 30: flint; 15, 22, 25, 28: flinty shale; 1, 3–7, 9–11, 13, 16–18, 23, 27, 29, 31: quartz.

Figure 1.4.4 Stone tools of the Late Mesolithic. Sites: 1–23: Kharlovka IX; 24–33: Morskaya I. Stone tools: 1, 15: flint; 2: flinty shale; 3–14, 16–33: quartz.

The *microlithic component* is well represented in the Mesolithic assemblages in Northern Scandinavia. Commonly found are small bladelets thought of as beveled points (lanceolate forms) or microretouched pieces (such as in the Tollevik locality). Such forms are present in small numbers in many settlements and are sometimes classified as microliths judging mainly from their size.

The subdivision of the Komsa culture into chronological periods was raised in archaeological literature more than once. K. Odner (1966) suggested its division into three phases based of the presence or absence of the key tool types and the development of the material inventory from tanged forms to transverse points. Those phases were distinguished on the basis of materials from the Varanger Fjord taking into consideration the altitudes of the marine terraces. P. Simonsen (1975) distinguishes, on

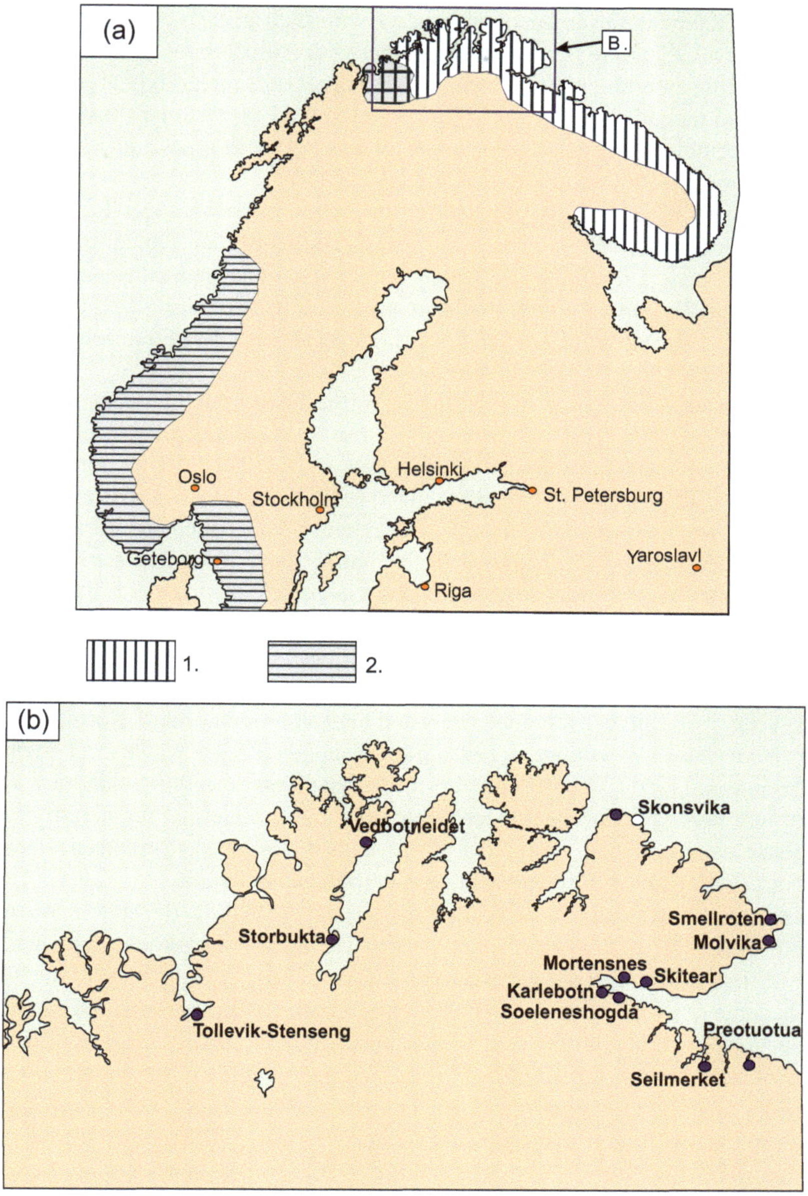

Figure 1.4.5 Mesolithic cultures in Fennoscandia (a) and the best studied Komsa sites in Finnmarken, Northern Norway (b). 1: Komsa; 2: Fosna-Hensbaska.

the basis of the same materials, two phases, the main Komsa and Sæleneshøgda Phase, transitional to the Neolithic.

Having studied early settlements in Tromsø, A. Sandmo (1989) suggested three-phase divisions, namely the following: I, 10,000–9700 BP; II, 9700–9300 BP; III,

9300–9000 BP. Not all the archaeologists agree with such an early age, but the division put forward by A. Sandmo is in good agreement with radiocarbon dates. According to her data, the tool assemblage from the early Preboreal sites includes blades, small single points, and tanged points presumably related to the Fosna cultural tradition. Later Preboreal assemblages reveal disappearance of some of those types and an appearance of more pronounced "northern" types (Sandmo, 1989).

Judging from the studies of the Mortensnes sites at the Varanger Fjord (eastern Finnmark) (Schanche, 1988), microliths, borers, and blades are typical of the earlier phase of the Fosna Culture, while single points and transverse-edge points appear somewhat later. The same types of tools are present at interior sites dated to later time (Helskog, 1974; Simonsen, 1975) and occasionally at Early Neolithic settlements, as relics (Simonsen, 1961, Schanche, 1988).

Another subdivision of the Mesolithic in Arctic Fennoscandia was suggested by Woodman (1993). In his opinion, the term "Komsa" may be applied to a certain assortment of tools including stemmed points, points with truncated back, end scrapers, and flake axes. That Mesolithic phase is dated to 9500–8500 BP. The previous phase termed tentatively "Prestemmed" is still insufficiently represented and should be dated older than 9500 yr BP and attributed to the initial stage of human settlement. The Komsa phase technology was aimed at making large-size blades/flakes using percussion technique. The fact of the Komsa technocomplex absence, even at such an early site as Mortensnes, suggests, in Woodman's opinion, a short duration of that phase: it came presumably to its end ~8500 BP. The phase could develop from the Fosna culture (as seen from Simavic site materials) and expand to the north of Norway.

The next phase (Woodman's (1993) Sæleneshøgda Phase) was earlier identified by P. Simonsen (1961) who attributed it to the Neolithic. It is quite clear now that it is older and belongs entirely to Mesolithic. The basic materials for identifying the later phase were obtained from excavations in Sæleneshøgda (Simonsen, 1961), in Karlebotn (Engelstad, 1989), and Mortensnes (Schanche, 1988).

These assemblages are dated by radiocarbon at 7710 ± 480 BP/8680 ± 540 cal. yr BP/ and 8500 ± 120 BP/9490 ± 150 cal. yr BP/. The Sæleneshøgda Phase inventory is slightly different from that of the earlier phase. Large end scrapers typical of the Komsa phase become rare and are replaced by series of small end scrapers and flakes. Similar tools occur at some Mesolithic sites of the same age in Nordland. Most of Mesolithic sites in Western Lapland are confined to the sea coast, so their oldest inhabitants were supposedly adapted to marine resources. On the whole, the Mesolithic people settlements, and maybe those of later inhabitants, concentrated in the maritime zone.

In the past the presence of typical dwelling structures was considered to be a characteristic attribute of Neolithic settlements. Recently found Mesolithic coastal sites, however, appeared to include both open sites and remains of small pit-houses and tents (Helskog, 1974; Engelstad, 1989; Schanche, 1988). In Finnmark, for example, there are dwelling areas with shallow (5–15 cm) open pits. A few of them excavated in the Varanger Fjord region are dated by radiocarbon as follows: Karlebotn at 7710 ± 480 BP/8680 ± 540 cal. yr BP/ (Engelstad, 1989) and Mortensnes at 8500 ± 120 BP/9490 ± 150 cal. yr BP/ (Schanche, 1988). The presence of such complexes suggests a sedentary lifestyle in those regions.

In the Late Mesolithic, some sites appear in the inner region. Usually, they are small short-term camps of hunters and fishermen. In spite of intensive survey of Finnmarksvidda and Troms mountain regions, only a few localities were discovered with poor inventory including rare transverse-edge points. The only exception is the Devdis site (Engelstad, 1991) with several pits, one small stone construction of unknown purpose, and a series of beveled, transverse, and single points, retouched flakes, end scrapers, and cores. The only ^{14}C date obtained for that assemblage is 6575 ± 150 BP/7460 ± 130 cal. yr BP/.

The location of the sites on the sea coast and a related economy based on fishery and sea mammal hunting was characteristic for all the regions of Mesolithic sites. There are 13 Preboreal open-type sites known in the region varying in size from tens to a few thousands of square meters. The large sites were most probably repeatedly visited camps. One of such settlements—Simavik, in the vicinity of Tromsø—occurs at an altitude 28–30 m a.s.l. It corresponds to the Younger Dryas sea level and is dated by ^{14}C at 9200 ± 200 BP/10 420 ± 300 cal. yr BP/ (Sandmo, 1989). All the settlements dated at 10,000–9000 BP belong to the oldest stage of settlement and show dependence on the marine resources (most probably, marine mollusk gathering on littoral). The stone inventory includes few typologically identifiable tools: there are only blades, various stemmed (tanged) points, and cores. Microliths are completely absent. The raw materials are mostly quartzite, diabase, and chert, mostly of local provenance (immediately near the sites or at a small—no more than 20 km—distance).

When excavating shell deposits of the Mortensnes site (the Varanger Fjord) (Schanche, 1988), some faunal remains were recovered, which shed light on the Komsa people's diet. Fish bones—mostly of cod—account for about 70% of the bone material; a few bones of seal are present, but not one bone of deer is found nor a piece of processed antler. The archaeologist having excavated the site supposed it was seasonally (in spring and summer) inhabited. The presence of cod, however, does not rule out a possibility of human habitation in other seasons. The data, however, are still insufficient for a final conclusion.

According to Odner (1966), the Mesolithic people of the region could migrate from winter settlements in the inner parts of the fjord to the open coast in summer. However, no unambiguous proof of such migrations has been found as yet. The polyvariant analysis of implements from Mesolithic settlements at the Varanger Fjord (Bolviken et al., 1982) does not agree with this model either.

Some archaeologists (Welinder, 1981; Siiriainen, 1981; Sandmo, 1989) were of the opinion that at the first stage the northern coasts of Norway were seasonally used by groups of gatherers coming from the south for a short time. It is difficult, however, to find reasons for such long travels from the regions hardly poorer in natural resources than the northern territories. Long migrations were possible in the Neolithic and proved to be practiced in the Iron Age (Schanche and Olsen, 1985), though the reasons for them were different. There is little doubt, though, that the contacts existed with southern groups as well as with the population of eastern regions (Russian Lapland). Stone rings at the tent periphery and pit-houses are often found in Western Lapland in groups (up to five constructions). Unlike the older localities, they suggest greater and more stable population.

Though the dwellings could be constructed and used at different times, still those were long-term base habitation sites and not seasonal camps. It should be noted that since the Middle Mesolithic stage the sites occur in various environments: on islands, on fjord coasts, along rivers and on lakes, and even far from the sea coasts, in the inner mountains and valleys. It is still uncertain what kind of relations existed between coast dwellers and inland inhabitants. It might be some systems of integrated economy, from a permanent residence with seasonal migrations to the beginning of colonization of various ecological niches.

The oldest people groups that reached the extreme north of the continent in the Preboreal succeeded in adapting themselves to the severe environments of the arctic zone and colonized the region widely using the natural resources.

In the first millennia of their development, they traveled a long way having changed the initial culture appearance. In the process of adaptation the Mesolithic people mastered techniques of splitting quartz and developed rational procedures for toolmaking. As follows from the studied archaeological materials and data on paleoenvironments, the earliest assemblages are confined to the western and northern sea coasts and may be dated to the middle of the 8th millennium BC at the latest. In their gradual movement eastward onto the ice-free territories on the arctic coasts the human groups moved to a greater distance from the places of their former habitation, their lifestyle being changed accordingly.

In the western sector of Northern Fennoscandia the population kept to the sea coast over the entire Mesolithic. Feeble attempts at penetrating inland had no visible effect. In Eastern Lapland the process was more successful at the Middle (7th–6th millennia BC) and particularly Late (6th–5th millennia BC) Mesolithic. That was presumably due to specific characteristics of the environments (Shumkin, 1984).

An analysis of the tool assemblages did not reveal essential differences in the tool assortment, nor in their types, the latter gaining gradually in accomplishment and diversity. Differences in the raw materials used depend mostly on the local sources and do not reflect on the culture appearance. Considering topography of the coastal sites and spatial distribution of the findings, we may suppose that the inhabitants at that early stage were gatherers of seafood who constantly wandered along fjords and open coasts. This strategy was dominant in Western Lapland; it led gradually to the appearance of permanent dwellings, toward a more sedentary lifestyle, and a control over a certain territory. In the regions farther east, along with that type of subsistence (though it retained longer its mobility), existing groups specialized in forest animal hunting and occasionally practiced primitive fishing; in all probability, techniques of the latter did not differ much from that of hunting.

More detailed analysis of the archaeological and environmental data suggests that the oldest groups on the territory under a consideration descended from the people, bearers of the Ahrensburgian culture sites, and were related to their southern neighbors, people of the Fosna culture.

A distinct continuity in the technology, a relative isolation of individual sites in the Arctic, and the fact that changes in various parts of the region tend to occur simultaneously and in a similar way all give grounds to consider the Mesolithic sites in Northern Fennoscandia as belonging to a special archaeological culture and to keep the name it had been known under: the Komsa culture. The data available at present permit one to distinguish local variants of the culture: western, northern, and southeastern (Shumkin, 1984, 1990). The differences between them begin to be noticeable toward the latest stage and appear mostly to be a result of inner evolution and only partly of external contacts.

1.4.2 Eastern sector

The detailed studies of the Stone Age in the north of Scandinavia (Finnmarken) were started in 1925 by the curator of National Museum in Oslo, A. Nummedal, on the initiative of A. Broger. The studies proved successful. Their results, published as a monograph (Nummedal, 1929), aroused a considerable interest among archaeologists and geologists, as the author's conclusions contradicted the common notions of the delayed colonization of the European North. The fieldwork begun by A. Nummedal was soon continued on adjoining territories by Bøe and Nummedal (1936), Gjessing (1948), Freundt (1948), Luho (1956, 1967), Odner (1966,1970), and Simonsen (1961). Later the investigations were successfully continued by specialists from Norway, Sweden, and Finland, such as Gaustad (1969), Siiriainen (1971, 1972, 1973, 1974, 1978), Helskog (1974), Indrelid (1976, 1978), Broadbent (1979), and many others.

In 1935–37 a team of Soviet specialists headed by B.F. Zemlyakov carried out an archeological survey on the Rybachiy Peninsula in search of the so-called "Arctic Paleolithic" sites already known from Northern Norway. The works resulted in a discovery of 12 sites of the "Arctic Paleolithic" (later attributed to the Komsa Mesolithic culture). The sites were described in a series of papers (Zemlyakov, 1937a,b, 1940) where, among other things, a relative chronology of the Rybachiy Peninsula sites based in their altitudes was given.

The studies of the Mesolithic site were continued in 1965 and in 1969–82 by an expedition of the Institute of Archaeology (Leningrad Division), the USSR Academy of Sciences, under the supervision of N.N. Gurina (1971, 1973). Several sites in the southern and central regions of the Kola Peninsula were excavated by archaeologists from the Karelian Department of the Institute of Language, Literature and History (Pesonen, 1978).

At present, there are more than 70 Mesolithic sites in Eastern Lapland (Fig. 1.4.2). They occur near the Russian–Norwegian boundary, on the Rybachiy Peninsula, on the northern and northeastern coasts of the Barents Sea, on the Terskiy and Kandalaksha coasts of the White Sea, and in the inner part of the Kola Peninsula.

The sites are usually situated on high marine and river terraces and occupy a relatively small area. Because of severe climate and weakly pronounced soil formation, the finds occur on the surface or—rarely—under a thin layer of soil (10–12 cm). Only occasionally the archaeological material is concentrated in rounded spots about 3–5 m in diameter. Twenty sites of that kind were excavated, but the excavated assemblages were similar to the pieces collected at the surface both in quantity and quality. That may be attributed to the fact that the cultural layer of all the excavated sites had been repeatedly redeposited by wind. The excavations result as a rule in obtaining several tens or hundreds of stone objects and very seldom of hearth remains. No evidence of dwellings or domestic areas has been found except for the aforementioned concentrations of lithics (Pesonen, 1978).

All the preceding does not mean that excavations are useless. We would only underline that both surface finds and excavated ones may be considered as archaeological sources of equal value. It should be also noticed that the absolute majority (91%) of the Mesolithic sites and localities yielded materials belonging solely to that epoch; in a few sites with mixed materials the younger objects are few in number and belong to the early age of metals, so they may be easily recognized and separated from the Mesolithic assemblage.

The Kola Peninsula is deficient in silicified rocks, so the Mesolithic man made his tools primarily from quartz; the latter occurs here in abundance both in solid rock outcrops and as debris or pebbles and cobbles.

As has been stated earlier, there are more than 70 Mesolithic sites known on the Kola Peninsula. All the objects are divided into four groups according to their geographic position, as follows:

1. northern region: the Rybachiy Peninsula and adjoining northern coast of the Barents Sea;
2. western region: river and lake drainage basins adjoining the Russian–Norwegian boundary;
3. eastern region: northeastern coast of the Barents Sea;
4. southern region: the Terskiy coast of the White Sea and the upper reaches of the Voronya River.

The regions were outlined on the basis of data on their geology, topography, and soils geography (Atlas of the Murmansk Region, 1971; Rikhter, 1936; Belov and Baranovskaya, 1969). In the northern region the Mesolithic sites are located on the sea coast, in bays, and coves of the Barents Sea. They do not show any relation to river mouths and occur mostly on old marine terraces. The sources of fresh water were probably small lakes and streams. The archaeological materials are usually scarce: no more than a hundred stone objects, mostly quartz flakes. The area of the distribution of finds is small (40–100 m^2), which suggests a short-term stay of limited groups of inhabitants.

The earliest sites are typically positioned on marine terraces 38–91 m high; according to the geological data, the terraces formed no later than the Preboreal (Armand et al., 1963; Nikonov, 1966; Armand and Armand, 1965; Dolukhanov, 1971). The raw material was mostly quartz of good quality, often milk-white or transparent, less common were silicified shale or quartzite, and rarely, flint of average quality. Comparatively scarce tools found on those sites include oblique end scrapers or end scrapers with a curved outline, scrapers with fine nibbling retouch at the very edge; dihedral angle burins, multiple burins on flakes; rare borers, drills, knives, adzes, and large cutting tools on quartz pieces. Noteworthy are relatively abundant multipurpose tools, most often a combination of scraper and burin. In spite of a general scarcity of inventory, the number of burins exceeds, or at least is equal to, that of scrapers. A small number of chisel-like tools were found, treated on one end only at one surface. More diversified are arrowheads. The most common are asymmetrical tanged; subtrapezoidal with transverse edge; transverse on elongated flakes; rhomboidal; and symmetrical tanged (stemmed). All the types are noted for the most thorough treatment of stem and less careful of blade and tip (point). Most often, they are treated by abrupt nibbling retouch, but sometimes, in case of quartz arrowheads in particular, the burin blow technique was used. The pieces with burinated edge are typical of the entire complex on the considered sites. Several amorphous cores are found, along with prismatic ones. The tools made of siliceous and quartzite rocks seem to be treated more carefully than those made of quartz. Such is a brief characteristic of the most ancient complexes of stone tools that we attribute to the initial stage of the Mesolithic on the Kola Peninsula.

The Middle Mesolithic sites occur usually at the same elevations as the aforementioned or somewhat lower: on marine terraces 33–45 m a.s.l. In the northern part of the region the sites are also confined to the sea coast and occupy a small area. The sites are similar to the previously described in the types of tools, raw materials, and technique of treatment. As for distinctive features, the tools are more accurately treated (in particular, end and side scrapers), chisel-like tools are more common, and the knapping seems to have been more accurate to obtain blanks of more regular form: blade or blade-like flake.

The more important feature of that stage was a territorial expansion of the archaeological site occurrence. Besides the north, they appear also in the western and eastern regions, and they occur not only on the sea coast but in river valleys as well. That is probably indicative of partial change and expansion of the economic activities of the site inhabitants. Some of the sites attributed to that stage are noted for a larger area, up to 300–500 m^2, and a corresponding increase in quantity of finds. Having moved into new regions, the Mesolithic people sometimes found new raw materials (such as silicified shale in the area near the town of Nickel, Western region, Shuoniyoki-Kolasyoki group of settlements). That encouraged development of a new technique of knapping and trimming and resulted in the appearance of new types of tools.

Thus, new types of scrapers appear, six of them, instead of two at the early stage: end scrapers with a wide edge, nosed, subrectangular, and subtriangular. The proportion of scrapers increases also; they dominate over burins, the latter becoming more diversified. Dihedral burins (type I) are not so definitely prevalent; the proportion of angle burins (on flakes, blades, and blade-like flakes) is growing.

Along with arrowheads of old types, there are subleaf-shaped and those of irregular (casual) configuration. Though asymmetrical tanged arrowheads are still dominant, others also increase in number; that is true of rhomboidal, subtrapezoidal transversal, and transversal arrowheads on elongated flakes. An amount of cutting tools, knives, drills and borers, and chisel-like notches also grows. Some sites display trapezoid blades, along with irregular blades and prismatic cores increasing in number; dominant though is still flake technique.

At that time, there were known specialized workshops for initial splitting of silicified quartz (Shuoniyoki VI) at the outcrops of that valuable raw material. At the same time, some traditions in stone toolmaking are visibly inherited from the previous stage; among those traditions are the usage of the burin blow technique and partial "occasional" retouch. The raw material mainly used is good-quality quartz, which is typical for most of the sites.

So, the sites of the Kola Peninsula attributed to the Middle Mesolithic stage are noted, on one hand, for keeping many features of the earlier stage in toolmaking, and on the other for the development of new procedures and new types of tools that would be further improved in the subsequent period.

Sites of the late stage of the Mesolithic are recorded in all the regions mentioned earlier and occur both on the sea coast and on the coast of lakes and river banks. Along with small-sized, short-term camps, typical in particular to the Rybachiy Peninsula (20–40 m a.s.l.), there are larger (up to 800–1000 m^2), presumably permanent habitations (Figs. 1.4.6 and 1.4.7). Inventory collected from them amount in some cases

Figure 1.4.6 Typical location of Mesolithic sites on the Kola Peninsula. Kil'din Strait coast. The Late Mesolithic site of Zarubikha 2.

Figure 1.4.7 Typical occurrence of Mesolithic artifacts in the absence of soil. The Late Mesolithic site of Zarubikha 2.

to a few thousands, with the tool number exceeding 300–400 pieces. By that time the Mesolithic people on the Kola Peninsula had mastered the technique of quartz processing and made it as close as possible to processing flint and siliceous shale.

A new type of arrowheads appears: triangular, straight-based ones. The arrowheads of the Late Mesolithic are distinguished for a greater degree of retouching of the entire tool, and of its stem and point in particular. On the whole, a distinct progress is seen in the technique of raw material knapping (as seen from increasing proportion of prismatic and conical cores), in the lithic technology and the typological diversity. At the same time, many types of implements show inherited features that suggest a gradual evolution of the stone industry, without any special impact from the outside. It should be noted that sites on the sea coasts still existed at the same time as those on rivers and lake coasts; in all probability, the inhabitants of the marine sites retained then their lifestyle from the previous stage. The results of archaeological studies of the stone industry obtained from the Mesolithic sites on the Kola Peninsula and the data on the site topography provide a reconstruction of subsistence of the Mesolithic people in the region.

At the end of the Preboreal to the beginning of the Boreal (transition from the 8th to 7th millennia BC), small groups of early people arrived to the north of the Kola Peninsula (Rybachiy). They founded their settlements on the high marine terraces dated to the earlier Late Glacial transgression (Dolukhanov, 1971). Small size of the sites and the fact that they are confined to marine terraces in bays of the Barents Sea suggest that they were inhabited by very small mobile groups (hardly more than 10–15 persons) of hunters and gatherers of seafood. It seems that the principal means of subsistence was collecting marine fauna found dead during storms and brought into bays, as well as gathering sea mollusks on littoral at low tide. In addition, the inhabitants of the sites could gather eatable things on land or hunt wild reindeer. However, the latter could not be the main resource, as reindeers were migrating to the north only in summer, and the sites were inhabited all the year through.

The lack of the early sites in other parts of the Kola Peninsula suggests the early people to be rather well adapted to their chosen way of living in that region. The archaeological materials, location of the sites in a certain altitudinal interval (38–91 m a.s.l.), and the presence of similar dated sites in Northern Norway (Freundt, 1948; Odner, 1966; Dolukhanov, 1971; Siiriainen, 1971; Indrelid, 1976) allow dating the oldest Mesolithic settlements to the 8th–7th millennia BC (Boreal).

Later, in the 7th–6th millennia BC the Mesolithic people spread toward the southwest and southeast. Part of the groups continued to practice mainly marine food gathering and so kept to the sea coast (sites Chernaya I; Lumbovka III, Ambarnaya I, II, 22–45 m a.s.l.), while another part began to colonize river valleys in the interior regions and proceeded to hunting and fishing (Kolasyoki III, Shuoniyoki I–VI). It should be noted that the groups considered had similar types of tools and used similar procedures in their making. It is quite possible that the population did not grow in number, and its expansion onto a larger territory may partly account for the reduced number of sites on the Rybachiy Peninsula.

The Late Mesolithic sites are considerably more widespread. They are found in all the regions of the Kola Peninsula, both on sea coasts (at 20–40 m a.s.l.) and in river valleys and on lakes. An individual site area is growing (which suggests more stable settlements); the inventory and types of tools increase in number. So, by the end of the Mesolithic (the 6th millennium BC), various regions of the Kola Peninsula had been successfully colonized along with an increase in number of sites, which suggests the growing population.

A specific feature of the distribution of the earliest sites (their concentration in the extreme north and the absence farther south), as well as the direction of the archaeological objects' advancement from north to south, are in good agreement with recent geological and geomorphological data on the history of the last glaciation, late glacial, and postglacial time on the considered territory and adjacent regions (Koshechkin et al., 1973; Armand and Nikonov, 1963; Nikonov, 1966; Armand and Armand, 1969; Dolukhanov, 1971; Holtedal, 1960; Strelkov, 1971; Lavrov, 1977; Vakorin and Kuptsova, 1975; Blazhchishin and Kvasov, 1980; Elina, 1981; Armand, 1961).

The problem of the ways and time of the earliest people coming to the Kola Peninsula is closely related to the Komsa culture origin and ways of the human penetration into the north of Fennoscandia.

From the very moment of the discovery of the Komsa culture, many specialists proposed various hypotheses aimed at elucidating this question (Nummedal, 1929; Freundt, 1948; Luho, 1956, 1967; Odner, 1966, 1970; Gaustad, 1969; Zemlyakov, 1940; Gurina, 1961, 1971; Pesonen, 1978; Meinander, 1982; Clark, 1975; Pankrushev, 1978, 1980). Some of them are mostly of historical interest [in particular, the hypothesis of the preglacial age of the culture and the people migration from the interior of Asia (Nummedal, 1929)]. At present, there exist two points of view. First is the Zemlyakov's hypothesis elaborated by V. Luho and G.A. Pankrushev and supported by Meinander (1982). According to the hypothesis, the Mesolithic people came here by way of Karelia and Finland. It should be noted, however, that the author of the idea, Zemlyakov, put forward his suggestion with caution and did not exclude a possibility of another way. His followers are much more categorical in their statements and confidently draw the early humans' way to Karelia, Finland, and Kola Peninsula from the east, from the northern Urals starting from the assumption of the kinship between the peoples living in those regions (Pankrushev, 1978, 1980).

There are many vulnerable points in this hypothesis, such as the absence of sites not only older, but even of the same age east of Karelia; the absence of intermediate early sites between the Karelia and Finland on one side and the Kola Peninsula on the other; and differences in the stone inventory types and categories (let alone the inhospitable environments that could prevent humans coming from the south).

More justified is the hypothesis advanced by Freundt (1948), who considered it possible that the Komsa people came from the south by way of the western coasts of Norway. At present, many West European scientists share that opinion. In Russia, similar ideas have been put forward by Gurina (1961, 1971, 1973) and Khlobystin (1998), though in their opinion the people moved from two regions: from the northwest and from the south. A similar concept was advanced by A. Hagen.

Taking the hypothesis advanced by E. Freundt as a basis, we can reconstruct the process of evolution of the Late Paleolithic and Early Mesolithic cultures in Northern Europe. It is along that line that investigations are carried out at present in archaeological institutions of Norway, Finland, and Sweden. In the process, some items formerly contradicting the theory got a plausible explanation. To take but one example, the main objection against the hypothesis was the Fosna culture (neighboring the Komsa on the south) being younger than the Komsa. Later, however, some newly obtained data proved the Fosna culture had come into being at the beginning of the 8th millennium BC at the latest (Mikkelsen, 1975); settlements in both cultures are mostly on the sea coasts, and the stone inventories have much in common.

At the same time, some difficulties are still to be overcome. In particular, the chronological analysis is hindered by the fact that many sites were occupied repeatedly; in the Fosna culture, it is not easy to distinguish assemblages of different age; chronological stages in the culture of Northern Norway are not distinct as yet (Odner, 1966).

The comparison between typology of the two cultures is also hampered by the different sources of raw materials: the Fosna culture abounds in flint tools due to its location close to the southern flint-containing regions of Scandinavia. Taking all the preceding in consideration, it is still possible to suggest that the Komsa and Fosna cultures are related and correspond to two stages ("waves") of migrations from a single region: most probably from the area of Ahrensburgian tradition sites (Pinneberg–Lyngby).

That suggestion agrees well with the presence of tanged and transverse-edge points and scrapers with beveled (oblique) edge. It is quite possible that the people began to move northward and eastward in the Preboreal–early Boreal. The reason for that movement was most probably related to the change in environments and crisis of the reindeer hunter economy. The resulting situation was not unlike to that of the Swiderian culture bearers.

It is quite possible that the migrations resulted in the formation of the cultures Fosna and Suomusjärvi north from the Komsa area. The difference between those consists in that the Fosna people came across the former strait from the southern coast of the Baltic Sea to Western Scandinavia, while the Suomusjärvi group traveled along the southern Baltic Sea into Finland.

The data available by now show that the Fosna and Komsa people radically changed their economy having turned from reindeer hunting to the exploitation of marine resources or even to sea mammal hunting. It is not inconceivable that the new way of getting sustenance appeared due to influence of prehistoric inhabitants of Denmark and of the North Sea islands who practiced marine food gathering. Some way or other, the people gradually colonizing the coasts of Norway and later the Kola Peninsula were already well acquainted with advanced marine gathering. Some common features in the stone industries of Komsa and Suomusjärvi cultures may be explained by the fact that peoples of both cultures had similar sources of raw material and became adapted to the new conditions.

Very soon, however, noticeable differences appeared in the technique of stone processing and in the use of new materials. The population of Karelia and Finland, for example, early came to usage of schist as a raw material and applied a grinding technique when making axes and adzes. An analysis of Mesolithic and Neolithic sites of the Kola Peninsula has shown that the Mesolithic industry was mainly based on quartz, while the usage of slate or shale for making tools involving an adequate technique was a Neolithic innovation. The regional differences may be attributed to a relatively isolated position of the Kola Peninsula, a specific orientation of the economy, and to characteristics of environments.

The Mesolithic sites of the Kola Peninsula form, therefore, a single unit and a constituent of the Komsa culture. The changes observable in the transition from the earlier to subsequent stages result from the internal evolution of the culture, its expansion onto new areas, and the usage of more advanced means of subsistence. On the whole the industry seems to evolve gradually; no obvious trace of the influence from southern or southeastern population groups has been recorded. The lithic industry is based almost entirely on quartz, while quartzite-sandstone, silicified shale, and chert are rarely used. The first two types of rocks are undoubtedly of local sources; the chert, however, is often supposed to be brought from afar, in particular from the eastern coast of the White Sea (near Arkhangelsk) (Gurina, 1961, 1971, 1973).

Meanwhile, the analysis of material has shown that the Mesolithic people used mainly good-quality quartz present in pebbles (found in abundance in beach ridges on terraces) and silicified shale found both as pebbles and in solid rock outcrops. An analysis of archaeological complexes with a large number of flint flakes and tools shows that 35%–70% of the entire inventory have typical pebble cortex and are very small in size. That suggests its local provenance. One of earlier investigators of the region, Zemlyakov (1940), mentioned such a possibility when considering the occurrence of flint in till on the northern coast of the peninsula (Zarubikha Bay).

As the culture evolves, there is a noticeable tendency toward better splitting technique to obtain more perfect blanks for tool production. As the quartz, when used as material in toolmaking, is hardly fit for making blades, the Mesolithic people continued to apply the flaking technique, though the flakes became longer. As for making the tools themselves, it may be concluded that the burin blow technique (that had come into being rather early) was gradually replaced by retouch (at the Late Mesolithic assemblages in particular), though it did not disappear completely.

On the Kola Peninsula the transition from the Mesolithic to Early Neolithic period was gradual and dated by radiocarbon to the beginning of the 4th millennium BC at the latest. The transition itself came about without any new groups arriving to the region.

Many things that had already been present in the Mesolithic industry evolved further during the Neolithic, though some new features also appeared. This is especially true in regard to the appearance of new raw materials, particularly slate, the latter never being found at the Mesolithic sites. Some new techniques appear in stone working, such as sawing, grinding, pecking, along with improved methods of getting sustenance.

Eastern Fennoscandia and the adjacent regions of the northwestern East European Plain

1.5

Paul M. Dolukhanov[1], Elena A. Kosheleva[2], Sergey N. Lisitsyn[3], Dmitriy A. Subetto[2]
[1]Newcastle University, Newcastle upon Tyne, Great Britain; [2]Herzen Russian Pedagogical University, St. Petersburg, Russia; [3]Institute for the Material Culture History, Russian Academy of Sciences, St. Petersburg, Russia

Chapter Outline

1.5.1 Introduction

The term "Fennoscandia" has been put forward by the Finnish geologist Ramsay (1898). It was chosen to designate a relatively isolated part of Europe bounded by isthmuses between the Gulf of Finland, Lake Ladoga, Lake Onega, and the White Sea, and characterized by the uniformity of geological structure and environments (Ramsay, 1902). Therefore, this unit includes denudation plains and uplands of the Baltic Shield, the Scandinavian upland of block type, and the structural Skåne plain in the south of the Scandinavian Peninsula.

The Baltic Sea basin developed against the background of the ice sheet retreat, isostatic uplifts, and the eustatic sea level fluctuations. The central part of the region

Human Colonization of the Arctic: The Interaction Between Early Migration and the Paleoenvironment
http://dx.doi.org/10.1016/B978-0-12-813532-7.00005-X

is occupied by the Baltic Sea itself, with two large gulfs: the Gulf of Finland and the Gulf of Bothnia. The region is unique in that its colonization of the prehistoric people has been represented since the Mesolithic.

The objectives of the present review may be stated as follows:

1. to reconstruct the process of the initial settlement of the region in changing environments;
2. to elucidate the general factors in the Fennoscandia area colonization.

1.5.2 General characteristics of the changes in environment during the Late Glacial

The development of the present-day landscapes of Fennoscandia depended to a great extent on the natural conditions in the Late Glacial and the Early Holocene, since the final stage of the last (Late Valday, Late Weichselian) ice sheet decay.

At the Last Glacial Maximum the entire region was covered by the Scandinavian Ice Sheet (Svendsen et al., 2004). In the interval 18,000–16,000 BP a warming of climate induced a gradual reduction of the ice volume and a consequent isostatic rebound. Resulting ice-free areas were suitable for colonization by plants, animals, and finally, by man.

The deglaciation, most of which fell on the Dryas, was not instantaneous, though rather short (in terms of geological time). The retreat of ice and disappearance of ice-dammed lakes proceeded at a varying rate in different regions. The ice retreated from the southwestern part of Norway as early as 18,000–16,000 BP; the northern coasts were ice-free between 13,000 and 12,000 BP (Anundsen, 1996). Southern Scandinavia was completely ice-free 13,500 BP (Björck et al., 1988; Larsson, 1996). In Finland the ice sheet receded northwestward, so the southern, southeastern, and eastern regions became ice-free about 10,000 BP (Matiskainen, 1996). Judging from the ^{14}C dates obtained on the peat and lacustrine deposits, the north of Sweden was free of ice approximately 8700 yr BP (Hörnberg et al., 2006).

An important feature of the postglacial time was a change of the zonality structure (Velichko et al., 2002a,b), which was gradually approaching its present-day latitudinal position. The process, however, was occasionally interrupted by the Late Glacial oscillations and periodic "cold waves," so the zone migration was a kind of reciprocating process. Among the factors that exerted control over environmental changes were growing mean annual temperatures along with a fast increase in moisture supply and a gradual afforestation of the territory (Khotinskiy, 1977). As seen from pollen assemblages, the areas getting free of ice became occupied by tundra plant communities with shrubs and the most easily adapting tree: birch (Spiridonova, 1991; Spiridonova and Aleshinskaya, 1999).

Shrub tundra with *Salix*, *Betula nana*, *Lycopodium*, *Sphagnum*, *Bryales*, and small spots of open pine and birch forests occurred at that time on the Scandinavian Peninsula north of 68°N. The south of the peninsula, between 61° and 67°N, was covered with a forest-tundra combination of pine and birch forests with *Populus*, *Juniperus*, *Hippophaë rhamnoides*, and tundra plant communities.

An arctic tundra-steppe with periglacial complexes and communities of herbs and dwarf shrubs (Elina et al., 2000) could exist on the ice-free area in the southeast of Karelia. The White Sea stood 65 m above the present-day level, but the water body occupied only the southern part of the modern sea basin (Devyatova, 1976). Ice-dammed lakes persisted in the southern parts of the Onega and Ladoga Lake basins (Kvasov, 1975). On the Kola Peninsula, tundra plains alternated with mountain and periglacial complexes. In Karelia, dominated by tundra plant communities, there were isolated birch open groves south of 65°N. Periglacial complexes occurred on the lake coasts. It is possible that birch forest-tundra occurred south of 62°N in combination with dwarf shrub tundra (Elina et al., 2000).

Since the Late Glacial time the Baltic depression has been occupied successively by a series of water basins. First, there were separate lakes including the arctic Lomma Sea (13,200–12,700 BP). In the course of time, some of the basins were drained, while others merged together and formed the South Baltic Lake (12,700–10,800 BP).

A single freshwater basin, Baltic Ice Lake (BIL), existed from 10,800–10,200 BP. It was fed by meltwater from the degrading Scandinavian Ice Sheet. The lake occupied the central part of the Baltic depression up to the Bornholm Island in the south. The Gulf of Riga and the Gulf of Finland extended eastward, and the lower part of the Daugava valley and eastern and northern Estonia were flooded. The northern coastline passed along the Salpausselkä morainic ridge, southern Finland and Karelian Isthmus were under water, and Ladoga Lake became a gulf of the BIL. In the west the water flooded central Sweden east of Billingen Mount. Occasionally the lake discharged through a channel that opened into central Sweden or through the Danish Straits. In the north of the Karelian Isthmus the BIL level was 70–80 m above the present-day sea level.

1.5.2.1 Holocene

The Preboreal period (10,200–8000 BP) was marked by the maximum of the Early Holocene transgression of the White Sea and of large lakes in Karelia. It is to that interval that high terraces of Onega Lake (60–85 m a.s.l., that is, 40–50 m above the lake level of today) are dated. The greater part of the Kola Peninsula was occupied by dwarf shrub tundra with *Ericales* and *B. nana*, and mountain tundra on uplands. Periglacial complexes were still widely spread. South of 66°N, there were open forests of birch with occasional pine trees. On the Maanselka Upland, forest-tundra alternated with mountain tundra (Fig. 1.5.1).

The rest of the region was covered with open birch forests, not unlike the northern taiga in appearance. In southern Karelia, pine-birch forests were dominant (Elina et al., 2000). At the interval corresponding to Pereslavl cooling (10,000–9300 BP), periglacial plant communities were partly restored (Khotinskiy, 1977). At the same time an activity of Icelandic volcanoes was recorded, ash layers Vedde, Borrobol, and Saksunarvatn having marked significant climatic events at the end of Pleistocene and in early Holocene.

The ocean water first penetrated into the Baltic basin in the Preboreal (10,200–9300 BP). The BIL discharged through channels that opened into central Sweden near Billingen Mount. about 8500 yr BP (9500 cal. yr BP) (Björck, 1995). As a result of the Baltic basin

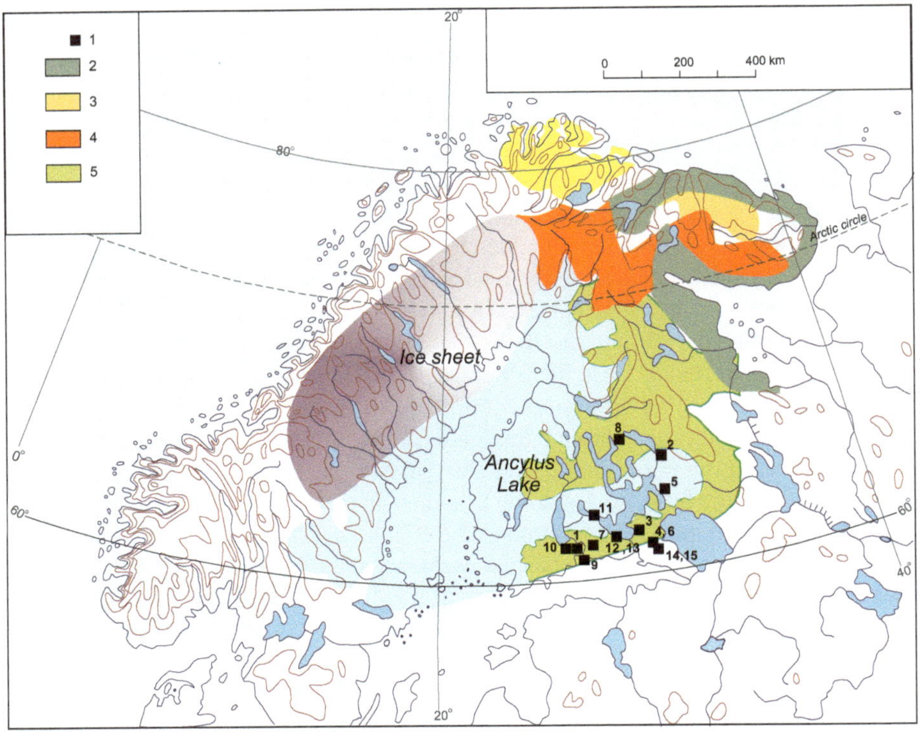

Figure 1.5.1 Paleogeographic map and Mesolithic sites in southeastern Fennoscandia. 1: Mesolithic sites; 2: areas of ancient seas and lakes; 3: shrub tundra with *Betula*; 4: birch forest-tundra; 5: northern taiga forest. Numbers on the map mark sites: 1, Orimattila Myllykoski; 2, Eno Rahakangas; 3, Joutseno Kuurmanpohja-Saarenoja 2; 4, Antrea Korpilahti; 5, Kirkkolahti 1; 6, Borovskoye Lake (Antrea Suuri Kelpojärvi); 7, Lahti Myllyoja; 8, Akunpohja Helvetinhaudanpuro; 9, Mäntsälä Pukinkallio; 10, Lahti Ristola; 11, Heinola Viikinainen; 12, Protochnoye 4 (Kaukola Rupunkangas 1a); 13, Protochnoye 5 (Kaukola Rupunkangas 3); 14, Veshchevo 10 (Heinijoki Valklampi 1); 15, Veshchevo 11 (Heinijoki Valklampi 2).

connection with the World Ocean, the level of the former dropped by 25–27 m, and the Preboreal Yoldia Sea came into being. The rapid discharge of the lake resulted in the emergence of a large area. The cold and rather dry climate of the Late Glacial changed to a warmer and wetter one, the mean July temperature rising from 4 to 10–12°C (Wohlfarth et al., 2006).

The climate warming, however, was a rather irregular process. The geological record shows a so-called "Preboreal oscillation" between 9300 and 9150 cal. yr BP: a short cooling supposedly resulted from large masses of cold water coming from the BIL into the Atlantic, which interfered with the air mass circulation.

The ice sheet persisted in the western part of Fennoscandia (west of the Gulf of Bothnia) for almost a thousand years (Khotinskiy, 1982); it created a high pressure area and blocked the transfer of the warm air mass eastward.

At ~9500–9300 BP the Närke Strait (central Sweden) that connected the Yoldia Sea with the ocean became shallow due to the isostatic rebound and was replaced by the so-called Svea River flowing from the Baltic basin. Having lost the two-way connection with the World Ocean, the Baltic basin gradually freshened and turned into a freshwater water body known as Ancylus Lake. It reached its maximum level in the Boreal, ~8700–8500 BP. At present, that level occurs in northern Estonia at ~35 m a.s.l., while near Rugen Island, its elevation is only 10 m a.s.l.

The Ancylus Lake level soon rose above that of the World Ocean, and a new spillway developed in the south. After 8300 cal. yr BP the lake regressed, and Lake Ladoga became isolated from the main water body. The buried peatbogs related to the regressive phase of Ancylus Lake have been thoroughly studied in many regions of the Baltic coasts; their dates suggest that the process of water level lowering was gradual and prolonged.

As the Baltic basin level became close to the level of the World Ocean (since 7200 cal. yr BP), salt water flowed into the eastern part of the basin, so came the stage of the *Litorina* Sea (Dolukhanov, 1969; Miettinen, 2002, p. 14; Subetto et al., 2002). Bottom sediments of the lakes on the Karelian Isthmus recorded the salt water penetration about 6500 cal. yr BP. The *Litorina* transgression related to the eustatic rise of the World Ocean level reached its maximum about 5600–4500 cal. yr BP (Miettinen et al., 2007).

1.5.3 Specific characteristics of environments and the initial colonization of the territory

During the Late Valday the ice sheet was the dominant factor controlling the environments of the East European Plain. The marginal zone of the ice sheet (Gerasimov and Velichko, 1982) occupied the drainage basins of the Neman and Zapadnaya Dvina (Daugava) rivers and the upper reaches of the Volga River. The climate was severe, with mean annual temperatures more than 10°C below those of present, and noted for its extraordinary aridity, strong winds, and winters with little snow (Velichko and Nechaev, 2005). The solar activity, however, was close to the present-day level, which was favorable for increasing phytomass volume. The dominant landscape was tundra-steppe: a hyperzonal type of landscape unknown at present.

The tundra-steppe was dominated by herbs and grass typical for steppe, though dwarf shrubs and other tundra plant communities were permanently present, along with mixed forest communities in places protected from wind: in balkas and on river floodplains. The fauna composition was also specific and known as the mammoth assemblage (Upper Paleolithic); it included mammal species typical to different natural zones: tundra (mammoth, wooly rhinoceros, reindeer, lemming, polar fox, musk ox), forest (hare, red deer, moose, beaver), and steppe (horse, bison, aurochs, saiga) (Agadjanian, 2001; Markova, 2005).

At the second half of the Upper Paleolithic the plain was inhabited by groups of mammoth hunters (Anikovich and Anisyutkin, 1995; Anikovich, 1998; Soffer, 1985). At the end of the glacial time the people found themselves in drastically changed environments: not only temperatures changed, but also the duration of cold and warm seasons, the regime of rivers, and number of rainless days.

No less (probably, more) important for the early humans than the environmental restructuring at the Pleistocene/Holocene boundary were changes in the fauna. The most conspicuous was the disappearance of the hyperzonal species—members of the mammoth assemblage—and their substitution for animals adapted to the new ecological niches in forests, forest-steppe, and steppe formed in place of former periglacial landscapes (Fig. 1.5.2).

The general increase in moisture supply and afforestation of the former tundra-steppe was the main cause of the mammoth extinction. After the mammoth—their principal competitor—had disappeared, the hoofed herd animals were able to sharply increase their population. At the first stage of afforestation, when open pine and birch forests developed along with meadow communities replacing tundra-steppe ones, the dominant species became reindeer and horse.

Judging from rare identifiable fragments of calcinated bones recovered from Early Mesolithic sites in southeastern Fennoscandia, there were typical forest animals, such as moose (Antrea, Helvetinhaudanpuro, Ristola, Veshchevo 10 sites), beaver, and hare (Ristola). Noteworthy is the constant presence of fish bones practically in all the settlements; fish remains are dominated by pike and pike perch, with perch and dace occurring in smaller quantities.

The fauna similar in composition is known from the Mesolithic sites farther south in the forest zone, such as Pulli, Kunda, and Sokolok (Timofeyev, 1993; Zhilin, 2004). The findings of the forest-tundra animals, and first of all of reindeer, in Early Mesolithic sites of southeastern Fennoscandia are still open to question; the reindeer has not been identified with confidence in any cultural layer there (unlike Lapland, where the reindeer was one of the main objects of the hunt) (Halinen, 2005).

Similarly, in the Mesolithic sites in the central East European Plain the reindeer occurs only occasionally. The earliest finds of its remains recovered from the third cultural layer of the Stanovoye 4 and the lower layer of the Ivanovskoye 7 sites were attributed to Mid-Preboreal time on the basis of palynological and radiocarbon dating results (Zhilin et al., 2002; Zaretskaya, 2005). It noteworthy in particular that the recovered remains of reindeer belonged to a forest variety of the reindeer, not related to a larger Late Glacial form (*Rangifer tarandus tarandus*), which became extinct at the end of the Pleistocene. The later form belongs to another subspecies (*R. tarandus fennicus*) that spread over the forest zone from the Urals westward and reached Fennoscandia by the beginning of Atlantic (Ukkonen, 2001).

The aforementioned archaeological and ecological data prove that the early man's dispersal northwestward was a natural consequence of the expansion of the ecological niche for boreal forest hunters and did not involve a colonization of a new region with completely different environments.

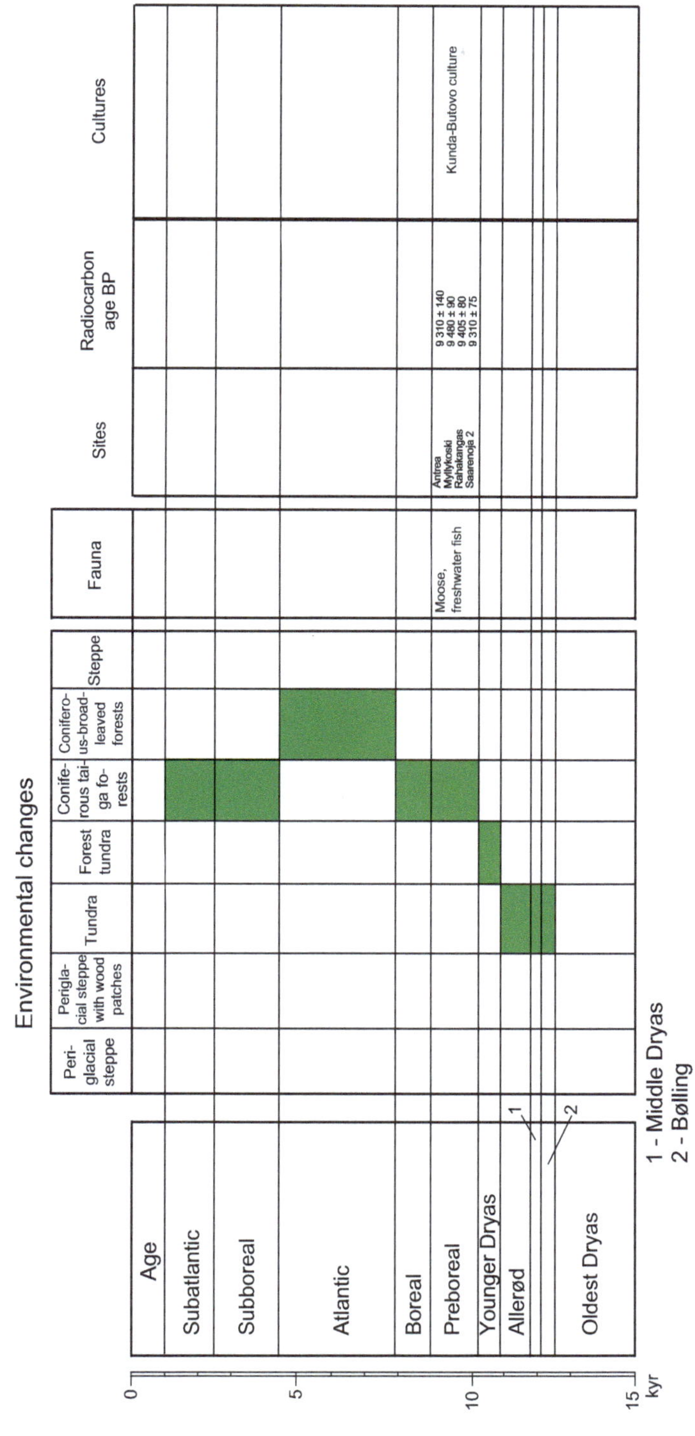

Figure 1.5.2 Environmental changes and archaeological cultures from the oldest Dryas to the Subatlantic period. 1: Middle Dryas; 2: Bølling.

1.5.4 Initial colonization of southeastern Fennoscandia

The time of the initial colonization of that region (Table 1.5.1; Fig. 1.5.1) is still a matter of argument. As noted by some authors (Bjerck, 1994, 1995; Bang-Anderson, 2003), the regions first having become ice-free remained uninhabited for a long time. The coastal zone in southwestern Norway became suited for human habitation a few thousand years before people first appeared there. A similar interval between the ice retreat and human arrival was proven for Finland (Matiskainen, 1996). Once the colonization process began, however, the hunter-gatherer communities settled the new lands promptly. The coastal regions of Norway were settled in 200–300 years (Bjerck, 1994, 1995; Thommesen, 1996); the arrangement of settled areas closely to each other suggests numerous small and highly mobile groups coming over the new land at a considerable rate (Bjerck, 1995). The rapid colonization may be attributed to the development of arctic marine technology and enhanced adaptivity.

The settlement and stable habitation of humans in those regions depended on a number of factors that distinguished the Early Mesolithic of southeastern Fennoscandia from the Mesolithic cultures in other regions. The great Ancylus Lake exerted, undoubtedly, a great influence on the coastal areas. Vast wetlands and dammed lakes account for the climate humidity during the Holocene. During the entire postglacial period the region was exposed to humid air masses and to arctic air coming from the retreating, but still close, ice sheet. The Ancylus Lake basin was gradually filling and subjected to isostatic tilting (due to isostatic rebound). That resulted in a long-term transgression and indentation of the coast line abounding in skerries and islands. Besides, southeastern Fennoscandia was separated from the continent by Heinijoki Strait (in place of the modern Karelian Isthmus), which prevented the inhabitants from a direct communication by land with the regions farther south.

Quite possibly, it was for that reason that the first settlers had to use water-borne transportation to a much greater degree than the dwellers of the forest zone on the East European Plain. In Scandinavia, boats were actually the only means of quick penetration into new lands. The wide usage of boats by those people is confirmed by findings from Antrea, which also indicate an importance of fishing in the Early Mesolithic of the region.

The constant presence of ichthyofauna on the excavated settlements suggests a well-developed fishing industry. According to the geochronology of the Early Mesolithic sites developed for southern Finland (Jussila and Kriiska, 2004), almost all the sites under consideration were related to the Ancylus Lake coastline, and half of them were positioned on islands (Fig. 1.5.2). In all probability, such sites were related to regular (likely seasonal) fishing; that conjecture is corroborated by a thick cultural layer with fish bones and remains of permanent pit-houses found on the sites. At the considered time interval the coasts of northern Norway were actively colonized from the west, and Southern Finland and Karelia adjacent to Ladoga Lake from the east (Dolukhanov et al., 2009).

Table 1.5.1 Absolute chronology of the early Mesolithic sites in southeastern Fennoscandia

No.	Name	Dates			Dated material	Location	References
		Uncal. yr BP	Lab. index	Cal. yr BC			
1	Orimattila Myllykoski	9480±90	Hela-552	9200–8550	Burnt bone	Skerry, Ancylus Lake terrace	Takala (2004)
2	Eno Rahakangas	9405±80	Hela-882	9150–8450	Burnt bone	Inland lake	Pesonen (2005)
3	Joutseno Kuurmanpohja-Saarenoja 2	9310±75	Hela-728	8750–8320	Burnt bone	Skerry, Ancylus Lake terrace, island	Jussila and Matiskainen (2003) and Takala (2004)
		7720±115	Hela-470	6721–6468	Charcoal		
		9477±57	Hela-2488		Burnt bone		Jussila et al. (2012)
		9425±56	Hela-2489				
		9438±56	Hela-2490				
		9411±56	Hela-2491				
		9431±58	Hela-2492				
		9385±57	Hela-2493				
		9350±75	Hela-758				
4	Antrea Korpilahti	9310±140	Hela-1303	8779–8379	Pine bark	Bottom sediments of the Ancylus Lake	Jungner (1979), Matiskainen (1996) and Takala (2004)
		9230±210	Hela-296	8795–8238			
		9140±135	Hela-404	8534–8223	Fabric of fishing net		
5	Kirkkolahti 1	9300±85	Ua-24774	8534–8223	Burnt bone	Island, Ancylus Lake terrace	Forsberg (2006) and Shakhnovich (2007)
6	Borovskoye Lake 1 (Antrea Suuri Kelpojärvi	9275±120	Hela-931	8675–8357	Burnt bone	Island, Terrace of the Ancylus Lake maximum	Takala (2004)
	Borovskoye Lake 2	9273±59	Hela-3163		Burnt bone		Unpublished
		9336±58	Hela-3164				

Continued

Table 1.5.1 **Continued**

No.	Name	Dates			Dated material	Location	References
		Uncal. yr BP	Lab. index	Cal. yr BC			
7	Lahti Myllyoja	9265±120	Hela-544	8612–8355	Burnt bone	Skerry, Ancylus Lake terrace	Takala (2004)
8	Akunpohja Helvetinhau-danpuro	9200±75	Hela-918	8610–8280	Burnt bone	Coast, ter-race of the Ancylus Lake maximum	Jussila et al. (2007)
9	Mäntsälä Pukinkallio	8960±65	Hela-706		Burnt bone	Island	Takala (2004)
10	Lahti Ristola	8880±75	Hela-727		Burnt bone	Skerry	Takala (2004)
11	Heinola Viikinainen	8840±90	Su-1710		Sled runner	Bottom sediments	Matiskainen (1996)
12	Protochnoye 4 (Kaukola Rupunkangas 1a)	8770±85 8130±65 7550±75 6595±55	Hela-1182 Hela-1187 Hela-1196 Hela-1195		Charcoal	Island	Mökkönen et al. (2007)
13	Protochnoye 5 (Kaukola Rupunkangas 3);	8740±80	Hela-1165		Burnt bone	Island	Mökkönen et al. (2007)
14	Veshchevo 10 (Heinijoki Valklampi 1)	8765±65 8720±70	Hela-748 Hela-744		Burnt bone	Island	Takala (2004)
15	Veshchevo 11 (Heinijoki Valklampi 2)	8720±70	Hela-744		Burnt bone	Island	Takala (2004)

1.5.5 Main archaeological sites of Southern Finland and of the Ladoga region of Karelia

1. *Orimattila Myllykoski* is a site in Orimattila municipality, at the outskirts of Virenoja village; it was discovered in 1998. It is located on the Ancylus Lake terrace associated with the transgression maximum (or slightly lower). In a test pit, H. Poutiainen found scarce quartz debitage, including cores, and calcinated bones. One of bones was dated by ^{14}C at 9480±90 BP (Hela-552)/9200–8550 cal. yr BC/, which is in good agreement with hypsometric position of the site (Poutiainen, 2002; Takala, 2004).
2. *Eno Rahakangas* is one of five points found in 2004 in the easternmost part of Finland, in Eno municipality, not far from Sarvinki village, on the coast of an ancient inner lake not connected in any way with the Ancylus basin. On the plowed surface of the site, beside a depression marking a former dwelling, there were collected quartz flakes and calcinated bones of animals. Among the faunal remains, pike bones were identified and those of a large, hoofed animal, moose, or reindeer. The date obtained on bone—9405±80 BP (Hela-882)/9150–8450 cal. yr BC—corresponds to Early Mesolithic. The same age may be seemingly attributed to a micro-sized end scraper on a flint blade found on the surface of the same terrace and described by Pesonen (2005).
3. *Joutseno Kuurmanpohja-Saarenoja 2* is one of 16 sites discovered in 1999–2000 in Joutseno municipality, not far from the town of Imatra (Jussila and Matiskainen, 2003). The site occurs on the Ancylus terrace, 8 m above the uppermost level of the transgression. A well-preserved cultural layer was found in a few test pits by T. Jussila in 2000. The findings include calcinated bones and quartz artifacts, as well as a few pieces of flint flakes (12 pieces), blades, and microblades (11 pieces), among them a retouched inset, a core, and a burin on blade (Takala, 2004). The cultural layer also yielded a broken point made on bone and belonging, in the opinion of T. Jussila, to the Shigir type (Jussila and Matiskainen, 2003).

   Two ^{14}C dates were obtained on a fragment of calcinated bone, 9350±75, Hela-728 (8750–8320 cal. yr BC), and on charcoal, 7720±115, Hela-470 (6721–6468 cal. yr BC) (Takala, 2004); the latter is considered to be invalid. New excavations were performed in 2008–2010 on an area of 55 m^2. The number of artifacts recovered was 1702, including 792 pieces of quartz, 870 of flint, and 40 of other rocks. Among the flint pieces, no less than 65% were made from rocks of Carboniferous age, the rest were of Cretaceous (Jussila et al., 2012). Small-sized flint debitage included three arrowheads of the Pulli type, insets on microblades with marginal retouch, end scrapers on flakes, and an adze (Jussila et al., 2012). A series of eight dates on calcinated bones give the age of the Saarenoja 2 site between 9310±75 BP (Hela-728) and 9477±57 cal. yr BP (Hela-2488).
4. *Antrea Korpilahti* is a site on the Karelian Isthmus that has been known since 1914 (when the territory belonged to the Antrea municipality of Finland). The discovery was made in the process of swamp melioration, the unique finding occurred in the Ancylus bottom sediments at a depth slightly more than 1 m. The excavated objects included floats made of pine bark and sink stones (pebbles) together with remains of fishing net made from willow bast (Fig. 1.5.3) (Palsi, 1920). At the same place, some stone and bone artifacts were recovered (mostly from the dump piles of a recently dug ditch), among them quartz flakes, five partly ground hand axes and adze-like tools made of diabase and slate, and a few fragments of grinding plates. There were also 23 objects (including fragments) made of moose bone and antler (dagger with one slot, chisel-like and adze-like pieces, a willow-leaf point) (Palsi, 1920; Clark, 1975; Zhilin, 2002). At present, most of the specialists agree that the found objects

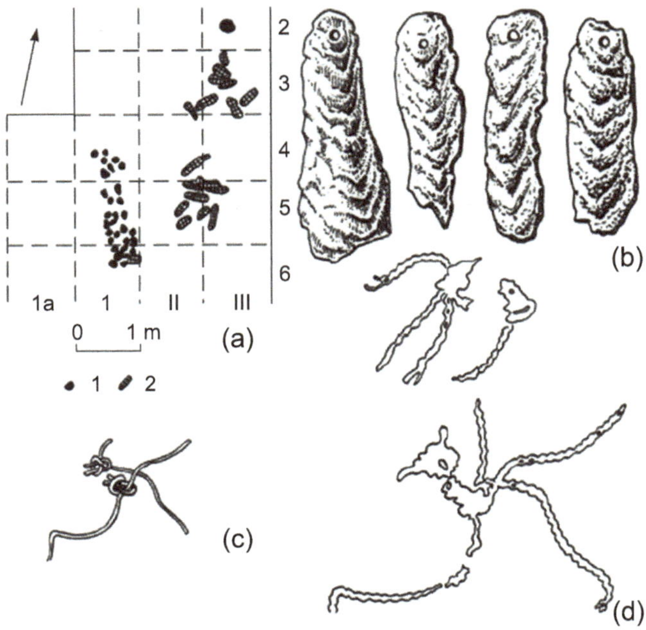

Figure 1.5.3 Fishing tackle remains from the Antrea Korpilahti site (Palsi, 1920): (a) excavation plan (1, sink-stones, 2, floats); (b) floats; (c) knot reconstruction; (d) fragments of a fishing net.

were from a shipwrecked boat of a Mesolithic fisher (Carpelan, 1999). The radiocarbon dates obtained on the bark that the floats are made of are as follows: 9310±140 BP, Hela-1303 (8779–8379 cal. yr BC), 9230±210 BP, Hela-269 (8795–8238 cal. yr BC) (Jungner, 1979; Matiskainen, 1996). The dates are in good agreement with that obtained on the fiber of the fishing net: 9140±135 BP, Hela–544 (8534–8223 cal. yr BC) (Takala, 2004).

5. *Kirkkolahti 1* is the site where a thick Early Mesolithic cultural layer was discovered in 2005 on Jänisjärvi Lake (Sortavala district of Karelia) (Forsberg, 2006; Shakhnovich, 2007a,b). The site is related to the terrace dated to the Ancylus transgression maximum. Excavations by M.M. Shakhnovich in 2005–06 over an area of more than 40 m^2 yielded a collection of about 3000 calcinated bone fragments and more than 27,000 stone artifacts including 26,881 pieces of quartz, 106 of flint, 62 of slate, and 17 of lydite. Among quartz pieces with traces of secondary treatment, there are abundant bipolar cores and those with a striking platform, end scrapers (including large and carinated ones), and knives. Burins, points, and arrowheads made of quartz are rare and typologically amorphous. Objects made of soft stones are diversified. Among tools made of slate, there are an axe and chisel with ground cutting edge, preforms and fragments of chopping tools, eight grinding stones, and one "hoe" or icepick. There are four perforated objects of sandstone: clubs, flattened in profile, processed to a different degree. Of flint objects, there are flakes and chips (94 pieces), broken atypical side scrapers (two pieces), two blades, blade sections including retouched ones (four pieces), one cutting tool, and three flakes bearing traces of usage (Shakhnovich, 2007a,b). Faunal remains are tentatively determined as belonging to large mammals and fish. One radiocarbon date obtained on the calcinated bone from the cultural layer is 9300±85 BP, Ua-24,774 (8534–8223 cal. yr BC) (Forsberg, 2006).

6. *Lake Borovskoye I* (*Antrea Suuri Kelpojärvi*) is the site on the Karelian Isthmus near Lake Borovskoye, Krasny Sokol village, Leningrad Region. The quartz debitage locality was discovered in 2000, on the lake terrace attributable to the Ancylus transgression maximum (Jussila et al., 2007). The site was seriously damaged by gutter trenches. In 2004, H. Takala, H. Poutiainen, and S.N. Lisitsyn cleared up the wall of a trench and made an excavation of the poorly preserved cultural layer, collecting also the surface material. The surface finds included 11 quartz flakes and small-sized calcinated bones (Takala, 2004). One of the bones was dated by ^{14}C to 9275 ± 120 BP, Hela-931 (8675–8357 cal. yr BC).

   In 2009, S.V. Aleksandrov found yet another locality with quartz inventory, Lake Borovskoye 2, on the same Ancylus terrace 30 m high on the northern coast of the lake. As the locality was in the zone of a railroad under construction and had been damaged already, S.N. Lisitsyn and A.A. Bessudnov carried out salvage excavations in 2012. An area of 1216 m^2 was opened at the northern periphery of the site damaged by construction equipment. Several concentrations of findings related to hearths (fire places) were studied, and 3536 quartz artifacts, mostly debitage, were collected. Among the tools, there were end scrapers and one chopping tool. Radiocarbon dates on calcinated bones are 9273 ± 59 yr BP, Hela-3163 (8675–8357 cal. yr BC) and 9336 ± 58 yr BP, Hela-3164 (8765–8425 cal. yr BC).
7. *Lahti Myllyoja* is a multilayered site 5 km south of the city of Lahti (Finland) that was discovered in 1998 at the level of the Ancylus transgression maximum. The first excavation on a small area performed by H. Poutiainen in 2001 showed the presence of Early Neolithic materials (Sperrings ceramics), Corded Ware Ceramic culture, and the Mesolithic culture. The inventory includes cores, flakes, and quartz pieces with secondary treatment, as well as grinding material and flakes of other rocks. Of several ^{14}C dates, one belonged to Iron Age, two were Neolithic, and one was Early Mesolithic: 9265 ± 95 yr BP, Hela-544 (8612–8355 cal. yr BC) (Takala, 2004).
8. *Akunpohja Helvetinhaudanpuro* is a site in Juankoski municipality, Savo Region, eastern Finland, which was discovered in 2003. In 2004–06 an area of more than 60 m^2 was excavated under the direction of Jussila et al. (2007). In common with the preceding, the site is positioned on the terrace related to the maximum stand of the Ancylus transgression. It contains a single Early Mesolithic cultural layer and remains of at least one rectangular pit-house.

   The collection (from excavations of 2003–05) consists of at least 10,880 stone artifacts: 10,859 are of quartz, 6 of flint, and 15 of other rocks. Dominant among quartz pieces with traces of secondary treatment are bipolar cores and single-platform cores (173 pieces); of 133 tools, 129 are end scrapers and side scrapers, almost in equal proportion. Burins are found occasionally. There are also five anvils made of anisotropic rocks, an adze-like piece, and a hammer stone of the same material. Flint pieces include two flakes, one scraper fragment, and two fragments of retouched insets made from flakes (Jussila et al., 2007). Faunal remains are rare: 56 calcinated bone fragments of which 12 were identifiable and attributed to moose, pike, perch, and dace. The ^{14}C date of 9200 ± 75 yr BP, Hela-918 (8610–8280 cal. yr BC) is in complete agreement with the site altitude (Jussila et al., 2007).
9. *Mäntsälä Pukinkallio* is the site in the south of Finland, in the Mäntsälä municipality, northeast of the city of Helsinki, on the bank of the Ohkolanjoki River, which was discovered by T. Jussila in 2002. Judging from its hypsometric position, the site was founded on the northern coast of a large island that existed at the Ancylus time. A compact concentration of artifacts was excavated on an area of 39.5 m^2 by H. Seppä (Takala, 2004). The collection consists of 504 objects, mostly calcinated bones and 289 quartz artifacts. Among the latter, there are flakes, cores, and other artifacts, made mostly

using the bipolar technique (Takala, 2004, p. 160; Jussila et al., 2007). Of the faunal remains, fish bones were identified visually. The date on bone, 8960 ± 65 yr BP, Hela-706 (8229–8008 cal. yr BC), corresponds to the Ancylus transgression peak and beginning of regression (Takala, 2004).

10. *Lahti Ristola* site is situated in Renkomaki at the southern margin of the city of Lahti. Hypsometrically, it is at the maximum level of the Ancylus transgression. The site was discovered in 1970 in the process of a road construction. The salvage excavations carried out in 1970–71 over an area of more than 2250 m^2 yielded a collection of 23,500 items attributed to the Early Metal Age (Corded Ceramics culture) and to the Mesolihic (Takala, 2004).

In 1995–99, H. Takala reopened excavations over an area of 500 m^2 and recovered a collection of more than 33,500 stone artifacts and only 25 ceramic fragments (while their number excavated in the 1970s was 1569). In the opinion of Takala (2004), the recently obtained materials (unlike those of the earlier collection) are mostly Mesolithic, with some admixture of younger pieces, mainly in quartz inventory. Quartz artifacts account for the greater part of 1995–1999 findings (33,083 pieces), while flint artifacts are small in number (282 pieces); so are tools made of other rocks (276 pieces).

H. Takala analyzed all the stone findings, obtained from the old and recent excavations together. Quartz artifacts include 1746 cores, 51,516 flakes and chips, 2113 angular fragments, 27 insets, 1470 scrapers, 35 burins, 445 retouched pieces, 105 points, 33 knives, and 5 drills. The composition of artifacts made of flint is as follows: 181 flakes and chips, 81 blade and bladelets (including 49 retouched ones), 1 residual core, 16 scrapers, 14 burins, 1 point, 1 drill, 4 knives, 3 insets, 5 arrowheads, and 8 composite tools (Takala, 2004). Besides, nine ground chopping tools made of soft rocks were found. Of flint arrowheads (both intact ones and fragments) made on blades, noteworthy are forms typical of Butovo and Kunda cultures, including those of Pulli type (two pieces in fragments). Insets on microblades and end scrapers, as well as angle burins on truncated blades, also have analogs in the mentioned cultures (Zhilin, 2002). Unlike flint inventory, made mostly using platform technique, quartz pieces are usually made using the bipolar technique of knapping.

Faunal remains recovered in abundance from the Ristola site are mostly calcinated bones. Among them there were identified moose, beaver, deer (not to a species level), hare, partridge, duck, pike, pike perch, and a fish of the cyprinoid family. An absolute age of the Ristola Mesolithic assemblage is based on the date of calcinated bone from the 1971 excavation: 8880 ± 75 yr BP, Hela-727 (8172–7884 cal. yr BC). Besides, there are 10 more ^{14}C dates made on charcoal from excavations of different years. The age thus obtained varies from middle and late Neolithic to the Middle Ages and the radiocarbon "present days" (Takala, 2004).

11. *Heinola Viikinainen*. A unique artifact was found in 1934 when mining for phosphate-bearing silt from Lake Ala-Rieveli: a sled runner almost 2.5 m long carved of pine was taken from the water. Pollen analysis of the Ancylus bottom sediments sampled at a depth about 1.2 m permitted the attribution of the finds to the Preboreal/Boreal boundary. The radiocarbon age of the wood the sled runner is made of —8840 ± 90 yr BP, Su-1710 (8146–7804 cal. yr BC)—is in a reasonable agreement with the stratigraphic context of the find (Matiskainen, 1996; Takala, 2004). The found sled runner suggests a high level of transportation development in the Early Mesolithic; evidently, the pioneer settlers of the region could travel over great distance in winter.

12. *Protochnoye 4* (*Kaukola Rupunkangas 1a*) is a multilayered site near Protochnoye village, Leningrad Region, Karelian Isthmus, discovered in 2004. It is located on the terrace corresponding to the Ancylus transgression maximum in Heinijoki Strait (Mökkönen et al., 2007). In 2005, S. Bel'skiy, T. Mökkönen, and K. Nordqvist studied remains of a pit-house over an area of $6\,m^2$. There were identified several horizons with findings attributed to Early Metal Age (Textile Pottery culture), Late Neolithic (Pit-Combed Ware culture), and below them, Early and Late Mesolithic. Among artifacts (223 pieces) dominant are those of quartz, with some pieces of quartzite and rock crystal also present. The only tool found is a rounded scraper of quartz. Cores are represented by bipolar and platform varieties. There are quartz blades in debitage. The fauna composition (by identifiable bones, 115 of the total of 380, only nine fragments being recovered from Mesolithic horizons) is published *en masse*, without subdivision into horizons, and it includes a large, hoofed mammal (moose or reindeer, one bone collected on the surface), perch, lake herring, fish of salmon and cyprinoid families, pike, and dace (Mökkönen et al., 2007). The radiocarbon dates on charcoal from lower horizons of the site cover the entire Mesolithic and the beginning of the Neolithic period: 8770 ± 85 yr BP, Hela-1182 (8000–7650 cal. yr BC), 8130 ± 65 yr BP, Hela-1187 (7190–7040 cal. yr BC), 7550 ± 75 yr BP, Hela-1196 (6480–6350 cal. yr BC), and 6595 ± 55 yr BP, Hela-1195 (5570–5480 cal. yr BC). In the authors' opinion, there possibly are errors due to a partial mixing of samples (Mökkönen et al., 2007).

13. *Protochnoye 5* (*Kaukola Rupunkangas 3*) is the site 0.5 km south of the Protochnoye 4. It was also found in 2004 by a concentration of quartz debitage on the surface of Ancylus Lake terrace disturbed by forest cutting. Two pit-houses have been identified on the surface, one rectangular, another rounded in outlines. Stone implements were mostly of quartz and, in a lesser quantity, of quartzite, rock crystal, and flint. There were quartz flakes, cores, blades, and scrapers, with only one burin and several flakes being made of flint (Mökkönen et al., 2007). Of the calcinated bones collected on the surface, there have been identified fish of the salmon family and three bones of seal. A ^{14}C date obtained on the bone fragment, 8740 ± 80 yr BP, Hela-1165 (7880–7610 cal. yr BC), permits one to assign the site to the regression that immediately followed the Ancylus maximum (Mökkönen et al., 2007).

14. *Veshchevo 10* (*Heinijoki Valklampi 1*) is the site on the Karelian Isthmus near Lake Belen'koye and Veshchevo village, Leningrad Region. It was discovered in 2004 on the Ancylus terrace at a level 3–4 m below the transgression maximum. The site was studied by H. Takala, H. Poutiainen, and S.N. Lisitsyn in 2004. Though they could not identify a distinct cultural layer, an assortment of pieces was recovered from 11 test pits and collected on the surface. The total collection consists of 257 fragments of calcinated bones and 102 stone artifacts, including a broken adze (partly ground) of slate and a fragment of the same tool, 93 flakes of quartz and 2 of slate, 3 flakes of flint, 1 core, and 1 end scraper of quartz. There were several crumbs of unidentifiable ceramics in the surface collection. Among diagnosable bone fragments, there were identified remains of moose, seal, pike perch, and a fish of the cyprinoid family. The calcinated bones yielded the ^{14}C date of 8765 ± 65 yr BP, Hela-748 (7955–7706 cal. yr BC), which agrees reasonably well with the hypsometric position of the site (Takala, 2004).

15. *Veshchevo 11* (*Heinijoki Valklampi 2*) is the site discovered at the same time as Veshchevo 10 (2004) and positioned on the same lake terrace about 200 m west of it. In all probability, both sites belong to a single inhabited zone at the ancient coastline. No cultural layer was found in seven test pits (1 × 1 m). The finds included 93 fragments of calcinated bones and 125 stone artifacts, among them 113 quartz flakes, 5 those of slate, and 7 quartz cores. Of the identifiable calcinated bones, three were determined as belonging to seal and one as hare. The ^{14}C date on bone of 8720 ± 70 yr BP, Hela-744 (7904–7656 cal. yr BC) coincides with the date of Veshchevo 10 and indicates the time of the Ancylus regression (Takala, 2004).

1.5.6 Discussion

The Early Mesolithic sites discovered in the southeast of Fennoscandia at the turn of the century contributed noticeably to the general picture of the initial colonization of the region (Lisistyn, 2006b). The materials obtained showed that the process began earlier than it had been assumed–the first settlements appeared here at least in the middle–or in the second half of the Preboreal. That put an end to the chronological gap between the Mesolithic chronologies in southeastern Finland and the adjacent Baltic regions on one hand and those in the upper reaches of the Volga on the other; that inconsistence between two regions had compelled the specialists to invoke various ecological models to explain the time lag and local distinctions in Mesolithic cultures (Matiskainen, 1996; Schultz, 1996; Carpelan, 1999). As it turned out, the fluctuation of the Baltic Sea and formation of a large water body—the Ancylus Lake—was no barrier to the peopling of the vast northern areas; on the contrary, the transgression was actually synchronous to the active colonization of the Karelian-Finnish lake district.

The Early Mesolithic sites appeared at the time of the cardinal climate change and a complete disappearance of relics of the tundra-steppe vegetation. It is at that time that dense boreal forests began to spread over the region. The rare fragments of calcinated bones suggest the presence of typical forest animals being killed by the Early Mesolithic hunters, such as moose, beaver, and hare. Fish bones (pike, pike perch, as well as perch and dace) are also commonly found.

The environments of the Early Mesolithic sites reconstructed on the basis of the shore displacement chronology developed for southern Finland (Jussila and Kriiska, 2004) indicated that almost all the sites under consideration were related to the Ancylus Lake coastline, and almost half of them were on islands. A similar situation is characteristic of the Mesolithic in the basin of a nearby freshwater lake, Onega, where no early Holocene site has been found as yet (Filatova, 2006).

The active exploitation of water resources in the Early Mesolithic seemingly had not yet resulted in forming an ecological niche of specialized coastal hunters and fishers. Unlike the Late Mesolithic of the coastal Estonia where some sites of seal hunters are known (Kriiska and Lougas, 1999) and the coastal Suomusjärvi culture of western Finland (Matiskainen, 1989), in the Mesolithic of the considered region the seal bones have been found until now only in surface collections, or in a multicultural context (Veshchevo 10, Protochnoye 5).

It should be taken into account that at the Mesolithic-Neolithic boundary the people lived on the coasts of the *Litorina* Sea, and not of the freshwater Ancylus Lake. Seals inhabiting the region in Early and Late Mesolithic belonged to different species and were not equally valuable as an object of hunt (Ukkonen, 2002).

The seal penetrating into the Baltic most early was the ringed seal (*Phoca hispida*); it came at the Yoldia Sea stage and inhabited the Ancylus basin. Its descendants still persisted in Ladoga Lake and in Saimaa lakes in Finland after connections with the sea were broken due to the isostatic uplift of the land.

The Greenland seal (*Phoca groenlandica*) replaced the ringed seal and inhabited low-salinity water of the *Litorina* Sea. It died out in the Subboreal after the environments had changed once more.

Both species were not very valuable objects of hunt as they are rather small-sized animals inhabiting shallow water and not forming large seal rookeries. Hunting those seals may be effective in winter at ice holes (polynyas). The bones of both species (ringed and Greenland seals) are found in the archaeological context in the northern Baltic region; they are confined to layers dated not earlier than the end of Mesolithic–beginning of Neolithic (Kriiska and Lougas, 1999; Ukkonen, 2002; Mökkönen et al., 2007; Knutsson, 1999). The main object of hunt could only be the large-sized gray seal (*Halichoerus grypus*), the species forming large rookeries. It did not inhabit the northern Baltic in the Early Holocene and became widespread there in the *Litorina* time only. The gray seal bones are known from the Estonian settlements dated to the Early Neolithic to the Iron Age inclusively, as well as from some sites on the Aland Islands (Lougas, 1997; Ukkonen, 2002).

Another distinctive feature of the Mesolithic in southeast Fennoscandia is a radically different source of raw materials for the stone industry. Unlike sites in the forest zone dominated by flint pieces, the Mesolithic inventory consists here almost entirely of quartz. That was due to the almost complete absence of high-quality isotropic stone raw material in the region, while quartz is commonly found as veins in the granite-gneiss solid rocks and in the form of pebbles and cobbles in glacial deposits. As quartz (in distinction to flint) is not a metamorphic rock and may be broken down only in accordance with its crystal lattice, it is practically unfit for treatment using traditional techniques applicable to flint (prismatic knapping, pressure flaking). Quartz debitage is hardly like regular spalls and in most cases looks more like a result of crushing than that of splitting; it bears all the characteristics of the natural fragments, such as indistinct flake typology and poorly pronounced attributes of a preliminary preparation and faceting (Callahan, 1987).

The only effective way of quartz treatment is splitting on a hard anvil; a widely used technique derivative from such a treatment was bipolar splitting, when a flake is detached by mutually opposite impulses from hammer stone and anvil. Bipolar flakes, cores, and their fragments are quantitatively prevalent in any quartz industry, which results from an uncontrolled splitting of the crystalline raw material. The mentioned specific features of quartz impose restrictions on the typology of pieces subjected to secondary treatment. It is impossible to obtain regular bladed blanks for tools. One cannot apply such a technique as burin blow, pressure retouch, and bifacial technique to quartz (except maybe the rarest variety of quartz rock crystal). As a result, quartz tools are mostly different from those of flint; the only tools that coincide typologically with their analogs made of flint are scrapers, while all the rest of the tools (knives, points, burins, insets, chisel-like, and heavy-duty tools) are identifiable only using functional and technical-morphological criteria.

Nevertheless, the drawbacks of the quartz usage for the stone inventory would be balanced out for the early people in southeast Fennoscandia by the availability of this raw material, so they could reject faulty pieces.

Flint as material imported from afar was highly valued. The nearest outcrops of flint are at a distance of 300–900 km south and southeast of the northern coasts of Ladoga Lake and the Gulf of Finland (Jussila et al., 2007). There are but single tools of flint recovered from the Early Mesolithic sites. Varieties of imported flint that they are made of are rather easily classified according to their origin and may be traced to certain sources: outcrops of solid rock in the northwest of the East European Plain. There are at least three groups of flint distinguishable that may be named by convention: Belorussian, the Upper Volga, and Estonian.

The Belorussian flint originates from Cretaceous rocks; their outcrops are found in central and western Belarus and in the south of Lithuania (Kovnurko, 1963; Baltrunas, 2006). It is noted for a high quality, luster, and relatively transparency; in alkaline medium, it is subjected to the appearance of a milky patina. The color of the Cretaceous flint varies from black to brown and gray.

The flint from the Upper Volga originates from Carboniferous layers. The closest outcrops of the rocks are found on the Valday, in the upper reaches of the Daugava, and occasionally on the Vepsovo Upland, south and east of the Ladoga and Onega lakes (Kovnurko, 1963; Zhuravlev, 1982). The Carboniferous flint is highly diversified in appearance, from black and brown color to ocherous, magenta, and bright red, often with a spotty and stratified coloration. The quality also varies, from relatively homogeneous to fractured, with caverns and a considerable content of fossilized microfauna. The Carboniferous flint, as distinct from the Cretaceous one, is almost never transparent, and it is not subjected to patina.

Estonian flint originates from the rocks of which the Paleozoic limestone plateau (central Estonia) is composed. Most probably, it is derived from Silurian rocks (Jussila et al., 2007; Yurgenson, 1958). Its color is dark gray, occasionally with a greenish hue. The Estonian flint is of mediocre quality, absolutely opaque, and contains many micro-inclusions.

There are also, though exceedingly rare, artifacts made from silicified rocks attributable to the glacier activity. Some of them are small-sized pebbles brought by the glacier from the inner regions of Scandinavia. Others resulted from the scouring or plucking of rock fragments from the glacier bed in the zone of flint-bearing rocks, which then were redeposited in the end moraines.

An analysis of ratio between various sorts of flint in the Early Mesolithic site inventory, as well as studies of the artifacts made of that flint, are of fundamental importance in the search for cultural and economic relationships of the early settlers of southeastern Fennoscandia. The Carboniferous flint is most commonly found in the collections of flint pieces. For example, the artifacts recovered from Kirkkolahti 1 and Veshchevo 10 are made from the Upper Volga flint only. Most probably, the scraper collected from the surface at the Rahakangas site was of the same material. Objects made by primary splitting occur extremely seldom.

For example, of 106 flint objects found in Kirkkolahti 1 (excavations of 2005–06), only four flakes have traces of the surficial cortex typical of flint nodules, all the rest being of secondary removal (Shakhnovich, 2007a,b). Flint cores have not been found in Kirkkolahti 1 nor Veshchevo 10, and in Saarenoja 2, they are present as fragments.

In the richest of the Finland flint collection from the Ristola site (314 pieces), 85.7% of the finds are made of Carboniferous flint, and only 14.3% of Cretaceous; meanwhile, if we take tools only, 63% of them are of Carboniferous flint and 37% of Cretaceous. As to the arrowheads found there, they are mostly of Cretaceous flint. Among the rest of the tools, the ratio between tools made of Cretaceous and Carboniferous raw materials is approximately 2:3, the latter being prevalent (Takala, 2004). It should be taken into consideration that the collection includes also some later debitage attributed to the Corded Ware culture when only Carboniferous flint was used. In common with other sites, there is scanty evidence of primary knapping in the Ristola collection. Of the flint findings, tools and by-products (fragments) amount to 32.5% of the total, micro-debitage to 36.6%, flakes to 14%, blades and microblades to 9.9%, and technical scales and spalls to 7% (Takala, 2004). Only one residual core was found; it is less than 3 cm in size and bears traces of splitting on an anvil. In the entire flint collection from the site, the cortex fragments are found on three flakes and on one side scraper. All the features testify to the fact that the site inhabitants did not practice the toolmaking at the settlement, and all the objects had been brought there ready for use.

Out of six flint objects found in Helvetinhaudanpuro, four are made from Carboniferous flint, one from Estonian Paleozoic rock, and another one from Cretaceous flint. Except for the latter (which was possibly a secondary core rejuvenation flake), there is no evidence of primary knapping in the site: there are three fragments of tools, one flake, and one chip (Jussila et al., 2007).

So, as follows from the analysis of flint pieces recovered from the Early Mesolithic sites in southeast Fennoscandia, the pieces of that rare material were brought to the sites as blanks or tools, rather than as nodules or cores ready for knapping. The flint knapping on the place was not characteristic. A recovery of the flint tools was rare, and reshaping of tools was hardly used.

In Ristola, for example, only one burin spall was found, while 14 ready burins are present; intact or partly fragmented objects are dominant among tools, while composite forms are rare, and there are only six cases when broken tools were refitted (Takala, 2004). Characteristically, the first settlers could use the flint sources from all the three nearest flint-bearing regions. Most often used was a Carboniferous flint, while flint from Estonia was the rarest; it should be taken into account that the latter was identified in the archaeological materials only recently, and it is often mistaken for gray Carboniferous flint derived from glacier deposits.

The Cretaceous flint was found on two sites only. Besides, it is known from two finds of blade fragments excavated from multilayered site Veshchevo 2 (Tarhojenranta) on the Karelian Isthmus; its hypsometric position on the site permits one to date it to the time interval from the Mesolithic to the end of Neolithic. The rest of the flint inventory at Veshchevo 2 is of Carboniferous flint, though the presence of some objects made from Estonian flint is quite possible (Takala, 2004).

1.5.7 Archaeological cultures and local specific features

The sources of the imported flint are indicative of the vectors of cultural impulses in the context of the origin and dispersal of the first settlers in southeast Fennoscandia; that is especially interesting as the quartz materials are not actually so informative.

At least at three sites relatively abundant in flint pieces (Ristola, Kirkkolahti 1, and Saarenoja 2), distinct analogies may be traced to inventories of the adjacent regions. Judging from published blade-like flakes, all of them were obtained from prismatic cores, had thoroughly prepared striking platform, regular faceting, and straight profile (Takala, 2004; Shakhnovich, 2007b; Jussila and Matiskainen, 2003). As follows from the technological analysis of the flint materials from Ristola and Kirkkolahti 1, the blades had been removed from single-platform (unidirectional) cores.

Besides, blanks for microblades show traces of pressure flaking (Takala, 2004; Tarasov, 2007). All the mentioned characteristics are equally typical of the early Kunda and Butovo flint industry. The same features of cultural similarity are seen in the tool assortment.

The arrowheads from Ristola include both Kunda and Butovo types; the insets made of retouched microblades (both Kirkkolahti 1 and Saarenoja 2 have one such inset each) and angle burins on truncated blades are also typical of both cultures. Scrapers are both shortened end and side scrapers; in Takala's opinion, they are closer quantitatively to Kunda than to Butovo types (Takala, 2004).

It should be taken into account, however, that among scrapers recovered from Ristola, there are those from the later assemblage (mostly side scrapers). So it is quite possible that the flint pieces recovered from all the Early Mesolithic sites in southeast Fennoscandia may be related to the Kunda culture as soon as to the Butovo one. It is all the more so as the findings of characteristic Kunda points of the Pulli type made of Cretaceous flint are known from some Preboreal Butovo sites at the Volga-Oka interfluve (Zhilin, 2001), while pieces made of the Upper Volga flint are often present at the Kunda sites. Cutting tools of slate, partly ground, have their Early Mesolithic analogs in the Butovo culture, e.g., at Stanovoye 4 and Ivanovskoye 7 sites, as well as at the oldest site of the Veretye culture: Veretye 1 (Averin and Zhilin, 2001; Zhilin et al., 2002; Oshibkina, 1997).

The findings of perforated heads of clubs (bulawa) from the Kirkkolahti 1 site are the oldest in the region. For a long time, they were considered to be a specific type of pieces characteristic of the Suomusjärvi culture of Finland (Matiskainen, 1989). Such pieces were occasionally found on the Karelian Isthmus as well (Huure, 2003); besides, they occurred in situ in Ristola (Takala, 2004), and they were recovered from the Late Mesolithic layer of the Bolshoye Zavetnoye 4 site not far from Priozersk town on the Karelian Isthmus (Huure, 2003).

The club heads are distinct from small perforated stones that were more common in the Late Stone Age; the latter served most probably as sink stones for fishing nets. The club heads are rather massive (up to 20cm in diameter) and have a large, biconically drilled hole designed for setting on a club, and not for being tied. The pieces are various in outline, from spherical to flattened oval or amorphous, though figured items occur sometimes. They are known from the Mesolithic in Norway (as occasional findings), and they occur in materials recovered from coastal sites (Ballin, 1999). Outside Fennoscandia the perforated club heads (as fragments) were recorded in Butovo materials at Stanovoye 4 (excavation 3, cultural layer III) and Ozerki 17 (cultural layer IV), as well as in Veretye 1 (Zhilin, 2004, 2006; Oshibkina, 1997).

What the clubs were intended for is still not quite clear. As some of them are polished (ground) or even have carved ornaments, they were sometimes thought of as having a sacral function or usable as a weapon. However, the fact that they are found only at the coastal or island sites suggests rather their usage in fishing.

It seems quite possible, in particular, that they could be used for increasing the weight of wooden boat anchors. Under this assumption, it becomes quite understandable why they are concentrated in Fennoscandia: in this region known for repeated water level oscillations the boats were the most efficient means of transportation for a long time.

The quartz industry of the Early Mesolithic sites in southeast Fennoscandia is noted for some specific features giving thought-provoking information on its origin. When considering abundant quartz material from Ristola, Kirkkolahti 1, and Helvetinhaudanpuro, there is a noticeable proportion of single-platform prismatic cores of quartz and flakes detached from them. In Ristola, prismatic cores amount to 59% and bipolar to 41% (though most of the bipolar cores were recovered from that part of the settlement where there are no later findings).

Unfortunately, there are no published reports on the relationship between flakes from single-platform cores and those from bipolar cores, nor on the quantity of blade blanks (Takala, 2004). Judging from the prevalence of bipolar quartz cores in Mesolithic materials, the flakes made in bipolar technique are dominant there. Even in assemblages with a noticeable presence of prismatic cores, flakes without platform are quantitatively dominant.

In Kirkkolahti 1, prismatic cores have not been considered separately from bipolar ones, though there is a group of the latter (2.9% from excavation 2005) distinct for elongated proportions and bearing blade removal scars (Tarasov, 2007).

In the collection of 2005 analyzed by A. Yu. Tarasov, the relation between flakes obtained using bipolar technique and those produced by applying prismatic technique is 3:1 (50% and 17.3%, respectively, from the total number of flakes, except for micro-debitage). Among flakes more than 2cm in size, there is a group of blade-like quartz flakes (7.2%) mostly obtained by bipolar technique (Tarasov, 2007). The Helvetinhaudanpuro assemblage is comparable with Kirkkolahti 1 in relationship of the used techniques of quartz knapping: 56.6% of cores are bipolar, and 43.4 are prismatic single-platform ones; while in debitage, 69.9% of flakes bear signs of bipolar flaking and only 30.1% of prismatic. Blade-like flakes amount to 6.5% of the knapped quartz (Jussila et al., 2007).

In the quartz industry of the Early Mesolithic sites considered, the bipolar knapping is prevalent on the whole. The prismatic technique of flaking constitutes a small, though steady, percent of the total. About 7% of quartz flakes from Kirkkolahti 1 and Helvetinhaudanpuro have features on the blanks for blades: their length is two times greater than their width. Specific features of the quartz industry, such as the presence of prismatic knapping elements together with those of blade-producing technique, make the Early Mesolithic assemblages distinct against the background of the Late Mesolithic ones (without regard for micro-debitage comparison).

In the quartz collections of the Late Mesolithic, more than 85% of cores are treated using bipolar technique. Findings of rocks other than quartz are extremely rare (Schultz, 1990). Blades on quartz are not typical of the late sites in Finland, flake technique being dominant. The same trend is noticeable in the Late Mesolithic sites with quartz inventory in northern Karelia (Shakhnovich, 2007a). That may be indicative of a gradual—throughout the Mesolithic—degradation of large blade reduction technique.

So the presence of prismatic technique elements and morphologically definite blade-like flakes in the quartz inventory of the postglacial sites in southeast Fennoscandia suggests relic techniques that were more typical for flint treatment in the Mesolithic industries of the central East European Plain (Jussila et al., 2007). The assemblages with stone industry of definitely transitional appearance from "flint" assemblages to "quartz" ones have not been found as yet.

1.5.8 Conclusions

As follows from the preceding, the postglacial colonization of southeast Fennoscandia took place no later than the third quarter of the Preboreal, which coincides with the beginning of the boreal forest expansion over the region. The first settlers were typical hunters of the forest zone, with fishery playing a noticeable part in their economy. They were originally related to the Kunda-Butovo technocomplex, the latter having been a single cultural community through the Boreal, as indicated by the cross-cultural flint import and similar types of implements recovered from various sites.

Large distances could not prevent people from the northward expansion from habitable forested lands. They could move fast enough using boats in summer and sledges in winter, which is confirmed by findings in Antrea and Heinola Viikinainen. The sites closest to southeast Fennoscandia and noted for their similarity to the early Butovo in appearance are found in the extreme east of the Leningrad Region and in the west of the Vologda Region, including the Sheksna drainage basin (Lotova Gora and Listvenka 3b) (Kosorukova-Kondakova, 1991; Kosorukova, 1996, 2000) and the Mologa drainage basin–Kurevanikha V–VI settlement, the Mesolithic complex (Kosorukova, 2005). They yielded artifacts made completely of the Carboniferous flint; all the sites are within 400–500 km of the northern coast of Ladoga Lake and the Gulf of Finland. The closest Kunda sites are known in the south of the Leningrad Region–Sokolok site at Lake Zalustezhskoye (Timofeyev, 1993; Galibin and Timofeyev, 1993), and in Estonia–Kunda site near the southern coast of the Gulf of Finland (Jaanits, 1965; Gurina, 1989), and also Pulli on the Pärnu R (Jaanits and Jaanits, 1975, 1978; Kriiska et al., 2011). There are quartz as well as flint varieties (Estonian, Cretaceous from Byelorussia, and that from the Upper Volga region) present in the early Kunda site inventory; the distance between them and the Early Mesolithic sites in Fennoscandia is 200–350 km.

The first settlers of the region became soon adapted to the local environments and did not feel a shortage of raw materials; that is indicated by a relatively small proportion of pieces of flint and slate (mostly imported ready for use), as well as the dominance of quartz, which was of primary importance in the stone industry. The permanent habitation of people even at early stages of the territory colonization is suggested by the presence of sites different in economic specialization: seasonal sites (including those on islands) and long-term settlements with a thick cultural layer and dwellings.

North and North-East of the East European Plain

North of the East European Plain

1.6

Alexander V. Volokitin[1], *Yuriy N. Gribchenko*[2]
[1]Institute of Linguistics, Literature and History, the Komi Centre, the Urals Division, Russian Academy of Sciences, Syktyvkar, Russia; [2]Institute of Geography, Russian Academy of Sciences, Moscow, Russia

Chapter Outline

1.6.1 Evolution of landscapes and climate during the Late Glacial and the Holocene

Spatial distribution of the Mesolithic sites in the north of the East European Plain depended not only on the general trend toward warming at the Late Pleistocene–Holocene boundary, but also on the regional specific features of the landscape evolution. At present the East European North is characterized by a complicated mosaic-like pattern of landscapes where tundra and forest-tundra communities are interdispersed with middle and southern taiga forest stands. This inhomogeneity was also typical for many regions during the Late Pleistocene and Holocene (Grichuk, 1982; Velichko et al., 2008a,b; Khotinskiy, 1977; Khotinskiy and Klimanov, 2002; Spiridonova, 1983; Nikiforova, 1982; Devyatova, 1982).

The earliest settlement and active colonization of the European North is dated to the Mesolithic. Judging from the spatial distribution of the regions initially inhabited by prehistoric communities in the East European North, one could identify the northwestern (including Lake Onezhskoye and the Onega and Sukhona drainage basins) and the northern (the Severnaya Dvina, Vychegda, Mezen', Pechora drainage basins) regions. The regions differ both in the landscape structure and in the

Human Colonization of the Arctic: The Interaction Between Early Migration and the Paleoenvironment
http://dx.doi.org/10.1016/B978-0-12-813532-7.00006-1

pattern of the Mesolithic sites distribution (Neustadt et al., 1982; Dolukhanov and Khotinskiy, 2002; Kol'tsov, 1989; Oshibkina, 1983, 2006; Sorokin et al., 2009; Volokitin, 1997, 2006).

The Late Valday ice sheet expanded from the Scandinavia onto the northwestern regions making them practically inaccessible for Paleolithic communities not only during the Last Glacial Maximum, but also during the Late Glacial (Velichko, 2002a,b).

The relief and environments of the Northern regions were mostly predetermined by the Early and Middle Pleistocene ice sheets. These ancient glaciers were responsible for complicated transformations of the river valleys; they also controlled cycles of glacial, lacustrine, and marine deposition. Most important were processes related to the large continental ice sheets of the late Middle Pleistocene glaciation (named Dnieper-Moscow, Babushkin or Vychegda). As for the Early Valday (Kalinin) ice sheet, it is not inconceivable that it could expand over the region, though the chronostratigraphic data available are insufficient for identifying the glacial formations of that time with confidence (Moskvitin, 1976; Zarrina, 1991; Mangerud et al., 2004; Svendsen et al., 2004).

Reconstructions of the last Late Pleistocene glaciation (centered at Scandinavia) are based on the studies of the glacial formations, their stratigraphy, and chronology (Chebotareva and Makarycheva, 1982; Velichko, 2002a,b). Boundaries of ice sheets identified as belonging to different stages of the Late Valday (Ostashkov–MIS 2) glaciation included a considerable portion of the Severnaya Dvina drainage basin.

The formation of the modern relief in the northern region was laid at the time of Middle Pleistocene: the expansive Dnieper ice sheets covered the entire area; so the most widely spread glacial formation in the region is the till attributed to the maximum stage of that glaciation. Fluvioglacial and limnoglacial formations are mostly related to the last (Moscow, Babushkin) stages of the Dnieper glaciation. As a rule, they are found in deeply incised valleys and closed depressions. Lakes and wetland landscapes in the valleys and basins persisted for a long time, and genetically diversified deposits accumulated there serve as the main sources of information on the Late Pleistocene environments (Devyatova, 1982; Gey and Kotlukova, 1982; Khomutova et al., 1998).

Modern landscapes of the region are dominated by spruce taiga forests (northern and middle taiga) giving way to pine forests on sandy surfaces. The light coniferous forests of pine are most common in the east of the region. Its northernmost portion belongs to tundra and forest-tundra zones. Paleogeographic data on the Late Pleistocene and Holocene formations of the East European Plain provide the basis for estimating environmental prerequisites to the initial colonization of the northern plains by the early human communities (Grichuk, 1982, 2002; Khotinskiy and Klimanov, 2002; Nikiforova, 1982; Devyatova, 1982).

Over most of the region, large lakes interfered with the long-distance migrations of the Paleolithic communities during the Late Pleistocene. At the Late Valday time the Scandinavian ice sheet penetrated into the lower parts of the Severnaya Dvina and Mezen basins, with vast ice-dammed lakes at its periphery (Kvasov, 1975; Aseyev et al., 1982) (Fig. 1.6.1).

In the southern part of the Severnaya Dvina drainage basin, and in particular in the valleys of its tributaries, the Vaga, Vychegda, and Sukhona rivers, numerous

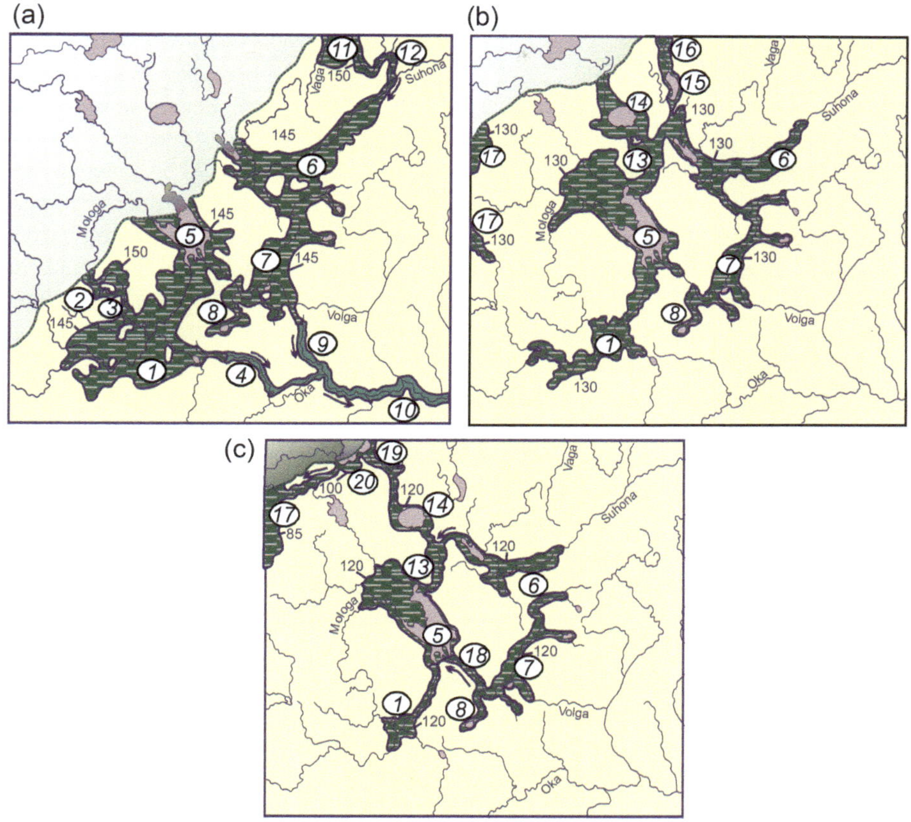

Figure 1.6.1 The ice-dammed lakes at the Late Pleistocene (Kvasov, 1975). (a) the maximum stage (20–18 ka BP); (b) Vepsovo stage (about 16 ka BP); (c) Luga stage (14–13 ka BP). *Numbers in circles*: 1, Lake Tverskoye; 2, Lake Upper Mologa; 3, spillway from Lake Upper Mologa; 4, spillway from Lake Tverskoye lake; 5, spillway from Lake Mologa-Sheksna; 6, Lake Sukhona; 7, Lake Kostromskoye; 8, Lake Rostovskoye; 9, spillway from Lake Kostromskoye; 10, drainage ways along the Klyazma, Oka, and Volga valleys; 11, Lake Volzhskoye; 12, drainage by way of the Sukhona valley; 13, Lake Upper Sheksna; 14, ancient Lake Beloye; 15, drainage way from Lake Sukhona; 16, Lake Vozhe-Lacha; 17, Lake Privaldayskoye; 18, Tutayev water gap; 19, drainage by way of Vytegra valley; 20, Lake Vytegra. *Numbers*, absolute altitude m a.s.l.

localities of large fauna have been discovered; the faunal remains belong mostly to the mammoth complex and suggest abundant herds of the mammals inhabited the territory or migrated there during the Middle and Late Valday (Smirnov, 1937; Devyatova, 1982).

However, only a few localities are known at present in the Sukhona and Vaga drainage basins where some Paleolithic stone tools occur together with large mammalian remains (Devyatova, 1982; Praslov et al., 2011). We may suppose that the Late Paleolithic man could occasionally penetrate into those regions, but at present, we have no factual materials indicative of the precise time and possible routes of his migrations.

The time interval close to the Late Pleistocene–Holocene boundary is noted for a series of climatic fluctuations and related changes in landscapes: relative warmings of Bølling and Allerød gave place to cooler intervals of Dryas (Dryas I – Oldest, Dryas II – Older, Dryas III – Younger) (Velichko, 1973, 1982; Grichuk, 1982). Later, forest formations began to develop on the East European Plain along with the process of permafrost degrading. At that time, considerable areas in the north of Eastern Europe were covered by shrub tundra including some periglacial elements and open woodlands of pine and birch with some steppe communities (Grichuk, 1982). Forests of spruce and those of pine and birch were confined to regions farther south; in the Sukhona, Yug, and Kama drainage basins, they penetrated as far north as 62°N.

Within the considered time interval the changes in landscapes and activation of depositional relief-forming processes were essentially influenced by the permafrost degradation. Polygonal ice wedge systems were widespread in the Severnaya Dvina drainage basin. The ice thawing at the Late Glacial and the Holocene resulted in development of residual polygonal block microrelief and active processes of eolian and deluvial deposition (Velichko, 1965, 1973; Velichko et al., 1982). This time interval (11–9 ka BP) was marked by a noticeable reduction in area of stable topographic elements (river terraces and interfluves) favorable for settlement and migrations of the prehistoric hunters. It is possible that it was due to the land surface destabilization that some of the northern regions have no signs of the early man's presence.

Certain changes in vegetation are recognizable in the Holocene pollen assemblages obtained from various regions of the North, in particular from the Mesolithic sites (Oshibkina, 1983; Volokitin, 1997, 2006; Nikiforova, 1982; Golubeva, 2008). In spite of a complicated mosaic structure of the northern landscapes and their changeability at the earlier part of the Holocene, one can easily see regularity in those changes well observable at the regional scale (Nikiforova, 1982; Golubeva, 2008). Taking into consideration chronological data available on the main Mesolithic sites, we can trace the main phases of the environmental evolution at the Early Holocene.

The Mesolithic people advanced into the region in the Holocene (Fig. 1.6.2), when forests had already covered the area (Dolukhanov, 2000, 2008; Dolukhanov and Khotinskiy, 2002). The Mesolithic communities extended considerably the area of their habitation and penetrated still farther in their migrations. The latter followed mostly traditional routes of the Paleolithic migrations northward along the Kama and Pechora valleys, though some new ways were laid in the Severnaya Dvina and Onega drainage basins as far as the forest-tundra and tundra boundaries. The harsh environments in the north did not present an insurmountable obstacle to the Mesolithic people migrations. The humans mostly used river valleys with stable terrace surfaces fully developed by the end of the Pleistocene–beginning of the Holocene.

The low terrace surfaces (the first and second) with meadow landscapes provided the Mesolithic communities with favorable habitats. During the Holocene, as the landscapes became dominated by forests, the Mesolithic people still preferred open habitats, though forest mammals (elk, deer, hare, etc.), as well as birds, occupied a prominent place among the objects of hunting. Fishing also gained in

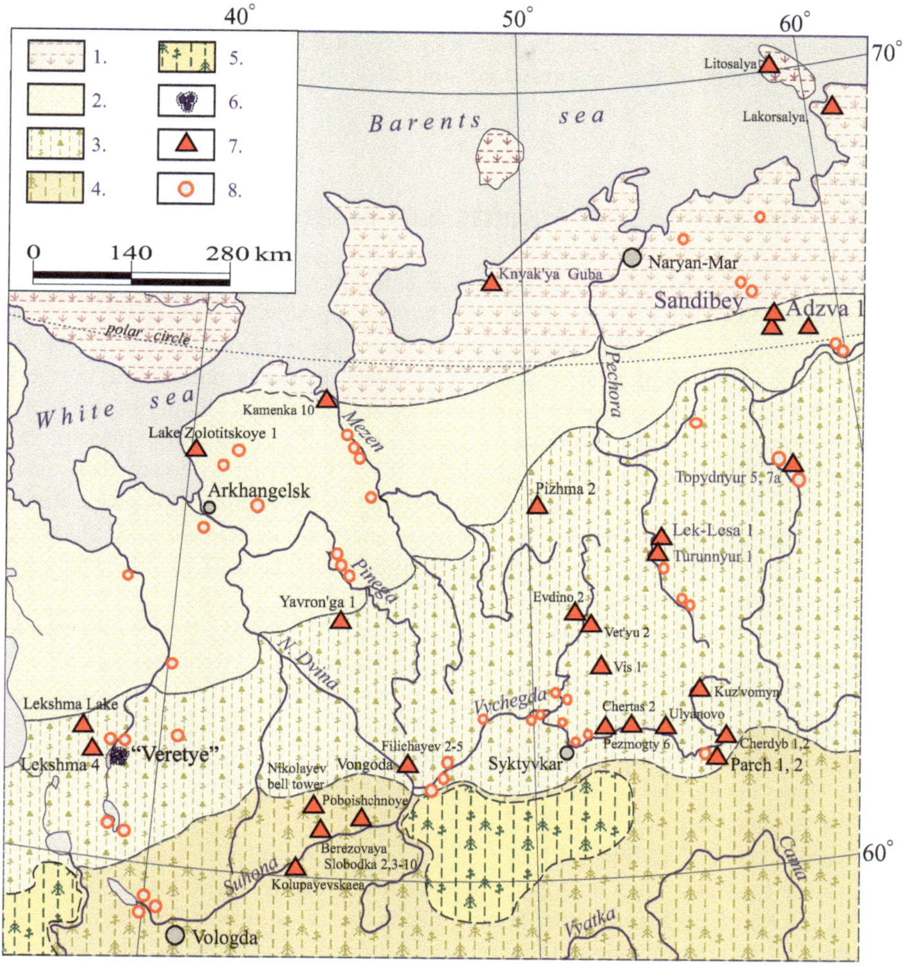

Figure 1.6.2 Mesolithic sites in the north of the East European Plain (according to A.V. Volokitin). The landscape reconstructions are shown according to (Grichuk, 1982, 2002; Khotinskiy and Klimanov, 2002). 1, tundra; 2, forest-tundra; 3, taiga; 4, southern taiga; 5, open forests; 6, the Veretye culture settlements (Veretye 1, Sukhoye, Nizhneye Veretye, Popovo 1, Lukinchikha, Peschanitsa and Popovo burials); 7, Mesolithic sites; 8, Mesolithic localities.

importance. Further north (in tundra and forest-tundra), humans hunted mostly herds of reindeer.

Some changes took place in the choice of places of long-term sites and habitations. Besides traditionally chosen river terraces, coasts of large lakes were also used (Veretye 1, Sukhoye, Yavronga 1, Pizhma 2). Cultural layers in the majority of sites occur close to the surface, at the base of modern (mostly acidic) forest soils; that led to a poor preservation of cultural remains and absence of bones and other organics (Volokitin, 1997). The best materials for paleoenvironmental reconstructions were obtained from the sites discovered in the context of peatlands, in lacustrine and floodplain deposits. Though few in numbers, they provided important information on the environment of the Mesolithic communities in the north of Eastern Europe (Table 1.6.1).

1.6.2 The oldest settlements and environments

The Mesolithic settlements are distributed equally enough over the entire territory under consideration (Fig. 1.6.2). Their total number amounts to 300 (Kuratov, 1978; Stokolos and Korolev, 1984; Chernov, 1985). There are localities discovered on the Arctic Ocean coasts (Lakorsalya, Knyak'ya Guba), including one on the northern point of Vaigach Island (Litosalya), and a few in the immediate vicinity of the Arctic coast: Kamenka 10, Lake Zolotitskoye, and others (Edovin, 2009; Edovin and Belichenko, 2009; Ivanov, 1993; Murygin et al., 2012). Only a few of them have been excavated, and in most cases, only prospecting and collecting took place. Better studied are Mesolithic sites located farther south and lying in the valleys of Sukhona, Severnaya Dvina, Vychegda, and Izhma and in the lake basins. It was there that the excavations were performed on large areas.

The most informative are sites at the Lacha and Vozhe lake basins referred to the Veretye culture (Veretye 1 and Sukhoye habitation sites; Nizhneye Veretye and Pogostishche sites; Peschanitsa and Popovo 1 burials) (Fig. 1.6.3). The presence of burial places (special places for burials outside the living sites) deserves a special attention as it is a rather rare occasion in Northern Eurasia. Both settlements and burials have been dated. The earliest dates (about 10 ka BP) were obtained on a skeleton from the Peschanitsa burial complex.

According to S.V. Oshibkina (1983, 1997, 2006), the typical Veretye culture is restricted to the lands lying east from the Onezhskoye Lake. As follows from materials on the Veretye sites, the area around Lacha and Vozhe lakes was permanently inhabited. In the rest of the region, there were only seasonal campsites attributable to various Mesolithic cultures.

1.6.3 The Veretye culture sites

The Veretye culture belonged to the Baltic group of archaeological cultures (Oshibkina, 1983, 1997). Its distribution was mostly restricted by the eastern coasts of Ancylus Lake and smaller water basins detached from it. The Veretye culture kept contacts with other cultures of this group (early phase of the Maglemöse, Kunda, early phase of the Onezhskaya, Suomusjärvi, stage Askola). The bearers of this culture occupied the same territory for a long time, and only at final stage, it expanded westward and southward.

The most important–and the earliest–settlement of this culture is Veretye 1 (see Fig. 1.6.3). Somewhat younger are Nizhneye Veretye, Pogostishche, and Sukhoye.

Table 1.6.1 Main Mesolithic sites in the north of the East European Plain

Site		Position					
Name	Type	Latitude	Longitude	^{14}C dates	Material of tools and implements	Faunal remains	References
Dated sites							
Veretye I	Long-term settlement	61°13′ N	38°58′ E	7.7–9.6 ka BP (11 dates)	Stone, bone, antler, wood	Abundant bones of animals, birds, and fish	Oshibkina (1983, 2006)
Peschanitsa	Burial complex	61°14′ N	38°53′ E	9.89 ka BP (1 date)	Flint and bone	Bones of animals, birds, and fish	Oshibkina (2006)
Popovo	Burial	61°13′ N	38°57′ E	7.62–9.52 ka BP (6 dates)	Flint and bone	Bones of animals, birds, and fish	Oshibkina (1983, 2006)
Visskiy 1 peat bog	Peat bog near settlement	62°48′ N	51°59′ E	7.09–8.08 ka BP (5 dates)	Bone and wood		Burov (1967b, 1993, 2009)
Parch 2	Seasonal site	61°36′ N	54°39′ E	9–9.5 ka BP (2 dates)	Flint, other rocks, bone	Bones of animals, birds, and fish	Volokitin (2006)
Sites with insufficient data							
Sukhoye	Seasonal site	61°12′ N	38°53′ E	The first half of the seventh millennium BC (?)	Flint, other rocks, bone, antler, wood	Animal bones	Oshibkina (1983, 2006)
Nizhneye Veretye	Multilayered site	61°13′ N	38°55′ E	The second half of the 7th millennium BC	Flint, other rocks, bone, antler, wood	Animal bones	Oshibkina (2006)
Pogostishche	Seasonal site	60°32′ N	39°00′ E	Transition from 8th to 7th millennium BC	Flint	Animal bones	Oshibkina (2006)

Continued

Table 1.6.1 Continued

Site		Position		^{14}C dates	Material of tools and implements	Faunal remains	References
Name	Type	Latitude	Longitude				
Yavron'ga 1	Multilayered site	63°24′ N	44°41′ E	8.5 ka BP (1 date)	Flint		Burov (1974a)
Lek-Lesa 1	Seasonal site	64°37′ N	53°21′ E	9 ka BP (1 date)	Flint	Animal bones	Volokitin (2005)
Pezmogty 6	Workshop site	61°48′ N	51°45′ E	~9 ka BP	Flint		
Cherdyb 1, 2	Seasonal sites	61°39′ N	54° 56′ E	7.5 ka BP	Flint	Animal bones	Pavlov et al. (2010)
Poboishchnoye 1	Seasonal site	60°28′ N	44°35′ E	9.5 ka BP	Flint		Adrianova and Vasil'eva (2000)
Berezovaya Slobodka 2-3	Seasonal site	60°22′ N	44°9′ E	9–9.5 ka BP	Flint		Ivanishcheva (2002)
Undated sites							
Filichayevski 2-5	Seasonal sites	61°28′ N	46°18′ E		Flint		Vereshchagina (2010)
Kolupayevskaya	Site with two Mesolithic layers	59°60′ N	43°00′ E		Flint		Oshibkina (1983)
Nikolayev "bell tower"	Seasonal site	60°30′	43°32′ E		Flint		Nedomolkina (2006)
Pizhma 2	Multilayered site	64°59′ N	50°14′ E		Flint		Luzgin (1973)
Adzva 1	Seasonal site	66°37′ N	59°26′ E		Flint		Kanivets (1973)
Turunnyur 1	Seasonal site	64°35′ N	53° 23′ E		Flint		Volokitin (1997)

Table 1.6.1 **Continued**

Site		Position		^{14}C dates	Material of tools and implements	Faunal remains	References
Name	Type	Latitude	Longitude				
Topyd-Nyur 5, 7a	Seasonal site	65°01′ N	57°20′ E		Flint		Volokitin (1997)
Sandibeyu 1	Locality	67° 05′ N	57°47′ E		Flint		Chernov (1985)
Chertas 2	Seasonal site	61°52′ N	51°60′ E		Flint		Vereshchagina (2010)
Ulyanovo	Seasonal site	61°50′ N	53°31′ E		Flint		Burov (1967b)
Kuzvomyn	Seasonal site	62°19′ N	54°13′ E		Flint		Burov (1967b)
Evadino 2	Seasonal site	63°03′ N	50°53′ E		Flint		Kosinskaya (2007)
Vetyu 1	Seasonal site	62°56′ N	50°49′ E		Flint		Kosinskaya (2007)

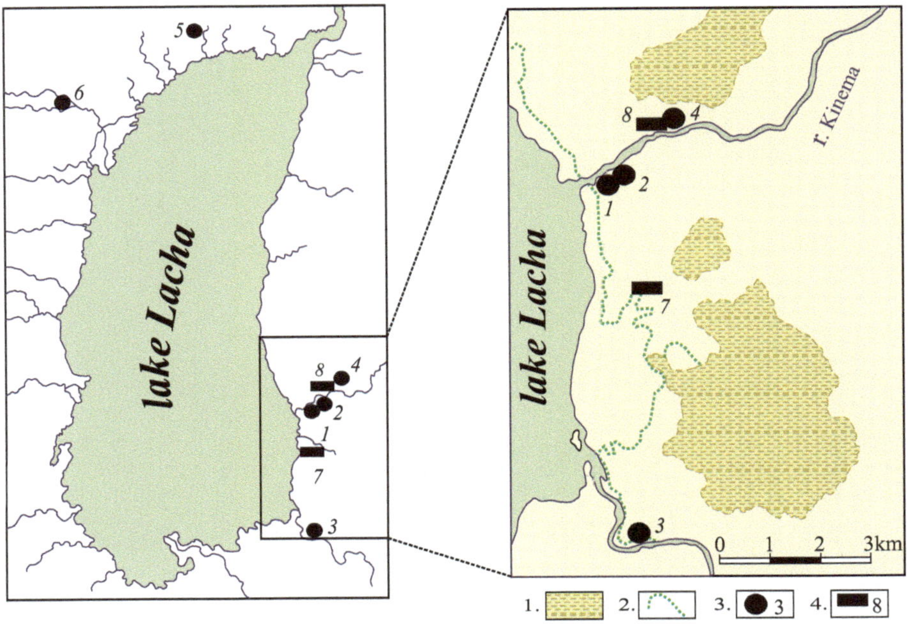

Figure 1.6.3 The Veretye culture sites (Oshibkina, 2006). 1, wetlands; 2, Lacha paleo-lake coastline; 3, the main sites; 4, graveyards. *The Veretye culture sites*: 1, Veretye; 2, Nizhneye Veretye; 3, Sukhoye; 4, Popovo site; 5, Lukinchikha; 6, Siyanga; 7, Peschanitsa; 8, Popovo graveyard.

The Veretye 1 is located near the Kinema River mouth, entering Lacha Lake, the latter having been much larger than at present. The cultural layer occurs in the pebble and sand context overlaid by a peat. The peat accumulation began after the settlement had been abandoned. Locally, there is a thin sand layer separating the cultural layer from the peat. It could result from a period of high stand of the water level in the lake when the people had to leave the settlement and to move to other places. Stratigraphic position of the remains suggests the people stayed here for a long period (Oshibkina, 1997).

The coastal part of the Veretye 1 excavated area has been sampled for pollen analysis. The composition of pollen spectra led N.A. Khotinskiy to the conclusion that the horizon embedding the cultural layer may be dated by the Boreal. It is noted for dominance of birch and pine pollen, a complete absence of broadleaf trees, and a small admixture of *Artemisia* and *Chenopodiaceae* (Fig. 1.6.4.). Noteworthy is the birch pollen dominance and pine pollen proportion steadily growing through the entire period of the site existence. Broadleaf pollen (*Ulmus*, *Corylus*, and later, *Quercus*) appears only in the deposits overlying the Mesolithic cultural layer. The Mesolithic findings occur usually at the contact of sand or gravel with peat, which is typical for the forest zone, as well as for the entire Baltic region (Oshibkina, 1983).

The pollen analysis of samples from the second section has been performed by E.A. Spiridonova (1997). The sediments underlying the cultural layer are distinct by the dominance of arboreal species, mostly of birch, and less by pine and spruce. Such

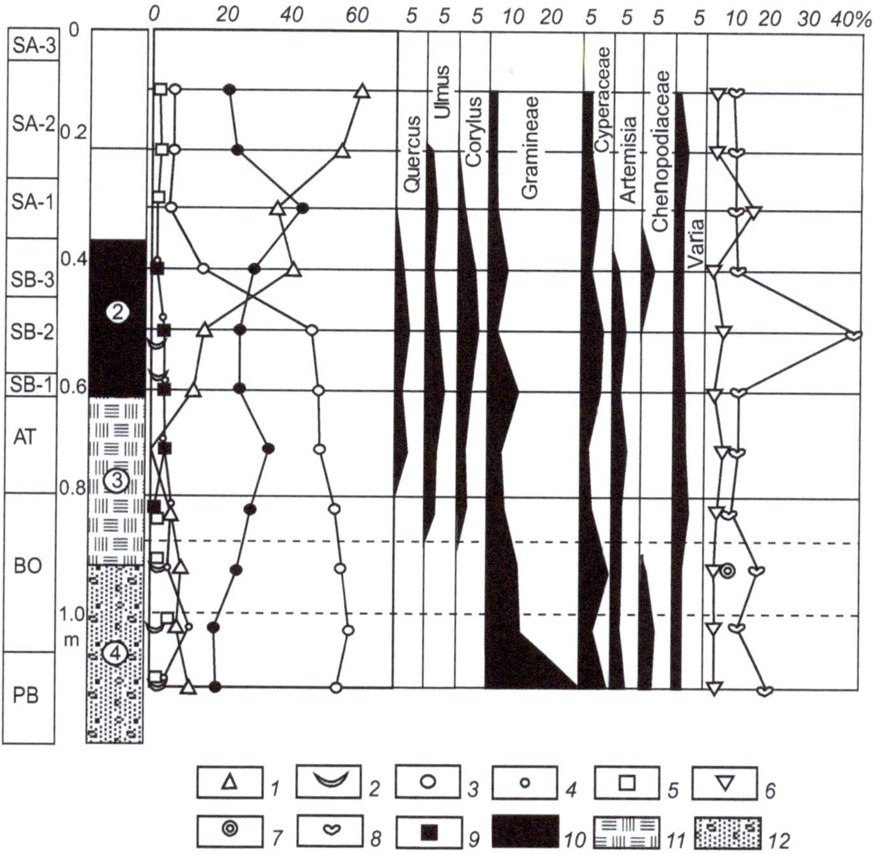

Figure 1.6.4 Pollen diagram of the section at the Veretye I site (compiled by N.A. Khotinskiy (Oshibkina, 2006)). 1, spruce; 2, pine; 3, birch; 4, alder; 5, oak, elm, lime; 6, hypnum moss; 7, equisetum; 8, ferns; 9, spores; 10, peat; 11, sapropel; 12, gravel and pebble.

composition of pollen indicates the dominance of boreal (taiga) forests of pine and birch, with sedge and grasses near the river's forests, gave place to wet meadows with *Spiraea*. Some periglacial relicts persisted locally. Such characteristics of pollen assemblages suggest the late Preboreal age of the sediments.

The samples from the cultural layers are characterized by the absence of periglacial relicts and a greater proportion of herbs and grasses, though arboreal species are still dominant (60%–70%). Among the latter, birch and pine prevail, while spruce is reduced. In the NAP group, grasses gain in importance, though the proportion of sedges is also considerable. The pollen assemblage suggests boreal forests (birch and pine with admixture of spruce and smaller of larch) prevailed in the site vicinity. River valleys featured meadows and swamps along with forests and willow coppices.

Pollen assemblages derived from sediments overlying the cultural layer demonstrate a slow change in vegetation, the dominant position belonging first to pine and later to spruce. According to E.A. Spiridonova, the pollen composition corresponds to the late phase of the Subboreal, when the site region was dominated by middle taiga forest. The cultural layer of the site is thought to be associated with the Preboreal/ Boreal contact (Oshibkina, 1997).

A series of radiocarbon dates is mostly in agreement with the time of the site inferred from the palynological data (Table 1.6.2). If the last two dates are excluded from the consideration, the habitation may be dated to the first half of the Boreal, which agrees with the age inferred from archaeological data, the pollen assemblages, and the composition of flora and fauna remains found. The earliest dates correspond to the end of the Preboreal.

There are three complex wooden constructions (dwellings) and a shrine identified on the excavated area. The shrine presented a wooden pole with a young elk skull on the top. Archaeological finds in the immediate vicinities of the pole suggest it to be the place of magic rites performed before a hunt, using a wooden imitation of the real weapon.

The stone inventory, and particularly that of bone and wooden tools, is one of the best collections of Mesolithic implements in Northern Europe. It includes numerous specimens of the portable art objects.

The inhabitants of Veretye 1 widely used various kinds of rocks (flint, slate, crystalline rocks) for making tools and weapons; the proportion of the stone implements is about 68.8% of the total number of artifacts (Oshibkina, 2006).

There have been documented 372 (6.6%) objects made of wood, bark of birch, and other trees, not including numerous wooden pieces and fragments of dwelling structures (Oshibkina, 2006). There have been used various trees, most often *Pinus silvestris* and *Pinus sibirica* (in a ratio of 8:1) (Table 1.6.3). Among wooden tools and implements, there are bows, arrows, spears, harpoons, axes, etc. (Fig. 1.6.5).

There have been also found a great number of objects made of bone and antler: 1386 items (24.6% of the total), including arrow points, harpoons, knives and daggers, points and needles, needle cases, fishhooks, hoes, and blade insets for large cutting tools (Fig. 1.6.6).

Table 1.6.2 Veretye 1. Radiocarbon dates

No	^{14}C date, yr BP	Laboratory index	Dated material	Calibrated age, yr BC
1	9600±80	LE-1469	Charcoal	8954–8602
2	9370±80	GIN-4833	Modified antler	7690–6870
3	9050±80	GIN-4031	Charcoal	8122–7994
4	8750±70	LE-1472	Wood	7899–7698
5	8560±120	GIN-2452	Charcoal	7690–7440
6	8552±130	GIN-2452	Wood	7490–7450
7	8520±80	GIN-4030	Charcoal	7572–7494
8	8340±120	GIN-4832	Modified antler	7490–7100
9	8270±100	LE-1470	Charcoal	7430–7090
10	7960±100	LE-1471	Charcoal	7000–6660
11	7700±80	LE-1473	Wood	6554–6222

After Oshibkina, S.V., 2006. Mesolithic of the Eastern Onezhskoye Region. Veretye Culture. Institute of Archaeology RAS 322 pp. (in Russian).

Table 1.6.3 Veretye 1. Tree species used for making wooden implements (identified by V.N. Chistyakova)

Tree taxa	Number of objects	%
Pine	56	65
Spruce	7	8.1
Larch	6	7.1
Birch	5	5.8
Willow	4	4.7
Juniper	3	3.5
Alder	3	3.5
Aspen	2	2.3
Total	86	100.0

Table 1.6.4 Veretye 1. Mammalian fauna

No	Mammal taxa	Number of bones/individuals	Percentage of individuals
1	Beaver	477/194	59.6
2	Elk	1449/66	20.3
3	Dog	342/41	12.6
4	Mustelids	14/7	2.1
5	Marten	23/5	1.5
6	Brown bear	62/3	0.9
7	Wolf	7/3	0.9
8	Reindeer	8/2	0.6
9	Hare	6/2	0.6
10	Red fox	1/1	0.3
11	Polar fox	1/1	0.3
12	Otter	4/1	0.3
Total		2394/326	100.0

Rich faunal material has been collected, including mammal and fish remains (Table 1.6.4). The total number of identified remains far exceeds 2000. Fish remains were studied using only the bones from the excavation of 1989. The list of the identified fish remains is given in Table 1.6.5. Among the preferred objects of hunting and fishing there were elk, beaver, ducks, swans, divers (loons), and pikes.

1.6.4 The burial complex of Peschanitsa

The object is located at a distance of 0.8 km from the present-day coast of Lake Lacha, on a sandy hill at the ancient Kortysh River mouth (see Fig. 1.6.3). The hill is composed of glacial deposits; there are varved clays exposed at its base and

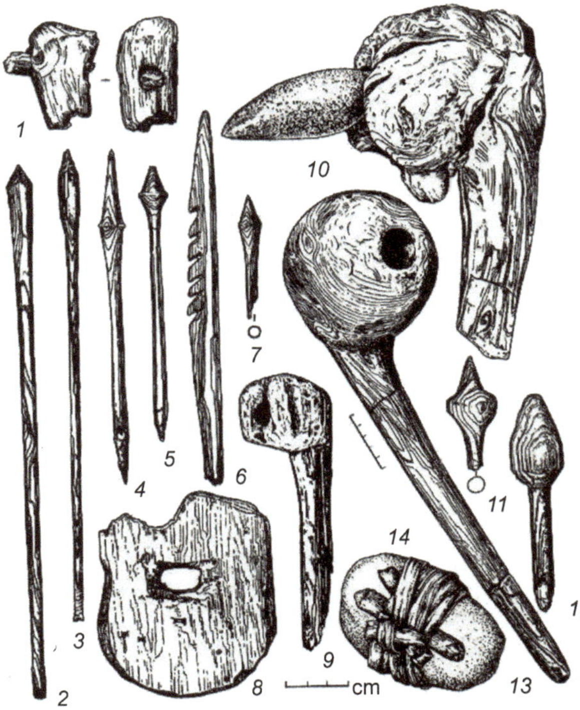

Figure 1.6.5 Veretye 1 site. Wooden tools (Kol'tsov, 1989). 1, 9, 10, 13, foreshafts; 2–5, 7, 11, 12, arrow points and their fragments; 6, harpoon; 8, float; 14, sink stone.

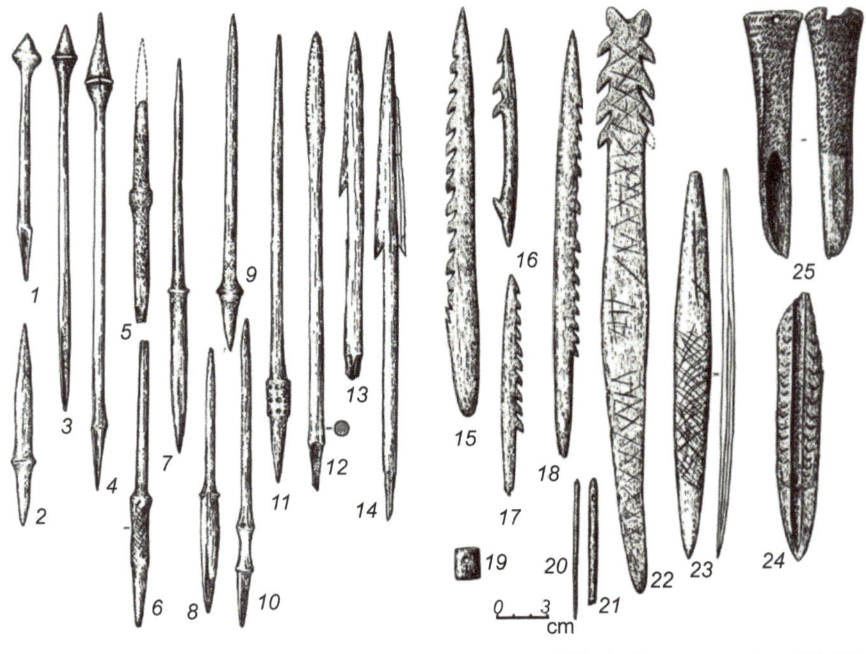

Figure 1.6.6 Veretye 1 site. Bone implements (Kol'tsov, 1989). 1–14, arrow points; 15–18, harpoons; 19, pendant; 20–21, needles; 22–24, daggers; 25, icepick.

Table 1.6.5 Veretye 1. Fish remains recovered in the excavation of 1989 (identified by E.A. Tsepkin)

No.	Fish taxa	Size (cm)	Number	Percentage
1	Pike	50–115	98	74.2
2	Perch	25–38	18	13.6
3	Bream	25–40	14	10.6
4	Tench	49	1	0.8
5	Burbot	93	1	0.8
Total			132	100.0

extending toward the lake. The burial ground was partly destroyed by a quarry. The excavations exposed four pits and a series of animal bone piles surrounding the destroyed burials. Besides those features, there were found two ocher spots, one being partly destroyed by the quarry and the other concealed in the wall of the excavation. Animal bones from destroyed piles are scattered around. The presence of bones in piles and pits may indicate some specific features of the burial rites at this complex structure.

The main object of the Peschanitsy site, with all the pits and bone piles being grouped around, was a damaged burial of a man. Skeletal remains (hip bones and ribs) were preserved between the quarry and pit No. 1. No leg bones have been found there. The first object to be found was a human skull covered with white-colored matter. Leg bones buried separately over pit No. 1 had been believed to belong to the same person; later, however, the suggestion has not been confirmed. Several specialists (including paleoanthropologists, forensic experts, pathologists) carefully examined the skull and the leg bones; some bone defects have been noted on the left facial part of the skull, presumably inflicted while the person was alive. The leg bones revealed small spots indicative of a disease (osteoporosis). Besides the mentioned defects, there are scratches on the bone surface made by a flint blade. It follows from the aforementioned that the leg bones were devoid of soft tissues when buried. The skull belongs to the archaic Europoid anthropological type, similar to those from the Popovo burial. The leg bones were dated by ^{14}C by 9890 ± 120 yr BP, GIN 4858 (9500–9010 cal. yr BP).

The faunal remains recovered from pits and bone piles include bones of several species of mammals, birds, and fish (Table 1.6.6). Elk bones (*Alces alces*) were found in pits 1 and 4 only, and scattered bones of fox (*Vulpes vulpes*) in pits 2 and 3. Almost every pit contained bones of hare (*Lepus timidus*), birds, and fish, the hare being found in various quantities, though usually as separate parts of the carcass. Among large animals, apart from elk, a reindeer (*Rangifer tarandus*) was found in pile 2. Besides, there were found some bones of marten (*Martes martes*) and a hedgehog (Erinaceidae), as well as one bone of lynx (*Lynx lynx*). No remains of animals typically living in mixed forests (with some broadleaf species), nor thermophilic fish have been recovered from the Peschanitsa site.

Table 1.6.6 Peschanitsa burial complex. Faunal remains (mammals identified by V.P. Danilchenko, and birds by E.N. Kurochkina)

Pits	1	Elk, birds, fish
	2	Fox, fish
	3	Fox
	4	Elk
Concentrations	1	Hare (1 individual), birds, fish
	2	Hare (6 individuals), reindeer, fish, hedgehog
	3	Hare (3 individuals), birds: wood grouse (*Tetrao urogallus*), duck
	4	Elk, hare, birds, fish
	5	Fish (*Tinca tinca*)
	6	Birds, fish
	7	Hare (1 individual), birds, fish
	8	Birds, fish
	9	Hare (7 individuals), birds, fish
	10	Hare (1 individual), birds, fish
	11	Hare (4 individuals), birds,
	12	Hare (1 individual), birds, fish
	13	Fish
	14	Elk, hare, marten, birds (*Anas*), fish
	15	Birds, fish
	16	Hare, bird

The bird remains were identified by E.N. Kurochkin in 2000. There are wood grouse (*Tetrao urogallus*), loon (Gaviiformes), hazel hen or grouse (Galliformes order), woodcock (Limicolae), and various ducks (Anas), both small (garganey) and middle-size (Fuligulinae). The duck remains are most abundant, wood grouse ranks next (7 individuals), followed by hazel hen (2), loon (1), and woodcock (1). The bird bones are all scattered, except for a duck skeleton found in pile 10.

Fish remains from Peschanitsa (identified by E.A. Tsepkin in 1990) are typical of the modern ichthyofauna of Lacha and other lakes in the region. Bones found at the site in abundance belong to river and lake species: pike (58.7%), perch (7.2%), bream (8.7%), burbot (13.8%), ide (6.5%), and goldfish (5.1%). The calculations did not include a single specimen of tench (*Tinca tinca*), 42 cm long, found in pile 5. The pike bone dominance in the fish remains agrees with data from other settlements of the Veretye culture and suggests the pike having been the main object of fishing.

The flint tools are mostly made of large blades, and the inventory is similar to that found at the Veretye 1 settlement. There are some flakes bearing use-wear traces.

As to bone tools, they could have been much more abundant if it were not for sandy soil and a high stand of groundwater. A wide knife made from an elk scapula (pit No. 1) is a piece characteristic for the Veretye culture. Similar tools are reported from other Mesolithic cultures in the forest zone where elk was the main object of hunt, and its bones were widely used for making various tools. A wide knife presents a simple form following a natural preform of the animal shoulder blade and requires only the working side sharpened to transform it into a cutting edge.

Table 1.6.7 **Radiocarbon dates from the Popovo burial ground**

Number of burial	Laboratory index	^{14}C dates, yr BP	Calibrated age, yr BC
I	GIN-4447	9430±150	9200–8300
II	GIN-4446	5430±150	4600–3950
III	GIN-4442	9520±130	9300–8500
VI	GIN-10211	7620+160	7050–6050
VIII	GIN-10210	9520±140	9300–8450
IX	GIN-10212	9520±100	9250–8600

1.6.5 Popovo burial ground

The site is located on the right bank of the Kinema River flowing into Lake Lacha, 2.5 km upstream from its mouth (see Fig. 1.6.3). Mesolithic graves are at the top of the hill and arranged in two lines, one of them having been disturbed by Medieval tombs and digging. A series of radiocarbon dates confirms its Mesolithic age (Table 1.6.7). There is a certain system in the grave arrangement, located 10–15 m from each other. Neither graves nor accompanying pits have been disturbed. It is possible that the graves formerly had some wooden constructions (or tree stump with roots) above or were otherwise marked on the land surface, so the later burials could be placed aside from earlier ones. It seems possible that all the graves belong to a relatively short time interval. That is supported by a similarity in details of the burial rite, materials used, and the radiocarbon dates.

All but one grave at Popovo are oriented toward the east, and burial pits are rather shallow (0.3–0.4 m) and narrow; the flat bottom is often powdered with red-colored matter (ocher). The bodies were probably bound or wrapped into soft tissue, the upper side being covered with red powder, less common with white powder (marl). The implements found in graves correspond to those of the Veretye culture. In the graves of men, there are stone axes at the head or at the foot, as well as bone knives or harpoons, necklaces, and belts made of animal teeth. No ornaments have been found in the woman's grave except a small knife. The graves are surrounded by ritual pits with broken tools or remains of the funeral feast. Often they are also painted red. Two complete skeletons of dogs were found in the soil covering one of the graves. Separate phalanges of human fingers are found in some ritual pits. All the pits belong to the Mesolithic burial ground and are overlain with sand and gravel soil.

The burial ground with graves surrounded with pits present specific structures that represented the funeral rites of the ancient bearers of the Veretye culture. Popovo is similar to Peschanitsa, with graves encircled with pits filled with animal bones, fragments of ocher-painted implements, and small charcoal spots (Fig. 1.6.7). Aside from areas with graves and pits, the locality is devoid of the cultural layer.

The animal bones were discovered in the pits associated with graves; individual bones were found scattered. No bones were found in the graves themselves or close to them except two finds of small undisturbed fish skeletons in graves. In grave VII the

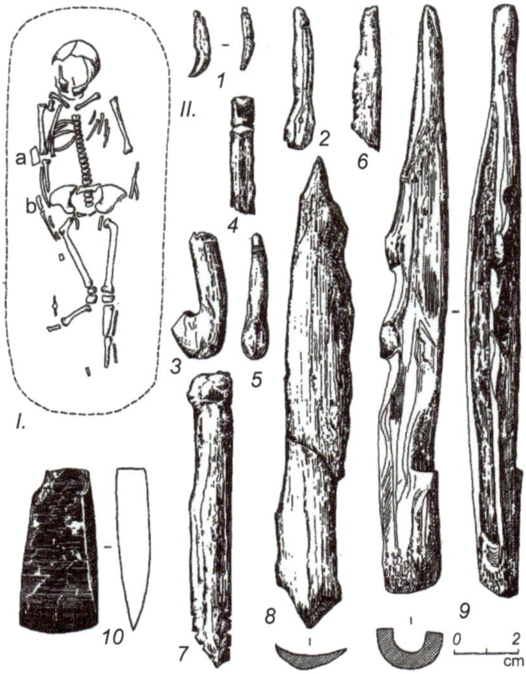

Figure 1.6.7 The Mesolithic graveyard of Popovo. Burial no. VII (I) and grave goods from it (II) (Kol'tsov, 1989; Oshibkina, 2006). a, hand adze, b, dagger, c, ocher, d, marl. 1, 2, pendants made of animal teeth; 3, fragment of a bone fishhook; 4, 5, pendants from elk incisors; 6, 7, fragments of tools; 8, dagger; 9, harpoon preform; 10, hand adze (schist).

fish was put close to the right hand of the child, and in grave V, near the left hand of the buried man. Putting an entire animal carcass or part of it into piles or pits at the burial ground was definitely related to the custom of gift offering to the dead tribesman.

The faunal remains recovered from the Popovo burial ground have been identified by E.G. Andreeva and fish remains by E.A. Tsepkin. Part of the collected material (fish, birds) have not been analyzed.

The species composition of animals recovered from Popovo is rather limited and includes reindeer (*R. tarandus*), elk (*A. alces*), beaver (*Castor fiber*), brown bear (*Ursus arctos*), and dog (*Canis familiaris*). Fish remains were determined in one case: the bones of eight pikes 40–80 cm long were recovered from pit No. 6 (excavation 4). According to E.A. Tsepkin, all the bones except one vertebra belonged to fish heads. Pit No. 6 filled with ocherous sand did not contain kitchen debris and is most probably related to the adjacent burial, its contents having been seemingly associated with burial rites, in common with other pits.

In the process of the analysis, there have been counted bones and number of bones and individuals, both with and without taking into account burial VIII. The latter proved to contain 58 lingual bones of horse having belonged to no less than 29 individuals. The faunal remains are absolutely dominated by elk, and the second place belongs equally to beaver and dog. Reindeer bones are present in noticeable quantities, though it is not typical of the fauna recovered from the Veretye culture. If we take into consideration pendants made of the elk lingual bones, the prevalence of elk would be still more substantial. There is also an uncountable amount of the elk teeth and bone fragments used for pendant making. The total amount of identified faunal remains presented as a percentage of the total is as follows (quantities of bones/individuals are given in brackets): reindeer – 8.3% (5/4); elk – 39.6% (227/19); beaver – 27.1% (25/13); bear – 4.2% (2/2); and dog – 20.9% (37/10). The total amount of identified remains is 48 (100%). In case burial VIII is taken into account, the statistics are somewhat different: reindeer – 5.2%, elk – 62.3%, beaver – 16.9%, bear – 2.6%, and dog – 13.0%.

The stone and bone implements recovered from the Popovo burial ground include objects used in the burial rites. Among them, there are many ornaments, including some unconventional and broken objects, while domestic utensils are completely absent. On the whole, the inventory from this burial ground matches up with materials recovered from other sites and settlements of the Veretye culture.

1.6.6 North

Some sites farther east of the typical Veretye culture range display features similar to the latter in the colonization strategy and in inventories. Among them, there are Yavron'ga 1 site (Krasny Okun' Lake, the Pinega River drainage basin), Pizhma 2 site (Lake Yamozero, the Pechora drainage basin), Vis 1, and Visskiy 1 peatland (Lake Sindorskoye, the Vym' River, Vychegda River drainage basin) (Fig. 1.6.2).

1.6.7 Visskiy 1 peatland

The most complete data have been obtained on the Visskiy 1 peatland, the easternmost site in this group. It is located 0.2 km southeast of the Sindor village. The cultural remains occur in peat deposits of the former crescent-shaped lake adjoining the residual hill of the river terrace 2–4 m above the floodplain. There is a multilayered (from the Mesolithic to the Bronze Age) site of Vis 1 on the residual hill (Fig. 1.6.8).

In the Visskiy 1 peatland the Mesolithic cultural remains occur at a depth from 1.2 to 2.1 m in the layer of hypnum-sedge peat and sapropel; they are partly embedded into underlying sand. According to G.M. Burov (1986), the accumulation of cultural remains took a rather long time. Some wooden objects were carried away from the human dwelling place on the hill, fell into the lake, and later were buried in peat. Later, it was ascertained by botanists that the cultural remains were confined to sapropel layers varying in composition (Zaretskaya et al., 2007). Palynological data give grounds for dating Visskiy 1 peatland to the end of the Boreal–beginning of the Atlantic. Wooden items excavated by G.M. Burov were dated by radiocarbon (Table 1.6.8). The ^{14}C age of sediments with cultural finds is 8480–7330 years, which is

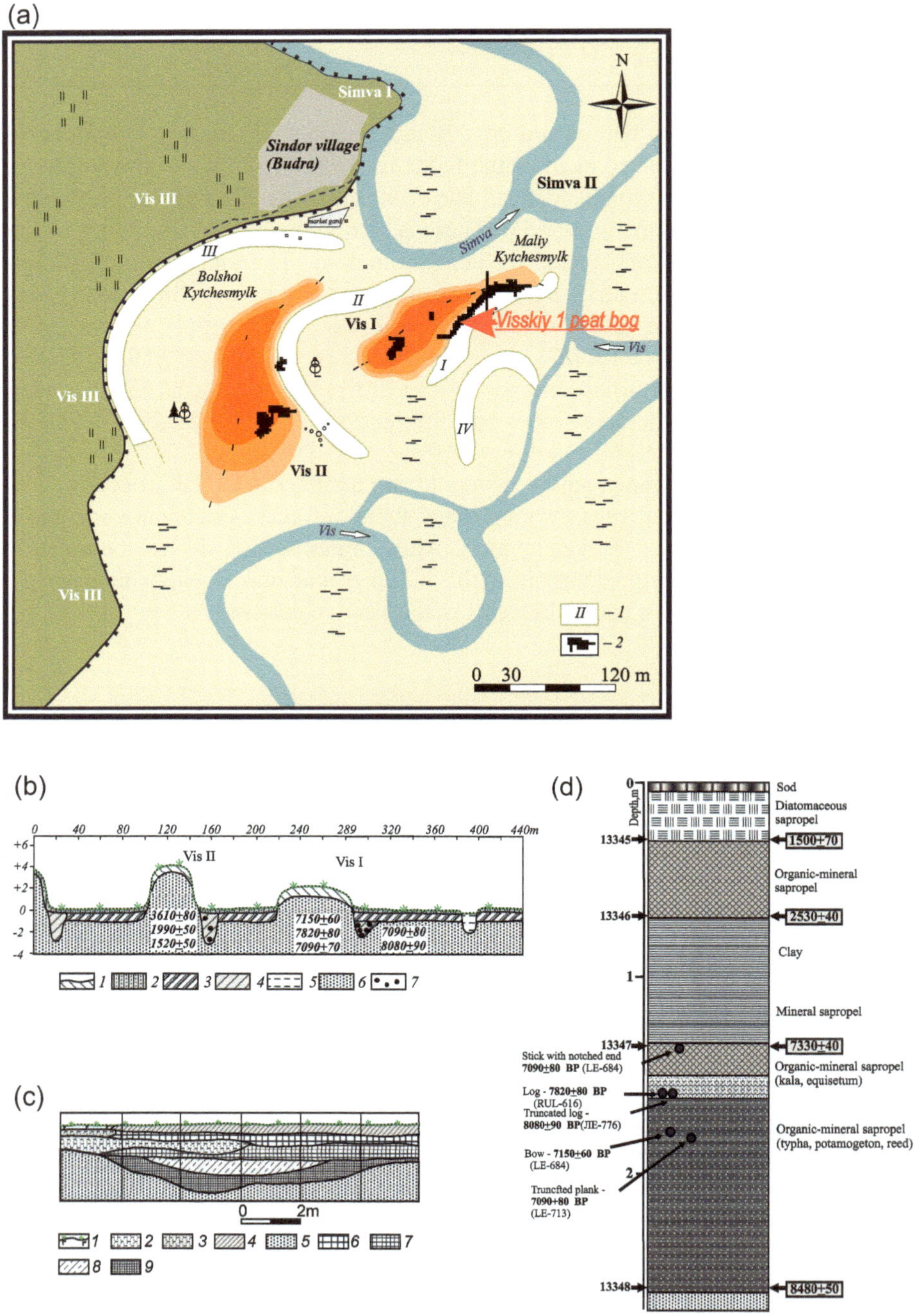

Figure 1.6.8 Visskiy 1 peatland–position and geological context of the site. **A**, topographic plan of the sites in the vicinities of Sindor village (after Burov (1967a,b)), with Additions made by V.N. Karmanov in 2007 (personal communication): *1*, numbers of peat bogs in former oxbow lakes; 2, excavations. **B**, Occurrence of cultural layers in Mesolithic sites of the Visskiy peat bog: 1, cultural layer of the terrace fragments; 2, peat; 3, lacustrine clays; 4, oxbow lake deposits; 5, water; 6, sand; 7, samples for 14C dating. **C**, Visskiy 1 peatbog. Stratigraphy: 1, humus horizon; 2, brownish sandy loam; 3, grayish sandy loam; 4, peat; 5, gray sand; 6, brownish clayey sapropel; 7, olive clayey sapropel; 8, sedge peat; 9, detritus sapropel (layers 8, 9, and partly 5 contain Mesolithic artifacts) (Burov, 2009). **D**, section of the Visskiy 1 peat bog with dates obtained on artifacts and deposit samples (Zaretskaya et al., 2007).

Table 1.6.8 **Visskiy 1 peatbog. Radiocarbon dates (Burov et al., 1972)**

Year of excavation	Material	Depth, m	^{14}C date (yr BP) and laboratory index	Calibrated ^{14}C age, yr BP
1967	Log with an end cut	1, 7	8080±90 LE-776 (478)	8980±160
1964	Log	1, 75	7820±80 RUL-616	8650±130
1966	Crossbow	1, 9	7150±60 LE-684	7970±60
1966	Stick with a notch	1, 9	7090±80 LE-685	7910±80
1966	Truncated plank	1, 9	7090±70 LE-713	7910±70

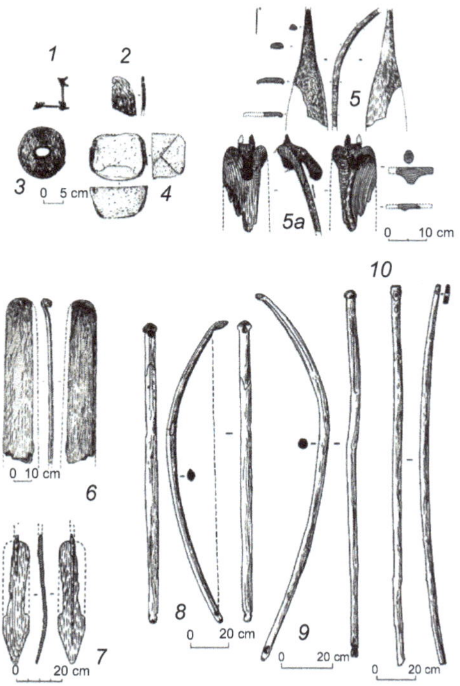

Figure 1.6.9 Visskiy 1 peatland. Items made of grass (1), bark (2), birch bark (4), and wood (3, 5–10). 1, fragment of a fishing net; 2, float; 3, disk of "botalo" (fishing facility); 4, box; 5, 5a, fragments of skis; 6, sledge runner; 7, paddle; 8–10, bows.

in reasonable agreement with earlier dates obtained for wooden objects (Zaretskaya et al., 2007).

A distinctive feature of this site is a large number of tools (more than 200 items, including objects bearing traces of treatment) made of wood, pine and birch bark, and sedge grass (Fig. 1.6.9). Among them there are hunting weapons and fishing tackle, means of transportation, both winter (ski, sledge runners) and summer (scull),

Figure 1.6.10 Mesolithic hunter equipment (based on materials from Visskiy 1 peat bog). Exposition of the Archaeological Museum, Institute of the Language, Literature and Art, Komi Research Center, Urals branch of Russian Academy of Sciences.
Drawing by S.K. Mikhailov and A.V. Punegov.

domestic utensils, construction details, etc. The assortment suggests the site to be inhabited the year round (Burov, 1993, 2009, 2011).

Among hunting weapons, there are bows, arrows, and spears. The hunter bows are about 1.5 m long, with an almond-shaped projection at the end and a hole for bowstring fixing. Besides, there are gigantic crossbows about 3.5 m long; they were made of young tree trunks bearing a hole, a constriction, or a knob on the butt end and a notch on the head. Arrows are all wooden, with spindle-shaped heads. There are some shafts with sharpened ends, as well as those with split for stone inset, or with notches for inserted pointed blades. Spears 1.5–2 m long are made of tree trunks with a flame-shaped blade on the butt end.

There are also tools for fishing, including a fragment of fishing net (made of a two-strand string of sedge), a float for net made of pine bark; a fragment of fishing trap of fine sticks interweaved with plant fibers; and bands for a scoop nets and trap nets, with holes along the whole length. There are also wooden elements of the fish spears.

Of the transportation means, there are sledge runners, skis, and paddle (oar). The means for transportation in winter are especially numerous (Fig. 1.6.10). There are sledge runners, trough-like, with horizontal openings for tying them in pairs; another kind is flattened, with vertical holes for crossbars joining two runners. The skis are distinct for a pointed upward-bent forepart; the most noteworthy specimen has a rounded rear part with a carved elk-head below the sliding plane; the latter would presumably hinder the backward motion of the ski. The paddle blade is pointed; supposedly, it could be used in shallow water as a punt pole.

Among tools intended for woodworking, there is an axe handle with a massive clamping sleeve and an arc-shaped scraper with slots for inserted blades.

There are small bows 0.55 and 0.26 m in size, parts of devices intended for drilling holes and making fire, as well as a square box made of birch bark held together with wooden pegs.

Some of the listed objects (bow, small drilling bows, skis, sledge runners) bear engraved ornaments in the form of zigzags, straight lines, oblique nets, crosses, or hatching.

According to identification of wood by V.E. Vikhrov and E.G. Melnikov, bows, handles, and fishhooks were mostly made of spruce wood; pine wood was used for small bows, and birch for sledge runners. Besides, there is noted the presence of Siberian pine, and a few implements of deciduous trees (Burov, 1986).

1.6.8 Parch 1 and Parch 2

The sites are located on the left bank of the Vychegda River, 2.5 km downstream from the Parch village (the Ust'Kuloma district), near the Parch River mouth. The sites are located at two segments of the Vychegda floodplain about 5 m above the water (Fig. 1.6.11). The Parch 1 site is located on a floodplain ridge and Parch 2 in an interridge hollow.

The finds recovered from the Parch 1 site occurred at a depth 0.6–1.0 m, in the lower horizon of the alluvial soil. Such a great thickness of soil is found only on higher ridges in the studied region. Except for a few tools made of quartzite pebble and one sandstone pendant, all the tools are made of flint. A considerable part of knapped flint fragments and broken tools appear to match each other, so the entire process of stone treatment may be reconstructed. Among the tools, there are stemmed arrowheads on blades. Numerous are points, reamers, piercing tools, borers, drills, end scrapers, burins, and insets for holders in composite tools. There are some flake tools and a series of large, massive scrapers.

At the Parch 2 site the cultural remains are found in the context of sandy loam and sand at a depth of 2 m, in the transitional facies from the channel to floodplain deposits. The finds occur as separate concentrations (presumably corresponding to dwelling remains) associated with coaly interlayers and lenses of calcined soil.

The analysis of pollen and spores obtained from the section of the Parch 2 site (dwelling 2) carried out by L.A. Konovalenko revealed several pollen assemblages (Fig. 1.6.12).

Pollen assemblage I obtained from deposits lying immediately above the dwelling remains suggests a rather limited afforestation of the land in the upper reaches of the Vychegda River under conditions of a climate colder than at present. Grass communities were dominant on the open landscapes, while trees were only beginning to colonize the region.

Assemblage II shows a wide distribution of spruce forests with some fir and common pine, that is, typical middle taiga forests indicative of the climate similar to the modern one.

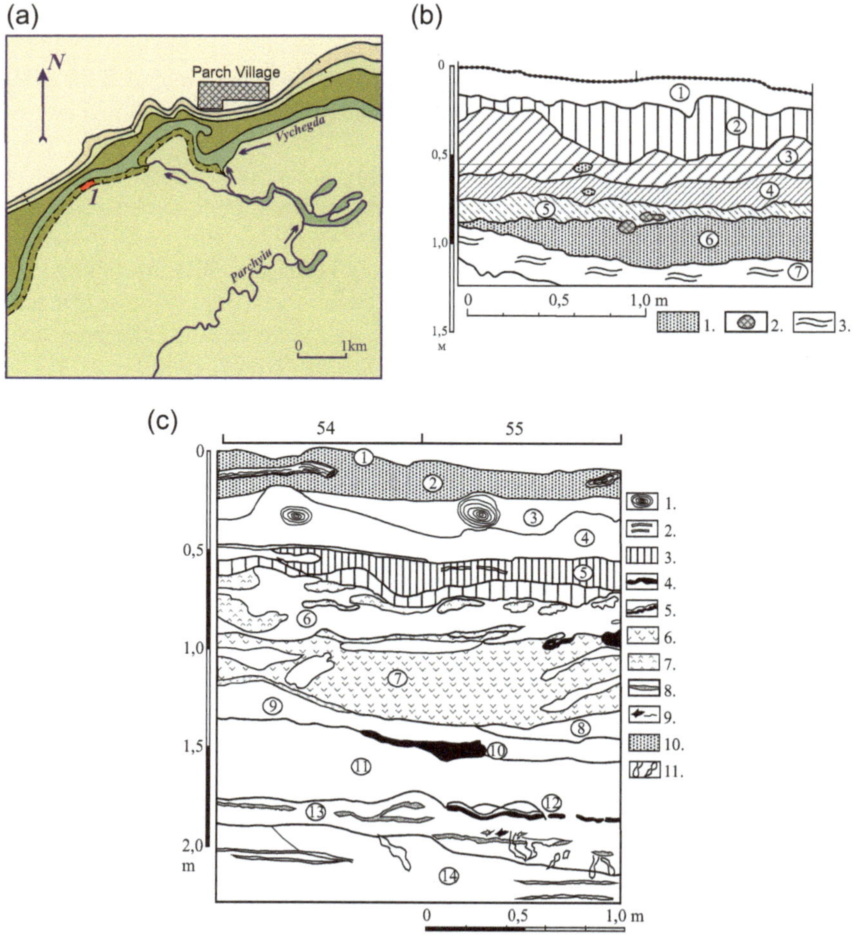

Figure 1.6.11 Parch 1 and Parch 2: location and stratigraphy. (a) Excavated areas of Parch 1 and 2 (1). (b), Parch 1. 1, sand; 2, mole burrows; 3, clay. Stratigraphic sequence from up to bottom (numbers of layers are shown in *circles*): 1, sod; 2, modern soil, dark-brown loam; 3, brown sandy loam; 4, dark-brown loam, buried soil; 5, brown sandy loam, colored by the humus penetrated from the overlying layer; 6, interstratification of sand and light brown clay. Cultural components occurred close to the base of layer 4 and in the uppermost part of layer 5. (c), Parch 2. 1, wood; 2, birch bark; 3, clay with peat; 4, manganese beds; 5, ferruginous crust; 6, ferruginous beds; 7, brown beds; 8, charcoal beds with stone artifacts; 9, fired sediments with animal bones; 10, sand; 11, root channels. Stratigraphic sequence (numbers of layers are shown in *circles*): 1, 2, sand and sandy loam (levee); 3, gray-brown loam with wood fragments, soil; 4, peat loam; 5, blue-gray clay; 5a, gray loam; 6, reddish brown loam; 7, dark-brown loam with manganese concretions; 8, brownish-gray sandy loam with cultural remains; 9, stratified grayish-brown sand; 10, gray loam; 11, brown loam; 12, bluish clay; 13, brownish-gray sandy loam; 14, brownish sand alternating with thin sandy loam beds.

At the time of assemblage III formation, there were spruce forests with participation of oak, elm, lime, and hazel. Dwarf birch occurred on wetlands, and thickets of European alder (*Alnus glutinosa*) in river valleys. The climate in the Vychegda River drainage basin at that time was distinctly warmer than at present.

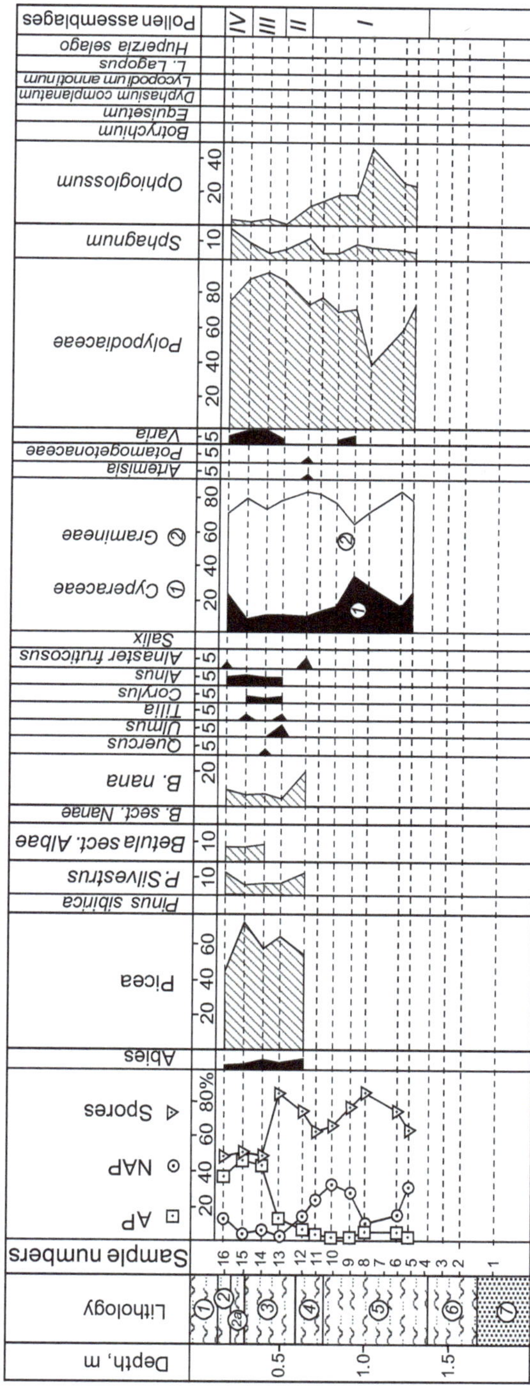

Figure 1.6.12 Pollen diagram of the Parch 2 section, dwelling no. 2 (Konovalenko, 2006); see Supplement 3 in (Volokitin, 2006). Stratigraphic sequence (numbers of layers are shown in *circles*): 1, gray sand; 2, loam, sandy loam; 3, sandy loam; 4, loam; 5, sandy loam; 6, yellow sand; 7, gray sand.

Table 1.6.9 **Radiocarbon dates from the Parch 2 site**

Laboratory index	Dated feature	Material	^{14}C date, yr BP	Calibrated age, yr BP
GIN-11911	Dwelling 3, hearth	Small-sized charcoal	7,800±300	8,710±350
GIN-11913	Dwelling 6, hearth	Small-sized charcoal	9,100±250	10,270±360
GIN-11912	Dwelling 5, hearth	Small-sized charcoal	9,500±250	10,830±350

Forests dominated by spruce persisted in the region at the time of assemblage IV deposition, though a disappearance of broadleaf species suggests a somewhat colder climate.

When compared with palynological data on the Northwest of European Russia in the Holocene (Nikiforova, 1982), pollen assemblage I may be tentatively associated with the Boreal, while assemblages II and III are attributable to the Atlantic. The considered assemblages are close to those described in the Parch 3 sequence (Volokitin and Konovalenko, 1988) and dated to the Boreal and Atlantic.

Few radiocarbon dates have been obtained for the site (Table 1.6.9).

There are remains of eight light surface dwellings studied at the Parch 2 site; two of them have been excavated entirely, while the rest have been partly destroyed by erosion. As follows from the analysis of the geological context and morphology of the site, it was positioned on a narrow projection of the river bank, where light surface dwellings, "chums" (tipi), were put up on a meadow (Volokitin, 2002). The dwellings were about 4–5 m in diameter, with a hearth in the center and trampled soil around it (Zaliznyak, 1991). The lithic assemblage is close to that of the Parch 1 site. A concentration of large tools (end scrapers and burins) intended for cutting, sawing, and scraping of bone and antler materials—probably the workshop of a prehistoric carver—was found in one of the fully excavated dwellings. Another dwelling yielded a lot of small-sized chert and flint debitage indicative of a lithic workshop. Tanged points of arrows were found in three dwellings. There are also fragments of an artifact (needle case) made of wood grouse bone.

Faunal remains seldom occur outside hearths due to the high acidity of the soil. Small fragments of calcined bones of mammals, birds, and fish were found in the hearths of the Parch 2 site and attributed to beaver (*C. fiber*), elk (*A. alces*), reindeer (*R. tarandus*), wolf (*Canis lupus*), dog (*C. familiaris*), Mustelide family (*Martes* sp.), and bear (*U. arctos*). Bird remains were identified as wood grouse (*T. urogallus* L.), hazel hen (*Tetrastes bonasia* L.), duck (Anatinae), and a bird of the Corvidae family. Besides, there were found fish bones: pike (*Esox lucius*), perch (*Perca fluviatilis*), and those of *Cyprinidae*. The remains were identified by P.A. Kosintsev (mammals) and A.E. Nekrasov (birds and fish). The faunal remains found cannot be identified with the diet of the Mesolithic inhabitants. Still the list of fossil fauna seems rather impressive and close to that of the present-day inhabitants of the region. It is typical of the taiga forest.

Figure 1.6.13 Paleolandscapes and living conditions of the Mesolithic man (reconstructed from materials from Parch 1 and Parch 2 on the Vychegda River). Exposition of the Archaeological Museum, Institute of the Language, Literature and Art, Komi Research Center, Urals branch of Russian Academy of Sciences.
Drawing by S.K. Mikhailov and A.V. Punegov.

The Parch 1 site was located in close vicinity to Parch 2, on another, more elevated, segment of the alluvial surface. Neither fire places nor dwelling traces were found there. It is quite possible that the Parch 2 dwellers chose that drier and better ventilated place for their outdoor activities. It is not inconceivable, however, that the two sites were inhabited subsequently during a season, depending on the rising water level in the river (Volokitin and Tkachev, 2004).

The inhabitants of the site used the resources of river and lakes on the floodplain as well as those of taiga forest. The water basins were used for fishing and hunting beaver and water fowl, while the forest was a source of upland fowl and most mammals. The most plentiful were open pine forests confined to outwash sandy terraces, more extensive at that time than at present. The area used for hunting and fishing was probably within 1–2 km from the sites (Fig. 1.6.13).

Both Parch 1 and Parch 2 bear an evident resemblance to the Mesolithic Butovo culture of the Volga-Oka interfluve (Volokitin, 2006).

The majority of the Mesolithic sites discovered on the Severnaya Dvina, Vychegda, Sukhona, Vym', Mezen', Pechora, and Izhma rivers are located on sandy river terraces, and their cultural components occur close to the surface. There are a few independent dates and finds of datable faunal material. No cultural layer as such can be found in most cases, which may be due to its high soil acidity. Most artifacts are found at the base of forest (podzol) soil or near the top of the illuvial horizon, often being mixed with later materials. In this case, we speak not of cultural layers, but of deposits embedding cultural components, or horizons with artifacts and some elements of prehistoric settlement structures. One of the sites most representative in this regard is the Lek-Lesa 1 site.

The excavated sites are typically characterized by lithic industry either West Uralian in appearance, that is, so-called Kama Mesolithic (the Ural tradition) or by lithic industries comparable with the Butovo culture typically distributed throughout the Volga drainage basin and the Volga-Oka interfluve in particular (the Western tradition). Some of the sites combine characteristics of both traditions. There are few dates and hardly sufficient faunistic material at the discussed sites (Volokitin, 1997). Next, we give a list of the best studied sites belonging to the three groups:

Ural tradition: Lek-Lesa 1 on the Izhma River; Ulyanovo and Kuz'vomyn on the Vychegda River; Evdino 2 on the Vym' River

Western tradition: Parch 1 and 2 on the Vychegda River; Adz'va 1 on the Usa river; Topyd-Nyur 5 and 7a on the Pechora river; Chertas 2 on the Vychegda River; Sandibeiyu on the Kolva river (Pechora drainage basin)

Sites of *mixed tradition*: Filichaev 2–5 on the Severnaya Dvina River; Turunnyur on the Izhma River; Pezmogty 6, Cherdyb 1 and 2 on the Vychegda River; Vet'yu on the Vym' River (see Fig. 1.6.2; Table 1.6.1).

1.6.9 Lek-Lesa 1

Lek-Lesa 1 is one of the most informative Mesolithic sites. It is located on a sandy terrace of the Izhma Rver on the right bank of the latter, 1 km upstream of the Lek-Lesa River (left tributary of the Izhma River) and 14 km downstream from Tom settlement (the Izhma district, Komi Republic). The terrace surface 17 m high descends gradually to 14–15 m westward, and it borders the floodplain to the southwest. The northern part of the site is a bog massif on the terrace. The findings were recovered from the white forest podzol and from the top of the illuvial horizon. As the majority of the archaeological sites of the region, we are dealing with a horizon with cultural remains or with culture-bearing deposits, and not with a cultural layer *sensu stricto*. In spite of unfavorable geochemical environment (typical for sandy terraces) and bioturbation, some elements of the cultural layer may be reconstructed.

There are remains of two light aboveground dwelling structures, one of them (dwelling no. 1) having been studied completely, and the other (dwelling no. 2) only partly (Volokitin and Kosinskaya, 1999).

The entirely excavated dwelling was represented by a colored charcoal spot 2–3 m, the long axis being oriented along a west–east line. Most of the finds were confined to that spot, including flint pieces, small charcoal fragments, fragments of burnt birch bark, small fragments of calcinated bones, and pieces of quartzite-sandstone cobble. Thirty pieces of the latter occur both in the podzol layer and in the illuvial horizon of the soil (mostly at a depth of 0.8 m), but within the limits of the coal spot. All cobble fragments were put together and formed a complete boulder about 20 cm in diameter. It should be noted that the fragments of quartzite-sandstone do not bear traces of fire; therefore they were not used for the hearth lining.

P.A. Kosintsev identified several animal species, such as beaver (3 fragments of a mandible), elk (part of a metacarpus), and bear (a jaw fragment) (Volokitin, 2005). A bulk sample of charcoal collected over the entire surface of the coal spot marking a surface dwelling was dated at 9010 ± 70 (LE-3607).

The tool inventory included burins, end scrapers, side scrapers, micro-scrapers, notched tools, large knives, fragments of retouched tools, points, truncated blades, and microblades. Among micro-scrapers there are so-called small chisels. Such implements are typical for the Middle Vychegda culture (Burov, 1967a). They are made on thick blades (occasionally crested blades) with curved profile. There is a point made on a large blade with a slightly asymmetrical "beak" (the piece was found in fragments, but completely reconstructed). Narrow blades and truncated bladelets are found with broken tips (usually distal). The truncation was made (with one exception) by dorsal retouch. The same series includes also bi-truncated blades with dorsal retouch; two of them are truncated transversely on both ends, while the other two are truncated transversely on one end and obliquely on the other. It seems possible that the blades truncated on both ends served as inserts in bone points, and the obliquely truncated blades were inserted at the end of the point and formed a cutting edge (spike).

Both the lithic technology and tool inventory indicate that the site could be attributed to the Mesolithic Ural culture tradition (Volokitin, 1997).

1.6.10 Conclusion

The true Paleolithic sites known in the north and northeast parts of European Russia are confined to a narrow zone lying immediately west from the Ural Mountains (Pavlov et al., 2010). On the rest of the territory the oldest sites are referred to the Mesolithic. They have been discovered both along large rivers and on the coast of the Arctic Ocean (see Fig. 1.6.2). Unfortunately, most of them are still insufficiently known, only prospecting and reconnaissance works having been performed. Southern regions are better studied; among these are the area lying east from the Onezhskoye Lake, the vicinities of the Vozhe and Lacha lakes, Lake Sindor, upper and middle riches of the Vychegda River, etc.

The archaeological sites in the region are mostly temporary camps and sites (Kosinskaya and Volokitin, 1993), except for sites attributed to the Veretye culture. The latter is a rare case of sites widely varying in type from hunting outlooks and seasonal camps to permanent habitations, as well as burial grounds located in close vicinity. The Visskiy 1 peatland near Lake Sindor yielded a large collection of pieces made of wood, bark, and grass, including wooden artifacts related to summer and winter transportation means. According to G.M. Burov, those materials are similar to those attributed to the middle stage of the Veretye culture.

Two sites located at floodplain on the Vychegda—Parch 1 and Parch 2—are dated by the Preboreal/Boreal boundary. Sites belonging to the same culture are discovered as far north as the Arctic Circle (Adzva 1) and display a distinct resemblance to the Butovo culture. That suggests a migration (possibly seasonal) of small groups from the Volga-Oka region. The latest sites—Chertas 2, Topyd-Nyur 5, and Topyd-Nyur 7a—are dated to the beginning of the Atlantic. Even wider distributed are sites with flint industry bearing a resemblance to the Mesolithic assemblages from the middle reaches of the Kama River. It seems that their penetration into the North of European Russia took place at the same time and much in the same way as the first group. The materials at hand suggest that the mutual influence of the two groups (traditions) resulted in the appearance of new industries and their subsequent spread over vast spaces. These new industries (Filichaev 2–5, Cherdyb 1 and 2, sites on the Sukhona

River) were similar to the industries of the Kama basin in lithic technology and to the industries from the Volga-Oka basin in the tool inventories.

It seems that the archaeological sites of the three groups belonged to migratory people with long seasonal migrations from south to north and backward (Gordon, 2003). It is possible that during the Mesolithic there was no permanent population on most of the East European North.

We argue that the northern part of the East European Plain was colonized by small mobile groups of Mesolithic hunters at the Preboreal–Boreal boundary. And the only really populated area seems to be eastern part of the Lake Onezhskoye basin (the Veretye culture).

Environments of the Northeastern East European Plain at the time of the initial human colonization

1.7

Yuriy N. Gribchenko
Institute of Geography, Russian Academy of Sciences, Moscow, Russia

Chapter Outline

1.7.1 Introduction

The Northeast of the East European Plain deserves a special attention when the primary colonization of the European North is considered. It is in this region that not only Late Paleolithic sites and localities were found and studied to a variable extent, but the Middle Paleolithic sites have been also discovered. The northernmost and earliest archaeological sites attributable to Middle and Late Paleolithic occur in the Pechora and middle Kama drainage basins (Gromov, 1948; Kanivets, 1976; Guslitser and Pavlov, 1993; Pavlov, 1997, 2002, 2004, 2008a,b; Pavlov and Makarov, 1998; Pavlov et al., 2010). All of the sites are highly variable not only in age and in the level of our knowledge, but they also differ in type. It should be specially noted that the sites in Northeastern Europe characterize three consecutive steps of the initial human colonization of the region (35–28 ka BP; 18–17 ka BP; and ~11 ka BP).

The studied sites are highly variable in the occurrence of cultural layers, in their geological context, age and other characteristics. The most important information on the human dispersal process was obtained from the sites of Byzovaya, Zaozerye, and Garchi I.

Human Colonization of the Arctic: The Interaction Between Early Migration and the Paleoenvironment
http://dx.doi.org/10.1016/B978-0-12-813532-7.00007-3

The integrated archaeological and paleogeographic studies of those sites provided a great volume of data on the conditions of the region colonization by the early man at the Middle Paleolithic and during the early and middle epochs of the Late Paleolithic. The Paleolithic chronology and ways of the Paleolithic man migrations obtained from the named above sites may be complemented by data from other Paleolithic sites.

The principal features of the colonization process in the studied region are predetermined by its considerable extension from north to south along the western slope of the Urals. That accounts for complicated landscape patterns, with plant communities varying from forest-tundra to middle and southern taiga types. Their natural variability may be traced at different epochs of the Late Pleistocene (Nemkova, 1976; Yakhimovich et al., 1988; Andreicheva and Marchenko-Vagapova, 2003; Andreicheva et al., 2006). It is the spatial diversity in the Late Pleistocene environments in the region that makes so difficult the comparison and correlation between Paleolithic sites distributed over the region.

The relief formation in the Middle Pleistocene was strongly influenced by the large ice sheets attributed to the last glaciation of the Middle Pleistocene (known as Vychegda or Moscow glaciation) (Fig. 1.7.1). The glacier activities resulted in deposition of thick tills, fluvioglacial sand and gravels, as well as in accumulation of periglacial loess-like loams of various thickness. Processes related to the erosional activities of ice sheets and meltwater resulted in development of large river valleys and wetlands in depressions filled mostly with sandy and clayey deposits. The latter formed the basis of many landforms and geomorphic levels (Lider, 1976; Lavrov et al., 1986; Lavrov and Potapenko, 2005; Yakovlev, 1956).

The Boreal transgression that occurred at the beginning of the Late Pleistocene, besides changing the East European coastlines, had a pronounced effect on relief-forming processes over a vast area. The subsequent cold epoch left various glacial and periglacial formations related to different stages of the Valday glaciation. Their facial relationships are still insufficiently known, so there is still no unambiguous data on the last glaciation age and extension.

There are noticeable differences between drainage basin of the Pechora River and that of the middle reaches of the Kama. The chronostratigraphy of the Pechora basin is based on the studies of glacial, fluvial, marine and lacustrine sediments (Andreicheva, 2002; Andreicheva and Konovalenko, 1989; Andreicheva et al., 1999; Guslitser and Loseva, 1979; Guslitser et al., 1985; Lavrov et al., 1986; Lavrov and Potapenko, 2005; Ryabkov, 1976; Yakovlev, 1956). The middle reaches of the Kama are more diversified in the Late Pleistocene sediment facies due to a wide occurrence of the loess-paleosol sequences (Butakov et al., 1981; Gribchenko, 2008; Vereshchagina, 1965; Yakhimovich et al., 1987; Lider, 1976).

The age determinations based on archaeological datable materials from the Paleolithic sites of the region vary over wide range, from the Early and Middle to the Late Paleolithic (Gromov, 1948; Guslitser and Kanivets, 1962b, 1965a; Guslitser, 1976; Kanivets, 1976; Pavlov, 1997, 2008a,b). The dating of the sites is usually based on finds of ancient fauna and associated stone tools, as can be shown by the example of Peshcherny Log, Elniki, Kharutin and other localities (Gromov, 1948; Bader, 1960; Pavlov, 1997, 2008b). Unfortunately, in most cases the archaeological material was redeposited, so that the initial occurrence of cultural layers became unrecognizable. The stone artefacts are rather scarce.

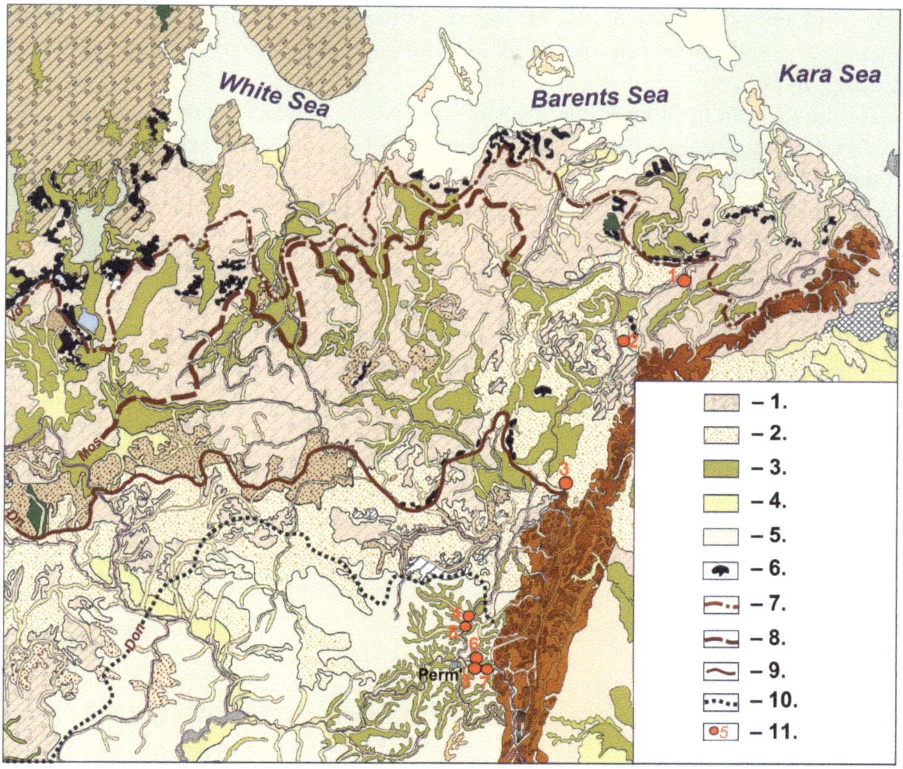

Figure 1.7.1 Middle and Late Pleistocene deposits in the north of the East European Plain (based on the Map of the Quaternary deposits of the USSR, sheets 2, 5, 6, scale 1:2 500 000 (Ganeshin, 1976)). 1 – Middle and Late Pleistocene moraines a and b; 2 – glaciofluvial and outwash deposits; 3 – glacio-lacustrine deposits; 4 – fluvial deposits; 5 – cover loams; 6 – end moraines; 7–10 – boundaries of ice sheets: 7 – Valday (Vd), 8 – Moscow (Mos), 9 – Dnieper (Dn); 10 – Don (Don); 11 – the most important sites: *1* – Mamontova Kurya, *2* – Byzovaya, *3* – Medvezhya, *4* – Garchi I, *5* – Shirovanovo; *6* – Talitskogo, *7* – Romakhino, *8* – Zaozerye.

Mousterian finds were documented in drainage basins of the Kama River (Peshcherny Log, Ganichata, Elniki, Garhi I), and Pechora River (Kharutin, Krutaya Gora) (Gromov, 1948; Kanivets, 1976; Guslitser and Kanivets, 1965b; Pavlov and Makarov, 1998; Pavlov, 1997, 2004, 2008a; Guslitser and Pavlov, 1993). No undisturbed Mousterian cultural layers have been found in the northern regions. According to various authors, the earliest sites may be attributed to different stages of the Middle Pleistocene or to the beginning of Late Pleistocene. The only exception is the unquestionable Mousterian site Garchi I in the middle reaches of the Kama where stone tools and their fragments have been found in abundance at the base of a high bank of the river (Pavlov and Makarov, 1998; Pavlov, 2002, 2008a,b). According to P. Yu. Pavlov, the cultural layer could be confined to the middle part of the thick Mezin paleosol complex where a few flint fragments had been found. However, neither series of analyses of the entire sequence, nor lithological studies of individual horizons revealed any signs of a cultural layer.

It follows from the above, that the Mousterian communities undoubtedly inhabited the Northeastern region; it is still uncertain, however, when the early man first came to the North. Consequently, we can hardly speak with certainty of the climate and environments at the time of the first human penetration into the middle Kama drainage basin and farther north. In the absence of factual materials on the cultural layer structure, we cannot estimate how long the Mousterian men stayed here and what kind of economy they practiced.

It was not until the second half of the middle Valday mega-interstadial that the early man dispersed widely over the northeastern part of the East European Plain. That time was marked by the appearance of Paleolithic sites belonging to various stages of the region colonization in the Pechora and Kama drainage basins. There are a few sites attributed to the early stage of the Late Paleolithic (Mamontova Kurya, Byzovaya, Garchi I) on the basis of radiocarbon dates on large mammal bones lying in the range between 36 000 and 28,000 BP (Pavlov, 2004; Svendsen and Pavlov, 2003; Pavlov and Indrelid, 2000). There are also some much younger sites known in the drainage basins of the upper Pechora and middle Kama rivers (Medvezhya, Talitskogo), both dated at approximately 18,000 BP.

1.7.2 Paleolithic sites in the Pechora drainage basin

The main cycles recognizable in the regional relief evolution are related to the expansion of ice sheets in the Early and Middle Pleistocene. High river terraces and interfluvial areas are mantled with till and glaciofluvial deposits. At the Late Pleistocene beginning (during the Mikulino Interglacial) thick series of lacustrine sediments accumulated in the central and southern part of the region; in the north they gave place to the marine series of the Boreal transgression. Large areas in river valleys are occupied by the first and second river terraces dated to the Late Pleistocene, along with widely spread floodplains. Of older terraces there are only fragments preserved, the corresponding fluvial series being found in the overdeepened segments of the valleys (Fig. 1.7.2). Specific features of the river valleys in the Pechora drainage basin made it possible for the Paleolithic communities to migrate over a long distance and to choose the most comfortable dwelling places.

There are several horizons recognized in the stratigraphic sequence of the Upper Pleistocene, including: Sulinskiy–interglacial (Mikulino, Eemian, MIS 5e); Layskiy–glacial (Podporozhskiy, MIS 4), Byzovskiy–interstadial (Lenindradskiy, MIS 3), and Polyarny–glacial (Ostashkov, MIS 2) (Andreicheva, 2002; Andreicheva and Marchenko-Vagapova, 2003). No deposits attributable to the Mikulino interglacial have been identified as yet in the southern part of the Pechora drainage basin.

Alluvial and subaerial formations in the region are mostly sand in composition, the source of the material having been glacial series of different age (Middle and Late Pleistocene). Alluvial sands of the terraces include a considerable proportion

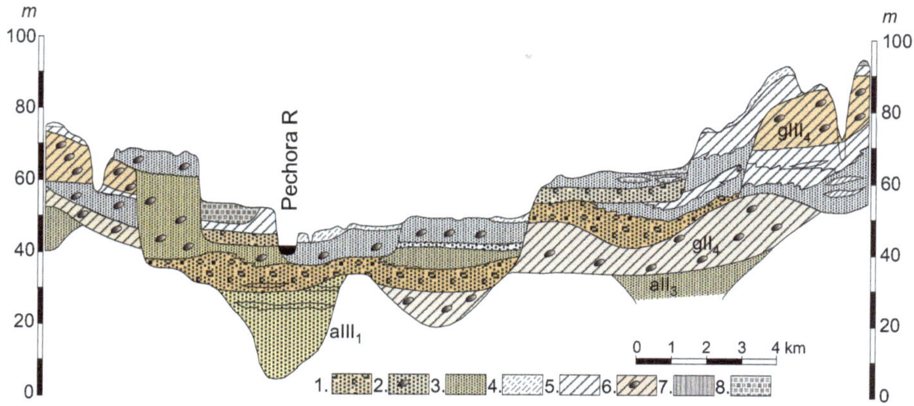

Figure 1.7.2 Cross-section of the Pechora valley near the town of Pechora (Rozanov, 1977; Lavrov and Potapenko, 2005). *1* – pebbles and cobbles; *2* – sand and gravel; *3* – sands; *4* – loess-like loam; *5* – clay loam; *6* – till; *7* – modern soil; *8* – peat.

of redeposited cobbles and pebbles derived from various glacial formations. The terrace composition and topography were essential for the Late Paleolithic man when he chose routes of his migrations northward.

Of special interest are factual data on the history of the Late Pleistocene glaciations in the NE of the East European Plain and on the climate and environments of the Middle Valday (Byzovsky) mega-interstadial (named after the Late Paleolithic site of Byzovaya). It should be noted that no proved interstadial deposits have been found as yet in the sequences at the site or nearby. The sequence at this site was chosen as the Middle Valday stratotype because of the chronological position of the cultural layer defined by numerous ^{14}C dates on bone material.

It has been established by the archaeological-paleogeographical investigations of the Late Paleolithic sites in Eastern Europe that the absolute majority of the early sites were not inhabited until the second half of the complex Middle Valday mega-interstadial. A single older site is known from the extreme north (Mamontova Kurya). There were found occasional stone tools and an ornamented mammoth tusk together with numerous bones of large mammals (Pavlov, 1997, 2002). The information available at present is not, however, sufficient for paleoenvironmental reconstructions (Svendsen et al., 2008).

The data on the vegetation of the period suggest forest formations of the northern taiga type existed in the north at the warmer intervals (Andreicheva and Marchenko-Vagapova, 2003; Andreicheva et al., 2006, 2008; Duryagina and Konovalenko, 1993). During the colder phases the northern part of the region was dominated by tundra and forest-tundra communities, though dark coniferous forests could persist farther south. Certain warm phases of the mega-interstadial could be close to interglacial conditions (see Chapter 11).

Numerous localities of large and small mammal fauna are known in the region, including those dated to middle Valday and to the last glaciation. It seems, however, that climate-controlled variations in the fauna composition are much less pronounced

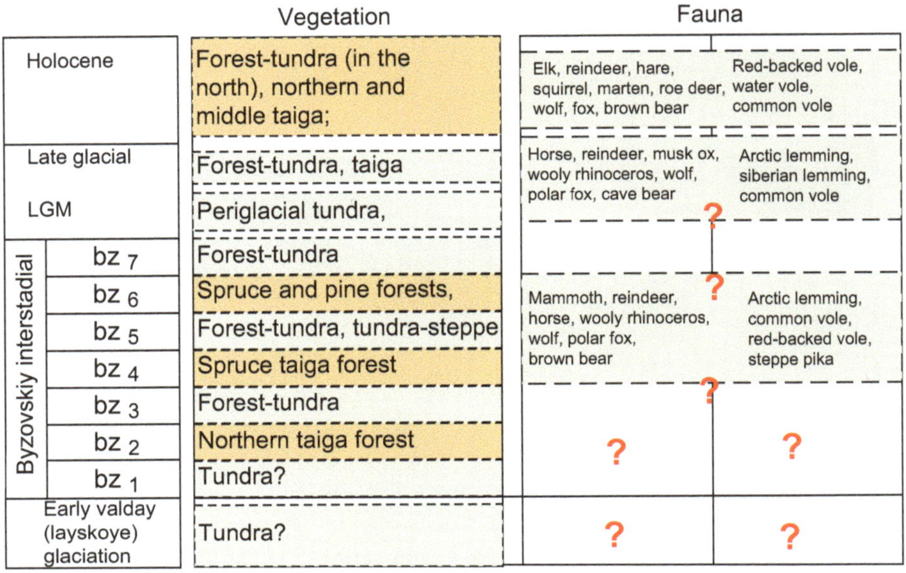

Figure 1.7.3 Changes in vegetation and fauna composition in the Late Pleistocene in the north of the East European Plain (Andreicheva and Marchenko-Vagapova, 2003; Andreicheva et al., 2006, 2008; Duryagina and Konovalenko, 1993; Kochev, 1991; Ponomarev, 2001).

than changes in vegetation reconstructed from palynological data (Fig. 1.7.3). Such a discrepancy may be assigned to the lack of radiometric dates and to the absence of precise correlation between the distinguished pollen zones and mammal fauna localities.

The last Valday (Polyarny) glaciation had a pronounced effect on the river valleys not only in the Pechora drainage basin, but also in that of the Kama River flowing southward. An integrated analysis of the glacial formations widely occurring in the north of the region and their relationships with fluvial and periglacial complexes widespread farther south permits to make a sound estimate of the duration and expansion of ice sheets in the Northeastern region (Yakovlev, 1956; Chernov, 1974; Lavrov and Potapenko, 2005; Andreicheva, 2002). Those data permitted to assign the Late Valday age to the moraines and marginal formations of the Polyarny ice sheet that occurred in the northern part of the region. Lithologically, the debris of the glacial deposits is of the Novaya Zemlya and the Polar Urals provenance. The accuracy in the last glaciation dating is of special importance in reconstructing natural environments and processes at the LGM and the Late Glacial time, when the Paleolithic man migrated into the region and colonized it.

There are numerous radiocarbon dates obtained on mammal bones for the main Late Paleolithic sites in the European Northeast (Pavlov, 2002, 2004; Pavlov and Indrelid, 2000; Svendsen and Pavlov, 2003). The dates vary considerably (within a few thousands of years). The spread is too great to correspond to the real duration of the man habitation at the sites and seemingly calls for further investigations. Stratigraphic criteria of the site dating (applied, for example, to sites of Mamontova Kurya, Byzovaya, Medvezhya, etc.) are often insufficient as the cultural components have been redeposited in many cases. Besides, the bones could be taken from earlier natural burials of animals (Pavlov, 1997, 2002). Under conditions of permafrost the mammoth cemeteries could persist for

Table 1.7.1 Radiocarbon dates on the materials from Mamontovaya Kurya and Byzovaya sites

Site	Laboratory index	^{14}C date, yr BP	Calibrated age	Dated material
Mamontova Kurya	LU-4008	12,380±60	14,450±180	Mammoth bone
	LU-3994	34,920±1040	39,540±1150	Mammoth tusk
	LU-4001	37,360±630	41,790±490	Mammoth bone
	T-15727	31,880±390	35,790±430	Mammoth bone
	T-15728	33,340±460	37,580±620	Mammoth bone
	T-15726	33,440±710	37,730±880	Mammoth bone
	T-11403	36,630+1310/−1130	41,180±1220	Mammoth tusk
	T-11503	36,770+2620/−1980	42,020±2870	Horse tooth
	TUa-3524	**34,655±570**	39,260±640	Mammoth tusk
	TUa-3525	**40,035+825/−1485**	43,840±700	Wolf bone
Byzovaya	LE-3048	14,150±150	17,210±230	Mammoth bone
	TA-121a	18,320±280	22,150±320	Mammoth bone
	TA-121б	25,450±380	29,640±470	Mammoth bone
	LE-3047	25,740±500	29,930±530	Mammoth bone
	LU-3995	27,110±240	31,110±150	Bone
	LU-3989	27,490±330	31,430±320	Bone
	T-11498	27,740±480	31,840±540	Bone
	T-13441	27,915±365	31,900±460	Bone
	LU-4010	28,230±920	32,460±910	Bone
	T-13439	28,485±290	32,460±450	Bone
	LU-3992	28,510±310	32,490±470	Bone
	LU-3979	29,160±430	33,240±500	Bone
	LU-3983	29.170±340	33,300±390	Bone
	LU-4007	33,180±2020	38,200±2410	Bone

AMS dates are shown in bold.

a long time, new animal carcasses being occasionally added to already existing ones. If such was the case, bones of different age could occur together, which would necessarily result in chronological discrepancies. The range of dates was considerable both on the Mamontova Kurya (31,880–40,035 BP) (Svendsen and Pavlov, 2003), and on Byzovaya (25,450–33,180 BP) (Pavlov and Indrelid, 2000; Svezhentsev, 1993; Sinitsyn et al., 1998; Velichko et al., 1999) (Table 1.7.1). As has been noted above, so prolonged existence of the Paleolithic sites is unlikely.

1.7.3 Mamontova Kurya

Mamontova Kurya located on the left bank of the Usa River (a right tributary of the Pechora River) is the earliest Late Paleolithic site in the Northeast of the East European Plain (Pavlov, 1997, 2002). Archaeological and paleogeographic information on it is rather scarce due to a complicated geomorphic and stratigraphic situation in the region

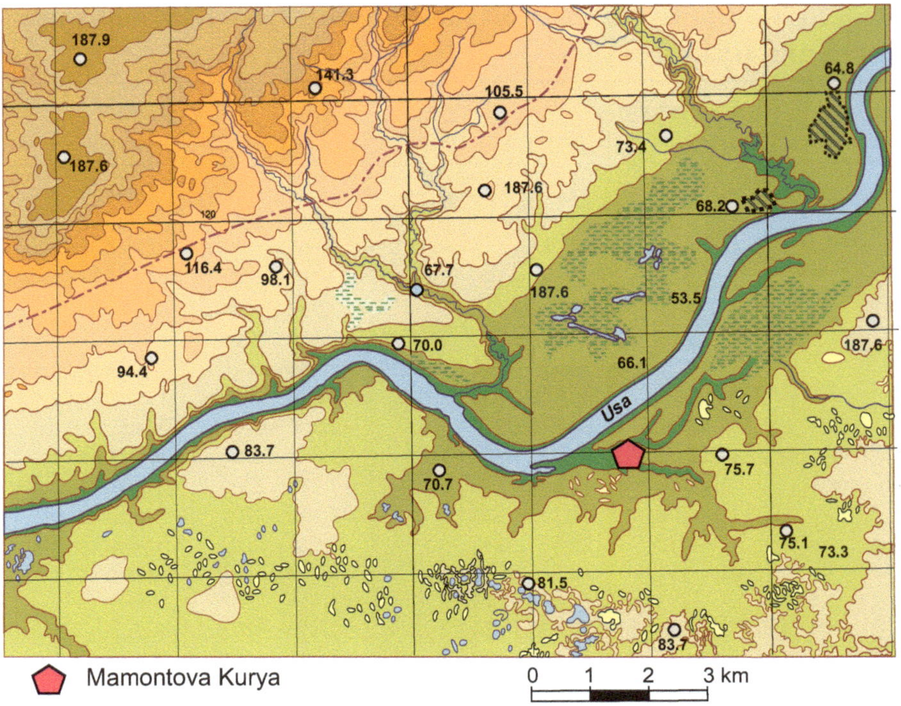

Figure 1.7.4 Location map of Mamontova Kurya (1 – Mamontova Kurya).

(Fig. 1.7.4). There are two levels of the third terrace composed of fluvial, lacustrine, and glaciofluvial facies dated to the Middle Pleistocene (Lider, 1976; Guslitser, 1960; Generalov, 1965; Kaletskaya, 1965; Rozanov, 1977; Lavrov and Potapenko, 2005). The available paleobotanical data suggest a relatively warmer climate recorded in the basal horizons of the fluvial formations presumably attributable either to early, or to middle Valday. There are numerous finds of the mammoth fauna related to those horizons.

The concentrations of bones and stone artefacts are confined to an enlarged segment of the Usa River valley. According to recent works (Svendsen et al., 2008; Mangerud et al., 1999), two geomorphic levels occurring at about 12–17 m above the river are in fact a single terrace built of a single alluvial series (Fig. 1.7.5). The fossiliferous cobble and pebble series are considered to be fluvial, though they are essentially different in lithological properties and mode of occurrence from alluvial formations described in the sections of the third and second terraces of the Usa River and its tributaries (Generalov, 1965; Lider, 1976; Rozanov, 1977). So, the chronostratigraphic context and genesis of the pebble bed with animal bones and artefacts call for further investigation. The studied section has no analogs in the Usa River terrace sequences. It is not inconceivable that its specific features are related to the bone deposition as an alluvial fan of balka (small tributary valley lacking perennial stream). A similar sequence was described at another site in the region, Byzovaya (Guslitser and Kanivets, 1965a; Guslitser and Liyva, 1972; Pavlov, 1997).

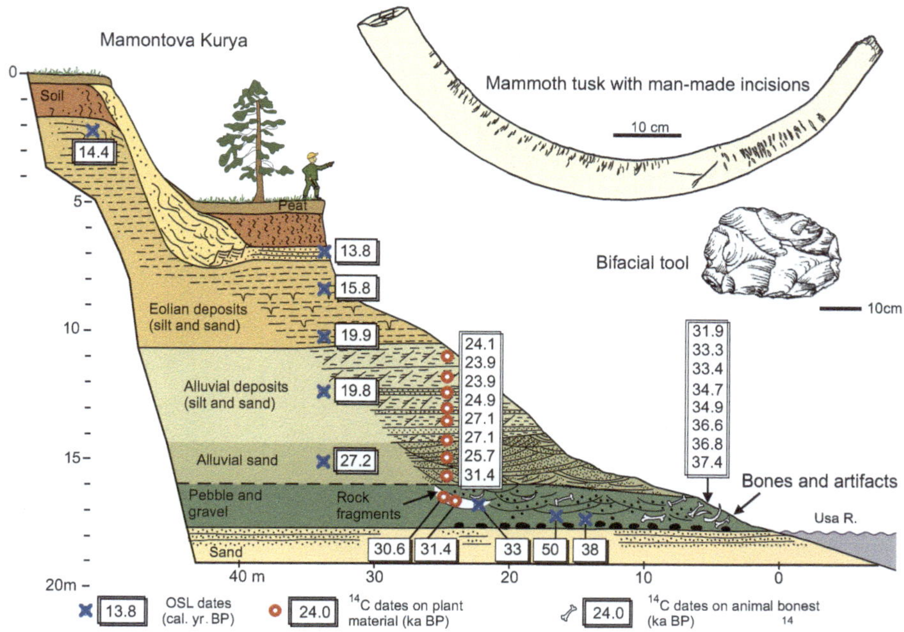

Figure 1.7.5 The section of the sediments at Mamontova Kurya (Svendsen et al., 2008). *1* – OSL dates, cal. yr BP; *2*–^{14}C dates run out on plant material, ka BP; *3*–^{14}C dates run out on animal bones, ka BP.

The fossiliferous horizons with artefacts, as well as overlying deposits, were dated using ^{14}C (including AMS) and OSL techniques (Svendsen and Pavlov, 2003; Svendsen et al., 2008). Ten ^{14}C (including two AMS) dates on the animal bones vary within a wide range–from 31 to 40 kyr BP. Such discrepancies are hardly attributable to errors in dating only.

The fauna is dominated by mammoth. The excavations conducted by P. Yu. Pavlov produced 28 large mammal bones. The bones identified by Ponomarev (2001) appeared to belong to the following mammal species: mammoth – 7 bones, wolf – 1, horse – 1, reindeer – 7, undetermined – 12 bones. Taking into consideration the limited quantity of bone remains in the sample, the faunal composition is comparable with that recovered from the Byzovaya.

1.7.4 Byzovaya

The Late Paleolithic site of Byzovaya discovered by E.M. Timofeyev has been studied by many specialists for many years (Guslitser and Kanivets, 1965a,b; Guslitser and Liyva, 1972; Kanivets, 1976; Pavlov, 1997, 2002). The site is located on the right bank of the Pechora River in its middle reaches (Fig. 1.7.6). It is positioned at the mouth of a large gully ("balka") cut into the steep slope of high terrace and of the watershed. Near Byzovaya, about 1 km downstream

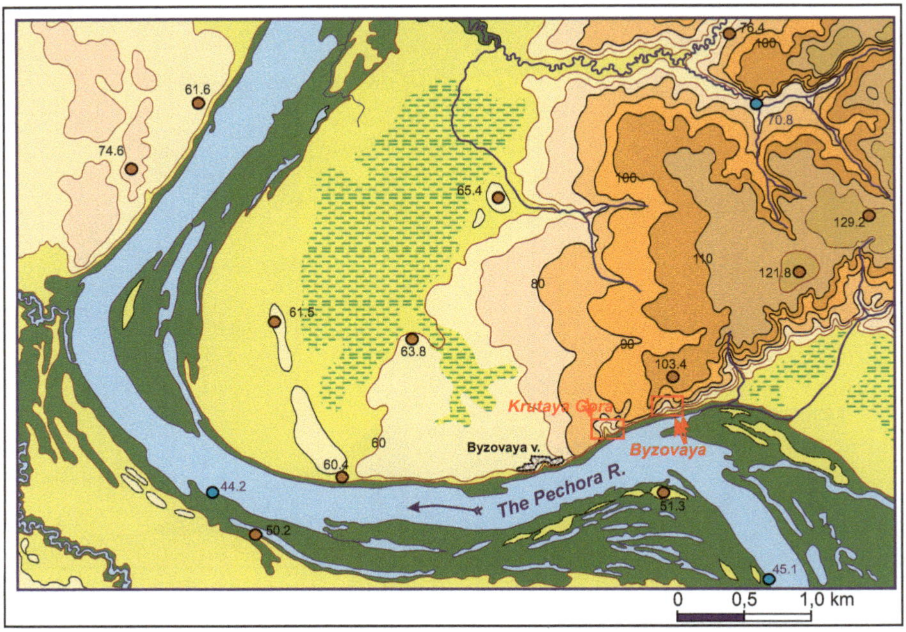

Figure 1.7.6 The Pechora River valley near the sites of Byzovaya and Krutaya Gora.

the Pechora River, there are Middle and Late Paleolithic sites (Krutaya Gora) and those attributed to Mesolithic (Topyd-Nyur) and Neolithic (Sirpach-Shor) (Fig. 1.7.7) (Kanivets, 1976). From all appearances, that wide segment of the river valley provided favorable environments for the Early Man both in the Late Pleistocene and in the Holocene.

The sediments enclosing the redeposited material of Byzovaya cultural layer are cobble and pebble in silt matrix (Guslitser and Liyva, 1972; Guslitser, 1976; Kanivets, 1976; Pavlov, 1997, 2002). Unlike Mamontova Kurya, this site is rich in stone tools and faunal remains. Various geomorphic levels of terraces (attributable either to the main valley or to its tributaries) are distinguished at the site (Fig. 1.7.8). The original occurrence of the cultural layer is not quite clear. Judging from the large-size enclosing material, it is unlikely that the cultural components were transported for a considerable distance. The stratigraphic position and genesis of the enclosing deposits are somewhat peculiar and have no analogs in other numerous excavations and drilling cross-sections of the Pechora terraces in the site vicinity.

The collection of lithics from excavation by Kanivets (1976) contains 136 items, 76 (56%) of them being stone tools. More than 2000 identifiable bones of animals were obtained from the same excavation. Such an abundance of stone tools and animal bones in a cultural layer is rarely found not only in the European Northeast, but in most of Paleolithic sites on the plain. The bones and stone objects are irregularly distributed over the excavation area, as was recorded during the excavations (1963–68) carried out by Kanivets (1976) (Fig. 1.7.9).

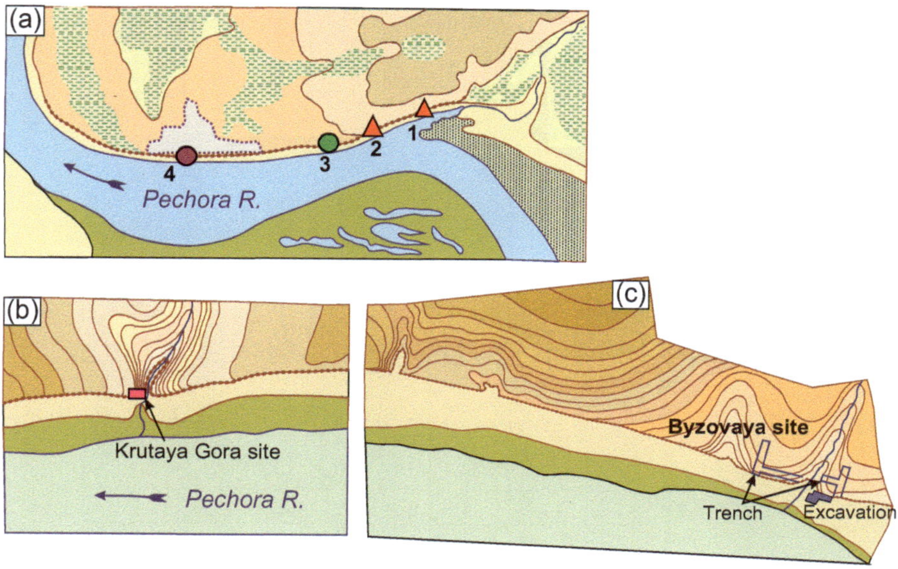

Figure 1.7.7 Position of archaeological sites near the Bysovaya village (Kanivets, 1976). (a), Krutaya Gora (b), and Byzovaya (c) *1*, *2* – Paleolithic sites: *1* – Byzovaya, *2* – Krutaya Gora; *3* – Neolithic site of Sirpach-Shor; *4* – Mesolithic site of Topyd-Nyur.

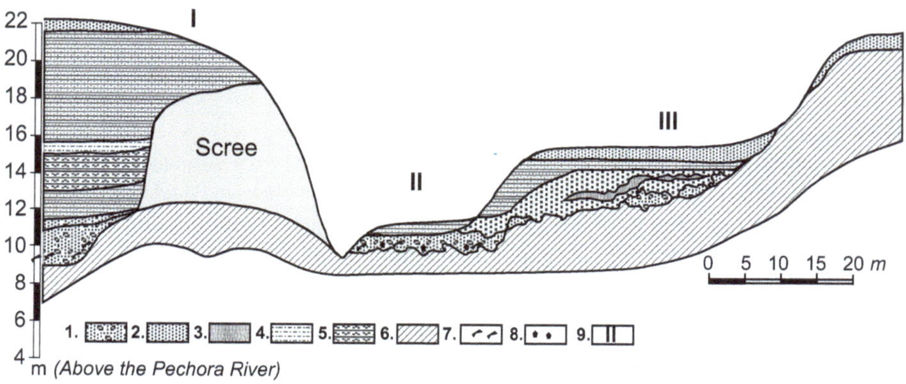

Figure 1.7.8 Geological section at Byzovaya (Guslitser and Liyva, 1972). *1* – pebbles and cobbles; *2* – sands with gravel; *3* – sand; *4* – sandy loam; *5* –loam; *6* – Triassic bedrock; *7* – mammal bones; *8* – stone tools; *9* – number of section. The elevations (in the left) are counted above the Pechora channel.

It seems possible that the abundance of bones in the layer bearing also cultural remains is due to the presence of a natural burial of large mammals at the same place (Pavlov, 1997, 2002). Excavations performed before 1997 revealed the absolute dominance of mammoth (2347 bones), with a much smaller proportion of other animals. There is, however, an essential difference in paleontological materials obtained from different areas of the excavation (Table 1.7.2). There are some areas marked by an absolute prevalence of horse or reindeer bones.

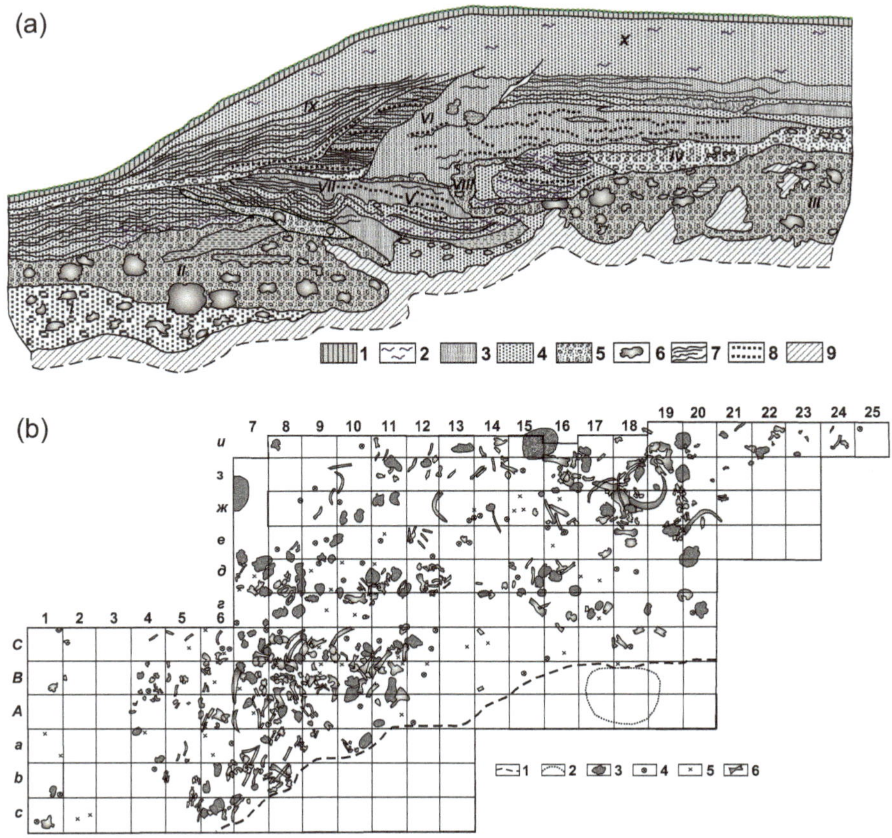

Figure 1.7.9 Byzovaya (Kanivets, 1976). (a)–cross-section of the excavation (squares И to К). *1* – modern soil; *2* – clay; *3* – sand; *4* – gravel; *5* – pebble; *6* – cobbles and boulders; *7* – lamination; *8* – ferruginization; *9* – Triassic bedrock. (b)–plan of the excavated area (from 1963 to 1968). *1* – boundary (the scarp); *2* – outlines of the bone concentration discovered by E.M. Timofeyev in 1962; *3* – cobbles and boulders; *4* – tools; *5* – flint fragments; *6* – large bones.

The balka deposits studied on several geomorphic levels and enclosing cultural layers of Byzovaya contain clastic material of various size redeposited from glacial formations of the Middle Pleistocene (see Fig. 1.7.9).

Aside from rock fragments, there are entire blocks of till found in undisturbed occurrence at greater altitude in higher terrace sequences (Ponomarev, 2001). There are still some uncertainties in the development of the small valleys (balkas) and therefore in the initial position of the site. Up to now the reconstructions of the Late Pleistocene environments in the Byzovaya vicinities and of the regional colonization process are based on archaeological data and on numerous radiocarbon dates on mammal bones and enclosing deposits (Pavlov, 2002; Svendsen et al., 2008).

Table 1.7.2 Faunal temains from the Byzovaya site (Ponomarev, 2001; Kuzmina, 1971; Pavlov, 1996)

Species	1963–66		1983–94	1997	Total number of bones
	Bones	Individuals			
Wolf	2	2	–	–	2
Polar fox	–	–	1	–	1
Brown bear	–	–	1	–	1
Mammoth	–	–	–	1	1
Horse	2051	21	296	100	2447
Woolly rhinoceros	1	1	–	–	1
Reindeer	1	1	–	2	3
Bison	12	3	14	4	30
Musk ox	–	–	–	2	2
Unidentifiable mammals	–	–	1	–	1
Total	2067	28	312	191	2570

1.7.5 Medvezhya Cave

A younger Late Paleolithic site–Medvezhya cave–is situated in the upper reaches of the Pechora River. The site is stratigraphically complicated due to specific features of sedimentation in caves or nearby (Guslitser and Kanivets, 1962a,b, 1965b; Pavlov, 1997; Svendsen et al., 2008) (Fig. 1.7.10). A correlation of the site with the Late Pleistocene formations on plains presents considerable difficulties. The radiocarbon dates vary over a wide range (which might result from a multistage human habitation at the site). Fossil mammal fauna vary only slightly from one layer to another (Kuzmina, 1971; Kochev, 1991), and the diagnostics of a warmer stage by the presence of fossil soils require a series of geochemical and paleopedological analyses. Pollen and spore analyses revealed a high proportion of fern spores, while arboreal pollen content is low enough (Guslitser and Kanivets, 1962b).

Medvezhya cave is noted for a complex stratigraphical context of artifacts and bones. Apart from the principal cultural layer of the site, the overlying beds contain also more than a 1000 of animal bones (Guslitser and Kanivets, 1962b; Guslitser and Pavlov, 1988; Ponomarev, 2001; Kochev, 1984; Svendsen et al., 2008). The determination of the site age was based on general concepts of the regional stratigraphy and changes in the small mammal fauna along the sequence of the enclosing deposits (Guslitser and Kanivets, 1962b). The main part of the 8-m sequence bearing cultural finds was conventionally dated by B.I. Guslitser to the Middle Valday.

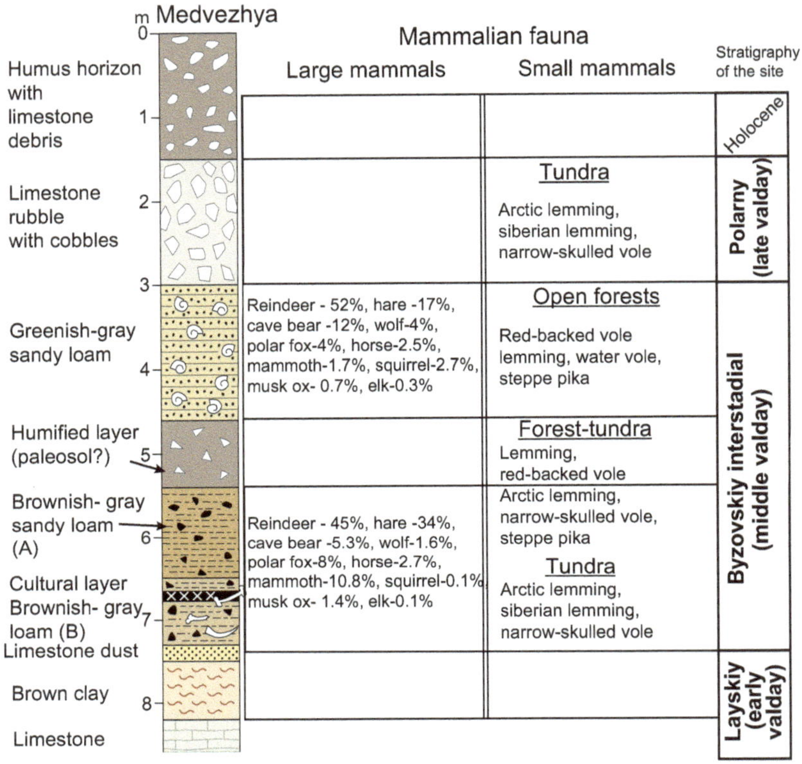

Figure 1.7.10 The sedimentary sequence of the Medvezhya Cave (Svendsen et al., 2008), mammal fauna from the cultural layers (Ponomarev, 2001; Kochev, 1991; Guslitser and Kanivets, 1962b; Guslitser and Pavlov, 1988), and the site stratigraphy (Guslitser and Kanivets, 1962b; Guslitser and Pavlov, 1988).

The species composition of the large mammal fauna is almost the same all over the sequence; the changes are noted only in proportion of individual species. There are species found together that occur originally in different faunal assemblages: polar fox, musk ox, elk and squirrel (Kuzmina, 1971) (see Fig. 1.7.10). The reasons for that are not quite clear, particularly taking into account that the radiocarbon age of the cultural layers (determined on bones) lies in the range of 18–12 kyr BP, and therefore they cannot be attributed to the Byzovskiy Interstadial (Table 1.7.3).

In the some specialists' opinion (based of the small mammalian fauna), the site sequence includes two cultural layers of different age dated to the last glacial age (18–16 kyr BP) and Late Glacial time (13–12 kyr BP) (Ponomarev, 2001).

The archaeology and paleogeography of the Medvezhya cave is still insufficiently studied. There is evidence for the site having been inhabited repeatedly, though nothing points to prolonged periods of human stay there. The cultural finds occur not only in the sheltered zone close to the entrance, but also in the open, just outside the cave. A few stone tools and animal bones were found there at a depth of 2 m below the surface (Fig. 1.7.11).

Table 1.7.3 **Radiocarbon dates on mammal bones from the Medvezhya Cave (Svezhentsev, 1993; Pavlov and Indrelid, 2000; Sinitsyn et al., 1998; Velichko et al., 1999)**

Laboratory index	^{14}C date, yr BP	Calibrated age
GIN-8400	11,840±50	13,660±60
LE-3059	12,230±100	14,200±220
GIN-8398	12,670±90	15,030±190
T-13476	13,260±230	15,930±340
LE-3060	16,130±150	19,460±200
LE-2876	17,980±200	21,780±270
LE-3059	17,960±200	21,750±270
GIN-8399	18,700±180	22,600±200

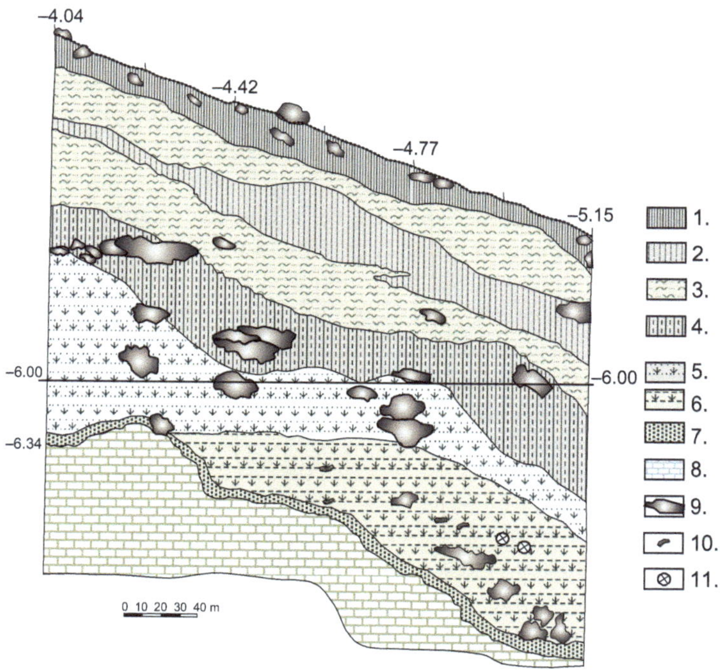

Figure 1.7.11 The cross-section of the Medvezhya Cave (Pavlov, 1997). *1* – modern soil; *2* – humified layer; *3* – greenish sandy loam; *4* – humified layer; *5* – brown loam, horizon A of a buried soil; *6* – brown loam, horizon B of a fossil soil; *7* – sandy loam; *8* – solid rock (limestone); *9* – limestone blocks; *10* – bones; *11* – flint tools.

1.7.6 Paleolithic sites in the middle reaches of the Kama River

The most significant Late Paleolithic sites in the southern part of the European Northeast are situated in the valleys of the Kama and its tributary Chusovaya River. Numerous sites of the Paleolithic, Mesolithic and Neolithic age known from this region

(Gromov, 1948; Bader, 1974; Denisov, 1976; Talitskiy, 1940a,b; Butakov et al., 1992; Galimova, 2001; Pavlov, 2002) suggest environments favorable for human habitation and long-distance migrations over the most part of the Valday. Mousterian sites in the middle reaches of the Kama (Peshcherny Log, Ganichata, Bolshaya Glukhaya, Garchi I) have no distinct chronostratigraphic reference marks and are tentatively attributed to the early part of the Valday on circumstantial evidence (Gromov, 1948; Pavlov and Makarov, 1998; Guslitser and Pavlov, 1993).

There are several sites in the region attributable to different stages of the Late Paleolithic (Gromov, 1948; Pavlov, 2002, 2008a,b); the best studied of them are Zaozerye and Garchi I located on the third terraces of the Kama and Chusovaya rivers. Another well known site–Talitskogo (Ostrov) – is noted for abundant archaeological and paleontological materials (Talitskiy, 1940a,b; Gromov, 1948), though not adequately dated (there is a single radiocarbon date of 18,700 yr BP). Stratigraphically, the cultural layer in this site occurs at the base of deluvial deposits overlying the second terrace of the Chusovaya (Gromov, 1948; Goretskiy, 1964; Vereshchagina, 1965).

Judging from geomorphic and stratigraphic position of the Late Paleolithic sites in the middle reaches of the Kama, it may be supposed that the most favorable conditions for the migrations of the prehistoric communities into the region developed there by the second half of the middle Valday interstadial (Velichko et al., 1999, 2005; Gribchenko and Kurenkova, 2006, 2007; Pavlov et al., 2005). Cultural layers of the sites attributed to the early part of the Late Paleolithic age (Zaozerye and Garchi I) occur within subaerial series on the third terraces of the Kama and its tributary–the Chusovaya River. The Talitskogo site is seemingly related to a fragment of the second terrace. Still uncertain is the geomorphic position of the Shirovanovo (Makarov and Pavlov, 2007) and Romakhino (discovered by P. Yu. Pavlov).

It is clear enough that the northward migration routes of the early communities depended heavily on physiographic characteristics of the region and specific features of the relief-forming processes. The latter were predetermined to a considerable degree by the natural processes at the epoch of the last Middle Pleistocene glaciation when the third terrace was formed in the Kama drainage basin. The terrace development was related to the meltwater flows from the last Middle Pleistocene ice sheet that expanded over the Pechora and Kama water divide, so the terrace is composed mostly of glaciofluvial material (Goretskiy, 1964; Obedientova, 1977; Lider, 1976). As for higher terraces (the fourth and fifth), they occur rather rarely and mostly as fragments. The surfaces of older terraces and interfluves are heavily dissected and therefore could hardly be used by prehistoric communities for traveling or inhabiting.

1.7.7 Talitskogo (Ostrov)

This Late Paleolithic site is located on a rather small fragment of the second terrace (Fig. 1.7.12). This geomorphic level is of rare occurrence, mostly as small fragments varying in altitude and sedimentary sequence. The terrace sediments are dated to the Mikulino interglacial.

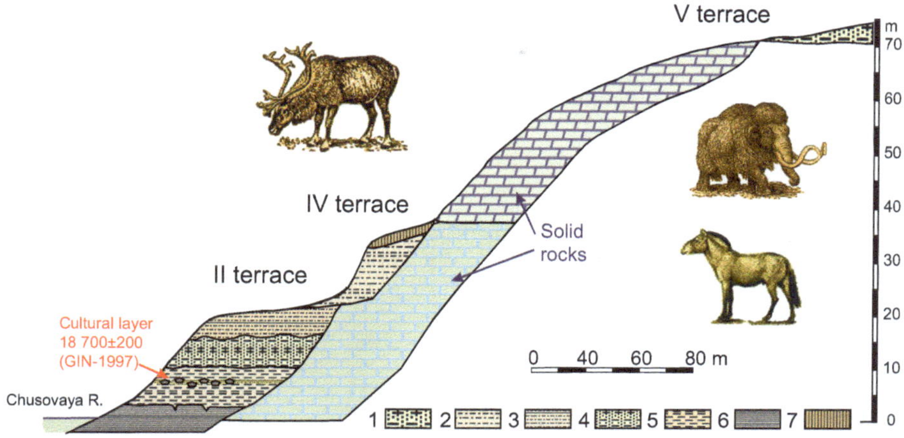

Figure 1.7.12 Schematic profile of the Chusovaya valley (right side) across the Talitskogo Paleolithic site (Gromov, 1948). *1* – loam with rock debris; *2* – deluvial loam; *3* – brown loam; *4* – laminated sands; *5* – interstratification of sands, sandy silt and clay; *6* – gray clay; *7* – modern soil.

The mode of the cultural layer occurrence and its ^{14}C age – 18,700 ± 200 (GIN-1997)/22,600 ± 230 cal. BP/– are seemingly not inconsistent with determination of this level as the second terrace adjoining the high scarp of higher (fourth and fifth) terraces or of the interfluve (Fig. 1.7.12).

Another fragment of the same terrace level was studied in the site vicinity (Vereshchagina, 1965). The pollen assemblages recovered from peaty deposits at the base of the section seemingly indicative of the Mikulino interglacial age of this horizon (Giterman, 1953; Vereshchagina, 1965). The correlation is not quite clear, however, between the cultural layer of the site and the alluvial sequence of the second terrace.

Along with the stone tools, the cultural layer yielded more than 1000 bones, only 270 of them turn out to be identifiable (Gromov, 1948), the reindeer being absolutely dominant (222 bones belonging to 5 individuals). There were also found 11 bones of mammoth belonging to 2 animals, 15 bones of horse (1 individual), 2 bones of woolly rhinoceros (1 individual), 15 – of polar fox (1 individual), and 7 – of hare (1 individual).

Most of small bones are calcined (Gromov, 1948). The structural characteristics of the cultural layer at the Talitskogo are still insufficiently known. The data of previous studies of the site archaeology and stratigraphy are scarce.

There is a noticeable similarity in the depositional sequences of the third terrace fragments housing Zaozerye and Garchi I (Figs. 1.7.13 and 1.7.14). There is a series of silts rich in plant detritus at the base of the sequence. Bones of large mammals are found in abundance. Pollen data are scarce. They suggest, anyhow, the landscape change from forests to forest-tundra (in the upper part of the sequence) (Goretskiy, 1964; Vereshchagina, 1965; Lider, 1976; Kuzmina, 1971; Yakhimovich et al., 1988) (Fig. 1.7.15).

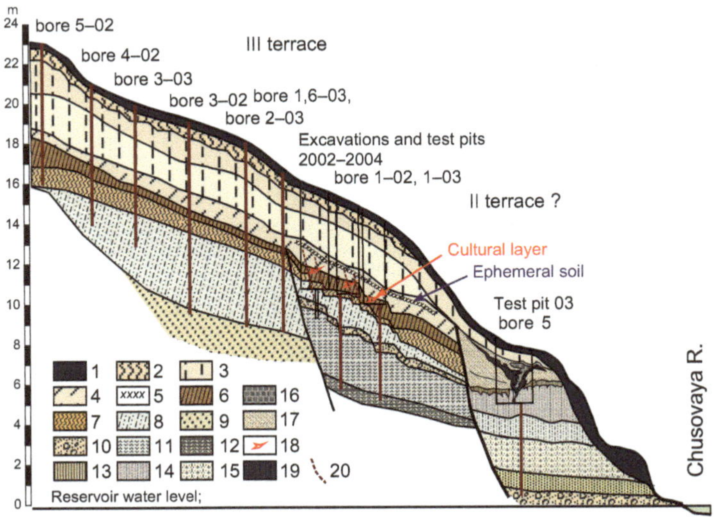

Figure 1.7.13 Geological profile across the second and third terraces of the Chusovaya River valley at Zaozerye. *1, 2* –modern soil: *1* – humus horizon, *2* – illuvial horizon; *3* – loess loam; *4* – brown loam; *5* – ephemeral soil formation; *6* – Bryansk and Mezin buried soils formation levels; *7* – illuvial horizon of the Mezin paleosoil; 8 – sandy silts of the Dnieper age; *9* – sand; 10 – gravel; *11* – gray laminated silt; *12* – brownish laminated silt; *13* – humified loam; *14* – gray loam; *15* – sandy loam; *16* – peat silt; 17 – brownish loam; *18* – cultural layer; *19* – peat; *20* – fault lines.

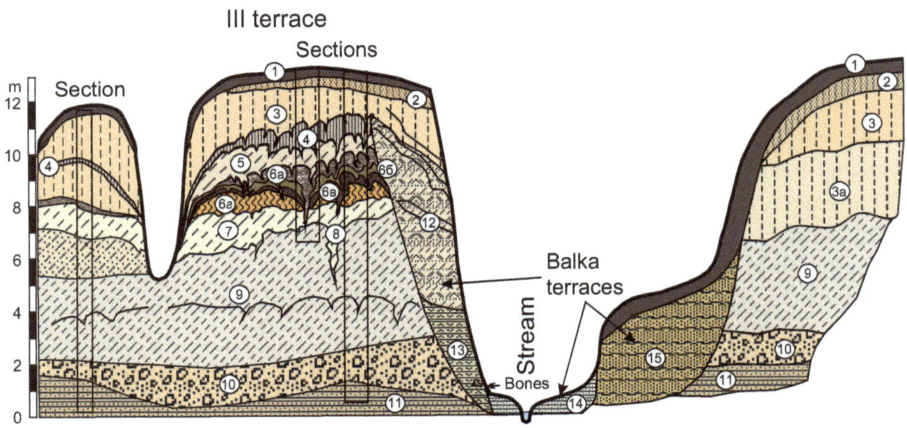

Figure 1.7.14 Geological profiles of the third terrace of the Chusovaya River and terraces of balkas at Zaozerye. *1, 2* –modern soil: *1* – humus horizon, *2* – horizon B; *3* and *3a*–loess-like loams; *4* – Bryansk buried soil; 5 – loess loam; *6* – Mezin pedocomplex (*6a*–Krutitsa humus horizon, *6б*–intra-Mezin loam, *6в*–Salyn humus horizon, *6г*–horizon B); *7* – late Dnieper loam; *8* – cryogenic wedge of the Dnieper age; *9* – gray silt loam of the Dnieper age; *10* – early Dnieper pebble and gravels; *11* – pre-Dnieper lacustrine silt; *12–15* – balka alluvial formations: *12* – thin laminated silt, *13* – greenish-gray silt with faunal remains, *14* – peat silt, *15* – laminated sand and silt.

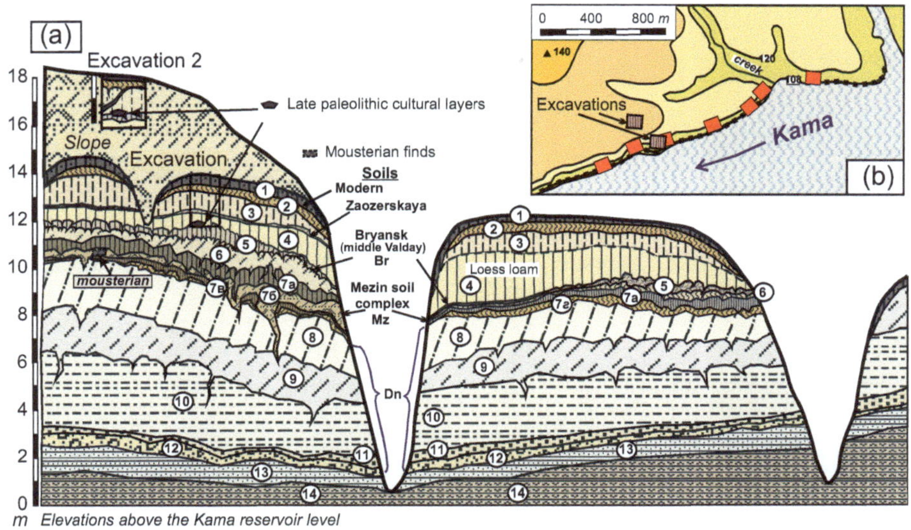

Figure 1.7.15 Sedimentary sequence exposed in the third terrace of the Kama River bank (a) and the plan of excavations and sections (b) at Garchi I. *1*, *2* –modern soil: *1* – humus horizon, *2* – horizon B; *3* – Late Valday loess; *3a*–Zaozerskaya buried soil; *4* – Late Pleistocene loess; *5* – Bryansk buried soil; *6* – Early Valday loam; *7* – Mezin pedocomplex: *7a*–Salyn humus horizon, *7б*–intra-Mezin loess, *7в*–Mikulino (Krutitsa) humic horizon, *7г*–Horizon B of the Mikulino buried soil; *8–10* – Dnieper glaciofluvial deposits: *8* – Late Dnieper loam, *9* – gray loam of Dnieper age, *10* –Early Dnieper silt with gravel; *11* – glaciofluvial sand; *12* – glaciofluvial gravel; *13* – lacustrine silt; *14* – gray lacustrine clay.

The section studied by Yakhimovich and her colleagues (Yakhimovich et al., 1988) is in the Kama valley downstream from Garchi I at the Obva River mouth. In common with the third terrace at Garchi I and Zaozerye, the terrace sediments here are diversified (Fig. 1.7.16). The upper part of the sequence–silts and sands with gravel lenses–corresponds to glaciofluvial facies of the Dnieper glaciation. At the contacts of sand and silt members of this series there are cryogenic deformations (large ice-wedge casts). Specific lithological properties of those formations could exert an essential effect on the Late Pleistocene sedimentation processes (Gribchenko, 2008). The pollen assemblages provided the basis for estimating the environment evolutions throughout the Late Pleistocene including the time of the Paleolithic site functioning in the Kama drainage basin (Yakhimovich et al., 1988).

Another Late Paleolithic site studied in the same region is Shirovanovo II at the Inva River mouth (Makarov and Pavlov, 2007). From the position of the site it may be concluded that it was initially positioned on the third terrace 19–22 m above the river channel, now it is at about 7–10 m above the Kama reservoir surface. In common with Zaozerye on the Chusovaya River, the cultural layer of Shirovanovo is overlain with Late Pleistocene loess-like loam 3 m thick.

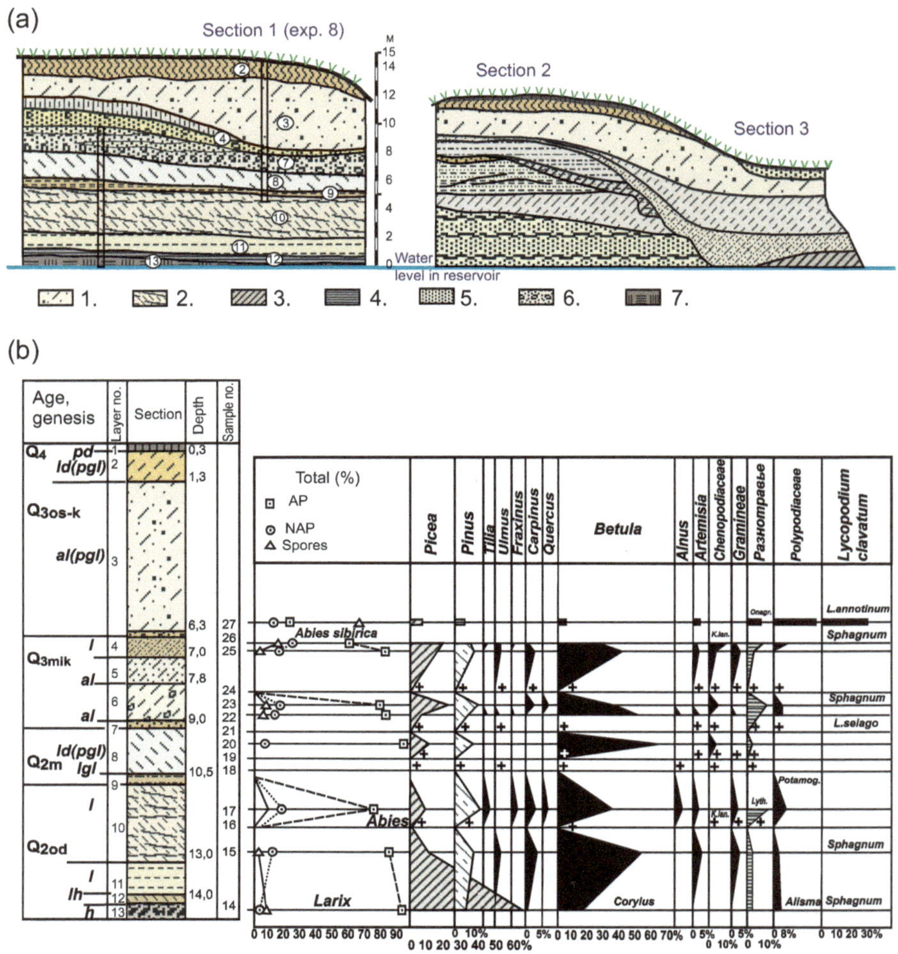

Figure 1.7.16 The Pleistocene sequence at the Obva River mouth (the right tributary of the Kama River) (a) and pollen diagram of the third terrace of the Kama near the Obva River mouth (Yakhimovich et al., 1988) (b). Section 1 – the Obva River mouth (exp. 8, the third terrace (Vereshchagina, 1965)). (b) Sections 2 and 3 – the Kama River valley about 1 km downstream from the Obva River mouth (Yakhimovich et al., 1988). 1 – loess loam; 2 – silt; 3 – loam; 4 – clay; 5 – sand; 6 – gravel; 7 – peat. Description of layers (figures are shown in *circles*): 1 – humic horizon of modern soil; 2 – Horizon B; 3 – loess; 4 – Mezin buried soil; 5 – sand and gravels; 6 - sand; 7 – pebbles; 8 – sandy loam; 9 – ferruginous loam; 10 – laminated loam; 11 – greenish loam; 12 – gray clay; 13 – peat silt.

1.7.8 Garchi I

The Mousterian artefacts from the Garchi I might be related to the Mezin paleosol complex. Though stone tools are found in abundance at the base of the exposure, their original position may be only inferred from the rare occurrence of flint or chert pieces in the paleosol (Pavlov and Makarov, 1998). Fig. 1.7.16 gives a comprehensive idea of the Kama third terrace, with the Late Pleistocene loess-soil formation overlying a glaciofluvial facies complex dated to the Dnieper glaciation. The Mousterian artefacts in Garchi I most probably correspond to the sandy loam horizon separating the Mikulino interglacial soils from those of the Early Valday interstadials (Brörup-Amersfoort). The soils overlying the Dnieper series form a part of a paleosol complex correlatable with the Mezin paleosol complex known from the central and southern regions of the East European Plain.

In the overlying series of the Valday loess there is a horizon of the Middle Valday (Bryansk) soil, the upper part of which contains the cultural layer of the Late Paleolithic site of Garchi I. The site age is determined by the radiocarbon date on charcoal at 28,750 ± 795, TUa-941/32,800 ± 830 cal. yr BP/and by a series of OSL dates on samples taken from soil and loess horizons (Svendsen et al., 2008). The Late Pleistocene loess-soil formations are noted for a complicated structure related to the multiple phases and cyclicity in the climate and landscape evolution during the early and middle Valday. Particularly distinct variations in the lithology of the uppermost horizons have been identified at the sites in the levels immediately underlying the habitable area.

The sequence exposed at Garchi I displays cryoturbations not only traceable in the Middle Pleistocene glaciofluvial silts and sands, but also disturbing the above-lying soil profiles (Fig. 1.7.17). A similar picture is observed in the scarp of the lowered third terrace near Zaozerye: faults of noticeable amplitude are distinctly seen in layers lowering towards a wide balka opening into the Kama valley. In common with the area of the site itself, the time of the deformations coincides with the permafrost degradation at the end of the Late Valday, when massive ground ice melted in local depressions.

Strange as it may seem, no signs of polygonal ice wedges has been recorded in close vicinities of Late Paleolithic sites and localities in spite of continuous permafrost present there in the Late Pleistocene (Velichko, 1973; Velichko et al., 1982, 1997; Butakov, 1986).

It seems possible that under conditions of increasing aridity towards the end of the late Valday glaciation the formation of polygonal ice wedge systems in the middle reaches of the Kama depended mostly on the local hydrological characteristics, so that large cryogenic wedge-like structures could develop both on lower geomorphic levels (Zaozerye) and on elevated surfaces (Garchi I) (Fig. 1.7.18). At the Zaozerye site there are cryoturbations distinctly seen in the system of faults on the side of the Chusovaya valley.

Noteworthy is the presence of several superimposed paleosols in the geological context of the Zaozerye cultural layer. That may indicate a prolonged period of low sedimentation rate (including the Mikulino Interglacial, Early and Middle Valday). As a result of that superposition, the cultural layers of the considered sites occur on the top of humified loams that are actually fragments of partly redeposited paleosols dated to the Mikulino Interglacial and to warmer phases within Early and Middle Valday.

The sedimentation rate increased considerably after the early man had left the region in the Late Valday. That acceleration resulted in accumulation of loess loam 5 m thick including two soil horizons, rather weakly pronounced. The horizons presumably correspond to intervals of a relative stability of the paleosurface.

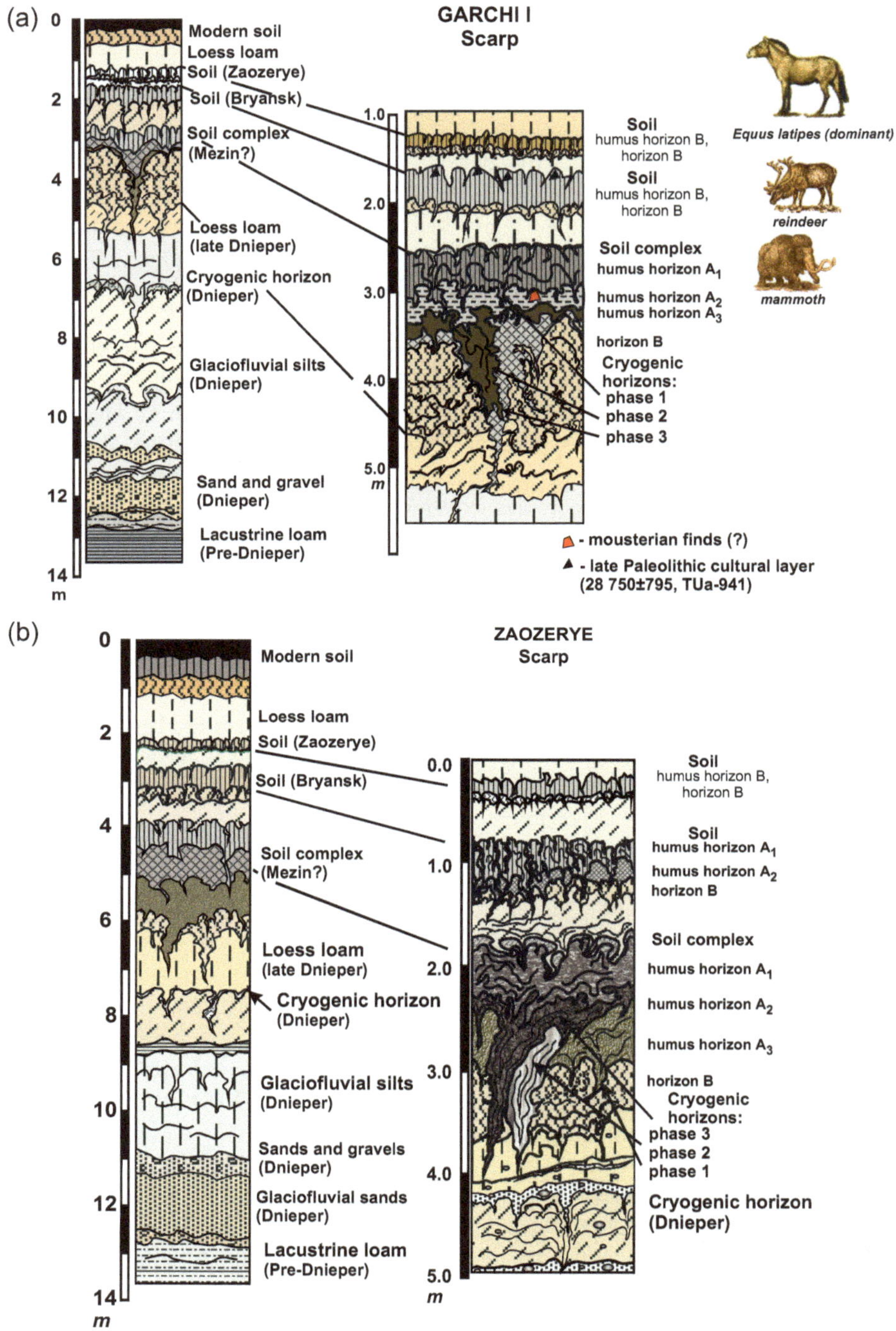

Figure 1.7.17 The Pleistocene sequences at Garchi I and Zaozerye. Vegetation at the Late Upper Paleolithic is reconstructed as open forest of spruce, pine and birch.
[After Nemkova, V.K., 1978. Stratigraphy of the late glacial and postglacial deposits Preduralye. In: Yakhimovich, V.L. (Ed.), On the History of the Late Pleistocene and Holocene of the Southern Urals and Preduralye. Bashkir Branch of the USSR Academy of Sciences Publ., Ufa, pp. 4–44 (in Russian).].

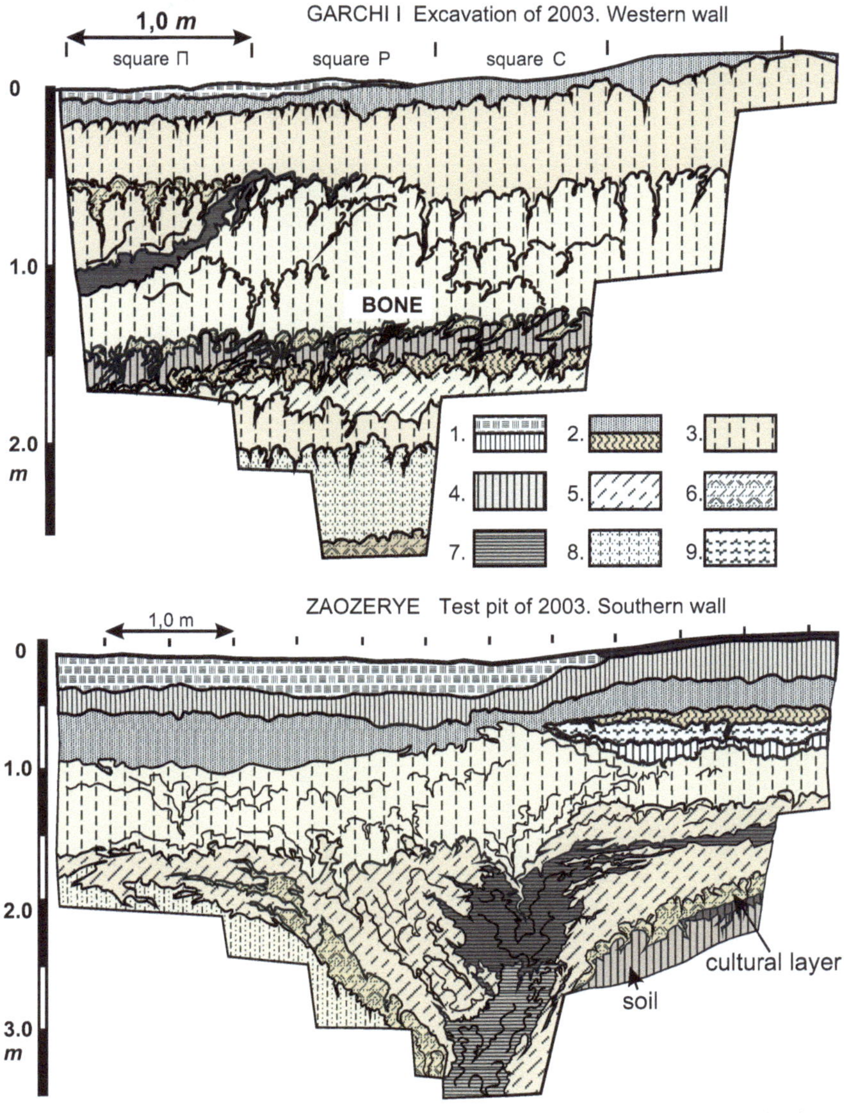

Figure 1.7.18 Wedge-like structures at Garchi I and Zaozerye. *1* – sod and humus horizon; *2* – podzols and horizon B of the soil; *3* – loess loam; *4* – signs of soil formation; *5* – sandy loam; *6* – gleying; *7* – charcoal; *8* – silt; *9* –loam.

1.7.9 Zaozerye

The archaeological and paleogeographic characteristics permit to attribute Zaozerye to the early part of the Late Paleolithic. Sites attributable to that epoch are rather scarce in the regions farther south. The site under consideration is noted for a complicated chronology (in spite of numerous radiometric dates (Pavlov, 2004)) (Table 1.7.4) and

Table 1.7.4 Radiocarbon dates (conventional and AMS) from the Zaozerye site (Pavlov, 2004)

Laboratory index	^{14}C date, yr BP	Calibrated age, cal. yr BP	Dated material
GIN-11499	31,000±400	34,970±400	Bone
GIN-11500	31,000±500	35,000±490	Bone
GIN-11501	31,000±700	35,100±720	Bone
GIN-11498	31,500±500	35,470±520	Bone
Poz-5124	30,700±400	34,680±380	Bone (AMS)
GrA-22751	30,140±190	34,190±190	Charcoal (AMS)
Poz-5075	33,150±410	37,380±580	Bone (AMS)
Poz-5076	33,450±420	37,700±580	Bone (AMS)

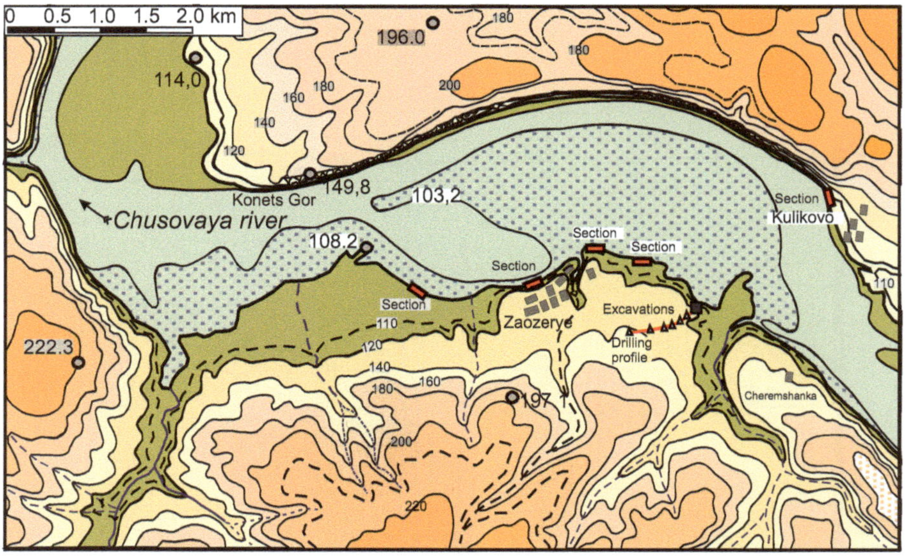

Figure 1.7.19 Topography of Zaozerye vicinities.

a specific stratigraphic context of the cultural layer. The site is situated on the third terrace of the Chusovaya River (Fig. 1.7.19). The cultural layer has been studied over a considerable area (Pavlov, 2004).

The site location is bounded on two sides by large gullies with steep slopes. The main excavations carried out by P. Yu. Pavlov (who had been working there since 2001) were concentrated near the outer edge of the terrace surface. The surface of the site itself rises gradually from 15 m above the reservoir to 18–22 m (Fig. 1.7.20). The loess loam overlying the cultural layer forms a mantle varying in thickness from 4.5 to 3.0 m.

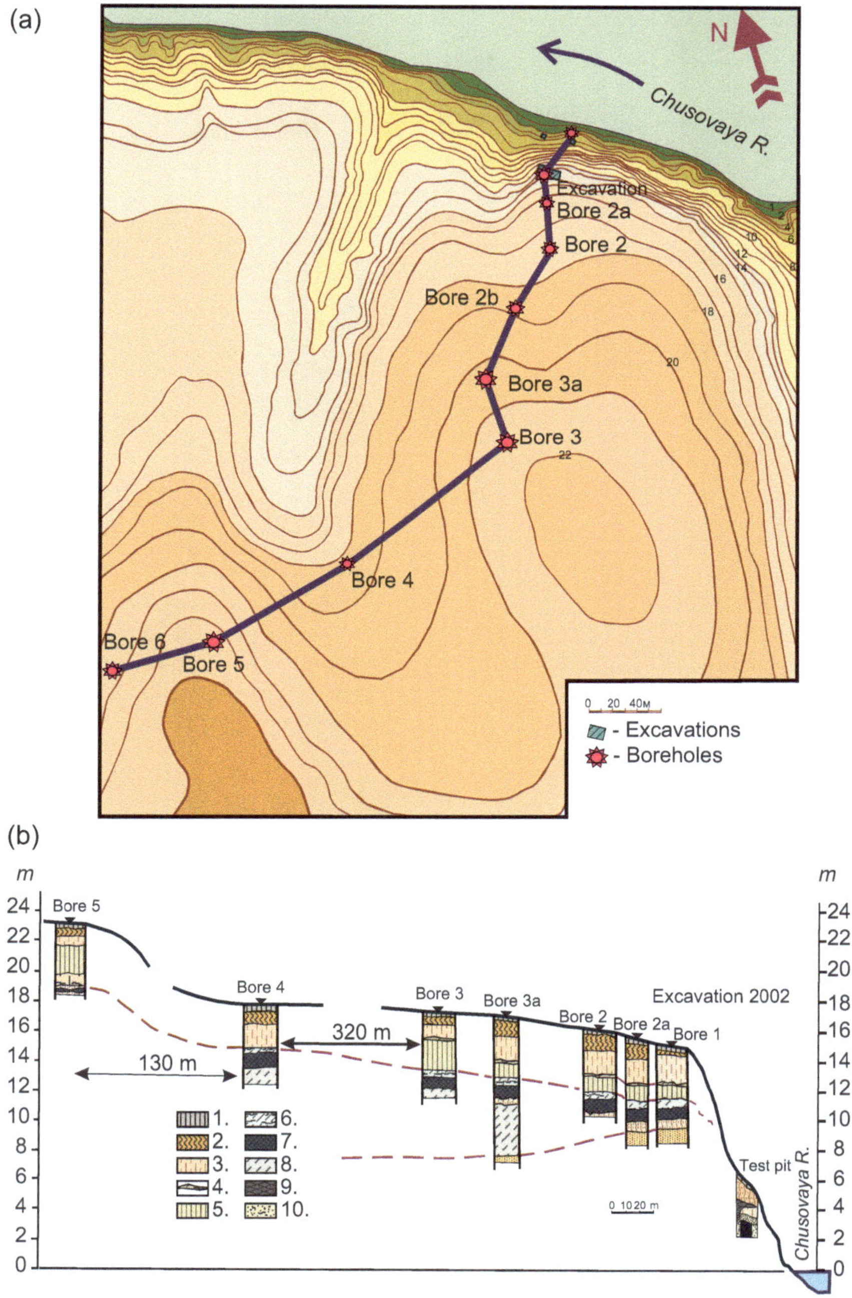

Figure 1.7.20 Topographic plan of Zaozerye (a) and the drilling profile across the site (b). *1*, *2* –modern soil: *1* – humus horizon, *2* – horizon B; *3* – loess loam; 4 – soil formation level; 5 – loam; *6* – gray gleyed loam; *7* – humified loam; *8* – silty loam; *9* – peaty loam; *10* – sand and gravels.

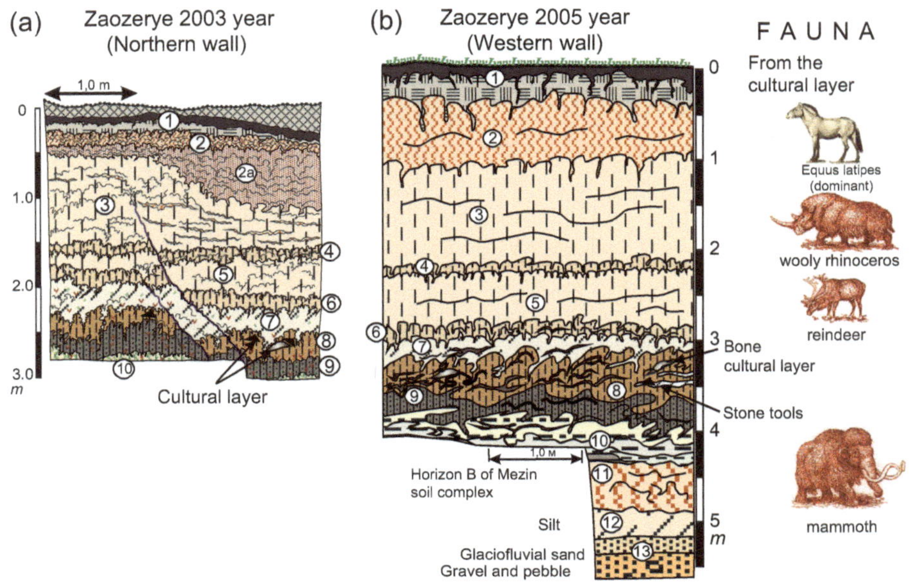

Figure 1.7.21 Geological context of the cultural layer at Zaozerye. (a)–northern wall of excavation of 2003; (b)–western wall of excavation of 2005. *1* – modern soil; *2* – illuvial horizon; *2a*–podzolic horizon; *4* – Zaozerye buried soil; *5* – loess; *6* – horizon of ephemeral soil formation; *7* – loess loam; *8, 9* – humified loams of the Bryansk and Mezin buried soils; *10* – gleyed loam; *11* – illuvial horizon of the Mezin buried soil; *12* – Dnieper silt; *13* – Dnieper sand and gravels.

The cultural finds were mostly recovered from a humified horizon with distinguishable fragments of multiple-aged soil formation (Fig. 1.7.21). Stone implements and animal bones are often confined to gleyed lenses indicative of excessive moistening periods at the time of human habitation.

By analogy with other sections studied, such formations may be related to buried seasonal-thawing layer dated to the end of middle Valday interstadial–beginning of the Late Valday glacial epoch. A considerable number of ^{14}C dates on horse bones from the cultural layer (33–30 ka BP) strongly suggest the soil formation to have been finished at the time of human habitation at the latest.

After that an active eolian and eolian-deluvial accumulation of loess started that was interrupted at least twice by short periods of the surface stability; those episodes were recorded by two levels of ephemeral soil formation and the development of large fault systems.

The occurrence of the soil complex in the site sections displays signs of weak downslope movements. There are lenses of darker organic-rich material related genetically to an earlier soil formation (presumably dated to the Early Valday interstadial). The underlying horizon displays morphotypical characteristics of horizon B of the interglacial (Mikulino) fossil soil. The entire series is no more than 0.5 m thick. Unlike the soil complexes and horizons (Mezin and Bryansk) exposed in the lowered terrace scarps, the excavations at the site itself showed them as a single series.

A distinctive feature of the sediments enclosing the Zaozerye cultural layer is a superposition of the several fossil soil horizons indicative of a prolonged period of a reduced subaerial sedimentation (the Mikulino Interglacial, Early and Middle Valday). That may also account for a spatial inhomogeneity of the soil formation (variations in structure, granulometry, humus and carbonate contents) found by laboratory analyses of deposits sampled in the course of excavations. It maybe safely supposed that the diversified composition of pollen spectra obtained from deposits enclosing the cultural layer and those underlying it resulted from superposition and distortion of soils attributed to different soil formation epochs.

Palynological analysis of samples taken from the cultural layer and overlying loam was performed by E.M. Zelikson and revealed assemblages mostly of the forest type dominated by tree species (Velichko et al., 2008b). Most common are pollen of hazel, hornbeam, oak, elm, lime, birch and common pine. Spruce is present in a smaller proportion. There have been identified also pollen of cold-tolerant dwarf birch, shrub birch, and *Alnaster*, as well as spores of fern *Botrychium boreale* (Fig. 1.7.22). So the presence of broadleaf trees makes the vegetation similar to a certain extent to that of the Mikulino Interglacial optimum, as there were no intervals warm enough for such thermophilic species during the Valday. That leads us to the only possible conclusion that the thermophilic pollen was redeposited from the Mikulino layers. As to birch and pine, they could persist in the region throughout the glacial–interglacial cycle.

The associated occurrence of ecologically-different pollen may be attributed to the mixing of humus material of the Mikulino soils and of those belonging to the mid-Valday mega-interstadial, the mixing taking place within the limits of the site on the habitable paleosurface itself. Such a conjecture is substantiated by the palynological data obtained from another section on the same terrace level at Zaozerye by V.V. Pisareva.

The reconstructions based on the samples from the cultural layer and enclosing deposits reveal a different type of vegetation (Velichko, 2009b). The spectra are dominated by arboreal pollen (40% of the total), while non-arboreal species account for 33% and spores–for 27%. In the AP group dominant are spruce (*Picea* sp.) and pine (*Pinus sylvestris* and *Pinus sibirica*). Of deciduous species there are birches (*Betula alba*, *Betula humilis*, *Betula nana*) and alder including shrub forms (*Alnus incana*, *Alnaster fruticosus*). As follows from the analysis, the deposition period was marked by forest communities of spruce, pine, birch and alder coexisting with cryophilic plants–dwarf birch, shrub alder, clubmoss *Selaginella selaginoides*, and fern *Botrychium boreale* indicative of a rather cold climate.

At the level of the cultural layer pollen of tree species is typical along with that of dwarf shrubs and herbs. There are present *Licopodium appressum* spores–the plant now occurring in tundra and forest-tundra. The presence of fern spores identified as *Pteridium aquilinum* and *Polypodium vulgare* suggests the mixed forest communities in combination with meadows and wetlands.

Pollen spectra retrieved from the deposits lying immediately above the cultural layer are dominated by tree species–mostly pine and birch, with occasional grains of dwarf birch pollen. Forest communities coexisted there with meadows and bogs. On the whole, the landscapes were not unlike those of the site functioning time.

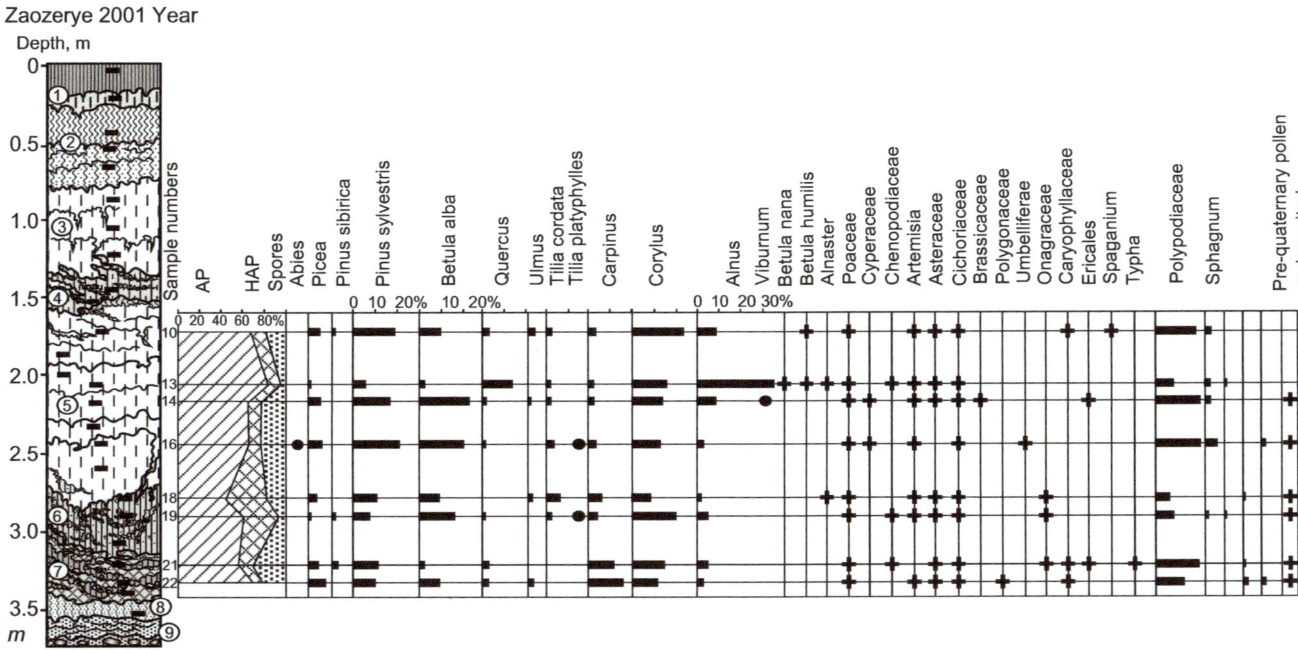

Figure 1.7.22 Pollen diagram of the Zaozerye site section sampled in test pit C in 2001 (analyses by E.M. Zelikson). Numbers in *circles* designate layers: *1* – modern soil; *2* – illuvial horizon of the modern soil; *3* – loess loam; *4* – Zaozersk buried soil; *5* – loess loam; *6* – Bryansk buried soil with the cultural layer; *7* – distorted fragments of Mezin buried soil; *8* – horizon B of the Mezin pedocomplex; *9* – sand and clay glaciofluvial sediments with pebbles; *10* –gravels and pebbles.

A conspicuous change in environments is recorded in the loess loam overlying the above described. The pollen assemblages are dominated by pollen of grass and dwarf shrubs (53%). Among them, there are Cyperaceae, Chenopodiaceae, Asteraceae, Polygonaceae, *Cichoriaceae*, Fabaceae. Of tree species, there are tree birch, Alnaster, and common pine, while neither spruce, nor Siberian pine have been found. According to the conclusion by V.V. Pisareva, during the period of the loess loam deposition the local vegetation consisted mostly of wetland, aquatic, and ruderal communities typical of floodplains and low terraces in river valleys.

In the mode of the cultural layer occurrence there are noticeable variations in altitudes due to a complicated system of faults. The vertical displacements at the level of the cultural layer amount to 10–20 cm. Individual fault lines are 2–4 m apart and oriented towards a large balka east of the site. The faults developed after the cultural layer had been overlain with loess loam about 0.5 m thick. In the overlying loess series the fault deformations cannot be traced. The faulting resulted in the gradual lowering of the cultural layer towards the river by a few meters.

The Late Pleistocene loess-paleosol series reflect multi-phase and cyclic character of the landscape and climate variations during the early and middle Valday. The comparative litho-stratigraphic analysis performed on deposits enclosing the Zaozerye cultural layer and on coeval deposits in vicinity of the site revealed noticeable differences principally at the level of human habitation.

Stratigraphic position of the cultural layers at the long-term settlements (sites) is inferred from their correlation with alluvial (terrace) series and the overlying loess-paleosol series. The fossil soils identified in the sequences are highly variable in morphotypical characteristics, which hinders spatial correlations of the soil horizons. Because of that, it becomes difficult to correlate fossil soil horizons in the sections widely apart from each other. Undoubtedly, further multidisciplinary investigations are required in order to trace a relation between the early man migrations and colonization of the European Northeast on one hand and the cyclic changes of climate and environment in the Late Pleistocene on the other.

The earliest Late Paleolithic sites are related to the geomorphic levels of the third terraces of the Pechora and Kama rivers. That is true of Byzovaya, Garchi I and Zaozerye, positioned on the terrace surfaces underlain with thick glaciofluvial series attributed to the last Middle Pleistocene glaciation. The surfaces inhabited by the Paleolithic people had been finally formed at the end of the Middle–Early Late Pleistocene.

The middle reaches of the Kama River are distinct for lower surfaces of the third terrace somewhat different in structure, the Late Valday loess series being reduced in thickness and fossil soil horizons absent. The local specific features, and most of all the long periods of surface stability, offered the prospect of human habitation.

1.7.10 Conclusion

A comparison of radiocarbon dates obtained for cultural layers shows a noticeable similarity in different Paleolithic sites, such as Mamontova Kurya and Byzovaya (with the only exception of the Medvezhya cave), while dates obtained for a single site often vary over a wide range. Taking this into account, it seems reasonable to consider them together with sites farther south, in the middle reaches of the Kama (Zaozerye, Garchi I, Talitskogo).

In spite of scarcity of sites known in the Northeast of Europe and insufficient paleogeographic information, it seems reasonable to distinguish, as a matter of convention, three main stages in the initial human penetration and colonization of this region (Fig. 1.7.23):

- the early stage–Mousterian (beginning of the Late Pleistocene);
- the main stage–early stage of the Late Paleolithic (end of the middle Valday Interstadial);
- the late stage – 18,000–17,000 yrs BP (LGM).

Supposedly, the periods of relatively warmer and milder climate at the second half of the middle Valday mega-interstadial permitted the Early Man communities to move to the Far North. The preferred routes, however, were confined to the Kama and Pechora valleys, along the Ural Range. The valleys, with wide and continuous terraces (the third in particular) composed of clastic material provided a means for colonization of the region. Other routes were much less suitable for traveling northwards.

The peopling and colonization of the northern plains of Eastern Europe proceeded in several steps throughout the Late Pleistocene and Holocene against the background of changing landscapes and climates.

Throughout the Middle Pleistocene and at the beginning of the Late Pleistocene a considerable part of the East European Plain (in particular, its northwestern regions) was practically unsuitable for human habitation, mostly due to specific features of the Scandinavian ice sheets at the various stages of the maximum–Dnieper–glaciation (Middle Pleistocene). In the Late Pleistocene the Scandinavian ice sheet expanded over the northwestern regions, with large ice-dammed lakes formed on the ice periphery.

The peopling and colonization of the northeastern regions of Europe began at the second half of the middle Valday mega-interstadial, after the Bryansk paleosol had been formed. The northwestern regions of the plain were gradually covered by the growing Scandinavian ice sheet. Both ice-covered areas and periglacial zones of meltwater drainage, ice dammed lakes, outwash plains, and waterlogged areas did not provide environments suitable for humans.

The most favorable conditions for man's migrations to the north existed in the eastern part of the region, in the Kama and Pechora drainage basins, where the people could use wide terrace surfaces for hunting and making dwellings. The sites are usually confined to the areas of the Late Pleistocene loess, most of cultural layers being directly related to the loess-paleosol formations. In many cases the sites were really hunting camps, with cultural layers containing but scanty artefacts. In the northernmost sites (Byzovaya, Mamontova Kurya) all the material had been redeposited, which makes the reconstructions hardly possible.

A relatively early penetration of man into the northern latitudes of the East European Plain was to a considerable degree due to the regional landscapes in the zone fringing the western slope of the Urals. The zone is noted for a mosaic pattern of landscapes, with vast plains and river valleys alternating with hilly forelands and low mountains. That provided favorable conditions for various herbivorous animals belonging to the so-called "Mammoth faunal assemblage" and therefore attracted Paleolithic hunters.

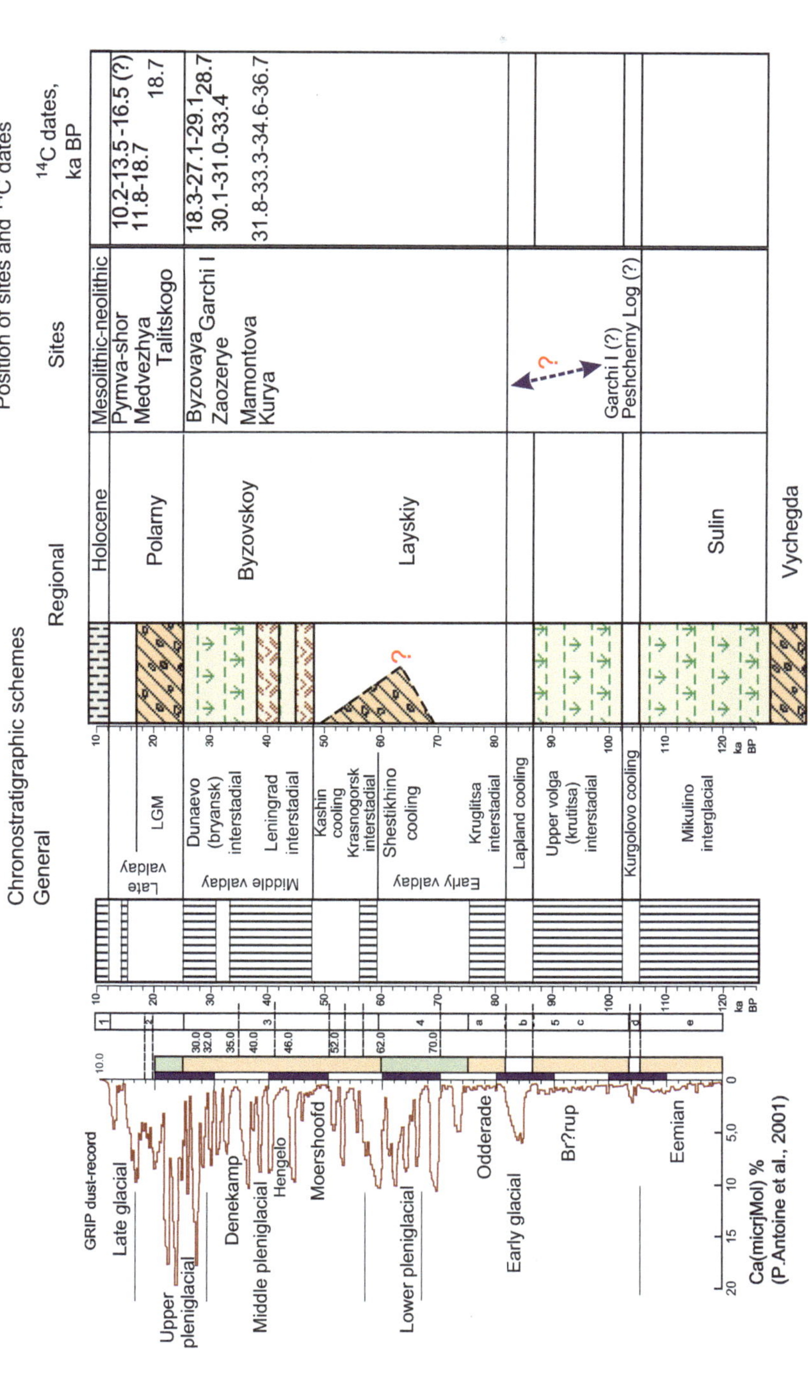

Figure 1.7.23 Chronostratigraphic position of the Paleolithic sites in the northeast of the East European Plain.

At present the interrelations between the fauna composition and the hunters' settlements or specific feature of hunting are not quite clear. Fluctuations of climate over the limited chronological interval (33,000–28,000 yr BP) were relatively insignificant and could only slightly change the migration routes of large animals. No sign of drastic changes in the sedimentation and relief-forming processes at the time of human habitation on the sites has been recorded in the sedimentary sequences close to the cultural layers. A distinct cyclicity of natural processes is easily traced only in alternating horizons of loess, paleosol, and cryogenesis lying above the cultural horizons of the sites.

It may be considered an established fact that the northeastern margin of the East European Plain and the western forelands of the Urals belong to the regions with sites dated to the early stage of the Late Paleolithic (Early Upper Paleolithic) in Eastern Europe. At the second half of the middle Valday the regional climate, a high diversity of biotopes, and relatively stable environments offered the favorable conditions for the Paleolithic hunters even at times of drastic changes of landscapes during the late Valday Ice Age.

It was only after the Scandinavian ice sheet had decayed that the Mesolithic and Neolithic groups began to colonize actively the north of Eastern Europe (including the Lower Pechora drainage basin) holding still to the Late Paleolithic traditions. They penetrated into the taiga and tundra zones and adapted themselves to the changing environments.

Middle and Upper Paleolithic sites in the Northeast of the East European Plain

1.8

Pavel Yu. Pavlov
Institute of Linguistics, Literature and History, the Komi Centre, the Urals Division, Russian Academy of Sciences, Syktyvkar, Russia

Chapter Outline

Most of sites attributed to the early stage of the Upper Paleolithic on the East European Plain are concentrated in the zone just west of the Urals (Fig. 1.8.1). Among them, there are such widely known sites as Mamontova Kurya, Garchi, and Zaozerye. There are also a few sites presumably of the Middle Paleolithic age.

1.8.1 Middle Paleolithic

1.8.1.1 Elniki II locality

The locality was discovered by A.F. Melnichuk in 1983 (Korenyuk and Melnichuk, 1985) and explored by the author together with B.I. Guslitser in 1985 (Guslitser and Pavlov, 1987). Its position is 58°01′ N and 56°45′ E. The locality is 0.9 km west of Elniki village, Perm' Region. It is exposed in the scarp of the third terrace of the Sylva River about 17–20 m high near its contact with a higher (up to 70 m) interfluve surface (the elevation of the terrace before the river damming and reservoir filling was 29–32 m).

Human Colonization of the Arctic: The Interaction Between Early Migration and the Paleoenvironment
http://dx.doi.org/10.1016/B978-0-12-813532-7.00008-5

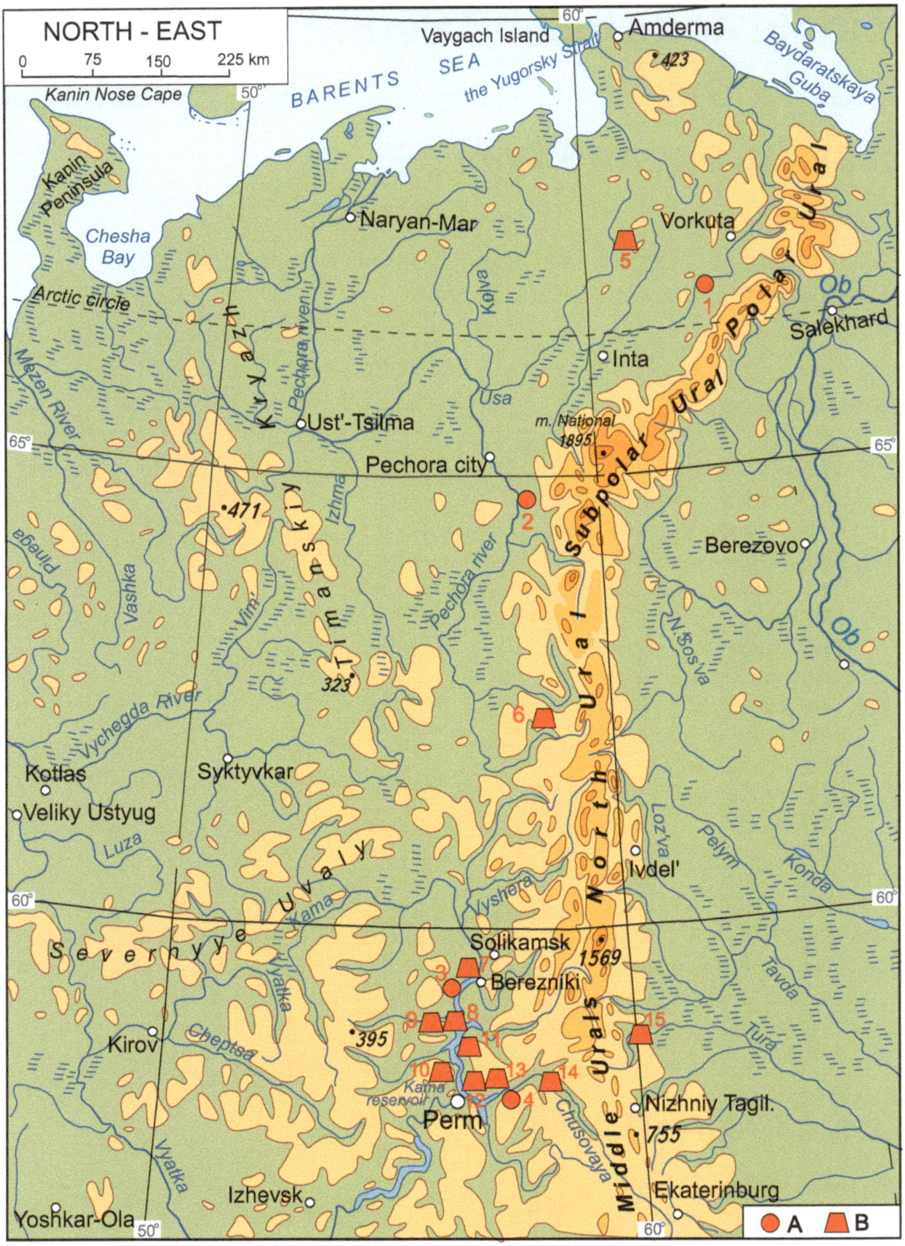

Figure 1.8.1 The main Paleolithic sites in the Urals region dated to the Early Upper Paleolithic (*A*) and the Late and Final Upper Paleolithic (*B*). Sites designated by numbers: 1: Mamontova Kurya; 2: Byzovaya; 3: Garchi I; 4: Zaozerye; 5: Pymva-Shor; 6: Medvezhya Cave; 7: Ust'-Pozhva; 8: Gorka; 9: Shirovanovo I; 10: Ganichata II; 11: Stolbovoy Rockshelter; 12: Talitskogo; 13: Gornaya Talitsa; 14: Bolshoy Glukhoy Rockshelter; 15: Gari.

In the spring of 1983, A.F. Melnichuk found in situ the lower mandible of an elephant with well-preserved teeth and fragments of the skull. The bones were recovered from the lower part of the terrace sequence cut across with an erosional rill. The poorly preserved elephant foot bones were found on the beach near the rill. When excavating the lower mandible, a quartzite chopping was found in the top of bluish-gray clays, at the contact of the clays with overlying loess-like deposits. Another object—a flake—was found directly under the exposure, together with fragments of the elephant foot bones. The teeth from the lower mandible were given to V.E. Garutt (Zoological Institute, RAS) and determined as belonging to the elephant *Archidiskodon trogonterii* Pohl.

In 1985 the author continued the work together with B.I. Guslitser. The sequence of the terrace deposits was described, and some additional paleontological material, including a large tusk, was collected. Unfortunately, neither that year nor in the subsequent works on the site could we open the top of bluish-gray clays as the base of the exposure had been overlaid with a large landslide.

1.8.1.2 Garchi I site (lower cultural layer)

The site is situated on the right side of the Kama reservoir 3 km south of Ust'-Pozhva settlement (Yusva district of the Komi-Permyak Autonomous Area, Permian Region). Its position is 59°04′ N and 56°07′ E. The site was discovered by E.Yu. Makarov in 1989 and was investigated by the author in 1990–91, 1995, 2003, and 2004 (Pavlov and Makarov, 1998; Pavlov, 2002; Pavlov et al., 2004). Two cultural layers have been identified in the section, the upper belonging to the Upper Paleolithic Kostenki-Streletskaya culture, while the lower one was dated to the Middle Paleolithic.

The sediments exposed in the scarp of the third terrace of the Kama are ~20 m thick. Bedrock at the base of the sequence is overlain with a member of lacustrine silts and clays a few meters thick. At the top of the member, there is a distinct horizon of cobbles, pebbles, and gravel indicative, in all probability, of erosion. At the base of the exposure, on the beach surface, bones of mammoth, bison, and other mammals have been collected; they probably were washed out of the lacustrine sediments. Yu.N. Gribchenko dated that member to the Middle Pleistocene (Gribchenko, 2008, 2006; Vereshchagina, 1965; Yakhimovich et al., 1987; Goretskiy, 1964).

There are yellow-brown silts and sands with a characteristic wave-like bedding that occur above the lacustrine layers. This member up to 5 m thick was interpreted as eolian deposits reworked by slope processes. They are overlain by a dark-brown paleosol displaying a typical granular texture and supposedly correlatable with the Mezin soil complex (Gribchenko, 2006). The dark soil underlies light-gray silt about 1 m thick, presumably, gley soil. The soil complex yielded a few stone implements datable to the Middle Paleolithic.

The foreconsidered soil complex is covered with a 4- to 5-m loess member that is yellow-brown loam (silt) with massive structure and typical columnar jointing. Variations in color make it possible to distinguish three horizons of loess: lower of yellowish-brown, middle of grayish-brown, and upper of light brown loess.

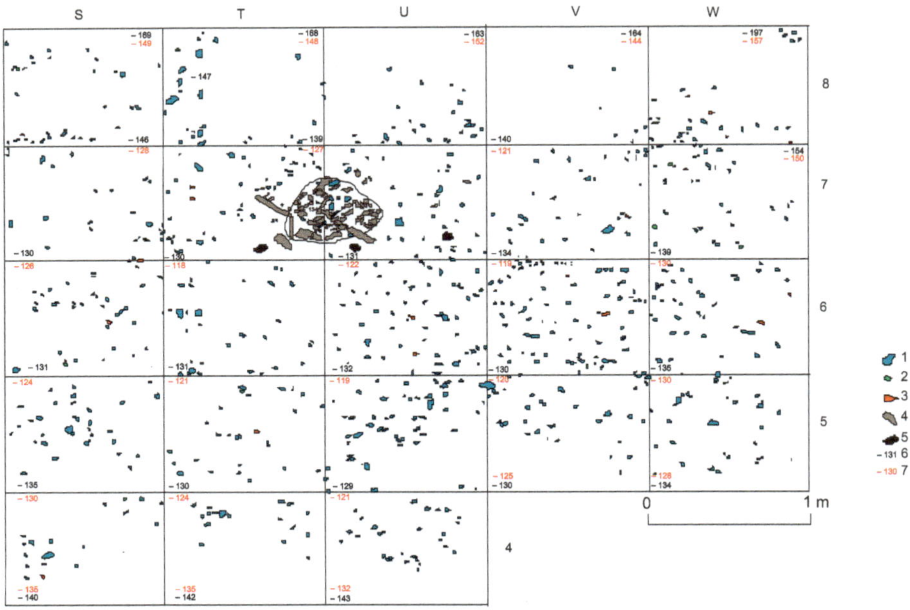

Figure 1.8.2 Plan of Garchi I excavation. 1: lithics; 2: end scrapers and side scrapers; 3: points, 4: bones, 5: charcoal spots; 6: altitude marks of the cultural layer base measured from the bench mark; 7: the same, of the top of the cultural layer.

In 1990, three stone tools and a fragment of bone were recovered from the Mezin soil complex (Fig. 1.8.2). Immediately below the excavation of 1990 there was found a concentration of artifacts on the beach surface; it consisted almost entirely of objects morphologically similar to the Middle Paleolithic inventory. The objects occurred closely to each other near the base of a large landslide over an area of $10\,m^2$. Many objects recovered from the concentration were covered with gray-greenish incrustation (sinter).

Similar sinter was found on artifacts recovered from the excavation. Some of the objects collected from the beach still bear traces of ocher and coaly particles preserved in the cavities on the stone surface. All those observations permitted the attribution of the surface artifacts to the Middle Paleolithic.

The lithic assemblage recovered from the lower cultural layer of the Garchi I site includes 71 objects. Three of them are found in the Mezin paleosol complex; the rest are artifacts collected on the beach surface. It should be noted that small flakes found on the surface had not been included into the collection, as they could be easily moved in and out within the surf zone and therefore could belong initially to the cultural layer of the Upper Paleolithic (Fig. 1.8.3).

The industry recovered from the lower cultural layer of the Garchi I site is fully consistent with typological characteristics of the eastern Micoquian industries (Chabay, 2004). It bears an undeniable resemblance to the materials of the second Middle Paleolithic site in the northeast of Europe: the Peshcherny Log locality.

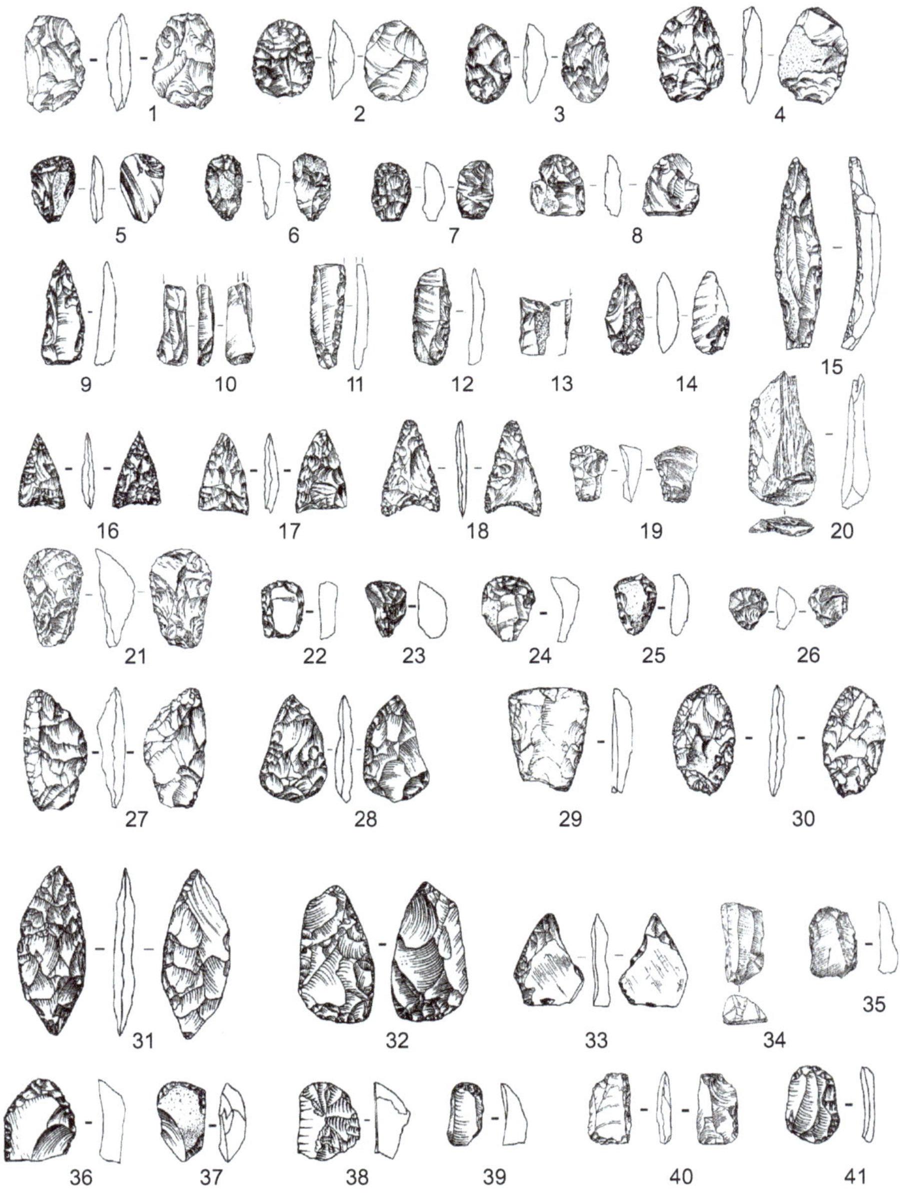

Figure 1.8.3 Lithic assemblage of the sites attributed to the Initial and Early Upper Paleolithic of the European Northeast. 1: Mamontova Kurya; 2–15: Zaozerye; 16–26: Garchi I; 27–41: Byzovaya.

1.8.2 Early Upper Paleolithic

1.8.2.1 The Mamontova Kurya site

The site was discovered by the author in 1992 (Pavlov and Indrelid, 2000; Pavlov et al., 2001). It is situated 7 km south of the Meska-Shor settlement (Vorkuta district, Komi Republic) on the left bank of the Usa River (right tributary of the Pechora River) near the western forelands of the Polar Ural Range (66°34′ N, 62°25′ E). The site is confined to the deposits of the 12-m terrace of the Usa River at the backwater pool of the same name as the site. The locality has been known since the late 18th century for numerous mammoth bones (Kanivets, 1976).

Excavations made over an area of about 54 m^2 revealed the fossil-bearing horizons to be continued upstream and downstream (westward and eastward) along the river bank and also southward, away from the river channel. Judging from the occurrence of faunal remains along the terrace scarp, the area of the site may be as great as 4000 m^2.

There is no cultural layer at the site. The finds are mostly redeposited and occur in the fluvial deposits. The distance of their transportation, however, was not long, as may be concluded from the lack of the traces of rounding on all the bones and stone tools recovered from the tool-bearing layer, while in the overlying alluvial series, all the bones are perfectly rounded.

A small assemblage of artifacts collected at the Mamontova Kurya site consists of five stone objects and an engraved mammoth tusk (see Fig. 1.8.3). The most interesting piece is a fragment of a double-convex biface retouched on both sides. There is a single side scraper made on a primary flake of what appears to be a greenish-gray andesite. The working edge is straight, treated by a large-faceted retouch. The presence of biconvex biface argues for the lithic assemblage belonging to the so-called seletoid technocomplex industries (Pavlov et al., 2001; Djindjian et al., 2003).

The mammoth tusk recovered from the cultural layer is unique in its ornament. The latter consists of two rows of twin incisions oriented at an angle to each other. The most distinct is a parallel series of 23 incisions. The second row is at a certain angle with reference to the first. The incisions are made at a right or oblique angle to the surface and are V-shaped in profile. In the opinion of F. D'Errico (personal communication), the incisions were made by different tools. Similar ornaments, though not so distinct, are found on other parts of the tusk.

The topographic position of the Mamontova Kurya site, as well as some specific features in the fauna composition, make the site somewhat different from others. First, noteworthy is the fact that the site is located at the place of a large concentration of mammoth bones known since the late 18th century. We do not know how much ivory was collected there for more than 200 years; in the 1990s, mammoth bones were absolutely dominant in the collections (88.7%). Other animal species are mostly represented by one or a few bones. The fauna composition and topographic position of the site are typical for natural concentrations of large mammal remains: the so-called "mammoth cemeteries."

1.8.2.2 The Zaozerye site

The one-layer Zaozerye site attributable to the Upper Paleolithic is at the left side of the Chusovaya River valley (at present, on the coast of the Kama reservoir) in the Ural Range foothills (58°09′ N, 56°59′ E). The site was discovered by V.P. Denisov (1976) when surveying coasts of the Kama reservoir and studied by the author in 2001–08.

The site is situated 1.5 km east of the Zaozerye village (Chusovskoy district, Permian Territory) on a protruding part of the third terrace of the Chusovaya River. Its elevation above the reservoir water's edge is 15–18 m; before reservoir filling, it was 27–30 m above the river channel.

The cultural layer in the Zaozerye site occurs at the base of the cover series on the third terrace. It is confined to the middle part of the humified layer overlain with loess-like loam 2.5–3.5 m. The radiocarbon (AMS) age of the site amounts to 33–35 ka BP.

In the course of rather numerous excavations (~1500), animal bones were recovered from the cultural layer, most of them well preserved. The bones are in small fragments and apparently are typical kitchen refuse. Among the identified bone fragments, most common are limb bones and teeth of *Equus latipes*. Some of the bones bear distinct traces of butchering distinguishable as thin parallel grooves mostly grouped at joints of the carcass.

There are three concentrations of cultural remains identified in excavations of 2002–05, all of them being in the preserved initial context (Fig. 1.8.4). The concentrations may be interpreted as short-term camps of hunter groups. The main human activities there included probably butchering of horse carcasses (hind and fore legs) brought from killing places; food preparation and consumption (and probably its preservation for keeping in store); sewing clothes from hides of fur-bearing animals; and making bone tools and ornaments.

We do not know stone inventory completely analogous to that described at the considered site. The group of blade tools from the Zaozerye site might be compared with implements of the oldest Upper Paleolithic sites on the East European Plain attributable to the Aurignacoid technocomplex, in particular, with the second layer of the second layer of Kostenki XVII site (Praslov and Rogachev, 1982; Anikovich and Rogachev, 1984). Some tools analogous to convex-plane bifaces and carinated (keeled) end scrapers with ventral trimming (low-angle partial retouch) are known from Middle Paleolithic sites ("Kielmessergruppe") in the south of the East European Plain (Kolesnik, 2003) and the North Caucasian foothills (Golovanova and Hoffecker, 2000).

So, as follows from the preceding, the Zaozerye site may be counted among the oldest sites of the Upper Paleolithic in Eurasia. The site is recognized as a specific cultural type of the Early Upper Paleolithic of Eastern Europe.

1.8.2.3 Garchi I site (upper layer)

The site is located on the right bank of the Kama water reservoir 3 km south of the Ust'-Pozhva settlement (Komi-Permyak Autonomous Area, Permian Region) at 59°04′ N, 56°07′ E. The site was discovered in 1989 by E.Yu. Makarov and studied in 1990–91, 1995, 2003, and 2004 by the author (Pavlov and Makarov, 1998; Pavlov, 2002; Pavlov et al., 2004).

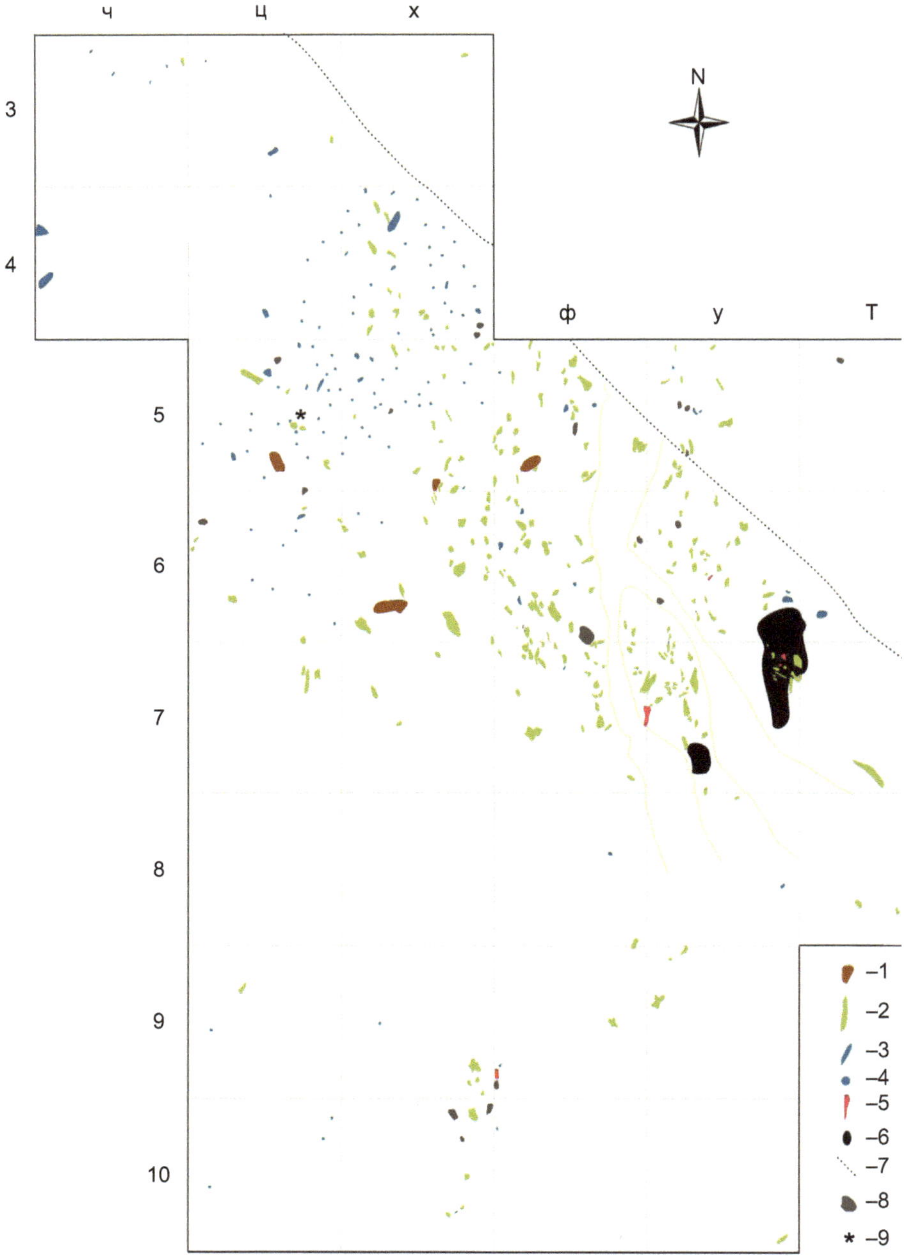

Figure 1.8.4 Plan of Zaozerye excavation. 1: "anvils," large pebbles of quartzite-sandstone; 2: bones; 3,4: lithic pieces more than 20 mm (3) and less than 20 mm (4) in size; 5: bone implement; 6: charcoal spots (several fireplaces or one large hearth); 7: fissure; 8: fragments and plates of limestone; 9: oval pendant made of *Unio* shell.

The site is positioned on a fragment of the third terrace of the Kama River. The fragment width is minimal (1–2 m) in the northern part of the site and increases toward the southern part up to 60 m. The northeastern part of the terrace fragment about 80 long is greatly damaged by the coastal erosion.

As follows from the data of prospecting pits, the area of the upper cultural layer in the southern part of the terrace fragment is at least 2000 m^2. The cultural layer is confined to a fossil soil; in the opinion of Yu.N. Gribchenko, the soil has morphotypical characteristics similar to those of the Bryansk paleosol typical to the central East European Plain. That conclusion agrees well with the results of the layer dating. At present, there is an AMS date 28,750 ± 795, TUa-941/32,800 ± 830 cal. yr BP/ on charcoal (Pavlov and Indrelid, 2000) and an OSL-date of 33 ka BP.

Faunal remains recovered from the upper cultural layer are poorly preserved. Among identifiable fragments (32 pieces), there were recognized some belonging to horse (20), reindeer (11), and mammoth (1) (identification by A.K. Hufthammer, Bergen University, Norway). Pollen analysis of samples from the cultural layer and overlying deposits appeared to be ineffective.

Two concentrations of cultural remains on the site are, most probably, remnants of surficial dwellings. It is not quite clear whether they existed at the same time, but in any case, it may be safely assumed that it was within a rather narrow time interval (as indicated by similar stratigraphic context of the cultural layer, similar structure of the concentrations, and identical typology of stone industry).

The collection from the upper layer of the site includes about 9000 objects and is the most representative one not only in the northeast of Europe but in the entire Ural-Volga region.

The cultural attribution of the site is quite clear: it belongs to the Kostenki-Streletsky culture of the Early Upper Paleolithic of Eastern Europe. The stone inventory recovered from the site displays clearly all the typical characteristics of that culture. In the technology and morphology of the inventory, and particularly in the presence of specific type of artifacts, the upper cultural layer of the Garchi I site is close to the collections from layer V of Kostenki I (upper reaches of the Don River), and even to the much earlier industry recovered from layer III of the Kostenki XII (Rogachev, 1957; Anikovich, 1977).

Judging from the cultural layer occurrence and the stone inventory composition, the site may be considered as a repeatedly occupied hunters' camp with the complete working cycle of the stone implement production. In all probability, there were also surface dwellings on the site.

1.8.2.4 Byzovaya site

The Byzovaya site is one of the northernmost Early Upper Paleolithic objects in Europe (65°01′ N, 57°25′ E). It was discovered in 1962 by E.M. Timofeyev. The site was studied by V.I. Kanivets together with B.I. Guslitser and E.M. Timofeyev in 1963–68 (Guslitser et al., 1965; Kanivets, 1969, 1976). The author worked on the site in 1983, 1985, 1992, 1997, and 2000 (Pavlov, 2008a,b). The area of the site excavated since the beginning of the investigations is 551 m^2, that is, ~14% of the total (supposed) area of the site (Fig. 1.8.5).

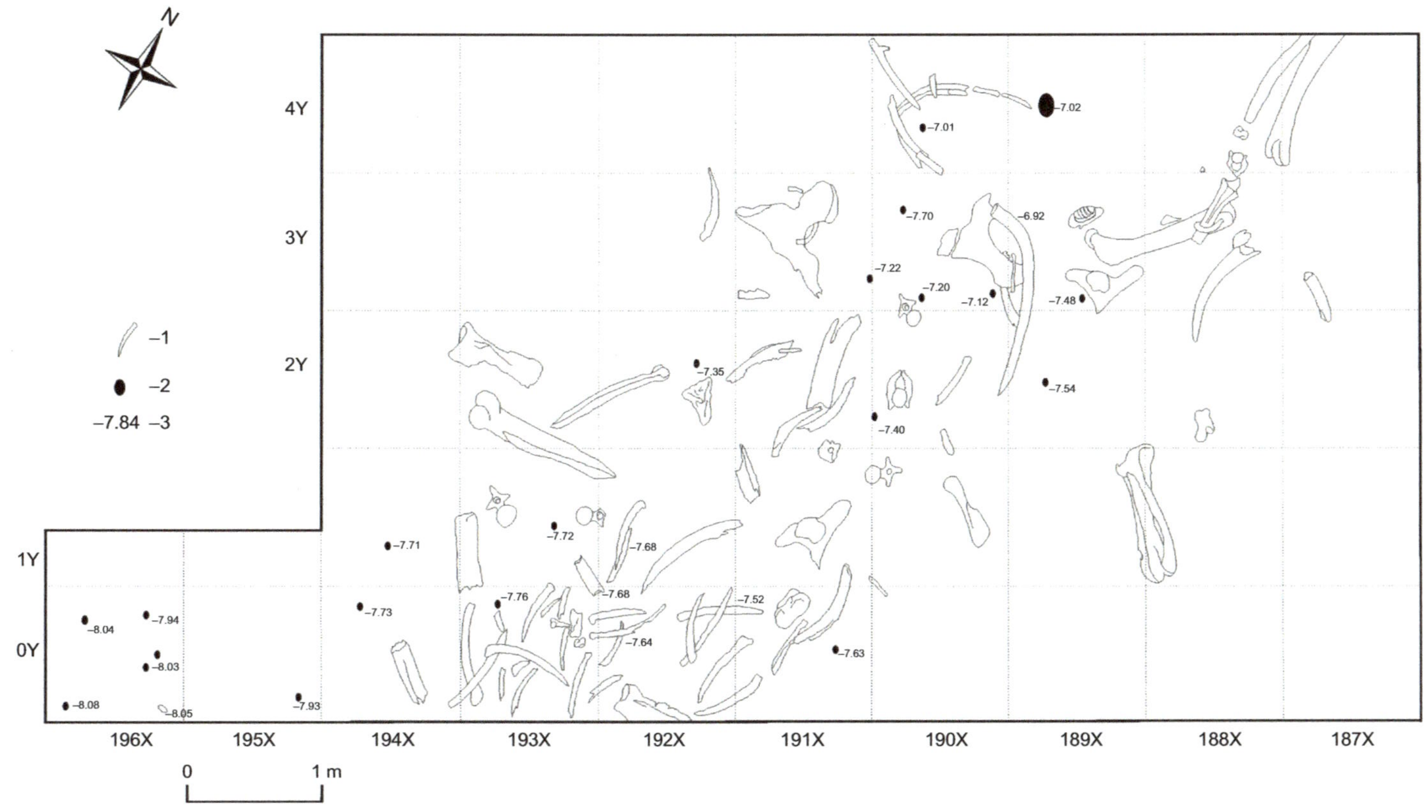

Figure 1.8.5 Plan of Byzovaya excavation. *1*: bones; *2*: lithic artifacts; *3*: altitudinal marks of cultural layer with reference to the bench mark.

The site is situated on the right bank of the Pechora R, at the bend of the river channel known as Byzovskaya Kurya Bay. The cultural remains are confined to the pebble horizon at the base of the second terrace sequence, at ~12 m above the river channel edge at the time of low water.

The deposits enclosing bones and stone artifacts up to 2 m thick occur immediately on the surface of Triassic siltstones, at the base of the deposits infilling an ancient gully cut into the alluvial deposits of the Pechora (the second terrace). The deposits infilling the gully are sands and gravel abounding in boulders and angular rock fragments. The presence of the latter suggests the source of the material to be glacial deposits occurring on the upper slopes of the valley. The layer enclosing cultural remains is overlain with a thick series of sands and silts, mostly of eolian origin. The sands form a mantle covering bones and tusks lying on the top of gravels. Most of the bones bear no marks of weathering or destruction: a fact indicative that the burial process was fast.

The series with cultural remains is weakly cemented and consists of rock fragments widely varying in size and roundness (including boulders up to 0.5 m in diameter), as well as in petrographic composition (quartzite sandstones, flints, sandstones, clay slates, etc.), with sandy or clayey matrix.

The cultural layer has been dated by radiocarbon (24 conventional ^{14}C dates giving averaged age of ~29 ka BP). That site may be confidently attributed to the second half of the Bryansk (Dunaevo) interval of the Middle Valday (Pavlov, 2008a,b; Svendsen et al., 2008).

The absolute majority of the bones recovered from the site belong to mammoth. Bones of other animals occur only occasionally. Bones are mostly unbroken, only a few of them being split. It seems the bones were somewhat sorted. On the left bank of the stream, prevalent are leg bones, skulls and their fragments, and lower jaw bones. On the right bank, there are mostly flat bones: pelvis bones, shoulder blades, vertebrae; rather abundant are tusks; some flat mammoth bones with incisions are also found. Leg bones are scarce, and lower jaw bones are absent completely. There are also rare skull fragments and single teeth. Even so, the found bones form a practically complete mammoth skeleton.

The pollen assemblage is dominated by shrub and grass pollen, with occasional pollen grains of spruce and birch.

The stone inventory typology reflects the functional specialization of the site as the place of primary butchering of mammoth carcasses. So the stone inventory at the Byzovaya site is a specific single-purpose assortment of tools; it is likely to reflect only partly the principal cultural characteristics. That specificity should be taken into consideration when studying tools recovered from the Byzovaya site.

The stone and bone inventory recovered from the Byzovaya site differs from other sites both in its composition and in the typology of stone tools (see Fig. 1.8.3). Both volumetric and flat splitting are detectable in the knapping technology. Two technical-morphological groups—Middle Paleolithic and Upper Paleolithic—are distinctly distinguishable in the tool set. The first one is represented by convex-plane bifaces, including backed segment-like knives of the Kielmesser type and various side scrapers, those of Quina type among them.

The Upper Paleolithic group consists of end scrapers on blades, carinated and ogival (pointed) end scrapers, angle burins on blades, pièces esquilles, points, and includes one thin foliate point.

On the whole, the Byzovaya stone inventory has no direct analogs in sites of the Early Upper Paleolithic of Eastern Europe and may be recognized as a specific Paleolithic industry. Some analogs to the Byzovaya inventory structure may be found in the materials from the Zaozerye site.

Distinctive features of the Byzovaya recognizable in stone inventory and fauna composition, as well its topographic position, suggest the site to be of a specific type. One of the earlier investigators of the site, V.I. Kanivets (1976), drew attention to an unusually high proportion of tools in the site inventory (56% of the total number of the found items). Our data gave somewhat lower, though still considerable, value: 40%. Such a proportion is most characteristic of short-term (seasonal) hunter camps (Kozlowski, 1980). Noteworthy is an almost complete absence of hunters' weapons; dominant are tools for carcass butchering: end scrapers, side scrapers, knives, and flakes with retouch bearing traces of usage (more than 80% of the total number).

Neither the presence of unbroken bones in such abundance (more than 4000) is typical for hunter camps, particularly considering that only one-fifth of the site has been excavated. The fauna is absolutely dominated by mammoth (97.6% of all identified bones).

All the mentioned confirms our opinion that the site was positioned in immediate vicinities of a natural mammoth burial (Pavlov, 1996). The stone tool inventory, as well as geological and archaeological context of the findings, identifies the locality as a butchering site.

1.8.3 Final Paleolithic

1.8.3.1 Talitskogo site

The site is situated on the right bank of the Chusovaya River in its lower reaches, 3 km east of the Krasnaya Sludka village in the Dobryansk district, Perm' Territory (58°10′ N, 56°31′ E). The site was flooded after the dam of the Kama Water Reservoir had been built.

The site was discovered by M.V. Talitskiy in 1938 (Talitskiy, 1939, 1940a,b). In 1945–52 the site was thoroughly studied by O.N. Bader (1960). The cultural layer was studied over the ~300 m^2 area during the years of excavation (Shcherbakova, 1994).

The site was positioned in the river valley, on a projection of the valley slope between the Chusovaya channel and that of a small stream (Bader, 1960), entering the central part of a large meander of the Chusovaya channel.

V.I. Gromov studied the Chusovaya valley in the site vicinity and identified there the second and third terraces (Gromov, 1948). The cultural layer of the site is related to a gleyed interlayer at the base of the cover mantle on the well-pronounced second terrace 20–25 m above the river channel.

In the early 1980s, L.D. Sulerzhitskiy obtained a ^{14}C date of 18,700±200, GIN-1997/22,600±230 cal. yr BP/ that confirmed the Late Valday age of the site. The stratigraphic position of the cultural layer, the ^{14}C date, and the presence of cold-tolerant mammal species permit one to assign the site to the beginning of the second half of the Late Valday.

In the faunal remains recovered from the site, V.I. Gromov identified the following species: reindeer, mammoth, wooly rhinoceros, horse, polar fox, hare, lemming, and field mice. A considerable proportion of identifiable bones—about 220 bones—appeared to belong to reindeer (five animals) (Gromov, 1948). Mammoth bones belonged to two animals; the rest of the remains belong as a rule to a single animal of the species and are represented by a much lesser number of bones.

The Talitskogo site is one of key sites representative of the Urals Late Paleolithic culture of the northeast of Europe and of the Ural region (Pavlov, 2007, 2009). On the whole, the stone inventory shows all the characteristic features of the Afontova technocomplex, most typical for the Late Upper Paleolithic of Siberia (Anikovich, 1998). Stone and bone tools of the site have their analogs in many Late Paleolithic sites of the northeast of Europe, such as Shirovanovo II and Ganichata II. All the investigators determined the type of the site as a seasonal camp of reindeer hunters inhabited by a relatively small group (Talitskiy, 1940a,b; Bader, 1947). No trace of long-term houses was found on the site, though some surface dwellings could exist.

1.8.3.2 Medvezhya Cave site

The site was discovered by B.I. Guslitser and V.I. Kanivets in 1960 (Guslitser and Kanivets, 1962a,b). It is situated in the upper reaches of the Pechora River (Troitsk-Pechorskiy District, the Komi Republic) on the right bank of the Pechora River, 17 km upstream of the Bolshoi Shezhim River mouth, within the limits of the Pechora-Ilych Biosphere Reserve (62°05′ N, 58°05′ E).

The cave site is in the dry valley of the Iordansky Log separated from the Pechora channel with a fragment of floodplain 2 m above the channel and 300–500 m wide. The cave itself is at a distance of 500 m north of the Pechora channel and 200 m east of the log mouth, in the rocks on its right side. The cave entrance is 40 m above the Pechora channel and 24 m above the log floor.

The excavations were performed in 1960–1962 by V.I. Kanivets and B.I. Guslitser (Guslitser and Kanivets, 1962a,b, 1965a,b; Kanivets, 1969). In 1982–84 the excavations were continued by the author with the participation of B.I. Guslitser (Guslitser and Pavlov, 1988, 1989, 1990).

The cultural layer is confined to the slope deposits near the cave entrance (brown loam with rubble). Stone tools were recovered from a thin (3–5 cm) cultural layer. The age of the latter in the Medvezhya Cave was determined reliably enough at 18–16 ka BP.

The faunal remains collected in 1982–84 include 2124 items (aside from bones of birds and small rodents that amount to a few thousands). When excavating the cave site in the early 1960s, a vast collection (about 60,000 bones) was gathered (Guslitser and Kanivets, 1965a; Kuzmina, 1971). The bones from the main horizons of the cave

Table 1.8.1 Species composition of fauna from the cultural layer of the Medvezhya Cave site

Species	Number of	
	Bones	Animals
Reindeer (*Rangifer tarandus* L.)	136 (including 60 shed horns)	5
Hare (*Lepus tanaiticus* Gureev)	64	8
Mammoth (*Mammuthus primigenius* Blum)	5	1
Cave bear (*Ursus speleaeus* Ross)	3	1
Rhinoceros (*Coelodonta antiquitatis* Blum)	3	1
Polar fox (*Alopex lagopus* L)	9	1
Undeterminable	590	

deposits were studied together; the material from the cultural layer had not been specially considered. This drawback, however, was corrected in excavations in the early 1980s. Special investigations of the cultural layer resulted in 810 bones recovered from the layer. The species composition is shown in Table 1.8.1.

Noteworthy is a sizeable amount of shed horns of reindeer: about 50% of the total number of identifiable bones are attributed to that animal. A similar proportion in the reindeer bone remains was noted when the monument was first studied in the early 1960s (Guslitser and Kanivets, 1965b).

The stone and bone inventory recovered from the Medvezhya Cave site shows an undoubted similarity to the materials from the Talitskogo site and other Paleolithic sites (Shirovanovo II, Ganichata II).

1.8.3.3 Pymva-Shor site

The site discovered by the author in 1994 is situated in the Bol'shezemelskaya tundra at 67°11′ N, 60°51′ E (Pavlov et al., 1996). It is in the middle reaches of the Adzva River, 110km northeast of Kharuta settlement (Nenets Autonomous Area, Arkhangelsk Region) in the valley of a small stream 2km east of its mouth. The valley of the stream is bounded by high rocky sides composed of massive Lower Carboniferous limestones; the cave itself is deepened in the limestones broken with bedding joints dipping toward the north-northwest.

Only side tunnels of the cave are preserved at present, as the rock shelter at the entrance is completely destroyed. The site itself is confined to the area (~150 m^2) before the former entrance gently sloping toward the southwest. The area excavated in 1994–95 amounted to 32 m^2.

The fifth (lower) cultural layer of the Pymva-Shor I site is attributed to the Final Paleolithic. Three radiocarbon dates obtained for this layer fall within an interval 11,460–10,255 yr BP, which is the very end of the Pleistocene.

The cultural layer occurred in the lower part of the gray fine sand at a depth of 1.10–1.20 m below the surface. It yielded 26 flint objects, all of them confined to a horizon no more than 0.05 m thick. Among the findings, there were flakes (14 pieces), small irregular blades (6 pieces), and 5 chips. The only tool found in the layer is a side scraper on quarzitic sandstone pebble with a high working edge.

Besides the stone pieces, the layer contained a few intact bones and fragments of bones belonging to large mammals: reindeer, musk ox, bison, horse, polar fox, and hare. The stone implements attributed to the Early and Middle Upper Paleolithic of the region are shown in Fig. 1.8.3.

Environments of the Paleolithic sites in the Northeast of the East European Plain

1.9

Pavel Yu. Pavlov
Institute of Linguistics, Literature and History, the Komi Centre, the Urals Division, Russian Academy of Sciences, Syktyvkar, Russia

Chapter Outline

1.9.1 The Late Pleistocene Environments in the Northeast of the East European Plain

The data on the environments of the Early and Middle Paleolithic sites are rather scarce and cannot serve as a basis for valid conclusions. More information is available on the sites of the initial and early Upper Paleolithic sites attributable to the Middle Valday mega-interstadial.

The earliest site of that chronological group is Mamontova Kurya located just north of the Arctic Circle. In the period of human habitation the climate of the region was cold and dry, with periglacial tundra-steppe on interfluvial surfaces and small willow groves in the Usa River valley (Svendsen and Pavlov, 2003). According to pollen data, at the time of the culture-bearing layer deposition (38–34 ka BP) the region was dominated by open landscapes of tundra-steppe with herbs and dwarf shrubs. No tree species pollen could be found in the overlying fluvial deposits, presumably as a result of progressing cooling and aridity 27–25 kyr BP.

The cultural layer of Zaozerye, slightly younger and located considerably farther south than Mamontova Kurya, was studied palynologically by V.V. Pisareva. The pollen assemblage appeared to be dominated by tree pollen (AP) –40%, with proportion of herbs and dwarf shrubs (NAP) about 33% and that of spores – 27% (Pavlov, 2004). AP group includes mostly spruce (*Picea abies*) and pine (*Pinus sylvestris*, *Pinus sibirica*), with some deciduous trees: birch (*Betula alba*, *Betula humilis*, *Betula nana*), alder (*Alnus incana*) and shrub alder (*Alnaster fruticosus*). Herbs are represented by an abundant group of weeds and ruderal species of *Artemisiae*, *Cichoriaceae*, *Onagraceae*, *Lamiaceae*.

Human Colonization of the Arctic: The Interaction Between Early Migration and the Paleoenvironment
http://dx.doi.org/10.1016/B978-0-12-813532-7.00009-7

The regional vegetation at the time of the Zaozerye cultural layer formation belonged to boreal and northern boreal type (taiga). Local vegetation consisted of wetland and aquatic communities with some ruderal ones, the latter being typical of terrace and floodplain surfaces still in the process of formation. Marshes occurred locally.

The composition of faunal remains (horse, woolly rhinoceros, reindeer and hare) suggests the presence of vast open landscapes in the vicinities of the site. Both pollen data and the fauna composition suggest the site existence under conditions of temperate cold climate and forest-steppe landscapes.

Another site dated to the second half of the Middle Valday – Garchi I (the upper cultural layer) – has not been studied palynologically. The fauna, however, includes horse, reindeer and mammoth and strongly suggests the presence of open forest-steppe landscapes in the vicinities of the site.

The site of Byzovaya (as indicated by pollen analysis data and composition of fauna) also was inhabited under conditions of temperate cold climate. Dominant landscapes were tundra-steppes with occasional forest patches in the Pechora River valley.

In general, the climate in northeastern Europe at the second half of the Middle Valday may be described as moderately cold. The upper reaches of the Kama River featured forest-steppe landscapes, with tree assemblages confined mostly to river valleys and open landscapes of steppe widely spread on the interfluves. The Pechora drainage basin was noted for a colder climate and domination of tundra-steppes.

Landscapes similar to the above were widespread all over the East European Plain at certain intervals of the Middle Valday (Spiridonova, 1989), which partly accounts for the early man migrations.

The first half of the Late Valday is noted for an extremely severe climate in the Northeast of Europe. Permafrost became widely spread and initiated various cryogenic processes (polygonal ice wedge formation, solifluction, etc.) (Velichko, 1973). Marks of those processes have been traced in the horizons overlying cultural layers at several sites, including Zaozerye, Garchi I, Byzovaya. The data obtained on Mamontova Kurya strongly suggest scanty vegetation in the site vicinities and landscapes similar to those of arctic deserts (Svendsen and Pavlov, 2003; Astakhov and Svendsen, 2008). Such environments were not very comfortable for the Paleolithic man to live in. So it is hardly surprising that no sites attributable to the first half of the Late Valday have been found as yet.

The data on the environments of northeastern Europe at the Late Paleolithic time indicate a colder climate as compared with the Middle Valday. Thus, small mammal remains recovered from the cultural layer of the Medvezhya cave (the Pechora River basin) are mostly those of tundra rodents – collared lemming (*Dicrostonyx* sp.) and Siberian lemming (*Lemmus sibiricus*) (Guslitser and Pavlov, 1990). It may be assumed that typical tundra landscapes existed at that time in the neighborhood of the site.

The Late Paleolithic cultural layers studied in the Bolshoy Glukhoy rockshelter yielded small rodent fauna closely resembling in composition the above except for somewhat greater proportion of steppe and forest species (Guslitser and Pavlov, 1987). The pollen assemblages are dominated by herb and grass pollen, with a negligible presence of dwarf arctic birch and shrub birch (*Betula nana*, *B. fruticosa*).

Similarly, the mammal fauna found at the site of Talitskogo includes cold-tolerant species of carnivorous (polar fox) and rodents (collared lemming). The environments of the Late Paleolithic sites may be reconstructed as tundra and tundra-steppe with cold climate.

The inhabitants of Garchi I and Zaozerye attributed to the Early Upper Paleolithic used wood as fuel for their hearths, while at Talitskogo large mammal bones were used for that purpose (Shcherbakova, 1994). That suggests a limited presence of trees in the vicinity of the site.

We have data on the environments for two Final Paleolithic sites only, both of the cave type: Pymva-Shor I and Bolshoy Glukhoy (cultural layer III). The composition of mammal fauna recovered from Pymva-Shor I is noted for presence of animal remains indicative of severe cold and dry climate (small mammals being dominated by collared lemming and the musk ox bones identified in the large mammal group) (Smirnov et al., 1999).

The small rodent fauna excavated from the cultural layer III of the Bolshoy Glukhoy rock shelter is dominated by steppe and forest inhabitants, with tundra species present in limited quantities. Judging from the species composition, dominant at that time were cold temperate forest-steppes, which are typical of the final interval of the Late Pleistocene in the Upper Kama drainage basin (Nemkova, 1978).

1.9.2 Subsistence of prehistoric people

By subsistence systems the author means a complex of ways and methods the early humans used in their adaptation to the environments and ensured their survival. The principal characteristics of the systems describe the key economic activities as reflected in types of the sites, animal species being objects of hunt, as well as the strategy of the stone raw material use.

Lithic raw material used. One of the most important feature in northeastern Europe is the abundance and availability of diversified isotropic rocks that served as a source of raw materials for the Paleolithic people. In the western foothills of the Urals Mountains there are numerous outcrops of siliceous rocks; they are found as chert inclusions in the Paleozoic limestone, as well as occur in fluvial and glacial (moraine) deposits (Gerasimov et al., 1968). Undoubtedly, that promoted colonization of the region in the Paleolithic.

Hunting game, types of sites, and economy. The composition of the hunted game was practically the same throughout the Upper Paleolithic; it included horse, reindeer, mammoth, woolly rhinoceros, bison, musk ox, and hare.

In the species composition of the animals and particularly in the dominance of horse remains, the Initial and Early Upper Paleolithic sites in the northeast of Europe are analogous to those in the central East European Plain dated to the same time (Praslov and Rogachev, 1982). As for the Late Upper Paleolithic, the sites in northeastern Europe are noted for the prevalence of horse and reindeer among the hunted species and differ markedly from sites of the central East European Plain dominated by mammoth (Praslov and Rogachev, 1982).

Topographic position and spatial organization of the open-air sites dated to the Initial and Early Upper Paleolithic are somewhat different from the characteristics of later sites. The Early Upper Paleolithic sites are usually located at the mouth of relatively large dry valleys ("logs"). This position is equally typical of the sites of the same age in the central East European Plain, in the

Kostenki-Borshchevo region in particular (Rogachev, 1957; Praslov and Rogachev, 1982). Later Paleolithic sites of the open-air type (Talitskogo, Shirovanovo II, Ganichata II) are mostly found near meandering river channels and confined to the central parts of the meander.

The sites associated with natural concentrations of large mammal remains (such as Mamontova Kurya and Byzovaya) are topographically different from the above and occur mostly at the base of high and steep slopes rising above sharp bends of the river channel.

Typically, the sites attributed to the Early Upper Paleolithic are greater in area than those of the later Paleolithic age. We have no data on the area of spatial distribution at Mamontova Kurya, though indirect indicators allow estimating it at a few thousands of square meters. The cultural layer occupies an area of 2500 m^2 at least at Zaozerye, no less than 4000 m^2 at Garchi I, and about 3000 m^2 at Byzovaya.

The younger sites feature much less expansive cultural layer – no more than 300 m^2 at Talitskogo, and ~150 m^2 at the Medvezhya cave. There are two concentrations of cultural remains about 200 m^2 in area recorded at Shirovanovo II. Other Late Upper Paleolithic sites in the region appeared to be small in area (a few tens of square meters) and scanty in finds.

The open-air sites of the Early Upper Paleolithic and those of the later interval of the Upper Paleolithic display an unquestionable similarity in the distribution of cultural remains over the cultural layer. In every case the artefacts occur as distinctly localized concentrations, most probably corresponding to the habitation units. Outside the concentrations the cultural layer is deficient in finds or absent.

There are two concentrations of cultural remains revealed at Garchi I. Both are oval in shape and one of them covers no less than 20 m^2 and has a line of hearths in the central part. Spatial organization and the cultural layer occurrence within the concentrations suggest them to be remains of relatively large living structures. The finds are mostly debitage – remains of making triangular points for arrows or spears. The site, therefore, served most probably as a hunting base camp.

The concentrations of cultural remains at Zaozerye are smaller (about 10 m^2) and somewhat different in structure. The living areas at the site were seemingly left by a small group of hunters having repeatedly visited the hunting place.

The difference in size of living areas at Garchi I and Zaozerye may be attributed to different size of the groups having inhabited the sites. The hunter groups at Zaozerye were smaller than those at the Garchi I.

The cultural remains at Talitskogo occurred within an irregular oval about 22 m long and 13 m wide. The remains occurred mostly around the hearths forming several large concentrations. The hearths themselves are pits round or irregular in outlines filled with burnt or charred bones. Close to the hearths, there were found numerous narrow holes, occasionally with remains of vertically arranged large tubular bones. According to Bader (1947, 1950), those were remnants of wind break constructions. All the investigators agree in that it was a seasonal base camp of the reindeer hunters inhabited by a relatively small group (Talitskiy, 1940a,b; Bader, 1947).

Shirovanovo II has some features in common with the Talitskogo, particularly in the distribution of cultural remains over the habitation space.

At the Medvezhya cave the artifacts also formed concentrations, no traces of cultural layer found between them. At this site, however, neither hearths, nor fireplaces have been found as yet, nor did even occasional charcoal occur. That may be indicative of the former existence of a sacral object at the site (Pavlov, 1997).

At the Final Paleolithic site of Gornaya Talitsa the cultural remains are grouped into distinctly localized concentrations (Melnichuk and Pavlov, 1987; Pavlov, 1988). There are short-term camps left by small hunter groups in the north of the region (north of the Arctic Circle). They are attributed to the Final Paleolithic. Such camps were typical for the preceding Late Paleolithic.

It should be noted that the Late Paleolithic site of Talitskogo has dug out hearths, while only flat fireplaces are found at the sites of the Initial and Early Upper Paleolithic. It is also worthy of note that both at Zaozerye (Initial Upper Paleolithic) and at the younger Paleolithic sites of Talitskogo (Shcherbakova, 1994) and Shirovanovo II the remains of the animals – principal objects of hunt – are present almost always as fragmented leg bones and skulls. I argue that those facts point out directly to the fact that the Paleolithic people of the region kept their ways of life unchanged throughout the middle and late Valday, the economy being based on hunting large herd animals.

So the most part of Upper Paleolithic sites in the region are seasonal hunter camps. No long-term habitation sites with thick and rich cultural layer has been found as yet in the northeast of Europe (as well as in the entire territory of the European North).

Supposedly some sacral objects existed at sites like the Medvezhya cave. The Final Paleolithic is noted for an appearance of a new type of sites in caves and shelters (Bolshoi Glukhoi rockshelter).

There is a notable difference between the Upper Paleolithic sites in the north and those in the south of the region. The sites in the upper reaches of the Kama River are mostly hunter camps with dwelling structures, while some of the sites in the Pechora drainage basin (Mamontova Kurya, Byzovaya) represent a specific type of sites associated with natural mammoth deathsites.

A certain relationship may be seen between the faunal composition and the type of sites. The faunal remains found at the hunter camps are typically dominated by large ungulates (horse and reindeer as much as 90%), with minimal admixture of other species. At the sites located at the natural deathsites of large animals in the north of the region the dominant species is mammoth (more than 90%), with occasional bones of other animals.

It follows therefore from the analysis of the Paleolithic site types and hunted animals that a single cultural-economic type – large herd animal hunters – persisted through the entire Upper Paleolithic, which was reflected in the various culture traditions. It should be stressed that, unlike the Final Paleolithic inhabitants of Western Europe, the Late Paleolithic people here were not specialized reindeer hunters (Zaliznyak, 1989). Specific features of the Late Pleistocene fauna in the Northeast of Europe (in particular, the abundance and diversity of large ungulates) permitted the ancient hunters to hunt different animals depending on the season. In summer, for example, when the reindeer herds moved northward, the horse became the main object of hunt.

So, as follows from the above, the data available on the Upper Paleolithic environments in the middle Valday suggest a cold temperate climate. There existed forest-steppes in the upper reaches of the Kama River. In the Pechora drainage basin the climate was colder and the dominant landscapes were tundra-steppes.

At the later stage of the Upper Paleolithic the environments of Northeastern Europe were much colder that in the Middle Valday. Cold tundra-steppes existed in the upper reaches of the Kama River and typical tundra landscape – in the Pechora drainage basin at that time. The Final Paleolithic sites in the upper Kama basin existed, most probably, under conditions of cold temperate forest-steppe, while cold tundra-steppe still persisted in the Pechora drainage basin.

As an element of the Paleolithic adaptation to the northern environments, noteworthy is an appearance of a specific type of sites at the natural cemeteries of mammoth, which are unknown outside the subarctic zone of Eurasia. Significantly, this type of sites persisted in the northeast of Europe, in West and East Siberia throughout the Upper Paleolithic epoch (which in all probability is due to specific characteristics of the subarctic environments of Eurasia); the Upper Paleolithic sites on the natural mammoth burials are found exclusively in the arctic and subarctic Eurasia (Zenin et al., 2006; Derevyanko et al., 2000a,b; Pavlov, 1994, 2007, 2008a,b; Petrin, 1986; Bader and Serikov, 1981; Okladnikov et al., 1971). That could be related to a diversified usage of food and raw material resources available in different natural zones at the Upper Paleolithic time.

Arctic Islands

Arctic islands of the Barents and Kara seas

1.10

Vadim F. Starkov
Institute of Archaeology, Russian Academy of Sciences, Moscow, Russia

Chapter Outline

1.10.1 Early settlements on the Spitsbergen archipelago

The Spitsbergen Archipelago is the best archaeologically studied region of the High Arctic. A study of the history of its colonization, including archaeological excavations, has been under investigation since the second half of the 18th century, and the studies are still in process. Expeditions from many countries (including Sweden, Norway, Russia, France, and Great Britain) took part in the search for historical places on the islands in the 18th and 19th centuries. Since the mid-1950s, Scandinavian archaeologists have explored the region professionally; expeditions from other countries joined them later.

As a result, at present, there is information about 88 settlements of Russian hunters and West European whalemen dated to the interval from the middle of the 16th to 19th centuries (Fig. 1.10.1). The multidisciplinary studies of that material, including chronometric dating and geomorphological analysis, provided data on the environmental changes in the 16th–18th centuries.

The scheme of the Russian settlements given in Table 1.10.1 clearly displays that the settlements dated to the 16th and 17th centuries (all but two of them) are confined to the coastal zone below the scarp of the first marine terrace, while later sites (the 18th to the first half of the 19th century) are at the edge of the first terrace, about 5–6 m a.s.l.

Considering the settlements dated to the 16th–17th centuries, some regularity is noticeable in their spatial distribution. The sites dated to the 16th century (Stabbelva and Sørneset) are rather far from the shoreline and positioned on the surface of ancient beach ridges (Fig. 1.10.2). They are hardly damaged by wave erosion, though the wooden constructions suffered heavily from water flowing from the land. The Stabbelva settlement dated at 1589 is at a distance of 100 m from the shoreline, behind an ancient beach ridge (Fig. 1.10.3). The ruins of the dwelling were repeatedly flooded with meltwater from the land and were partly eroded (Fig. 1.10.4).

Human Colonization of the Arctic: The Interaction Between Early Migration and the Paleoenvironment
http://dx.doi.org/10.1016/B978-0-12-813532-7.00010-3

Figure 1.10.1 Russian settlements on the Spitsbergen Archipelago. 1: settlements.

Two sites of the 16th century are found (as a part of mixed assemblages) at a level of the first marine terrace. One of them—Mosvattnet-2—was immediately at the water's edge. It consisted of three house units spaced some distance apart. All of them were destroyed by the sea waves (Fig. 1.10.5). A relatively well-preserved construction erected 10 m from the shoreline was completely buried with marine sediments. The time of its functioning is determined as the 16th–17th centuries.

The second group of sites, which are now below the terrace level and unsuitable for living, is dated to the late 16th–17th centuries. They occupy the lowest stretches of beach. All of them bear traces of destruction by waves. Some of them (Mosvattnet-3, Langstranda) are completely eroded, and the logs of the former house are scattered on the beach over a distance of 200 m. The lower parts of other constructions (Mosvattnet-2, Renardodden-1, Revelva) were overlain by younger marine deposits. The only exception is the Dunderbukta settlement on the edge of the first marine terrace at 5 m a.s.l.

Many whaler bases on the sea coast were also destroyed; in particular, the stoves for blubber rendering were destroyed on the Amsterdam island (Smeerenburg Dutch

Table 1.10.1 Correlation between age of settlements and their geomorphic context

Site	At the level of the first terrace	At the level of the beach	At low shores (below the first terrace)
Brĕgger	1780		
Russepynten	1775		
Kinghamna	1770		
Imerbukta	1763		
Recherche	1763		
Russkeila	1762		
Sletneset	1761		
Farnhamna-1	1747		
Ingerbrigtsenbukta	1747		
Kap-Li	1746		
Orvinelva	1717		
Mosvatnet		1648	
Dunderbukta	1647		
Revelva		1639	
Langstranda		1621	
Gravshön (bathhouse)			1593
Van-Muydenbukta			1593
Stabbelva	1593		
Mosvatnet-2		1588	1589
Sørneset			1563
Gravshön (living quarters)			1563 1548

settlement), on the Renardodden Point (in Recherche Bay, Western Spitzbergen), and other objects.

Particularly interesting information was provided by investigations of the Renardodden settlement in the Recherche Bay where both older Russian construction (the early 17th century) and structures of the Dutch whaler base (mid-17th century) were destroyed by natural processes.

The ruins of the Russian construction are located on a flattened surface of an old beach ridge armored with a compact layer of dark gray pebble on the surface. The remains are 47 m from the water edge and at 1.9 m a.s.l. (Fig. 1.10.6). Such position accounts for the site destruction and specifics of its stratigraphy. Only two posts fixed in the ground are preserved, along with remains of a brick stove destroyed by water. The cultural layer within the dwelling is eroded almost completely; only its basal horizon is preserved in its natural state (black in color, with chipped wood in abundance).

At 36 m south of the construction, there is a crest of another ridge, its seaward slope being eroded by waves. In the hollow between the two ridges, there are found numerous artifacts brought by waves from the destroyed settlements.

Figure 1.10.2 Ruins of the Sørneset settlement (16th century).

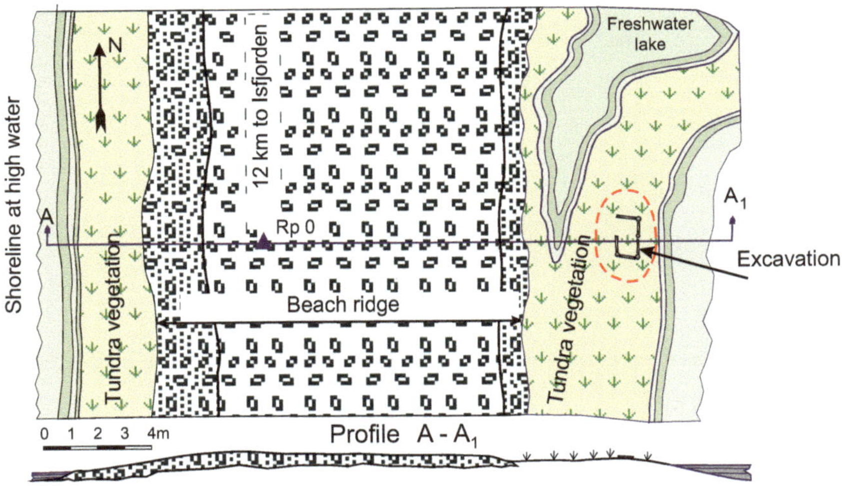

Figure 1.10.3 Plan of the Stabbelva settlement (16th century).

The remains of the Dutch whaler base situated farther from the beach are better preserved, except for stoves for blubber rendering that were positioned closer to the water. It is beyond question that the present state of the settlement is due to an active impact of natural processes (mostly coastal ones); that is supported by a number of facts, as follows:

- a low position of the site with reference to the sea level;
- the presence of distinct traces of beach ridge erosion by waves;
- the concentration of the washed-out finds on lower stretches of the coast, at the back of the site, at the base of the partly eroded beach ridge;
- the presence of younger marine deposits overlying the remains of the cultural layer.

At the same time, the performed studies indicated that even before being destructed by the natural erosional processes the Russian construction was partly dismounted by Dutch whalers.

Figure 1.10.4 The Stabbelva settlement. Remains of a house.

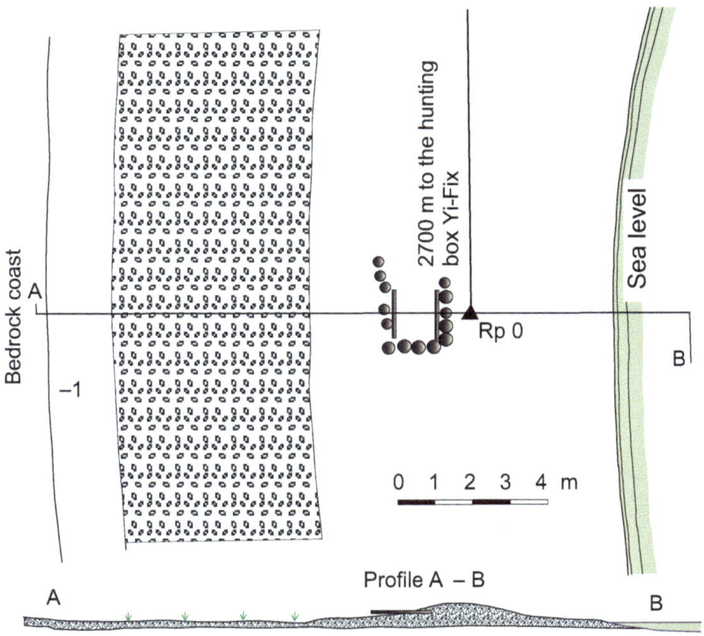

Figure 1.10.5 The Mosvatnet-2 settlement (16th–17th centuries).

A similar chronological relationship between Russian and Dutch contexts on the Renardodden Point may be inferred from the description of the cultural layer on the *pomors*' settlement. The layer consists of two horizons: the lower one, which is well preserved, black in color, and up to 19 cm thick, and the upper one, which is heavily eroded, red in color (due to redeposited material of the eroded stove), and no more than 10 cm thick.

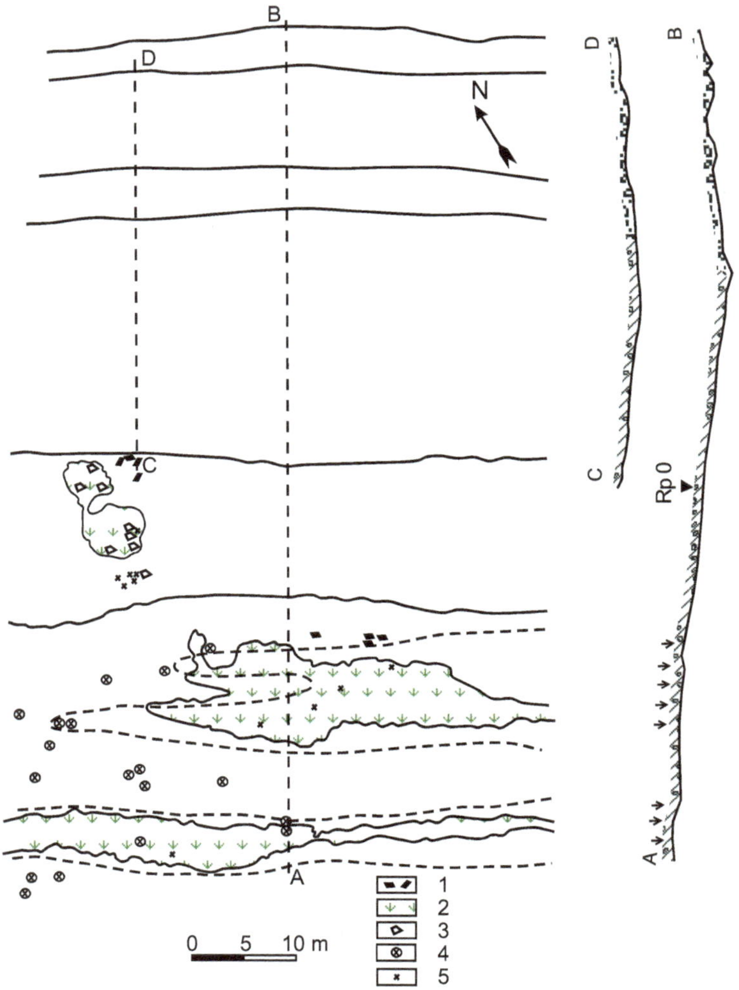

Figure 1.10.6 Plan of the Renardodden settlement (17th century). *1*: wood; *2*: tundra vegetation; *3*: brick; *4*: walrus bone; *5*: archaeological finds.

Within the lower—undisturbed—layer, there were found artifacts of exclusively Russian origin. Among them were fragments of gray clay ceramics (dominant by far) and black polished ware, gun flints, cramp irons, a door hood, a chisel-like tool, iron nails, and animal bones. In the upper—eroded—layer, there were items of Dutch production along with those of Russian making.

It follows from the facts that the Dutch whalers began to colonize the region later than the Russian hunters. The former could not appear there before the 1630s; earlier the region of the Bellsund was the zone of English whaler activity. This suggests that the Russian construction on the Renardodden Point functioned in the early 17th century and was abandoned after 1612 because of the appearance of English whaling

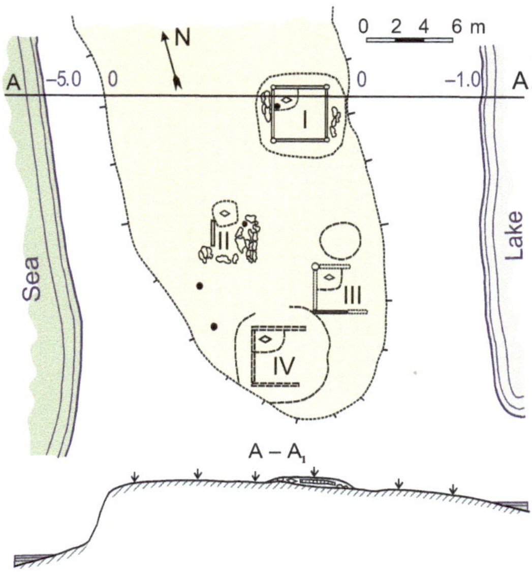

Figure 1.10.7 The Lognedalen settlement (18th century). Plan and profile of the marine terrace.

vessels in Bellsund. The demolition of Russian and Dutch settlements took place much later, at the final stages of the West European whaling activities, most probably at the end of the 17th century.

As to the 18th century settlements, they occur typically in places inaccessible to the action of marine erosion, most often just on the edge of the first marine terrace. Usually, they are at an elevation of 3–8 m a.s.l., though sometimes the sites are found at 15–20 m a.s.l. (Figs. 1.10.7–1.10.9). The house remains are fairly well preserved and contain many objects of the material culture.

The second half of the 17th century is estimated as one of the peaks of the so called Little Ice Age (the 17th–19th centuries) marked by a sharp decrease in temperature. As shown by isotopic measurements of the ice caps on Spitsbergen, the most severe cooling fell on the 1660–1770 interval (Kotlyakov and Troitskiy, 1985; Vaikmae and Punning, 1982).

That interval (the second half of the 17th century) is discernible also in the temperature curves based on isotopic measurements of ice cores from the Greenland ice cap (Fredskild, 1981). A growth of glaciers in northern Scandinavia was also recorded at the same time.

Besides the glaciological data (Hacquebord, 1983), the cooling and growth of the glaciers are known from written sources. When describing his voyage to Spitsbergen in 1767, F. Martens wrote: "There are the greatest icebergs here that never melt and grow every year due to snow, rains and sleet. It should be noted that the land was covered with snow in summer – which had never been observed previously." As follows from historical sources, the years 1660–65 were a period of extremely heavy sea ice. The boundary of the pack ice shifted so far south that many areas of whale hunting were blocked with ice, so the whalers had to leave their coastal bases and go away from Spitsbergen. The same is seemingly true of

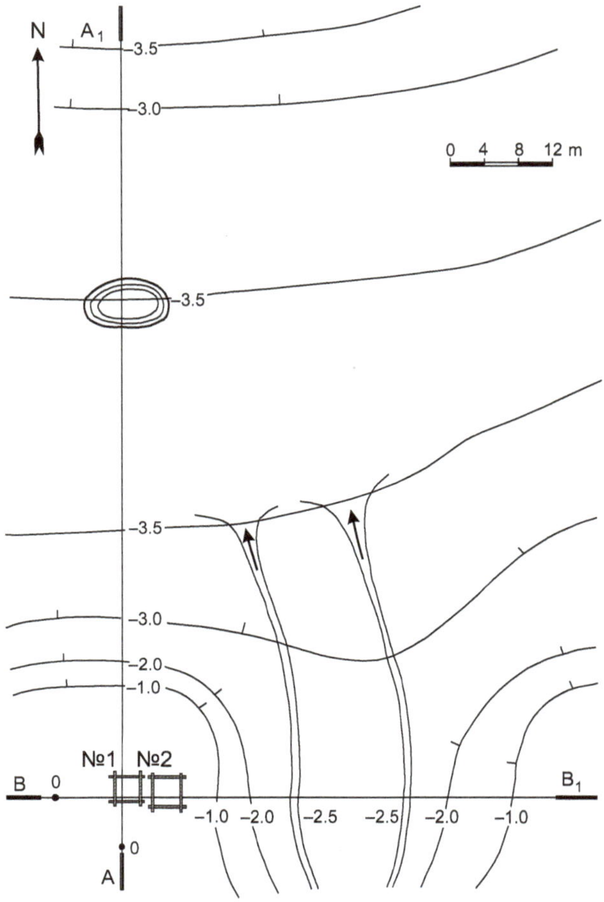

Figure 1.10.8 Geomorphological position of the Dunderbukta settlement (18th century). A–A_1 and B–B_1: profiles (see Fig. 1.10.9). Isohypses are designated in meters.

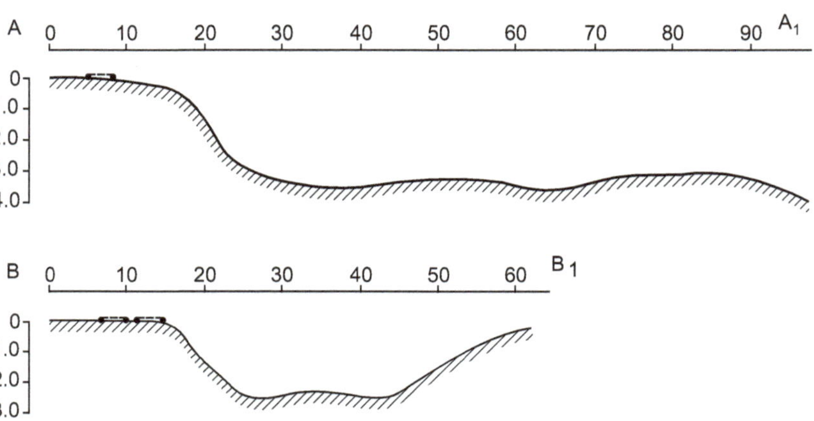

Figure 1.10.9 The Dunderbukta settlement. Profiles A–A_1 and B–B_1 (see Fig. 1.10.8).

the Russian hunters, as not one site dated to the second half of the 17th century has been found on the islands.

It may be suggested that the increasing ice load could exert some influence on the local land downwarping; though, glaciologists do not think it possible for the coastal zone to subside so much. In all probability, the process could be promoted by the ocean level fluctuations, one of maximums of the latter falling on the beginning of the 16th century. The subsequent transgression, with sea level not quite high enough to reach the terrace scarp, induced the hunters to build their houses near the new sea edge on the beach. When the sea level rose again in the early 18th century, the old constructions were destroyed, and new ones were built at a level of the first terrace. At that period the coastal deposition prevailed, so the remains of settlements were buried under the littoral deposits. To take but one example, the deposition zone on the Renardodden Point continues for about 70 m inland, and the remains of dwelling are overlain with sand (prevalent) and gravel (Figs. 1.10.10 and 1.10.11).

Figure 1.10.10 Ruins of a dwelling dated to the 18th century in Erdmann's tundra.

Figure 1.10.11 The Russkeila 2 settlement. Remains of house construction.

Figure 1.10.12 Cutlery (fork: *1*, knife: *2*) and points of hunting spears (*3*, *4*).

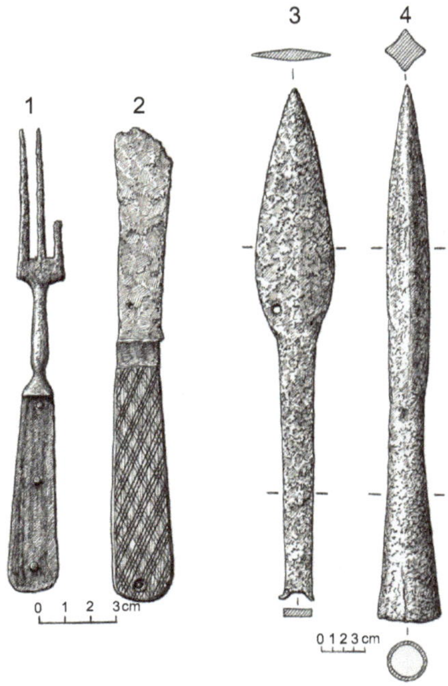

Extremely important for dating the Russian settlements on Spitsbergen are the excavated objects of material and spiritual culture of *pomors*. They include numerous household and hunting objects (Figs. 1.10.12 and 1.10.13), means of transportation (both by sea and by land), and navigation instruments. Among the findings, there are carved objects of art datable with a reasonable degree of accuracy. They are texts and inscriptions; one of them has a date of 1593 (the Van-Muydenbukta settlement).

Of other objects of the *pomors*' spiritual culture dated to the 16th century, there are two undoubtedly rare objects, namely an alphabet carved on a trifacial plank and a cross for keeping relics (Fig. 1.10.14). The first of them was found at the Sørkapvatnet settlement dated to 1563 (by tree rings). In the opinion of T.V. Nikolayeva, a leading specialist in the Old Russian script, the object was made in the 15th century. The cross intended for relic keeping and found at the same site is completely covered with ornamental carving and canonical inscriptions. The early date and highly artistic design make the object an outstanding specimen of the Old Russian art.

A high cultural level of the population of Northern Russia in the 16th–18th centuries is evidenced by numerous other art objects: carved drawings, chess pieces, calendars, and wooden pipes (Figs. 1.10.15–1.10.17).

Other vast regions of the *pomors*' activities in the 16th–18th centuries were lands farther east: Novaya Zemlya and Vaigach islands, the northernmost part of West Siberia (from the Yamal Peninsula to the Yenisey River mouth). There also occasional traces of the Taymyr Peninsula visiting by the Russian hunters in the 17th century.

Speaking of Vaigach Island, it should be noted that it played an important role in colonizing the eastern lands by *pomors*, as it was a border point in the navigation

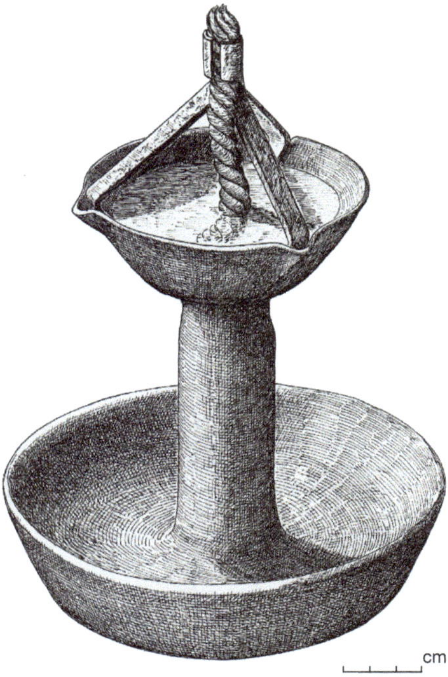

Figure 1.10.13 Oil lamp.

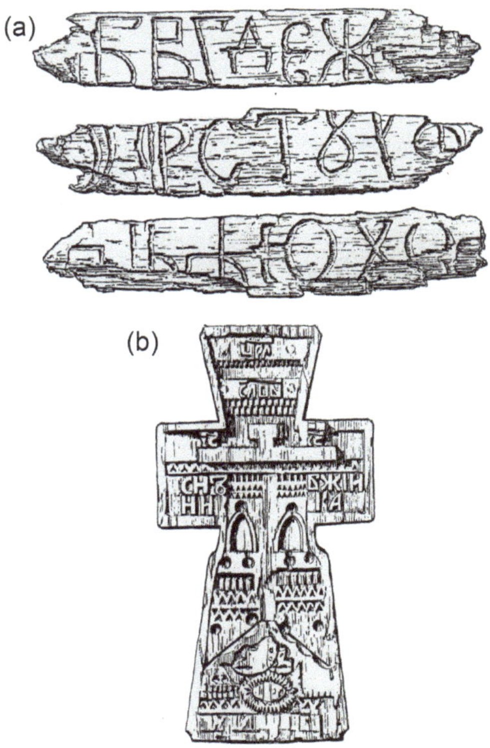

Figure 1.10.14 Finds attributed to the 16th century. (a) Old Russian alphabet, (b) relic-keeping cross.

Figure 1.10.15 Pieces of artistic carving (wood).

through the straits into the Kara Sea. It is not surprising therefore that during the survey on the island by the Marine Arctic Complex Expedition (MACE, the Russian Institute of Cultural and Natural Heritage), there were found rather numerous remains of coastal navigational marks: noticeable crosses. They are also found in abundance on other nearby islands (Matveyev, Golets, Dolgiy). Those marks were first recorded in 1594 by the members of a Dutch expedition that searched for the Northeast Passage into the Pacific Ocean (van Linschoten, 1611).

1.10.2 Novaya Zemlya

The first references to Novaya Zemlya in historical literature about the Russian voyages are dated to the 16th century. At that time the English and Dutch seamen looked for the so-called Northeast Passage to the Pacific Ocean. It was then that the outlines of the Vaigach Island and of the southern island of Novaya Zemlya were put on charts. In 1598, W. Barents published a chart where almost the entire archipelago of Novaya Zemlya was shown, except the eastern coast. W. Barents was also the first of West European seamen who noted the traces of the Russian presence in the north of the Novaya Zemlya western coast.

Figure 1.10.16 Objects of the *pomors'* spiritual culture: chess figures and a fragment of a calendar.

The Novaya Zemlya islands are especially rich in historical objects related to the *pomors'* hunting activities, as was established by surveys performed by MACE in the 1990s. In the published map (Boyarskiy, 1994), there are 65 historical objects including settlements, crosses, and burials (Fig. 1.10.18).

The Novaya Zemlya islands were the hunting ground for *pomors*, who actively hunted there in the 16th–18th centuries. That may be attributed to two main factors: a relative proximity to the Russian North territory on the continent and the presence of rich hunting grounds for killing both marine (walrus, seals, white whales) and terrestrial (polar fox, reindeer) animals (Koryakin, 1991). That is confirmed by some written sources (Krestinin, 1805), as well as by archaeological surveys. In particular, MACE performed fieldwork on the islands under the leadership of P.V. Boyarskiy for many years.

The survey carried out by the expedition covered a large area of the Novaya Zemlya Archipelago and discovered both recent objects of historical interest and remains of *pomors'* settlements dated to the 18th and early 19th centuries. They occur in the southern and southeastern parts of the archipelago, as well as in the southwest of the Northern (Severnyy) Island. Numerous sites of interest were recorded on the Southern Island coasts and on the northern coast of the Northern Island (Boyarskiy, 1994).

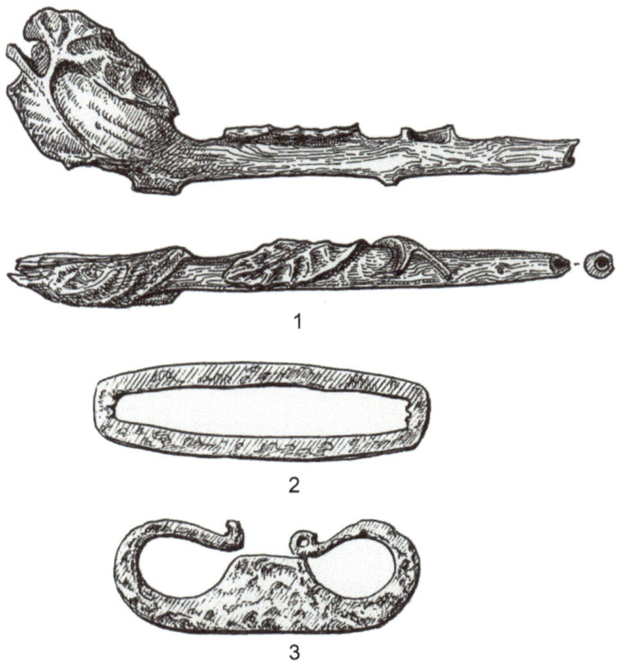

Figure 1.10.17 Smoking pipe (*1*) and fire strikes (*2*, *3*).

The *pomors* sailed to the Novaya Zemlya islands using a special route known as "*Novozemel' skiy Khod*" described in detail in a kind of a pilot's sailing direction. Such descriptions were written down from questioning of the *pomors*' helmsmen ("*kormshchiks*") performed by order of the local Siberian administration ("*voyevods*") in 1616 (Russian Historical Bibliotheca, 1875, p. 570).[1]

It should be noted that the way to the Novaya Zemlya was termed "*Khod*" (course), which was usually applied by *pomors* to a steady, constantly used route. It is also interesting that it is described as the way by the open sea.

Indeed, the *Novozemel' skiy Khod* was an open route across the sea. It started from the Kuloi River mouth falling into the Mezen Bay. In all probability, the Kuloi mouth was the first transit station in the *pomors*' traveling in the European Arctic seas. From there the boats sailed to the Novaya Zemlya as far as the Maksimkov Island (now the Severniy Krestoviy Island), one of the Krestovye Islands in the Archangelskaya Guba Bay (Vize, 1949). That island (located at ~76°N) was the extreme point of the *pomors*' activities on the Novaya Zemlya by the end of the 16th century (Starkov, 2001).

[1] The materials were published in the 19th century in a series of volumes under general name of the Russian Historical Bibliotheca (40 volumes altogether). The edition was performed by the Archaeographic Commission set up with the aim of publishing historical and ethnographic materials.

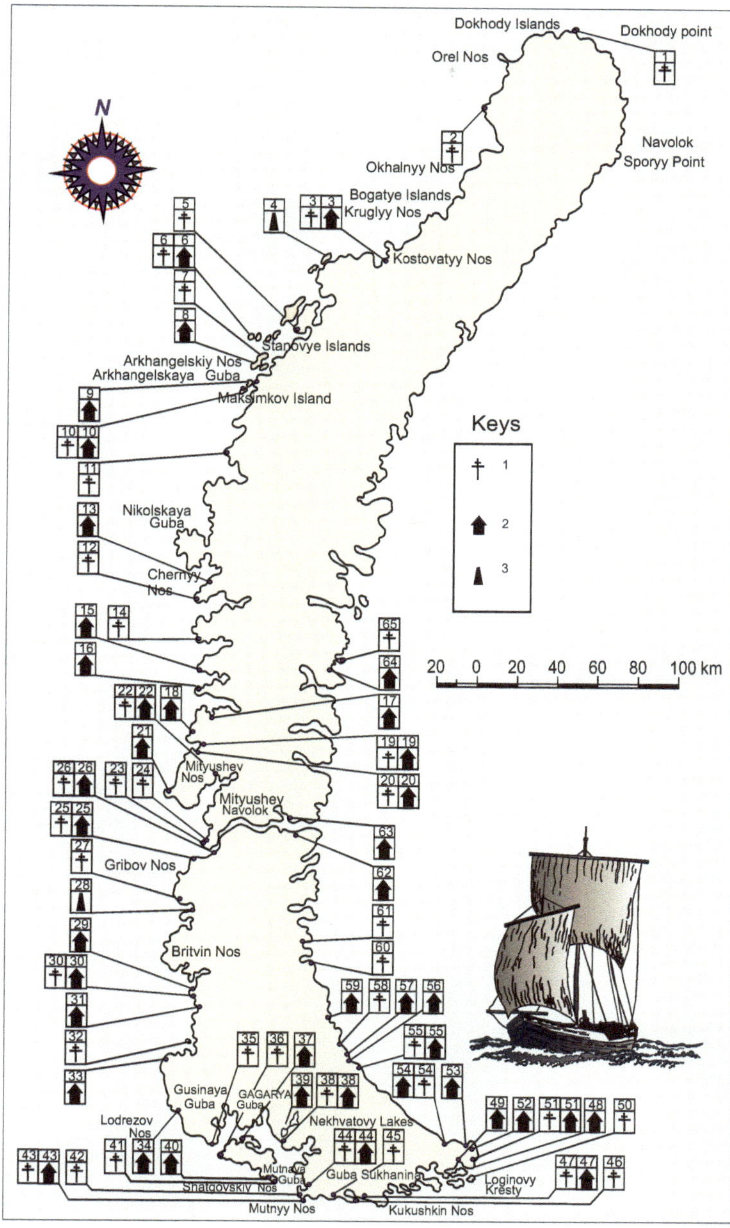

Figure 1.10.18 Traces of Russian sailors visiting the Novaya Zemlya Archipelago (Boyarskiy, 1994). *1*: crosses mounted by Russian sailors; *2*: ruins of houses and settlements; *3*: other traces of *pomors*' habitation, including burials.

Section Two

Siberian North

(photo by V.A. Sarana)

West Siberia

2.1

Vladimir S. Zykin[1], *Valentina S. Zykina*[1], *Sergey V. Markin*[2], *Liybov A. Orlova*[1]
[1]Institute of Geology and Mineralogy, the Siberian Division, Russian Academy of Sciences, Novosibirsk, Russia
[2]Institute of Archaeology and Ethnography, the Siberian Division, Russian Academy of Sciences, Novosibirsk, Russia

The colonization of West Siberia by the early people began with its southern margin, from the Altay Mountains and the adjoining Kuznetsk Depression. As the oldest hominids started their movement from Africa, pebble tools of the Oldowan type appeared in Eurasia. They are identified, in particular, at the Karama site in the Altay, formed under conditions of the warm climate of the first half of the Middle Pleistocene at least as early as 0.8–0.6 Ma BP (Derevyanko et al., 2005) or in the Early Pleistocene between 1.95 and 1.77 Ma BP (Zykin, 2012; Zykin et al., 2005). There is another variety of industry described in the Denisova Cave as well as in the Altay Mountains. The artifacts occur in the basal (layers 22.3–22.1) and overlying them (layer 21) deposits in the cave; their deposition corresponds to the second half of the Middle Pleistocene, beginning from the Tobolsk Interglacial, as well as in loose deposits (layer 19) on the Ust'-Karakol 1 site correlatable with the final part of the Middle Pleistocene. Those are advanced technocomplexes having some Mousterian traits (Derevyanko et al., 2003a,b).

Admissibly, the culture of the Altay sites that existed under conditions of warm and cold moderately wet environments is representative of another migration wave of hominids (at the stage of the late *Homo erectus*); those hominids, having already lost the most archaic forms of inventory, moved into Eurasia from the Near East (Derevyanko, 2005). The site MK 1 in the Kuznetsk Depression yielded few artifacts dated to the end of the Middle Pleistocene (Derevyanko and Markin, 2007).

The next Paleolithic phase in the south of West Siberia is represented by the Middle Paleolithic cave and open-air sites located in the Altay Mountains. Among them are Denisova Cave, Ust'-Karakol 1, Strashnaya (with many cultural layers in the middle part of the sequence), Okladnikov, Chagyrskaya, Ust'-Kanskaya, Tyumechin 1 and 2, and the lower layers of Kara-Bom. All of them were presumably inhabited within the time interval of 100–44.8 kyr BP (Derevyanko et al., 2003a,b). The latest Middle Paleolithic assemblages in the region (the upper layer in the Okladnikova cave) are dated by radiocarbon at 33.5 kyr BP.

The earliest appearance of the Upper Paleolithic culture in Altay is documented in the materials from layers 7, 8, and 11 of the Denisova Cave. The base of layer 11 is dated by ^{14}C at 48,650+2380/-1840 BP/KIA 25285 SP553/D19 (Derevyanko, 2009b). The stone industry of the cave demonstrates development of Aurignacian type characteristics.

Human Colonization of the Arctic: The Interaction Between Early Migration and the Paleoenvironment
http://dx.doi.org/10.1016/B978-0-12-813532-7.00011-5

The assemblage similar to that of Denisova Cave was discovered in layers 11–8 of the Ust'-Karakol 1 site, dated in the range from 50 to 29 kyr BP; on the sites of Anuy 3 (layers 11, 12) and Ushlep 6; layer 8 of the latter being dated to older than 42 kyr BP (SOAN-5045) (Derevyanko and Shun'kov, 2004; Baryshnikov et al., 2005). This tradition at the initial stage of the Upper Paleolithic is known under the name of Ust'-Karakol tradition. Another group of the Altay sites attributed to the early stage of the Upper Paleolithic is named the Kara-Bom tradition, which demonstrates developed blade industry lacking Aurignacian elements (Derevyanko et al., 2000a,b). The first to be mentioned are sites of Kara-Bom with six levels dated from 43,300 ± 1600 BP, GX-17596-AMS/46,970 ± 1420 cal. yr BP/ to 30,990 ± 460 BP, GX-17593-AMS/34,980 ± 450 cal. yr BP/and Kara-Tenesh dated in the range between 42,165 ± 4170, SOAN-2485/45,910 ± 2330 cal. yr BP/and 28,875 ± 625 BP/32,880 ± 710 cal. yr BP/. In the materials from Kara-Bom and Maloyalomanskaya cave dated to older than 33.5 kyr BP, there are known traces of mineral colors (Derevyanko, 2001; Derevyanko and Rybin, 2003). Objects attributed to the early Upper Paleolithic are found also in the inner regions of the Kuznetsk Depression; among them, there is the Mokhovo 2 dated to 30,330 ± 445 BP, SOAN-2861/34,390 ± 380 cal. yr BP/ (Derevyanko and Markin, 1998).

It follows from the preceding that the southern regions of West Siberia presented a high potential for human colonization at every stage of the Paleolithic (noteworthy that besides the aforementioned sites, there are those attributed to the middle and final stages of the Upper Paleolithic). Migrations of the early man northward, to the West Siberian plains, and the onset of the northern region colonization may be dated to the Karginsky time of the Late Pleistocene, as seen from the studies of sites (rather scarce) in the region.

The climatic fluctuations during the Late Pleistocene differed considerably from the earlier stages in the range and frequency. The Late Pleistocene warmings were smaller in amplitude and shorter as compared with previous interglacials. The soils had a humus horizon half as thick as that in older paleosols; their profiles are poorly differentiated, with insignificant degree of claying. Those characteristics suggest a shorter time of the soil formation and indicate the environmental trend toward increasing aridity and cooling.

Palynological data provided evidence of boreal cool climate in West Siberia at the Karginsky time. According to V.S. Volkova (Arkhipov and Volkova, 1994; Volkova, 2001), the forest zone boundary shifted south by 5 degrees of latitude.

Two buried soils (known as the Iskitim soils) developed at that time in the Gorny Altay Mountains, on the hilly plains adjoining the Ob' River (Priobye), and in the Kuznetsk Depression. Both paleosols are noted for a small thickness of humus horizon, as well as of the entire profile. They are separated by a loess horizon formed during an interval of colder and dryer climate.

The Upper Iskitim soil was dated by radiocarbon to the interval of 24–35 kyr BP (Zykina et al., 1981). The thermoluminescent dating permitted finding the age limits of the lower Iskitim soil and to ascertain the chronological boundaries of the Karginsky Interstadial as 24 ± 4–53 ± 4 kyr BP (Zander et al., 2003; Frechen et al., 2005); that gave us grounds to correlate the Iskitim pedocomplex with the MIS 3 stage. Automorphic soil formation was represented by weakly developed chernozems indicative of forest-steppe and steppe landscapes widespread in the south of West Siberia (Zykina, 1990).

Morphotypical characteristics of the Iskitim soils are similar to modern soils of the chernozem type formed under conditions of the temperate continental climate. The profile of buried soils, however, is less mature than those of modern soils, presumably due to a lesser duration of their development.

The Karginsky climatic optimum fell most probably on the time of the lower Iskitim soil formation. That soil formation took longer than the upper one, as suggested by its greater humus content and more distinct differentiation of its profile into genetic horizons. It is safe to say that the early Karginsky warming was wetter and, probably, somewhat warmer than the late Karginsky one. Besides the chernozems, there was soil formation of meadow and meadow-chernozem type.

Two buried soils have been also found in the drainage basin of the middle Irtysh River; they are also noted for shortened immature profile of soddy-gley type. Peatbogs were widespread. Pollen data from the lower soil suggest its development proceeded in forest or forest-steppe environments, cooler and wetter than at present. Pollen spectra from the upper soil permit tracing the vegetation evolving from steppe grass and sage communities to forest-steppe and then to pine and birch forests with grasses and finally to forests of the middle taiga type (Volkova and Zykina, 2000).

It is seen from the preceding that a sizeable portion of the West Siberian Plain at the Karginsky time was not a very suitable region for the Paleolithic human settlement. The paleoenvironments of that time differed from those of today: the climate was cooler, and wetlands were dominant in the northern and central parts of the West Siberian Plain. The sites, rather few in number, were concentrated in the southeast of the plain, in the Chulym River basin. The Aryshevo 1 site yielded scarce artifacts (scraper-like tools, denticulated and notched pieces) associated with the buried soil of the age older than 40,000 BP (SOAN-4178 and SOAN-4179) and to the overlying layer with ^{14}C date of 33,630 ± 995 BP, SOAN-4180/38,050 ± 1200 cal. yr BP/.

Few artifacts of quartzitic sandstone were found together with bones of moose and woolly rhinoceros at the Voronino-Yaya site below the deposits dated at 28,450 ± 850 BP/32,590 ± 860 cal. yr BP/ (Zenin, 2002). A small-size core and a backed bladelet recovered from layer 8 at Shestakovo on the Kiya River (right tributary of the Chulym) are attributable to the end of the Karginsky on the basis of radiocarbon dates: 24,590 ± 110 BP/28,630 ± 130 cal. yr BP/ and 25,660 ± 200 BP/29,870 ± 300 cal. yr BP/ (Derevyanko et al., 2003a,b).

The Karginsky warm interval was followed by the Sartan Glaciation (24–11.5 kyr BP), the traces of which are found in the extreme north of West Siberia (Arkhipov et al., 1980). In the extraglacial regions at the Sartan, highly arid climate favored accumulation of loess (known under the local name of Yeltsovka loess). Besides, under condition of aridity the tractional eolian load was accumulated as ridges and other eolian landforms (Volkov, 1973). At the same time the wind erosion formed vast deflation plains and closed deflation basins.

The Sartan was marked by the shift of the landscape zones by 700–1100 km southward; the Arctic flora and fauna migrated to the southern regions of West Siberia. Pollen assemblages display the dominance of the cold vegetation indicative of a severe continental climate. The period was noted for a great aridity and repeated waves of cold that were responsible for a wide occurrence of specific landscapes: tundra and open woodlands, locally with wetlands. The mean annual temperatures (as inferred from pollen data) were 8–10°C below those of today (Volkova, 1991; Arkhipov and Volkova, 1994).

A spatial distribution of subaerial (mostly eolian) deposits and landforms over the West Siberian Plain at the last glacial time suggests the plain to be enormous cold desert with typical environments of sedimentation (Velichko and Timireva, 2005; Velichko et al., 2007; Zykin et al., 2003; Dobretsov et al., 2006).

The dust content in the atmosphere was much higher than at present; that may be attributed to a greater climate aridity, stronger winds, desertification of vast areas, and the deflation product removal from those areas (Broecker, 2000).

The climate of the Sartan was highly variable. As follows from the analysis of the Sartan sequence in a closed deflation basin, both temperature and humidity varied over wide limits (Zykin et al., 2003). There were at least nine short-term phases of warming when permafrost degraded, ice wedges melted out, and a lake formed in the basin, and nine cold phases when the lake dried out, its bottom sediments froze, and ice wedges developed. The intense cold and arid phases with mean annual temperatures as low as −12 to −20°C and annual precipitation less than 100 mm alternated with cold temperate climate with temperatures about −3°C and less aridity.

At the coldest intervals the mean annual temperature was 13–21°C below that of today. The reconstructed values are in good agreement with the paleoclimatic simulation results (Kutzbach et al., 1998); according to the latter, mean annual temperatures about 21 ka BP were 10–15°C below the present-day values at that latitude.

At the time of deglaciation (Late Glacial) the southern West Siberia underwent short warmings when buried soils were developed (Volkov, 1971; Zykina et al., 1981). The earliest of the soils known as Suma soil was ^{14}C dated at 14,200 ± 150 BP, SOAN-78/17,270 ± 220 cal. yr BP/ (Firsov and Orlova, 1971); the date was obtained on the mammoth bones that occurred on the buried soil surface in Volchya Griva. Poorly developed chernozems were widespread in the south of the plain (Zykina et al., 1981).

Soils developed at that time farther south, in the dry steppe environments, were similar to the chestnut-like soils in their profile, in the microstructure of the horizons, and physical and chemical properties (Zykin et al., 2003; Zykina, 2006). The soil was dated by ^{14}C at 16,210 ± 850 BP, SOAN-3891/19,800 ± 1040 cal. yr BP/.

During the Sartan, man colonized a greater part of the West Siberian Plain, and first of all, its southeastern and central regions. That may be concluded from locations of isolated archaeological sites attributed to different stages of the Sartan on the basis of ^{14}C dating (Table 2.1.1).

A series of cultural layers (7a, 7–5) described at the Shestakovo site (in place of an animal salt lick) is dated to the interval from 23,330 ± 110 BP, GrA-13235/27,560 ± 100 cal. yr BP/to 18,040 ± 175 BP, SOAN-3610/21,860 ± 240 cal. yr BP/. The inventory is dominated by small-size cores and different flakes and blades, including irregular bladelets. Most common among tools are points; there are also denticulate pieces, retouched notches, retouched blades, atypical burins, and backed bladelets. Segmented micropoints are most representative (Derevyanko et al., 2003a,b).

The site of Tomskaya located on the Tom' River terrace 40 m high is a short-term hunting camp at the place of animal killing or butchering; it is dated by radiocarbon at 18,300 ± 1000 BP, GIN-2100/22,290 ± 1240 cal. yr BP/. Among mammoth bones scattered around, there are few artifacts, including angle burins, scrapers, retouched bladelets, and chisel-like tools of uncertain appearance (Abramova and Matyushchenko, 1973).

Table 2.1.1 **Conventional and calibrated radiocarbon age of the Late Paleolithic sites of West Siberia (Orlova et al., 2004)**

Name of the site	Radiocarbon date (yr BP)	Lab index	Calendar age (yr BP) (±1 σ)
Mogochino	20,150±240	SOAN-1513	22,478–21,877
Shikaevka-2	18,050±95	SOAN-2211	19,666–19,162
Tomskaya	18,300±1,000	GIN-2100	20,903–18,434
Troitskoye-1	16,300±300	IERZH-165	17,858–17,229
Chernoozerye-2	14,500±50	GIN-122	15,738–15,299
Volchya Griva	17,800±110	GIN-11463	19,285–18,852
Volchya Griva	14,450±110	SOAN-111	15,703–15,154
Volchya Griva	14,280±285	SOAN-4292	15,686–14,722
Volchya Griva	14,200±520	SOAN-78	15,931–14,367
Volchya Griva	13,600±230	SOAN-111	14,574–13,864
Volchya Griva	12,520±150	SOAN-4293	12,964–12,362
Lugovskoye	18,250±1,100	SOAN-3838	20,041–19,570
Lugovskoye	15,420±215	SOAN-5065	16,979–16,640
Lugovskoye	13,490±155	SOAN-4942	14,348–13,799
Lugovskoye	13,720±160	SOAN-4940	14,656–14,106
Lugovskoye	13,465±50	KIA-19643	14,233–13,859
Lugovskoye	13,455±60	OxA-12030	14,224–13,839
Lugovskoye	13,450±50	OxA-12029	14,216–13,832
Lugovskoye	13,205±60	OxA-12031	13,859–13,498
Lugovskoye	12,970±160	SOAN-5063	13,616–13,134
Lugovskoye	12,830±350	SOAN-4754	13,690–12,541
Lugovskoye	11,840±95	SOAN-4753	11,859–11,643
Lugovskoye	11,310±380	SOAN-4755	11,616–10,927
Lugovskoye	10,820±170	SOAN-4943	11,019–10,697
Lugovskoye	10,210±135	SOAN-4752	10,214–9,660

The Achinskaya site on the Chulym River is admittedly of the Early Sartan age; it displays large fireplaces, surrounded with concentrations of various cores, end scrapers, and notched or denticulate tools on blades. There were reported also side scrapers, chisel-like tools, points, and bone pieces, including a rod with dotted ornament (Lisitsyn, 2000).

An industry similar to that of the Afontova culture sites on the Yenisey River was discovered at Mogochino on the Ob' River bank, not far from the Chulym River mouth. The industry found together with remains of mammoth fauna consists of various cores (including microcores), side scrapers, end scrapers, chisel-like pieces, occasional burins, and retouched blades. The site is dated by ^{14}C at 20,150±240 BP, SOAN-1513/24,270±320 cal. yr BP/ (Petrin, 1986).

Another site belonging to the type of short-term hunting camps is Shikaevka II in the Tobol River drainage basin. It is dated by ^{14}C at 18,050±95 BP, SOAN-2211/21,870±150 cal. yr BP/. Two almost complete skeletons of mammoths have been found there, together with 35 pieces on blade-like flakes, kind of regular geometric forms (mostly asymmetric trapezes). The raw material seems to be of the Southern Urals provenance (Petrin, 1986).

Jasper from the Southern Urals was also used in the industry from the site of Gari on the right side of the Sos'va River valley. The industry included parallel cores, retouched blades, scrapers, and burins. The artifacts were supposedly left at the place of a natural death site of animals, mostly of mammoths. The age of the site is determined by the radiocarbon dates: 15,150±280 BP, SOAN-4462/18,390±310 cal. yr BP/ and 16,320±450 BP, SOAN-4461/19,750±540 cal. yr BP/ (Serikov, 2000).

There are sites attributed to the late Sartan, such as Berezovyy Ruchey 1 in the Chulym drainage basin. The lower cultural layer of the site is dated at 15,310±560 BP, LE-895/18,620±660 cal. yr BP/. The industry is described as a blade assemblage with prismatic cores, retouched blades, scrapers, burins, and other tools (Makarov, 2008).

The sample taken from layers 2 and 3 of the Chernozerye II site at the Irtysh River yielded the date 14,500±50 BP, GIN-622/17,680±100 cal. yr BP/. The faunal remains recovered from the site are dominated by bison. The cultural layers contain various structures, such as hearths, pits, caches, and workshops: areas for knapping and treatment of organic materials. The blade industry was based on exploitation of prismatic cores and large pebble cores. It includes end scrapers, bladelets, burins, notched pieces, and occasional pebble tools. Among bone pieces, there are needles, smoothers, and fragments of diadems; of particular interest is a slotted dagger with insets of flint, covered with geometric ornamentation (Gening and Petrin, 1985).

Several localities are known in the Barabinskaya Steppe (Vengerovo V, Novy Tartas). Of particular interest is Volchya Griva, a short-term camp at the place of an animal salt lick. The faunal remains were dated in the range between 17,800±110 BP, GIN-11463/21,550±170 cal. yr BP/ and 11,090±120 BP, SOAN-4291/12,940±110 cal. yr BP/. Among the bones (dominated by mammoth), there were found more than 30 stone pieces, mostly bladelets and some burins. That suggests occasional human visits to the salt lick (Zenin, 2002).

The northernmost Upper Paleolithic site in West Siberia is Lugovskoye located in the middle reaches of the Ob' River, in the Khanty-Mansi Autonomous Area (Zenin et al., 2006; Leshchinskiy, 2006; Leshchinskiy et al., 2006). The site is located near the edge of the first terrace of Maramka, a lateral channel connecting the Ob' and Irtysh rivers (Fig. 2.1.1).

Bones and cultural remains are related to the bottom deposits of a small stream cutting the terrace and falling into the river channel. The sequence of the bottom sediments includes a silty layer (layer 1) and stratified sands and clays (layer 2) that comprise two bone-bearing horizons with remains of large mammals (mostly of mammoths). A series of radiocarbon dates on the mammal bones from that horizon lie between 16.5 and 9.5 kyr BP. The materials from the site indicate the temperate cold climate, somewhat colder than at present, and forest-tundra landscapes with wetlands and lakes.

The lithic inventory (271 items) that occurred together with the remains of megafauna in layer 2 displays the use of variable rocks, including quartz, chalcedony, quartzite, jasper, chert, slate, and magmatic rocks. There are found parallel-sided

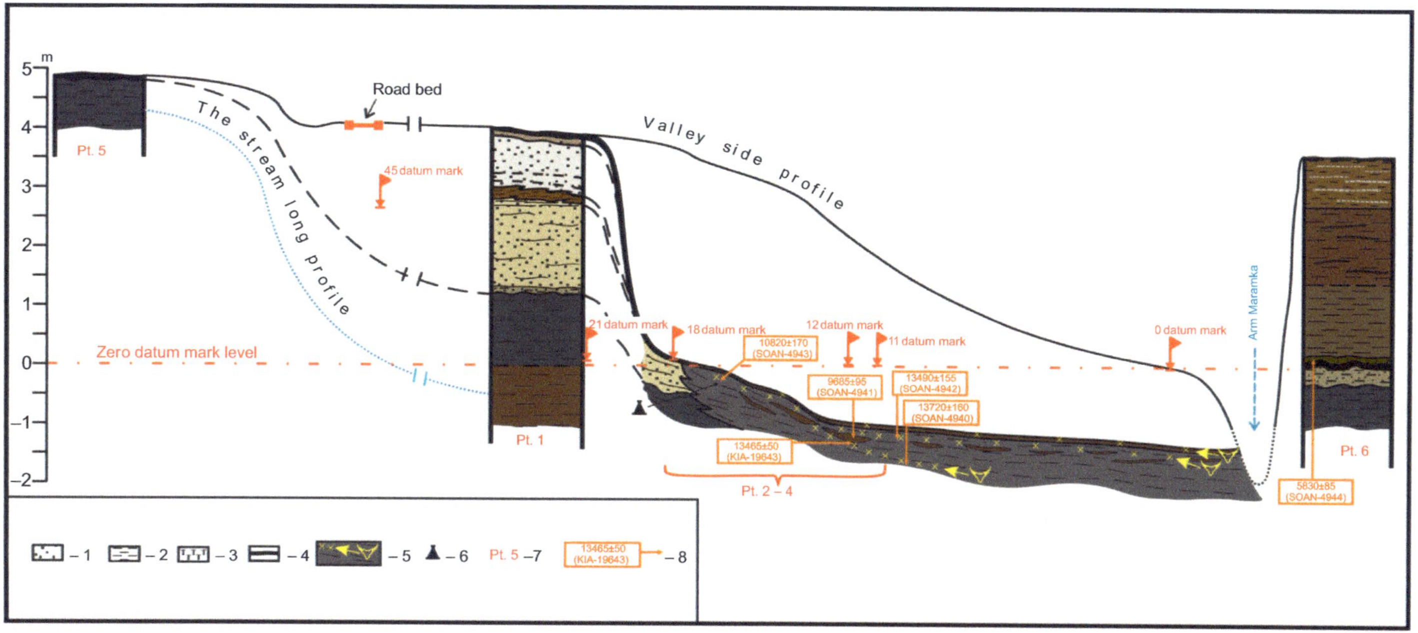

Figure 2.1.1 Generalized section of the quaternary deposits at Lugovskoye vicinity (Leshchinskiy et al., 2006). 1, sand; 2, clay; 3, peat; 4, modern and buried humus horizons; 5, bone-bearing horizons; 6, Paleolithic artifact occurrence; 7, observation point; 8, ^{14}C date and Lab index.

cores, microcores, retouched bladelets, beaked pieces, end scrapers on retouched blades, pieces with burin spalls, and a retouched notched piece (Fig. 2.1.2). The industry as a whole was considered to be a variant of bladelet technocomplex.

The site yielded the first (in Asia) real evidence of the Paleolithic hunting the mammoth. The most remarkable find was the thoracic vertebra of an adult mammoth with a broken acanthus; it was found at a distance of 60 m from the center of artifact

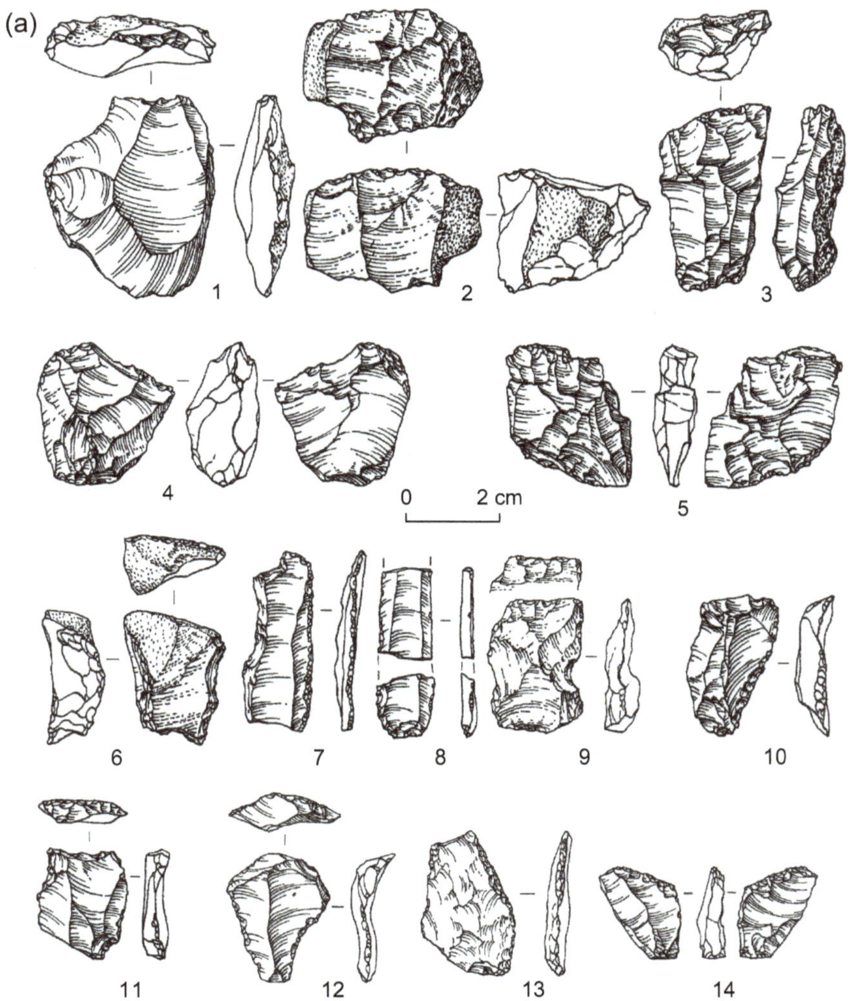

Figure 2.1.2 Stone inventory of the Lugovskoye. (a): 1–3, cores; 4, 6, 11, beaked tools; 5, 9, chisel-like tools; 7, 8, 10, blades with dorsal retouch; 12, end-scraper; 13, retouched flake; 14, notched tool. (b): 1, retouched chip; 2–4, 9, 11–13, 19, 23–25, 27, 28, blades with dorsal retouch; 5, 10, pieces with burin blows; 6, 15, beaked tools; 7, 8, 22, end-scrapers; 14, blade with alternate retouch; 16, 20, 21, retouched flakes; 17, 18, chisel-like tools; 26, blade with alternate retouch.

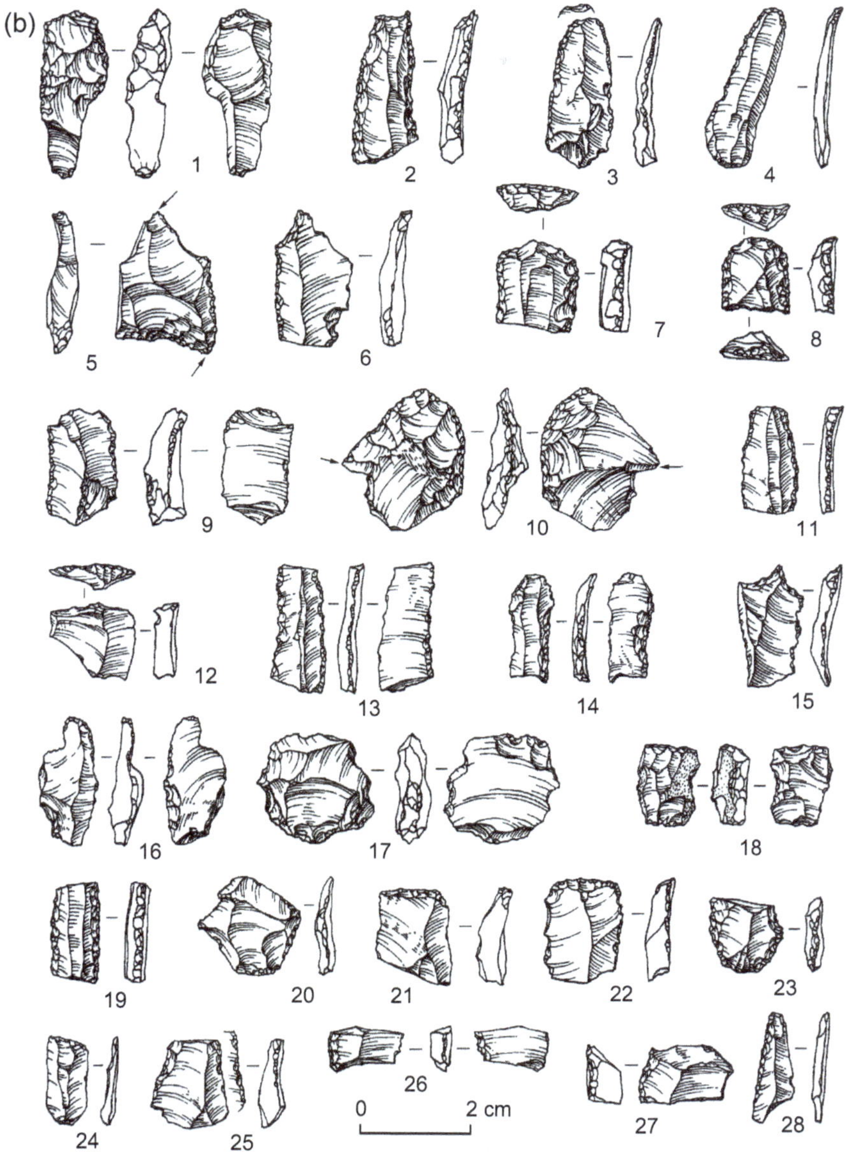

Figure 2.1.2 cont'd.

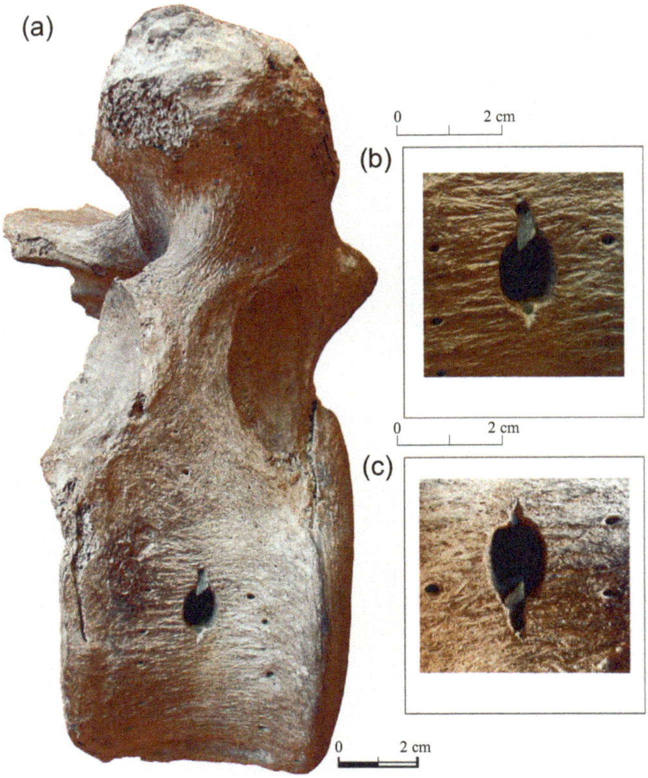

Figure 2.1.3 Mammoth's vertebra with a hole. Lugovskoye. (a) right lateral side of the vertebra; (b) view of the hole with a natural position of the vertebra; (c) view of the hole with the vertebra turned about (Zenin et al., 2006).

concentration, among other bones. In the lateral surface of the vertebra, there is a penetration hole with quartzite insets stuck in the bone (Fig. 2.1.3). The insets were presumably part of a distant hunting weapon made of organic materials.

The position of insets suggests the hunter used a double-slotted point. According to a graphic model of the spear flight, the hunter could hit the animal from a distance of 2–5 m. The radiocarbon date of the vertebra is 13,465±50 BP, KIA-19643 (the average calculated calibrated age is 14,225 BC). It follows that the animal was killed approximately 16,200 BP, which corresponds to the beginning of the second half of the Sartan ice age.

North of West Siberia

2.2

Evgeniy M. Besprozvanny[1], *Pavel A. Kosintsev*[2], *Andrey A. Pogodin*[3]
[1]AV-Kom Nasledie Ltd., Yekaterinburg, Russia; [2]Institute of Floral and Faunal Ecology, the Urals Division, Russian Academy of Sciences, Yekaterinburg, Russia; [3]The Urals Federal University, Yekaterinburg, Russia

Chapter Outline

The north of West Siberia is the northernmost zone of the West Siberian Plain between the Urals Mountains and the Yenisey River; it extends from 59–60°N up to the Kara Sea coast. Physiographically, it is a sedimentation plain with hills and ridges in meso-relief, dissected by rivers belonging to the Ob', Nadym, Pur, and Taz drainage basins.

2.2.1 General traits of the landscape system evolution

The available documentary evidence of the initial colonization of the West Siberian North permitted dating the beginning of the process to the second half of the Late Pleistocene. According to the regional stratigraphic scheme, the period of colonization corresponds to the Karginsky and Sartan horizons of the Late Pleistocene and the Holocene (Volkova, 2000). Chronologically, it corresponds to the middle and late Valday in Eastern Europe,

Human Colonization of the Arctic: The Interaction Between Early Migration and the Paleoenvironment
http://dx.doi.org/10.1016/B978-0-12-813532-7.00012-7

the middle and late Weichselian in Western Europe, and the Holocene, or to MIS 3, MIS 2, and MIS 1. The reconstructions of natural events differ considerably in the accuracy and the level of details in various regions and different time intervals. Considering the Karginsky (55,000–24,000 BP), there have been identified three relatively warm intervals and two coolings (Laukhin et al., 2006). Within MIS 2, there is the period of the maximum Sartan cooling (LGM, 24,000–17,000 BP). The Late Glacial interval (17,000–10,200 BP) has not been subdivided into smaller climatic-stratigraphic units.

Late Pleistocene

Karginsky megainterstadial

MIS 2 (50,000–24,000 $^{14C\ BP)}$

There are palynological data on the landscapes of that period obtained from localities in the northern part of West Siberia and the eastern slope of the Urals.

Two landscape zones are recognized in the *north of West Siberia* from that time, namely, the tundra-steppe with occasional trees and forest-tundra-steppe; the boundary between the two units was, in all probability, near the Polar Circle. The vegetation of the tundra-steppe was dominated by grasses and sedges, sagebrush plants were few, and other nonarboreal plants were rare. The arboreal group was represented by spruce, fir, larch, birch, and shrubs including dwarf birch, willow, and alder (Andreev et al., 2006).

The zone south of its boundary was forest-tundra-steppe; it persisted practically throughout the entire interval, with a somewhat higher proportion of forest cenosis at the warmings (Laukhin et al., 2006). The Siberian pine (*Pinus sibirica*) was always dominant among the trees; less was the participation of birch, spruce, and larch. In the nonarboreal group, grasses were dominant almost constantly, with a lower proportion of herbs and sagebrush, and *Asteraceae* and *Caryophyllaceae* families were almost permanently present. Other plants appeared only occasionally. In the south of the zone, *Ephedra* was occasionally present.

The paleoenvironments of the *eastern slope of the Northern Urals* were reconstructed on the basis of pollen assemblages recovered from the Karginsky deposits in the Cheremukhovo 1 cave site. The age was attributed on the basis of six radiocarbon dates in the range from 31,500 to 25,150 BP. The palynological data indicate the dominance of steppe vegetation with some birch and cold-tolerant shrubs (Strukova et al., 2006).

It should be noted that the results of pollen analysis of the Karginsky deposits in the north of West Siberia are somewhat controversial. To take but one example, the arboreal pollen (AP) proportion in the pollen spectra from the Zolotoy Mys locality (64°52′ N, 65°33′ E) is twice as great as their proportion in the synchronous layers sampled in the Kiryas locality, which is much farther south (60°51′ N, 75°45′ E) (Laukhin et al., 2006). The reason for such a discrepancy is still unclear.

The Sartan cold phase

MIS 2 (24,000–17,000 $^{14C\ BP)}$

The palynological results suggest the existence of a periglacial type of landscapes (Kuzmin et al., 2004). At the maximum of the Sartan cooling (20,000–18,000 BP), four

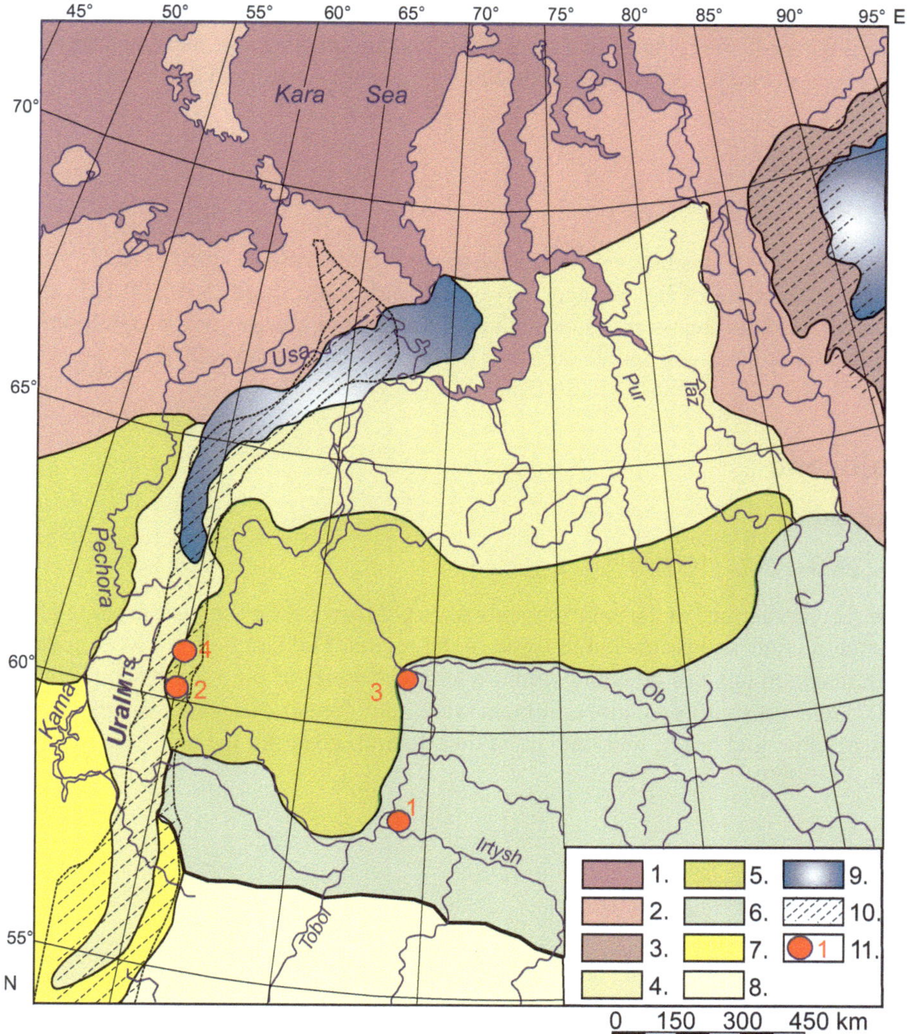

Figure 2.2.1 Types of landscapes in West Siberia in the Late Pleistocene (Late Valday, Sartan) glacial epoch (Grichuk, 2002). *1* – arctic deserts in combination with herb and moss tundra on the exposed shelf; *2* – arctic deserts and dwarf shrub-moss tundra; *3* – mountain dwarf shrub and moss tundra; *4* – moss and dwarf shrub tundra on plains and in mountains with birch, spruce, and larch open forests; *5* – spruce and birch open forests; *6* – meadow steppe alternating with birch, pine, and spruce forests; *7* – forests of pine and Siberian pine; *8* – sagebrush-grass steppe; *9* – glaciers; *10* – mountains; *11* – prehistoric sites (1 – Baigara; 2 – Gari, 3 – Lugovskoye, 4 – Cheremukhovo 1).

zones are identified in the north of West Siberia (Grichuk, 2002) (Fig. 2.2.1). In the extreme north, there was a *zone of arctic deserts* in combination with herb and moss tundra and halophilic communities; the southern boundary of the zone was approximately at 70°N. South of it, there was a *zone of moss and dwarf shrub tundra* with patches of birch and spruce open forests, extending southward up to 64°N. Farther south, it gave way to the *zone of spruce and birch open forests* that reached 61°N. Southeastern and central parts of the region were occupied by *meadow steppe zone*: meadow steppes in combination with forests of pine (including the Siberian stone pine) and birch.

Late Glacial

(17,000–10,200 ^{14}C BP)

The vegetation of this interval was reconstructed for the eastern slope of the Northern Urals. At the time of the Younger Dryas (Noril'sk stage, 11,000–10,200 BP), there were periglacial forest-steppes dominated by herbs and grasses; psychrophile shrubs and forests of birch and spruce occurred in river valleys (Lapteva, 2007).

Holocene

Preboreal

(10,200–9300 ^{14}C BP)

The reconstructions of the environments were put forward for the eastern slope of the Northern Urals and plains in the south of the region. The Urals slopes were covered with northern taiga of pine and spruce and open forests of pine and birch (Lapteva, 2007). On the southern plain, dominant were open forests of spruce with an admixture of larch and birch, with moss and dwarf shrub cover on the soil (Antipina and Panova, 2009).

Boreal

(9300–8000 ^{14}C BP)

There are four landscape zones reconstructed at that time in the West Siberian North (Fig. 2.2.2):

1. tundra zone extending southward to 69°N;
2. spruce and birch forest-tundra between 69°N and 68°N;
3. zone of open spruce forests (northern taiga) between 68°N and 64°N;
4. dark coniferous forests of spruce and larch (middle taiga) (Khotinskiy and Klimanov, 2002).

Atlantic

(8000–4500 ^{14}C BP)

Three landscape zones (and five types of vegetation) have been reconstructed in the north of West Siberia at the considered period (Fig. 2.2.3):

1. tundra, with southern boundary approximately at 70°N;
2. forest-tundra, with its southern boundary at about 67°N;

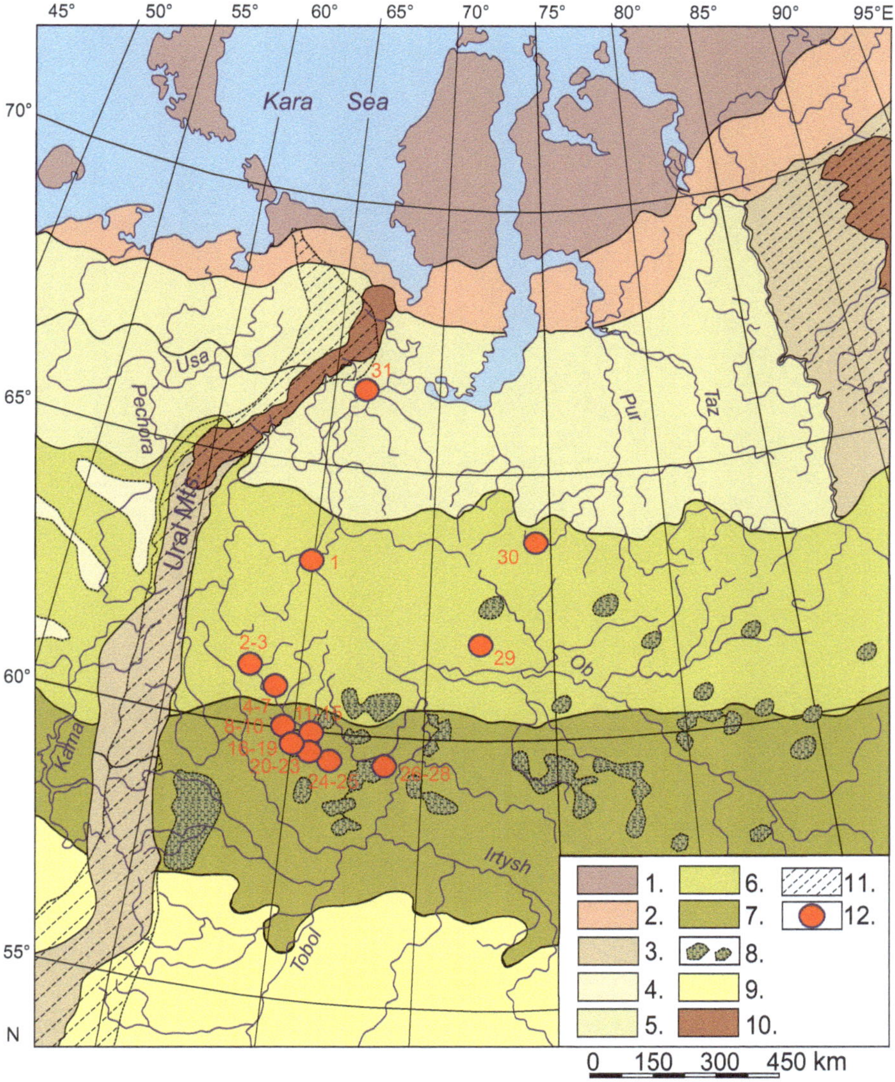

Figure 2.2.2 Types of landscapes in West Siberia in the Boreal (Khotinskiy and Klimanov, 2002). *1*– tundra; *2* – forest-tundra; *3* – birch open forest; *4* – pine forests; *5* – open forests of spruce and birch; *6* – dark coniferous spruce forests (taiga); *7* – birch forests; *8* – wetlands; *9* – steppe; *10* – mountain tundra; *11* – mountains; *12* – sites. Mesolithic sites: 1 – Smolyanoy Bor settlement; 2 – Geologicheskoye III settlement, 3 – Geologicheskoye XVI settlement, 4 – Arantur 16 settlement; 5 – Enyya 18 settlement; 6 – Maly Akh 3 settlement; 7 – Bolshaya Umytya settlement; 8 – Ushakhanskoye settlement; 9 – Uchinya XXIII settlement; 10 – Uchinskiy Mys site; 11 – Satyga XVIb site; 12 – Satyga XVIa site and graveyard; 13 – Satyga XV settlement; 14 – Satyga XVIIa site; 15 – Satyga XXIII site; 16 – Satyga 6 settlement; 17 – Satyga 3 site; 18 – Karagevo I settlement; 19 – Ryabinovaya 2 settlement; 20 – Leushi III settlement; 21 – Leushi IX settlement; 22 – Leushi XIII settlement; 23 – Leushi XIV settlement; 24 – Var-Bor IX site; 25 – Var-Bor XI site; 26 – Kondinskoye I site; 27 – Kondinskoye II site; 28 – Kondinskoye III site; 29 – Amtunyukh 1 settlement; 30 – Pyamaly-yakha IV site; 31 – Korchagi I-B site.

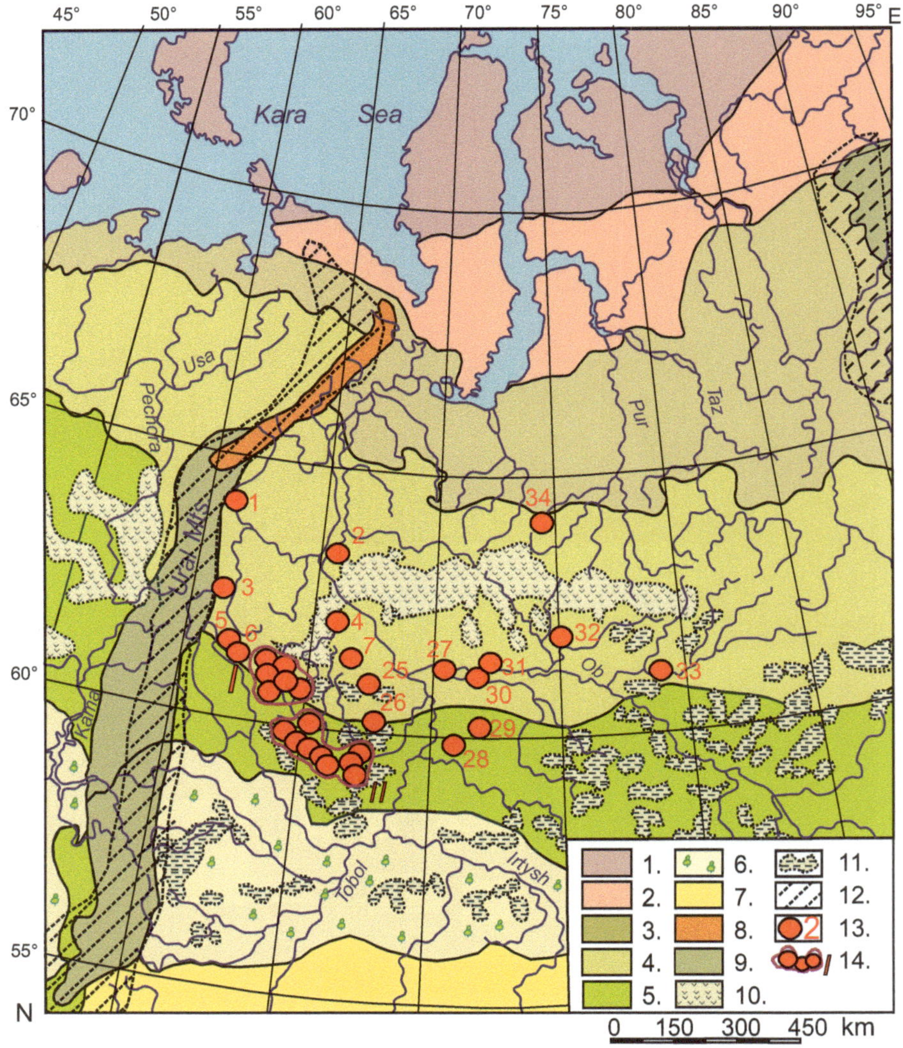

Figure 2.2.3 Types of landscapes in West Siberia in the Atlantic (Khotinskiy and Klimanov, 2002). *1*– tundra; *2* – spruce and birch forest-tundra; *3* – dark coniferous forests (northern taiga); *4* – spruce and coniferous-broadleaf forests; *5* – dark coniferous southern taiga; *6* – birch forests and forest-steppe; *7* – meadow steppe in combination with pine and larch forests; *8* – mountain forest-tundra; *9* – mountain forests; *10* – pine forests; *11* – wetlands; *12* – mountains; *13* – Neolithic sites; *14* – groups of sites (single sites are designated by Arabic, groups of sites by Roman). **Sites**: 1 – settlements of Saranpaul 1, Ches-tyy-Yag, and Iilpaul; 2 – Amnya 1 fortified settlement, and Kirip-Vis-Yugan settlement; 3 – Nyaksimvol settlement; 4 – Maloatlym fortified settlement, and Maly Atlym 1 settlement; 5 – Saylakhaldalansuy 2 settlement; 6 – Yalpyn-Yaki 1 settlement; 7 – Paginskoye settlement. **Group of sites I (8–14)**: **8** – Geologicheskoye 3, 7, and 21 settlements; **9** – Lemya settlements (1.1, 1.3, 1.4, 1.5, 1.8, 1.9, 1.14, 2.6, 2.7, 5.4, 5.5, 5.7–5.9, 5.12, 5.14–5.16, 6.2, and 16.2); **10** – Rangetur settlements (1/2 (2), 15, 17, and 20); **11** – Aran-Tur settlements (1, 2, 7/1, 9, and 17), Bolshaya Umytya (1, 2, 5, 8, 9, 13, 14/3, 15, 46, 51/1, 56, 57, 60, 64, 65/1, 68, 69, 89/1 (89),

3. taiga zone subdivided into three subzones:
   a. dark coniferous spruce forest (northern taiga); south of it,
   b. dark coniferous forests of Siberian stone pine and spruce (middle taiga); and between them, on the Sibirskiye Uvaly Hills,
   c. West Siberian pine forests.

Mountain tundra with brushwood is reconstructed in the polar and near-polar Urals and mountain dark coniferous forests in the Northern Urals (Khotinskiy and Klimanov, 2002).

2.2.2 Geological and archaeological context of the sites

Paleolithic

The Paleolithic man seemed to come to the southern boundary of the West Siberian North at the transition from the Middle to Upper Paleolithic. That is sustained by the find of a human bone at the Baigara settlement in the lower reaches of the Irtysh River at 58°N (Kuzmin et al., 2009). The bone was dated by AMS ^{14}C at >40,300 BP (AA-61831). Neither a cultural layer nor artifacts have been found near it.

The find could be correlated with the onset of the Karginsky (MIS 3) Interstadial (Volkova et al., 2003; Volkova, 2000) or maybe with the Zolotoy Mys warming (Laukhin et al., 2006). The Zolotoy Mys deposits are noted for buried peat layers indicative of relatively warm and humid climate of that time. The middle part of West Siberia featured partly open landscapes with patches of forests (fir, spruce, Siberian pine, and birch) and ground vegetation of herbs, grass, and ephedra.

Cheremukhovo 1 locality is on the right bank of the Sos'va River (the Tavda River drainage basin), 6 km from the village of Cheremukhovo (60°24′03″ N, 60°03′26″ E) on a solid rock massif. The massif is composed of limestone forming scarps along the right side of Sos'va valley over a length of 0.5 km. The rocky outcrops abound in caves, rock shelters, microcavities, and sinkholes. The Cheremukhovo 1 cave locality occurs in the central part of the rock massif composed of compact cross-stratified

90, 100, 110, and 114); Enyya 7, 12, 20, and 26; Maly Akh 3; **12** – fortified settlement Malaya Umytya (1/1, 7, 16A, 17, 19, 20, 24, 28, 32, 36, 37, and 41); **13** – fortified settlement Arpavla 2; **14** – settlements Supra 4, 9, 19, 22, and Ubinka 2. **Group of sites II (15–34): 15** – Evra 5 fortified settlement, and settlements Indra 3, Evra 9, and 12, 17, 23; **16** – Bolshaya Uchinya 14, 17, 20, 23, and 25 settlements, Vysokaya Gora 2, 4, 5 settlements; **17** – settlements Satyga 9, Sumpanya 6, 8, Leushi 21; sites Sumpanya 3, and 4; **18** – Satyga 6, Popuya 2 settlements, and Satyga 5 site; **19** – settlements Leushi 9 (loc. 5), 13 (loc. 4), 19 (loc. 2); **20** – Talya 3 settlement; **21** – Chertova Gora settlement; **22** – Katysh settlement; **23** – Alymskiy Bor settlement; **24** – Landina settlement; **25** – Mikishkin Bor 5 settlement; **26** – Kama 5 settlement; **27** – Kushnikovo 1, 2, and 8 settlements; **28** – Kygyn-Ega settlement; **29** – Kayukovo 2 fortified settlement, and Kayukovo 1 settlement; **30** – settlements: Bystryy Kulyegan 26, 32, 64, 66, 67, Serebryanka 8, Chernaya 3, Pykhty 1; **31** – settlements: Barsova Gora 2/8a, 2/10, sites: Barsova Gora 11, 2.19a, 2.42a, 1.2a, 1.3a, 2.6a, 2.16a, 2.17, 4.5, 1.8a, 2.8a; **32** – Mamontovo 1, 2 settlements; **33** – Chekhlomeyevskoye III settlement; **34** – Et-To 1 settlement.

Devonian limestone. The cave entrance is 8–8.5 m in size and faces northwest 5.5 m above the floodplain. From the entrance inward the ceiling descends gradually to 4 m, the width decreases to 4–7 m, and it passes gradually into the main corridor extending toward the southeast along a fissure. At a distance of 11 m from the entrance (from the drop line) a roof collapse occurred leaving the back part of the cave open at a length of 12 m. Besides, there is a series of small open-end holes in the ceiling and in the upper part of the northeastern wall (Borodin et al., 2000).

The cave deposits were studied in excavation no. 1 at the cave entrance (on both sides of the drop line) over the area of 9 m^2. From layer 10 (brown sandy loam with debris, up to 1.1 m thick, overlain with the Holocene deposits) a stone knife was recovered, broken to a form of asymmetrical segment. That is the oldest archaeological find (Borodin et al., 2000) from the Cheremukhovo 1. The layer it was recovered from was dated by ^{14}C on large and small mammal bones to an interval between 31,500 and 25,150 BP (Strukova et al., 2006), that is, the second half of the Karginsky (Volkova et al., 2003; Volkova, 2000) (Table 2.2.1).

The pollen assemblages from the deposits enclosing the find indicate cryoxerophitic vegetation with sagebrush and herb communities (Strukova et al., 2006). The proportion of nonarboreal plants varies from 67% to 91%; dominant in the group are xerophytes (*Artemisia* 59%–84%; *Chenopodiaceae* from 6% to 21%) and mesophytes (*Asteraceae* 3%–13%; *Caryophyllaceae* up to 10%; *Poaceae* up to 8%; *Brassicaceae* up to 6%). Other herbs do not exceed 1%. AP proportion varies from 9% to 23%. Among tree species, there are birch (8%–24%), pine *Pinus silvestris* (4%–20%), and spruce (up to 4%). Of shrubs, there were noted willow, alder, shrub birch, and *Alnaster*. The proportion of spores (mostly those of ferns) is not too large: 1%–12%. The landscapes may be interpreted as meadow-sagebrush steppe with trees and shrubs on river floodplains and in intermountain basins.

The dominant large mammal was reindeer (*Rangifer tarandus*), with horse (*Equus caballus*) and hare (*Lepus tanaiticus*) being also abundant. Besides, there are present woolly rhinoceros (*Coelodonta antiquitatis*), bison (*Bison* sp.), saiga (*Saiga* sp.), wolf (*Canis lupus*), polar fox (*Alopex lagopus*), cave lion (*Panthera spelaea*), and marmot (*Marmota bobak*). Small mammals are dominated by collared lemming (*Dicrostonyx torquatus*) at 77%; common are Siberian lemming (*Lemmus obensis*) at 14%, and narrow-skulled vole (*Microtus gregalis*) at 6%. Other species include steppe pika (*Ochotona pusilla*), ground-squirrel (*Citellus* sp.), gray hamster (*Cricetulus migratorius*), steppe lemming (*Lagurus lagurus*), root vole (*Microtus oeconomus*), and Middendorff's vole (*Microtus middendorffii*) (Strukova et al., 2006).

Other Paleolithic sites of the West Siberian North appeared at the end of the glacial time, not earlier than 16–12 ka BP.

Gari. The site is located on the right bank of the Sos'va River, at the southern margin of the village of Gari (59°N, 62°E) (Bader and Serikov, 1975, 1981). At the base of the steep and high (about 25 m) side of the valley, there are outcrops of white tufaceous limestone. They are exposed also at the base of the second terrace 16 m high and overlain by dark-brown loam. The first terrace is absent, though the floodplain 7–8 m high is developed. The site is located at a rectilinear segment of the terrace bounded on the north and south by deep (up to 5 m) ravines; the latter drain a swamp located

Table 2.2.1 Regional archaeological and paleoenvironmental characteristics of the Paleolithic sites

Stratigraphy	Environments	Fauna	Sites and localities	^{14}C date /cal. yr BP/
Sartan 24,000–10,200	Tundra-steppe with wetlands and low forests of pine, shrub birch and alder	Mammoth	Gari	15,150 ± 280 (SOAN-4462)/18,390 ± 310 cal. yr BP/ 16,320 ± 250 (SOAN-4461)/19,700 ± 310 cal. yr BP/ 16,700 ± 240 (SOAN 4843)/20,160 ± 300 cal. yr BP/
		Mammoth	Lugovskoye	18,250 ± 1100 (SOAN-3838)/22,270 ± 1370 cal. yr BP/ 12,830 ± 350 (SOAN-4754)/15,220 ± 580 cal. yr BP/ 11,840 ± 95 (SOAN-4753)/13,660 ± 110 cal. yr BP/ 11,310 ± 380 (SOAN-4755)/13,230 ± 420 cal. yr BP/ 10,210 ± 135 (SOAN-4752)/11,900 ± 280 cal. yr BP/
Karginsky 50,000–24,000	Meadow and sagebrush steppe with trees on river floodplains	Reindeer, horse, woolly rhinoceros, collared lemming	Cheremukhovo 1	25,150 ± 500 (SOAN-5302)/29,340 ± 580 cal. yr BP/ 28,520 ± 840 (AA-36469)/32,630 ± 850 cal. yr BP/ 29,120 ± 230 (OXA-10894)/33,310 ± 270 cal. yr BP/ >34,140 (GIN-101152)

upslope 300–350 m from the terrace. There are five more ravines cutting the bank line at the site area. The age of the site was determined by ^{14}C (see Table 2.2.1). According to Serikov (2007), the site may be placed into the second chronological group of the Paleolithic sites (dated approximately 18–15 ka BP).

Both archaeological finds and the Pleistocene mammal bones occur in two concentrations on a narrow strip of the beach at the base of the terrace at a level of 7–8 m above the river. The northern concentration is about 15 m in diameter, the southern one about 20 m. Occasional finds occur outside the concentrations, within 200–300 m from them. More than 600 bones have been collected, with 98% of them being identified as belonging to mammoth. Dominant are intact large bones of young animals; bones of large animals are rare. Among other animal remains, there were identified bones of woolly rhinoceros, bear, reindeer, saiga antelope, wild horse, cave lion, wolverine, and polar fox. A single bird bone was also found.

Gari was visited by the Paleolithic hunters under conditions of harsh climate of the Sartan. Serikov (2007) described the landscapes of that time as open tundra-steppe and wetlands, with numerous lakes and streams, and marshes alternating with patches of low pine, birch, willow, and alder bushes.

Gari is considered to be a Final Paleolithic hunter site at a natural mammoth death site. According to Serikov (2007), however, the bones could have been brought to the site from elsewhere. He also admitted a possibility that there could have been dwellings at the site constructed using mammoth bones.

The lithic collection from Gari comprises 599 pieces. The lithic technology was mostly prismatic: blades were obtained from butt cores (17), prismatic and conical cores (1 core of each type). Dominant are blanks 0.8–1.7 cm length (78.7%). The inventory includes retouched blades (53), blades with use-wear retouch (29), burins (12), points (14), end scrapers (8), side scrapers (4), retouched flakes (13), and chisels (7). There are also an adze, an abrader, a pebble tool, and a hammer stone.

Among the bone pieces (99), there are ivory flakes, bones with traces of grinding, mammoth shoulder blades with holes, and bones with incisions, with pocket holes and traces of splitting. Burned bones are also present.

P.Yu. Pavlov attributes Gari to the Middle Urals Paleolithic culture he first identified and dated to the chronological interval of 17–13 ka BP (Pavlov, 1996). According to Serikov (2007), the cultural affiliation of the site is not yet clear because of the insufficiently advanced Paleolithic studies in the Urals region.

Lugovskoye is a place of concentration of the Late Pleistocene mammal remains 25 km west from the city of Khanty-Mansiysk (60°57′28.8″ N, 68°32′17.4″ E) (Pavlov et al., 2002). It is confined to the eroded lateral part of the first terrace on the left side of the channel connecting Ob' and Irtysh rivers. The terrace surface is 5–5.5 m above the water. Finely laminated clays lying at the base of the terrace sequence are gray in color, with bluish, greenish, or (rare) brownish bands, very dense and viscous, and of apparent thickness more than 2 m. They are tentatively dated by the beginning of the Late Pleistocene glacial: the Ermakovo cooling; it is possible, however, that their age corresponds to Middle Pleistocene: the Samarovo–Taz. The basal clays are overlain with clayey sands horizontally or wave-like bedded, locally massive, with occasional interbeds of brown clays, up to 3 m thick. In the southern part of the locality the sands form the entire upper part of the sequence, up to the present-day soils (the latter are 0.05–0.1 m thick). In the central and particularly in the northern part of the area the geological sequence is more complicated, as the floodplain deposits are included in the sediments of the first terrace.

Five radiocarbon dates have been obtained on mammoth bones (see Table 2.2.1) and two on woolly rhinoceros bones (*C. antiquitatis* Blum.): 30,090 ± 800 Yr BP, SOAN-4756/34,240 ± 830 cal. yr BP/ and 10,770 ± 250 yr BP, SOAN-4757/12,630 ± 300 cal. yr BP/.

The paleontological uniqueness of Lugovskoye consists of abundant mammoth remains distinct for a high degree of preservation and exclusively young radiocarbon dates, not typical for the center of West Siberia. There are representatives of almost all principal species of small and large mammals belonging to the "mammoth complex": Rodentia of *Microtus* sp.; Lagomorpha of *Lepus* sp.; Proboscidae of *Mammuthus primigenius* Blum.; Carnivora of *A. lagopus*, *C. lupus*, *Ursus arctos*, *P. spelaea*; Perissodactyla of *C. antiquitatis* Blum., *E. caballus*; and Artiodactyla of *R. tarandus*, *Alces* sp., *Bison* sp., and *Ovibos moschatus*. Prevalent are bones and teeth of mammoth (more than 98%). No less than three fragments of mammoth skeletons are found in situ. The woolly rhinoceros ranks next to mammoth (about 20 bones). Other mammals are represented by a few bones or teeth (from 1 to 20). Bones of wolf and polar fox are dominant among the carnivores, while only a single tooth fragment was found of brown bear and cave lion.

Remains of at least 27 mammoths have been recovered from Lugovskoye, both adult males and females, and young (including newborn) animals are present. Of great interest is a unique finding of two mammoth embryos at late stages of prenatal development. Among other animals, there were recovered remains of at least seven individuals of wooly rhinoceros (including one young animal), four individuals of reindeer (one young animal and three adults), one moose, one bison, one musk ox, three or four adult wolves, one brown bear, one cave lion, and two or three polar foxes.

The human presence at Lugovskoye is ascertained by the stone inventory recovered by the washing of the sediments forming layer 2 and bottom of the stream. The high level of the groundwater table inhibited excavations, so the samples were washed on a local plot 7 × 1 m in the area in a stream.

The stone industry recovered from the site amounts to about 300 pieces. It is mostly blade inventory with a high proportion of formal tools, up to 15%; that includes retouched blades, end scrapers, burins, and chisel-like tools. One of the most interesting finds in 2002 was a dorsal vertebra of a female mammoth pierced by a point (Mashchenko et al., 2006).

The locality formed due to regular visits of mammals during the second half of the Sartan cryochron. The reason for such visits was seemingly the presence of an animal salt lick of hydromorphic origin. In all probability, the salt-containing sediments are the highly plastic bluish and brownish-gray clays enriched with scarcely found minerals. The clays occur at the base of the first terrace fluvial sediments and, due to its high hygroscopic capacity, they easily formed mud baths, the favorite places of salt-licking animals. The combined conditions of salt lick place and trap at the locality were undoubtedly compelling for the Paleolithic humans, both with a view of scavenging and easy hunting on weak or ill animals trapped in the quagmire.

It may be supposed that the first attempts at the colonization of the north of West Siberia took place 40–30 ka BP. Visits occurred periodically, at the intervals of a milder climate. Precursors of the "permanent colonization" were identified on the basis of places of accumulation of the Pleistocene mammal bones.

Mesolithic, Neolithic

At the Holocene the north of West Siberia was colonized nonuniformly. There are at least three regions different in the concentration of sites and settlements:

1. *Southwestern part* of the West Siberian North between 59°N and 63°N and from 60°E to 68°E. It includes the drainage basin of the Konda River and interfluves of the Konda–Tavda and the Konda–Severnaya Sos'va rivers, and it is rich in archaeological sites. In particular, a series of long-term settlements were found on the banks of the Konda tributaries.
2. *Central* region approximately coinciding with the Sibirskiye Uvaly upland. There are a few Mesolithic sites in this region, mostly of seasonal camp type.
3. *Northern* ("near-polar") region.

A general archaeological-paleogeographic characteristic of the Mesolithic and Neolithic sites is given in Table 2.2.2.

Geologicheskoye III settlement (61°23′ N, 63°13′ E) is on the right bank of the Ess River (right tributary of the Konda River). It is located on a flat surface of a peninsula protruding into the waterlogged floodplain and connected with the river terrace by a narrow isthmus.

Five Mesolithic constructions have been studied in the settlement, among them one aboveground dwelling and four semisubterranean structures. Three of them are two-chambered.

The chambers in one structure are rectangular in plain view, measuring 40 and 30 m^2, and connected by a corridor-like passage. The entrance of the construction is facing the river. The depression varies in depth from 70–75 cm (chamber 1, northeastern) to 30–35 cm (chamber 2, southwestern). There is a step about 35–40 cm high in the passage between two chambers. Walls of the structure are vertical; the floor is flat. The dwellings had a wooden roof and floor (traced by charcoal layers), as well as a post construction (no less than 200 posts fixed in postholes 8 to 30–35 cm in diameter). The bottom of the construction (under the floor) was covered with ocher. Both chambers had storage pits. Deep niches were recorded in the corners and in the wall in the upper part of the foundation pit. The indoor hearth was a hollow more than 3 m long filled with a layer of charcoal and burned ground. Under the floor of the construction there was found a round pit 2.5 m in diameter and 1 m deep (below the floor). Several tens of brown quartzite flakes occurred on its bottom.

The second construction consisted of two rectangular hollows measuring 21 and 29 m^2. The entrance was formed as a rectangular projection in the wall. The passage between the chambers was arranged along the same axis. The depth of chambers was 0.5 and 0.8 m. The floor was flat, with postholes along the walls.

The third construction was partly destroyed. There is one chamber and a passage into the second chamber preserved, as well as parts of the northern and western walls

Table 2.2.2 Regional archaeological and paleoenvironmental characteristics of the Mesolithic and Neolithic sites

Period	Landscape	Fauna	Sites	^{14}C dates and lab index/calibrated
Atlantic	Broadleaf-coniferous and broadleaf-small-leaf taiga forests	Reindeer, moose	Amnya 1	6900±90, LE-4973/7750±90 cal. yr BP/
			Chestyy-Yag	6130±50, LE-2711/7030±80 cal. yr BP/ 6150±40, LE-2713/7060±70 cal. yr BP/ 6455±40, SOAN-2718/7370±40 cal. yr BP/ 6165±35, SOAN-2717/7070±60 cal. yr BP/ 6150±40, SOAN-2721/7060±70 cal. yr BP/ 6095±30, SOAN-2720/6970±60 cal. yr BP/ 5785±40, SOAN-2716/6580±50 cal. yr BP/ 5750±60, SOAN-2719/6550±70 cal. yr BP/
			Et-to 1	6230±280, LE-4975b/7090±290 cal. yr BP/ 6110±150, LE-4975c/6990±180 cal. yr BP/ 5670±80, LE-4975a/6470±90 cal. yr BP/
	Pine West Siberian forests	Reindeer, moose	Geologicheskoye VII	
			Bolshaya Umytya 57	6030±85, SOAN-7639/6890±120 cal. yr BP/ 6030±80, SOAN-7640/6890±110 cal. yr BP/
	Dark coniferous spruce, pine and fir forests (southern taiga)	Reindeer, moose	Sumpanya IV	6850±60, LE-1440/7690±60 cal. yr BP/ 6590±70, LE-1814/7490±60 cal. yr BP/ 6520±70, LE-1813/7430±70 cal. yr BP/
			Sumpanya II	6530±70, LE-1818/7440±70 cal. yr BP/
		Reindeer, moose	Bystryy Kulyegan 66	6150±200, LE-5680/7020±220 cal. yr BP/ 5930±90, LE-5335/6770±120 cal. yr BP/ 5910±130, LE-5336/6750±170 cal. yr BP/ 5780±130, LE-5689/6590±150 cal. yr BP/ 5725±130, LE-5689/6540±140 cal. yr BP/ 5725±70, LE-53370/6530±80 cal. yr BP/

Continued

Table 2.2.2 Continued

Period	Landscape	Fauna	Sites	^{14}C dates and lab index/calibrated
Boreal–Preboreal	Spruce and birch forest-tundra		Yuribey 1	
	Open spruce northern taiga		Korchagi 1-B	7260 ± 80, LE-1376/8090 ± 80 cal. yr BP/
	Pine West Siberian forests	Beaver, reindeer	Geologicheskoye III	8380 ± 710, LE-4220/9580 ± 920 cal. yr BP/
			Bolshaya Umytya 74	
	Birch forests (Ural-Siberian formation)	Beaver, reindeer, moose, fish, birds	Satyga XVIIa	9520 ± 60, LE-5692 (10,870 ± 140 cal. yr BP/ 9330 ± 110, LE-6581 (10,550 ± 170 cal. yr BP/ 9280 ± 25, LE-5686 (10,470 ± 50 cal. yr BP/
			Leushi IX	6680 ± 70, LE-2241/7550 ± 60 cal. yr BP/ 6720 ± 80, LE-2242/7580 ± 70 cal. yr BP/ 7030 ± 80, LE-2243/7850 ± 80 cal. yr BP/ 7240 ± 80, LE-2244/8070 ± 80 cal. yr BP/ 7380 ± 80, LE-2245/8200 ± 90 cal. yr BP/ 7420 ± 80, LE-2246/8240 ± 90 cal. yr BP/ 7560 ± 80, LE-2247/8360 ± 80 cal. yr BP/ 7590 ± 80, LE-2248/8400 ± 80 cal. yr BP/ 7430 ± 80, LE-2249/8250 ± 80 cal. yr BP/ 7510 ± 80, LE-2250/8310 ± 80 cal. yr BP/ 6890 ± 70, LE-2726/7740 ± 70 cal. yr BP/
			Tukh-emtor IV	8360 ± 1510, SOAN-1476/10,240 ± 2110 cal. yr BP/

of the latter. The hollow unearthed is subrectangular in plain view, measuring 26 m^2 and at least 0.6 m deep. Below the floor of the construction, there were found storage pits and postholes (post structure) along the walls. The aboveground dwelling and one semisubterranean construction were destroyed during late construction works and cannot be reconstructed.

The bone material from the settlement includes calcinated bones of reindeer, moose, and beaver. A charcoal sample from dwelling no. 5 (excavation III) yielded the date 8380 ± 710 Yr BP, LE-4220/9580 ± 920 cal. yr BP/.

The raw materials for toolmaking were local pebbles, mostly siliceous rocks, with slate, silica clay, and quartz also present. The raw material was knapped using prismatic (normal pressure) and counterblow technique. The main type of blanks was blade, mostly 0.6–1.0 cm width. The tools made of those blades were pieces with marginal retouch and end scrapers; there are also found points, burins, cutting tools, and inserts.

There are various end scrapers (circular, rectangular, etc.) on flakes, chisel-like tools, burins, etc. Pebbles were used as hammer stones, anvil stones, and chopping tools.

According to Koksharov and Pogodin (2000), the appearance of two-chamber constructions (dwellings) indicates the transition to a sedentary way of life and the growth of the group size. The settlements with such constructions functioned seasonally, most probably in winter, as base camps.

Leushi IX settlement is located on a fragment of a terrace 4–5 m above the floodplain bounded by a river secondary channel (59°39′ N, 65°39′ E). The site consisted of several constructions. The construction no. 1 measured about 30 m^2 within the upper outlines of the subrectangular pit. Its southeastern portion about 20 m^2 was slightly deepened. The floor is inclined toward the southeast (by 0.07–0.12 m). A niche up to 1 m wide is found in the northwestern corner at 0.2–0.3 m above the floor. At the floor the foundation pit was rounded, with two rows of postholes (more than 20) along the outer boundary.

Constructions nos. 2 and 3—one of them possibly with two chambers, the other with only one—were built on the flat surface of the terrace. Both the foundation pits are rectangular, 50–80 m^2 in size, with vertical walls and flat floors. In both constructions the entrance was on the side opposite to the river. Deep niches are found in the corners and on the walls of the pit. The floor in the constructions was covered by ocher. Remains of post construction traceable as two rows of holes are noted along the pit walls, part of them being tilted toward the center of the construction. A hearth on an artificial ground pad was found in the two-chamber construction. Storage pits were found under the construction floor.

^{14}C dates run on charcoal samples from the floor and bottom part of constructions nos. 1 and 3 gave the age of the settlement approximately between 6700 and 7500 BP (Table 2.2.2).

The lithics recovered from the settlement include more than 30,000 pieces. Small pebbles (3–5 cm) of high-quality rocks—flint, jasper, as well as quartz, quartzite, tuffite, and limonite—were used as raw materials. The stone knapping was performed using pressure technique, percussion, and counterblow technique. The cores for blade making—prismatic, conical, and edge-faceted (end) ones—were made on pebbles and cobbles, flakes, and flakes detached from polished (ground) tools. Flake component dominates over the blade one, though the blade is the prevailing type of blanks; it is microlithic in size (80% of blanks are 0.5–0.6 cm wide). The inventory includes inserts; angle, lateral, and dihedral burins; cutting tools; points; and end scrapers. There are a number of tools made on flakes and pebbles, including end scrapers, chisel-like tools, burins, cutting tools, points, adzes, and various combination tools. A special group is formed by adzes with "ears."

Leushi III settlement is located 3–3.5 km northwest from Listvennichny village (the Konda district, Khanty-Mansi Autonomous Area). The settlement sits on a protruding point on the right bank of the river 200 m southwest of its entrance to the Olupya River. The excavation was carried out near the river bank where a Mesolithic structure was studied.

The structure was subrectangular in plain view (Besprozvanny, 1997); the size of the preserved part is (8–8.5) × (4.5–5) m. At present, southern, western, and eastern walls are preserved, while the northern one was destroyed by a landslide. There is a slightly deepened (by 0.1 m) step along the western wall, about 2.6 m wide. The central and eastern parts of the construction are embedded into the sand by 0.2–0.3 m. The bottom descends gradually toward the northeast. In the southeastern and southwestern corners of the construction the niches are identified. There are postholes (remains of a post construction) arranged along the western and southern walls tilted toward the center. The postholes are noted also at the center of the structure. In the latter case, there are holes 0.3 m in diameter each arranged in two rows. Storage pits are found at the bottom of the construction. No traces of hearth or a supposed dwelling entrance were detected.

The number of the lithics from the construction exceed 5700 items. Local pebbles, mostly siliceous and jasper-like rocks, as well as shale and salica clay, were used as raw material. Knapping was performed by normal pressure, the main types of cores being prismatic, conical, and pencil-shaped. The blade preforms do not mostly exceed 0.8 cm in width. There are retouched and truncated bladelets; points with symmetrical tips, and end scrapers. Flakes have been transformed into end scrapers and chisel-like tools, as well as retouched flakes. Among the pebble tools, there are a hammer stone and an anvil. Small-size flakes and broken pebbles are waste.

Bones of reindeer and beaver are found in kitchen refuse.

Seasonal camps of Mesolithic people are widely distributed in the Konda drainage basin. Besides, there was a Mesolithic graveyard found at one of the sites.

Kondinskoye III site is situated near the Lenino village on the Konda River, 100 km downstream from the Kondinskoye village. The site sits on the river terrace 15 m high (Serikov, 1998), at its southwestern slope about 14 m above the water.

The cultural remains are found on the eroded surface of the terrace slope over a distance of about 200 m. The excavation 135 m^2 in area placed at 12–13 m above the water revealed a concentration of 4340 findings. The concentration was oval in plain view, 7.4 m from north to south and 6.4 m from west to east. In its southwestern part facing the river, there were 10 postholes. One more hole was found in the center of the concentration. The sand within the limits of the object concentration was dust-like and loose when dry, while outside the concentration, it stayed firm in the dry state. Serikov (1998) suggested that the concentration formed in place of the aboveground dwelling. About 200 lithics were collected between dwellings. The fireplace was reconstructed from the dark spots of sand found in the southwestern part of the concentration. It may be supposed that the fireplace was located near the entrance, and the latter faced the river, oriented toward the southwest.

The collected items include more than 8000 pieces of stone and 6 fragments of bone items, 5 of them being cases for insets. There are various types of cores (butt-ended, prismatic, conical) and core trimming flakes (core tablets, core rejuvenation flakes, ridge flakes). The blanks are mostly blades (predominantly 0.4–0.8 cm wide), flakes, and split pebbles. The tool inventory includes retouched bladelets, burins (angle, dihedral, truncation burins), cutting tools, points, end scrapers, and flakes with marginal retouch.

The dwelling studied is not the only one at the site. At least three more structures supposedly existed as indicated by the concentrations of stone pieces. Concentration no. 2 is at a distance of 15 m from the excavated dwelling, the distance between concentrations nos. 2 and 3 and that between 3 and 4 amounts to 18 m. According to the author of excavations, there existed a seasonally functioning settlement of four dwellings arranged in a single line along the bank with a lithic processing area in front of every dwelling.

It is suggested that the settlement was inhabited by a group of hunters close to the route of the seasonal migration of animals (moose, reindeer); men waited for the end of the spring flood preparing for the hunt.

The materials from the Kondinskoye III site, according to Yu.B. Serikov, differ from those from other sites at the Konda River. It may be assigned to a difference in age or to local context. It is not inconceivable that the site under consideration provides evidence of the people coming to the lower reaches of the Konda River from the territory located to the east from the Middle Urals.

Leushi XIV settlement is on the right bank of the middle Leushinka River, 1.5–2 km from its flowing into a lake. It is positioned on an elevation rising about 3 m above the river and covered with coniferous forest.

The settlement was excavated over an area of 146 m^2. The Mesolithic finds were recovered from the light-brown and gray-brown sandy loam at the living floor and dark brown in pits. The remains of the construction were oval in plain view, its long axis extending from the southeast to the northwest. The size of the construction was 6–6.5 × 4–4.5 m. There was a subrectangular pit 6.2 × 2.3 m in size in the center of the construction. Its bottom is slightly inclined toward the southeast, with 0.3 m difference in elevation. Accordingly, its depth with reference to the main floor of the construction is 0.1 m in the southwest and 0.4 m in the southeast.

Sixteen postholes are recorded along the eastern, northern, and western walls of the construction, six of them with axes inclined toward the center. A hearth is found 2.4 m south of the construction, with a pit 0.6 m deep in front of it. The deposits within the hearth are 0.1 m thick.

The osteological material consists of very small calcinated bones of fish, birds, and animals. One bone belonging to reindeer was identified.

The construction studied may be considered as the remains of an aboveground dwelling with a deepened area of indoor activities. The entrance, in all probability, was on the southern side, opposite to the river. The hearth was outside the dwelling. The upper cover (roof) consisted presumably of inclined toward the center and slightly embedded poles, probably rested upon the ridge oriented along the long axis of the construction. The construction could be open on the south; in that case, sloping walls could serve as reflectors of the heat of the outer hearth.

Satyga XVIa site and graveyard. The site is situated on the northern coast of the Satyginskiy Tuman drainage lake (the basin of the Evra River, left tributary of the Konda River) near the Gorogishche River mouth (approximately 59°58′ N, 64°50′ E) (Besprozvanny and Pogodin, 1998). The site sits on a triangular cape pointing toward the southwest. The height depends on the level of water in the lake, about 2 m on the average. East of the point, there is a low hill bounded by a short linear hollow. The site and the graveyard occupy its central area and southern slope facing the lake.

The sandy terrace deposits appeared to contain remains of a surface construction and a hearth, as well as three concentrations of calcinated bones that yielded stone and bone artifacts. The cultural remains within the concentrations occurred in the layer of light-brown sandy loam at its contact with clay (basal) bed or sandy horizon; the artifact-bearing stratum was separated from the bedrock with a sterile interbed.

The surface structure is subrectangular in plan, measuring at least 5.2×2.8–3 m, its long axis oriented from west to east. On the west, in front of the entrance, there are remains of a hearth 0.7 m in diameter, with the fired ground up to 0.14 m thick. There are small spots of ocher south from the hearth.

The concentrations of the calcinated bones and microinventory form a burial assemblage dated by the Mesolithic. Concentration no. 1, rounded in plain view, 0.24 m in diameter, and 0.05 m thick, turn out to be devoid of inventory. The bones were unidentifiable due to the small size of fragments. Concentration no. 2 is oval in plain view (1.6×1.2 m) and segment-like in cross-profile, with a bone-bearing layer 0.38 m thick. There was a spot of fired ground 0.76 m in diameter within the concentration, 0.1 m below its top. Among the bones, two pieces were identified, a fragment of a human skull and a proximal phalanx of hand, all the rest being unidentifiable fragments of tubular bones and ribs. The microinventory included 63 items, 67% of them bearing traces of thermal action. There are cores, bladelets with and without marginal retouch, and end scrapers on flakes and points, as well as debitage.

Concentration no. 3 is oval with the long axis oriented from south to north (with a negligible deviation toward the east). The concentration dimensions are 1.6×1.1 m; the bone-bearing layer is at least 0.18 m thick and contains human and animal bones. The human bones are represented by fragments of the proximal phalanx of hand, a fragment of skull, a wrist bone, a foot phalanx, fragments of humerus and radial bones, and fragments of thoracic and cervical vertebrae. According to D.I. Razhev, the remains belong to different parts of a single skeleton; the bones underwent thermal action when soft tissues was still preserved on them. Among the animal bones, there were identified moose bones, a phalanx of wolf or dog, and some bird bones.

Satyga XVIIa site is on the northern coast of the Satyginskiy Tuman Lake 10.5–11 km to the southwest of the village of Yagodny (the Konda district, Khanty-Mansi

Autonomous Area), ~59°58′ N, 64°51′ E (Pogodin and Besprozvanny, 1999). The coast at the site is 5 m high; it is subjected to erosion by waves and wind. The surface of the terrace is overgrown with coniferous forest gradually replaced with mixed one away from the coast.

As indicated by the distribution of finds over the beach, the cultural layers bearing the Mesolithic pieces occur over a length of 100 m. The eastern part of the Mesolithic site is overlain with the cultural remains of the Satyga VII Gorodishche (fortified settlement) of the Kulminsky culture (Early Iron Age). As the western part of the site has no distinct features on the surface, the excavations were mostly performed close to places of maximum concentration of finds in the beach zone. The age of the site (seemingly, of the second habitation horizon) was dated by radiocarbon at 9520–9280 BP (see Table 2.2.2).

The excavation opened 150 m^2 of the cultural layer. Its thickness is 0.9 m on the average and up to 1.45 in the pits (holes). There is a charcoal dark-gray interbed noted within the pits. The cultural layer is underlain by an ortsand interlayer 0.05–0.1 m thick (within the pits the interlayer occurs at the same depth and cuts through the pit filling). Some lithics were found in the ortsand. The Mesolithic artifacts form concentrations in places of former surface structures (at least two structures have been identified). There are also areas of maximal concentration of lithics. The pits are deepened into the underlying strata, cut horizontally by interbeds of ortsand; the latter included occasional lithic pieces. Three cultural (habitation) horizons were identified; the lower of them is tentatively dated to the Early Holocene or Late Pleistocene. Occasional white inclusions are seen in the deposits filling the pits; in all probability, they developed in places of former ice wedges. The sandy horizons display faults and depressions that could be due to solifluction.

The artifact collection includes more than 2500 stone pieces and one piece of bone. The stone inventory is uniform both typologically and technologically. Dominant is prismatic technology with normal pressure knapping. Local pebble deposits were used as a source of raw material; most often used was red and green jasper as well as black flint. Among cores, there are prismatic single-platform and double-platform, conical, and pencil-shaped ones. The most common blanks were blades, most often of microlithic proportions (bladelets) used as inserts in composite tools. There are also found burins, cutting tools, and end scrapers made on blades. Some end scrapers were made on flakes and pebbles. The only bone piece presents a fragment of a case for an inset tool.

The osteological material includes small calcinated bones of moose, beaver, and fish.

Bolshaya Uchinya XXIII settlement is located on the right bank of the river of the same name, 530 m from its mouth, 200 m west-southwest from the Polovinka village (Besprozvanny and Pogodin, 2006). The settlement sits on the river terrace 11 m above the river. In the central part the fragment of the terrace is cut by a ravine up to 4 m deep; the alluvial fan of the ravine is well pronounced in the relief. The northern slope of the terrace gently descends toward the Konda River; the western slope facing the Bolshaya Uchinya River is steep due to its being undercut by the flood flow.

In the site area, there is a concentration of finds 3.2 × 2 m in size, with its long axis oriented southwest to northeast. Outside the concentration, within 3–6 m from its boundaries, Mesolithic pieces are found in small number. The western margin of the concentration has been destroyed by a bulldozer.

The total number of the Mesolithic pieces is 251. Pebbles were used as raw material for making cores, tools, and preforms. Most of the pieces (61.75%) are made of siliceous gray-green rock, 20.32% of black flint, 7.97% of gray flint; other rocks were used only occasionally (quartzite 2.79%, quartz 2.39%, silica clay, limonite, brown-red chert 1.59% each).

The knapping technology was mostly blade removal. The cores are prismatic, single- or double-platform, and the knapping performed circle-wise or from the end and front face. The recovered cores are up to 32 mm high, with straight faceted platforms. The core-like forms are residual cores utilized by counterblow knapping. The elements of the core trimming are core tablets removing part of the platform or the entire platform; one crested flake has been documented.

The blanks are mostly blades 9–26 cm long (unbroken). Among the blanks and waste, there are flakes, chips, and intact and split pebbles. The tool assemblage includes retouched and truncated blades, an angle burin, end scrapers on medial parts of blades and on flakes, and points on flakes.

The use-wear analysis revealed the dominance of tools used in the game processing (meat, hide, and bones). In all probability, the site was functioning as a seasonal camp (during spring and summer) used for processing hunting game.

The Mesolithic sites in the *Central* and *Northern* (near-polar) regions of West Siberia are rather scarce and indicative of a gradual colonization of the northern latitudes.

In the *Central region* (approximately coinciding with Siberian Uvaly Uphills) the Mesolithic sites are mostly of seasonal camp type, such as Pyamaly-yakha IV at the interfluve of the Nadym and Pur rivers and Amtunyukh 3 in the Pim River basin (left tributary of the Ob' River).

Pyamaly-yakha IV site is located on the northwestern coast of Lake Pyaku-to, on the left bank of the Pyamaly-yakha River, swamped at present (Pogodin, 1998). There is a low (4–5 m) ridge on the coast of the lake; the site occurs on a gentle, smooth, southern slope of the ridge, close to its base at about 1 m above the water. Most of the lithics were collected from the disturbed surface, except one finding recovered from a test pit; it occurred within an interlayer of gray sand at a depth of 0.2 m from the surface. The collection includes 308 items. Among them, there are prismatic cores and core rejuvenation flakes, blades, end scrapers on flakes, and plaque, chisel-like tools, hammer stones. The blade-blanks are 3–8 mm wide.

The site was the place of seasonal stay of hunters and fishers who visited the lake in spring and summer. The lack of visible remains of foundation pits and the stone tools occurrence in the upper sandy layers suggest the dwellings were light, transportable tents ("chums"). The lithic industry has some features in common with the Mesolithic assemblages of the Konda lowland.

The northernmost Mesolithic sites are found in the near-polar ("*circumpolar*") zone of West Siberia: Korchagi I-B, Yuribei I, and Nulmayakha.

Korchagi I-B locality (site) is on the right bank of the Ob' River downstream from Salekhard city, on a fragment of a high (15–20 m above the river) terrace bounded by a gully on the south and by a linear hollow on the north (Khlobystin, 1987). A concentration of lithics was discovered in a layer of light-yellow loam; they occurred in a shallow pit (0.15–0.2 m) measuring 0.3–0.4 m. There have been found 38 lithics: 6 cores, 3 end scrapers, a side scraper, flakes, and chips. A ^{14}C date of 7260 ± 80 yr BP, LE-1376/8090 ± 80 cal. yr BP/ has been obtained on a charcoal interlayer overlying the loess-like loams with the finds.

According to Khlobystin (1987), the concentration was a cache of lithics buried for better preservation. The cultural attribution has not been ascertained, though some resemblance (in the morphology of end scrapers and cores) may be traced to the materials of the Mesolithic sites known from the Middle Urals.

The *Yuribey I site* is on the left bank of the river of the same name in the middle part of the Yamal Peninsula (68.90196° N; 70.18770° E) (Brusnitsyna and Oshchepkov, 2000). The site is positioned on a point formed by an erosional terrace of the Yuribey rising 8–13 m above the floodplain. The slopes of the terrace are steep and covered with grass. The flat upper surface descends gradually eastward. The central part of the surface is heavily eroded.

A collection of 56 artifacts has been recovered from a deflation basin; it includes 31 blades, 2 retouched blades, 4 crested flakes, 14 flakes, 3 chips, 1 hammer stone, and an intact pebble.

The surface material was gathered from an area of 1080 m^2 (27 × 40 m). The site territory included the terrace surface and its steep slope (about 3500 m^2).

Central Siberia (the Yenisey-Lena-Yana Region)

2.3

Yaroslav Y. Kuzmin
Institute of Geology and Mineralogy, the Siberian Division, Russian Academy of Sciences, Novosibirsk, Russia

Chapter Outline

2.3.1 Introduction

The chapter deals with the environmental changes and the oldest human settlements in the north of Central (Eastern) Siberia, within the drainage basins of the main rivers of the region: the Nizhnyaya Tunguska, the Lena and its large tributaries the Aldan and Vilyuy rivers, and the Yana River; some traces of the human presence were also found on the Taymyr Peninsula (Fig. 2.3.1). There will be also discussed the process of the initial colonization against the background of the changing environments and specific features of the dispersal of Paleolithic man over the Siberian sector of the Arctic.

As known from the earlier studies (Kuzmin, 1994; Kuzmin and Tankersley, 1996; Vasil'ev et al., 2002), the north of Central Siberia was colonized by the early humans in the Upper Paleolithic. It should be mentioned, however, that there are two archaeological localities that may be older than others discussed in this chapter, namely Diring-Yuryakh and Mungharyma.

Diring-Yuryakh in the middle reaches of the Lena (61°12′ N, 128°28′ E) is one of the most debated archaeological localities in Siberia, in particular regarding the age and the type of its lithic industry. According to Mochanov (1992), the artifacts

Human Colonization of the Arctic: The Interaction Between Early Migration and the Paleoenvironment
http://dx.doi.org/10.1016/B978-0-12-813532-7.00013-9

Figure 2.3.1 The oldest archaeological sites of the Yenisey-Lena-Yana region.

are very old (3.2–1.8 Ma BP) and belong to the Lower Paleolithic; the hypothesis has practically no supporters. In particular, thermoluminiscence dating of the culture-bearing layer yielded the values not exceeding 267 kyr BP (Waters et al., 1999). Some specialists consider the main part of the Diring-Yuryakh collection to be even younger (Alekseev et al., 1990; Ranov and Tseitlin, 1991); see also an overview by Kuzmin and Krivonogov (1999).

As to the Mungharyma site in the lower reaches of the Vilyuy River (64°01′ N, 123°25′ E), the publications are few in number as yet (Mochanov, 2007; Mochanov and Fedoseyeva, 2001). The site is tentatively attributed to an initial stage of Middle Paleolithic/pre-Dyuktay quartzite industry of the Mousterian type; see Mochanov (2007). The overlying deposits were dated by radiotermolumininscence (RTL) to 150 ± 38 BP (RTL-958) and underlying layers to 600 ± 150 BP (RTL-957) (Mochanov and Fedoseyeva, 2001). There is a single ^{14}C date obtained on a mammoth bone from the cultural layer (>41,310 BP). As follows from the preceding, the two objects are still insufficiently studied with a view to their age or archaeological context, so they have not been included into the present review.

At present the earliest find indicative of the early man presence in the central and northern parts of Siberia is that of a femoral bone attributed to the modern man (*Homo sapiens sapiens*) near the Ust'-Ishim village (57°42′ N, 71°10′ E). The radiocarbon date on the bone is 41,400 ± 1300 BP (Oxa-25516) (Fu et al., 2014). Those data indicate unambiguously the human presence in the center of West Siberia long before the LGM.

2.3.2 Environments of northern Central Siberia in the second half of the Late Pleistocene (30,000–10,000 BP) and in the Early Holocene (10,000–6000 BP)

The present chapter is mostly based on the work performed by a number of international expeditions in the region (MacDonald et al., 2000a,b; Pisaric et al., 2001; Schirrmeister et al., 2002; Andreev et al., 2002, 2003, 2004a, 2006; Sher et al., 2005), as well as on integrated studies of Russian investigators in the 1950s–90s (Velichko, 1993). The entire period under study may be divided into three parts:

1. the second half of the Karginsky (MIS 3) interval: 32,000–24,000 yr BP;
2. Sartan stadial (MIS 2): 24,000–10,000 yr BP;
3. Early–Middle Holocene (MIS 1): 10,000–6000 yr BP.

During the second half of the Karginsky (32,000–24,000 BP) the Central Siberian North presented favorable environments; it was the so-called Lipovka-Novoselovo warming, according to Kind (1974). The central part of the region was covered by open forests (Isayeva et al., 1993), and farther north the forests gave way to landscapes of tundra/tundra-steppe with isolated forest massifs (Kienast et al., 2001; Andreev et al., 2002, 2003, 2006). Still farther north, at the Laptev Sea coast (the Mamontova Khayata section near Tiksi (Sher et al., 2005)), the vegetation at the Karginsky and Sartan time (50,000–14,000 BP) was mostly tundra-steppe dominated by herb and grass formations that provided the sustenance for various mammal species (mammoth, horse, musk ox, bison, etc.). The climate was cold and dry. About 14,000 BP the dominant landscape became that of forest-tundra that existed under conditions of more humid climate about 14,000–12,000 BP (Andreev et al., 2002; Sher et al., 2005).

At the Sartan maximum (c.20,000–18,000 BP) the region was occupied by a cold steppe (or tundra-steppe) (Grichuk and Borisova, 2009; Tarasov et al., 2000). The recently obtained data (Svendsen et al., 2004) proved the absence of an ice sheet in the north of Central Siberia, including the Taymyr Peninsula.

The Late Glacial (c.15,000–10,000 BP) was marked by a gradual replacement of the vegetation typical of the dry and cold Late Pleistocene climate by the formations growing under more humid and mild conditions. In the south of Taimyr, for example, tundra-steppe was dominant until 12,000 BP and was replaced by tundra during the interval 12,000–9700 BP (Andreev et al., 2003, 2004b). A similar trend is traceable in the region of the Lena River mouth; forest communities settled there about 8500 BP and persisted until 3500 BP when they gave way to the present-day tundra (Pisaric et al., 2001; Andreev et al., 2003; Sher et al., 2005).

As early as the Early Holocene (c.8900–8600 BP) the birch grew on the Yamal Peninsula and in the Lena delta (Forman et al., 2002); on Taimyr, birch appeared about 9200 BP and larch at 8600 BP (Andreev et al., 2002). In other regions of the arctic high latitudes, trees appeared approximately at the same time or even earlier, c.11,000–10,000 BP (MacDonald et al., 2000a,b), and in greater number, c.9500 BP (Andreev et al., 2004a,b). So in the Early Holocene, some arboreal species penetrated to the north, a few degrees beyond the present-day tree line. To take but a few examples, the south of Taymyr (the Lake Lama vicinities) was occupied by forest-tundra with spruce, birch, and shrub varieties of alder and birch (Andreev et al., 2004b). At the Holocene optimum (Atlantic period (c.6000–5000 BP)), there were open forests of larch and birch on the Taymyr Peninsula (Andreev et al., 2002).

2.3.3 Principal regions of the initial settlement

2.3.3.1 The Yana and Yenisey drainage basins

The Yanskaya site (RHS) is at present the oldest reliably dated Paleolithic object in the polar regions of Eurasia (70°43′ N, 135°25′ E, Fig. 2.3.1) (Pitul'ko 2006; Pitulko et al., 2004, 2007; Pitul'ko and Pavlova, 2007, 2010). The first artifacts in the vicinities were found in 1993; the site itself was discovered in 2001 and has been studied methodically since 2002 under the direction of Pitul'ko and Pavlova (2010).

The site is positioned on the second terrace of the Yana River, on its left bank, 16–18 (up to 20 m) above the river. The cultural layer occurs at a depth of 7–8 m from the surface, in the context of the sandy-clayey deposits of the "*yedoma*" type (Pitul'ko, 2008b; Pitul'ko et al., 2007; Basilyan et al., 2009). The sediments are frozen, with a high content of ice (up to 50%–70% of the volume), and they contain ice wedges. The artifacts occur in situ in the layer composed of sandy silt with plant remains; when viewed from above, the cultural layer displays a polygonal pattern of thick, syngenetic ice wedges (Pitul'ko, 2008b). The excavated area is about 200 m^2, and the cultural remains occur within the "ground columns" confined between ice wedges; the total area of the excavated artifact-bearing ground is about 90 m^2. Near the water's edge, there are alluvial sands (Fig. 2.3.2).

The sediment sequence at the Yanskaya site may be schematically presented as follows (Pitul'ko and Pavlova, 2010):

Strata	Depth (m)
1. Red-gray clayey silt interlayered with peat	0.0–0.15
2. Gray clayey silt with high content of ice, horizontally and wave-like laminated, with interbeds of sand	0.15–4.05
3. Gray silt, with numerous roots and plant detritus	4.05–4.85
4. Gray silt, sandy-clayey, horizontally laminated	4.85–6.65
5. Silty sand, gray-brown, horizontally laminated	6.65–7.45
6. Sandy silt and clayey-sandy silt, gray-brown, with thin interbeds of sand and inclusions of artifacts (cultural layer)	7.45–7.65
7. Gray sandy-clayey silt interlayering with fine sand	7.65–13.15 (visible thickness)

The radiocarbon dates on the objects from the cultural layer of the site are given in Table 2.3.1. Besides, there are several ^{14}C dates obtained on bone remains collected on the beach immediately below the exposure, on plant remains from the terrace deposits near the site, and on humic acids extracted from the deposits enclosing the plant remains (Pitulko et al., 2004; Pitul'ko and Pavlova, 2007, 2010). The dates most closely related

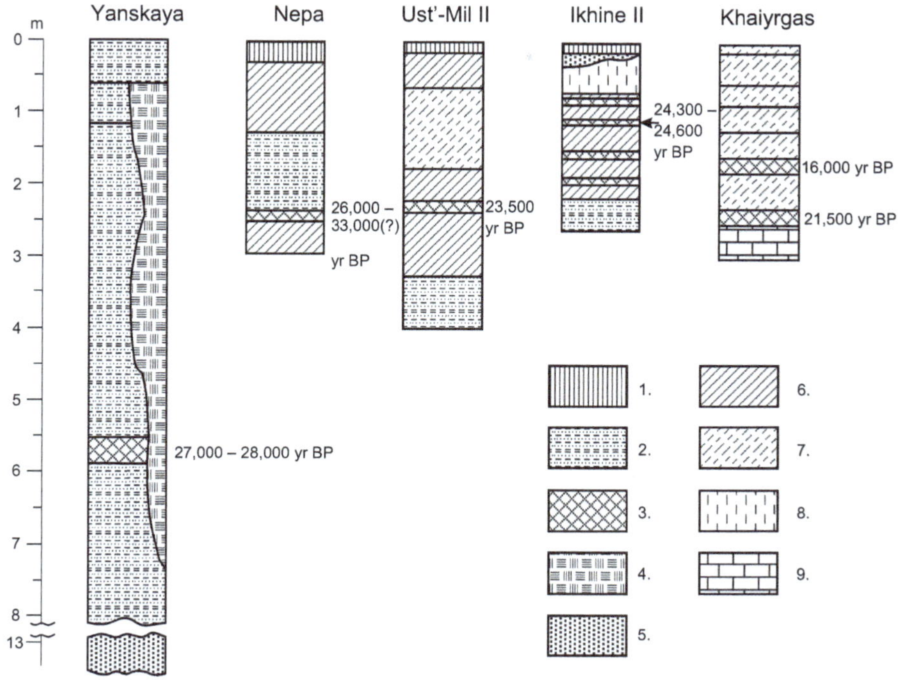

Figure 2.3.2 Geological context of the Paleolithic sites of the Yenisey-Lena-Yana region (Yanskaya, Nepa, Ust'-Mil II, Ikhine II, and Khaiyrgas). *1* – humic horizon of soil; *2* – interstratification of sand and loam; *3* – cultural layer; *4* – ice wedges; *5* – sand; *6* – silt; *7* – sandy loam; *8* – loess-like loam; *9* – limestone.
Modified from Mochanov, Yu.A., 1977. The Most Ancient Stages of the NE Asia Colonization by Humans. Nauka, Novosibirsk, 263 pp. (in Russian); Stepanov, A.D., Kirillin, A.S., Vorobyev, S.A., Solovyeva, E.N., Efimov, N.N., 2003. The Khaiyrgas cave site in the middle reaches of the Lena River (results of the survey in 1998–1999). In: Alekseyev, A.N. (Ed.), Old Cultures of the Northeastern Asia. Astro-Archaeology. Paleo-Informatics, Nauka, Novosibirsk, pp. 98–113 (in Russian); Goebel, T., 2004. The search for a Clovis progenitor in sub-Arctic Siberia. In: Madsen, D.B. (Ed.), Entering America: Northeast Asia and Beringia Before the Last Glacial Maximum. Univ. of Utah Press, Salt Lake City, pp. 311–356; Pitul'ko, V.V., 2008b. The main scenarios of excavations under conditions of permafrost an experience of works on Zhokhov and Yanskaya sites, Northern Yakutia. Archaeology, Ethnology and Anthropology of Eurasia 2 (34), 26–33.

to the human presence (those on the objects from the cultural layer and on bone artifacts) permit one to place it at about 28,500–27,000 BP (Pitul'ko, 2010; Pitul'ko and Pavlova, 2007). The end of the site existence coincided approximately with the boundary between the Karginsky and Sartan stages; the boundary is drawn at c.27,400 BP (Lozhkin et al., 2000; Lozhkin et al., 2008a,b; Anderson and Lozhkin, 2001).

Noteworthy is a series of very old dates on the wood fragments from the cultural layer: 44,010±1100 BP, Beta-216803/47,500±1100cal.yr BP/; >45,200 BP, Beta-216802, and >47,000 BP, LE-7446 (Pitul'ko and Pavlova, 2007; Pitul'ko, 2010). According to Pitul'ko and Pavlova (2007) (see also (Pitul'ko, 2010)), this case shows that the ^{14}C

Table 2.3.1 Radiocarbon dates of the samples from the cultural layer of the Yana site[a]

Sample number	Material dated	^{14}C date (yr BP)	Lab. index	Calibrated date (cal. yr BP)	References
1259	Musk ox bone	27,140 ± 180	Beta-191321	31,120 ± 120	Pitul'ko et al. (2007)
YA06/3W17/FP	Mass filling the hearth	27,200 ± 2400	ЛЕ-7668	32,930 ± 3530	Pitul'ko and Pavlova (2007, 2010)
LY06-L23/2FP/8	Charcoal	27,250 ± 230	Beta-223411	31,190 ± 150	Pitul'ko and Pavlova (2007)
Y02TUMS1HL	Horse bone	27,300 ± 270	Beta-173067	31,230 ± 190	Pitul'ko and Pavlova (2007)
2001/3	Bone	27,400 ± 600	ГИН-11466	31,610 ± 640	Pitul'ko and Pavlova (2007)
–	Point made of rhinoceros horn[b]	27,440 ± 210	Beta-162233	31,300 ± 150	Pitul'ko and Pavlova (2007, 2010)
1411	Plants remains	27,510 ± 180	Beta-191332	31,330 ± 140	Pitul'ko and Pavlova (2007, 2010)
2001/2[c]	Horse bone	27,600 ± 500	ГИН-11467	31,720 ± 550	Pitulko et al. (2004)
LY04-NP/W-ES/F16/15b	Bone	27,620 ± 240	Beta-204863	31,460 ± 240	Pitul'ko and Pavlova (2007, 2010)
40275-YB	Bone	27,670 ± 210	Beta-250634	31,470 ± 220	Pitul'ko and Pavlova (2010)
907	Plants remains	27,820 ± 190	Beta-191328	31,600 ± 250	Pitul'ko et al. (2007)
907	Humic acids	28,520 ± 200	Beta-191594	32,490 ± 370	Pitul'ko and Pavlova (2007)
1385	Plants remains	27,890 ± 190	Beta-191335	31,680 ± 270	Pitul'ko and Pavlova (2007)
1386-1	Hearth filling	27,900 ± 200	Beta-191333	31,710 ± 290	Pitul'ko and Pavlova (2007, 2010)
APA 799	Bone	27,910 ± 280	Beta-191323	31,810 ± 380	Pitul'ko and Pavlova (2007, 2010)
977	Plant remains	28,000 ± 190	Beta-191329	31,820 ± 300	Pitul'ko et al. (2007)
1150	Bone	28,090 ± 200	Beta-191327	31,940 ± 330	Pitul'ko and Pavlova (2007)
40345-YB	Bone	28,210 ± 200	Beta-173064	32,090 ± 340	Pitul'ko and Pavlova (2010)
38670-YB	Bone	28,250 ± 200	Beta-250633	32,130 ± 340	Pitul'ko and Pavlova (2010)
Y02FS	Ivory point[b]	28,250 ± 170	Beta-250635	32,120 ± 310	Pitul'ko and Pavlova (2007, 2010)
33BJ2/1	Bone	28,350 ± 250	Beta-204881	32,270 ± 400	Pitul'ko and Pavlova (2007)
K1093	Bone	28,500 ± 200	Beta-191326	32,460 ± 370	Pitul'ko and Pavlova (2007)
1241	Hare bone	28,570 ± 300	Beta-191322	32,560 ± 470	Pitul'ko et al. (2007)
LY04-NP/W-ES/F16/10	Soot	29,130 ± 410	Beta-204864	33,210 ± 490	Pitul'ko and Pavlova (2010)

[a]The data in the table were chosen as the closest to the time of the human habitation; for the complete list, see Pitul'ko and Pavlova (2007, 2010).
[b]The artifacts are collected on the Yana River floodplain.
[c]The sample (horse bone fragments) was collected from the beach, from a fallen ground block; a stone flake was found in one of the bone fragments ((Pitulko et al., 2004), Supplement, Table S1).

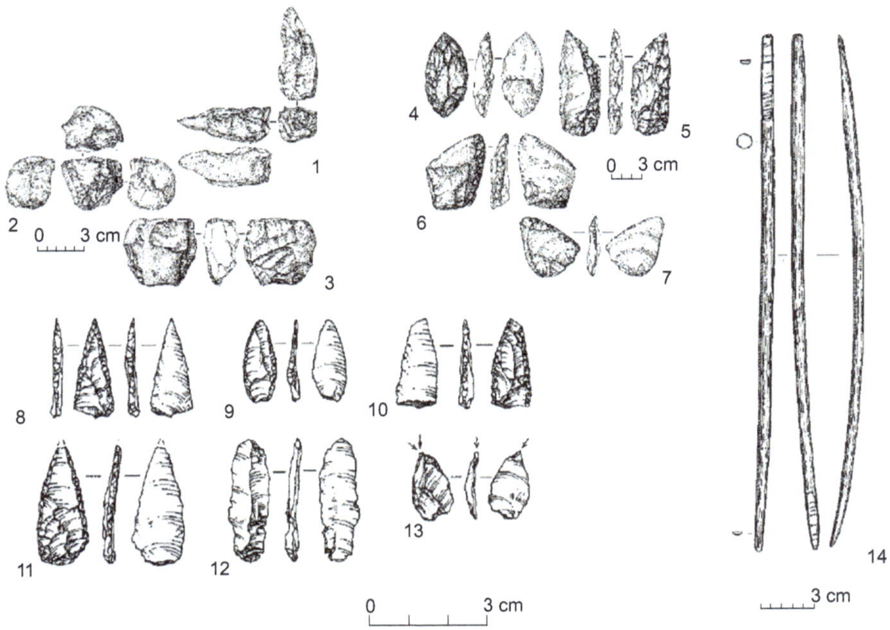

Figure 2.3.3 Lithic and antler tools from Yanskaya. *1, 2* – orthogonal cores; *3* – flat core; *4, 6, 7* – side scrapers; *5* – point; *8–11* – small points; *12* – blade; *13* – burin; *14* – foreshaft made from rhinoceros horn.
Modified from Pitul'ko, V.V., 2010. The Late Neopleistocene dispersal of humans in the Northeast of Asia and adaptation. In: Derevyanko, A.P., Kudelin, A.B., Tishkov, V.A. (Eds.), Adaptation of Peoples and Cultures to Changes of Environments, Social and Technogenic Transformations. Russian Political Encyclopedia, Moscow, pp. 38–46 (in Russian).

dates obtained on the wood recovered from alluvial deposits may be unreliable in the Arctic environments; that has been repeatedly noted in the studies of the oldest sites of the Dyuktay culture on the Aldan River (see, for example, (Abramova, 1979)).

The lithic industry of Yanskaya (Fig. 2.3.3) indicates the overwhelming prevalence of unifacial tool production on pebbles and flakes (Pitul'ko, 2010), with extremely rare occurrence of blade reduction (see, e.g., (Pitul'ko, 2006b)). The stone tools include side scrapers, end scrapers, composite tools, chisel-like, notched, and large cutting tools, and occasional choppers and choppings (Pitul'ko, 2006b). In 2002–07 the cultural layer (excavated over the area of about 400 m^2) yielded more than 7000 stone artifacts (more than 5000 flakes, 465 cores, 685 side scrapers, 37 chisel-like tools, and 47 microtools), more than 1500 bone pieces, and numerous (more than 21,000) animal bones, as well as artifacts from wood, ocher, and burnt stone (Pitul'ko, 2010; Pitul'ko and Pavlova, 2010).

Yanskaya yielded also a rich assortment of bone objects. In particular, there have been found a point made of rhinoceros horn (Fig. 2.3.3, № 14) and two ivory points (Pitulko et al., 2004). Various bone awls, borers, and needles were recovered from the northern segment of the site. There are also spearheads, chisel-like tools made of a large mammal bone, and ivory tools (Pitul'ko, 2006b, 2010). Among ornaments, there are pendants made of animal teeth and numerous (about 100) beads (Pitul'ko, 2006b).

Of the faunal remains recovered from the cultural layer (2380 bones) (Pitul'ko and Pavlova, 2010), there were identified mammoth (*Mammuthus primigenius*), woolly rhinoceros (*Coelodonta antiquitatis*), bison (*Bison priscus*), reindeer (*Rangifer tarandus*), horse (*Equus caballus*), polar fox (*Alopex lagopus*), wolf (*Canis lupus*), and hare (*Lepus tanaiticus*), along with a few bones of other mammals (beaver, lemming, red deer, moose, musk ox), and bones of birds (Pitulko et al., 2004; Pitul'ko, 2006b, 2010; Pitul'ko and Pavlova, 2010). In the cultural layer, dominant are bison (38.9%), reindeer (23.4%), hare (21%), and horse (6.8%); there are also remains of forest species: beaver and red deer (Pitul'ko, 2010). Many bones bear traces of human actions: incisions (Pitulko et al., 2004) and puncture holes (Pitul'ko, 2010). Brown bear bones (*Ursus arctos*) were found on the riverside downstream from the site (Pitulko et al., 2004). A bone of cave lion (*Panthera spelaea*) found on the Yana bank in autumn (during the low water level) is dated by radiocarbon at 26,050 ± 240 BP (Beta-173066) (Pitulko et al., 2004). That may be considered as an evidence of cave lion habitation in the site vicinities simultaneously with the humans (Pitulko et al., 2004).

At the time of the man inhabiting Yanskaya the environments were not unlike those of the present days: it was the continental climate with a dry and relatively warm summer and a winter with little snow (Fig. 2.3.4, zone YT2a) (Pitul'ko et al., 2007). The dominant tundra-steppe landscapes were characterized by a prevalence of xerophile species, a low proportion of sedges, and the dominance of sagebrush and *Selaginella sibirica*; shrub birch occurred on the Yana River floodplain. The palynological and paleo-entomological data (Pitul'ko et al., 2007) are in a reasonably good agreement with the mammal fauna composition dominated by tundra and steppe species.

The aforementioned specialists came to the conclusions that the early humans hunted primarily bison, reindeer, and horse (Pitul'ko, 2010). It is not inconceivable that the Yanskaya site was a hunting camp inhabited at least seasonally (no remains of long-term dwellings have been discovered). Based on studies of the stone and bone artifacts, Pitul'ko (2006b) identified an original culture, which he termed "Yana," and considered it to be older than the Dyuktay culture (Pitul'ko and Pavlova, 2010).

Nepa 1 site has not been adequately described in the literature. Here, we mostly use the work by Goebel (2004). The site is located in the upper reaches of the Nizhnyaya Tunguska River (59°10′ N, 108°18′ E), immediately south of the Nepa River mouth. It was discovered by M.I. Semin in 1983, and only a preliminary survey was performed (Semin and Shelkovaya, 1991).

The site is positioned on the fluvial terrace at 18–22 m above the river. The upper 2.5 m of the sequence is composed of sands and silt deposits (see Fig. 2.3.2). The section may be schematically presented as follows (Goebel, 2004):

Strata	**Depth (m)**
1. Recent soil	0.0–0.3
2. Fine silt, loess-like	0.3–1.25
3. Sand with silt interlayers, cryoturbated	1.25–2.35
4. Brown silt with artifacts and animal bones (cultural layer)	2.35–2.50
5. Sandy-silty member with pebbles and cobbles	2.50–3.00 (visible)

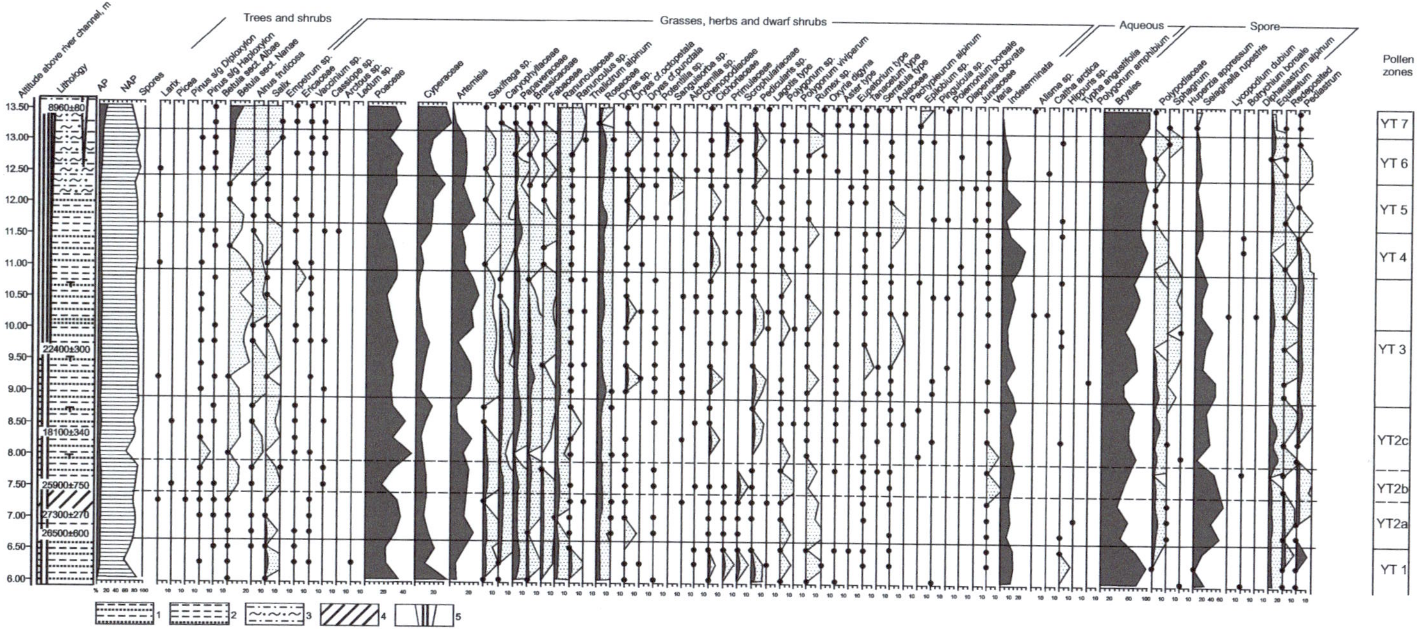

Figure 2.3.4 Pollen diagram of Yanskaya. *1–3* – silts: *1* – sandy silt, *2* – clayey-sandy silt, *3* – sandy-clayey; *4* – cultural layer; *5* – ice wedges. Modified from Pitul'ko, V.V., 2010. The Late Neopleistocene dispersal of humans in the Northeast of Asia and adaptation. In: Derevyanko, A.P., Kudelin, A.B., Tishkov, V.A. (Eds.), Adaptation of Peoples and Cultures to Changes of Environments, Social and Technogenic Transformations. Russian Political Encyclopedia, Moscow, pp. 38–46 (in Russian).

In the faunal remains recovered from the cultural layer, there were identified horse (*Equus* sp.), woolly rhinoceros (*C. antiquitatis*), aurochs (*Bos primigenius*), and roe deer (*Capreolus capreolus*). The artifacts are few in number and include a biconical core and some flakes. Bone samples of roe deer or reindeer were dated by radiocarbon as follows: vertebra 26,065±300 BP (AA-8885) and metatarsal bone 33,100±1500 BP (AA-27382) (Goebel, 2004). According to T. Goebel, those dates suggest the human presence in the region long before the Last Glacial Maximum (LGM) (20,000–18,000 BP); in all probability, the site presents, along with Yanskaya, one of the earliest evidences of the penetration of the modern man (*H. sapiens sapiens*) into the central Subarctic Siberia.

2.3.3.2 The Lena and Aldan drainage basins

The Ust'-Mil II site in the Aldan River valley at the Mil River mouth (59°37′ N, 133°04′ E) was discovered by Mochanov in 1966 (Mochanov et al., 1983) and excavated in 1968, 1970–73, 1980, and 1982; the total excavated area is 312 m^2. Three Paleolithic cultural layers (A, B, C) are distinguished in the sequence (Mochanov, 1977).

The site is positioned on the third terrace of the Aldan at 16–18 m above the river channel (Mochanov, 1977). The sequence from the top downward is as follows (Fig. 2.3.2):

Lithology	Thickness (cm)
1. Turf	0.0–0.3
2. Silt brownish-red	40–60
3. Sandy silt, grayish-yellowish, horizontally laminated, with interbeds of buried soils; earth veins penetrate into the underlying layer (traces of cryogenesis); the lower boundary is sharp and irregular (Paleolithic culture unit "A")	110–120
4. Silt brownish-grayish, horizontally laminated; at the top there is a thin (2–5 cm) interbed of sand at the contact with layer 3; Paleolithic culture unit "B" is identified in the upper part of the layer, and culture unit "C" in the middle part	150–160 (total)
5. Yellowish sand horizontally laminated, with thin interbeds and lenses of grayish sandy silt (bedrock)	200 (visible)

According to Tseitlin (1979), layer 3 belongs to the alluvial deposits of the second terrace of the Aldan superimposed onto the earlier deposits of the third terrace (layers 4 and 5); he also presumed that the uppermost part of layer 4 might belong to the Early Sartan deposits of the second terrace overlying the older sediments. Tseitlin considers the sequence to be a result of tectonics; the subsided surface of the third terrace appeared within the area of deposition at the time of the second terrace formation. That point of view differs considerably from Mochanov's opinion: the latter thinks the Paleolithic layers of Ust'-Mil II are confined to the third terrace of the Aldan exclusively (Mochanov, 1977) (for critical review of those points of view see (Pitul'ko and Pavlova, 2010)).

There are several ^{14}C dates obtained on wood from the Ust'-Mil II site (Mochanov, 1977): Component A – 12,200 ± 170 BP (LE-953)/14,230 ± 320 cal. yr BP/; Component B – 23,500 ± 500 BP (LE-999)/27,740 ± 470 cal. yr BP/; Component C – 33,000 ± 500 BP (LE-1000)/37,240 ± 660 cal. yr BP/, 30,000 ± 500 BP (LE-1001)/34,120 ± 450 cal. yr BP/; and 35,400 ± 600 BP (LE-954)/40,030 ± 660 cal. yr BP/. One more radiocarbon date was obtained on wood from layer 5 (bedrock), 55 cm from the top of the layer: 35,600 ± 900 BP (LE-955)/40,230 ± 890 cal. yr BP/.

There are not many artifacts recovered from the Ust'-Mil II (45 objects) (see Fig. 2.3.5). Component A yielded 21 flint, quartzite, and diabase flakes, 3 fragments of a flint blade, and a pebble core (Pitul'ko and Pavlova, 2010). Seven flint flakes were found in Component B. As to Component C, 12 stone pieces and 1 mammoth bone bearing traces of human treatment were found (Mochanov, 1977; Mochanov et al., 1983). Among the stone artifacts, there are flint flakes (5), flint blades (2), a core blank, a wedge-like core, a burin, an end scraper, and a blank of bifacial knife or a spear point. According to Mochanov (1977), the core blank could be considered as a preform of a bifacial knife or of a spearhead. Therefore, he believes it is possible that Component C bears evidence of bifacial technique and microblade technology. Kashin (2003) casted doubt on the wedge-like core really belonging to the same assemblage as the rest of materials from

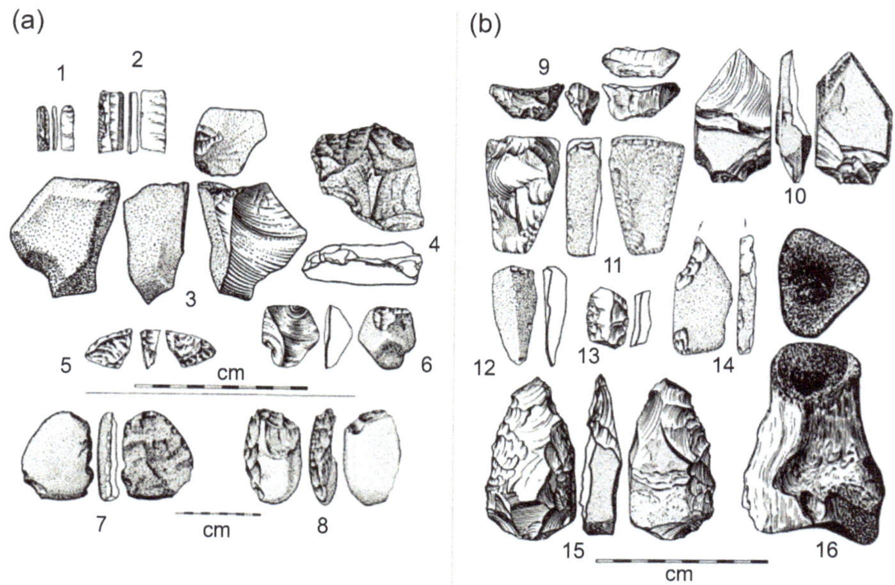

Figure 2.3.5 Stone and bone tools of the Dyuktay culture in Yakutia. *A* – Ikhine II: *1–4* – layer IIa; *5–7* – layer IIб; *8* – layer IIд; *B* – Ust'-Mil II (layer C). *1* – blade; *2* – knife; *3*, *5* – preforms of wedge-shaped cores on pebbles; *4*, *7*, *8* – end scrapers; *6* – flake; *9* – preform of a wedge-shaped core; *10*, *15* – performs of bifaces; *11* – core preform; *12*, *13* – blades; *14* – burin; *16* – processed mammoth bone.
Modified from Mochanov, Yu.A., 1977. The Most Ancient Stages of the NE Asia Colonization by Humans. Nauka, Novosibirsk, 263 pp. (in Russian).

Component C. Besides the materials recovered from the cultural layer, more than 4000 artifacts were collected on the surface near the site (Mochanov et al., 1983).

The data on the mammal fauna from the cultural layers of the Ust'-Mil II site are presented in Table 2.3.2; there are not many bones recovered, 64, with 45 of them being identifiable. Mammoth, bison, and horse are dominant in Components A–C; reindeer and woolly rhinoceros occur occasionally.

Data on pollen assemblages of the Ust'-Mil II site were published more than once (Tomskaya and Savvinova, 1975; Mochanov et al., 1983; Savvinova et al., 1996). According to the palynologists, layers 4 and 5 are noted for dominance of spores (as much as 88%–93% of the total), while the proportion of arboreal pollens (trees and shrubs) amounts to 11%–39% (Fig. 2.3.6). Those data suggest the early man lived there under conditions of a cold climate (Components B and C). According to Tomskaya and Savvinova (1975), the most cold conditions fell on 30,000–23,000 yr BP, while a warming occurred about 35,000–33,000 yr BP (according to ^{14}C dates run out on wood). They also drew an inference (see, for example, (Savvinova et al., 1996)) that layers 4 and 5 correspond to the Karginsky (warm enough) and that the vegetation at that time was dominated by pine and larch forests (though it decidedly contradicts the available factual material: see subsequent discussion).

Mochanov (1977) came to the conclusion that the inhabitants of the Ust'-Mil II site hunted mammoth, woolly rhinoceros, bison, and horse. On the grounds of the presence of the bifacial technology, he attributed the site to the Dyuktay culture.

Ikhine II site is located at the confluence of the Aldan River and its small tributary Ikhine (63°07′ N., 133° E) (Fig. 2.3.1). It was discovered in 1965 by Yu.A. Mochanov, who worked on it in 1965–66, 1968, 1973–74, and 1980 (Mochanov et al., 1983); the total excavated area is 216 m^2.

According to Yu.A. Mochanov, the site is positioned on a lowered segment of the third terrace of the Aldan River about 12 m above the water (Fig. 2.3.2). The sedimentary sequence is as follows:

Strata	Thickness (cm)
1. Sod horizon	10–15
2. Coarse light-gray sand (superimposed Holocene alluvial deposits)	4–50
3. Reddish-brown loam, broken with frost fissures, includes artifacts of the Early Iron age and some Paleolithic pieces (deluvial mantle, *cultural layer 1*)	35–45
4. Silt bluish-gray and yellowish-brown, stratified (individual layers are 0.5–2.0 cm thick), with thin interbeds (0.1–0.3 cm) of nonsorted sand and inclusions of plant remains; at a depth 0.75–1.95 m from the surface, there are five horizons (A–D) enclosing Paleolithic artifacts of the *cultural layer II* (sublayers IIa–IIe) (floodplain facies of the 3rd terrace alluvial deposits)	150–170
5. Alternating bands of yellowish-gray silt 1–2 cm thick and nonsorted gray sand 0.5–1 cm thick (alluvial deposits of the 3rd terrace)	25–45
6. Sand and gravel with pebbles, channel facies of the third terrace alluvial deposits	40 (apparent thickness)

Table 2.3.2 Mammalian fauna and number of bones recovered from Ust'-Mil II and Ikhine II (Mochanov, 1977)

	Site and cultural layer							
	Ust'-Mil II			Ikhine II				
Mammalian species	**A**	**B**	**C**	**IIa**	**IIb**	**IIc**	**IId**	**Total**
Mammoth (*Mammuthus primigenius*)	4	7	2	7	4	–	9	**33**
Wooly rhinoceros (*Coelodonta antiquitatis*)	–	3	1	4	5	2	1	**16**
Bison (*Bison priscus*)	3	–	5	50	58	14	29	**159**
Horse (*Equus caballus*)	9	–	7	71	50	14	43	**194**
Musk ox (*Ovibos moschatus*)	1?	–	–	–	–	–	–	**1?**
Reindeer (*Rangifer tarandus*)	3	–	–	16	6	2	10	**37**
Moose (*Alces alces*)	–	–	–	1	–	–	–	**1**
Red deer (*Cervus elaphus*)	–	–	–	–	–	–	1?	**1?**
Wolf (*Canis lupus*)	–	–	–	–	–	–	1	**1**
Fox (*Vulpes vulpes*)	–	–	–	–	1	–	–	**1**
Polar fox (*Alopex lagopus*)	–	–	–	–	1	–	–	**1**
Fish (Pisces)	–	–	–	–	2	–	–	**2**
Unidentifiable split bones	7	4	8	105	75	32	–	**231**
Total number of bones	**27**	**14**	**23**	**254**	**202**	**64**	**94**	**678**

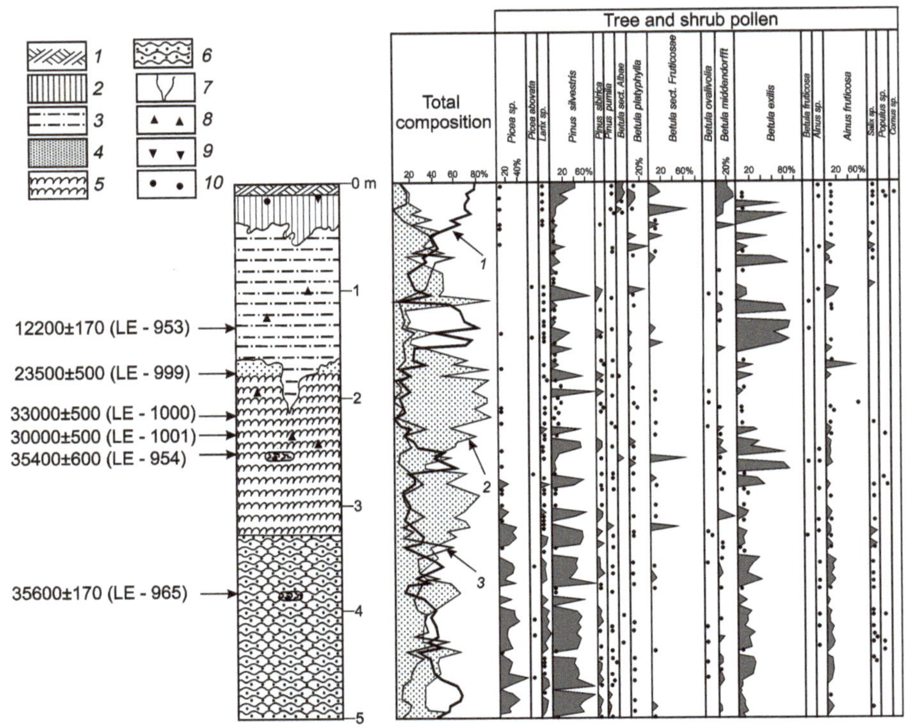

Figure 2.3.6 Pollen diagram of the Ust'-Mil II. *1* – sod; *2* – cover loam; *3* – horizontally laminated sandy loam; *4* – sand; *5* – laminated loam; *6* – thinly laminated loam; *7* – cryogenic structures; *8*, *9* – artifacts dated to the Late Pleistocene (*8*) and Early Holocene (*9*); *10* – Neolithic pieces; *11* – cultural layer; *12* – samples for pollen analysis. Total composition of pollen and spores: 1 – tree and shrub species pollen; 2 – herbaceous and dwarf shrub pollen; 3 – spores. Modified from Mochanov, Yu.A., Fedoseyeva, S.A., Alekseyev, A.N., Kozlov, V.I., Kochmar, N.N., Shcherbakova, N.M., 1983. Archaeological Sites of Yakutia. The Aldan and Olekma River Drainage Basins. Nauka, Novosibirsk, 391 pp. (in Russian); Fig. 12: only total pollen and spore composition and arboreal species are shown.

A series of ^{14}C dates has been obtained for the site, mostly on wood (Mochanov, 1977), and five of them are for the cultural layer IIb (from the depth 0.90–0.95 m from the surface): 24,330 ± 200 BP, LE-1131/28,360 ± 220 cal. yr BP/, 24,500 ± 480 BP, IM-203/28,620 ± 510 cal. yr BP/, 24,600 ± 380 BP, IM-153/28,670 ± 420 cal. yr BP/, 27,400 ± 80 yr BP, IM-205/31,250 ± 90 cal. yr BP/, and 30,200 ± 300 yr BP, GIN-1019/34,260 ± 260 cal. yr BP/.

Wood and rhinoceros bones from cultural layer IIc yielded four dates: two for the middle part of the layer (1.20 m from the surface): 26,600 ± 900 yr BP, IM-201/30,780 ± 980 cal. yr BP/ and 31,200 ± 500, GIN-1020/35,190 ± 500 cal. yr BP/, both dates run out on wood, and two more for the lower part (1.3 m from the surface): 26,500 ± 540 yr BP, IM-202/30,570 ± 510 cal. yr BP/ on wood and

26,030 ± 200 yr BP, IM-239/30,270 ± 290 cal. yr BP/ on bone. A wood sample taken from the upper part of layer IId (from 1.6 m depth) was dated at 27,800 ± 500 yr BP, IM-206/31,910 ± 560 cal. yr BP/. Mochanov (1977) considers the last date to be too young; in his opinion, the age of that layer should be about 31,500–35,000 yr BP.

Three radiocarbon dates were obtained on bones collected by Yu.A. Mochanov (presumably in 1980) on the site: 20,080 ± 150 yr BP, SOAN-3185/24,150 ± 190 cal. yr BP/ on rhinoceros bone, 19,695 ± 100 yr BP, SOAN-3186/23,720 ± 150 cal. yr BP/ on bison bone, and 15,780 ± 70 yr BP, SOAN-3187/19,040 ± 100 cal. yr BP/ on moose bone (Orlova, 1998).

Artifacts recovered from the Ikhine site are few (Fig. 2.3.5), made on flint, diabase, chert, and granite (Mochanov, 1977). There were 11 lithic pieces recovered from layer IIa (4 flakes, 1 knife-like blade, 3 worked pebbles, 1 wedge-shaped (?) core), two tools (a knife and a side scraper), and 254 animal bones. In layer IIb, there were found 6 lithic pieces (3 flakes, 1 wedge-shaped core, 1 chipped pebble, and 1 side scraper) and 202 animal bones. Kashin (2003) considered the identification of the wedge-shaped core to be doubtful. Two stone artifacts (a flake and a split pebble) were found in layer IIc along with 64 mammal bones. Layer IId yielded 2 lithic pieces (side scraper and chipped pebble) and one bone tool (mammoth rib with worn holes), along with 127 animal bones.

The mammal fauna composition at Ikhine is listed in Table 2.3.2. The species dominant in layers IIa–IId are horse (178 bones), bison (151 bones), and reindeer (34 bones); there were also mammoth and woolly rhinoceros remains, and a few fish bones (in layer IIb). The palynological analysis (Mochanov et al., 1983) provided data on the environments. The pollen assemblages recovered from deposits enclosing the cultural layers are dominated by spores and nonarboreal pollen (herbs, grasses, and dwarf shrubs), while tree pollen is of secondary importance. The floral reconstruction (Savvinova et al., 1996) suggests the prevalence of open larch and pine forests at the time when the lower part of cultural layer II was being deposited (horizon IId, 1.5–1.8 m from the surface); that phase is correlatable with the Malaya Kheta optimum of the Karginsky. The uppermost part of cultural layer II (at 1.0–1.5 m depth) is noted for the herb and grass pollen dominance, which suggests the forest degradation and steppe landscape development at that time.

In another work, Mochanov and Savvinova (1980) state that the pollen spectra from cultural layer II (samples nos. 7–11, from the depth of 0.8 to 2.0 from the surface (Mochanov et al., 1983)) are indicative of the forest degrading and the steppe vegetation spreading, only single grains of arboreal pollen being found. It should be noted that such an interpretation of the pollen data is in conflict with the modern views on the vegetation in Siberia at the Karginsky, when tree species were prevalent. The fauna composition, however (mostly a species of open spaces, bison, and horse; see Table 2.3.2), suggests rather open landscapes: steppes or, at least, forest-steppe.

Mochanov (1977) is cautious in his statements about the Ikhine II inhabitants as possible bearers of the Dyuktay culture; that conclusion calls for further investigations. He takes the presence of the site on the Aldan at about 25,000–30,000 yr BP as an ample evidence of the region having been settled at the pre-Sartan time.

2.3.4 Regions of the later colonization

2.3.4.1 The Lena drainage basin

Khaiyrgas cave site (Fig. 2.3.1) is in the middle reaches of the Lena River, 2 km downstream from the Maly Patom River mouth (59°56′ N, 117°28′ E). The site is located in a cave in Riphean limestone (Stepanov et al., 2003). The cave is 500 m from the Lena River. The total area of the cave is about 200 m^2, it is about 30 m long and 2–10 m wide, and the entrance is positioned about 10 m above the water's edge. The cave roof is 7 m high at the entrance and 2–3 m in the central part of the cave, and the entrance faces northeast. The site was discovered in 1984 and excavated in 1984–85 and 1998–99; the total excavated area is about 103 m^2 (Stepanov et al., 2003). There are seven cultural layers identified (Fig. 2.3.2): layers 1 to 4 attributed to the Holocene and layers 5–7 to the Late Pleistocene.

The general sequence of the cave deposits (based on the results of excavations in 1998–99 (Stepanov et al., 2003)) is as follows:

Strata	Thickness (cm)
1. Dump pile of the 1984–85 excavations	2–35
2. Whitish sandy loam	2–5
3. Stratified sandy loam, greenish-brown, with interbeds of plant detritus, abundant inclusions of rubble and small blocks, cultural layer 2	15–145
4. Sandy loam with a sizeable proportion of sand, greenish-gray, with a detritus interbed in the lower part, cultural layer 3	6–25
5. Rubble and blocks, cultural layer 4	Up to 35
6. Interlayered gray and black sandy loam, with a considerable admixture of black rubble covered with calcareous crust; there are two or three interlayers of black sandy loam rich in debris; cultural layer 5	Up to 50
7. Debris with a matrix of light-gray and dark gray loam; cultural layer 6	40–80
8. Debris with a matrix of light-gray and dark gray loam; cultural layer 7	45–60
9. Solid rock	

Two radiocarbon dates have been obtained on mammal bones from the Pleistocene layers of the Khaiyrgas cave: 16,000 ± 200 yr BP, IM-887/19,320 ± 240 cal. yr BP/ from layer 6 (animal not identified) and 21,500 ± 775 yr BP, SOAN-4249/25,820 ± 850 cal. yr BP/, layer 7, mammoth bone (Stepanov et al., 2003).

Numerous stone and bone artifacts were found in the Paleolithic cultural layers 5–7 (Stepanov et al., 2003). Five hundred and twenty-two pieces of stone and more than 75 pieces of bone, including ornaments, have been recovered from the *cultural layer 5*. Among lithics, flint flakes are prevalent (125), along with chipped diabase, quartzite, rock crystal pebbles (165), blades (73), pebbles (more than 50), nodules (14), and split pebbles (more than 22). There are also 12 cores (wedge-shaped, flat, and prismatic). Stone tools are represented by burins, one arrowhead, side scrapers and end scrapers, points, knives, and inserts. Among bone artifacts, there are a smoother, beads, awls and borers, pendants, arrowheads, a fishhook, a harpoon, and needles. Of special interest is a bone dagger (or spearhead) 34 cm long and 2.9 m wide made of mammoth rib. In *cultural layer 6,* there have been found 1203 lithic and 59 bone pieces. There are 889 flint flakes, 120 blades, 17 retouched blades, 14 cores (conical, edge-faceted, wedge-shaped), and 30 end scrapers. Among the bone artifacts, there are arrowheads, harpoons, pendants, slotted pieces (for stone inserts), and needles. From *cultural layer 7* were recovered 705 stone and 20 bone artifacts. The stone pieces are mostly flakes (588 items), blades (80), and scrapers (6); two wedge-shaped cores were also found. Bone artifacts are slotted points (4 items), needles and their fragments (9), and arrowheads (3).

In spite of faunal remains being found in abundance in the Paleolithic layers 5–7 (more than 4360 bone fragments in layer 5, 7600 in layer 6, and about 2000 in layer 7) (Stepanov et al., 2003), only 41 bones have been identified to a species level (Boeskorov, 2003). The results of the identification of the remains from layer 5 are as follows (numerals in brackets are number of bones versus number of animals): horse (*Equus* sp.) (4/1), reindeer (*R. tarandus*) (7/4), bison (*B. priscus*) (3/2), saiga (*Saiga* sp.) (1/1), bighorn sheep (*Ovis nivicola*) (3/2), cave lion (*P. spelaea*) (2/1), and wolf (*C. lupus*) (1/1). Among the remains recovered from layer 6 the following species were identified: woolly rhinoceros (*C. antiquitatis*) (1/1), horse (1/1), musk deer (*Moschus moschiferus*) (1/1), red deer (*Cervus elaphus*) (1/1), moose (*Alces* sp.) (1/1), cave lion (*P. spelaea*) (1/1), sable (*Martes zibellina*) (1/1), brown bear (*U. arctos*) (1/1), and *A. lagopus* (1/1). Besides, from undivided layers 6 and 7 were recovered hare (*Lepus timidus*) (1/1), reindeer (1/1), and ermine (*Mustela erminea*) (1/1). The following mammals were recovered from layer 7: hare (1/1), long-tailed Siberian souslik (*Spermophilus undulatus*) (1/1), moose (2/2), bison (1/1), ermine (1/1), and polar fox (2/2). A mammoth (*M. primigenius*) was also found in layers 5–7 (Stepanov et al., 2003).

Besides the previously listed animals, a number of mammals were identified by Kasparov (1998) in the collections of 1984–85 recovered from layers 5 to 6: hare (33 bones recovered from layers 5 to 6); pika (*Ochotona* sp.), 1 bone in layer 5; Siberian chipmunk (*Tamias sibiricus*), 3 bones in layer 5; squirrel (*Sciurus vulgaris*), 22 bones in layers 5–6; ground squirrel (*Citellus* sp.), 1 bone in layer 5; fox (*Vulpes vulpes*), 1 bone in layer 6; Siberian roe deer (*C. pygargus*), 6 bones in layer 5; and argali (*Ovis ammon*), 6 bones in layer 6. Besides the mammal remains, there were bird and fish bones noted in layer 5, as well as a human molar (unfortunately, lost) (Stepanov et al., 2003; Stepanov, pers. comm., 2009).

Layers rich in debris are associated with a cold climate (Stepanov et al., 2003). The cultural layer 5, however, displays black humified interlayers presumably related to warmer intervals when organic matter could be deposited on the cave walls; due to its flowing onto the cave floor, there could be formed interlayers imbued with organics (Stepanov et al., 2003).

As to functional attribution of the site, no suggestions had been advanced on the subject by the authors having performed the excavations. It is not inconceivable that the cave served periodically as a base camp for hunters and fishers. The presence of wedge-shaped cores in cultural layers 5–7 and the ^{14}C dates having fallen into the interval 16,000–21,000 yr BP permit a tentative attribution of artifacts from layers 6 and 7 in the Khaiyrgas cave site to the Dyuktay culture; layer 5 might belong to Sumnagin culture (Stepanov et al., 2003).

2.3.5 Taymyr Peninsula

The Tagenar 6 site (70°56′ N, 93°40′ E) is located on the bank of the Tagenar River (tributary of the Volochanka River) 5 km upstream from its mouth (Fig. 2.3.1). The site was discovered in 1967 and studied in 1968 and 1972. The description is given next according to Khlobystin (1998, 2005). The total area of the excavations is about 40 m^2.

The site is positioned on the surface of the first terrace of the Tagenar River, at about 5–7 m above the river. The sequence of the deposits from top to bottom is as follows (Khlobystin, 1998; Levkovskaya et al., 1972):

Strata	Depth (m)
1. Sod	0.0–00.2
2. Gray-yellow sand with thin humus interbeds, the number of the latter increasing downward; (ice-wedge casts begin from the layer base and go down to the depth of 1.2 m from the surface)	0.02–0.47
3. Brown indistinctly laminated sand with thin (0.5–2 cm) dark-colored (due to humus and charcoal) interbeds; one of the interbeds at a depth of 1.20–1.32 contains fire spots, ocher, stone artifacts, animal and bird bones (cultural layer)	0.47–1.32 (visible)

A radiocarbon date has been obtained on charcoal from the fireplace with an artifact concentration: 6020 ± 100 yr BP, LE-884/6890 ± 130 cal. yr BP/; the second date, 5160 ± 60 yr BP, LE-789/5910 ± 90 cal. yr BP/, was obtained on charcoal sampled on the periphery of the site and cannot be used in dating the cultural layer as it has been probably contaminated by younger material coming from the overlying coal-humus interbed (Khlobystin, 1998).

Cultural remains are not very numerous: there have been found 296 artifacts (the minimum number; see (Khlobystin, 1998)) made of multicolor flint fragments. Dominant are thin prismatic knife-like bladelets (215) and flakes (62). There are some other pieces: prismatic cores (3), micro-inserts (2), angle burins (2), cutting tools (at least 4), a retouched bladelet (probably, a fragment of arrowhead or a borer), end scrapers (at least 4), and a chisel-like tool. Some flakes from siliceous slate bear traces of polishing. According to Khlobystin (1998), the cultural attribution of the site is Mesolithic.

There are palynological and paleontological data on the site; the faunal remains, however, are rather scarce, only reindeer *R. tarandus* bones having been identified. The pollen analysis (Levkovskaya et al., 1972) revealed a distinct tree birch maximum in the cultural layer (at a depth of 1.10–1.20 m from the surface), along with a high concentration of spruce pollen and occasional grains of fir.

The assemblages are dominated by spores (mostly of *Lycopodium* and ferns) and NAP pollen (grasses and sedges). Alder and birch (both tree and shrub species) are dominant in the AP group. As suggested by the pollen data, the vegetation in the site vicinities was represented mostly by communities of alder and birch, with larch, spruce, and pine being of secondary importance (Levkovskaya et al., 1972). It may be concluded that the Tagenar 6 site existed in the northern taiga zone of larch forests (at present the region is occupied by the forest-tundra with open forests of larch in river valleys (Khlobystin, 1998)). At the time of the early man habitation the climate was warmer and wetter than at present (Levkovskaya et al., 1972).

The data obtained up to now suggest that the Tagenar 6 site was a short-term (seasonal) hunting camp; the people were hunting reindeer and birds in the vicinities and making tools.

2.3.6 Initial colonization of the north of Eastern Siberia: a synthesis

2.3.6.1 Unresolved problem

One of the key problems of the initial colonization of the East Siberian North is the age of the oldest sites in the Aldan valley: Ikhine I–II, Ust'-Mil II, and Ezhantsy. Mochanov (1977) dated them to ~35,000–30,000 yr BP (Ust'-Mil II and Ezhantsy) and to ~30,000–25,000 yr BP (Ikhine I–II). Many archaeologists and geologists disagree with this attribution (Abramova, 1979; Yi and Clark, 1985; Hopkins, 1985; Kuzmin and Orlova, 1998; Vasil'ev et al., 2002) (for summary see (Kashin, 2003)). Earlier, the author made an attempt to analyze pollen assemblages and radiocarbon data from the aforementioned sites in the Aldan River valley (Kuzmin and Orlova, 1998); here the results are presented in full for the first time.

The sections considered next are found in eastern and northeastern parts of Siberia; they have been studied palynologically and dated by radiocarbon. The paleoclimatic stratigraphy we are using is based on the monograph by Kind (1974) that has not lost its validity up to now. According to her, there were three main warmings identifiable within the Karginsky (~50,000–25,000 yr BP): 50,000–44,000 yr BP (earlier one), 42,000–35,000 yr BP (Malaya Kheta, the climatic optimum of the Karginsky), and 30,000–25,000 yr BP (Lipovka-Novoselovo); they are separated by two coolings: 44,000–42,000 yr BP (earlier) and 34,000–31,000 yr BP (Konoshchelye) (Kind, 1974).

To correlate the aforementioned sites with climatic phases, we consider the general pattern of the vegetation in Eastern and Northeastern Siberia during the Karginsky. In the alluvial sequence of the second terrace of the Markha River (the Vilyui drainage basin), there have been distinguished three phases of the forest dominance, the oldest being dated at about 50,000 ± 2000 yr BP (GIN-803) (Shofman et al., 1977), and the two younger being attributed to the Malaya Kheta and Lipovka-Novoselovo warmings. Between them, during the phases of relative coolings, steppe landscapes prevailed, the earlier dated at 43,100 ± 1800 yr BP (GIN-896) and the later presumably corresponding to the Konoshchelye cooling (Shofman et al., 1977).

The prevalence of forest vegetation is clearly seen from the data in other sections of Central and Northern Yakutia, such as a mammoth locality at Chekurovka, 26,000 ± 1600 yr BP, Mo-215/30,630 ± 1850 cal. yr BP/ (Korzhuev and Fedorova, 1962; Tomskaya and Savvinova, 1971); the Bolshaya Balakhnya River, Taymyr Peninsula, section K-54, 26,600 ± 1000 yr BP, GIN-999/30,830 ± 1100 cal. yr BP/ (Nikolskaya, 1980); the Bolshaya Balakhnya River, 29,700 ± 1000 yr BP, GIN-963/33,830 ± 1120 cal. yr BP/ (Andreyeva, 1980); the Molodo River, 36,100 ± 200 yr BP, GIN-1427/40,750 ± 260 cal. yr BP/ (Klimanov and Shofman, 1982); and the Vilyui River, 37,600 ± 800 yr BP, GIN-1103/41,970 ± 620 cal. yr BP/ (Klimanov and Shofman, 1982). Similar regularity was seen at the sites with the mammoth fauna at the Indigirka and Kolyma valleys dated around 41,000–39,000 yr BP (Arslanov et al., 1980; Ukraintseva, 2002).

The most complete record of the environmental changes during the Karginsky in the north of Siberia was inferred from the Molotkovskiy Kamen section, the Maritime Lowland in Yakutia (Giterman, 1985; Kaplina and Giterman, 1983). The warmer intervals recorded in the sequence at 11.5–12 m above the water's edge (Malaya Kheta) and at 17.5–19 m (Lipovka-Novoselovo) were noted for forests of larch and birch, while the Early Karginsky (6–11.5 m) and Konoshchelye (12–17.5 m) coolings featured open landscapes dominated by grasses and *Artemisia* with a small participation of light coniferous trees (larch).

It follows from the preceding that the warm phases within the Karginsky interval were marked by a noticeable proportion of tree species in the vegetation all over the Siberian North (including the Arctic coast). More than 90 ^{14}C dates (70 of which belong to palynologically studied sections) (Kaplina and Lozhkin, 1982) lend support to the statement that the transition from Karginsky to Sartan in the Siberian Northwest was marked by a reduction of the tree species proportion (Kaplina, 1979). The recently obtained results (Schirrmeister et al., 2002; Sher et al., 2005) confirm the data.

All the studied sites of the Aldan River valley attributed to the Early Dyuktay and Proto-Dyuktay cultures have a common characteristic in the total composition of their pollen assemblages, namely, the prevalence of spores. According to G.M. Savvinova and A.I Tomskaya (Tomskaya and Savvinova, 1975; Mochanov and Savvinova, 1980; Tomskaya, 1981), that indicates a prevalence of treeless landscapes and the accumulation of the culture-bearing deposits under conditions of a cold climate.

The assemblages attributable to the Karginsky, however, also contain a sizeable proportion of spores; that is true of the sections on the Markha River (Shofman et al., 1977), and of the Stanchikovsky Yar and Molotkovsky Kamen' (Giterman, 1985) on the Maritime Lowland. Neither of the mentioned authors, however, made a conclusion on periglacial conditions at the time of the deposition of the mentioned sediments.

2.3.6.2 An analysis of radiocarbon dates

Let us consider the data on the Paleolithic sites of the Aldan River valley against the background of the environmental evolution of the entire region. The treeless plant communities reconstructed from the pollen assemblages recovered from the Ikhine II cultural layer seemingly disagree with the environments of the Lipovka-Novoselovo warming (if the ^{14}C dates of the layer are taken to be correct). If we assume, however, that some of the dates partly correspond to the age of artifacts, we may take the age to be about 24,300–24,600 yr BP, which agrees better with the general paleoclimatic situation at the end of the Lipovka-Novoselovo warming and the transition to the cold Sartan stage. It should be noted that the continentality gradually increases up the Ikhine II sequence, which is not typical of the warm stages of the Karginsky time in Siberia.

It is necessary to stress that the radiocarbon dates of the Ezhantsy, 17,150 ± 395 yr BP, IM-459/20,760 ± 510 cal. yr BP/ (Kostyukevich et al., 1980), and Ikhine, I – 16,600 ± 270 yr BP, IM-452/20,040 ± 330 cal. yr BP/ (Kashin, 2003), suggest much later age of those objects, which is in better agreement with "cold" pollen assemblage and totally opposite to the Mochanov's conclusion about the age of the sites about 35,000–25,000 yr BP (see previous).

The pollen assemblages and radiocarbon dates of the Ust'-Mil II appeared to be most difficult to interpret (see Fig. 2.3.6). Cold environments have been reconstructed for the interval between 30,000 and 23,500 yr BP, though that chronological interval corresponds presumably to the Lipovka-Novoselovo warming. The lower lying series dated to 35,400–30,000 yr BP yielded pollen assemblages of forest type, though chronologically that time corresponds to the Konoshchelye cooling (Kind, 1974). If we assume, however, that the lower part of the Ust'-Mil II sequence lacking artifacts (at ~2.5–5 m depth) belongs to the Lipovka-Novoselovo warming (that is, ~30,000–24,000 yr BP), then the overlying artifact-bearing series occurring at a depth of 1.75–2.5 m and typically showing a lower tree pollen content may be attributed to the end of the Karginsky–early Sartan. If such is the case, the ^{14}C date ~23,500 yr BP (Component B) fits into the general picture of the environmental evolution and of the Paleolithic material culture in Siberia. The anomalous dates ~35,400–30,000 yr BP may be explained by the redeposition of the wood from older deposits, which often occurs in the permafrost regions, such as Siberia (Katasonov et al., 1979) and Alaska (Hopkins, 1985).

It should be noted that Tseitlin (1979) considered it possible to assign Component B of the Ust'-Mil II site to the Early Sartan. So it would not be inconceivable to take the age of the oldest sites in the Aldan valley to be around 24,000–17,000 yr BP (see also (Kuzmin and Orlova, 1998)).

Pitul'ko and Pavlova (2010) reviewed the geological and archaeological data on the Dyuktay culture sites in the Aldan River valley. They suggested that the artifacts from the Ust'-Mil II and Ikhine II sites had been redeposited, and the ^{14}C dates did not correspond to the real age of cultural remains. In their opinion, the lower chronological boundary of the Dyuktay culture in the Aldan valley is about 17–18 ka BP, though it might be older: 22–23 ka BP (Pitul'ko and Pavlova, 2010). They think also possible is the presence of the pre-Dyuktay in the Aldan River valley (Ust'-Mil II and Ikhine II) on the condition "…if there exists and could be proved an interrelation between archaeological materials and the dates 27 to 26 ka BP" (Pitul'ko and Pavlova, 2010, p. 194).

The data obtained in the studies of the environments and the chronology of the early human settlements in Eastern Siberia are shown schematically in Figs. 2.3.7–2.3.11.

2.3.7 Conclusion

The previously stated data make it possible to reveal the main trends in the initial colonization of the East Siberian North. In all probability, man had already penetrated into the southern part of the Eastern (Middle) Siberia as far as 59°N (Nepa) and into northeast of Siberia about 33,000 yr BP.

About 28,500 yr BP, man could visit the Arctic regions of East Siberia and stay there, at least seasonally (Yanskaya, 71°N, Fig. 2.3.7). According to Pitul'ko (2006, 2010), Yanskaya reflects the human migration northeastward from the Transbaikal region and South Siberia, and it could not be a result of the evolution of local culture; he places it (together with the culture of the same name, the Yana culture) before the Dyuktay culture identified by Mochanov (1977). So the eastern part of the Yenisey drainage basin (Nepa) and the Yana drainage basin (Yanskaya) could be considered as the regions of the initial settlement of northern Siberia.

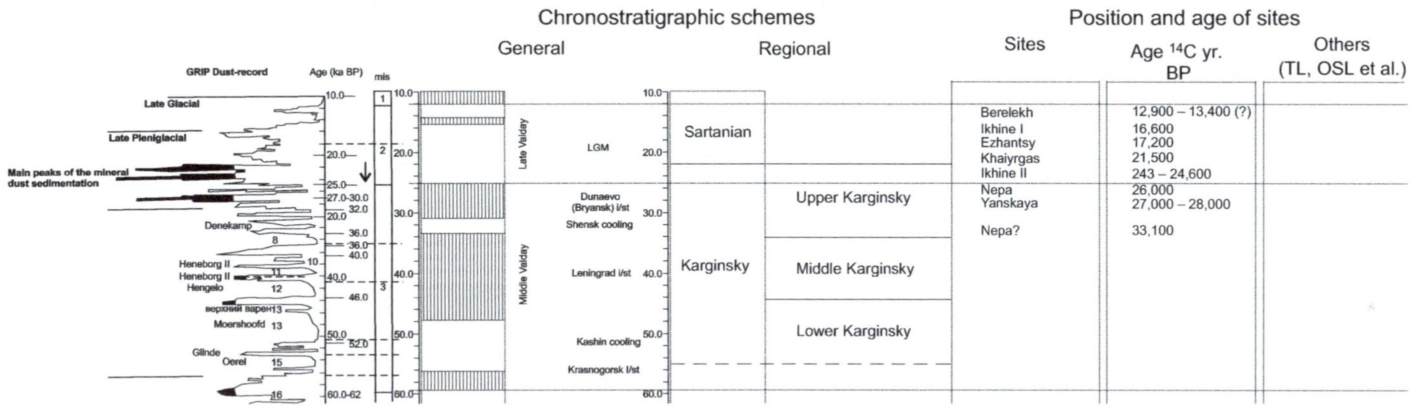

Figure 2.3.7 Chronostratigraphic position of the Upper Paleolithic sites of the Yenisey-Lena-Yana region and their correlation with stratigraphic schemes (regional scheme by Krasnov (1983)), boundaries of subhorizons are shown according to Volkova and Zykina (2000).

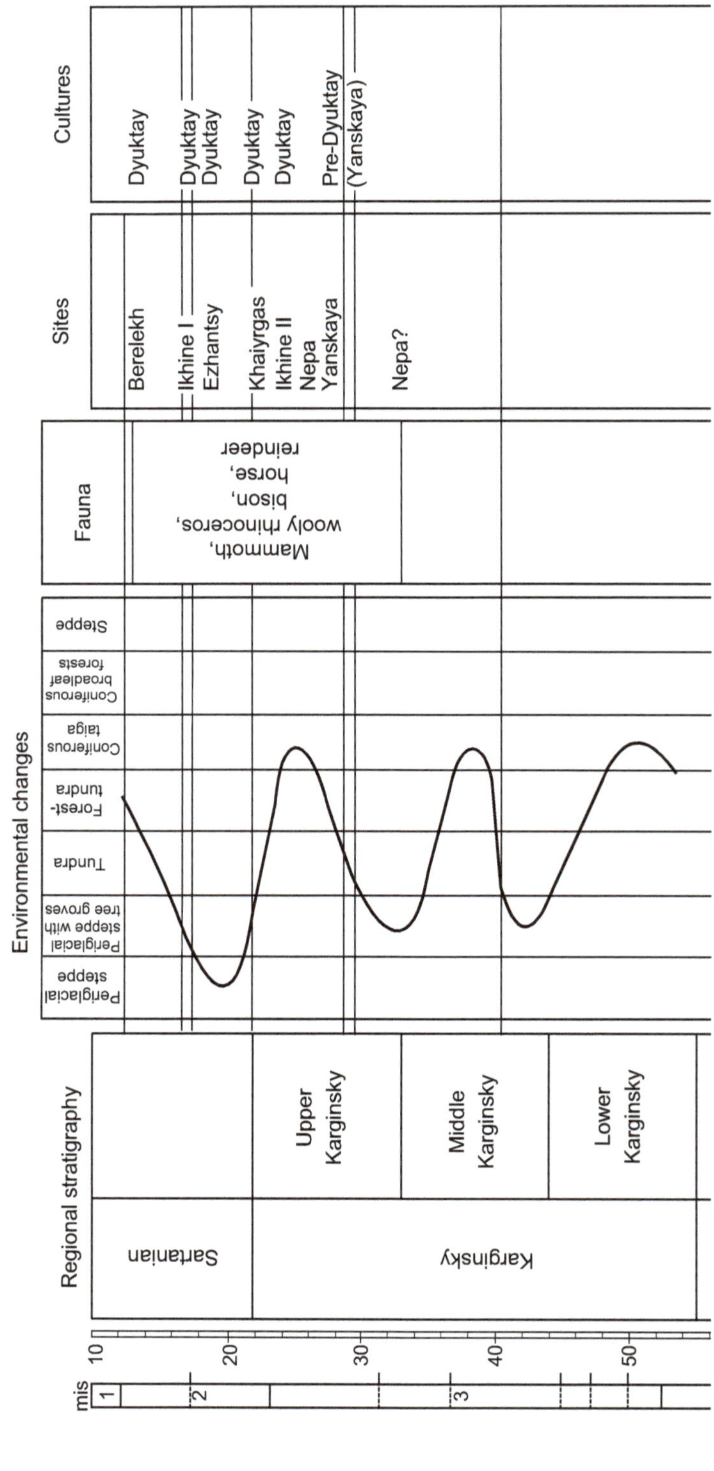

Figure 2.3.8 General archaeological and paleogeographic characteristics of the Paleolithic sites in the Yenisey-Lena-Yana region.

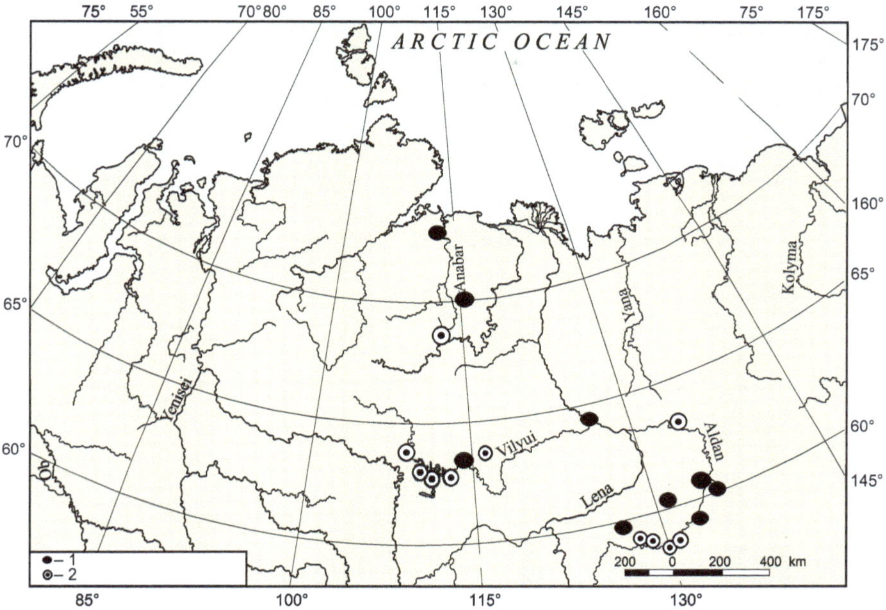

Figure 2.3.9 Position of the Dyuktay culture sites and Mesolithic sites (the Sumnagin culture) in the north of East Siberia west of 140°E (Mochanov, 1977; Mochanov et al., 1983, 1991; Argunov, 1990). *1*, *2* – cultures: *1* – Sumnagin, *2* – Dyuktay.

Though the data available on the Upper Paleolithic colonization of the East Siberian North are still fragmentary (Figs. 2.3.7 and 2.3.8), it may be stated with confidence that the central part of the region (valleys of the middle reaches of the Lena and Aldan) was constantly inhabited by early man since 24,000 yr BP. At the LGM, about 20,000–18,000 yr BP, the early humans were in close vicinity of the studied region, or even within its limits (the sites of Ikhine II, Verkhne-Troitskaya, and Ezhantsy dated by ^{14}C to the interval of 18,300–16,600 yr BP) (Vasil'ev et al., 2002; Kuzmin, 2008). After the LGM, man was constantly present in the north of East Siberia (Figs. 2.3.7 and 2.3.8).

In the Early–Middle Holocene (8000–6000 yr BP), man occupied the northernmost areas of continental Siberia (Taymyr, the Tagenar 6 site) and the Novosibirskiye Islands (Zhokhov) (Fig. 2.3.8). The concentration of animal bones on Taymyr (the eastern coast of Engelhardt Lake, ~75°05′ N, 100°24′ E) (Pitul'ko et al., 2003) dated at 10,020 ± 80 yr BP, LU-3152/11,540 ± 170 cal. yr BP/ and 9680 ± 130 yr BP, LU-3153/11,010 ± 190 cal. yr BP/ suggests the human appearance on Taymyr before the habitation of Tagenar 6. However, that cannot be taken as a final deduction until any unquestionable artifacts are found (Pitul'ko et al., 2003).

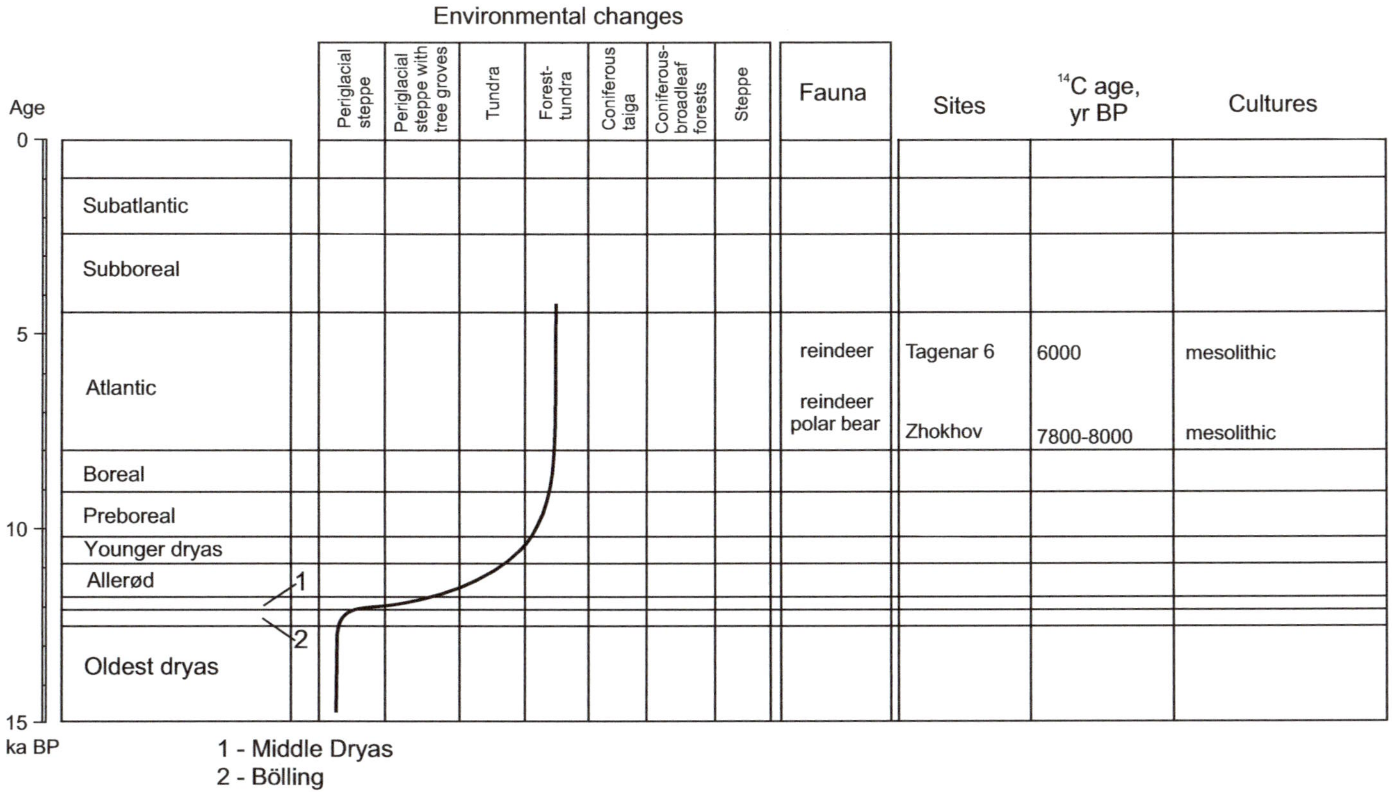

Figure 2.3.10 General archaeological and paleogeographic characteristics of the Mesolithic sites in the Yenisey-Lena-Yana region. *1* – Older Dryas; *2* – Bølling.

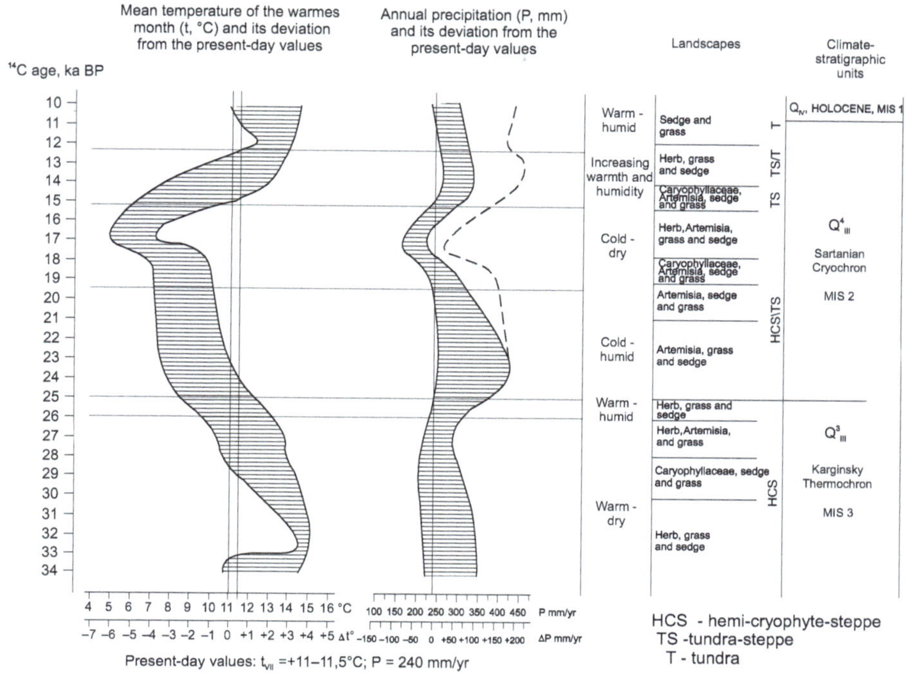

Figure 2.3.11 Temperatures of the warmest month and annual rainfall and environmental changes during MIS3–MIS1 stages in the west of the Yana-Indigirka lowland (Pitul'ko and Pavlova, 2010). *HCS*, hemicryophytic steppe; *T*, tundra; *TS*, tundra-steppe.

Therefore, the human presence has been known in arctic and subarctic East Siberia since 28,500 yr BP (and probably since 33,000 yr BP). His constant presence in the region in 17,000–6000 yr BP may be considered proven (Figs. 2.3.8–2.3.10). The general characteristics of the natural environment of the early man habitation in the East Siberian Arctic between 10,000 and 34,000 yr BP may be inferred from the paleoclimatic data on Yanskaya (Fig. 2.3.11) (Pitul'ko and Pavlova, 2010).

Section Three

Beringia (Northeast of Asia, Alaska, Yukon)

(photo by V.A. Sarana)

Western Beringia (Northeast Asia)

3.1

Sergey B. Slobodin[1], Patricia M. Anderson[2], Olga Y. Glushkova[1], Anatoly V. Lozhkin[1]
[1]Northeastern Interdisciplinary Research Institute, the Far Eastern Division, Russian Academy of Sciences, Magadan, Russia; [2]Washington State University, Seattle, WA, United States

Chapter Outline

Beringia includes the Arctic-Boreal region extending from the Verkhoyansk Range to the northwest of Canada. In the Late Pleistocene, when the northern regions were gradually colonized by early man, the Eurasian and North American continents were connected by the Bering Land Bridge.

The appearance of a dry land in place of shelf seas, increasing ice coverage over the Arctic Ocean (the latter being reduced noticeably in area), lower temperatures of the surface water in the Pacific, and marginal seas resulted in a dryer climate in Beringia, which led, in turn, to the formation of plant communities and animal assemblages fundamentally different from the present-day ones. The Bering land controlled dispersal of the species (both animal and plant ones) from one continent onto another and exerted an influence on migrations of human populations during the period of the initial colonization of arctic and subarctic regions.

Western Beringia, the region considered next in the paleogeographic reconstructions, includes the Northeast of Russia. It extends from the Lena River valley in the west to the Bering Sea in the east, and from the Arctic Ocean coast to the Sea of

Human Colonization of the Arctic: The Interaction Between Early Migration and the Paleoenvironment
http://dx.doi.org/10.1016/B978-0-12-813532-7.00014-0

Okhotsk, and includes also the Kamchatka Peninsula in the south. The recent archaeological (Pitul'ko, 2006a,b; Pitul'ko and Pavlova, 2010), paleoclimatic, and paleoenvironmental data (Velichko, 2002a,b) indicate the human colonization of the arctic and subarctic zones in the Northeast of Asia to begin not later than the second half of the Late Pleistocene. Numerous new facts on the time and ways of the human colonization of those regions have been accumulated. The potential presence of more ancient sites was suggested by the studies of the Diring-Yuryakh site in Yakutia; it follows from the data on this site that the Paleolithic men inhabited the Lena River valley up to 61°N as early as 200–300 ka BP, and probably even earlier (Mochanov, 1992; Ranov and Tseitlin, 1991; Waters et al., 1997; Derevianko et al., 2006).

The landscapes in the Northeast of Asia are highly diversified, so the humans colonized various parts of the region at different times. The most favorable conditions for the human migration eastward existed on the Maritime Lowlands fringing the Arctic Ocean coast; there was no expansion of glaciers onto the lowlands in the Late Pleistocene. We have no unambiguous evidence of human habitation on the Kamchatka and Chukchi peninsulas in the Karginsky. There are, however, materials from the Yana site (Indigirka-Kolyma region) dated at 28,000–27,000 BP (Pitulko et al., 2004) and indicative of man's penetration into Northeast Asia at the end of the Karginsky. Paleogeographic reconstructions also confirm the conditions to be favorable for humans' dispersal over the Maritime Lowlands (Gubin et al., 2008; Bezus'ko et al., 2008; Brigham-Grette et al., 2004).

No site dated to that time has been found in the mountains of Western Beringia. That may be attributed to the fact that mountain ecosystems take a longer time to recover after deglaciation. Harsh landscapes hindered the mountain crossing by men and animals and made more difficult their passage from one river system to another and the dispersal northward and eastward. Nevertheless, there are discoveries in the Western Beringian mountains typologically comparable with the oldest pebble tool assemblages of Siberia and Mongolia, about 30,000–50,000 BP (Dikov, 1979, 1993a,b; Kiryak, 2005; Vorobey, 1999).

More reliable evidence of the human presence in Western Beringia is dated to the Last Glacial and the beginning of the Holocene (Fig. 3.1.1; Table 3.1.1). As an illustration, we refer to the materials on the Bolshaya Khaya IV site near the divide between the Sea of Okhotsk and the Kolyma drainage basin (Slobodin et al., 2008); according to those data, the site vicinities were inhabited at early stages of the last glaciation. Much more is known about the human settlement of that region at the end of the Late Pleistocene; there are a number of sites of that age, such as Berelekh, Druchak-Vetreny, Kheta in the Indigirka-Kolyma region, Ushki I–V, Anavgay on Kamchatka, and Tytylvaam IV and others on the Chukchi Peninsula.

The available typological characteristics of the materials permitted distinguishing several stages in the distribution of different cultural traditions over Western Beringia in the Late Pleistocene; among the Upper Paleolithic traditions, there were the Dyuktay-Ushki culture noted for the presence of microblades and Early Ushki culture with stemmed points (Dikov, 1993a,b; Kiryak, 2005; Slobodin, 2000). The fact that similar tool assemblages appeared in Alaska (Holmes, 2011) and farther south in America as early as 12 ka BP (judging from ^{14}C dates) strongly suggests the

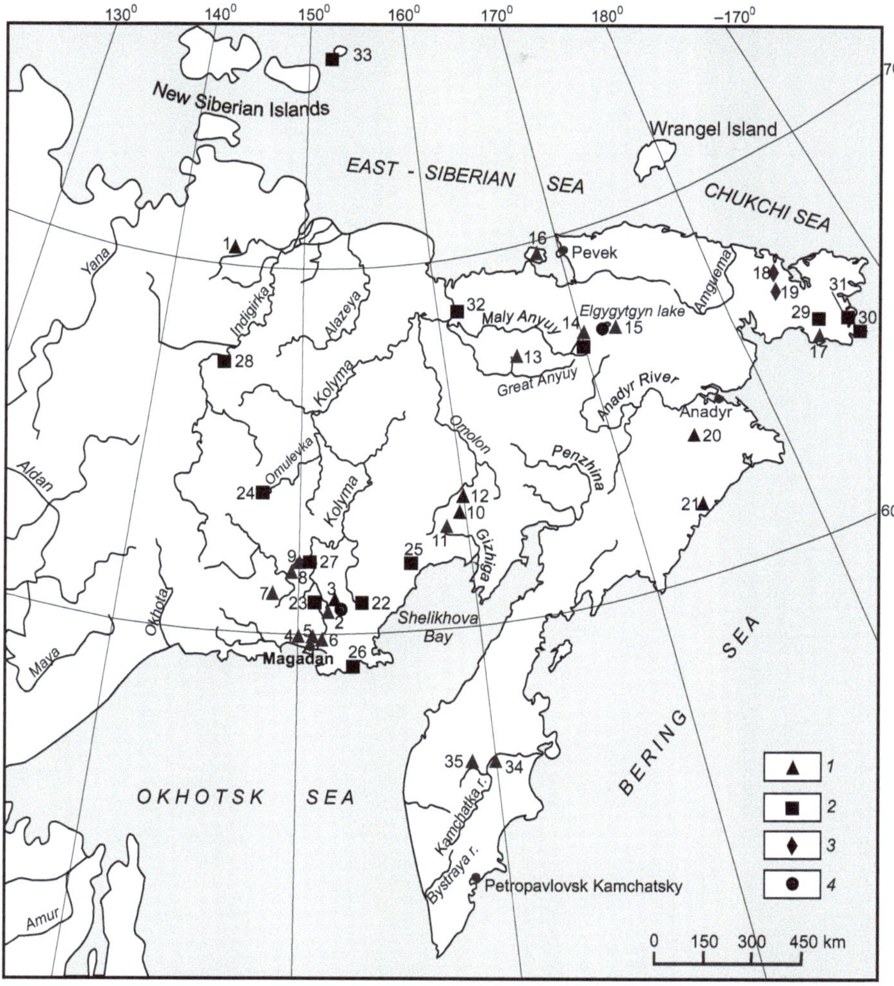

Figure 3.1.1 Paleolithic and Mesolithic sites of Western Beringia. *1–3* – sites: *1* – Paleolithic, *2* – Mesolithic, *3* – others; *4* – key sections. The sites: 1 – Berelekh; 2 – Bolshaya Khaya IV; 3 – Kheta; 4 – Ust'-Magadaven I; 5 – Uptar; 6 – Serdyak; 7 – Omchik II; 8 – Siberdik; 9 – Kongo; 10 – Bolshaya Avlondya; 11 – Druchak-Vetreny; 12 – Bolshoy Elgakhchan I; 13 – Orlovka II; 14 – Tytylvaam IV, Tytyl I, III, IV; 15 – Elgygytgyn II; 16 – Ayon; 17 – Kurupka I; 18 – Kymyneykey; 19 – Kymynanonvyvaam XII, XIII; 20 – Talyain; 21 – Inaskvaam; 22 – Buyunda III; 23 – Urtychuk IV, Azamat; 24 – Uy, Pridorozhnaya; 25 – Omsukchan III; 26 – BMA -3; 27 – Kongo -78; ING-78; 28 – Yubileyny; 29 – Chelkun IV; 30 – Nayvan; 31 – Puturack; 32 – Panteleyikha I–VIII; 33 – Zhokhova; 34 – Ushki I–V; 35 – Anavgay II.

human colonization of the Indigirka-Kolyma region at an earlier time. The hypothesis of several "waves" of the Eastern Beringia colonization is supported by the data on the earlier sites in America (Vasil'ev, 2004) and data of genetic reconstructions (Fix, 2005).

Table 3.1.1 **Paleolithic and Mesolithic sites in Western Beringia**

Site (period)	Longitude, latitude, elevation a.s.l.	^{14}C date/ marker beds	Lab index	Dated material	Geological context of cultural layer	Epoch	Cultural attribution	References
1	2	3	4	5	6	7	8	9
Berelekh Paleolithic	70°29′ N 144°02′ E 20 m	11,970 ± 50 11,450 ± 150 12,220 ± 40 19,820 ± 80 12,930 ± 80 13,420 ± 200	Beta 243748 Beta 190085 Beta-216796 Beta-243743 GIN-1021 IM-152	Hare bone Hare bone Mammoth bone Tusk spear Wood Wood		Late Pleistocene	Dyuktay (?)	Pitul'ko et al. (2009) and Mochanov (1977)
Druchak-Vetreny Paleolithic	63°01′ N 159°57′ E 435 m	7790 ± 250 5120 ± 180	LE - 4711 LE - 4712	Coal from soil	Soil, eluvial formation	End of the Late Pleistocene	*Cis*-Baikal	Vorobey (1996, 2003)
Kheta Paleolithic	61°04′ N 151°46′ E 812 m	Below ash layer dated at 7650 ± 50		–	Soil, eluvial formation	End of the Late Pleistocene	Dyuktay–Ushki	Slobodin (1999a,b) and Melekestsev et al. (1991)
Bolshaya Khaya IV Paleolithic	60°41′ N 151°34′ E 890 m	Below ash layer dated at 7650 ± 50		–	Paleosol	Late Pleistocene	Complex of Bolshaya Khaya IV	Slobodin (2005), Slobodin et al. (2008) and Melekestsev et al. (1991)
Uptar Paleolithic	59°54′ N 150°43′ E 145 m	Older than 8260 ± 330	MAG-1262	Coal from soil, above the finds	Paleosol	End of the Pleistocene	Uptar complex	Slobodin (1999a,b)

Siberdik Paleolithic (?) Mesolithic (?)	61°26′ N 149°24′ E 405 m	7865 ± 310 8020 ± 80 8130 ± 100 8480 ± 200 9700 ± 500 13,225 ± 230	MAG -184 Kril-250 MAG -606 Kril-249 MAG -1019 MAG -916	Coal Coal Coal Coal from hearth Coal Coal	paleosol	Early Holocene	Siberdik	Dikov (1979) and Lozhkin (2008)
Buyunda III Mesolithic	60°54′ N 153°20′ E 835 m	8135 ± 220 7510 ± 205 7790 ± 190 7620 ± 75	GX-17064 GX-17065 ЛЕ-3991 ETH-17512	Coal from hearth soot deposit on artifact	Alluvial sediments	The Holocene beginning	Sumnagin	Slobodin (1999a,b)
Urtychuk Mesolithic	60°38′ N 151°28′ E 1015 m	8285 ± 95 4920 ± 240	GX-17063 LE-3899	Coal from cultural layer	Paleosol	Early Holocene	Sumnagin	Slobodin (1999a,b)
Khurendzha (Azamat) Mesolithic	60°41′ N 151°15′ E 1040 m	8780 ± 70	Beta-56847	Coal from cultural layer	Paleosol	Early Holocene	Sumnagin	Slobodin (1999a,b)
Uy Mesolithic	63°50′ N 147°53′ E 1060 m	8310 ± 240 8370 ± 190 8695 ± 100 8810 ± 235	LE-4652 LE-3990 GX-17066 GX-17067	Coal from cultural layer	Paleosol	Early Holocene	Uolba	Slobodin (1999a,b)
Zhokhov Mesolithic	76°06′ N 152°42′ E 13–20 m	8563 ± 180 7910 ± 180 7810 ± 180 7930 ± 40	LE-3527 LE -3535 LE -3534 GIN-6400	Coal from cultural layer	Paleosol	Early Holocene	Sumnagin	Pitul’ko (1998)

The climatic and environmental changes in the Early Holocene initiated a wide dispersal of the people better adapted to the new conditions and provoked changes in the cultural processes. In the region the cultures dated to that time are Sumnagin, Uolba, and Siberdik (Mochanov, 1977; Dikov, 1979; Slobodin, 1999b).

Subsequently, we consider the large region of Western Beringia, the Indigirka-Kolyma region, Kamchatka, and Chukchi Peninsulas in detail.

3.1.1 Indigirka-Kolyma region

The area under consideration occupies the extreme eastern and northern (between 59°N and 73°N) territory of Eurasia. Its geographic position in high latitudes and close to the seas of the Arctic (Laptev, East Siberia, and Chukchi seas) and the Pacific (Sea of Okhotsk and Bering Sea) determined a somewhat specific response to the global climate changes in the second half of the Late Pleistocene and the Holocene.

At present the extreme continental climate dominates almost the entire region heavily influenced by the Siberian anticyclone. Generally, it is distinct for very low winter temperatures and rather high summer ones, with a relatively small amount of precipitation. The winter is long, up to 9 months; the summer is very short (1.5–2 months). Most of the region is highland country, with variously oriented narrow ranges 1000–3000 m high and plateaus 800–1000 m high. The Arctic Ocean coast is fringed with the Maritime Lowlands; similar lowlands are found at the sea coasts in the south. Because of such distribution of plains and mountains, there is a distinct differentiation of the region into three zones: arctic deserts and tundra on the northern lowlands; polar desert or stony tundra in mountains; and tundra and forest-tundra in the Kolyma and Indigirka drainage basins and on the mountain slopes facing the Sea of Okhotsk and in southern coastal regions. Except for lowlands, the landscapes are controlled by the altitudinal zonality superimposed on the latitudinal zones; the permafrost and related cryogenic and postcryogenic processes and phenomena also exert a noticeable effect on the landscapes.

During the Late Pleistocene and Holocene the vegetation underwent a series of transformations due to changes of the climate, the latter varying repeatedly from very cold and dry to very close to that of present day. The climate-induced changes in vegetation are recorded in pollen assemblages recovered from ^{14}C-dated sequences of lacustrine deposits. The longest and therefore most informative records have been studied in the deposits of the lakes Elikchan-4 (northern coasts of the Sea of Okhotsk) and Elgygytgyn (Chukchi Peninsula), and also in some natural exposures.

The so-called Karginsky interval of the Siberian chronological scale (Kind, 1974) covers the time between 64.0 and 27.0 ka BP and includes two noticeable coolings and three warmings (Anderson and Lozhkin, 2002). The Karginsky is distinguished for a high climatic variability resulting in repeated changes in vegetation (tundra communities being replaced by forest ones and vice versa). The changes might be differently expressed in different regions. On the Yana-Indigirka and Kolyma lowlands the high-amplitude fluctuations of climate were rather indistinct. Among the natural processes, dominant were soil formation, accumulation of peat with high proportion of wood, deposition of lacustrine silts, and accumulation of the so-called "yedoma" formation, the fine-grained frozen sediments being heavily disturbed by thermokarst. The World Ocean level was higher than during the preceding cryochron (Zyryanka), but still lower than at present.

Toward the end of the Karginsky, about 34 kyr BP (Hopkins, 1976; Lozhkin, 2002), the ocean level was about 50 m below that at present. The main components of the landscapes were tundra-steppe communities (Yurtsev, 1972) dominated by grass and wormwood, with rich and diversified herbs. Floodplain meadows and marshes were widely spread, which accounts for a high proportion of nonarboreal plants in the vegetation. As follows from the studies of the Zeleny Mys sequence in the lower reaches of the Kolyma, about 69°N, the winters were frosty, with a low snow cover, and the summers were warm and dry. The zone of open larch forests shifted northward from its present-day position. The Siberian shrub stone pine (*Pinus pumila*) was confined to the places of snow drift accumulation (Lopatina and Zanina, 2006).

The tundra-steppe vegetation prevailing on the northern Maritime Lowlands provided hospitable living conditions for a specific mammoth fauna at the second half of the Karginsky and during the subsequent cooling. Numerous faunal remains in that area (Pitulko et al., 2004; Sher, 1971) included the following species: mammoth (*Mammuthus primigenius),* horse (*Equus caballus* L.), reindeer (*Rangifer tarandus* L.), bison (*Bison priscus* Bojanus), hare (*Lepus tanaiticus* Gureev), woolly rhinoceros (*Coelodonta antiquitatis* Blumenbach), musk ox (*Ovibos moschatus* Zimmerman), wolf (*Canis lupus* L.), polar fox (*Alopex lagopus* L.), brown bear (*Ursus arctos* L.), cave lion (*Panthera spelaea* Goldfuss), wolverine (*Gulo gulo* L.), and red deer *(Cervus elaphus).*

The plant communities in the mountains were more diversified due to a superposition of latitudinal and altitudinal zones. Pollen analysis of lacustrine and alluvial sequences in the Indigirka and the Kolyma drainage basins gave a clue to a specificity of climate fluctuations and vegetation changes in the second half of the Late Pleistocene and the Holocene. In the higher mountains of the Indigirka–Kolyma interfluve (such as the Bolshoy Annachag Range) dominant during the Karginsky were wormwood tundras with shrubs of *P. pumila*, dwarf birch, and *Alnaster* on lower slopes. The comprehensive information so far have been obtained from the pollen record of the Elikchan-4 Lake sediments (60°44′ N, 151°52′ E, 810 m a.s.l.) located in a deep valley in the Maimadjin Mts, near the water divide between the Kolyma (Arctic Ocean) and the Yama Rivers (Sea of Okhotsk) drainage basins. The pollen record analysis permitted identifying the climate and vegetation changes over the last 70 ka BP. The record includes five pollen zones (Fig. 3.1.2). Pollen assemblages of zone EL1 correspond to MIS 4. Zone El2 displays repeated changes in vegetation during the late thermochron of the Pleistocene (the Late Pleistocene interstadial) correlated with MIS 3. That zone is divided into five subzones associated with warmings (subzones El2a, El2c, El2e) and coolings (subzones El2b, El2d). As compared with those of the present day, the climate of the warmings featured cooler and wetter summers. The landscapes were dominated by open larch forests with *P. pumila* and shrub birch and alder in the undergrowth. A relatively high proportion of NAP in the assemblages suggests a presence of tundra communities on mountain

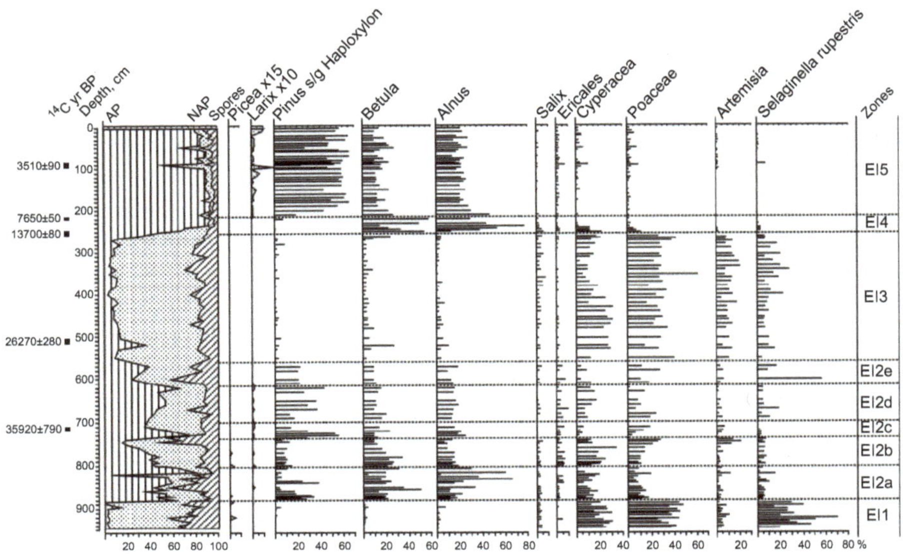

Figure 3.1.2 Pollen diagram of the Lake Elikchan-4 sequence (Okhotsk Sea–Kolyma River interfluve).

slopes. During the Late Pleistocene interstadial the colder intervals were marked by increasing areas of the dwarf-shrub and herbaceous tundra in mountains, while the open larch forests were confined to valley floors. The upper boundary of the El2 pollen zone is dated at 27,400 BP (Lozhkin, 1997).

The climate cooling started in the region about 27 ka BP; it was marked by the appearance of glaciers in mountains. The glaciers appeared in the higher mountains (the most elevated ranges and massifs) (Fig. 3.1.3). Except for the Chersky glacial region, most of glaciers were concentrated on the Pacific slope of mountains (Glushkova, 1984, 2001). In the high mountains the glaciation was of "reticulated" type or formed ice sheets. In the medium-height mountains, there were valley glaciers (15–25 km long) and cirque glaciers. The mountain slopes were modeled by nivation and cryogenic processes.

At the Last Glacial Maximum (LGM) time (20–18 kyr BP) the World Ocean level dropped by 100–110 m, which resulted in vast areas of the shelves being exposed. The sea receded about 1000 km from the coastline of today (Fig. 3.1.4) near the New Siberian Islands and 640 km near the Ayon Island (the Pevek Bay). At the Okhotsk Sea coasts the width of the exposed shelf did not usually exceed 40–60 km.

The LGM, which is considered to be the "main climatic boundary of the Pleistocene," is noted for glacio-eustatically lowered level of the World Ocean, low temperatures, and aridity (Velichko, 1968); in the Elikchan-4 pollen record, it corresponds to pollen zone El3. The vegetation was noted for a mosaic pattern and formed a discontinuous cover of *Artemisia*, various herbs, and Siberian spikemoss

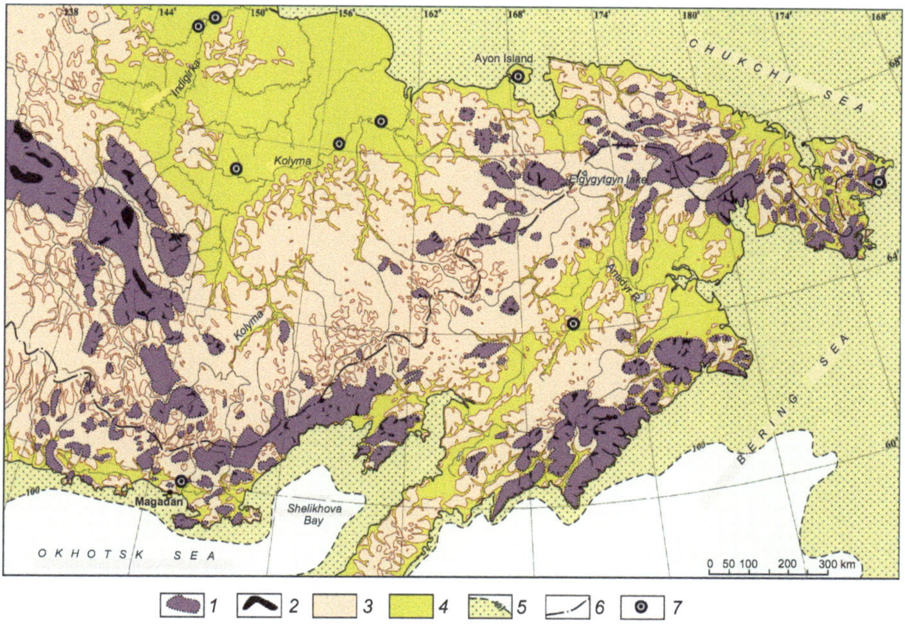

Figure 3.1.3 Paleogeographic scheme of Last Glacial Maximum (20,000–18,000 yr BP). *1* – glaciated areas in mountain ranges; *2* – the largest glaciers; *3* – mountains and foothills (more than 200 m a.s.l.); *4* – plains and lowlands (less than 200 m a.s.l.); *5* – boundary of the emerged shelf; *6* – the main divide between the Pacific and Arctic; *7* – localities of the ^{14}C-dated mammoth fauna.

(*Selaginella sibirica*) on dry stone slopes; on lower slopes, there were wetlands with communities of sedge and moss, or of sedge and grass, with creeping willows on the river and lake banks. Shrub birch and alder were present as small isolated populations. Noteworthy is a wide distribution of the Siberian dwarf pine (*P. pumila*), suggesting a deep snow cover in winter (as the species is highly sensitive to the seasonal soil freezing and thawing). At the LGM (about 21–15 ka BP) a relatively narrow zone of land along the northern coast of the Sea of Okhotsk served as a refugium not only for arctic herbs and shrubs, but also for some Boreal trees and bushes; besides *Larix dahurica* (*Larix gmelinii*), which is the main forest-forming species at present, there were other tree species: *Picea obovata*, *Chosenia arbutifolia*, *Populus suaveolens*, *P. tremula*, *Alnus hirsuta*, *Betula lanata*, *Betula platiphylla*, and *Betula Cajanderi*.

Conspicuous changes of climate took place about 12.5 ka BP in Western Beringia and resulted in a radical change of the vegetation. That sudden warming was recorded in the pollen assemblages recovered from the Elikchan Lake (El4 pollen zone). Similar zones have been identified in all the pollen records recovered from lacustrine sequences. They characterize the transitional interval from the Pleistocene to the Holocene.

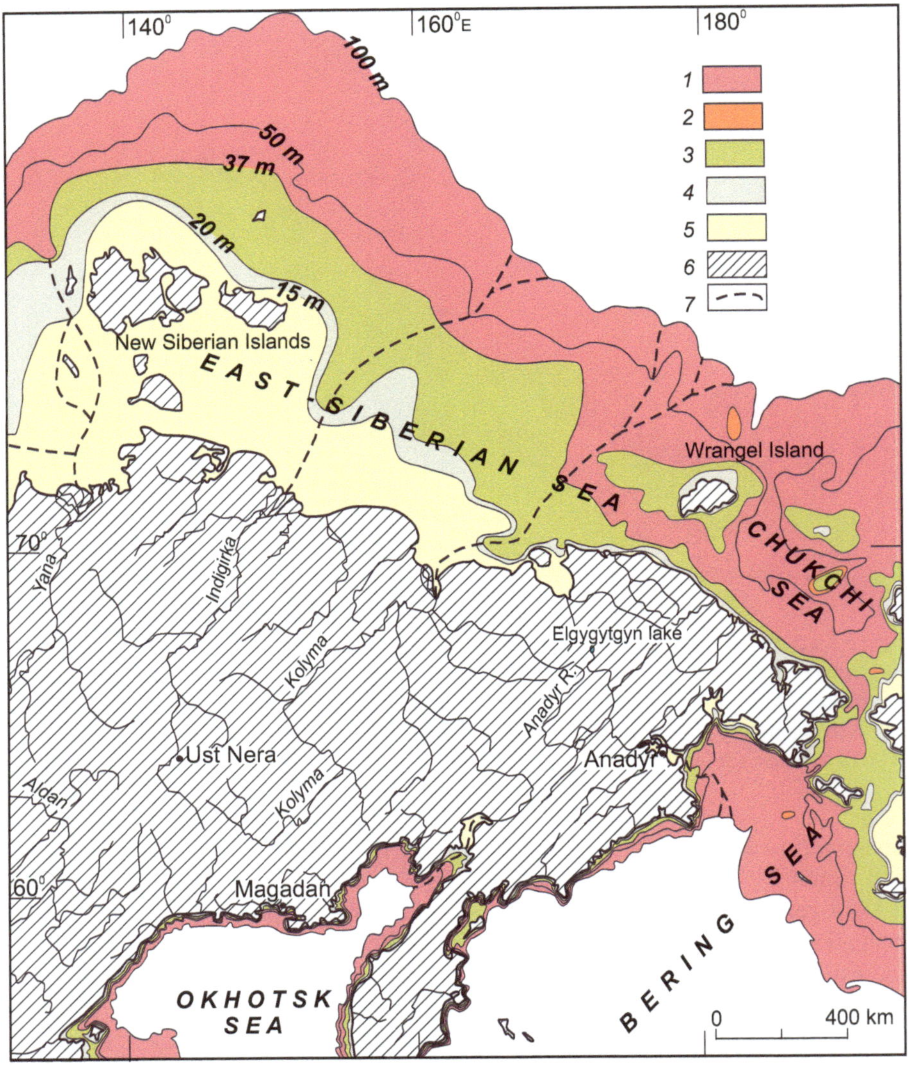

Figure 3.1.4 Ocean level changes between 18,000–8500 yr BP dated by ^{14}C. *1* – 18,000 yr BP, sea level at −100 m; *2* – 15,000 yr BP, sea level at −50 m; *3* – 12,000 yr BP, sea level at −37 m; *4* – 10,000 yr BP, sea level at −20 m; *5* – 8500 yr BP, sea level at −15 m; *6* – the present-day land; *7* – paleo-valleys.

The noticeable rise in the proportion of dwarf birch and *Alnaster* in the pollen spectra of the zone suggests an essential restructuring of the vegetation, the grass and bush tundra of the glacial stage being replaced by the shrub birch tundra, and later by tundra with large bushes of birch and *Alnaster*. It has been found that 11.6 ka BP open forests of larch occupied already considerable areas. The larch appearance, in particular, indicates the summer temperatures were not lower than 12°C. The later cooling (Younger Dryas) is noticeable only in pollen assemblages from the upper reaches of the Indigirka River.

About 8 ka BP the Siberian dwarf pine (*P. pumila*) became an important constituent of the larch forest undergrowth and formed an altitudinal zone of the pine shrub (pollen zone El5) above the tree line in mountains (Anderson and Lozhkin, 2002). Quite possibly the plant communities similar to those of today appeared there at that time.

On the Maritime Lowlands the postglacial rise of the sea level resulted in a gradual change of the landscapes, while many animals died out or migrated out of the region. The arctic tundra-steppe seems to have disappeared relatively fast. The vast land exerted a noticeable influence on the climate presumably until the beginning of the Holocene and in particular during the Preboreal and Boreal, when several species of trees and large shrubs grew within the modern zone of herbaceous and dwarf shrub tundra. There are several radiocarbon dates obtained on the tree birches within the interval from 9315±50 yr BP, MAG-123/10,510±80 cal. yr BP/ to 8000±30 yr BP, MAG-128/8880±70 cal. yr BP/, on alder (*Duschekia fruticosa*) 8620±200 yr BP, MAG-243/9700±260 cal. yr BP/, and large shrub *Salix* sp. from 10,730±230 yr BP, MAG-244/12,580±280 cal. yr BP/to 8450±160 yr BP, MAG-138/9430±210 cal. yr BP/.

Those data permitted considering the interval synchronous to the Boreal to be the thermal optimum of the Holocene (Kaplina and Chekhovsky, 1987; Lozhkin and Vazhenina, 1987).

3.1.1.1 Early man sites in the Indigirka-Kolyma region

3.1.1.2 Late Pleistocene sites

Berelekh belongs to the Berelekh complex of geoarchaeological objects. It is located in the upper reaches of the Berelekh River (left tributary of the Indigirka River, 70°29′ N, 144°02′ E, 20 m a.s.l.). It was discovered in 1970 by Vereshchagin (1977) during the study of the Berelekh "mammoth cemetery," at a distance of a few tens of meters from the latter. The cultural layer of the site was found in the upper part of the 12–13 m terrace section (Fig. 3.1.5), at a depth of 1.5–2.5 m from the top (Vereshchagin and Mochanov, 1972; Nikolskiy et al., 2010).

A thick (about 2 m) bone-bearing horizon occurred at 3.5–4 m depth from the top and was traceable for 70–80 m in the terrace deposits at the Berelekh "mammoth cemetery" (Vereshchagin, 1977; Mochanov, 1977; Pitul'ko, 2008a; Pitul'ko and Pavlova, 2010). The "cemetery" confined to an oxbow lake on the Berelekh River floodplain was dated to the period 14–12 ka BP on the basis of ^{14}C dates run out on bones (12,000±130 yr BP, LU-149/13,870±170 cal. yr BP/) and soft tissues of mammoth (12,000±130 yr BP, MAG-114/16,590±570 cal. yr BP/) (Lozhkin, 1998; Pitul'ko and Pavlova, 2010).

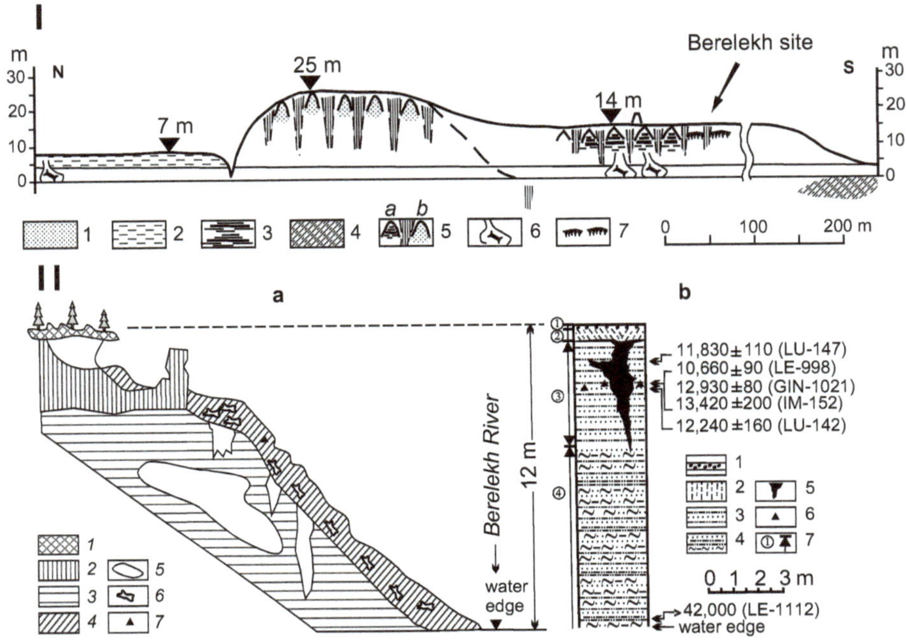

Figure 3.1.5 Paleolithic site of Berelekh. (a) –Profile across Berelekh (Nikolskiy et al., 2010). *1* – sand; *2* – silt; *3* – silt rich in organics; *4* – distorted silts; *5* – ice-wedge polygons: *a* – in the sediments forming the second terrace, *b* – in the sediments forming the third terrace; *6* – mammoth bones; *7* – artifacts. (b) – Schematic profile of Berelekh (Vereshchagin and Mochanov, 1972) (*a*) and generalized sequence of the Berelekh site deposits (Mochanov, 1977) (*b*). **a** – *1* – sod; *2* – loess; *3* – alluvial deposits; *4* – landslide; *5* – ice veins; *6* – animal bones; *7* – artifacts; **b**: *1* – sod; *2* – crumbly sandy loam; *3* – horizontally laminated sandy silt; *4* – horizontally laminated sand and sandy loam; *5* – ice vein; *6* – finds of the Paleolithic tools; *7* – numbers and boundaries of geological strata.

The archaeological materials were obtained from the excavation (more than 30 m^2 in area) and from exposures on the terrace slopes. The finds occurred in the upper part of grayish-brown sandy loam and nonsorted sand (4–4.5 m thick altogether) overlain with a yellow-brown crumbly sandy loam and the modern soil. The sequence is distorted by cryogenic processes, ice wedges, slumps, etc., so the finds might be partly redeposited. That conjecture is supported by a wide range of the radiocarbon dates and variations in the depth of the object occurrence from 1.5 to 2.5 m.

More than 300 stone artifacts were recovered from the site, among them flakes, retouched blades, bifacial knives, and foliated (leaf-shaped) points, burins on flakes (Fig. 3.1.6: *8*, *14–17*fn11[1]), chisel-like tools, and pendants made of small drilled pebbles (Fig. 3.1.6: *20*). There have been also found numerous mammoth bones and tusk fragments, split, burnt, and treated; some fragments were reworked into knives, scrapers, and one fragment of spear 94 cm long.

[1] Here and following, numbers in italic designate archaeological objects in figures.

The faunal materials recovered from the cultural layer of the Berelekh site (Vereshchagin, 1977; Mochanov, 1977) revealed the species composition as follows:

Fauna	**Number of bones**
Mammoth (*Mammuthus primigenius*)	256
Woolly rhinoceros (*Coelodonta antiquitatis*)	1
Reindeer (*Rangifer tarandus* L.)	2
Bison (*Bison priscus* Bojanus)	3?
Horse (*Equus caballus* L.)	1
Wolf (*Canis lupus* L.)	18
Musk ox (*Ovibos moschatus* Zimmerman)	1
Hare (*Lepus tanaiticus* Gureev)	850
Partridge (*Lagopus*)	92
Goose (*Anser*)	4
Fish (*Pisces*)	2

Based on the ^{14}C-dated plant remains and mammoth bones recovered from the cultural layer, the site age was first determined as 13–12.5 ka BP (Mochanov, 1977). Judging from other dates, however, obtained on samples from the sediments overlying the bone-bearing horizon—11,870 ± 60, MAG-117/13,680 ± 70 cal. yr BP/(Lozhkin, 1977) and 11,830 ± 110, LU-147/13,660 ± 120 cal. yr BP/(Arslanov et al., 1980)—the site cannot be older than 12,000 yr BP. Considering that the inhabitants of the Belelekh took the mammoth bones from the "cemetery," the main object of hunting was the hare, as its bones could hardly appear in such abundance at the site in a natural way. That being so, the most reliable date of the cultural layer is that on the hare bone, that is, 11,970–11,450 yr BP (Table 3.1.1) (Pitul'ko et al., 2009). As follows from the correlation between the site geological context and the "cemetery" sequence, man settled there soon after formation of the bone-bearing deposit and used the bones as a raw material for toolmaking.

Mochanov (1977) attributed the assemblage to the Dyuktay culture. However, the site did not yield any pieces considered typical of that culture: microblades, typical bifaces, and wedge-shaped cores (Slobodin, 1999a, 2000). The fact gave grounds to separate the Berelekh culture from that of Dyuktay sites and consider it as another microblade-lacking bifacial assemblage of Beringia (Goebel at al., 1991). Teardrop-shaped bifaces found at the site (Fig. 3.1.6: *17*) have analogs in Beringia (Nenana complex at Alaska) and in Siberia (Tarachikha) (Abramova, 1983).

It has been supposed that the cultural objects recovered from the site may be of different age and partly belonging to an older component, genetically related to the Yanskaya culture (Pitul'ko, 2006a,b). That agrees with the date—19,820 ± 80, Beta-243,743/23,860 ± 130 cal. yr BP/—obtained on a spear from mammoth tusk (Pitul'ko et al., 2009). A microblade wedge-shaped core characteristic of the Dyuktay tradition was reported to be found on the site near the excavation (Fig. 3.1.6: *18*) (Mochanov and Fedoseeva, 1996); its origin, however, as well as its relationship to the assemblage recovered from the excavation are still not clear.

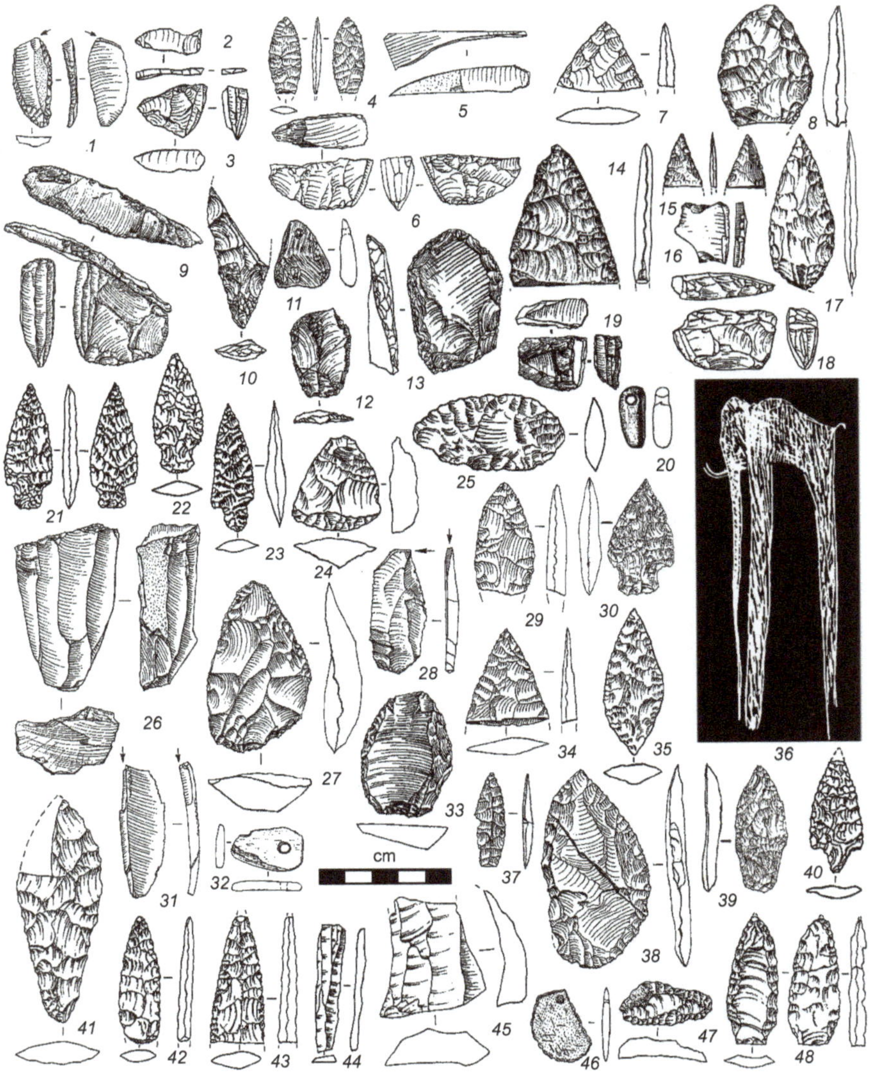

Figure 3.1.6 Artifacts from Paleolithic sites of the Indigirka-Kolyma region (Mochanov, 1977; Vorobey, 1996; Kiryak, 1993). *1–7, 11* – Kheta; *8, 14–18, 20, 36* – Berelekh; *9, 10, 12, 13* – Druchak-Vetreny; *19* – Ust'-Magadaven 1; *21* – Serdyak; *22* – Bolshaya Avlondya; *23–25, 35, 40* – Bolshoy Elgakhchan I; *26–29, 31–34, 37–39* – Bolshaya Khaya; *30* – Omchik IIб; *36* – engraved image of a mammoth on a tusk from the area near Berelekh; *41–48* – Uptar.

It is not inconceivable that the Berelekh inhabitants could meet with the remaining mammoth population; judging from the radiocarbon dates, the latter still existed on the coastal lowlands, as well as on Taymyr, Yamal, and New Siberian Islands, until the Early Holocene (Sulerzhitskiy, 1997), and on Wrangel Island up to the second half of the Holocene (Vartanyan, 2007). A tusk fragment found at a distance of 48 km from the Berelekh bone concentration bears an engraved image of a mammoth (Bader and Flint, 1977). Judging from the type of the image and technique of engraving (most probably, made with a stone burin), as well as from the cover of patina on the tusk, the picture was performed in high antiquity, possibly by a contemporary of mammoths (Fig. 3.1.6: *36*).

Bolshaya Khaya IV located in a small valley of the Bolshaya Khaya River on the Okhotsk-Kolyma Highlands (60°41′ N, 151°34′ E, 900 m a.s.l.) was discovered in 2001 (Slobodin et al., 2008). It sits on the surface of an erosional terrace 11.5 m high and occupies an area of more than 1000 m^2. The cultural pieces occur at the base of thin (no more than 35 cm) deposits with cryoturbations overlying the solid rocks. No radiocarbon dates are available as yet, but the stratigraphic position of the artifacts under the volcanic ash layer 7600 years old (Melekestsev et al., 1991), as well as technical and typological characteristics of the tools, suggest its Late Pleistocene age.

The tool assemblage is uniform in the raw material, typology, and technology. As to the raw material, it was mostly flint slates and chalcedony pebbles collected near the site. The primary knapping is represented by a few types of cores: flat cores for blade-like flakes; end cores on slates; flat subprismatic core; flat double-platform cores with retouched oblique platforms; and a fragment of the core front with traces of blade removals.

Among tools, there are bifaces of asymmetric oval, crescentic, or subtriangular outlines, flattened, lens-like in cross-profile; foliated and subtriangular points; lateral end scrapers on flakes and end scrapers on blades; and dihedral, lateral, and transversal burins on blade-like flakes (Fig. 3.1.6: *26–29*, *31–34*, *37–39*). A flat pendant with a hole and beads represent personal ornaments. A large area of the site, an abundance of raw material, and the presence of ornaments suggest the site had been repeatedly visited by the prehistoric hunters.

The cultural affiliation of the assemblage and its age have not been determined. It might be quite a new assemblage, previously unknown in the region. It may be confidently attributed to the Paleolithic judging from the technique of the core reduction; the latter shows transition from the subprismatic technique (when blade preforms are removed from the wide core front) to the edge-faceted core technique. Such a transition was recorded at the early stages of the Upper Paleolithic at the Altay (Kara-Bom) (Derevyanko et al., 2002a,b), and in the Middle Upper Paleolithic (25–18 ka BP) in the Transbaikal area (Konstantinov, 1994). The lithic technology described at Bolshaya Khaya IV has some features in common with the fourth component of the Abaykan dated at 21–15 ka BP (Derevyanko et al., 1998) and with the materials of the lower layer of Ustinovka I in Primorye (Russian Far East) dated at 22–14 ka BP (Vasilyevskiy and Gladyshev, 1989; Krupyanko and Tabarev, 2001) or even 33–30 ka BP (Kononenko, 2001). Supposedly, in the Far East, there is a pre-Ustinovka Paleolithic complex provisionally attributed to the Middle Sartan, with parallel-sided cores, blades and tools on blades, and bifaces, though lacking the microlithic technique (Dyakov, 2000). The characteristics of the tool assemblage of the Bolshaya Khaya IV site are in full accord with those criteria.

Kheta discovered in 1991 is in the upper reaches of the Kheta River close to the divide between the Sea of Okhotsk and the Kolyma drainage basin, 180 km southeast from the city of Magadan (61°04′ N, 151°46′ E, 800 m a.s.l.). It sits on a rocky surface of the erosional terrace 12–14 m above the river (Slobodin, 1999b; Slobodin et al., 2008). The cultural materials attributed to the Paleolithic and Neolithic occur in sediments from the day surface to a depth of 0.8 m. The sequence is as follows:

	Thickness (cm)
1. Sod	2–10
2. Brown sandy loam	3–10
3. Yellow-brown sandy loam rich in organic matter, with lenses and interlayers of volcanic ash	10–30
4. Medium-grained, yellow-brown sand with lenses of yellow sandy loam, pebbles, and angular debris	10–20
5. Pebble layer with sand and gravel	2–30
6. Bedrock	

The chemical composition of the volcanic ash permits attributing it to the so-called "Elikchan" type dated by radiocarbon to 7.6 ka BP. The lower (Paleolithic) cultural layer has been unearthed in the stratigraphic layer 4. It contains flakes, microblades, cores, blades, technical spalls, bifacial knives, arrowheads, end scrapers, burins, and pendants (Fig. 3.1.6: *1–7*, *11*). No faunal remains were found.

Wedge-shaped cores are prepared using the Yubetsu technique on bifaces and flint plaques with retouched margins; their pressure platforms are fashioned by ski spall removals. The blades are removed from subprismatic cores. Bifaces are thin, elongated, and oval in outline. Arrowheads and dart points are leaf-shaped and flat in cross-section. There are also end scrapers and angle and transversal burins of the Upper Lena type. Among the ornaments, there are flat, polished beads and a pendant made of agalmatolite.

The small size of the site (about 100 m^2) suggests it to be a seasonal camp of reindeer and wild sheep hunters. The technical and typological characteristics of the materials recovered from the site suggest it belongs to one of the Late Paleolithic traditions: the Selemdzha-Dyuktay-Ushki (Derevyanko et al., 1998) and/or Far Eastern Ustinovka one (Vasilyevskiy and Gladyshev, 1989); the lower complex of Kheta is attributable therefore to the late stages of the Upper Paleolithic (18–12 ka BP). In the southwestern part of northern coasts of the Sea of Okhotsk a similar industry was recovered from Ust'-Magadaven (Fig. 3.1.6: *19*) (Vorobey, 2008).

Bolshoy Elgakhchan I is in the upper reaches of the Omolon River on a high (37 m above the river) surface at the mouth of its tributary, the Bolshoy Elgakhchan River (64°14′ N, 16,106′ E, 407 m a.s.l.). It was discovered by Yu.A. Mochanov in 1979 (Kistenev, 1980). The Paleolithic materials were recovered from the lower step of the terrace headland (Kiryak, 1993). The excavated area amounts to 97 m^2.

	Thickness (cm)
1. Sod	4–14
2. Brown humified layer (locally absent)	3–22
3. Yellow sandy loam	5–24
4. Greenish sandy loam	30–50
5. Sandy loam	About 50
6. Bedrock	

Most of the findings occurred at a depth of 35 cm. As follows from the palynological results, the time of the upper layer deposition was marked by the dominance of open larch forests with birch (Titov, 1991). The pollen assemblages recovered from the base of the cultural layer (from a depth of 25–30 cm) suggest the communities typical of the end of the Late Pleistocene. The deposits bear traces of cryogenic processes (Kiryak, 1996).

The lower cultural layer of the site yielded 127 pieces, among them are pebble tools, points, scrapers, knives, and burins (Fig. 3.1.6: *23–25*, *35*, *40*) (Kiryak, 1993). The tools are typically treated using bifacial technique with covering retouch on their broad surfaces; there are also unifacial tools with a marginal retouch. Pieces on pebbles include choppers, adzes, hammer stones and cores for large flakes. The bifacial knives of oval shape are flattened and lens-like in cross-section. Bifacial stemmed points of the "Ushki" type (Slobodin, 2002), 3–5 cm long and 1.5–2.3 cm wide, are made on flakes. The working part (blade) is triangular, slightly asymmetric in plan, and flattened, lens-like in cross-section; there are distinct shoulders and a short rectangular or subtriangular stem. The assemblage is homogenous in its technical and typological characteristics and could be correlated with the assemblage of layer VII of Ushki at Kamchatka (the site having been given the name to the Late Upper Paleolithic Early Ushki culture dated by ^{14}C to 11,300–14,000 yr BP) (Dikov, 1979; Slobodin, 2010b; Goebel et al., 2010). Pebble tools form a sizeable (by weight and volume) part of the tool complex, indicative of an active and relatively long-term economic activities of the site inhabitants.

Apart from the Bolshoy Elgakhchan I, the stemmed points of the Ushki type were found in the surface collections from the sites of Omchik IIb on the Kolyma River, Serdyak on the northern coasts of the Sea of Okhotsk, and Boshaya Avlondya on the Okhotsk-Kolyma interfluve (Fig. 3.1.6: *21*, *22*, *30*) (Slobodin, 2002, 2010a).

Druchak-Vetreny in the upper reaches of the Gizhiga River, 140 km north of the Sea of Okhotsk coast (63°01′ N, 159°57′ E, 420 m a.s.l.), was discovered in 1988 by I.E. Vorobey (1996). It is positioned on the erosional terrace on the right side of the Druchak River valley (Fig. 3.1.7).

The cultural remains occur within the cryoturbated cover sandy loam at a depth 2–60 cm from the surface. The total thickness of the deposits varies from 0.1 to 1.6 m. There have been recovered about 1000 tools made of hornfels.

The primary knapping objects are represented by large cores for macro- and micro-blades and by wedge-shaped microcores (the Yubetsu technique) with ski spalls and microblades (Fig. 3.1.6: *9*). The tools are transversal burins of the Upper Lena type, cutting tools, notched tools, borers, end scrapers, bifacial pieces (an axe, an adze, core preforms, foliated points) (Fig. 3.1.6: *10*, *12*, *13*), and choppers.

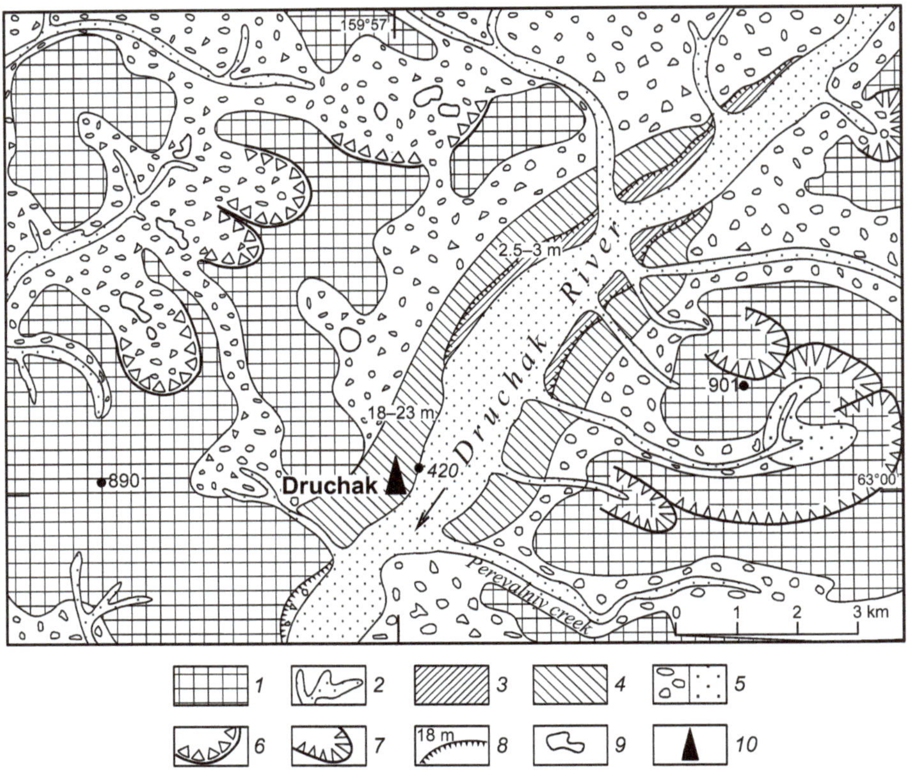

Figure 3.1.7 Geomorphological scheme of the Druchak-Vetreny site setting. *1* – mountain and uphill slopes; *2* – the river channel and floodplain; *3* – the first terrace; *4* – the second terrace; *5* – areas of glacial deposits: *a*–Zyryanka, *b*–Sartan; *6* – old glacial cirques; *7* – young cirques; *8* – scarps and their height; *9* – glacial lakes; *10* – archaeological site.

A charcoal from the upper sandy loam was dated by ^{14}C at 7790±250 and 5120±180yr BP (Table 3.1.1); in all probability, that may be explained by the action of cryogenic processes on the site; the latter transported the charcoal from the upper layers downward, into the cultural layer (Vorobey, 1996). The technical-typological characteristics of the tools from the site suggest its age to be about 14–13ka BP (Slobodin, 1999b, 2000). Morphological elements of the Druchak industry are comparable to those of the Middle Upper Paleolithic in Siberia, Far East, and the Baikal region dated at 15–11ka BP (Vorobey, 2003).

A great number and diversity of the artifacts suggest the site functioned in summer and autumn, at the time of hunting and salmon fishing, when the raw material sources were also available.

Uptar was discovered in 1984 in the valley of the Uptar River (the Khasyn River drainage basin), 45km from the Sea of Okhotsk coast (59°54′ N, 150°43′ E, 144m a.s.l.). It is positioned on the second terrace 4–6m above the river (Fig. 3.1.8) (Slobodin, 1999b).

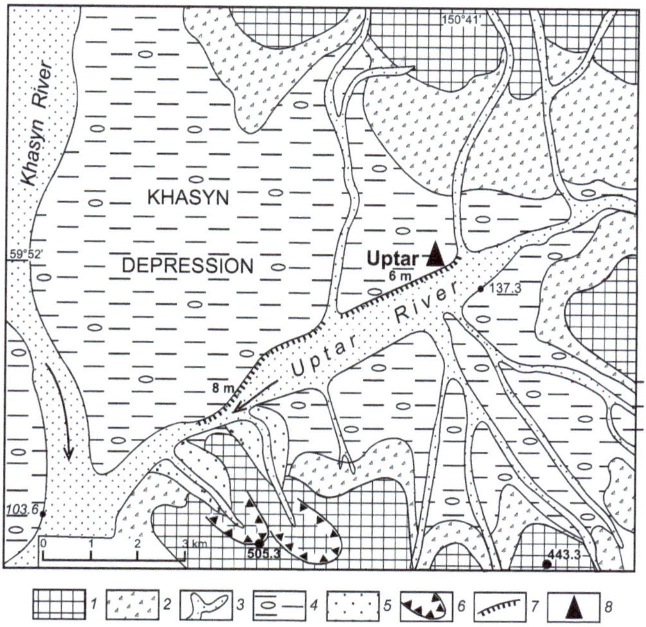

Figure 3.1.8 Geomorphological scheme of the Uptar site setting. *1* – mountain slopes; *2* – foothill apron; *3* – the river channel and floodplain; *4* – fluvial-lacustrine plain, locally overlain by glacial deposits; *5* – Sartan till; *6* – glacial cirques; *7* – scarps and their height; *8* – archaeological site.

The artifacts from the excavated area measuring about 50 m^2 occurred at a depth of 8–35 cm, in red-brown sandy loam overlain with a volcanic ash, fine sand, and sod. Below the cultural layer, there is a thick series of coarsely stratified gravel and pebbles with sand. The radiocarbon date—8260 ± 330 yr BP, MAG-1262/9220 ± 420 cal. Yr BP/—was obtained on charcoal from the layer above the culture-bearing stratum. Both patina on the surface of tools and the traces of eolian corrosion suggest that the findings rested on the surface for a long time, before they were buried under the ash.

The cultural layer yielded 45 tools, flakes from the frontal surface of subprismatic cores, bladelets, and a pendant made of tuffite and siliceous rocks (Fig. 3.1.6: *41–48*). Of the tools in the Uptar complex, 85% are bifacial oval knives, leaf-shaped arrowheads, or points of darts, as well as spear points with a narrow haft. One of the points bears the trace of a removed long channel flake, from the base to the end of the point, resembling typical Paleoindian points of North America. End scrapers (10% of tools), both typical end and lateral are made on flakes. Cutting tools are trimmed pebbles and an axe-like tool. Microliths (microblades?) are few, and no microcores (prismatic or wedge-shaped), have been found in the complex. A polished pendant made of slate has a hole and a series of marginal incisions. The assemblage has no distinct analogs among the known cultures of the northeast of Asia. The Uptar points with a narrow haft show a similarity to the tools belonging to the Osipovka Paleolithic culture in the Amur drainage basin (Slobodin, 1999b). The Uptar inhabitants hunted moose and reindeer in river valleys and probably practiced fishing at the season of salmon spawning.

3.1.1.3 Early Holocene sites

A group of sites (Buyunda III, Urtychuk, Khurendzha, or Azamat, Zhokhov, Panteleikha I–VIII, BMA-3, and others) are known to have existed in the Early Holocene in the Indigirka-Kolyma region. They are dated by ^{14}C to 9–7 ka BP and contain materials attributable by the technical-typological characteristics to the *Sumnagin* tradition of early reindeer and moose hunters in the Northeast of Asia (Mochanov, 1977). Another cultural tradition, *Uolba*, is represented by the sites of Uy, Kongo-78, ING-78, Pridorozhnaya, Agrobaza II on the Upper Kolyma and Yubileyny on the Indigirka (Slobodin, 1999b). The *Siberdik culture* (Siberdik and Kongo sites) presents another episode of the Upper Kolyma colonization in the Early Holocene (Dikov, 1979).

Buyunda III in the upper reaches of the river of the same name belonging to the Kolyma drainage basin is situated closely to the Okhotsk-Kolyma water divide (60°54′ N, 153°20′ E, 830 m a.s.l.) (Slobodin, 1999b). The site sits on the Buyunda River terrace 5–6 m high (Fig. 3.1.9). There is small fragment of the terrace preserved as an outlier composed of cross-stratified alluvial gravels and separated by a lateral channel from the valley. The cultural layer is practically undisturbed except for a few animal

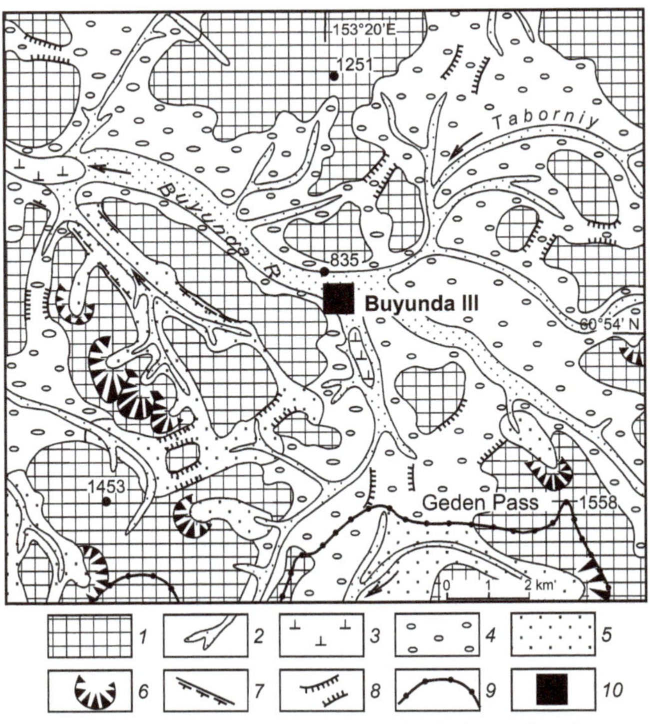

Figure 3.1.9 Geomorphological scheme of the Buyunda III site setting. *1* – mountain slopes with traces of glacial erosion; *2* – river channels and low terraces; *3* – large ice fields; *4, 5* – tills: *4* – Zyryanka, *5* – Sartan; *6* – glacial cirques; *7* – trough shoulders; *8* – water gaps (through valleys); *9* – Okhotsk Sea–Kolyma Riverdivide; *10* – archaeological site.

burrows. The finds were recovered from the terrace deposits at a depth from 20 to 60 cm. The section is as follows:

	Depth (m)
1. Sod and modern soil	0.0–0.18
2. Gray sandy loam with interbeds of sand and gravel, including artifacts	0.18–0.60
3. Compact reddish sandy loam with gravel and pebbles	0.60–0.70
4. Gray pebble and coarse gravel	0.7–1.20

Layers 2 and 3 (to a depth of 0.7 m) belong to the floodplain facies, and layer 4 belongs to the channel facies of fluvial sediments.

There is a hearth rounded in plain view, about 1 m in diameter, lined by pebbles, with a thick layer of charcoal. The finds occurred in abundance around the hearth, within 2 m from it; their concentration is lessened noticeably with distance. The objects from the hearth have traces of soot on the surface together with imprints of burnt grass. The charcoal from the hearth yielded radiocarbon dates of the following: 7510 ± 205 yr BP/8340 ± 220 cal. yr BP/ – 8135 ± 220 yr BP/9050 ± 280 cal. yr BP/ (see Table 3.1.1). The soot on a surface of a knife-like blade from the hearth was AMS-dated at 7620 ± 75 yr BP/8430 ± 70 cal. yr BP/. There was found a large burnt fragment of birch bark and a knife-like blade with soot bearing an imprint of a plant, presumably attributed to *Poaceae* or *Cyperaceae* (identified by Dr. L.B. Golovneva, Botanical Institute, St. Petersburg). There were numerous findings around the hearth, including prismatic and conical microcores; core platform removal spalls; microblades retouched along one or two margins; end scrapers, angle burins and gravers made on the microblades; and an oval-shaped bifacial cutting tool (Fig. 3.1.10: *6, 7, 9–11, 14, 16*). The maximum depth of the cultural layer (60 cm) is recorded in the center, near the hearth. Closer to the excavation margins, it occurs at 25–20 cm from the surface. That gives ground for interpreting the dwelling as a kind of tent ("chum") in a natural depression.

Urtychuk IV is at the head of the Urtychuk stream on the Kolyma upland (60°38′ N, 151°28′ E, 1015 m a.s.l.) (Slobodin, 1999b). In the excavation (7.5 m^2) the cultural layer occurs immediately under the turf, 1–2.5 cm thick. The artifacts occurred in sandy loam at depths of 3–22 cm. Charcoal from the cultural layer was dated at 8285 ± 95 yr BP, GX-17,063/9270 ± 120 cal. yr BP/. The excavations produced 11 prismatic cores and their preforms, more than 1000 microblades, several platform rejuvenation flakes, 30 crested blades, an end scraper on a blade, and flakes (Fig. 3.1.10: *5, 8, 12, 13, 15*). The cores present all degrees of their processing, from preforms to completely exhausted pieces. It may be suggested that the site was used as a workshop and a hunting lookout. The inhabitants of the site procured flint from the adjacent slopes, made cores, and obtained microblades used in making inset tools.

Khurendzha (Azamat) is on the Kolyma highland, and on the Ola and Khurendzha river interfluve (60°41′ N, 151°15′ E, 1060 m a.s.l.). It sits on a hilltop; the northern and western slopes of the latter are gentle and smooth, while the southern and eastern ones are steep, with local scarps (Slobodin, 2005). The top of the hill is hummocky and broken with cryogenic polygons. Stony areas devoid of vegetation alternate with small parcels covered with moss, lichens, and herbs.

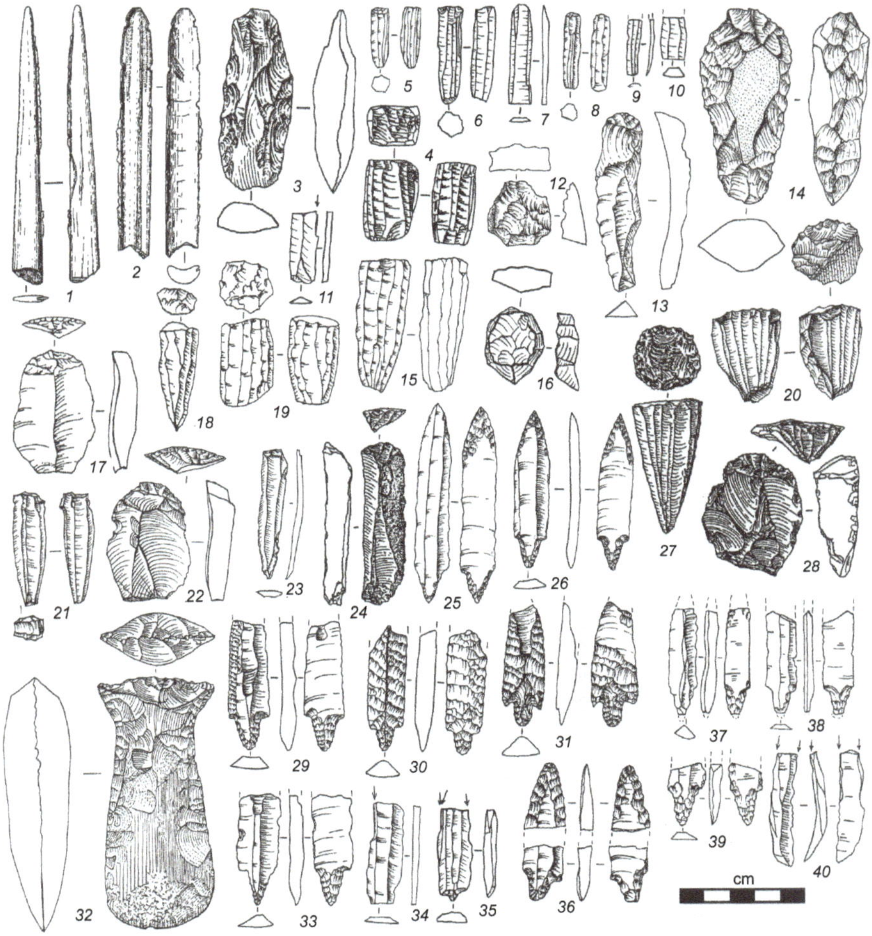

Figure 3.1.10 Artifacts from the Mesolithic (Early Holocene) sites in the Indigirka-Kolyma region (Pitul'ko, 1998; Mochanov, 1977; Kashin, 1983). *1–4* – Zhokhov; *5, 8, 12, 13, 15* – Urtychuk IV; *6, 7, 9–11, 14, 16* – Buyunda III; *17–19* – Panteleyikha I and II; *20–23* – Omsukchan III; *24, 27, 28* – BMA -3; *25, 32* – Kongo -78; *26, 34* – ING-78; *29, 30, 33, 35* – Uy; *31, 36* – Pridorozhnaya; *37–40* – Yubileyny.

The sequence exposed in the excavation (2.5 m^2) is as follows:

	Thickness (cm)
1. Underdeveloped sod layer	1–1.5
2. Gray sandy loam with debris	10
3. Sandy loam, with charcoal spots and artifacts (down to a depth of 20–27 cm)	10–15
4. Coarse stone fragments lacking matrix	

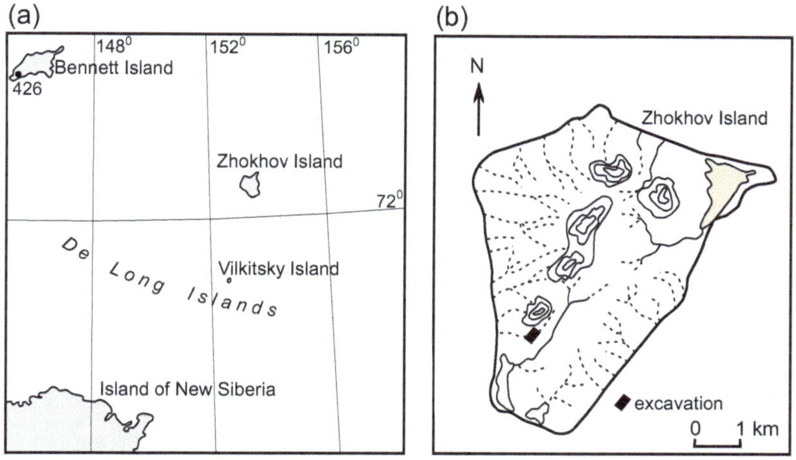

Figure 3.1.11 Position of the Zhokhov Island (Pitul'ko, 1998). (a) – The Zhokhov Island location in the East Siberian Sea; (b) – excavated area at the island.

The sample from the depth of 15–20 cm (layer 3) was dated at 8780 ± 70 yr BP, Beta-156,847/9830 ± 150 cal. yr BP/. The flakes and microblades bear traces of heat impact, so we may confidently associate them with the age of the dated charcoals. There were found about 150 microblades, along with ridge flakes. Almost all the microblades are broken; fragments are mostly medial segments. There are few opening flakes (cortex flakes); the microblades were removed from already prepared cores. Taking the site position into consideration, it is not inconceivable that the site was really a hunting overlook rather than a workshop.

The inhabitants of the Buyunda III, Urtychuk, and Khurendzha (Azamat) hunted reindeer migrating across the mountain passes or grazing on the mountain slopes. They made their tools and weapons from the raw material mined nearby.

Zhokhov Island was discovered by researchers working at the polar station on the Zhokhov Island (De Long Archipelago, in the group of New Siberian Islands) (Fig. 3.1.11) (Pitul'ko, 1998). It is situated in the southwestern part of the island, at the base of the hill slope 115 m high, at a distance of 1 km from the coastline (76°06′ N, 152°42′ E, 13–20 m a.s.l.). The site was a long-term hunter camp. The cultural layer area is estimated at about 8000 m^2. Within its limits, there have been identified 13 concentrations (dwellings?) marked by a high concentration of archaeological objects (animal bones, fragments of driftwood, artifacts).

The cultural layer, 8–40 cm thick, is distinguishable on the side walls of the excavations, in the perennially frozen, light-brown silt, overlain with light loam with interbeds of peat. The cultural layer has been heavily disturbed by cryogenic processes that resulted in the artifact redeposition, both laterally and vertically, over the entire upper layer of the deposits, from the surface to the underlying alluvial deposits. It hampered the reconstruction of the dwellings. The identified constructions of driftwood look not unlike the Eskimo lodgings with a roof and wood floor (Pitul'ko, 1998).

The cultural layer is rich in finds. Among the lithic pieces of silicified rocks, chalcedony, and obsidian, there are prismatic cores, retouched microblades, and adzes; the latter are oval in plain view and asymmetric lens-like in cross-section, with a grounded edge (Fig. 3.1.10: *3*, *4*). Among the bone tools, there are one- and two-slotted points made of antler, walrus tusk, and fossil mammoth tusk with microblade inserts (Fig. 3.1.10: *1*, *2*), hoes, scrapers, harpoon, awl, adze handles, etc. Numerous wooden articles are present (ladles, dishes, arrow shafts, sledge skid).

The faunal remains collected on the coastal plain near the site include the following species:

Fauna	Number of animals
Wolf (*Canis lupus*)	1
Polar fox (*Alopex lagopus*)	1
Polar bear (*Ursus maritimus*)	21
Reindeer (*Rangifer tarandus*)	20
Seals (*Phocidae* sp.)	1
Walrus (*Odobenus rosmarus*)	2
Dog (*Canis familiaris*)	2
Birds (*Aves*)	
Swan (*Cygnus bewickii*)	2
Bean goose (*Anser fabalis*)	6
White-fronted goose (*Anser albifrons*)	2
Gull (Laridae)	1

An analysis of bone remains shows the site inhabitants to hunt equally reindeer and polar bears. They could probably kill (or scavenge) beached marine mammals (walruses, seals). The presence of dog remains (*Canis familiaris)* suggests they kept domestic animals, and the sled runner indicates a possible usage of sleds for transportation, probably harnessing dogs.

There is a large series of ^{14}C dates obtained for the site (on charcoal, bone, and wood) in the range of 8740±190 yr BP/9830±230 cal. yr BP/ – 7810±180 yr BP/8690±220 cal. yr BP/ (Table 3.1.1). Taking into consideration possible errors due to the utilization of old bones or driftwood, the age was determined within the range of 8000–7800 yr BP (Pitul'ko, 1998). The technical and typological characteristics of the stone inventory permit attributing the site to the arctic variant of the Early Holocene Sumnagin culture (Mochanov, 1977; Pitul'ko, 2003).

There are other sites in the Indigirka-Kolyma region tentatively attributed to the same Sumnagin culture (tradition) (Fig. 3.1.1); those are Panteleikha I–VIII in the lower reaches of the Kolyma (Fig. 3.1.10: *17–19*) (Mochanov, 1977); BMA-3 on the northern coast of the Sea of Okhotsk (Fig. 3.1.10: *24*, *27*, *28*) (Vorobey, 2007); and Omsukchan III on the Kolyma Highland (Fig. 3.1.10: *2–23*) (Slobodin, 2011).

Uy discovered in 1987 is situated in the central part of the Chersky Range on the Lake Uy coast (63°50′ N, 147°53′ E, 1068 m a.s.l.) (Slobodin, 1999b). It sits on a gently sloping lacustrine terrace 3–4 m above the lake (Fig. 3.1.12.). In common with other lakes of the Okhandzha massif on the left side of the Momontay River the Lake

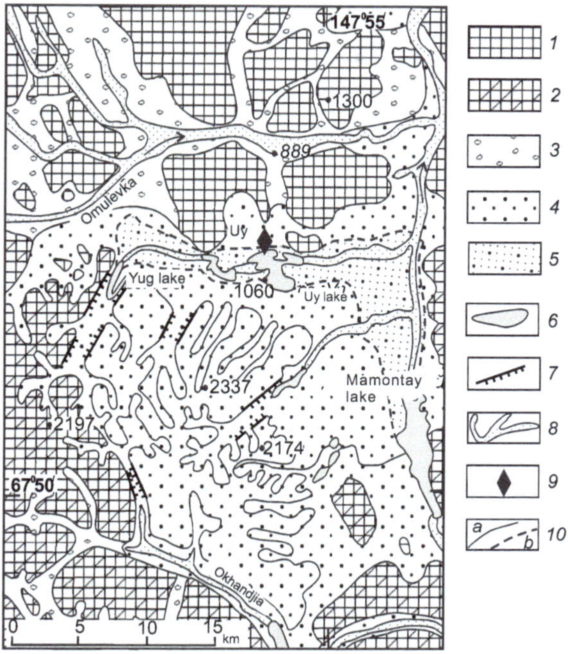

Figure 3.1.12 Geomorphological scheme of the Uy site setting. *1* – mountain slopes: *2* – slopes with traces of glacial erosion; *3* – till of the Zyryanka glaciers; *4* – Sartan till; *5* – outwash plains; *6* – ice-dammed lakes; *7* – glacial troughs; *8* – river channels and low terraces; *9* – archaeological site; *10* – geomorphological boundaries: *a* – proved, *b* – supposed.

Uy was originally ice-dammed and developed due to large blocks of the dead ice melting out. Morainic hills descend to the lake, forming low steps.

The finds recovered from the excavation (60 m^2) occurred under a thin soil and sod layer to a depth of 20 cm. Charcoals from the find concentrations yielded radiocarbon dates in the range 8810±235 yr BP/9910±280 cal. yr BP/– 8310±240 yr BP/9260±310 cal. yr BP/ (Table 3.1.1). The presence of the numerous ready-to-use tools suggests a long-standing residential complex with workshops.

The primary knapping is represented by conical and prismatic cores for macro- and middle size blades, as well as microblades; the blades of various size served as blanks for various tools, such as end scrapers, angle burins, borers, points, knives on blades, axe without grinding, and retouched inserts (Fig. 3.1.10: *34*, *35*). A specific feature of the latter was an oblique, retouched margin creating gradual transition from the bone point surface to the stone (flint) cutting edge. Large blades were used for making stemmed arrowheads or points of darts (Fig. 3.1.10: *29*, *30*, *33*). The raw material was of local provenance, though occasionally rare rocks were used, such as obsidian, its nearest outcrops being at a distance of hundreds of kilometers from the site (Slobodin, 2007). The site inhabitants hunted reindeer and moose and seemed to practice fishing.

Assemblages similar to the Uy industry were also identified at several other sites: Pridorozhnaya, Kongo -78, and ING-78 (Slobodin, 1999b), Yubileyny (Kashin, 1983)

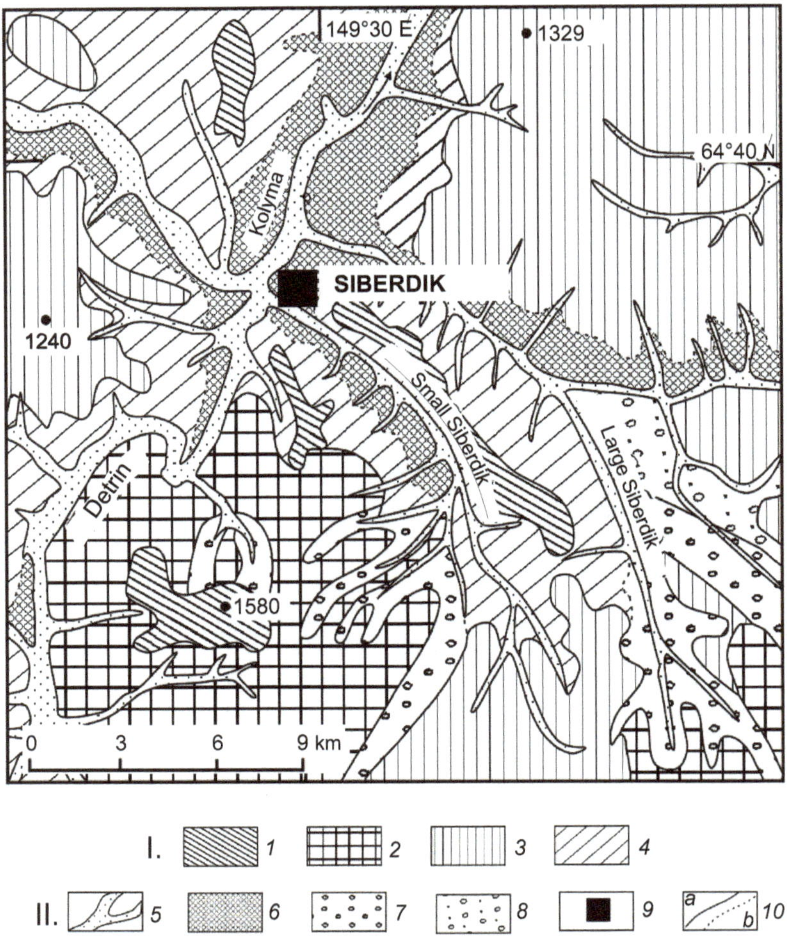

Figure 3.1.13 Geomorphological scheme of the Siberdik site setting. *1–4 – structural-denudational landforms*: *1* – flat or dome-like interfluves; *2* – steep slopes with rockfalls and screes; *3* – medium-steep scree slopes; *4* – gentle slopes dominated by deluvial and solifluction erosion and deposition; *5, 6 – fluvial landforms*: *5* – river channel and floodplain; *6* – fluvial terraces and terrace-uvals (terraces buried under the mantle of slope deposits); *7, 8 – glacial landforms*: moraines of the last (Sartan) glaciation; *8* – outwash trains; *9, 10* – other objects: *9* – archaeological site; *10* – geomorphological boundaries: *a* – proved, *b* – supposed.

(Fig. 3.1.10: *25, 32, 26, 34, 31, 36, 37–40*). Those materials contain specific stemmed points on blades (the culture-determinative feature) and gave grounds to recognize the Uolba culture widely spread over Yakutia, in the Kolyma drainage basin, and on Chukchi Peninsula in the Early Holocene (7–9 ka BP) (Slobodin, 1999b).

Siberdik (the site flooded at present) was discovered by N.N. Dikov in 1971 at the Detrin River mouth (tributary of the Kolyma River) (61°26′ N, 149° 24′ E) (Dikov, 1977, 1979). The site is positioned on the terrace 14 m high (Fig. 3.1.13). The terrace base is composed of clay slates overlain with covering deposits 3 m thick. The upper member of the latter contains cultural materials forming several cultural layers. The lower layer (III) occurs at a depth of 0.8–1.0m in peaty sandy loam; it is partly destroyed by water and wind erosion.

Layer III excavated over an area of 800 m^2 reveals lithic workshops and habitation units with hearths, tools, and poorly preserved bone remains (teeth, antlers, shoulder blades) of herbivorous animals (Dikov, 1977). In the center a "sacral complex" was identified, with a hearth surrounded with stone tools and bone fragments. One of the bone concentrations marked with spots of red ocher led the archaeologists (Dikov, 1977) to the conclusion (based on the bone arrangement) that the bones were the remains of human burial. No special study of the bones has been carried out, except for teeth, which were identified by N.K. Verehchagin as belonging to a horse (Dikov, 1977). The presence of horse teeth in layer III of the site at the Kolyma is in contradiction with the data on the species extinction at the end of the Pleistocene (Mochanov, 1977). It is not inconceivable, however, that some fossil teeth had been brought to the site.

Radiocarbon dates obtained on the charcoal from the layer range from 7865±310yr BP/8790±360cal.yr BP/to 9700±500yr BP/11,220±720cal.yr BP/ to a date of 13,225±230yr BP/15,870±340cal.yr BP/ (Table 3.1.1). The latter date does not agree with palynological data, nor with other ^{14}C dates, so it can be ignored (Lozhkin, 2008).

The tool assemblage from layer III includes choppers and adzes, typical and lateral end scrapers, oval bifacial knives, points, a fragment of polyhedral burin, and retouched blade-like flakes. The microblade components are represented by knife-like bladelets, some of them retouched, and one wedge-shaped micro-core; the latter, in the opinion of Dikov (1977), gives a "Paleolithic appearance" to the complex. The bifacial points are leaf-shaped with rounded base and lens-like in cross-section. The tool assemblage is dominated by choppers, adzes, and scrapers. Some of the working places yielded only choppers, anvils, and flakes. Judging from the spatial organization of the site, it was a long-term hunter camp with aboveground dwellings of the tent ("chum") type. Judging from the bone remains, the hunt objects were horse (?), reindeer, and moose. The site location in the Kolyma valley suggests fishing. Based on tool assemblage recovered from layer III of Siberdik and similar finds from Kongo, the latter site being dated to 9470±530yr BP, Kril-314/10,920±760cal.yr BP/and 8655±220yr BP, MAG-196/9740±280cal.yr BP/, Dikov (1979) identified the Early Holocene Siberdik culture of the Upper Kolyma drainage basin.

3.1.2 Kamchatka

The Kamchatka Peninsula is one of the key regions for getting an insight into the problem of the initial colonization of the northern part of the Russian Far East by the early man and his migration to Alaska by the Beringian Land Bridge. The multilayered archaeological complex of the Ushkovskoye Lake (Ushki I–V) on Kamchatka (Fig. 3.1.14) was discovered and investigated by Dikov (1977, 1979, 1993b). In the distinctly stratified sequence, there were found seven cultural layers

Figure 3.1.14 The Late Paleolithic sites of the Kamchatka Peninsula.

covering the Final Pleistocene (layers VII and VI) and the Holocene (layers V to I). The excavations of Ushki V had been continued later (Goebel et al., 2003; Kiryak, 2005; Ponkratova, 2007).

The excavations of Ushki I and V (cultural layers VI and VII) recovered remains of long-term settlements of the Paleolithic hunters with dwellings, various activity areas, and burials. The excavations produced extensive collections of artifacts and samples for radiocarbon dating. Recently additional investigations of layers VII and VI (Ushki I–V) were carried out with the aim of getting more precise dating (Goebel et al., 2003, 2010; Ponkratova, 2007; Slobodin, 2010b).

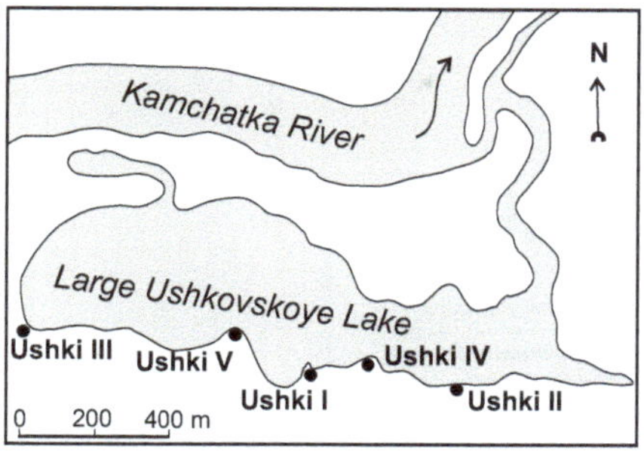

Figure 3.1.15 A cluster of the sites on the southern coast of the Ushkovskoye Lake (Dikov, 1977).

The archaeological discoveries in Kamchatka opened a new line in the Paleolithic investigations in the northeast of Asia and for a long time predetermined an approach to the problems of the initial colonization of that territory (Slobodin, 2000, 2010b; Kiryak, 2005; Goebel et al., 2003).

For a long time the sites on the Lake Ushkovskoye were the only ^{14}C-dated Paleolithic sites on Kamchatka with cultural layers distinctly stratified. When the stratified Anavgay II site (the Bystraya River valley, left tributary of the Kamchatka River, 65 km from the Lake Ushkovskoye) was discovered in 2008 (see Fig. 3.1.14), it became clear that the range of the Late Ushki culture (layer VI) had been larger than previously supposed (Ptashinskiy, 2010).

Both *Ushki I and V sites* are situated in the middle reaches of the Kamchatka River, on two caps, Kamenny and Bolshoy, 150 m apart, on the southern coast of the Lake Ushkovskoye (56°09′ N, 159°56′ E) (Fig. 3.1.15). The cultural layers occur in the upper part of the second terrace. The terrace surface is 4–6 m above the lake. The geological context of the sites is formed with sandy loams and silts interlayered with volcanic ash (Dikov, 1977) (Fig. 3.1.16). The sequence includes seven cultural layers. The layer attributed to the "retarded" Neolithic occurs at a depth of 8–50 cm from the surface. The lower lying layers II and III represent the Late (to a depth of 0.75 m) and Middle (to 1.17 m) Neolithic. Layer IV (at the depth of 1.5 m) is attributed to the Early Neolithic or Mesolithic. Cultural layer V bearing the Final Paleolithic assemblage is dated to the Early Holocene. Locally, in places of the deposit thickness increasing to 2.5–3 m, Paleolithic layers VI and VII are identified.

The Upper Paleolithic layer VI, attributed to the Late Upper Paleolithic Ushki culture by Dikov (1979, 1993a), is traceable in the lower horizons of the sequence at depths of 1.8–2.1 m to 2.0–2.05 m. Its age was determined by conventional ^{14}C dates at 10,860 ± 400 yr BP/12,670 ± 490 cal. yr BP/, 10,760 ± 110 yr BP/12,680 ± 100 cal. yr BP/, and 10,360 ± 350 yr BP/12,050 ± 480 cal. yr BP/. There is an AMS ^{14}C date of 10,240 ± 75 yr BP/11,970 ± 170 cal. yr BP/ obtained at the Ushki V on the sample from layer VI (Goebel et al., 2003) (Table 3.1.2). The cultural layer VI in Ushki I was investigated over an area more than 5000 m^2. It included dwelling constructions, burials, and numerous concentrations of artifacts. There were 40 dwellings found altogether, among them 13 pit-houses with hearths and corridor-like entrance passages,

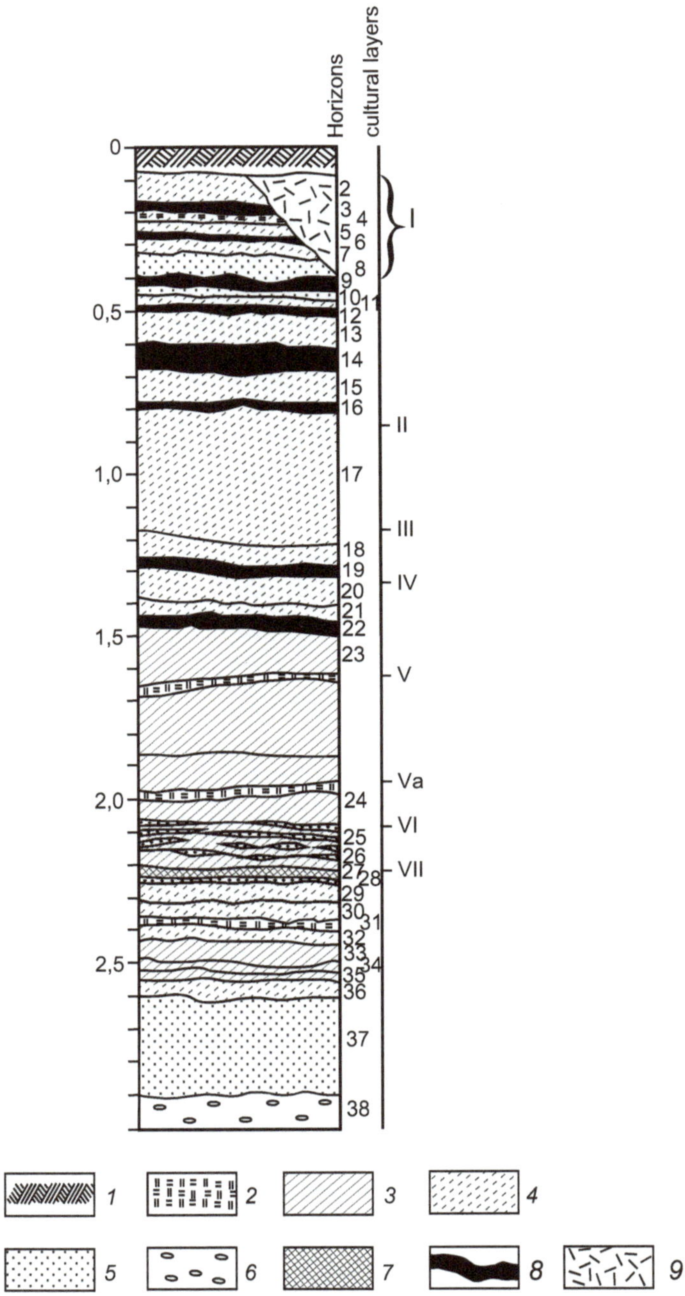

Figure 3.1.16 Generalized stratigraphy of the Ushki I site (Dikov, 1977). *1* – modern soil; *2* – charcoal beds; *3* – loam; *4* – sandy loam; *5* – sand; *6* – volcanic bomb fragments; *7* – ocher; *8* – volcanic ash; *9* – disturbed ground.

Table 3.1.2 Radiocarbon dates of layers VII and VI of the sites Ushki I and V and Anavgay II (lower layer)

^{14}C age, Yr BP (±1 σ)	^{14}C age, cal. yr BP	Complex, feature	Method	Lab. Index	References
Ushki I, layer VII					
13,600±250	16,270–17,032	Burial	Standard	GIN – 167/168	Cherdyntsev et al. (1969)
11,360±330	12,885–13,577	Hearth, dwell. 10	Standard	Le – 3695	Lisitsyn and Svezhentsev (1997)
11,120±500	12,432–13,600	Hearth, dwell. 11	Standard	Le – 3697	Lisitsyn and Svezhentsev (1997)
13,800±500	16,069–17,613	Unspecified	Standard	MAG - 522	Dikov (1979, 1993a,b)
14,300±800	16,588–18,558	Burial	Standard	MAG - 550	Dikov (1979, 1993a,b)
11,650±100	13,382–13,626	Hearth, dwell. 7	Standard	MAG – 594	Slobodin (2010a,b)
11,185±25	12,974–13,164	Hearth, dwell. 10	AMS	UCIAMS-53553	Slobodin (2010a,b)
11,210±25	13,084–13,183	Hearth, dwell. 11	AMS	UCIAMS-53554	Slobodin (2010a,b)
11,205±25	13,079–13,185	Hearth, dwell. 11	AMS	UCIAMS-53555	Slobodin (2010a,b)
11,220±25	13,091–13,184	Hearth, dwell. 8	AMS	UCIAMS-53556	Slobodin (2010a,b)
Ushki I, layer VI					
10,860±400	12,395–13,150	Unspecified	Standard	MAG-400	Dikov (1979, 1993a,b)
10,760±110	12,760–12,860	Unspecified	Standard	MAG-219	Dikov (1977)
10,790±400	12,290–13,100	Square. 20 e	Standard	MAG-218	Dikov (1979)
10,360±350	11,400–12,800	Depth. 1.7 m	Standard	Mo-345	Dikov (1977)
11,130±100	13,175–13,000	Test pit	AMS	AA-45717	Goebel et al. (2003)
Ushki V, layer VII					
11,005±115	12,743–13,067	Hearth, sq. B 1	AMS	AA-41388	Goebel et al. (2003)
11,050±75	12,843–13,086	Hearth, sq. B 1	AMS	AA-41389	Goebel et al. (2003)
11,320±50	13,156–13,271	Square A 15	AMS	CAMS-74639	Ponkratova (2007)
11,060±25	12,896–13,077	Square Γ (–3)	AMS	UCIAMS-32199	Ponkratova (2007)

Continued

Table 3.1.2 Continued

^{14}C age, Yr BP (±1 σ)	^{14}C age, cal. yr BP	Complex, feature	Method	Lab. Index	References
Ushki V, layer VI					
10,240±75	12,315–11,705	Excavation	AMS	AA-41386	Goebel et al. (2003)
Anavgay II, lower layer					
10,870±40	12,804–12,648	Excavation, hearth	AMS	IAAA- 80842	Ptashinskiy (2010)
10,600±50	12,612–12,530	Excavation, hearth	AMS	IAAA-92682	Ptashinskiy (2010)
11,060±60	13,090–12,872	Excavation, hearth	AMS	IAAA-907-72	Ptashinskiy (2010)

The dated material: coal.

14 slightly deepened dwellings with hearths, and 13 surface domestic units with fireplaces. The best distinguishable are rounded dwellings 0.5 m deep up to 48 m^2 in area, with a side passage-like entrance and a fireplace in the center, lined with 7–10 large stones (Fig. 3.1.17). There are three habitation horizons identified in the cultural layer; the dwellings occasionally overlap each other, as may be seen in some pit-houses where the stone linings of hearths lie above one another. Such superposition may be evidence of the site being repeatedly occupied in the course of long time periods.

In the Late Ushki culture the primary knapping technology is represented by wedge-shaped cores for microblades and large-size cores for flakes (see Fig. 3.1.17). The wedge-shaped cores are made on bifacial preforms. The tool assemblage includes leaf-shaped, crescentic, and oval bifacial knives; knives on slate plates and blade-like flakes with edge retouch; narrow, leaf-shaped arrowheads and dart points; end scrapers and circular scrapers; side-scrapers, burins, borers, grinding and sizing plaques; and crude choppers and scraper-like tools. They were made of chalcedony, obsidian, basalt, rocks of jasper type, and flint of various colors. Knife-like blades were used as segments-inserts of bone tools. Among ornaments, there are perforated flat or rounded polished pendants made of agalmatolite.

The dwelling interior may be inferred from the artifacts recovered from the charcoal floor around hearths, remains of burnt grass (seemingly used as a kind of matting), and caches. In one of the dwellings, small holes were found on the floor around the fireplace. They were left by sticks pushed into the soil and tilted toward the fire or outward. They could be used for frying fish and/or drying footwear. An anvil found in another dwelling was presumably used for stone tool processing. It consists of two small stones arranged vertically and supporting each other. Nearby, in a small pit near the fireplace, there was found a "cache" holding two grindstones, a slate knife, and a burin. One more "cache" with 13 crescentic bifaces —preforms of wedge-shaped cores—was discovered in a niche near a wall in the passage.

Near the dwelling, there was a working area where arrowheads were made and stuck into the ground forming a row. Another dwelling revealed a burial of a dog (Fig. 3.1.17) and a bison shoulder blade. According to Vereshchagin (1979), it was a dog of husky kind (wolf-dog); it was buried on its side, with paws close to muzzle. There was a spot of red ocher nearby, with an obsidian knife, scraper, and a grindstone within it. The dog burial suggests a very early (for Northeastern Asia) domestication of the animal at the end of the Pleistocene.

There are two child burials found with well-preserved teeth and decayed bones. One of them contains remains of two children. It is a small pit 0.7 m wide and 0.2 m deep, its bottom densely inlaid with numerous lemming incisors, with ocher spots and grave goods. Besides a baby's rootless teeth, there have been teeth with ground and perforated roots, probably for wearing as an amulet, found on one of the skeletons. The second burial contains as many as 150 teeth in five clusters of bones. The deceased were laid together with grave goods into a pit up to 1 m deep and 1 m wide, covered with a scapula of a large animal and ocher. The burials contained abundant inventory: arrowheads, bifacial knives, end scrapers, cores, blades, burins, grindstone plates, and pendants; among the latter there were some made of amber. Some

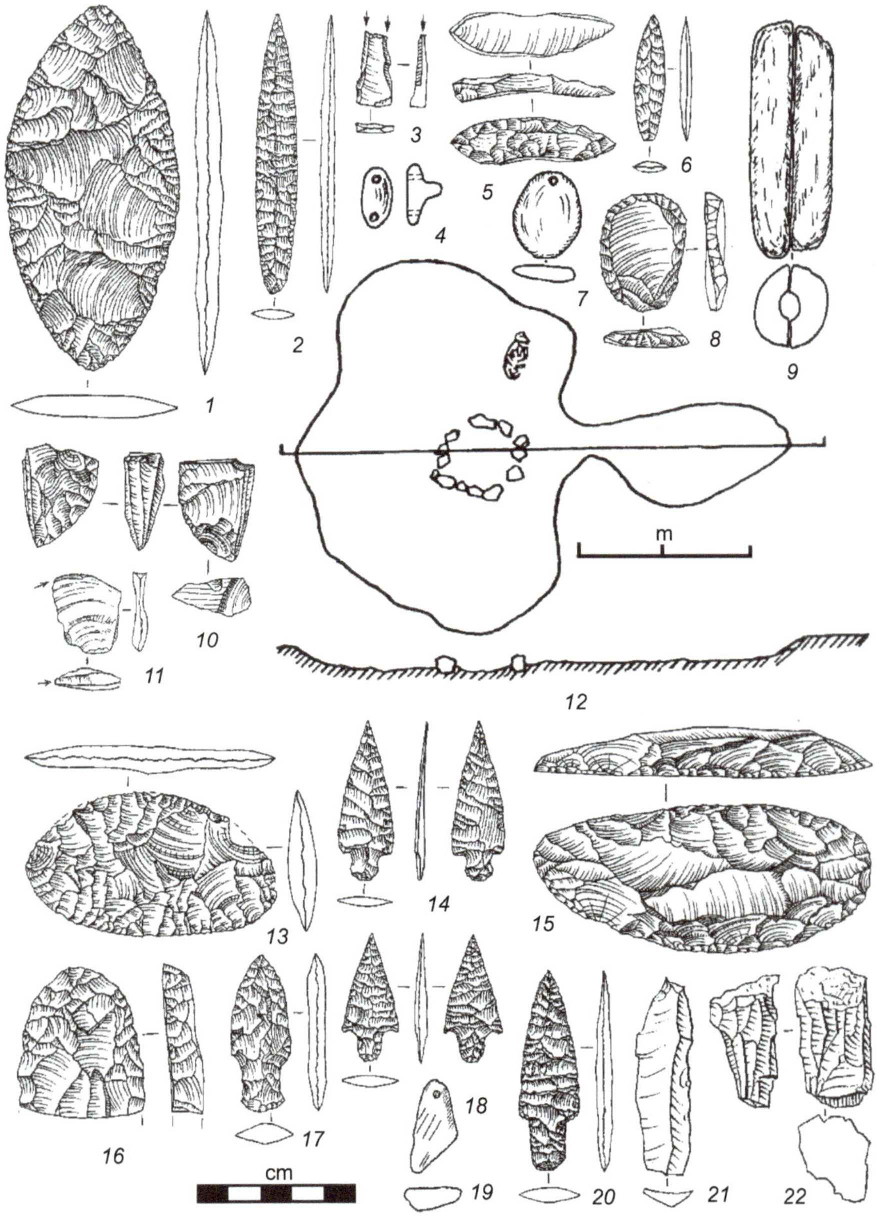

Figure 3.1.17 Stone tools and pendants from Ushki I. *1–11* – from layer VI; *13–22* – from layer VII; *12* – the sketch map of the dwelling with the dog burial (top) and its position on the profile (bottom).

pieces of walrus tusk have been also found. The items recovered from both burials are distinct for a high quality of work and a choice of rocks (brightly colored flint, agate, jasper, amber). The artifacts form clusters and probably were initially put in leather bags.

At Ushki V, cultural layer VI has been excavated over an area of more than 60 m^2. The excavated dwelling belongs to semisubterranean type with a hearth and elongated entrance passage; the passage yielded numerous wedge-shaped cores, microblades, scrapers, bifacial knives, and leaf-shaped points (Kiryak, 2005; Ponkratova, 2007).

The remains of decomposed animal bones and teeth in cultural layer VI of the sites suggest the inhabitants hunted bison, bighorn sheep, and horse. Burnt salmon bones in the layer and in particular within the hearths indicate fishery. Finally, judging from numerous finds of gastroliths from birds stomachs, birds were also hunted.

Originally, the materials from layer VI of the Ushki sites were considered to be evidence of the formation of the Proto-Eskimo–Aleutian culture in Western Beringia (Dikov, 1979); later, though, they were attributed to the ancestors of Athabascan Indians (Na-Dene) (Dikov, 1993a). The materials of the Late Ushki Upper Paleolithic culture belong to a wide range of Late Paleolithic microblade cultures (including Dyuktay, Selemdzha, Beringian) widely distributed in Beringia between 25 and 10 ka BP (Slobodin, 2000).

The assemblages recovered from a lower layer at Ushki I (210–220 cm) were attributed by Dikov (1977, 1993a) to the Early Ushki culture. The first ^{14}C scintillation dates were obtained on charcoal from layer VII in the burial: 14,300 ± 800 yr BP/17,360 ± 1050 cal. yr BP/ and 13,600 ± 250 yr BP/16,440 ± 370 cal. yr BP/ (Cherdyntsev et al., 1969; Dikov, 1993a) (see Table 3.1.2). Later, N.N. Dikov obtained ^{14}C scintillation dates on the charcoal from hearths in dwellings nos. 10 and 11: 11,120–11,360 yr BP (Lisitsyn and Svezhentsev, 1997) and 11,650 ± 100 yr BP/13,480 ± 110 cal. yr BP/. On repeated dating of charcoal from the hearth in dwellings nos. 10, 11 and 8, there were obtained ^{14}C AMS dates in the range 11,185 ± 25 yr BP/13,060 ± 20 cal. yr BP/ to 11,220 ± 25 yr BP/13,090 ± 20 cal. yr BP/ (Slobodin, 2010b; Goebel et al., 2010). Similar AMS dates were obtained on the samples from layer VII, Ushki V: 11,000–11,300 yr BP (Goebel et al., 2003), 11,320 ± 30 yr BP/13,160 ± 40 cal. yr BP/, and 11,060 ± 25 yr BP/12,920 ± 60 cal. yr BP/(Ponkratova, 2007).

The new series of dates make cultural layer VII much younger (by 2–3 ka) and probably mark the upper limit of the Early Ushki culture. The question of its lower limit (that is, before it became distributed over the Kamchatka Peninsula) still remains to be solved: it may be a few millennia older than the new dates have shown and close to the age estimated by N.N. Dikov.

The excavations of layer VII at Ushki I site carried out over the area more than 2000 m^2 discovered a long-term settlement with dwellings, a burial, and working areas with a high density of artifacts (Fig. 3.1.18). The dwellings marked with thick accumulations of charcoal and burnt bones with many finds were of aboveground type, rounded in outline (like a tent with a frame of poles covered with skins), and several hearths without slab linings in the center. Eleven dwellings have been found measuring from 16 m^2 to 100 m^2. Two of them were described by N.N. Dikov as double,

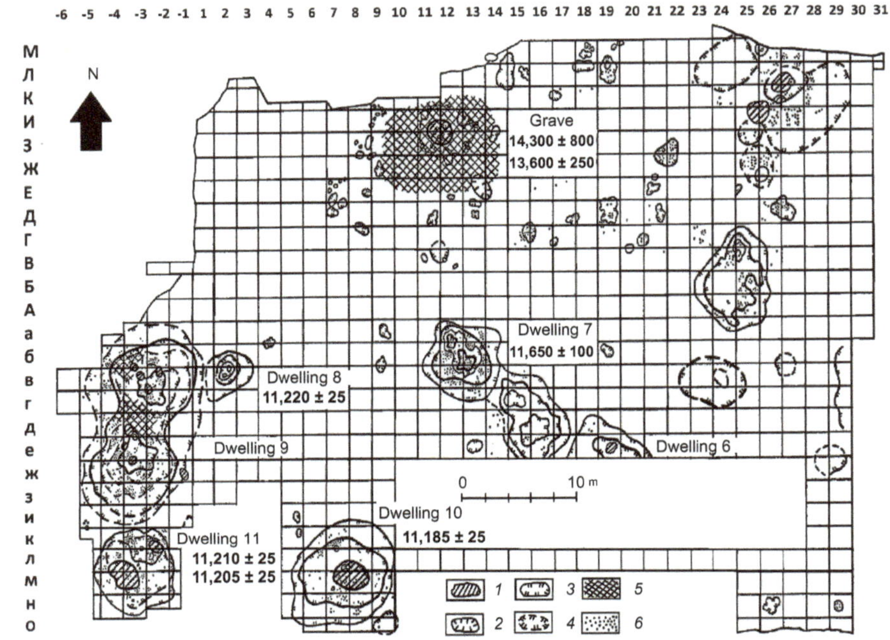

Figure 3.1.18 Ushki I: plan of the cultural layer VII. *1* – spots with abundant charcoal and burnt bones; *2* – charcoal spots of hearths and fireplaces; *3* – weak charcoal spots; *4* – outlines of dwellings; *5* – red ocher spots; *6* – lithics. Dwellings are designated by numbers (for indexes of ^{14}C dates see Table 3.1.2).
Modified from Dikov, N.N., 1993a. Asia at the interface of America in Ancient times. Nauka, St.-Petersburg, 304 pp. (in Russian).

though it could be a result of superposition of two charcoal concentrations belonging to two ordinary dwellings.

The dwellings were arranged around a burial discovered on the coast of Ushkovskoye Lake. The rounded in plain view grave 1.8 m in diameter and 0.7 m deep is covered with a layer of ocher with charcoal (presumably, remnants of a funeral fire). Judging from the position of bone fragments the dead was laid in a crooked position, his head pointed toward the lake. There were about 900 small beads of pyrophyllite found in the grave, together with pendants: supposedly, those were ornaments on the dead's clothes; besides, 26 chalcedony gravers and two stemmed points were recovered there.

The tools of the Early Ushki culture were made on flakes (Fig. 3.1.17). The primary knapping is represented also by subprismatic blade cores and blade-like flakes, though they were not the main types of blanks. The tool inventory included bifacial stemmed points (more than 50 pieces) up to 5–6 cm long, leaf-shaped and oval bifaces, oval typical and lateral end scrapers on flakes, and burin-like points. They were made of siliceous rock, chalcedony, and obsidian. The obsidian used for toolmaking originates from several volcanoes at a distance of 200–300 km from the site (Kuzmin et al., 2008).

An aboveground dwelling with a hearth excavated at Ushki I, layer VII, yielded teeth, vertebrae, and scales of silver salmon (*Oncorhynchus kisutch*), surrounded with stemmed arrowheads, scrapers, and pendants of pyrophyllite (Kiryak, 2005; Ponkratova, 2007). The settlement inhabitants practiced large animal hunting, as may be concluded from the presence of burnt and decomposed large bones and antlers. Gastrolith finds indicate the water fowl hunt, and the burnt salmon bones in hearths suggest fishing.

The unique characteristics of the culture recovered from layer VII makes it difficult to understand its origin and to place it in the range of the Late Pleistocene cultures of Siberia. No analog has been found yet among the archaeological cultures of Siberia or Alaska. It has some technical characteristics, however, in common with Paleoindian complexes of Alaska and of North America as a whole, in particular, the absence of microblade technology. Some artifacts similar to stemmed points from Ushki are present in the North American Paleoindian "Western Stemmed Point Tradition" (Stanford et al., 2005), such as at Lind Coulee site on the Columbia Plateau dated at 12,000–11,000 yr BP (Davis and Sisson, 1998).

The *Anavgay II site* (lower layer) is an analog of the Late Ushki Upper Paleolithic culture on Kamchatka. It enlarges considerably the area of the culture outside the Lake Ushkovskoye basin. The site is located on the right bank of the Anavgay River, approximately 650 m from its entering the Bystraya River, about 360 m a.s.l. The cultural layer of the site occurs within the upper portion of deposits exposed in the scarp of the glaciofluvial terrace 11 m high. It was excavated over an area about 40 m^2 (Ptashinskiy, 2010).

The excavation exposed 11 lithological horizons of sandy loams (paleosols) and volcanic ash. The maximum depth of the artifact occurrence does not exceed 66 cm. There were two cultural layers identified in the sequence. The first one (Early Neolithic or Mesolithic) lies under the ash of Hangar volcano, which erupted 6850 BP (Melekestsev et al., 1996). The lower cultural layer, 1–3.5 cm thick, is dark-brown, compact, sandy loam with calcinated bones. It is embedded in the sandy loam at a depth up to 66 cm underlying the ash layer of Shiveluch volcano dated to 8300 yr BP. In a charcoal spot, there was identified a group of stones arranged in an irregular oval not unlike the fireplaces at Ushki I.

There were 1120 artifacts recovered from the lower layer of the site including two wedge-like cores (on a flake and on a bifacial preform), 254 microblades, and about 80 tools. Among tools there are ovate bifaces, a leaf-shaped arrowhead, blades with traces of burin blows, retouched microblades, and several fragments of graphite. Most of the tools and most microblades on the Anavgay site are made of obsidian, its source being in the Payalpan Ridge, 70 km from the sites. That tool assemblage is identical to the finds from the Final Paleolithic layer VI of Ushki I and V.

The charcoal samples from the site hearths yielded the dates: 10,870 ± 40 yr BP, IAAA-80,842/12,740 ± 30 cal. yr BP/, 11,060 ± 60 yr BP, IAAA-907-72/12,920 ± 80 cal. yr BP/, and 10,600 ± 50 yr BP, IAAA-92,682/12,580 ± 60 cal. yr BP/ (Ptashinskiy, 2010). The date 11,060 yr BP is at present the oldest one obtained for the Late Ushki culture, though dates (in a somewhat uncertain geological context) older than 11,000 yr BP have been obtained already for layer VI of Ushki I and V (Goebel et al., 2003, 2010) (see Table 3.1.2).

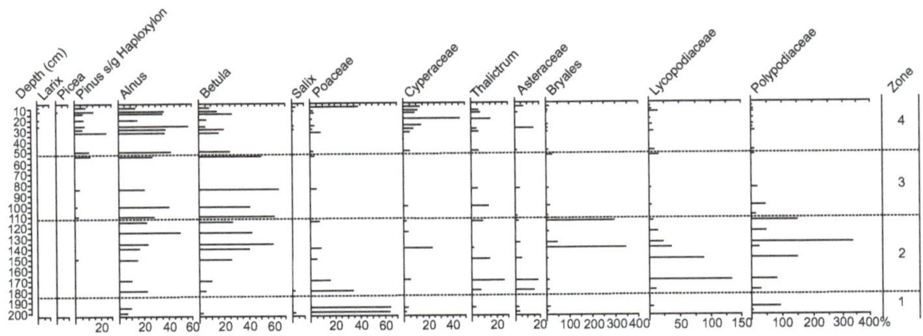

Figure 3.1.19 Pollen diagram from Ushki I.

Noteworthy is the absence of cryoturbations, frost wedges, or man-induced deformations in the sequence of Ushki V. Palynological analysis permits one to reconstruct the climate fluctuations at the end of the Late Pleistocene and the first half of the Holocene. The sediments enclosing Upper Paleolithic layers VII and VII (Dikov, 1977) (see Fig. 3.1.16) contain rare pollen grains of *Alnus*, *Betula*, *Thalictrum*, Rubiaceae, and rare spores of *Bryales* and *Polypodiaceae*; the scarce data cannot be used in vegetation reconstructions. The small quantity of pollen grains might be a result of low pollen productivity, which was characteristic of plants in the Kamchatka River valley at the Late Glacial. Four pollen zones may be recognized in the profile (Fig. 3.1.19).

Zone 1 is very distinct. It is dominated by *Poaceae*, pollen of *Cyperaceae*, *Caryophyllaceae*, *Asteraceae*, *Brassicaceae* are found in noticeable amounts, and pollen of *Alnus*, *Betula*, *Salix* are found occasionally. That is a typical herbaceous zone (not unlike a similar zone in the lacustrine pollen records in Beringia) indicative of dominance of tundra grass communities at the last glacial stage of the Pleistocene. The moderately diversified herbaceous pollen taxa of secondary importance correspond to various habitats: from a discontinuous cover of grass and dwarf shrubs (*Artemisia*, *Caryophyllaceae*, *Brassicaceae*, *Saxifragaceae*, *Papaveraceae*, *Rosaceae*) on dry, stony slopes to wetlands overgrown with sedges, moss, and grass communities with *Ranunculaceae*, *Primulaceae*, *Polemoniaceae*, *Pedicularis*, and dwarf willow. The presence of *Alnus* pollen produced by shrub species of *Alnaster*/*Duschekia* and *Betula* pollen suggests small near-coast groves of alder and dwarf birch.

Pollen zone 2 distinguished in the layers of the Late Paleolithic–Neolithic cultural layers is the zone of birch. In the pollen assemblage composition, it is close to birch zones identified in the sediments of Beringian lakes that developed during the last glacial stage and the transitional period from the Pleistocene to the Holocene.

The uppermost layers in the birch zone characterize the Early Holocene. A sharp increase of the *Betula* and *Alnus* proportion in the spectra suggests a noticeable restructuring of vegetation due to the climate warming. Shrub growths (of *Alnaster* and shrub birch) increased their area. The most noticeable change in landscapes was an expansion of birch forests with a rich herbaceous cover on soil and the species of *Thalictrum* (such as *Thalictrum contortum*, *Thalictrum sparsiflorum*) typical of floodplain forests and the riverine zone. Some *Lycopodiaceae* spores are mostly of forest species. An abundance of *Polypodiaceae* spores suggests a wide distribution of various ferns typical of different habitats, from riverine forests to stony slopes.

A high proportion of *Bryales* spores indicates wetlands on river floodplains. Also noteworthy is the presence of typical aquatic plants of *Typhaceae* (27%) growing near lakes on wet silt soils.

Pollen zone 3 recognizable in the deposits enclosing mid-Neolithic and late Neolithic cultural layers is the birch and alder zone. Distinctive features of the pollen assemblages in the zone are a high proportion of arboreal species (up to 74%), distinct maximums of *Betula* pollen (41%–75%), and an essential role of *Alnus* (19%–53%). Typically the content of *Pinus* subgen Haploxylon (*P. pumila*) pollen is rather low in zone 3. Pollen assemblages of the zone suggest forests of Erman birch (*Betula ermanii*) were widely spread on the slopes of mountain ranges and volcanoes (Klyuchevskaya, Bezymyanny, Tolbachinskiy). Those birch forests could be with a high grass layer, with shrubs and grass, *Alnaster* in undergrowth, and small reeds (*Calamagrostis* sp.) in the lower layer. On the valley floor in the center of the Central Kamchatka Depression the forests of Erman birch were replaced with those of Asian white birch (*Betula platyphylla*). No larch pollen has been found; considering its poor preservation in the fossil state, however, it might be supposed that *Larix kurilensis* formed the lower forest zone in mountains. The birch forests, in turn, were replaced upward with shrub alder. The *P. pumila* proportion increases sharply at the boundary of 3 and 4 pollen zones. Both in the Central Depression of Kamchatka and in glaciated regions in the south of the Chukchi Peninsula the maximum is dated at ~6–7 ka BP.

The warming recorded in pollen assemblages of zone 3 corresponds, therefore, to the hypsothermal interval of the Holocene. It may be inferred from the pollen characteristics of the Ushki site that the appearance of humans in the Kamchatka River valley took place at the last glacial stage of the Late Pleistocene, when the tundra grass and dwarf shrub communities were dominant in landscapes.

3.1.3 Chukchi Peninsula

The Chukchi Peninsula (Chukotka) played an important role in migrations of the early human populations at the time of the initial colonization of the arctic and subarctic regions. It was by way of Koryak Upland, Western Chukotka, and the Chukchi Peninsula that early man came to the North American continent.

The earliest Paleolithic localities of Chukotka attributed to the end of Karginsky (though there is no reliable evidence of their age as yet) are considered to be Orlovka II, Kymyneykey, and Kym'ynanonvyvaam XII, XIII (Dikov, 1996). Sites of the Dyuktay-Ushki Paleolithic tradition with microblades and wedge-shaped cores (most characteristic of them are Tytylvaam IV, Kurupka, Talyain, Inaskaam, Elgygytgyn, Ayon) existed during the Sartan (Late Pleistocene) and at beginning of the Holocene (Dikov, 1993a,b; Slobodin, 2011; Kiryak, 2005). The Early Holocene on Chukotka was marked by the sites of the Sumnagin and Uolba cultures (Kiryak, 2005) (Table 3.1.3).

The existence of the Bering "land bridge" between the Eurasian and North American continents along with continuously frozen Arctic Ocean and partly frozen

Table 3.1.3 **Paleolithic and Mesolithic sites of the Chukchi Peninsula**

Site Period	Coordinates, elevation a.s.l.	^{14}C date, yr BP	Laboratory index	Dated material	Genesis of sediments	Epoch	Cultural attribution	References
Orlovka II Paleolithic	66°57′N 165°05′E 250 m				Slope deposits	End of Karginsky int.	Orlovka complex	Kiryak (2005)
Elgygytgyn II Paleolithic	67°28′N 172°13′E 495 m				Lacustrine deposits	Late Pleistocene–Early Holocene	Late Dyuktay	Slobodin (2001a,b) and Kiryak and Makarov (2006)
Tytylvaam IV Paleolithic	67°19′ N 169°25′ E 530 m	9725 ± 45 9820 ± 40 9790 ± 60	CAMS-80788 CAMS-80789 AA-60203	Coal from hearth	Fluvial deposits	Late Pleistocene end to Holocene	Late Dyuktay	Kiryak (2005)
Chelkun IV Mesolithic	65°22′ N 173°55′ W 100 m	8150 ± 450	MAG-719	Coal from hearth	Fluvial deposits	Early Holocene	Sumnagin	Dikov (1993a)
Naivan Mesolithic	64°28′ N 172°23′ W 15 m	7700 ± 500 9000 ± 500 6990 ± 320 8160 ± 600	LE-5508 LE-5509 GIN-10474 GIN-10471	Coal from hearths	Marine deposits	Early Holocene	Sumnagin	Gusev (2002, 2006)
Verkhnetytylskaya VI, Mesolithic	67°21′ N 169°27′ E 507 m				Lacustrine deposits	Early Holocene	Uolba	Kiryak (2005)

Pacific promoted migration of various Late Pleistocene fauna by the "bridge" in both directions.

The Chukotka topography, in common with the Indigirka-Kolyma region, is dominated by mountains, mostly uplands 600–1000 m a.s.l. and linear ranges 1400–1800 m high. The Arctic Ocean coastline is fringed with Kolyma and Vankarem lowlands, and in the east, there is the vast Lower Anadyr lowland drained by the river of the same name. The modern climate is extremely severe, with very cold winters and cool summers.

Both in the Indigirka-Kolyma basins and in Chukotka the Karginsky was noted for a highly contrasting climate, the contrast having been aggravated by the ocean regression. Coastal regions adjoining the East Siberian and Chukchi seas featured very cold winters, relatively warm summers, and a small amount of rainfall. The Bering Sea coasts differed in lower summer temperatures and a greater moisture supply due to frequent rains and fogs. The climate continentality grew with a distance from coasts. Small glaciers persisted in higher mountains.

The maximum cooling of the 20,000–18,000 BP brought about radical changes in landscapes. Judging from the glacial landform assemblages well preserved in the modern topography, most of the valleys in the Koryak and Chukchi highlands, as well as the low mountains in Chukotka, were covered by ice (Glushkova, 2001). High interior mountains (Chantal and Ekityk ranges) were heavily glaciated, with glaciers up to 80–90 km long (20–30 km on average). In low mountains closer to the Bering Sea (Pekulney Ridge) the glaciers only occasionally exceeded the length of 20 km (Glushkova and Sedov, 1984). Thin ice caps were dominant in the Anadyr, Chukchi, and western Koryak uplands (all the regions distinguished for a wide occurrence of plateaus). In the vicinity of the Providence Bay the glaciers expanded onto the emerged shelf. The ice-free areas featured a periglacial regime with various cryogenic processes being extremely active. An intense frost weathering and sorting of debris resulted in development of vast rock fields (felsenmeer), rock streams (stone rivers, kurums), etc.

The Pacific influence did not practically reach the northern coastal lowlands (Chaun, Vankarem) protected by a wide belt of mountains. According to the climatic reconstruction based of the tundra-steppe distribution (Alfimov and Berman, 2004), the summer temperatures on the arctic coastal plains and the adjoining emerged shelf were 5–7°C above the present-day values, while in the inner regions, they were 2°C and more below those of the present day. The emerged shelf landscapes were low plains on ice-saturated soils, with scanty herbaceous vegetation. In the east, on the Lower Anadyr lowland, there were tussock tundra landscapes.

The paleoclimatic data inferred from the analysis of the upper 250-cm layer of the Lake Elgygytgyn bottom sediments (core PG-1351) (Minyuk et al., 2003) is correlated with marine isotopic stages (MIS) 3–1.

The pollen zones Eg11–Eg13 recognized in the pollen diagram (Fig. 3.1.20) reflect complicated climatic fluctuations during the second half of the Late Pleistocene and the Holocene. The pollen assemblages of zone Eg11 correlated with MIS 3 indicate the prevalence of xerophytic tundra communities: sagebrush and grass tundra with shrub birch. The end of the late thermochron and the late cryochron of the Pleistocene

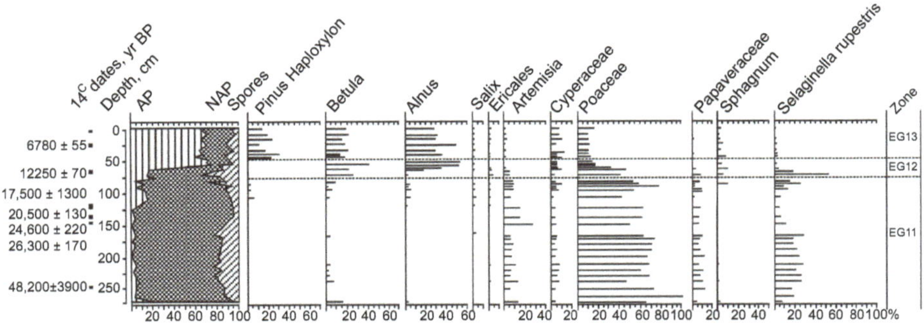

Figure 3.1.20 Pollen diagram of the upper 250 cm layer of the Elgygytgyn Lake sediments.

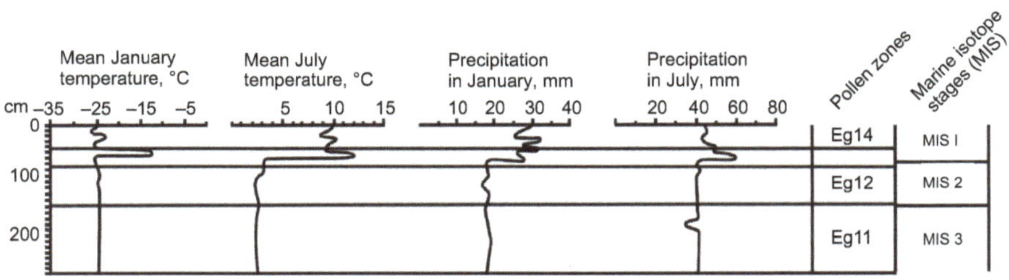

Figure 3.1.21 Changes in climatic indexes of the Northwestern Chukotka in the MIS 3–MIS 1 interval.

(pollen zones EG11–EG12, their boundary being at about 27 ka BP) were noted for July temperatures and seasonal precipitation considerably below those of today. The January temperatures were about −24 to −25°C. At the LGM the July temperatures were +2 to +3°C, which is 7–8°C below the present-day ones (Fig. 3.1.21). Reconstructed climatic characteristics at the late thermochron to late cryochron (based on the Elgygytgyn sequence) showed low winter precipitation (about 17–19 mm) (Anderson and Lozhkin, 2011) and a rather dry July, with rainfall of 28–40 mm. It may be safely concluded that low temperatures and precipitation in summer exerted a greater influence on the shrub tundra development than the extremely low temperatures in winter (Lozhkin et al., 2007).

The history of the early man dispersal is closely related to the glacio-eustatic fluctuations of the World Ocean level. Of the particular importance was the ocean level drop by 100–110 m (with reference to the present-day level) about 20 ka BP, which resulted in the appearance of the Bering Land Bridge. In place of the Bering Strait, there was lowland with residual massifs of St. Lawrence and Diomede islands (Svitoch, 2003) (see Fig. 3.1.4). In common with the Kolyma Lowland, that land was occupied by tundra-steppe and tundra landscapes. The width of the land bridge (along the meridian) was about 2000 km.

After 17 ka BP the ocean level began to rise, and finally the Bering Strait reopened. The shelf flooding led to increase in humidity not only on the coastal lowlands, but also over the entire region. About 15 ka BP the ocean level rose to −52 m, so the Chukchi

Sea and the Bering Sea were separated by a narrow rock threshold at a depth of 46 m (below the present-day sea level). About 14 ka BP the strait was completely opened; after 11.5 ka BP, when the sea level rose to –20 to –15 m, it resumed functioning normally. Since then the sea has exerted a constant warming influence on the coasts; the climate became considerably warmer and less continental. A noticeable change in the composition of pollen spectra recovered from the Elgygytgyn Lake sequence in the 47–75 cm interval (zone EG12) indicates the mosaic grass tundra replacement by dwarf birch tundra and large shrub tundra (with birch and *Alnaster*); such a replacement is typical of the period transitional from the Pleistocene to the Holocene.

The rise of the ocean level and the disappearance of the Bering Land Bridge led to changes in landscapes, and to migrations and extinction of the animals adapted to the tundra-steppe environments. The arctic tundra-steppe degraded relatively quickly (Svitoch, 2002). On the western side of the Bering Sea appeared the Anadyr River estuary, gradually filled with sea water of normal salinity. In the Lower Anadyr lowland the formerly dominant dry grass tundra with some creeping birch and willow was replaced by shrubs, including *Alnaster* (*Duschekia*). Both heat and moisture supply increased.

Palynological and carpological analyses of peats exposed in terrace scarps near Lake Elgygytgyn also display a wide expansion of the shrub alder (Shilo et al., 2008). Radiocarbon dates obtained on fragments of trunks and large branches of *D. fruticosa* are as follows: 9250 ± 90 yr BP, MAG-1477/10,430 ± 110 cal. yr BP/; 9125 ± 30 yr BP, MAG-994/10,270 ± 40 cal. yr BP/; and 8120 ± 25 yr BP, MAG -1478/9050 ± 40 cal. yr BP/. The dates suggest the peat accumulation to be related to the Boreal period of the Holocene. The larch needles present in the deposits indicate that 9–8 ka BP this tree grew near Lake Elgygytgyn, which is much farther north of its recent range.

It has been found that at the beginning of the Holocene mean July temperatures at the Polar Circle were up to 13°C, and mean January temperatures were –15°C; the precipitation amount was 60 mm in July and 25 mm in January (Anderson and Lozhkin, 2004; Overpeck et al., 1985).

3.1.3.1 Description of the sites

3.1.3.2 Late Pleistocene sites

The *Orlovka II site* was discovered by S.B. Slobodin in 1980 in the course of a field survey performed by the West Chukotka archaeological group (Northeastern Interdisciplinary Research Institute, the Far Eastern Division of the Russian Academy of Sciences). It is located in the lower reaches of the Orlovka River, right tributary of the Bolshoy Anyuy in the Kolyma drainage basin (66°56′ N, 165°08′ E) (Kiryak, 2005). It sits on a grass-covered slope of a low mountain ("sopka"), about 160 m above the river, at a distance of 1 km from the river (Fig. 3.1.22).

There are traces of solifluction noticeable on the slope surface that disturbed the cultural layers. Some of the finds (pebble tools and microblades) were recovered from the surface. In the excavated area of 23 m^2, there are two levels of the artifacts at depth from 0 to 22 cm.

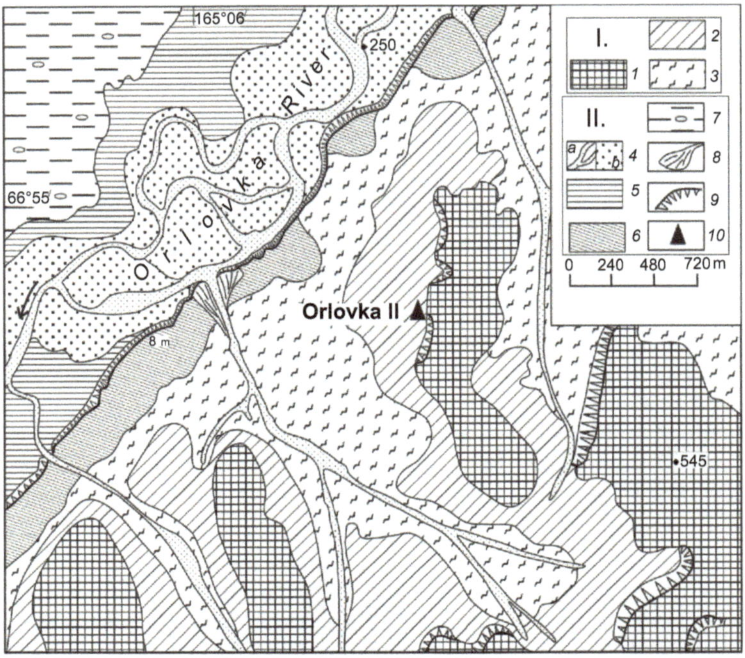

Figure 3.1.22 Geomorphological scheme of the Orlovka II site setting. *1–3 – structural-denudational landforms*: *1* – dome-like top surfaces bounded by scarps; *2* – gentle scree slopes; *3* – slope bases dominated by solifluction; *4–8 – fluvial landforms*: *4* – river channel (*a*) and floodplain (*b*); *5* – terrace 3–5 m high (Early Holocene?); *6* – terrace 8–12 m high (Karginsky?); *7* – lacustrine-fluvial plain; *8* – alluvial fans; *9*, *10* – other objects: *9* – erosional scarps and their height, *10* – archaeological site.

The upper level at a depth of 7–10 cm is embedded in the soil with rubble and gravel, and the lower one is in yellowish sandy loam with sand and gravel at 5–15 cm depth. The upper layer contains mostly flint microblades, while the lower one yielded 217 finds, including choppers, cores used for blades, flakes, blades, a side scraper, end scrapers, and burins (Fig. 3.1.23: *1–10*) (Kiryak, 2005). There are two clusters of finds measuring 6–8 m^2 and 5–7 m^2 in area in both layers. The presence of charcoal in the place of large pebble concentrations might be indicative of a fireplace and/or anvil (Kiryak, 2005).

The position of the site suggests it was a hunting outlook, though abundance and diversity of findings give evidence of rather long-term activities. The raw material, mostly large-size pebbles, was procured in considerable amounts from the Orlovka River and required considerable efforts to be delivered to the site. The dominance of side and end scrapers (about 50% of tools) is indicative of the hide processing.

The site is considered to be the earliest evidence of the colonization of the region by the Paleolithic man during the second half of the Late Pleistocene (Kiryak, 2005). The data available are not sufficient for its cultural attribution and dating. If we exclude the microblade component from the archaeological context of Orlovka II, it might be

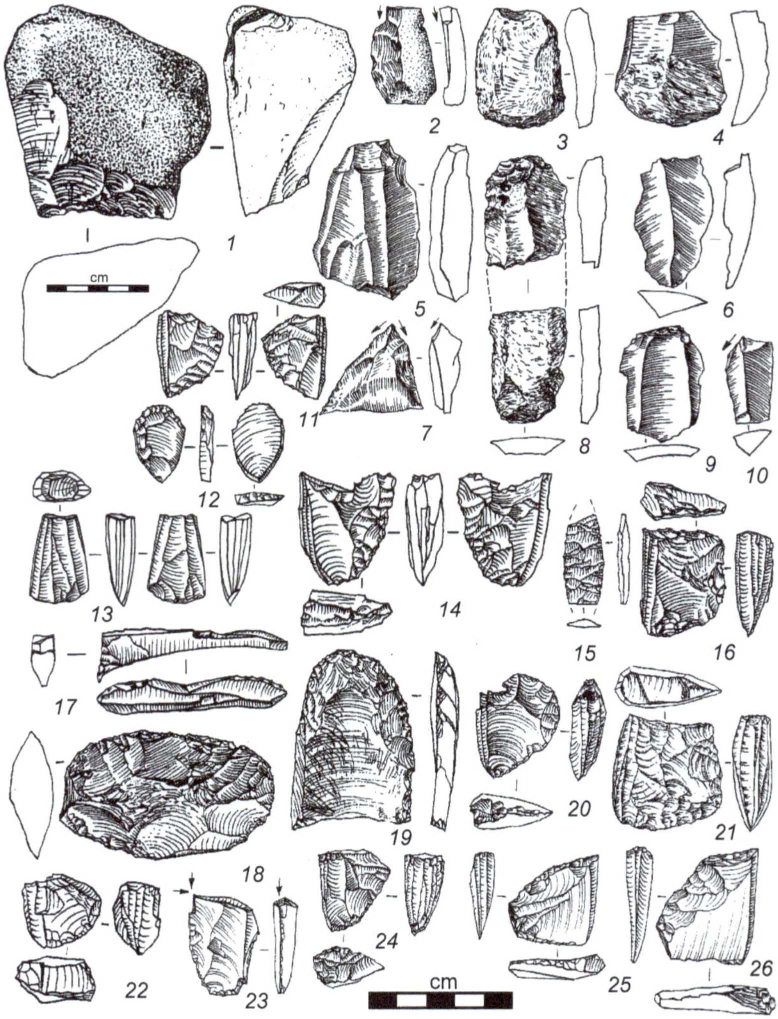

Figure 3.1.23 Artifacts from the Paleolithic (Late Pleistocene) sites of Chukotka.
1–10 – Orlovka II; *11*, *14–16*, *18*, *19* – Tytylvaam IV; *12*, *17* – Ayon; *13*, *23–26* – Kurupka I; *20* – Talyain; *21* – Elgygytgyn II; *22* – Inaskvaam II.
1–11, *14–16*, *18*, *19* – after Kiryak, M.A., 2005. Late Pleistocene assemblages of the Ushki V site: history of discovery and prospects for investigation. In: Goryachev, N.A., Kokorev, E.M., Mumortsev, Yu.I. (Eds.), Northern Pacific: Cultural Adaptations at the End of Pleistocene – In Holocene. Materials of International Conference, August 29–September 08, 2005. Northern International Univ. Publ., Magadan, pp. 62–66 (in Russian).

tentatively identified as the Orlovka complex of the pebble tool tradition of the time preceding the appearance of the microblade technology, that is, at least 25–30 ka BP.

Orlovka II has a number of features in common with pebble assemblages of Mongolia, the Far East of Russia, and Southern Siberia dated supposedly to the Late

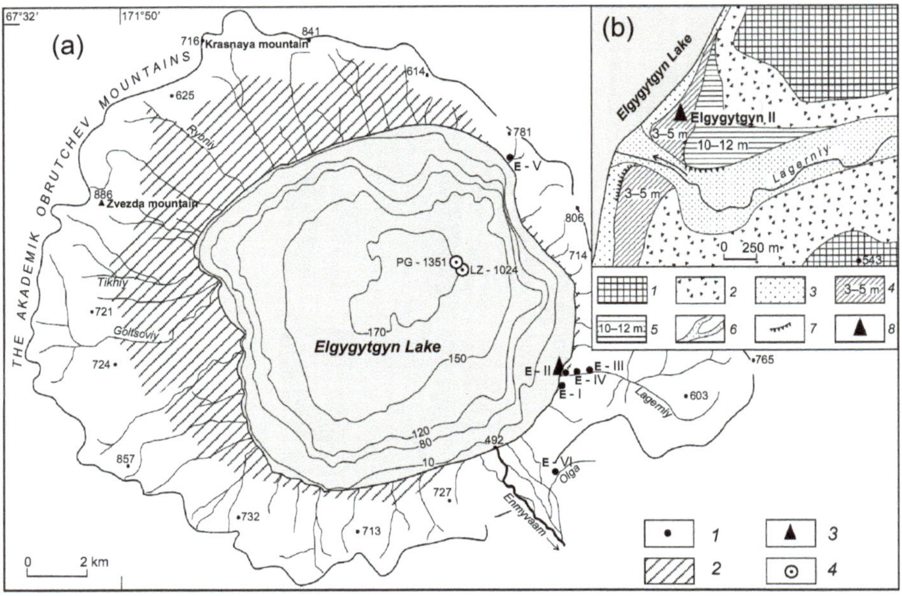

Figure 3.1.24 Geomorphological scheme of the Elgygytgyn II site setting. (a) – position of the site: *1* – Neolithic finds; *2* – upland; *3* – Paleolithic site; *4* – boreholes. (b) – geomorphological scheme: *1* – mountain slopes; *2* – aprons; *3* – beach; *4,5* – lacustrine terraces: *4* – 3–5 m high, *5* – 10–12 m high; *6* – river channel and floodplain; *7* – erosional scarps; *8* – archaeological site.

Pleistocene (30–35 ka BP) (Kiryak, 2005; Dikov, 1996), though they have neither reliable dates nor a developed typology of the tool assemblages. In particular, the materials of Orlovka II are thought to be similar to the tools recovered from Kumary II on the Amur River (Kiryak, 2005). There is a reason to agree with such correlation, if we exclude the Orlovka microblade component from the consideration. The age of Kumary II itself is questionable, and in spite of its being considered among the earliest complexes of the Amur region (Derevianko et al., 2006), there is an opinion that "…in that case one might confidently speak of the inventory appearance, and not of its age" (Derevyanko et al., 1994).

The *Elgygytgyn II site* is on the coast of the lake of the same name at the Lagerny Creek mouth (67°28′ N, 172°14′ E, 497 m a.s.l.) (Fig. 3.1.24). It was discovered in 1955 by the expedition of the Permafrost Institute, Siberian Branch of the Russian Academy of Sciences. The deep drilling performed in the central part of the lake gave information on the climate and environment changes in the northeast over the last 3.6 Ma.

As follows from the paleogeographic data, the Lake Elgygytgyn vicinities were not glaciated at the Sartan. That is corroborated by the presence of artifacts of the Paleolithic appearance at the eastern coast of the lake. The tools were found on the flat surface of the terrace 4–5 m above the lake, under the sod. The terrace deposits hardly differ from those on the beach: poorly sorted series of gravel and pebbles. The terraces around Elgygytgyn Lake abound in cryogenic landforms. More often found are tundra spot-medallions, mounds and ridges, stone rings, frost cracks, erosional grooves and furrows, solifluction terracettes, thermokarst depressions, gullies, etc. Ice-wedge polygons are often found. On the surface of the lake terrace, among objects of different age (mostly Neolithic), there were found two wedge-shaped cores on bifacial preforms (Slobodin, 2001a; Glushkova and Smirnov, 2008) (Fig. 3.1.4: *21*). The character of core preparation and use, the pressure platform formed by removal of short frontal flakes, was not unlike the core treatment technique used at Tytylvaam IV and Ushki I (layer V) dated by ^{14}C to about 10ka BP.

The *Tytylvaam IV site* was discovered by M.A. Kiryak in 1995 in the Tytylvaam River valley, 4km from its mouth (67°22′ N, 169°32′ E, 530m a.s.l.), on the surface of a glaciofluvial terrace 4–8m above the river (Kiryak et al., 2003). The surrounding landscapes are mountain tundra. The Tytyl Lake came into being after the Tytliutin River had been dammed by the end moraine ridge, the latter indicating the maximum expansion of the Sartan glacier about 34km long. The glacier was formed by confluence of several valley glaciers flowing down the valleys of Tytliutin, Tytylvaam, Lukovaya, and some others. The radiocarbon dates suggest the maximum advance of the Sartan glaciers in Chukotka to be 17–20ka BP (Glushkova, 2009).

The terrace is scarcely covered with sod, and cobbles and pebbles are often exposed on the surface. The deposits are underlain by a compact pebble bed with a sandy and gravel matrix; the deposits are presumably the remains of the Sartan till redeposited by the river (Fig. 3.1.25). There are two clusters of finds (localities 1 and 2) 75m apart, with a small stream between them, where the material was collected and excavations were performed.

In loc. no. 1 excavated over an area of 20m^2, finds were recovered from a layer of humified sandy loam with gravel at a depth of 2–5cm. There were found 25 large pebbles in the excavation, probably having formed a stone circle around a light dwelling (Kiryak, 2005). The concentration of artifacts in the center of the excavated area is confined to a pit 18cm deep; charcoal found in the pit yielded AMS radiocarbon dates of the following: 9725±45yr BP/11,150±80cal. yr BP/; 9820±40yr BP/11,230±20cal. yr BP/; and 9790±60yr BP/11,210±70cal. yr BP/ (see Table 3.1.3).

In loc. no. 2 the finds occurred over an area of 27m^2 in humified sandy loam, at a depth of 13–37cm under a layer of sod and white sandy loam. The cultural layer is broken by frost wedges and partly redeposited. A charcoal sample from the excavation (from a depth of 20–25cm) yielded a date of 8645±40yr BP/*9600*±*50*cal. yr BP/ (Table 3.1.1).

The tools from the site compose a homogenous assemblage including a few tens of subprismatic and wedge-shaped cores, macro-blades, knife-like bladelets, typical and lateral end scrapers, bifacial ovate knives, small bifacial arrowheads, burins on flakes, denticulate tools, choppers, and a flat small pebble with traces of polishing and ocher on the surface (Fig. 3.1.23: *11*, *14–16*, *18*, *19*). Pressure platforms of the cores were modeled by short frontal and side blows, without the use of ski spall removals.

In all probability, it was a working area where the site inhabitants prepared for hunting. The principal objects of hunt were reindeer migrating along the Tytylvaam River valley and wild sheep on mountain slopes. The technical-typological characteristics of Tytylvaam IV assemblage (localities 1 and 2) permit one to attribute it to the final phase of the Ushki-Dyuktay tradition of the Late Paleolithic.

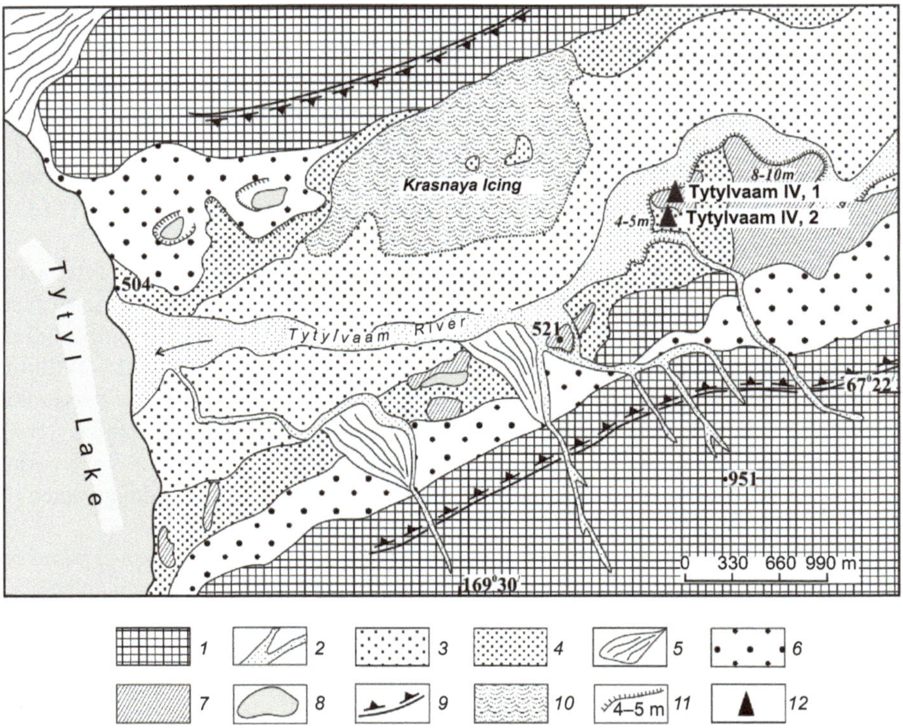

Figure 3.1.25 Geomorphological scheme of the Tytylvaam IV site setting. *1* – mountain slopes; *2* – river channel and low floodplain; *3* – high floodplain (1.5–2.0m); *4* – the first terrace (3–4m); *5* – alluvial fans; *6* – Sartan moraine partly overlain with slope deposits; *7* – fluvioglacial terrace (8–10m); *8* – lakes; *9* – shoulder of trough; *10* – ice; *11* – erosional scarps and their height; *12* – archaeological sites and their numbers.

That cultural tradition was widely spread in Chukotka (Fig. 3.1.1); finds attributable to it are found on the Ayon Island in the Chaunskaya Guba bay (Dikov, 1977; Slobodin, 2001a) (Fig. 3.1.23: *12*, *17*); at Kurupka I on the Chukchi Peninsula (Dikov, 1993a) (Fig. 3.1.4: *13*, *23–26*); at Talyain in the Velikaya River Valley, right tributary of the Anadyr River (Dikov, 1990) (Fig. 3.1.23: *20*); and at Inaskvaam in the Khatyrka drainage basin, 40km from the Bering Sea coast (Dikov, 1979) (Fig. 3.1.4: *22*).

At the Kymyneykey locality the findings were identified from the core of a borehole drilled near Kymyneykey Mountain in the Vankarem River valley (Laukhin and Drozdov, 1989). The pieces occurred at a depth of 33m; they were described as flakes and wedge-shaped cores and attributed to the microblade technique of the Northeast of Asia. The deposits at the locality are described as glacial till. The deposits underlying the till were dated at 39–40ka BP. The presence of artifacts in the till may be explained by their transportation by the glacier from a neighboring river valley in a block of frozen ground and the subsequent redeposition. The findings were tentatively dated about 30ka BP on the basis of their correlation with the Dyuktay culture in Yakutia (Laukhin, 1997). Both the characteristics of the finds and the circumstances of their discovery give rise to doubt as to their belonging to man-made artifacts (Goebel and

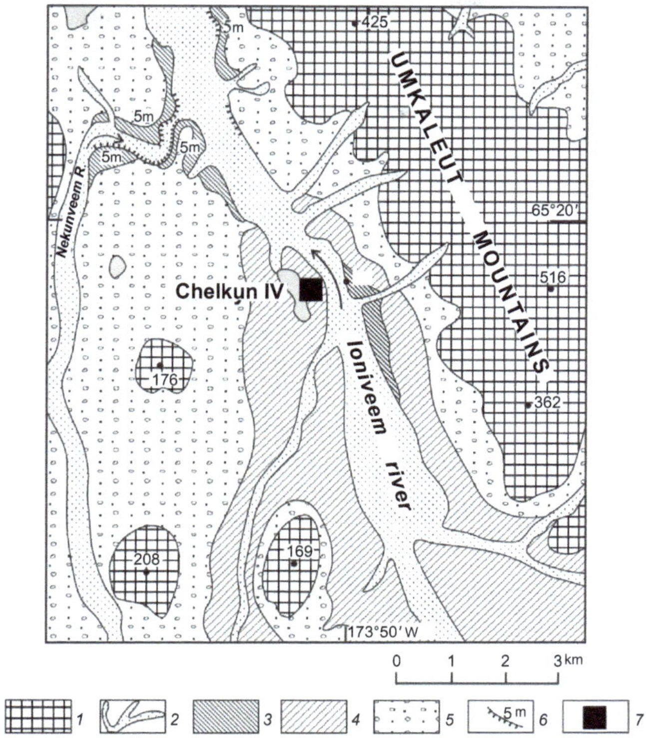

Figure 3.1.26 Geomorphological scheme of the Chelkun IV site setting. *1* – mountain slopes; *2* – river channel and floodplain; *3, 4* – terraces: *3* – 3–5 m high (Early Holocene?), *4* – 8–12 m high (Karginsky?); *5* – outwash plain (Zyryanka?); *6* – erosional scarps and their height; *7* – archaeological site.

Slobodin, 1999; Slobodin, 2000). Equally doubtful is the theory of the glacial origin of Kymyneykey Mt. (Kundyshev, 1992).

3.1.3.3 Early Holocene sites

There are two groups of sites in the Chukchi Peninsula dated to the Early Holocene (Fig. 3.1.1; Table 3.1.1). One of them includes Chelkun IV, Nayvan, Tytyl I, III, IV, Puturak, and others; materials recovered from those sites are dated by radiocarbon at 8–7 ka BP and may be attributed by their technological characteristics and typology to the *Sumnagin tradition* of reindeer hunters in tundra and moose (elk) hunters in taiga zone of the Northeast of Asia (Mochanov, 1977).

Another Early Holocene cultural tradition is *Uolba* (Slobodin, 2000) developed in the Kolyma and Indigirka basins. In Chukotka, it is represented by the materials from Verkhnetytylskaya VI; Tytyl IV, loc. 2; and Nizhnetytylskaya IV, loc. 3 (Kiryak, 2005).

The *Chelkun IV site* was discovered by Dikov (1993b) in 1979 in the upper reaches of the Ioniveem River, at the mouth of the Chelkun River (65°19′ N, 173°50′ W, 87 m a.s.l.). It sits at the 8-m scarp of the river terrace (Fig. 3.1.26). The excavations over an area of 31 m^2 revealed the cultural layer at a depth of 0.4 m; the layer contains numerous pieces embedded in gray sandy loam underlain by gravels and pebbles. A charcoal sample from the layer yielded a ^{14}C date of 8150±450 (see Table 3.1.3). Charcoal sandy loam forms a spot about 2 m in diameter. Its outlines coincide with the area of artifact occurrence. Within the spot, there are two elongated pebble ridges arranged in parallel (in the opinion of Dikov (1993a), they are artificial constructions), 2.5 and 1 m long, 0.5 m wide and 20–25 cm high. The fact that the outlines of the spot coincide with the area rich in finds makes possible identification of the construction as a light aboveground dwelling. The finds include two prismatic cores, flakes, microblades, and burins made on bladelets. Among the surface collections, there are prismatic double-platform cores, microblades, and marginal spalls (Fig. 3.1.27: *8*, *10*, *13*).

According to Dikov (1993b), the site was a lithic workshop. Taking into consideration the recorded elements of a dwelling, a small amount of core preforms, and debris by-products of their processing, the site may be considered as a point of various activities, where people made tools (e.g., inset bone points) and dealt with some house tasks. They could also hunt reindeer in the vicinities. The technical and typological characteristics and the dates suggest that the assemblage belongs to the Sumnagin culture.

The *Nayvan site* was discovered in 1997 by Gusev (2002) on the Chukchi Peninsula, on the Bering Strait coast between Chaplin and Mertens capes (64°29′ N, 172°23′ W, 25 m a.s.l.). It is situated on a flat top of rock remnant about 16 m high between lakes (the latter are inner lagoons separated from the ocean by Chaplin spit 200–300 m wide) (Fig. 3.1.28).

Excavations on an area of 86 m^2 revealed the cultural layer at a depth of 37–60 cm embedded into light-brown sandy loam. The cultural layer contains a complex of surface dwellings about 3–4 m in diameter, with hearths about 1 m in size in the center, as well as working areas and stone tools. Individual hearths, probably used at different episodes of habitation, consisted of small river pebbles and a few angular stones bearing traces of thermal action. Judging from the tool assemblage and the climatic conditions of that time, the site was most likely seasonal and was repeatedly (at least for a few years) visited during summer and fall. Charcoal samples from the hearths yielded ^{14}C dates in the range from 6990±320 yr BP/7870±*300* cal. yr BP/ to 9000±500 yr BP/10,260±690 cal. yr BP/ (Gusev, 2006) (Table 3.1.1). A wide range of the dates and a considerable statistical error may be related to the use of driftwood. The palynological characteristics of the sediments enclosing the cultural layer suggest the sediments formed during the Boreal and Atlantic periods of the Holocene (Gusev, 2002).

About three thousand lithic artifacts of siliceous rocks, chalcedony, and obsidian have been recovered from the site; among them, there are prismatic microcores, bladelets (including those with marginal retouch), a bifacial ovate cutting tool (axe), and end scrapers on microblades and on a large blade (Fig. 3.1.27: *4*, *5*, *15*, *16*). The use-wear analysis has shown traces of the tools being used to cut, plane, and saw bone and wood, as well as to cut fish and meat (Gusev, 2002).

The inventory recovered from the site permits reconstructing the lifeway of the Early Holocene nomadic reindeer hunters; the people inhabited mostly intracontinental regions, but they went to the coastal tundra and to the Bering Sea coasts

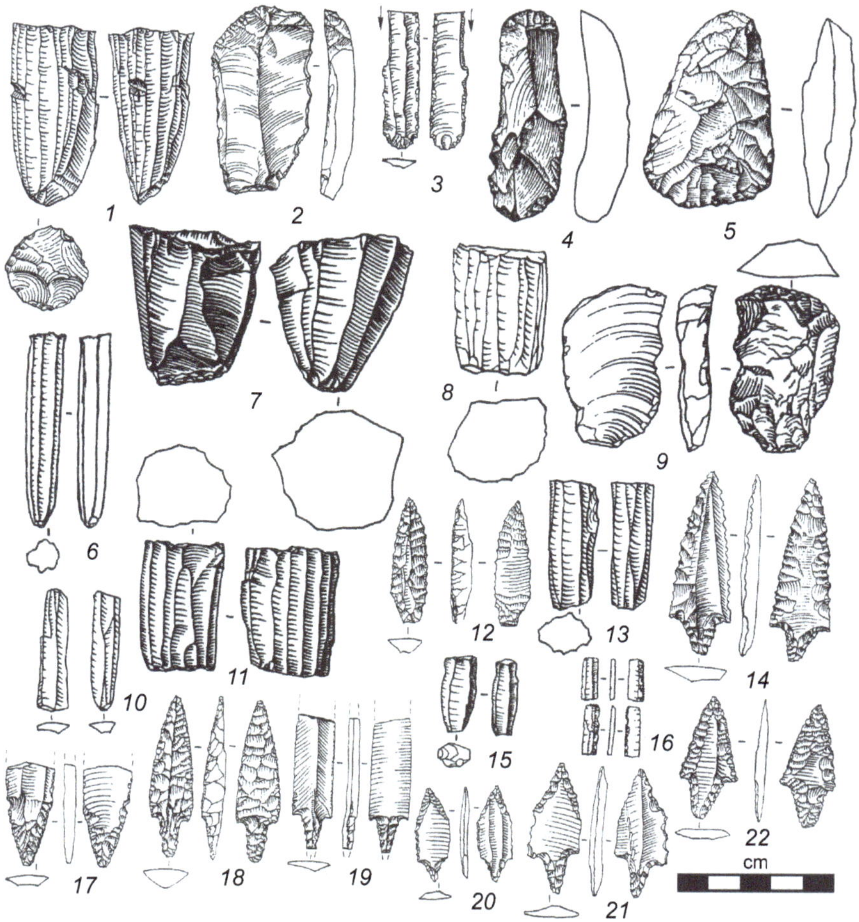

Figure 3.1.27 Artifacts from the Mesolithic (Early Holocene) sites of Chukotka.
1–3 – Tytyl I; *4, 5, 15, 16* – Nayvan; *6, 7, 9, 11* – Puturak; *8, 10, 13* – Chelkun IV; *12, 14, 17–22* – Verkhnetytylskaya VI.
1–6 and *11–22* – after Kiryak, M.A., 2005. Late Pleistocene assemblages of the Ushki V site: history of discovery and prospects for investigation. In: Goryachev, N.A., Kokorev, E.M., Mumortsev, Yu.I. (Eds.), Northern Pacific: Cultural Adaptations at the End of Pleistocene – In Holocene. Materials of International Conference, August 29–September 08, 2005. Northern International Univ. Publ., Magadan, pp. 62–66 (in Russian); Gusev, S.V., 2002. The early Holocene Naivan site in the Bering Strait (Chukchi Peninsula). In: Materials of Dikov's Conference II. Northeastern Interdisciplinary Research Institute, Far Eastern Division, RAS, Magadan, pp. 356–363 (in Russian); Dikov, N.N., 1993a. Asia at the Interface of America in Ancient Times. Nauka, St.-Petersburg. 304 pp. (in Russian).

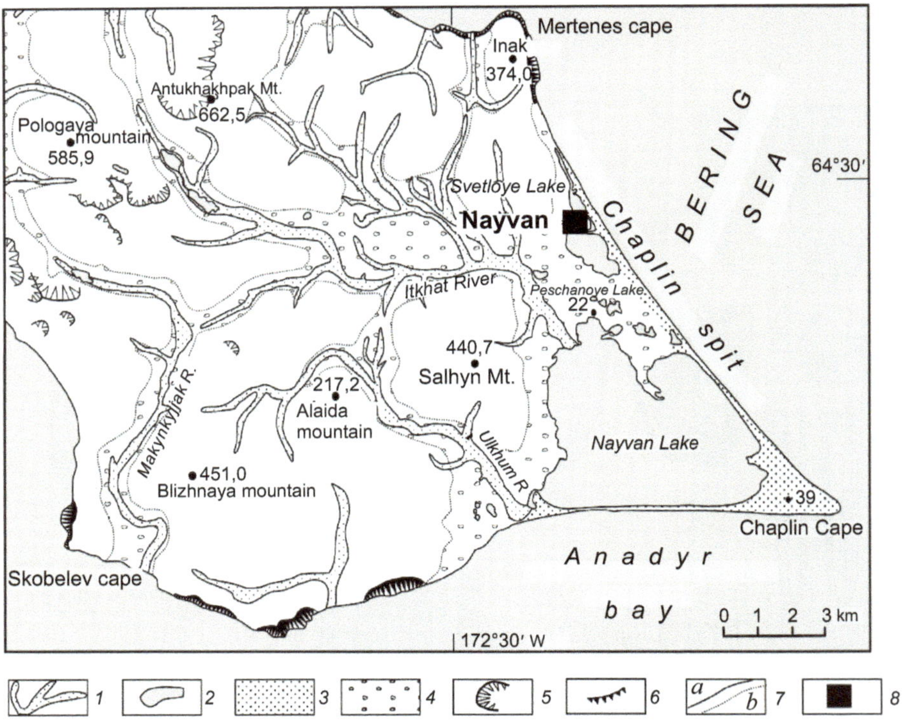

Figure 3.1.28 Geomorphological scheme of the Nayvan site setting. *1* – river channel and floodplain; *2* – lakes; *3* – beach ridges and spits; *4* – Sartan till; *5* – glacial cirques; *6* – coastal scarps; *7* – geomorphological boundaries: *a* – proved, *b* – supposed; *8* – archaeological site.

in summer. There is no direct indication (such as organic remains) of the marine resource usage by the site inhabitants, though they could practice marine animal hunting on the nearby ground or fishing salmon at the spawning season. They could also collect mussels, with small fragments of shells being found in the hearths. The technical and typological characteristics of the inventory and the age of the site are indicative of its belonging to the Sumnagin culture.

The *Puturak site* was discovered by N.N. Dikov in 1985 on the Chukchi Peninsula, on the Puturak pass between rivers Itkhat and Tkachen drainage basins, 15 km from the Bering Sea coast (64°31′ N, 172°39′ W, 207 m a.s.l.) (Dikov, 1993b). The excavation, 28 m^2 in area, yielded more than 10,000 artifacts belonging to a single cultural assemblage component; the latter includes prismatic (pencil-shaped) and conical cores at various stages of processing (from preforms to almost exhausted pieces), microblades, blades and large-size blade-like flakes, and end scrapers (Fig. 3.1.27: *6*, *7*, *9*, *11*).

Those materials permitted a new culture to be identified; known as "*Puturak culture*" it is noted for primitiveness of technology and the absence of burins and side scrapers (Dikov, 1993b). Those "specific features" may be probably ascribed to the poor quality of raw materials and to the type of the site: it was a camp-workshop where cores were prepared and a good deal of rough spalls and blades were rejected.

The technical and typological features of the materials permit one to attribute it to the Sumnagin culture (Slobodin, 2000).

The *Tytyl I, III, and IV sites* were discovered by Yu.A. Mochanov in 1977 (Mochanov et al., 1978) on the Tytyl Lake coasts in western Chukotka (67°17′ N, 169°15′ W, 499 m a.s.l.). Lithics collected on the terrace surface or retrieved from shallow excavations included conical and prismatic cores, microblades, burins, and end scrapers on blades (Kiryak, 1993, 2005) (Fig. 3.1.27: *1–3*). On the whole, the artifacts correspond to the Sumnagin culture of the Early Holocene, though differ from it by the larger size of the tools and the absence of bifacial axes.

The *Verkhnetytylskaya VI site* discovered by M.A. Kiryak in 1995 is situated on the eastern coast of Lake Tytyl 0.5 km south of the Tytylvaam River mouth, on a morainic hill 9–10 m high (67°21′30″ N, 169°27′10″ W, 507 v a.s.l.) (Kiryak, 2005). The cultural layer was excavated over an area of 24 m^2. The artifacts (more than 2000) are embedded in humified brown sandy loam at a depth of 3–25 cm.

The knapping products are conical and prismatic cores and microblades. The tool assemblage includes stemmed and leaf-shaped points on blades, end scrapers, angle and dihedral burins, bifacial pieces, and a blank of adze with "ears" (Kiryak, 2005) (Fig. 3.1.27: *12*, *14*, *17–22*). The site location and the tool assemblage suggest its inhabitants were mostly reindeer hunters. The assemblage has not been dated by radiocarbon; the technical-typological characteristics of the materials made possible their attribution to the *Uolba* tradition dated to the Early Holocene (Kiryak, 2005).

The *Uolba culture* is considered to be noted for the expansion of stemmed bladed points over the Chukchi Peninsula in the Early Holocene (they were found at several sites at Lake Tytyl, such as Tytyl IV, loc. 2, Nizhnetytylskaya IV, loc. 3). Within the succession of the archaeological cultures at Chukotka, they take place during the Late Mesolithic (Kiryak, 2005). By analogy with the Kolyma sites (Slobodin, 1999b), they are dated to the Early Holocene. The tool assemblage, besides blade points, includes prismatic and conical cores, end scrapers, and multifaceted burins.

3.1.4 Initial human colonization of Western Beringia: a synthesis

The archaeological and paleogeographic data (including those on paleontology and paleobotany) suggest that the initial colonization of Western Beringia (the Indigirka-Kolyma and Chukotka regions) proceeded by way of migrations from the southern and eastern regions of East Siberia, Far East, and Yakutia at the end of the Karginsky, the beginning of the Sartan (27 ka BP) at the latest (Fig. 3.1.3). The materials of the Yana site (Yanskaya culture) show that the man advanced to 70°N on the Yana-Indigirka Lowland as early as 27–29 ka BP (Pitul'ko, 2006a,b; Pitul'ko and Pavlova, 2010).

The Beringia land bridge persisted seemingly throughout the Karginsky (MIS 3). That is indicated unambiguously by paleontological data from Alaska: the earliest remains of the Pleistocene mammals that migrated from Siberia—mammoths, horses, and bison—are ^{14}C-dated at 35(40)–29 ka BP (Guthrie, 1982, 1990). Under conditions of tundra-steppe being dominant in Beringia at that time, man hunted mostly the animals of mammoth assemblage: mammoths themselves, reindeer, horses, and bison (Pitul'ko and Pavlova, 2010). The hunters lived in seasonal camps; some of the latter, like Yana and Ushki I and V, where the hunt was particularly successful, were used repeatedly for a long enough time (Table 3.1.2).

In the arctic and subarctic regions of the Kolyma and Chukotka regions the same age is tentatively ascribed to surface collections of pebble tools from Orlovka II (Kiryak, 2005) and a few sites on the Omolon River (Vorobey, 1999); the collections are comparable in typology with the oldest pebble tool assemblages in Siberia and Mongolia dated at 30–50 ka BP. On the whole, such an interpretation of those materials calls for further corroboration, and the age of those sites is still conjectural. Surface collections from Chukotka (Kymynanonvyvaam XII, XIII sites) (Dikov, 1993b) attributed to the same period are extremely questionable, and there is no corroborative evidence for dating them to that time (Slobodin, 2000). Equally controversial is interpretation of objects found in the borehole core from a depth of 33 m at Kymyneykey Mt., Eastern Chukotka (Laukhin and Drozdov, 1989). A featureless shape of the finds and a problematic genesis of the geological context (Kundyshev, 1992) make their cultural attribution difficult (Goebel and Slobodin, 1999; Slobodin, 2000).

More reliable data on the human colonization of the mountain regions in the Kolyma drainage basin at the first half of the Sartan were obtained from Bolshaya Khaya IV; the site produced an assemblage lacking microblades and comparable with the materials of the middle stage of the Upper Paleolithic in Eastern Siberia (Slobodin, 2005; Slobodin et al., 2008). There, in the Bolshaya Khaya River valley, man was attracted by high-quality flint outcrops and, probably, good prospects for reindeer hunting at the time of their seasonal migrations across the Okhotsk-Kolyma divide. A large area of the site, an abundance of stone material, and ornaments found there suggest repeated visits of the ancient hunters to the site.

During the entire Sartan and the Early Holocene, there were several waves of people migrations in Beringia reflected in the different archaeological cultures. When considering the problem of the initial colonization of the region, it would be important to take the data on Eastern Beringia into consideration, as the man could get into there only after he had colonized the western part of Beringia. It is no less important for the problem under discussion to make a comparison between the technical characteristics and typology of the tool assemblages of the considered region and the assemblages known from reliably stratified and radiocarbon dated sites that are key sites in the regional cultural-chronological schemes of eastern Siberia (Yakutia, Primorye, the Amur drainage basin, the Baikal region).

A wider presence of man in the Indigirka-Kolyma region and Chukotka at the second half of the Sartan is testified by the materials from the sites of the Okhotsk-Kolyma uplands (Kheta, Druchak-Vetreny, Omchik II, Ust'-Magadaven I) and Chukotka (Tytylvaam IV, Elgygytgyn, Ayon, Kurupka, Talyain; Inaskvaam); the sites are referred to the widespread microblade cultures and traditions: Dyuktay, Ushki, Beringian, Baikalian, and Selemdzha. Most of the sites have not been ^{14}C dated yet; considering, though, that the microblade Beringian tradition was reliably dated as older than 12,300 yr BP (Beta-209,882) (Holmes, 2011), it is not inconceivable that it had expanded over the Kolyma region 2000 or 3000 years earlier, that is 15–16 ka BP at the latest. In Kamchatka, it is dated at about 10,000–11,000 ^{14}C yr BP (Dikov, 1979). As to Chukotka, the tradition persisted there until the beginning of the Holocene (as indicated by the ^{14}C date on Tytylvaam IV).

The tool assemblages of the sites include wedge-shaped cores, microblades, leaf-shaped and ovate knives-bifaces, leaf-shaped arrowheads and projectile points, end scrapers, burins, and borers. Bifacial-type, wedge-shaped cores suggests those sites to be more closely related to the Selemdzha, Osipovka, and Ushki (Kamchatka) cultures than to the Dyuktay culture where cores of such type are unknown. The technological traditions of bifacial wedge-shaped cores expanded from the Upper Kolyma and Chukotka across Beringia onto Alaska; they were found at a number of sites: Red Dog (Gerlach and Hall, 1996), Broken Mammoth (Holmes, 1996), and Little Panguingue Creek (Hoffecker and Powers, 1996).

The sites of that period are mostly hunting camps with aboveground dwellings located in river valleys; the inhabitants hunted mostly reindeer and probably sheep. Salmon fishing could be practiced in the spawn season on Druchak-Vetreny, Ust'-Magadaven I, Ushki I, and Anavgay II positioned on spawning streams. Most of the sites do not display any system in spatial organization; the only exception is Ushki I in Kamchatka: a long-term settlement with slightly deepened dwellings, each for an individual family, with stone-lined rounded hearths, passage exits, burials of children and of a dog, and caches of blanks and end products. It is not inconceivable that a stone circle found at Tytylvaam IV was really a lining of a dwelling of the tent ("chum") type.

The human activity at the end of the Late Pleistocene (12,000–11,400 ^{14}C yr BP) is reconstructed from the materials recovered from the Berelekh site on the Indigirka Lowland: besides seasonal hunting, man exploited the bone-bearing horizon of a mammoth "cemetery" dated to the Sartan (Pitul'ko et al., 2009). Judging from the bones composition, the main object of hunt of the Berelekh site dwellers was hare; according to the ethnographic information, the northeast native people practiced hare hunting on a great scale (more than 12,000 animals per winter) in the recent past (Khersonskiy, 1898).

Another episode of the Western Beringia settlement at the end of the Sartan is presented by the Upper Paleolithic Ushki culture (Dikov, 1979), known from the sites Bolshoy Elgakhchan I on the Omolon River, Omchik IIb on the Kolyma, Bolshaya Avlondya on the Okhotsk-Kolyma Upland, and Serdyak in the Okhotsk Sea coastal region (Kiryak, 1993; Slobodin, 2002, 2010a). The lithic assemblages of the sites do not include microblades, though bifacial stemmed points of the Ushki type are usually present. Judging from small collections of tools from the sites mentioned, they were seasonal hunting camps. In Kamchatka, at Ushki I and V (layer VII) with dwellings and a burial, the culture was dated by ^{14}C AMS to 11,000–11,300 yr BP (Slobodin, 2010b; Goebel et al., 2010), and probably even to 14,000 yr BP (Dikov, 1979). N.N. Dikov stressed a similarity between the Ushki points and the Paleoindian stemmed points in America (from the Marmes rock shelter) and used the term "Americanoid" in reference to the Siberian objects. The Ushki points are comparable with those in the oldest in North American assemblage from the Lind Coulee site on the Columbia Plateau dated at 12,000–11,000 yr BP (Davis and Sisson, 1998; Vasil'ev, 2004).

The latest of the known episodes of human migrations into the regions north of the Sea of Okhotsk (presumably, from the Amur River basin) occurred at the very end of the Sartan. To that episode the Uptar site belongs, its tool assemblage being clearly oriented toward hunting. In summer the inhabitants could also practice fishing in the Uptar River during the season of salmon spawning.

About 12 ka BP the entire area of the Northeast of Asia underwent essential changes of climate, as well as of the fauna and vegetation. The higher mean annual temperatures induced ice melting, sea level rising, and finally the Beringian land submergence (within the limits of the modern shelf). The opening of the Bering Strait separated the Eurasian and American continents. The rising summer temperatures and increased moisture supply resulted in tundra-steppe transformation into wetlands and tussock bogs impassable for most of the mammals of the mammoth assemblage. In winter the access to fodder was hampered by a thick snow cover. The mammoth fauna partly died out and partly migrated to more hospitable regions. The most successful herbivorous species that were able to adapt themselves to the new environments were the reindeer in tundra and the moose in taiga. The environmental changes in the Early Holocene led to a crisis in the economy of the Paleolithic hunters; their culture was replaced at that time by new Mesolithic cultures distinct for a pronounced tendency for the microlithic tools to be in usage, such as meso- and microblades removed from conical and prismatic cores.

At that time, man dispersed over the entire Western Beringia (Northeastern Asia) from the Sea of Okhotsk coast to the northernmost margins washed by the Arctic Ocean (now the New Siberian Islands) at 72°N. He adapted to the changed environments and occupied all the ecological zones in the considered region: from mountain passes to coastal lowlands. At that time the most widely distributed—over the entire Kolyma drainage basin and the Chukchi Peninsula—were sites of the Sumnagin culture, which is dated by ^{14}C to 8800–7000 yr BP.

The distribution of this culture over the Sea of Okhotsk-Kolyma interfluve is marked by the sites Buyunda III, Urtychuk IV, Khurendzha, Omsukchan III, etc. (Slobodin, 1999b); the BMA-3 site on the Sea of Okhotsk coast is attributed to the same culture (Vorobey, 2007). The Zhokhov site on the island of the same name (the New Siberian Islands) belonging to the Sumnagin culture displays some unique characteristics of adaptation to the harsh environments of maritime Arctic (Pitul'ko, 1998). On the Chukchi Peninsula the Sumnagin culture materials were recovered from Chelkun IV, Nayvan, Tytyl I, III, IV, Puturack, etc., (Dikov, 1993a; Kiryak, 1993, 2005; Gusev, 2006) and from Panteleyikha I–VIII (Mochanov, 1977) in the lower reaches of the Kolyma River. It is noteworthy that the Sumnagin culture sites in the upper reaches of the Kolyma are mostly confined to mountain passes.

In the Holocene the changed environments of Beringia induced a change in the hunting specialization, and the primary object of hunt became reindeer in tundra and moose in the taiga. The Sumnagin culture people living close to the sea coasts (sites BMA-3, Zhokhov, Nayvan) could hunt marine mammals on the nearby beaches or catch salmon at the spawning season. The Zhokhov site inhabitants hunted also polar bear, reindeer, and birds, as seen from the faunal remains at the site.

The sites of this culture were mostly hunting camps with tent-like transportable dwellings: "*yaranga*." The findings are few, mostly small prismatic and conical cores, microblades, adzes, and flakes. This inventory corresponds to the nomadic lifeway of the inhabitants. The only exception is represented by the Zhokhov site: a large, probably repeatedly visited camp with slightly deepened dwellings of the Inuit pit-house type, with wooden roof and platform on the floor made of driftwood. There is evidence for a presence of dogs and, probably, dog-drawn sledges.

Another culture, less widely distributed in the Early Holocene in Western Beringia than the Sumnagin one, is the Uolba culture. Sites attributed to that culture were found in the Upper Kolyma region (Uy, Kongo-78, ING-78, Pridorozhnaya, Agrobasa II), in the upper reaches of Indigirka (Yubileyny), and in Chukotka (Verkhnetytylskaya VI, etc.) (Kiryak, 2005). Besides the typical stemmed blade points, the tool assemblages contain prismatic and conical microcores, end scrapers on blades, retouched inserts on blades, bifacial knives, adzes, some of them with edge grinding, and other tools. The assemblages include both meso- and microblade techniques.

Unlike the people of Sumnagin culture, the Uolba people lived primarily on coasts of large lakes (Uy) and river banks (on the Indigirka and Kolyma Rivers). The very distinctive stemmed points on blades typical of that culture originated presumably from regions located very far from Western Beringia, such as steppes of Eurasia (Okladnikov, 1950) and Northeastern Europe (Mochanov, 1977). From the Indigirka and Kolyma the culture expanded over Chukotka (Kiryak, 2005) and Kamchatka, where it persisted until 6200 yr BP (Ponomarenko, 2000). The Uolba culture preceding the wide occurrence of the Neolithic humans all over the area of the Kolyma and Indigirka drainage basins completes the period of the initial colonization of the territory.

The Early Holocene in the upper reaches of the Kolyma River was marked by another short episode: an appearance of the so-called "Siberdik" culture represented by Siberdik and Kongo sites (Dikov, 1979). Several ^{14}C dates obtained for layer 3 of Siberdik fall within the Early Holocene. Another date—13,225 ± 230 yr BP, MAG-916/13,870 ± 340 cal. yr BP/—does not agree with palynological data and has been rejected (Lozhkin, 2008). The tool assemblage of that culture includes pieces typical of both the Paleolithic (wedge-shaped cores, pebble tools) and Mesolithic (conical cores), as well of the Neolithic (polyhedric burins). The origin of this culture of Northeastern Asia, which combines Paleolithic, Mesolithic, and Neolithic features, is not clear (Slobodin, 1999b, 2001b).

3.1.5 Conclusion

It follows from the preceding that from the second half of the Karginsky and through the earlier part of the Sartan (at least since 27 ka BP), man colonized the arctic and subarctic regions of Northeastern Asia with increasing intensity. The colonized regions—the Indigirka-Kolyma drainage basins and Chukchi and Kamchatka peninsulas—were intracontinental areas of Beringia at that time with prevalent landscapes of tundra-steppe inhabited by the animals of the mammoth assemblage.

During the middle and late Sartan the colonization took place in several stages; there were different migration waves: humans of various cultures were coming from Yakutia, the Far East, and the Baikal region. Along with occupation of open landscapes in the lower reaches of large rivers, they settled mountain regions in the drainage basins of the Kolyma and Indigirka rivers and Okhotsk-Kolyma Upland. Having colonized those regions the people moved further to Chukotka, Kamchatka, and Alaska.

At the Late Pleistocene–Holocene boundary the environments were radically restructured, and both vegetation and animals changed, which induced a change of cultures and a complete reorganization of prehistoric lifeways and economy. At the thermal maximum of the Holocene the forest-tundra zone expanded almost to the Arctic Ocean coasts. The newcomers from Yakutia and still farther regions well adapted to the forest-tundra environments migrated north and northeast along the Kolyma and Indigirka rivers up to the margins of shrinking Beringia (near the New Siberian Islands). In the process, they mastered all the ecological niches (mountain passes, river valleys, coasts of large lakes, etc.). We have no evidence of an active settlement in maritime areas and a transition to a usage of marine resources in the Early Holocene, though the people presence has been recorded on the coasts of the Sea of Okhotsk and the Bering and East Siberian seas.

Another principal change in population took place about 6 ka BP. The Early Holocene cultures disappeared and were replaced by Neolithic ones, the permanent population of Northeastern Asia being formed on the basis of the latter. That population, according to the anthropological data, is related to the northern ethnic groups of today.

Eastern Beringia: climates and environments during the initial peopling

3.2

Patricia M. Anderson[1], *Anatoly V. Lozhkin*[2]
[1]Washington State University, Seattle, WA, United States; [2]Northeastern Interdisciplinary Research Institute, the Far Eastern Division, Russian Academy of Sciences, Magadan, Russia

The archaeological sites—the evidence of the first migrations of humans into Eastern Beringia—are mostly concentrated in three regions of Alaska:

- central Alaska (the Nenana River and Tanana River basins and the Tangle Lakes vicinities);
- north and northwest of Alaska (North Slope, the Noatak River valley, the Seward Peninsula);
- southwest of Alaska (lower reaches of the Kuskokwim River).

We will center our attention mostly on the history of environments in those regions; the sites in other regions, though interesting on their own accord, have not been properly dated by radiocarbon yet.

Based on the continuous pollen records from lacustrine sediments, the changes of vegetation and climatic fluctuations in the past were reconstructed. Of numerous archaeological sites in Eastern Beringia, we have chosen for the paleoenvironmental reconstructions the following: Tukuto Lake (north and northwest of Alaska, 68°30′ N, 157°03′ W, 541 m a.s.l. (Oswald et al., 1999)); Windmill Lake (63°50′ N, 148°49′ W, 640 m a.s.l. (Bigelow and Edwards, 2001)), and Harding Lake (62°26′ N, 146°52′ W, 217 m a.s.l. (Ager, 1983)) for the Nenana and Tanana regions in central Alaska; and Grandfather Lake (59°48′ N, 158°31′ W, 142 m a.s.l. (Hu et al., 1995)) and Nimgun Lake (59°33′ N, 160°46′ W, 320 m a.s.l. (Hu et al., 2002)) in the southwest of Alaska (Fig. 3.2.1).

The early human settlement in Eastern Beringia is dated at about 12.5–9 ^{14}C ka BP (Vasil'ev, 2004), or 14.7–8.14 cal. yr BP. That period of the early migration of man into Eastern Beringia was marked by noticeable changes in environments:

- The vegetation of the period has no analogs in that of today due to quite different climatic conditions.
- The first postglacial forests in various regions of Eastern Beringia were composed of deciduous tree species (such as poplar) and therefore presented a biome quite different from the modern coniferous forests.
- According to most of paleoclimatic models, the postglacial time in Eastern Beringia was marked by a thermal maximum, with the only exception of Southern Alaska where the Younger Dryas featured a cold climate.

At the LGM the climate and vegetation in Eastern and Western Beringia were much the same (a cold and dry climate, the dominance of herbaceous or herb and

Human Colonization of the Arctic: The Interaction Between Early Migration and the Paleoenvironment
http://dx.doi.org/10.1016/B978-0-12-813532-7.00015-2

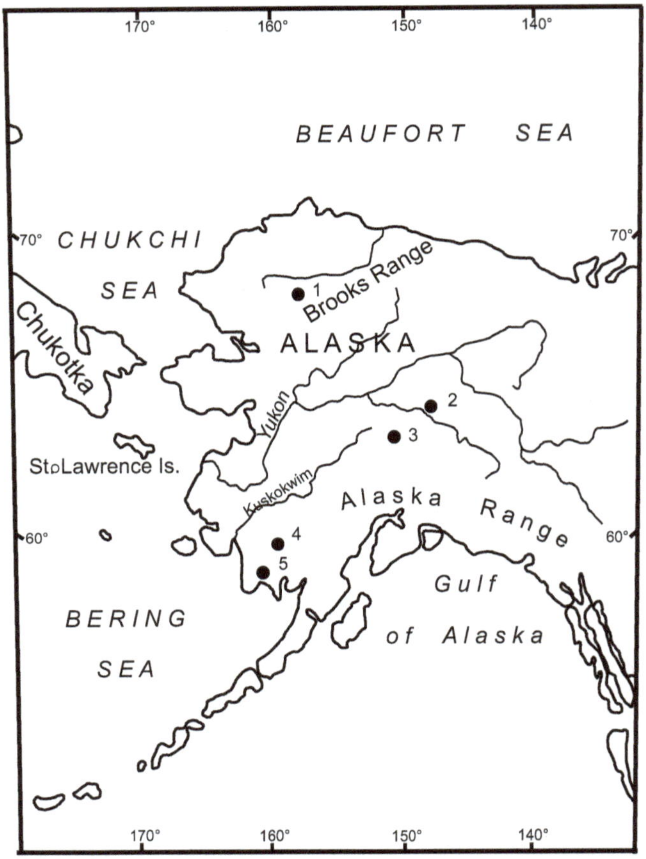

Figure 3.2.1 Geographic position of lakes with pollen records used in reconstructions of the Late Quaternary climates and vegetation in Beringia: 1, Tukuto Lake; 2, Harding Lake; 3, Windmill Lake; 4, Grandfather Lake; 5, Nimgan Lake.

shrub willow tundra). At the transition to the Holocene the climate became close to temperate, and the landscapes were dominated by the birch shrub tundra all over Beringia. It should be noted, however, that the appearance of the shrub birch tundra occurred not simultaneously in various regions of Eastern Beringia. The *Betula* percentage in lacustrine pollen records increases about 14–13 ka BP (^{14}C), or 16.8–15.44 cal. ka BP in western and central Alaska ((Anderson et al., 2004) zone HA2, Harding Lake) (Fig. 3.2.2).

The rate of pollen accumulation (pollen influx) in lake sediments in central Alaska suggests, however, that the shrub birch tundra did not become dominant the landscape until 12 ka BP (13.4 cal. ka BP), which agrees with the time of the birch shrub tundra appearance in Western Beringia. The aforementioned chronological speculations on its earlier appearance were based on the radiocarbon dating of the bulk samples of lacustrine sediments, and at present the dates are revised (Bigelow and Edwards, 2001; Abbott et al., 2010). To take but one example, the new data from the Noatak

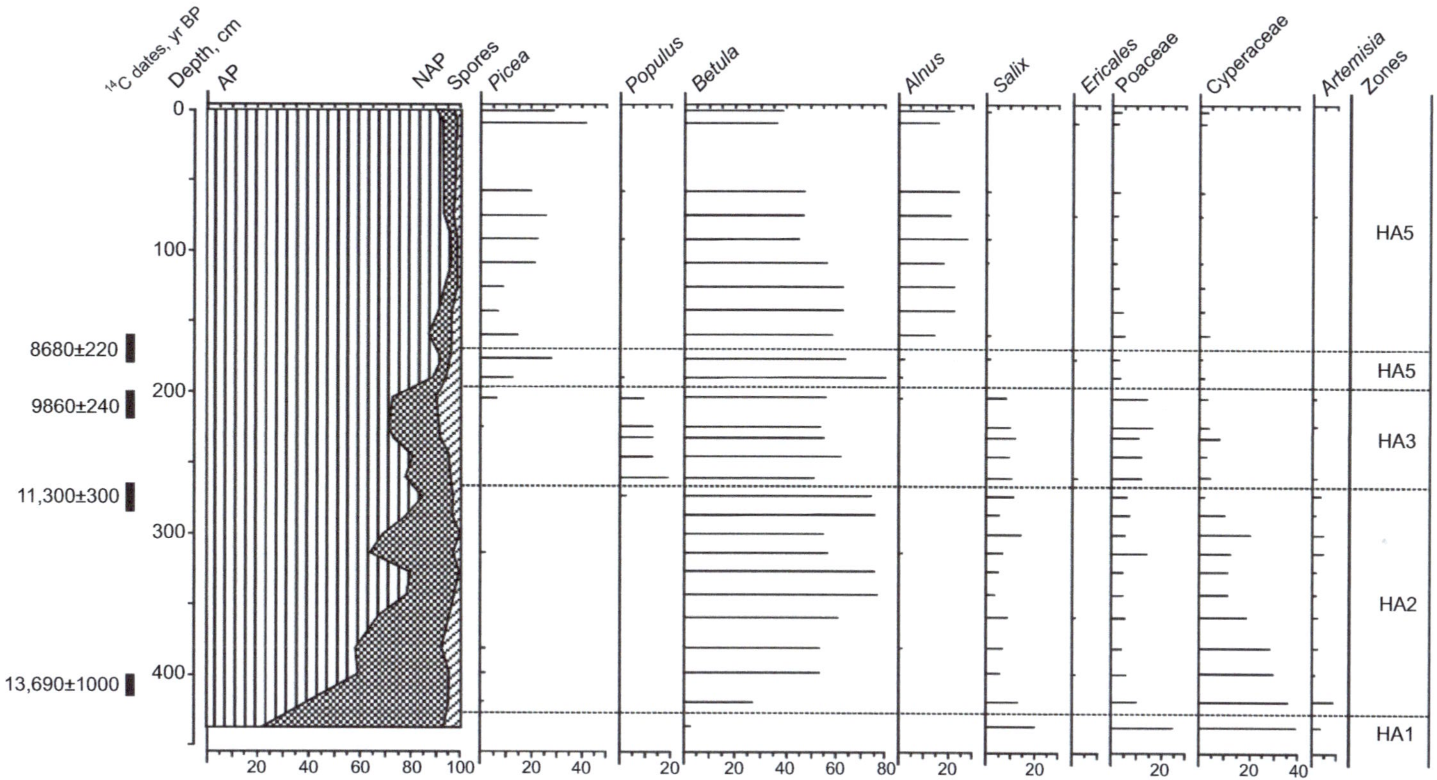

Figure 3.2.2 Pollen diagram of the Harding Lake sequence.
Modified from Ager, T.A., 1983. Holocene vegetational history of Alaska. In: Wright H.E. (Ed.), Late Quaternary Environments of the United States. The Holocene. Univ. of Minnesota Press, Minneapolis, pp. 128–141.

River basin (northwestern Alaska) based on the dating of plant macro-remains suggest the herbaceous tundra replacement by the shrub birch tundra could occur 12,000 cal. yr BP (Abbott et al., 2010).

Studies of the Windmill Lake sediments (Bigelow and Edwards, 2001) (Fig. 3.2.3) in the Nenana River valley also reject a wide occurrence of the shrub birch tundra at the end of the last glacial stage. According to the recent data, the shrub birch tundra appeared there about 11.8 ka BP (13,690 cal. yr BP) (pollen zone W12). Though local shrub birch communities could exist in the west and in the central part of the Brooks Range about 13–12.5 ka BP (15,440–14,680 cal. yr BP), most of the region was dominated by a herbaceous tundra similar to that of the Pleniglacial (Zone TU1, Tukuto Lake) (Fig. 3.2.4). The shrub birch tundra was of a restricted distribution until 10 ka BP (11,110 cal. yr BP); after that, it expands over most of the northern Brooks Range. In the southwest of Alaska the shrub birch tundra could appear in the region of the modern Yukon River mouth ~14 ka BP (16,800 cal. yr BP). On the Kuskokwim and Nushagak lowlands the shrub tundra did not exist until 12–11.5 ka BP (^{14}C). The vegetation in the region varied from the shrub tundra prevalent in the upper reaches of the Kuskokwim River to herbaceous (grass and sedge) communities dominant in the regions south of the Akhlun Mountains (Grandfather Lake, Nimgan Lake, pollen zones NU1, NU2) (Figs. 3.2.5 and 3.2.6).

A wide occurrence of the shrub birch tundra about 12 ka BP (^{14}C) suggests an increase in summer temperatures and moisture supply, though the vegetation as a whole indicates the environments to be colder and dryer than those of today. There are data on the rising lake levels in the east of central and northwestern Alaska, probably as a result of increasing rainfalls or more effective moisture supply as compared with the last glacial stage (Edwards et al., 2001; Abbott et al., 2010).

The climate reconstructions based on fossil beetles in Southeastern Alaska indicated the following mean seasonal temperatures ~12.5 ka BP (^{14}C): 13.5°C in summer and −30°C in winter (Elias et al., 1996). Between 11.5 and 11 ka BP (13,330–12,900 cal. yr BP) the data on fossil insects indicate the mean summer temperatures in the north and in the south of the Bering Strait to be 11.5–13.0°C and 9.25–10.25°C, respectively. The mean winter temperatures in the south of the strait varied from −28.5 to −23.5°C and in the north from −32.25 to −27.25°C. Those conditions were warmer than at present. They agree with the paleoclimatic scenario developed by M.E. Edwards et al. (2001) on a basis of the computer simulations of paleoclimates and the recent models of anomalies; the latter indicate that the west of Eastern Beringia was warmer than its east (unlike the modern temperature distribution, the present-day temperatures in the west of Eastern Beringia are lower than in the east of the region).

As has been shown by L.B. Brubaker et al. (1983), the shrub birch tundra was not equivalent to that existing at present in the north of the Brooks Range. Probably, there were higher shrub forms, not unlike those found now on the uplands bordering the range on the south. Studies of fires in Alaska (Higuera et al., 2008) showed a high frequency of charcoal occurrence in lacustrine sediments being related to landscapes abounding with easily flammable birch trees.

About 11 ka BP (12,920 cal. yr BP) the first postglacial forests appeared in many regions of Eastern Beringia (Anderson et al., 2004; Edwards et al., 2005). Unlike the

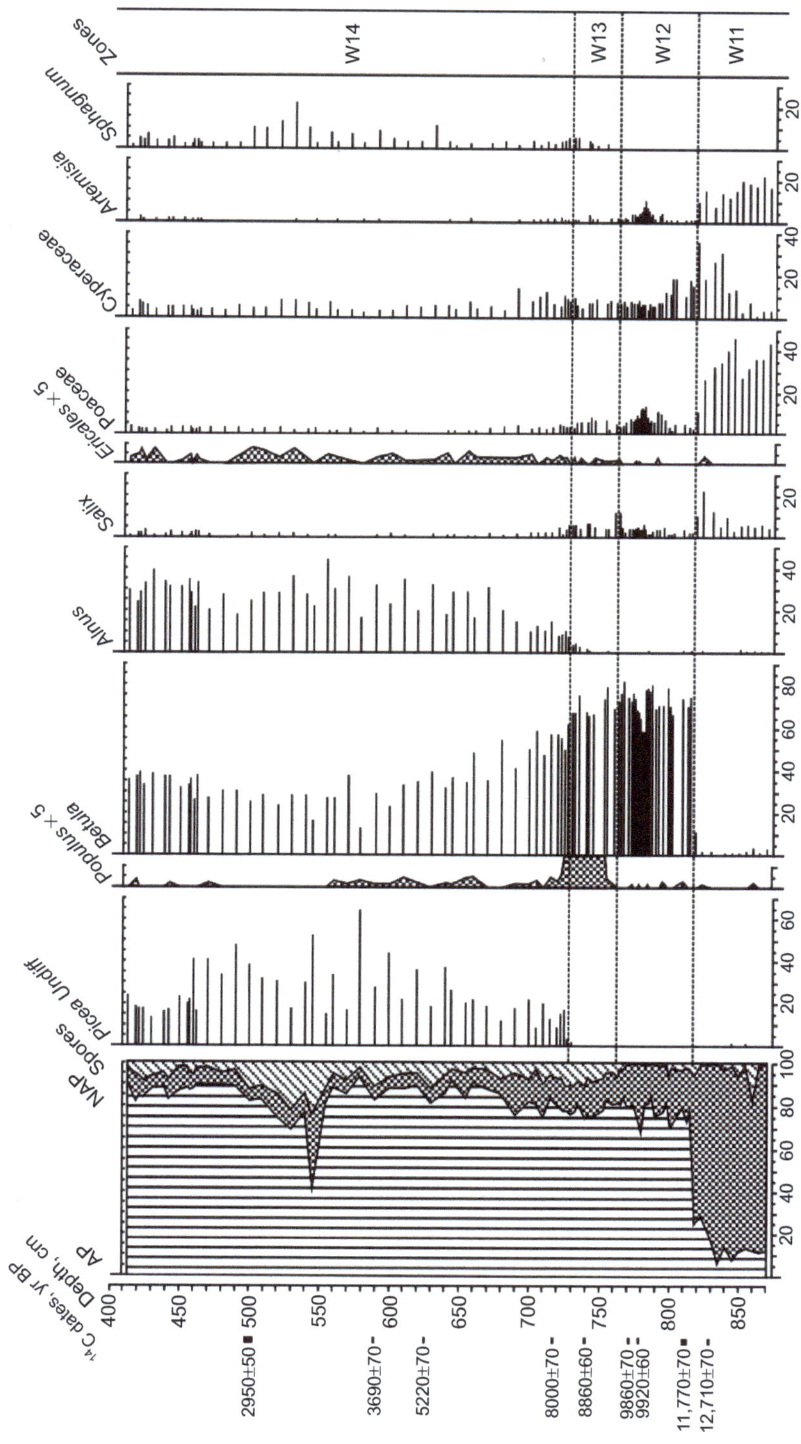

Figure 3.2.3 Pollen diagram of the Windmill Lake sequence.
Modified from Bigelow, N.H., Edwards, M.E., 2001. A 14,000 yr paleoenvironmental record from Windmill Lake, central Alaska: Late Glacial and Holocene vegetation in the Alaska Range. Quaternary Science Reviews 20, 203–215.

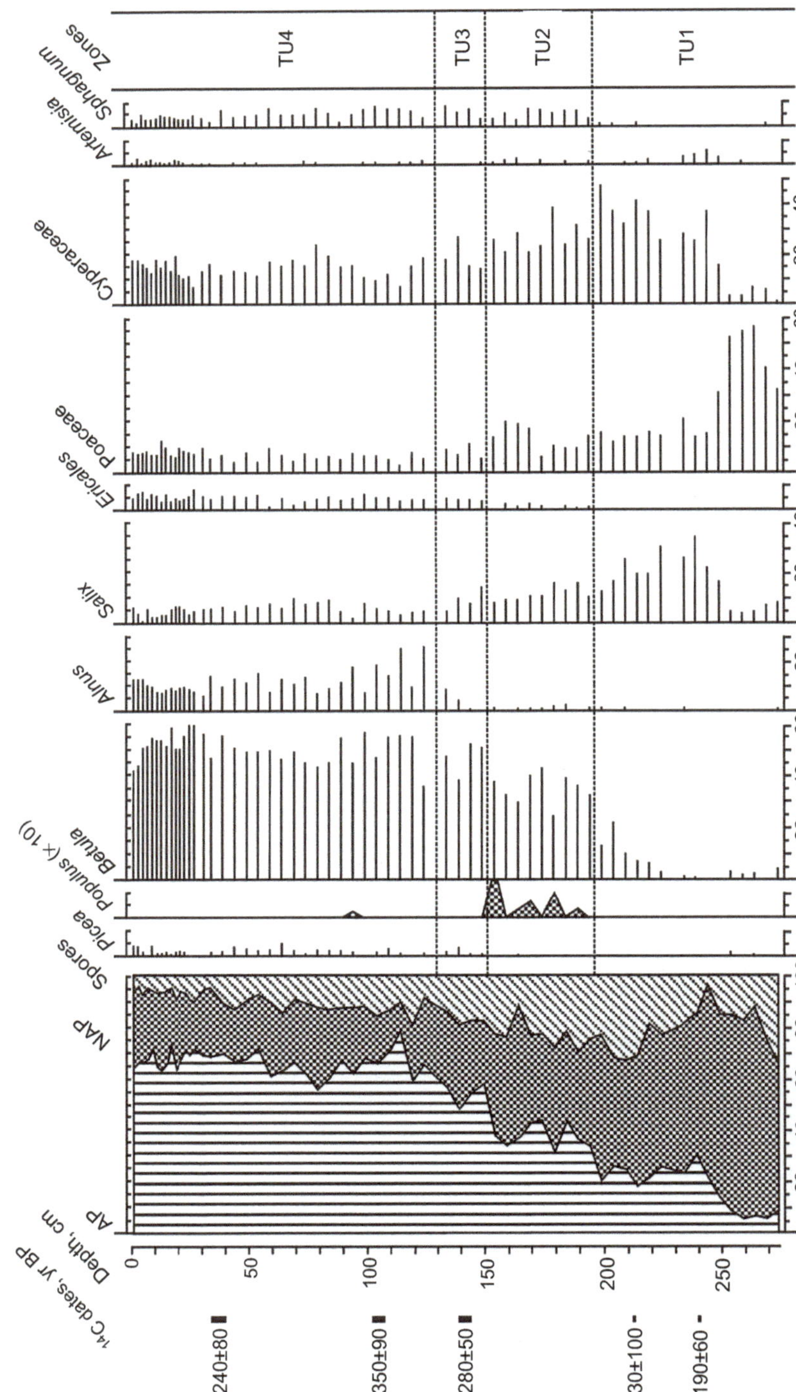

Figure 3.2.4 Pollen diagram of the Tukuto Lake sequence.
Modified from Oswald, W.W., Brubaker, L.B., Anderson, P.M., 1999. Late Quaternary vegetational history of the Howard Pass area, northwestern Alaska. Canadian Journal of Botany 77, 570–581.

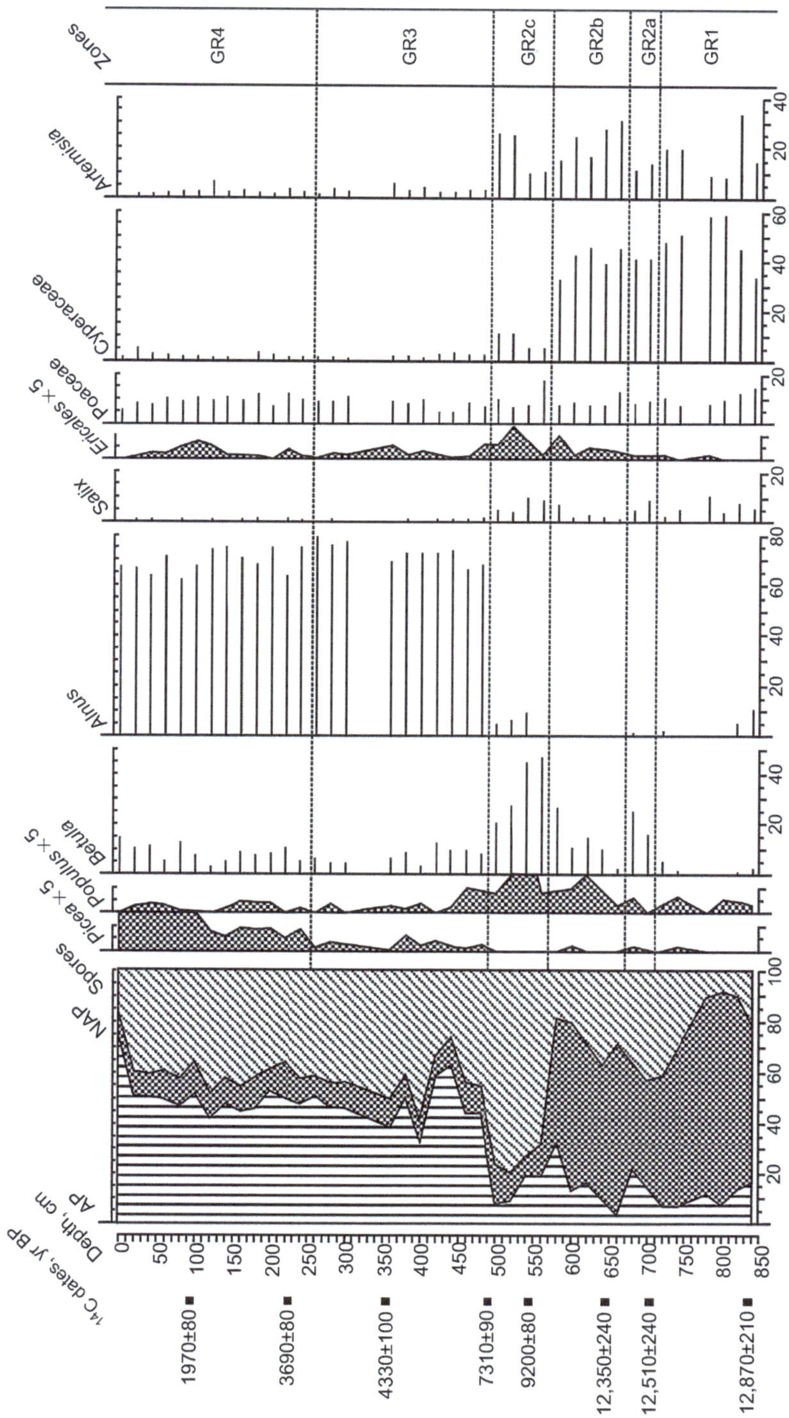

Figure 3.2.5 Pollen diagram of the Grandfather Lake sequence.
Modified from Hu, F.S., Brubaker, L.B., Anderson, P.M., 1995. Postglacial vegetation and climate change in the Northern Bristol Bay region, Southwestern Alaska. Quaternary Research 43, 382–392.

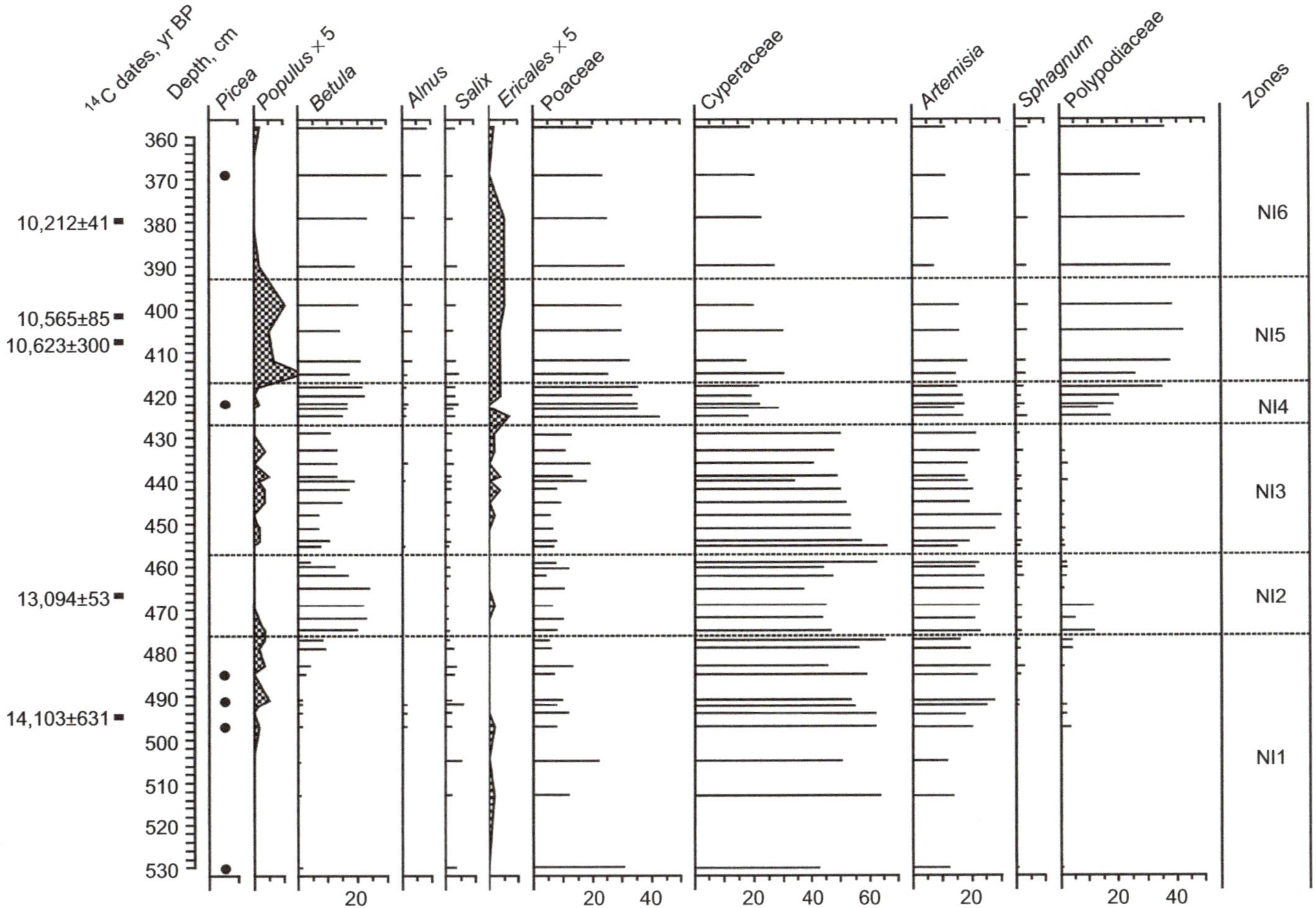

Figure 3.2.6 Pollen diagram of the Nimgan Lake sequence.
Modified from Hu, F.S., Lee, B.Y., Kaufman, D.S., Yoneji, S., Nelson, D.M., Henne, P.D., 2002. Responses of tundra ecosystems in southwestern Alaska to Younger Dryas climatic oscillations. Global Change Biology 8, 1156–1163.

present-day forests dominated by spruce, those were composed mostly of deciduous trees, the poplar being the main forest-forming species.

According to the pollen analysis of the lacustrine sediments, forests were widely spread in Eastern Beringia about 11–9 ^{14}C ka BP (12,920–9985 cal. yr BP). Quite possibly, thick forests could exist in central Alaska (zone HA3, Harding Lake; zone W13, Windmill Lake) and south of the Brooks Range. In the north of the Brooks Range (zone TU2, Tukuto Lake) and near the southwestern coasts of Alaska (zone GR2, Grandfather Lake) the forested areas were limited, and the treeless areas were occupied by grass-*Betula-Salix* tundra.

The wide distribution of poplar suggests a climate different from that of today. Most of the reconstructions of the Eastern Beringian paleoclimates suggest that 11–9 ^{14}C ka BP the summer was warmer and dryer than at present. Not only was a higher percentage of *Populus* in pollen spectra recovered from lacustrine sediments dated to that interval, but also macrofossils indicate the tree expansion was beyond their present-day northern boundaries. The climate warmer than at present is indicated by finds of *Castor* remains, by signs of intensified eolian activity, and the ice wedge melting out. Plant macrofossils and reconstructions based on fossil insects suggest the summer temperatures were 3.5°C above the present-day ones.

In central Alaska the lake levels were at an intermediate position: they were higher than at the time of the birch tundra dominance, though lower than at present (Edwards et al., 2001). Most of data on the south of Eastern Beringia indicate its cold climate at the Younger Dryas, while Central and Northern Alaska were dominated by poplar forests that suggest warmer conditions (Kokorowski et al., 2008).

In the Nimgun Lake sedimentary sequence, for example, those climatic changes are noted in pollen assemblages (zone NI5) for a reduced amount of *Betula* pollen (from 25% to 4%) and an increase of *Cyperaceae* proportion (from 37% to 63%). The data on northern regions are scarce or lacking completely. To take but one example, we consider the statistically processed pollen data of the Windmill Lake sequence dated by ^{14}C to 10.5–10.2 ka BP (12,400–11,970 cal. yr BP) (zone W12). As follows from the analysis, the pollen assemblages of the zone are closer in composition to those of the predominantly NAP zone of the last glacial stage than to any other pollen assemblage (Bigelow and Edwards, 2001). Such a specific reaction of the vegetation in certain geographic regions resulted, most probably, from a differences in the atmospheric pressure: the atmospheric high in central Alaska induced warm and dry conditions, while the Aleutian Low together with low temperatures of the sea surface account for cold summers in the south of Alaska (Kokorowski et al., 2008).

Further changes in the vegetation are recorded between 9.5 and 9 ^{14}C ka BP (10,740–10,970 cal. yr BP). Plant macrofossils show a development of small populations of spruce near Tangle Lakes about 9.1 ^{14}C ka BP (10,240 cal. yr BP) (Schweger, 1981) and in the lower reaches of the Tanana River 9.5 ^{14}C ka BP (zone HA4, Harding Lake sequence (Hu et al., 1993)). The data on macrofossils from the Windmill Lake sediments (zone W14) showed a relatively late appearance of spruce in the Nenana River valley: 8 ^{14}C ka BP (8980 cal. yr BP).

Pollen data obtained from bulk samples of the lacustrine sediments give evidence of spruce advance from east to west, from the Nenana River valley to the regions south of the central Brooks Range and the upper reaches of the Kuskokwim River: 9–8 ^{14}C ka BP (Brubaker et al., 2001).

Forests dominated by *Picea glauca* were mostly confined to river valleys and well-drained slopes. In the regions lacking in spruce forests, in central and south-western Alaska, poplar forests were gradually reduced (zone GR3, Grandfather Lake), probably persisting as separate populations or being replaced with shrub tundra (Brubaker et al., 2001; Anderson and Brubaker, 1994). Shrub birch tundra was dominant in the north of the Brooks Range (zone TU3, Tukuto Lake) and locally in northwestern Alaska. The Early Holocene climate was colder and more humid as compared with the postglacial thermal optimum, at the time of widely spread poplar forests (8.5 ^{14}C ka BP, 9500 cal. yr BP). The lake levels in central Alaska were close to the present-day ones due to rather humid climate (Edwards et al., 2001).

The archaeological investigations provided convincing evidence of an earlier appearance of humans in Beringia: ~27 ^{14}C ka BP in the west of the region, in the Yana River drainage basin. That time was transitional from temperate interstadial conditions to much more severe climates of the LGM. As follows from the paleobotanic data, the last glacial stage 27–12.5 ka BP was typically characterized by scanty vegetation and a harsh climate.

The remains of mammoth fauna, however, suggest more productive environments capable of supporting relatively rich megafauna of the Late Pleistocene. The absence of archaeological finds dated to the LGM in Beringia may be interpreted as evidence of human absence from the region during that time. But it is not improbable that people came to Western Beringia in small groups (for example, at the Berelekh mammoth cemetery in the lower reaches of the Indigirka).

As the Beringian prehistory indicates, the earliest and long-lasting penetration of human populations occurred at the transition from the Late Pleistocene to Holocene, approximately 12.5–9 ^{14}C ka BP (14,680–9985 cal. yr BP).

It was a time of radical changes in the Beringian environments, with the sea level rising, mountain glaciers shrinking, and plant and animal communities changing. In spite of some inconsistencies in the chronology of vegetation changes in different regions of Beringia, it is possible to identify four principal intervals, as follows:

1. the early postglacial climate mitigation with the climate warmer and more humid than before, though still colder and drier than at present (12.5–11 ^{14}C ka BP, 14,650–12,920 cal. yr BP);
2. postglacial thermal maximum (11–9 ^{14}C ka BP, 12,920–9985 cal. yr BP) as a result of higher summer insolation compared with that of today;
3. climatic fluctuations (most pronounced in the south of Alaska) associated with the Younger Dryas (11–10.2 ^{14}C ka BP, 12,920–11,970 cal. yr BP);
4. a new Early Holocene cooling since ~9 ^{14}C ka BP (9985 cal. yr BP) marked by an increase in precipitation.

The regional vegetation also displays a series of changes in response to changing climate:

1. shrub tundra;
2. deciduous forests;
3. shrub or herbaceous tundra (in some regions of Beringia);
4. mixed forests of deciduous and coniferous species.

There is a discernible difference between the plant communities in the western and eastern parts of Beringia. The birch shrub tundra replaced the grass and herb one in all the parts of Beringia. Shrub alder (*Duschekia fruticosa*) spread quickly over Western Beringia about 10 ^{14}C ka BP (11,115 cal. yr BP), while it was not typical of the Eastern Beringia until 8–7 ^{14}C ka BP (8960–7790 cal. yr BP). The deciduous forests in Western Beringia also differ from those in Eastern Beringia: the former were dominated by larch, and the latter by poplar. Finally, there was a difference between the regions in coniferous constituent: *P. glauca* is the principal component of forests in Eastern Beringia; the forests in Western Beringia are mostly of larch, *Pinus pumila* being one of the main elements in underbrush.

Considering the areas of early colonization in Beringia, grass tundra gave place to shrub tundra about 12–11 ^{14}C ka BP (14,000–12,920 cal. yr BP) in central and southwestern Alaska and about 10 ^{14}C ka BP (11,110 cal. yr BP) in the north of Alaska. In Western Beringia the same change took place about 12.4 ^{14}C ka BP (14,260 cal. yr BP). Neither wide occurrence of deciduous forests, nor warmer and dryer climates than at present were essential in the human expansion to the Arctic regions. In Eastern Beringia, such an expansion coincides with periods of warming (*Populus* zone) and periods of the climate instability (transition from grass tundra to birch shrub tundra; climate oscillations during the Younger Dryas; transition to mixed coniferous forests), as well as at the time when landscapes differed noticeably from those of today (large shrub tundra, poplar forests) (Fig. 3.2.7).

The studies were carried out with financial support of the FEB RAS (project 09-I-OH3-11), target program of the FEB RAS (project 09-II-YO-08-003).

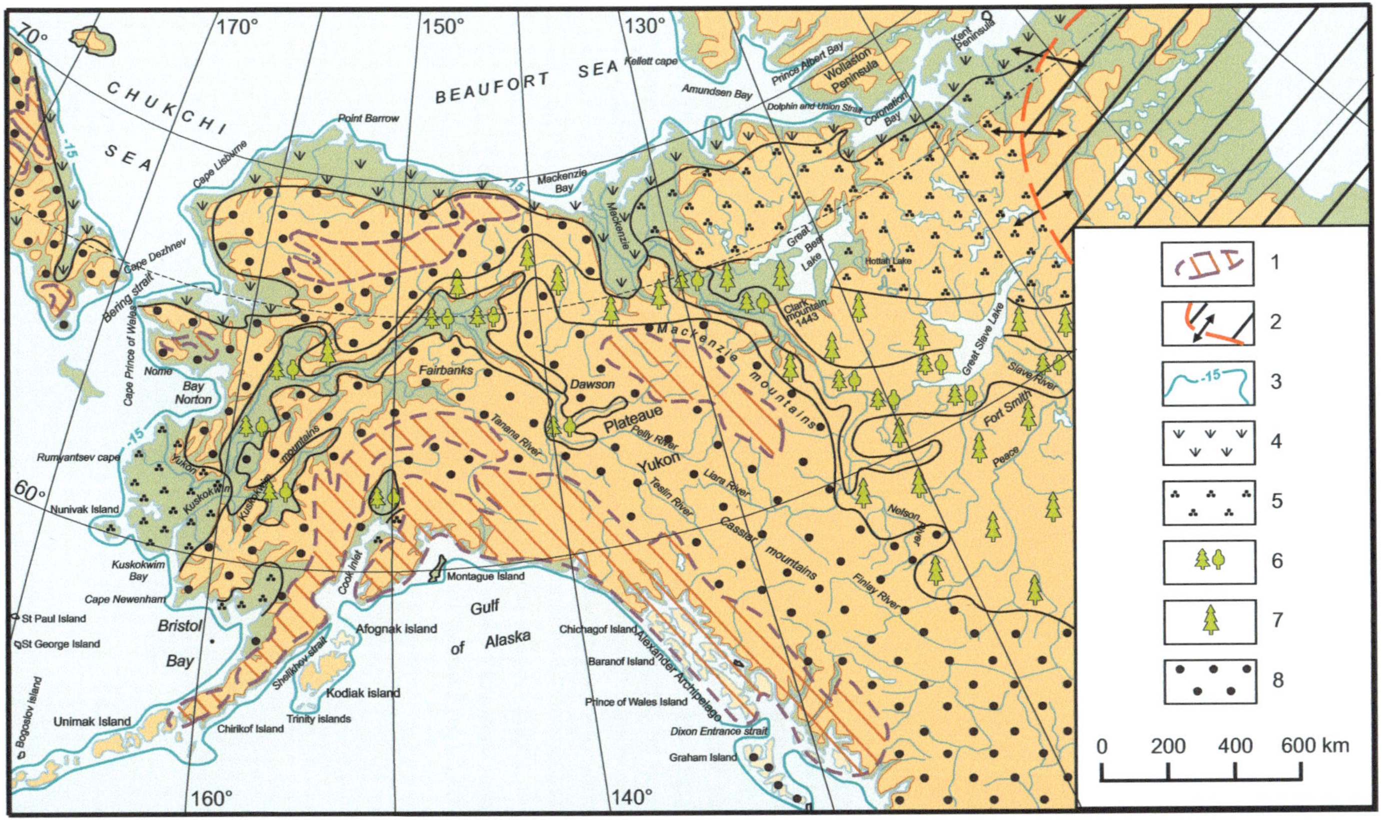

Figure 3.2.7 The landscapes of the subarctic zone in North America at the Late Glacial (about 12 ^{14}C ka BP). Compiled by O.Yu. Glushkova and A.V. Lozhkin. 1, areas of alpine glaciation; 2, presumed position of the Laurentian ice sheet; 3, supposed position of the World Ocean coastline; 4, sedge and cotton grass tundra, with dwarf shrub willow; 5, dwarf shrub tundra; 6, poplar forest with some dark coniferous trees along large river valleys (up to 500–600 m a.s.l.); 7, dark coniferous (spruce) forest (up to 1600–1800 m a.s.l.); 8, mountain forest-tundra and tundra.

The Paleolithic of eastern Beringia from western Alaska to Canadian Yukon

3.3

Kelly E. Graf, Ted Goebel
Texas A&M University, College Station, TX, United States

Chapter Outline

Human Colonization of the Arctic: The Interaction Between Early Migration and the Paleoenvironment
http://dx.doi.org/10.1016/B978-0-12-813532-7.00016-4

3.3.1 Introduction

In this chapter, we review current archaeological evidence of Paleolithic humans in Alaska and the Canadian Yukon. During the Late Pleistocene, the Bering Land Bridge connected this region with northeast Asia (western Beringia) to the west, and the vast Laurentide and Cordilleran ice sheets separated it from the rest of ice-free North America to the east. As ice sheets melted and sea level rose at the end of the Pleistocene, however, the land bridge became submerged and "ice-free corridors" opened in the interior and maritime regions of western Canada, effectively separating eastern and western Beringia but reconnecting eastern Beringia to the rest of North America. It was during this dynamic period of climatic and environmental change that Late Glacial humans appear to have dispersed into eastern Beringia.

At the time of this writing, at least 22 Late Pleistocene–aged Paleolithic sites were known from Alaska and the Yukon (Fig. 3.3.1). Conservatively, they range from 12,000 to 10,000 cal. yr BP. Older sites have been proposed, but these lack incontrovertible evidence of early humans. We thus focus attention on the last 2000 years of the Pleistocene, reviewing environmental conditions during this time as well as the archaeological evidence for early human cultures and adaptations. Paleoclimatically, this period can be conveniently separated into three chronozones: the Older Dryas stade, Allerød interstade, and Younger Dryas stade.

Most of the well-dated Paleolithic sites of eastern Beringia have been found in the Tanana River basin, a major tributary of the Yukon River that drains the north slope of the Alaska Range in east-central Alaska. Early sites, however, have also been found in the upper Kuskokwim valley, Seward Peninsula, and Brooks Range of northern Alaska. These sites are typically combined into a single archaeological complex, called the Beringian tradition; however, archaeologists have recognized technological, temporal, and geographic differences between the lithic assemblages, so more specific traditions or complexes have been defined. The Paleoarctic tradition or Denali complex represents a group of archaeological sites from across eastern Beringia that contain evidence of microblade and burin technologies as well as lanceolate bifaces and typically date to less than 11,000 cal. yr BP. The

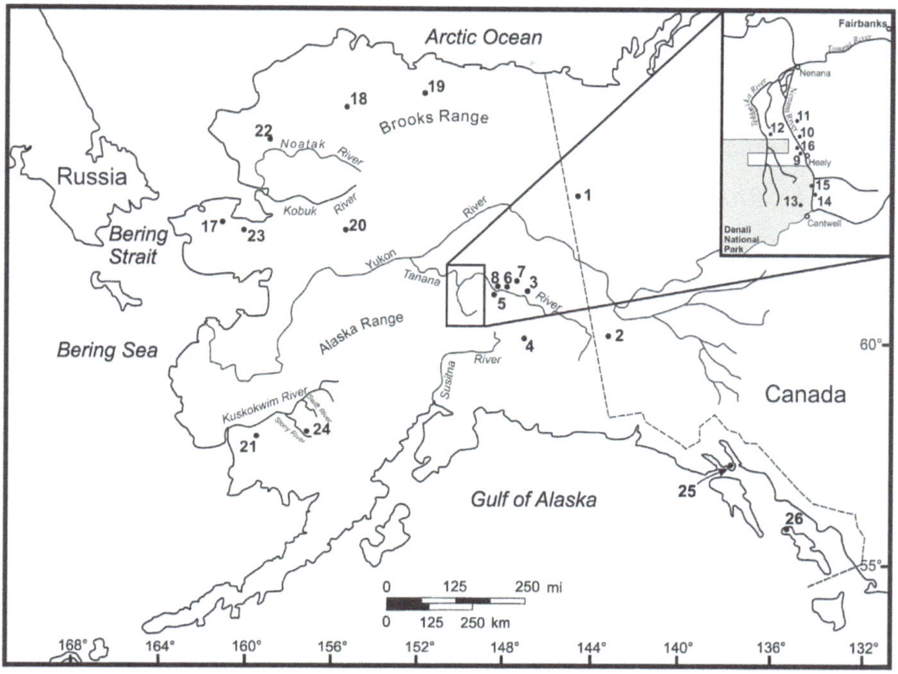

Figure 3.3.1 Map of eastern Beringia with locations of sites discussed in the text. Site numbers correspond to the numbers in the text.

Nenana complex represents a group of sites without microblade and burin technologies but with small teardrop-shaped and triangular bifaces, typically referred to as Chindadn bifaces or points, cobble tools, and unifaces made on large blades and flakes. Nenana complex occupations date to about 11,300–11,000 cal. yr BP at sites in the Nenana valley and 11,800–11,000 cal. yr BP at sites in the Tanana valley. The Northern Paleoindian tradition represents a group of sites in northern and western Alaska that also typically lack microblades but contain large lanceolate bifacial points (sometimes fluted) reminiscent of Paleoindian points of temperate North America. Although much interassemblage variability existed across eastern Beringia during the latest Pleistocene, understanding the causes of this variability are still debated. Some argue that the variation is due to site-specific activity or seasonality differences, while others argue that the sites without microblades are the product of restricted sampling, and that continued work at these nonmicroblade sites likely would minimize the differences. Nonetheless, there do seem to be some strong temporal and geographic patterns to the recognized variability, suggesting to some archaeologists that many of the differences between assemblages could reflect different Paleolithic populations—or cultures—that existed during the latest Pleistocene of eastern Beringia.

3.3.2 Characteristics of ancient-most archaeological sites of eastern Beringia

3.3.2.1 Sites from the Porcupine River and Upper Tanana River Valleys

3.3.2.1.1 Bluefish Caves

The Bluefish Caves are located about 50 km southwest of the town of Old Crow, Yukon Territory. The site includes three small karst cavities in limestone bedrock situated near the top of a ridge overlooking the Bluefish River (Morlan and Cinq-Mars, 1982). J. Cinq-Mars discovered the site in 1975 and excavated there in 1977–1987 (Cinq-Mars, 1979, 1990; Cinq-Mars and Morlan, 1999; Morlan and Cinq-Mars, 1982). Caves 1 and 2 yielded lithic artifacts and faunal remains in sediments dated to the Late Pleistocene.

The caves are relatively small in size (interior spaces <30 m^3) and contain loose deposits ranging from 30 to 200 cm in thickness (Cinq-Mars and Morlan, 1999). Both caves 1 and 2 have the same basic stratigraphic profile. A deposit of loess (reaching 1.4 m thick) lies unconformably on Devonian-aged limestone. The loess is overlain by a humus-rich layer of angular rock debris, and this is capped by a layer of humus. At least 27 ^{14}C dates have been reported for the site; most of these are on bones recovered from the loess, in association with the stone artifacts. The dates range from 33,550±350 to 11,570±60 cal. yr BP, but are concentrated between around 24,000 and 11,000 cal. yr BP (Cinq-Mars and Morlan, 1999; Morlan, 2003). In Cave 1, Cinq-Mars recovered a small assemblage of microblade fragments, biface thinning flakes, and microdebitage (Cinq-Mars, 1979), and in Cave 2, his team found a similar yet more expressive assemblage of microblade cores, burins, and debitage (Cinq-Mars and Morlan, 1999). Faunal remains were well preserved in the caves, too, and consisted of a number of species, some extinct, including mammoth, bison, horse, sheep, caribou, moose, wapiti, saiga, muskox, lion, cougar, bear, and wolf, as well as a variety of small mammals, birds, and fish (Burke and Cinq-Mars, 1996, 1998; Cinq-Mars, 1990, 2001; Cinq-Mars and Morlan, 1999; Harington and Cinq-Mars, 1995; Morlan, 1988; Morlan and Cinq-Mars, 1982). Some of the faunal remains are interpreted to have been modified culturally, either through butchery or through their use as tools (Cinq-Mars, 2001). The dated fauna indicate that the loess containing the lithic artifacts in Bluefish Caves 1 and 2 accumulated during full-glacial conditions of the Late Pleistocene, but a direct association between lithic artifacts and dated fauna has not been established. In addition to lithic artifacts, some fauna from the caves are interpreted as representing a "bone tool industry" with cores and flakes and other features suggestive of human modification. However, the bone tool component of the Bluefish Caves assemblage is fraught with controversy. Many archaeologists disregard the "bone tool industry" as nothing more than naturally fractured bones, possibly the work of carnivores (Beebe, 1983; Sattler, 1991).

3.3.2.1.2 Little John

The Little John site is located in the uppermost reaches of the Tanana River valley just east of the Alaskan-Canadian border, about 12 km north of the community of Beaver Creek, Yukon Territory. N. Easton has been excavating the site since 2002. To date, approximately 100 m^2 have been excavated (Easton et al., 2011). The site is situated in loess and colluvial deposits capping two glacial outwash lobes, possibly contemporaneous with the Healy and Delta glaciations in the North Alaska Range (Yesner et al., 2011; Easton et al., 2007, 2009; Hoffecker and Elias, 2007). Mantling deposits on the western lobe are thin (approximately 5–30 cm), but contain archaeological remains divided into two discrete cultural horizons: an upper microblade-bearing Denali complex component found in the B2 horizon of the modern soil and a lower Chindadn point-bearing Nenana complex component found in an underlying unweathered loess at the base of the profile (Easton et al., 2007). Two ^{14}C dates are reported from this area of the site and came from the B2 horizon of the modern soil. One date, 1740 ± 40 (Beta-182799) cal. yr BP, was obtained on charred material, while the second date on wood returned a modern result (Easton et al., 2011).

Covering deposits on the eastern lobe are over 1-m thick and contain an upper loess unit ranging from 40 to 60 cm in thickness, a 20–25-cm-thick paleosol complex found near its base, and a 10-cm-thick unweathered loess unit located near the base of the profile just overlying glacial outwash (Yesner et al., 2011). In this eastern lobe of the site, archaeological materials associated with the Denali complex are found within the paleosol complex, while a set of nondiagnostic artifacts generally thought to be associated with the Nenana component from the western lobe are found in the lower loess unit (Yesner et al., 2011; Easton et al., 2007, 2009). Seven AMS ^{14}C dates are reported from this eastern lobe profile. One date, 1620 ± 40 (Beta-231794) cal. yr BP, was obtained on hearth charcoal from the loess overlying the paleosol complex. Five dates, 8890 ± 50 (Beta-182798), 9530 ± 40 (Beta-217279), 9550 ± 50 (Beta-218235), 9580 ± 60 (Beta-241522), and 10,000 ± 60 (Beta-241525) cal. yr BP, were obtained on culturally modified faunal remains from the paleosol complex and found associated with Denali complex lithic materials. Faunal remains from this component were numerous and showed clear signs of human use (Yesner et al., 2011; Easton et al., 2011). The earliest date from the eastern lobe profile, 12,020 ± 70 (Beta-241523) cal. yr BP, was obtained on an unmodified bison bone from the lowest loess unit (Easton et al., 2007, 2009, 2011). To date, little details about the lithic and faunal assemblages from below the paleosol complex have been provided, especially with regard to their association with the dated bone piece described earlier.

3.3.2.1.3 Healy Lake

The Healy Lake site is located at an abandoned Athabaskan Village about 30 km east of Delta Junction. The site was initially located by R.A. McKennan during ethnographic research of the region. Healy Lake sits along the eastern margin of Healy Lake on a bedrock ridge projecting into the lake near its outlet. Healy Lake was initially tested in 1966 by McKennan et al. and further excavated between 1967 and 1972 by J.P. Cook, who exposed an area of about 400 m^2 (Cook, 1969, 1996). The bedrock ridge at Healy Lake is mantled by a gruss layer of weathered bedrock. Overlying the gruss is a set of sands and silts capped by the modern soil. A thick loess layer forms the majority of the profile and consists primarily of fine sands that fine upward to silts (Cook, 1969, 1996). Cook (1996) excavated the site in ten arbitrary, 5-cm-thick levels and recognized three cultural zones. Generally speaking, the lowest cultural zone or "Chindadn" component included materials from layers 6–10, and these were found associated with two paleosols. Sixteen conventional and six AMS ^{14}C dates were obtained mainly on charcoal from these lowest layers. Dates vary considerably both within each layer and between the layers. For example, dates range from 10,500±280 (GX-1944) to 8465±360 (GX-2175) cal. yr BP in layer 10, 9401±528 (AU-2) to 6645±280 (GX-2159) cal. yr BP in layer 9, 11,090±170 (GX-1341) for layer 8, 11,550±50 (CAMS-16523) to 8655±280 (GX-2171) in layer 7, and 11,410±60 (CAMS-15914) to 5110±90 (Beta-76064) in layer 6.

Artifacts associated with the Chindadn layers include a variety of materials such as wedge-shaped microblade cores, microblades, small, teardrop-shaped and triangular Chindadn points, bifacial knives, scrapers, and other flake tools. Hearth features and large concentrations of burnt bones were encountered during excavations of these lowest layers. The great diversity in ^{14}C dates coupled with the thickness of the Chindadn cultural component may indicate postdepositional mixing of the strata, so more than one cultural assemblage is misrepresented as a single cultural occupation (Cook, 1996; Hoffecker and Elias, 2007; Hoffecker et al., 1993).

3.3.2.1.4 Tangle Lakes

The Tangle Lakes archaeological district is riddled with sites that contain Denali complex artifacts. The area, located at the headwaters of the Delta River in the central Alaska Range, contained a large proglacial lake system that formed as mountain glaciers receded up the major river valleys of the central Alaska Range during Late Glacial times (Ten Brink and Waythomas, 1985). Geomorphic processes of both glacier retreat and high water produced a rather irregular landscape with sediment ridges and pocked kettle lakes. Between about 12,000 and 8000 cal. yr BP a high-level phase of the lakes produced lake features that lie about 30 m above modern lake levels, and several Denali complex sites are located on these geomorphic features (West, 1996). Initial archaeological discoveries in the region occurred in 1956 by W. Hosley. Subsequent investigations took place through the 1950s by I. Skarland and C.J. Keim, and beginning in 1964, F.H. West undertook more intensive investigations of the region until 1974 (West, 1996). Over 100 sites have been recorded in the region, and over 20 of these contain Denali complex artifact assemblages (Hamilton and Goebel, 1999; Hoffecker and Elias, 2007). Two of these sites, Phipps and Whitmore Ridge, may contain 10,000 cal. yr BP occupations and are outlined in more detail subsequently.

The Phipps site is located on a lacustrine feature composed predominately of sand. The site was discovered and excavated by West during the 1960s (West et al., 1996a). A single cultural component containing Denali complex artifacts was found associated with a paleosol in a sandy loess unit found about 30–40 cm below the surface of the site (West et al., 1996a) (Fig. 3.3.2). Radiocarbon dates indicate that the site was occupied between 10,230 and 10,150 cal. yr BP. The artifact assemblage numbers over 1600 pieces and includes wedge-shaped cores and core fragments, microblades, burins and burin spalls, gravers, modified flakes, and debitage. The lack of other characteristic Denali complex artifacts such as bifaces is interesting and might indicate

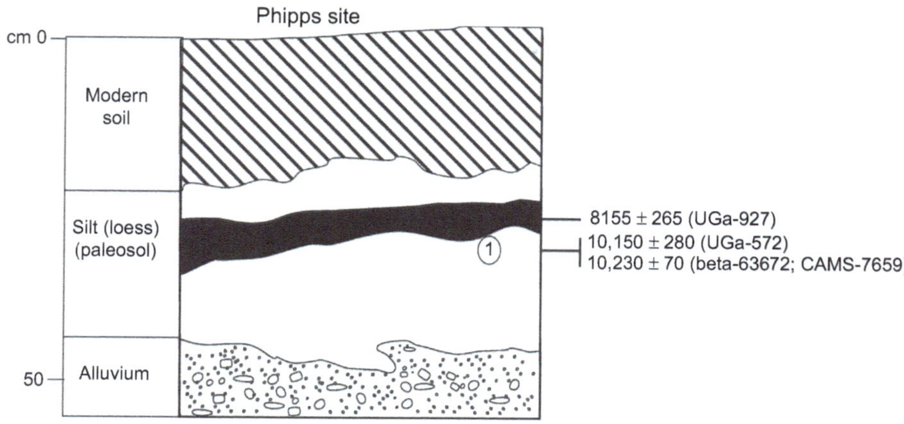

Figure 3.3.2 Stratigraphic profile of the Phipps Site with radiocarbon dates.

activities at the site were limited to slotted point production and maintenance. No faunal remains are reported from the site (West et al., 1996a).

The Whitmore Ridge site is located within 250 m of the Phipps site. Cultural materials were found in a shallowly buried context of a loess unit overlying glacio-fluvial sediments (West et al., 1996b). More specifically, two cultural components were found in a paleosol complex in the lower 10 cm of the loess unit. Four ^{14}C dates, ranging from 10,270–9890 cal. yr BP, obtained on materials from the lower component indicate a possible Younger Dryas age of for these occupation events. Lithic materials in the lower component were limited to two concentrations of activity. Locus 1 contains primarily wedge-shaped microblade cores, technical spalls, and microblades, representing a lithic workshop area devoted to microblade manufacture. Locus 2 reportedly contained bifaces, biface thinning flakes, large blades, burins, and scrapers. This second locus may indicate a camp where bifaces and other various tools were being manufactured and maintained. Lithic materials from the upper component were confined to a single locus. Blades, microblades, small conical blade cores, bifaces, and a burin comprise the assemblage (West et al., 1996b). Hoffecker and Elias (2007) claim this later assemblage is similar to early Holocene-aged component 2 at the Panguingue Creek site and, therefore, may represent an end-of-the-Pleistocene, interior Alaska industry.

3.3.2.1.5 Upward Sun River (formerly known as Little Delta Dune)

The Upward Sun River site is located in the Tanana River valley of central Alaska, near the confluence of Little Delta River and the Tanana River about halfway between Eielson Air Force Base and Delta Junction, Alaska. The site was discovered and initially tested by B. Potter and colleagues prior to 2007, and 16 m^2 have been excavated to date (Potter et al., 2008). Upward Sun River site is located in well-stratified, loess deposits that mantle a stationary sand dune in a large inactive dune field

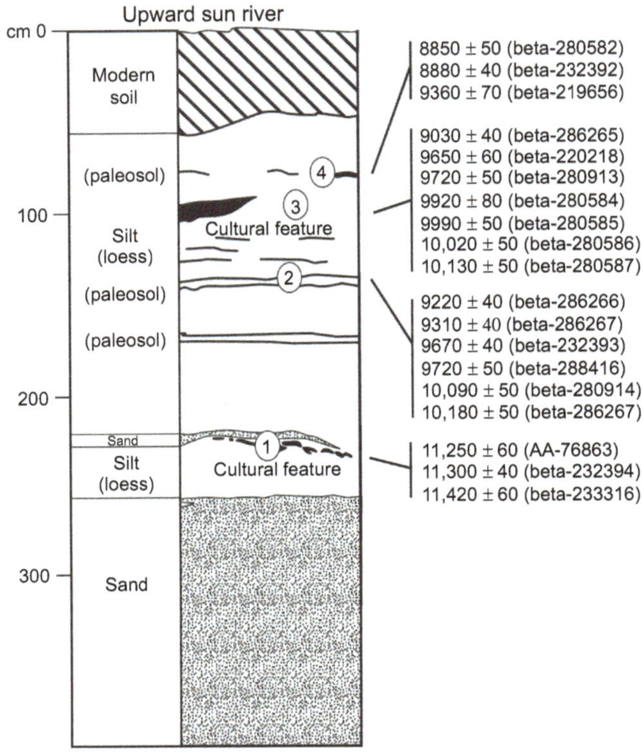

Figure 3.3.3 Stratigraphic profile of the Upward Sun River site with associated radiocarbon dates.

(Fig. 3.3.3). Four stratigraphic units mark the profile beginning with over 100 cm of laminated sands, followed by a 20–25-cm-thick silt zone, a 10–20-cm-thick sand unit, and finally a 160-cm-thick massive silt zone. Several paleosols characterize the upper silt, and the entire profile is capped with a modern boreal forest soil. Four cultural layers are present in the upper silt zone. The lowest cultural component comes from the bottom of the upper silt and has three ^{14}C dates from a single hearth feature of 11,250±60 (AA-76863), 11,300±40 (Beta-232394), and 11,420±60 (Beta-233316) cal. yr BP. The other cultural components are early Holocene in age with cultural component 2 dating from about 10,180–9670 cal. yr BP (paleosol and hearth charcoal), cultural component 3 dating from about 10,020–9360 cal. yr BP (hearth and burial pit charcoal), and cultural component 4 dating to about 8800 cal. yr BP (paleosol charcoal) (Potter et al., 2011). Seventeen lithic artifacts and 320 faunal remains have been recovered from cultural component 1; plus 28 carbonized bearberry seeds came from the hearth sediment (Potter et al., 2008). The faunal remains and seeds indicate an autumn habitation of the site during the terminal Pleistocene occupation event. Though it is early Holocene in age, cultural component 3 contained the burial remains of an infant, the first of its kind in eastern Beringia (Potter et al., 2011).

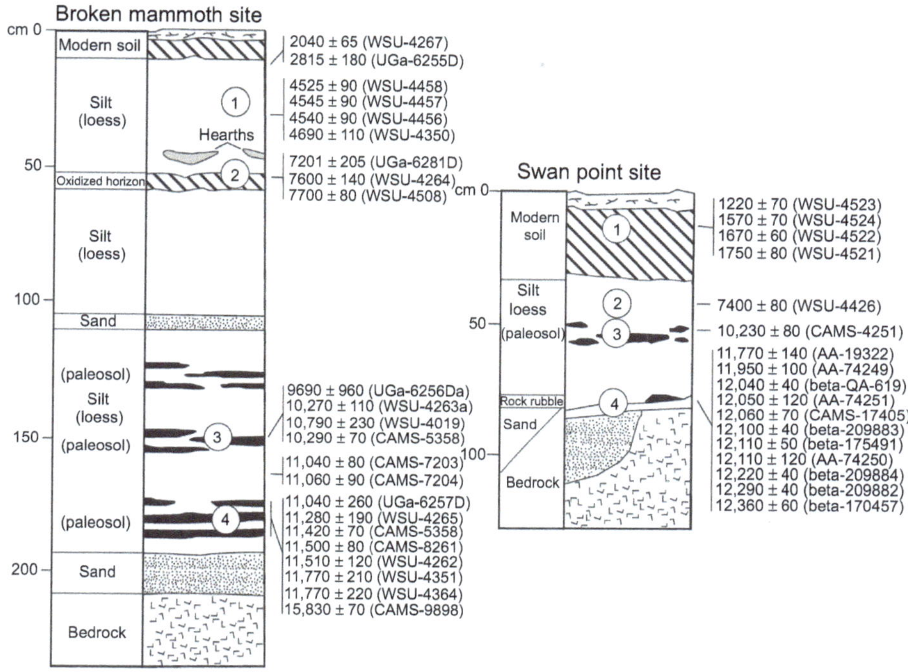

Figure 3.3.4 Stratigraphic profiles of the Broken Mammoth and Swan Point sites with associated radiocarbon dates.

3.3.2.1.6 Swan Point

The Swan Point site is located in the middle Tanana River valley of central Alaska, about 15 km northwest of the town of Big Delta. R. VanderHoek and T. Dilley discovered the site in 1991, and C. Holmes excavated there intermittently from 1992 through 2010 (Holmes et al., 1996; Holmes, 1996, 2011). The site is situated on a dome-shaped bedrock knoll about 25 m above the surrounding Shaw Creek Flats and Tanana River floodplain. On the site, bedrock is covered by about 1 m of unconsolidated sediment (Fig. 3.3.4). The base of the stratigraphic profile consists of up to 40 cm of eolian sand covered by a 3–5-cm thick band of colluvium (Hamilton and Goebel, 1999) or lag as Holmes (2011) suggests. Above this is 70–80 cm of loess; its lower part contains several weakly developed and intermittent paleosols, and its upper part has been masked by a strong B horizon of the modern soil. The profile is capped by a 10-cm thick modern A horizon. Two Late Pleistocene cultural components (called cultural zones, or CZ by the site's excavators) have been identified. The lower of the two, CZ4, occurs at the base of the loess, just above the thin band of colluvium, while CZ3 is associated with a weak paleosol about 20 cm above CZ4. (CZ2 and CZ1 date to later in the Holocene.) Eleven ^{14}C dates on a variety of materials (carbonized fat/grease residue, charcoal, mammoth dentine collagen, and ivory collagen) from CZ4 range from 12,360 ± 60 (Beta-170457) to 11,770 ± 140 (AA-19322) cal. yr BP (Holmes, 2011). Nine of the 11 dates fall between 12,300 and 11,950 cal. yr BP, and taken together, all

11 provide a weighted mean of 12,142 ± 17 cal. yr BP. CZ3has four AMS ^{14}C dates: 10,010 ± 90 (Beta-190578), 10,050 ± 60 (Beta-170458), 10,230 ± 80 (Beta-56666), and 10,570 ± 40 (Beta-209885) cal. yr BP; all are on charcoal from hearths (Holmes, 2011; Holmes et al., 1996).

Excavations at Swan Point continue, so published reports are preliminary. However, it is clear that CZ4 is a wedge-shaped core and microblade industry (Holmes, 2011; Holmes et al., 1996). Recent excavations have produced a representative sample of bifacial core performs, cores, and technical spalls indicating that core reduction followed northeast Asian Yubetsu technology, while preparation and use of end (or *tortsovyi*) cores made on thick flakes is uncommon. Besides microblades, other lithic tools from CZ4 include burins, bifaces, cobble tools, side scrapers, and marginally retouched flakes (Holmes, 2001, 2011). Worked fragments of mammoth ivory have also been found (Holmes, 2001). Other faunal remains include bones of grouse/ptarmigan as well as ducks/geese, teeth of horse, and cervid antler (Holmes, 2011). Hearths from CZ4 do not contain charcoal, but instead are made up of a residue of carbonized fat and grease, likely the product of burning bone in fires instead of wood (Kedrowski et al., 2009).

The appearance of the CZ3 lithic assemblage is very different from CZ4. Although it contains a small set of microblades, no microblade cores or expressive technical spalls have been found. Associated with the microblades are small bifacial points of various shapes: triangular, teardrop-shaped, and concave-based (Holmes, 2011; Holmes et al., 1996). Faunal remains from CZ3 include grouse/ptarmigan, ducks/geese, elk, bison, and fish (Holmes, 2011). Seven hearth features have been excavated; these are described as oval-shaped, unlined concentrations of charcoal, lenticular in cross-section (Holmes, 2011).

Until excavations and analyses have been completed, we cannot ascertain the function of the Swan Point site during the CZ4 and CZ3 occupations.

3.3.2.1.7 Mead

The Mead site is located in the middle Tanana River valley of central Alaska, about 2 km north-northeast of the Broken Mammoth site. T. Péwé (1965, pp. 48–49; Péwé and Reger, 1983, pp. 33–39) originally described it as a geological and paleontological locality; its archaeological significance was not evident until 1991 when a team led by C. Holmes conducted test excavations and found a mammoth ivory point and several mammoth tusk fragments (Yesner, 2000), as well as some stone flakes and tool fragments (Hamilton and Goebel, 1999, p. 167). Excavations at Mead in the early 1990s were limited to just a few square meters, but B. Crass and Holmes resumed excavations there in 2009 (B. Crass, personal communication, 2009). The 1991 finds came from near the base of a 2-m-thick section of loess, in association with a basal paleosol, and are referred to as CZ4. From CZ4 come three ^{14}C dates, two on charcoal of 11,560 ± 80 (CAMS-5198) and 11,600 ± 80 (CAMS-4877) cal. yr BP, and one on ivory collagen of 17,370 ± 90 (CAMS-17408) (Hamilton and Goebel, 1999; Holmes, 2001). Given the regional loess stratigraphy developed by Holmes (2001) for the middle Tanana valley, the date of 17,000 cal. yr BP obviously does not date the cultural component but instead is likely the product of the prehistoric collecting of fossil ivory (Hamilton and Goebel, 1999). More artifacts appear to have been recovered from a paleosol above CZ4 and ascribed to CZ3, dated to ~10,400–10,700 cal. yr BP (Hamilton and Goebel, 1999; Holmes, 2001); however, these materials have not been described in print.

3.3.2.1.8 Broken Mammoth

The Broken Mammoth site is located in the middle Tanana River valley, central Alaska, about 15 km northwest of the town of Delta Junction. C. Holmes and D. McAllister discovered the site in 1989, and Holmes and D. Yesner conducted large-scale excavations there through the 1990s and early 2000s (Holmes, 1996; Krasinski, 2005; Yesner, 1996). An area of ~250 m^2 had been excavated through 2002 (Krasinski, 2005). Results of the excavation have been presented in a series of publications (Holmes, 1996, 2001; Krasinski, 2005; Krasinski and Yesner, 2008; Yesner, 1994, 1996, 2001; Yesner et al., 1992, 2000). The site occupies a south-facing 30-m high bluff overlooking Shaw Creek Flats and the confluence of Shaw Creek and the Tanana River. Site stratigraphy consists of ~2 m of loess and eolian sand lying unconformably on frost-shattered and weathered bedrock (Fig. 3.3.4). The basal covering deposit is eolian sand likely deposited since the Last Glacial Maximum (Holmes, 1996), and this is overlain by 60–65-cm-thick loess deposit with three stratigraphically separated paleosol complexes. Two cultural components dating to the Late Glacial have been identified in these paleosols: CZ4 contains up to three living surfaces associated with the lowest paleosol complex, and CZ3 is associated with the upper paleosol complex (Holmes, 1996). (CZ2 and CZ1 date to later in the Holocene.) Lowest CZ4 has two AMS ^{14}C dates (on charcoal) of 11,770 ± 220 and 11,770 ± 210 cal. yr BP, middle CZ4 has five dates (on hearth charcoal, swan bone collagen, and ivory collagen) ranging from 11,540 ± 140 to 11,280 ± 190 cal. yr BP, and upper CZ4 has one date (on charcoal) of 11,040 ± 260 cal. yr BP. CZ3 has four dates (on charcoal) ranging from 10,790 ± 230 to 9690 ± 960 cal. yr BP (Hamilton and Goebel, 1999; Holmes, 1996). There is also a date of 15,830 ± 70 cal. yr BP on mammoth ivory from CZ4; it is discordant from the rest of the dates and suggests that the site's Late Pleistocene occupants mined ivory from a nearby paleontological locality (possibly Mead?).

Lithic assemblages from Broken Mammoth have been cursorily described (Holmes, 1996, 2001; Krasinski, 2005; Yesner, 1994, 1996; Yesner et al., 1992). CZ4 has yielded an assemblage of ~1300 flakes and 35 tools, including several scrapers, retouched flakes, and cobble scraper-planes (Holmes, 2001; Krasinski, 2005; Yesner, 2001; Yesner et al., 1992). Although bifaces are missing from the assemblage, biface thinning flakes do occur, indicating the reduction of bifacial points or knives on the site. Ivory artifacts were also recovered from the excavations. These include several scratched pieces of ivory (one with a chert flake embedded within it) as well as three worked ivory rods (two of which may have been projectile points and the third an atl-atl handle) (Holmes, 1996, 2001; Yesner, 1994; Yesner et al., 2000). Besides mammoth ivory, economically important fauna recovered through 1991 include steppe bison, wapiti, moose, bear, arctic fox, river otter, hare, hoary marmot, tundra swan, Canada goose, white-fronted goose, snow goose, mallard, pintail, gadwall, widgeon, green-winged teal, willow ptarmigan, and cycloid/salmonid fish (Yesner, 2007). Excavations revealed eight unlined hearth features in CZ4; some reflect generalized work areas, others food preparation areas. A cluster of hearths in the southwest corner of the excavation may represent a group work area, perhaps a set of hearths inside a dwelling (Krasinksi, 2005).

CZ3's lithic assemblage contains ~4000 debitage pieces, including 44 microblades, as well as 41 tools (Krasinski, 2005), though D. Yesner (personal communication, 2011) has stated that these were misidentified in the field. Among the tools are bifacial points (triangular or concave-based forms), bifaces, and scrapers (Holmes, 2001). Osseous artifacts include a beveled bone rod and bone, eyed needle (Holmes, 1996, 2001; Yesner et al., 2000). The list of fauna for CZ3 is very much the same as for CZ4, except that moose, brown bear, river otter, gadwall, and widgeon are not represented, while caribou, mountain sheep, and wolf are added to the list (Yesner, 2001). There are also differences in relative proportions of fauna. Specifically, CZ4 is dominated by avifauna (~60%), while CZ3 is dominated by large mammal fauna (~60%) (Yesner, 1996, 2007). Nine hearths were exposed in CZ3; these were unlined circular to elliptically shaped smears of charcoal. As in CZ4, these represent either general work areas or more specifically food preparation areas (Krasinski, 2005).

Based on ^{14}C dates and assemblage character, CZ4 may be assigned to the Nenana complex (Yesner, 1996, 2001, personal communication, 2011); however, the lack of diagnostic lithic artifacts makes this assignment provisional (Holmes, 2001). CZ3, with an age of ~10,500–10,000 cal. yr BP and interesting combination of artifacts (i.e., lack of microblades and slotted osseous points combined with the presence of both Chindadn/Nenana complex-like bifacial points and concave-based bifacial points), cannot be assigned to either Nenana or Denali complexes. Throughout the Late Pleistocene, the Broken Mammoth site appears to have been repeatedly occupied as a campsite, maybe even a short-term residential base. Based on faunal assemblages, however, CZ4 appears to have been occupied during late winter and spring months, while CZ3 was occupied primarily during the fall and early winter (Yesner, 2001).

3.3.2.2 Nenana Valley

3.3.2.2.1 Dry Creek

The Dry Creek site is located in the Nenana valley of central Alaska, about 1 km north of the town of Healy. C. Holmes discovered the site in 1973, and W.R. Powers excavated there in 1974–1978 (Hoffecker et al., 1996; Powers and Hamilton, 1978; Powers et al., 1983). In 1992, N. Bigelow and Powers (1994) reopened the excavation to collect ^{14}C samples. An excavated area of about 350 m^2 has yielded ~35,000 lithic articles (i.e., debitage and tools). Dry Creek is situated upon a south-facing bluff exposure of the Healy terrace, tentatively dated to the early Upper Pleistocene (oxygen isotope stage 4). Healy outwash is mantled by nearly 2 m of loess, windblown sand, and paleosols (Thorson and Hamilton, 1977) (Fig. 3.3.5). Two Late Pleistocene cultural components have been identified. The basal component 1 occurs in an unweathered Loess 2 near the base of the profile, while component 2 lies in above-lying Loess 3,

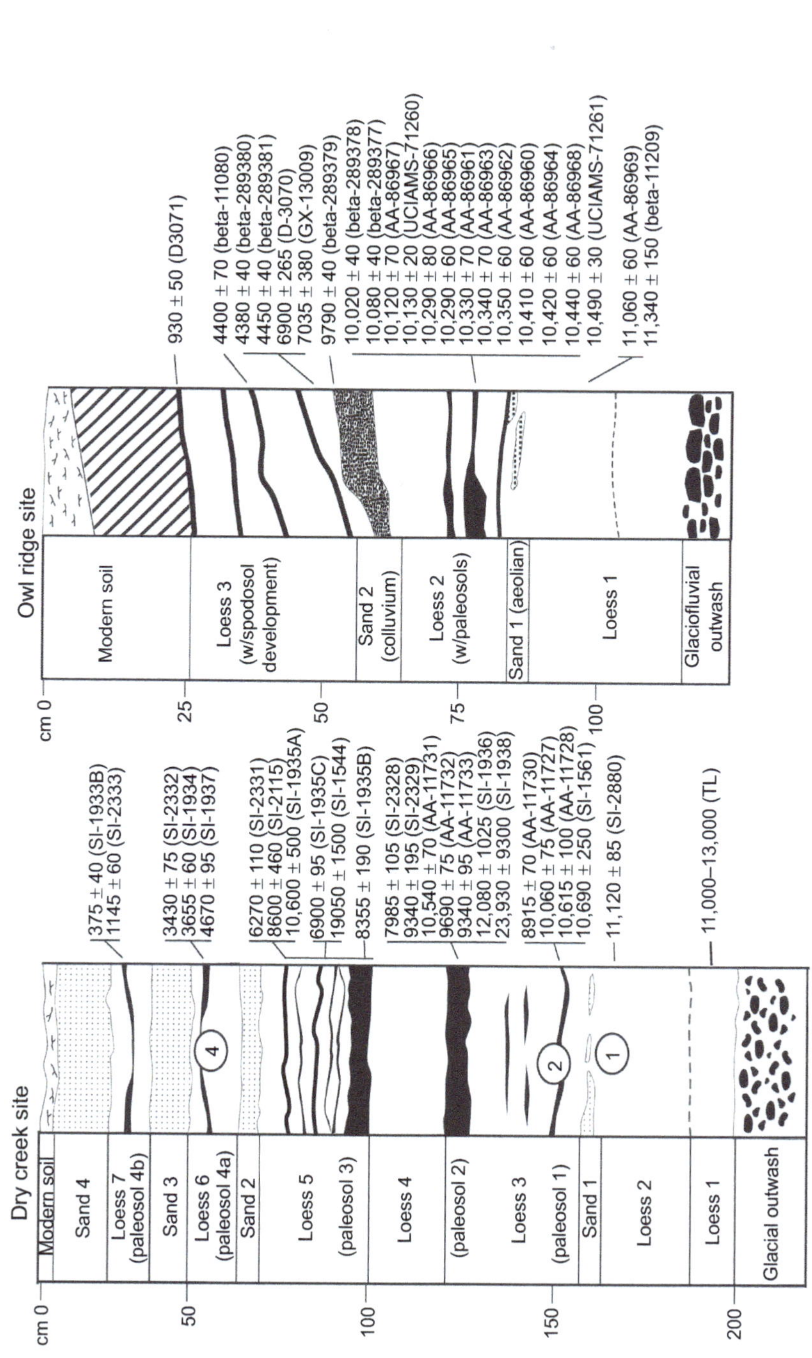

Figure 3.3.5 Stratigraphic profiles of Dry Creek and Owl Ridge with associate radiocarbon dates.

about 15 cm higher in the profile. Loess 3 also contains weakly developed and intermittent Paleosol 1. The two components are separated by Sand 1, a thin band of sand that is discontinuous in places across the site. Charcoal in component 1 was scarce; as a result, there is just one ^{14}C date for this occupation: 11,120±85 (SI-2880) cal. yr BP (Thorson and Hamilton, 1977). This was a conventional date on a bulk sample of charcoal collected from the site's bluff face early in the excavation. Three dates on nonhearth charcoal from Paleosol 1 provide estimates of the age of component 2: 10,690±250 (SI-1561), 10,615±100 (AA-11728), and 10,060±75 (AA-11727) cal. yr BP (Bigelow and Powers, 1994; Thorson and Hamilton, 1977). A seemingly aberrant date of 8915±70 (AA-11730) cal. yr BP was also recovered from charcoal loosely associated with component 2 (Bigelow and Powers, 1994).

Detailed descriptions of the components 1 and 2 assemblages were presented by Powers et al. (1983), with summaries in Powers and Hoffecker (1989) and Goebel et al. (1991). Briefly, the component 1 assemblage contains ~4500 debitage pieces and 56 tools, including three small triangular-shaped bifacial points and point fragments, several bifaces, end scrapers, side scrapers, cobble choppers, and a scraper-scraper plane, marginally retouched blades and flakes, gravers, and notches. One end scraper with a burinated edge also occurs. There are no microblades or microblade cores from component 1, but there are several nondescript flake core fragments and an unprepared bipolar core. Associated faunal remains from component 1 are poorly preserved; however, tooth fragments of Dall sheep and wapiti were recovered (Powers et al., 1983).

The component 2 lithic assemblage consists of >28,000 debitage pieces and 330 tools, including numerous wedge-shaped microblade cores and associated technical spalls, microblades, lanceolate bifacial points, bifaces, burins, side scrapers, marginally retouched blades, microblades, and flakes, cobble tools (including choppers, chopping tools, and hammerstones), gravers, notch-denticulates, and one end scraper. Faunal remains include a few teeth of Dall sheep and steppe bison (Powers et al., 1983). No hearth features were uncovered in either component, but artifacts were concentrated into numerous clusters (three in component 1 and 14 in component 2). In component 2, five clusters were rich in wedge-shaped cores, microblades, and burins, while nine were rich in bifacial points (Powers et al., 1983). Typically, these differences were mutually exclusive, leading some to suggest that, during component-2 times, the site was occupied by two different groups of people: those accustomed to making microblades and inset-projectile weapon tips and those accustomed to making bifacial weapon tips (Hoffecker and Elias, 2007).

Based on ^{14}C dates and assemblage characters, component 1 has been assigned to the Nenana complex and component 2 to the Denali complex (Powers and Hoffecker, 1989). Throughout the Late Pleistocene, the Dry Creek site appears to have been repeatedly occupied as a short-term "spike" camp (Powers and Hoffecker, 1989; Powers et al., 1983). The presence of Dall sheep in both components suggests repeated short-term autumn occupations of the site (Powers et al., 1983). Thorson and Hamilton (1977) and others (Dumond, 2001) expressed concerns regarding the geological integrity of the site and clear separation of components 1 and 2. During the summer of 2011, our team, led by K. Graf, revisited Dry Creek and found information that supports Powers' original interpretations. Geoarchaeological analyses from our investigations are underway, and we expect to have results to report in the near future.

3.3.2.2.2 Walker Road

The Walker Road site is located in the Nenana valley of central Alaska, about 16 km north of the town of Healy. J. Hoffecker discovered the site in 1980, and W.R. Powers and T. Goebel excavated there from 1985 to 1989, exposing an area of about 100 m^2 (Goebel et al., 1991, 1996; Powers et al., 1990). The site occupies an alluvial terrace surface just below the major Healy outwash terrace, thought to date to the early Upper Pleistocene (oxygen isotope stage 4). Terrace gravels are mantled by about 1 m of well-stratified loess and eolian sand, with at least two paleosols that date to the middle Holocene. In the lowermost pair of loess units is cultural component 1, ^{14}C dated to 11,820 ± 120 (Beta-11254), 11,300 ± 120 (AA-2264), 11,170 ± 180 (AA-1681), and 11,010 ± 230 (AA-1683) cal. yr BP. All of these dates were run on wood charcoal from two hearths located about 6 m apart. The earliest date was a conventional ^{14}C date, while the other three were AMS ^{14}C dates on individual charcoal pieces (two from the same feature as the conventional date); on this basis, the 11,820 cal. yr BP age has been rejected (Hamilton and Goebel, 1999).

Besides the two hearths, the site's excavators mapped two circular concentrations of lithic debris, which were tentatively interpreted to represent surface dwellings (Goebel et al., 1996; Powers et al., 1990). Lithic artifacts include 4762 pieces of debitage and 209 tools. Technological activities focused on the production of bifacial and unifacial tools from simply prepared flake and blade cores and unprepared bipolar cores. Elements of microblade technology are absent. Tool forms include small teardrop-shaped "Chindadn" bifacial points, end scrapers, side scrapers, gravers, notched denticulates, and cobble tools (including a set of large scraper-planes). No identifiable faunal remains were recovered. Component 1 has been ascribed to the Nenana complex, based on the character of its lithic assemblage and age (Goebel et al., 1991). The occupation represents a short-term base camp or residence.

3.3.2.2.3 Moose Creek

The Moose Creek site is located in the Nenana valley of central Alaska, about 24 km north of the town of Healy. J. Hoffecker discovered the site in 1978, and it was later excavated by Hoffecker in 1979 and 1984, and by G. Pearson in 1996. A total of ~70 m^2 have been excavated (Hoffecker, 1985, 1996; Pearson, 1999). The site is situated on a Tertiary-aged alluvial terrace, 210 m above the modern floodplain of the Nenana River. Terrace gravels are mantled by about 80 cm of loess and windblown sand; these contain two major paleosol horizons. During Hoffecker's (1985) initial excavations, artifacts were found within and below the lower paleosol, and were dated between 11,730 and 8160 cal. yr BP. These dates, however, were from bulk samples of soil organics and were not consistent with sample depths. Pearson (1999) opened a larger (~45 m^2) area, establishing that two Late Pleistocene cultural components were distinguishable in the site's lower loesses. Component 1 occurred in the basal loess, while component 2 occurred 10–15 cm higher, in loess associated with a weakly developed and discontinuous buried A horizon. Pearson (1999) reported two dates on charcoal from hearth features: 11,190 ± 60 cal. yr BP (Beta-96627) for component 1 and 10,500 ± 60 cal. yr BP (Beta-106040) for component 2. Given that these dates were run on culturally produced features, they are taken to represent the ages of the site's two cultural occupations (Hamilton and Goebel, 1999; Hoffecker and Elias, 2007; Pearson, 1999).

Both of the Moose Creek cultural components have yielded small, expressive assemblages of lithic debitage and tools in association with single hearth features. Debitage from component 1 has not been described, but Pearson (1999) noted that nearly all of the debitage and tools originated from the same nodule of basalt. Tools include two small bifacial points (one teardrop-shaped and the other subtriangular), a biface, side scraper, retouched flake, and scraper plane. Component 2 yielded 27 microblades, a flake core, transverse burin, three bifaces or biface fragments, and three bifacial point fragments (among them, bases of two lanceolate-shaped points). Although no microblade core was recovered from component 2, Pearson (1999, p. 338) pointed out that several microblade overshots bore remnants of a wedge-shaped core's bifacial keel. No identifiable faunal remains were recovered from either component. Based on the character of their lithic assemblages and ages, component 1 was ascribed to the Nenana complex and component 2 to the Denali complex (Hamilton and Goebel, 1999; Pearson, 1999). Both occupations seem to represent short-term spike camps probably used in both instances as a hunting overlook (Hoffecker and Elias, 2007; Pearson, 1999).

3.3.2.2.4 Owl Ridge

The Owl Ridge site is located along the Teklanika River, a major tributary to the Nenana River in central Alaska, and about 30 km west-northwest of the town of Healy. D. Plaskett and R. Thorson discovered the site in 1976, and W.R. Powers conducted test excavations there in 1977 and 1979 (Hoffecker et al., 1996; Phippen, 1988). P. Phippen, a student of Powers, returned to the site in 1982 and 1984, excavating an area of 22 m^2 (Phippen, 1988). The site is situated on a south-facing bluff exposure of the Healy-aged outwash terrace, 61 m above the modern river floodplain. On the site, terrace gravels are covered by a series of loess, eolian sands, and colluvial deposits together reaching about 130 cm in thickness (Fig. 3.3.5). Phippen (1988) identified three stratigraphically separated cultural components. Basal component 1 occurred in the lower loess deposits near the base of the profile, stratigraphically below a paleosol dating to the early Holocene. A sample of dispersed charcoal from component 1 produced a conventional ^{14}C date of 11,340 ± 150 cal. yr BP (Beta-11209).

The lithic assemblage from component 1 included 151 debitage pieces and tools; among them were three bifaces, a large cobble chopping tool, end scraper, and side scraper, and several utilized flakes (Hoffecker et al., 1996; Phippen, 1988). One of the bifaces is a nearly complete triangular point similar to Nenana complex points found at Dry Creek, Walker Road, and Moose Creek. No faunal remains or features were found during the excavations, and excavations were not laterally extensive to define prehistoric activity areas.

In 2007 and 2009–2010 the authors returned to Owl Ridge and expanded previous excavations to approximately 65 m^2 (Graf and Blong, 2009; Graf and Goebel, 2010; Graf and Bigelow, 2011). Several new AMS dates were obtained: one date of 11,060 ± 60 (AA-86969) cal. yr BP on wood charcoal (*Salix* sp.) for component 1 and 13 dates ranging from 10,490 ± 30 (UCIAMS-71261) to 10,020 ± 40 (Beta-289381) cal. yr BP all on wood charcoal (*Salix* sp.). Analyses of newly collected assemblages are still underway; however, we found an additional three thin, triangular points in component 1. In component 2, we found one concave-based bifacial point, one lanceolate bifacial point, and one microblade midsection. The character of the lithic assemblages and ^{14}C ages suggest that component 1 is a Nenana complex occupation dating to the end of the Allerød, and component 2 is a subsequent Denali complex occupation that occurred during the Younger Dryas interval. The character of the artifact assemblages, artifact numbers, and distributions of artifacts in both components indicate that Owl Ridge was repeatedly used as a short-term hunting camp/outlook site where bifaces were being manufactured and maintained.

3.3.2.2.5 Bull River II

The Bull River II site is located in the southeastern section of Denali National Park and Preserve in the mountains west of Broad Pass at an elevation of 1030 m (Wygal, 2009). Site stratigraphy is shallow (~30 cm thick) and interpreted to represent alluvial/colluvial processes. Cultural materials were found approximately 15 cm below the surface within a well-developed paleosol. Natural charcoal produced four AMS dates ranging from 10,490 to 10,310 cal. yr BP (Wygal, 2009). Artifacts include more than 500 debitage pieces and 13 tools (five bifaces, one chi-tho, one notch, and six utilized flakes). Based on ^{14}C dates and similarities with the artifacts in nonmicroblade workshop areas of component 2 at Dry Creek, the Bull River II occupation may be assignable to the Denali complex. The nature of the artifact assemblage implies that the site functioned as an upland hunting location where bifaces were manufactured and maintained (Wygal, 2009).

3.3.2.2.6 Carlo Creek

The Carlo Creek site is located in the southern reaches of the Nenana River gorge approximately 10 km south of McKinley Village and 20 km north of Cantwell in central Alaska. The site was discovered and excavated in 1976–1977 by P. Bowers, where he opened an area of about 45 m^2 (Bowers and Reuther, 2008). The site occupies a west-facing, bedrock ridge covered by fluvial, alluvial, and loess deposits on the east side of the Nenana River. Site deposits are rather thick (3.5 m in places) and divided into four stratigraphic units, including a lowermost fluvial gravel and sand unit, cross-bedded fluvial sand unit, alluvial fan sand and silt unit, and uppermost loess that contains lenses of the Mt. Hayes tephra and is capped by a modern boreal forest soil (Bowers and Reuther, 2008). Two cultural components were discovered in the profile with component 1 observed in a paleosol between stratigraphic units 2 and 3 and component 2 observed in the upper loess unit. Originally, four conventional

^{14}C dates, ranging from about 5000–10,000 cal. yr BP, were obtained on one soil humic acid sample and three hearth charcoal samples from two component 1 hearths. Recently, three additional charcoal samples from the two hearths and a bone sample from component 1 were submitted for AMS ^{14}C dating. Five dates were returned from the three charcoal samples. For two of the three charcoal pieces, the samples were split to provide dates on both the charred wood and humates from these samples. The six new dates of 9870±65 (AA-75049), 9650±60 (AA-75050), 9900±50 (AA-75051), 9760±50 (AA-75051), 9980±50 (AA-75052), and 10,040±50 (AA-75052) cal. yr BP indicate an age of about 9700–10,000 cal. yr BP for component 1 (Bowers and Reuther, 2008).

Lithic materials from component 1 include 12 bifaces and 8300 pieces of debitage. No microblade cores or microblades were found at the site. All stone tools were manufactured on local toolstones of argillite and hornfels. Other artifacts include one cobble anvil, two bone points, and one bone awl. The faunal assemblage included bones from one caribou, one Dall sheep, and nine arctic ground squirrel individuals. Based on faunal remains, the component 1 occupation of the site likely occurred during the autumn, and similar to Bull River II the Carlo Creek component I assemblage may represent a Denali complex occupation where hunters manufactured and maintained bifaces (Bowers and Reuther, 2008).

3.3.2.2.7 Eroadaway

The Eroadaway site is located at a roadside pullout on the west side of the George Parks Highway less than 10 km north of Carlo Creek. The site rests on an outwash plain of the Nenana River thought to correspond to the Carlo readvance and date to the Younger Dryas (Holmes, 1988). Several artifacts were found clustered around two possible hearths. Initial test excavations yielded bulk charcoal from one of the features and an age estimate of 8640 cal. yr BP (Holmes, 1988). A new AMS date reported for the feature, however, provided an estimate of 10,890 cal. yr BP, indicating human occupation of the site at the onset of the Younger Dryas. Lithics from the site are mainly bifaces with associated debitage (Holmes et al., 2010). In a presentation to the Alaska Anthropological Association, Holmes et al. (2010) presented metric data showing similarity of the Eroadaway bifaces with Chindadn points from other sites in the region. Possibly, Eroadaway represents a final Nenana complex occupation of the Nenana River valley. Meanwhile, we await a detailed report before making more definitive conclusions.

3.3.2.2.8 Panguingue Creek

The Panguingue Creek site is located in the Nenana valley, 5 km northwest of the village of Healy on a Healy-aged outwash terrace overlooking Panguingue Creek (Goebel and Bigelow, 1996). Site deposits are nearly 200 cm thick and interbedded with alternating loess and colluvium. The three cultural components occur in the upper 70 cm of the stratigraphic profile (Goebel and Bigelow, 1996). Component 1 yielded two AMS dates of 10,180 and 9840 cal. yr BP on natural charcoal. The small lithic assemblage included lanceolate bifacial points, unifacial flake tools, a subprismatic blade/flake core, and debitage. No faunal remains or features were found. This component probably represents a Denali complex occupation, presumably where hunters made a short-term camp (Goebel and Bigelow, 1996).

3.3.2.3 Northern and western Alaska

3.3.2.3.1 Serpentine Hot Springs

The Serpentine Hot Springs site is situated on a bedrock ridge overlooking an unnamed creek that forms part of the headwaters of the Serpentine River near Serpentine Hot Springs in the Bering Land Bridge National Preserve on the Seward Peninsula. The site was initially discovered and test excavated during the summer of 2005 by R. Gal, C. Young, and S. Gilbert-Young and occupies a granitic bedrock outcropping along a tributary of the Serpentine River. Hundreds of lithic artifacts, including one fluted point base made on brown chert, were found littered across the gruss-rich surface of the site in discrete blowout areas. Two 50 cm^2 test squares were excavated within 1.5 m of the location of the fluted point. These tests provided seven flakes and several pieces of charcoal. Most of the materials were found 25–47 cm below surface; however, a chalcedony channel flake was recovered from about 5 cm below surface. Four AMS ^{14}C dates, 9480 ± 40 (Beta-208369), 10,060 ± 40 (Beta-208367), 10,250 ± 60 (Beta-208368), and 10,250 ± 60 (Beta-208370) cal. yr BP were obtained from charcoal collected between 30 and 39 cm below surface (Young and Gilbert-Young, 2007).

In 2009–2011 the authors returned to Serpentine Hot Springs to systematically map and collect surface artifacts from blowout areas and expand previous test excavations to define the geoarchaeological context of the site, search for cultural features, and increase the sample of the buried fluted point assemblage. T. Goebel is directing this project. Analyses from our expedition are currently underway, and we expect to have results to report soon, but presumably this location was used as a short-term hunting camp.

3.3.2.3.2 Mesa

The Mesa site is located in the Arctic Foothills of the Brooks Range of northern Alaska, in the headwaters of the Colville River drainage. M. Kunz and colleagues discovered the site in 1978, tested its deposits in 1979–1980, and excavated there during 1989 and 1991–2002, exposing an area of nearly 300 m^2 (Kunz and Mann, 1997; Kunz and Reanier, 1994, 1995; Kunz et al., 2003). The site occupies a prominent igneous bedrock butte or "mesa" that rises about 60 m above the surface of Iteriak Creek and provides an impressive view of Iteriak valley and its surroundings. Bedrock is mantled by less than 35 cm of silt loess, and cultural materials are limited to the base of these eolian deposits (Fig. 3.3.6). Forty hearth features were recovered from four localities along the east face of the mesa. Twenty-eight of these provided wood charcoal for 51 ^{14}C dates, placing the age of site occupation between about 10,300 and 9300 cal. yr BP or at the very end of the Younger Dryas (Kunz et al., 2003; Mann et al., 2001).

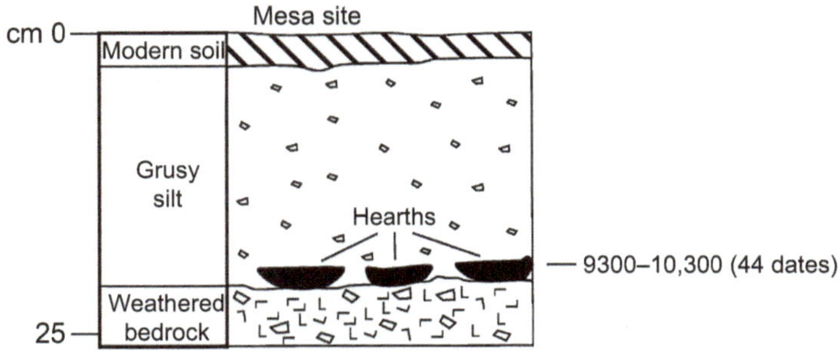

Figure 3.3.6 Stratigraphic profile of the Mesa site.

Besides the 40 hearth features, 121,466 lithic pieces were encountered at Mesa, and most of these were found concentrated around the hearths. Of the lithic materials, only 864 represent worked implements, and of these 43% are bifaces with over 150 of these pieces being projectile points or point fragments (Kunz et al., 2003). These projectile points are primarily lanceolate in shape with concave to straight bases, rather thick, lenticular cross-sections, and edge grinding on hafting elements. A few of the points were fluted; however, this does not appear to have been intentional and likely resulted from impact (Kunz et al., 2003). Other tools include side scrapers, end scrapers, and gravers. The Mesa assemblage has striking similarities with Plains Paleoindian assemblages, which has led M. Kunz and others to claim a possible direct cultural affinity with these groups (Kunz and Reanier, 1995). Interestingly, one of the site localities produced a wedge-shaped microblade core, microblade core fragments, and over 130 microblades that Kunz et al. (2003; Bever, 2006) contend represent a later occupation. No identifiable faunal remains were recovered from the Mesa localities; however, based on the nature of the Paleoindian occupation and the reported Younger Dryas increase in steppe bison in Alaska, Kunz et al. (2003) assume the site's occupants were specialized steppe bison hunters. The discrete clustering of lithic materials surrounding the numerous hearth features seems to provide compelling evidence for repeated occupation of Mesa by terminal Pleistocene–early Holocene foragers.

3.3.2.3.3 Putu-Bedwell-Hilltop

The Putu, Bedwell, and Hilltop sites are all three located in the upper Sagavanirktok River valley along the north flank of the Brooks Range. The Putu site was discovered and excavated during the early 1970s by H.L. Alexander (1987). Putu rests on a ridge about 215 m above the river valley and about 30 m below the top of a shale bedrock knob. Bedrock is mantled by 25–50 cm of loess that is divided into three stratigraphic units. The upper unit (5 cm thick) is characterized by a modern tundra soil, below is a 10–30-cm thick layer of tan loess, which contained cultural materials mainly concentrated near its base, and the lower portion of the profile is a gravelly mixture of rounded glacial stone clasts and decomposed shale bedrock (Alexander, 1987). Fragments of artifacts that refit were found dispersed through the profile. This coupled with the shallow nature of the lowest stratigraphic layers indicate the site witnessed considerable postdepositional disturbance. Six ^{14}C dates were obtained on charcoal from hearth features reported from all three stratigraphic units and range in age from 11,500–650 cal. yr BP (Alexander, 1987). Reanier's (1995) reinvestigation of field notes and ^{14}C lab records prompted him to submit a curated charcoal sample from the same hearth feature that produced the original date of 11,470 ± 500 (SI-2382) cal. yr BP for AMS ^{14}C analysis. This sample returned an age of 8810 ± 60 (Beta-69901 [CAMS-11038]) (Reanier, 1995). Lithic materials from the site include four fluted point fragments, four lanceolate points similar to those found at Mesa, large bifaces, side scrapers, end scrapers, gravers, retouched blades and flakes, blade and flake cores, and blades and blade-like flakes. No faunal remains are reported (Alexander, 1987). Given Reanier's reevaluation of site materials, initial occupation likely occurred during the early Holocene. Stratigraphic issues raised earlier coupled with a mixture of diverse lithic technologies and ^{14}C reversals suggests that the site's deposits are mixed, and clear understanding of the exact timing and number of occupations events will remain a mystery.

The Bedwell site is located approximately 100 m north of Putu along the same bedrock knob. The site assemblage came from a near-surface context of very shallow loess deposits, very similar to the context found at nearby Putu. Lanceolate projectile points make up the majority of the lithic assemblage, and they are thought to be very similar to Mesa points. No faunal remains were found. Reanier (1995) obtained a single AMS ^{14}C date of 10,490 ± 70 (Beta-69895 [CAMS-11032]) cal. yr BP, which suggests an occupation coeval with Mesa and dating to the Younger Dryas.

The Hilltop site is located on a similar bedrock knob about 17 km southwest of the Putu-Bedwell site complex. As with other sites in this area, the lithic assemblage was found in a near-surface context and contains lanceolate points similar to those found at Bedwell and Mesa. Other lithic materials include large bifaces, scrapers, and gravers. Originally, a conventional ^{14}C age of 6169 ± 130 (GaK-4924) cal. yr BP was obtained on a bulk charcoal and black organic material. Reanier (1995), however, obtained an AMS ^{14}C date of 10,360 ± 60 (CAMS-11034) cal. yr BP on similar materials, dating this Paleoindian assemblage to the Younger Dryas.

3.3.2.3.4 Nogahabara-1

The Nogahabara-1 site is located just north of Three Day Slough on the Koyukuk River in the Koyukuk Wilderness Refuge, western Alaska. The site was initially discovered by U.S. Fish and Wildlife Service biologists in 2001, subsequently visited and tested by D. Odess in 2003, and excavated by D. Odess and J. Rasic in 2004. The site is situated within a rather large sand dune field, but specifically the site rests in a flat, open area between two sand dunes. The majority of artifacts collected from the site were found in a surface context. Though nearly 112 m^2 were excavated, only about 20 pieces were recovered from excavation. Lithic materials number 484 and consist of bifacial cores, microblade cores, lanceolate and notched projectile points, bifacial knives, biface performs, scrapers, utilized flakes, and debitage, including microblades and blade-like flakes. The majority of lithic materials were manufactured from Batza Téna obsidian, which is located 140 km east of the site. Faunal remains are quite numerous, but not well preserved. Most pieces were less than 1 cm in dimension and therefore only taxonomically identifiable to either bird or mammal. Two of these bone samples, however, did yield enough collagen for AMS ^{14}C dates of 10,780±70 (CAMS-107314) and 11,815±40 (CAMS-112709) (Odess and Rasic, 2007). Some of the bones were found quite calcined, and the presence of three fire-cracked-rock pieces have led Odess and Rasic (2007, p. 696) to consider the burning of these bone as intentional. Because the Nogahabara-1 lithic assemblage is concentrated, seems to represent a few reduction trajectories, the artifacts of which have the same levels and patterns of wear, and is made primarily of the same raw material, Odess and Rasic (2007) contend it represents a single occupation event and quite possibly the toolkit of a single hunter. The presence of notched points (typically indicative of mid-Holocene occupation) coupled with the site's other lithics and 1000-year spread in ^{14}C dates has led others to question the integrity of single occupation at Nogahabara-1 (Hoffecker and Elias, 2007; Holmes et al., 2008). Likely, this site represents a palimpsest of activity possibly over thousands of years of time.

3.3.2.3.5 Spein Mountain

The Spein Mountain site is located on a basalt bedrock ridge between Spein Mountain and the Kisarilik River, a tributary to the lower Kuskokwim River in far southwestern Alaska. The site was initially discovered in 1979 during archaeological survey of the Kuskowkim Mountains via floating the river (Ackerman, 1996, 2001). The site contains four separate loci (062–065), but only loci 063 provided artifacts from a buried context. In 1992, Ackerman (1996) returned to Spein Mountain to conduct test excavations. Eighty-five square meters were excavated to bedrock, and deposits at loci 063 were found to be thin with a 3–6-cm-thick surface vegetal mat atop a 15–20-cm-thick dark brown loess that overlies weathered basalt regolith or degraded bedrock (Ackerman, 1996, 2001). Cultural materials were observed throughout the profile, but mainly limited to a depth of about 15 cm below the surface in the loess. Therefore, Ackerman (1996, 2001) suggests all materials from the site are from a single cultural component. One AMS ^{14}C date of 10,050±90 (Beta-64471 [CAMS-8281]) was obtained on charcoal from a fire pit feature. Lithic artifacts from both surface and excavation include 96 lanceolate points, one pentagonal point, two willow-leaf-shaped points, 28 bifaces, 32 unifaces, five hammer stones, one abrader, 25 split cobbles, and 4109 debitage pieces. A local argillite was the predominant lithic raw material used, though the presence of a few cherts and chalcedonies in the debitage and unifacial assemblages indicates tools from other regions were brought to Spein Mountain, refurbished there, and then carried away (Ackerman, 1996). Based on the morphology of projectile points and date from the fire pit feature, the Spein Mountain assemblage has been assigned to the Mesa Paleoindian complex, indicating the site served as a hunting camp at the very end of the Younger Dryas (Ackerman, 1996, 2001; Bever, 2001).

3.3.2.3.6 Sluiceway-Tuluaq

The Irwin Sluiceway and Tuluaq Hill sites are located along major tributaries of the Noatak River system on the western flank of the Brooks Range in the Noatak National Preserve. Artifacts found at both of these sites are noted for their morphological similarities to artifacts from Mesa complex sites located on the northern side of the Brooks Range (Bever, 2006; Hoffecker and Elias, 2007). Irwin Sluiceway is situated in the Anisak River valley and was discovered by R. Gal and T. Hamilton in 1992. Excavations of the site were performed in 1994 and 1998 during a joint effort by R. Gal of the National Park Service and D. Stanford of the Smithsonian Institution. A block excavation totaling 17 m^2 uncovered a centralized hearth feature and associated lithic artifact concentration in a near-surface (15–20 cm) depositional context. Three AMS ^{14}C dates, 9550 ± 50 (Beta-120696), 10,050 ± 70 (Beta-134677), and 10,060 ± 80 (Beta-131336) cal. yr BP, were obtained on charcoal from the hearth feature and date the occupation to the end of the Younger Dryas. Lithic materials from the site include a small debitage assemblage, 10 bifacial lanceolate points, other bifaces, and a uniface (Rasic, 2003, 2011). Sluiceway projectile points are broadly similar to Mesa points in that they are lanceolate-shaped, edge-ground, and lenticular in cross-section. Similar projectile points in the Noatak region have come to be known as Sluiceway Paleoindian points (Rasic, 2011).

Tuluaq Hill is situated on Wrench Creek, a tributary to the Kelly and Noatak Rivers, and was discovered by National Park Service archaeologists during regional survey in 1998. Just over 32 m^2 of the site was excavated between 1999 and 2001. The site is positioned on a large bedrock knob and, similar to other sites in the tundra environment of the Brooks Range, cultural materials come from both surface and near-surface contexts. During excavations, several hearth features were found in the shallow sediments of the site. Two of these features provide seven AMS ^{14}C dates. One hearth, feature C, has a date of 7950 ± 40 (Beta-133394) cal. yr BP. The other hearth, feature D, produced charcoal dates of 11,110 ± 80 (Beta-122323), 11,120 ± 40 (Beta-159913), 11,160 ± 40 (Beta-159916), 11,180 ± 80 (Beta-122322), 11,200 ± 40 (Beta-159914), and 11,200 ± 40 (Beta-133393) cal. yr BP (Rasic, 2011).

The lithic assemblage from Tuluaq Hill is quite extensive, covering an area of about 1000 m^2 and including 67 projectile points, over 300 bifaces, 23 unifaces, 1 hammer stone, 1 abrader, 17 cores, and "hundreds of pounds" of debitage pieces (Rasic, 2003, p. 24, 2011; Rasic and Gal, 2000). All artifacts are manufactured from the same, high-quality chert found only 5 km from the site. Nearly all of the projectile points (n = 65) from the site are Sluiceway type points, and Rasic (2011; Rasic and Gal, 2000) indicates that biface reduction materials found at the site are overwhelmingly linked to the manufacture and reduction of these Sluiceway points, which he most recently argues are associated with the hearth feature D dates of 11,100–11,200 cal. yr BP. Two of the artifacts, a fluted point and wedge-shaped microblade core, were found somewhat in isolation among the surface artifacts, but exactly how these implements relate to the other artifacts and hearth features remains ambiguous (Rasic, 2003, 2011; Rasic and Gal, 2000). Another problem is that in other places Sluiceway points are nearly

1000 years younger. Due to its location and proximity to a high-quality raw material, the Tuluaq Hill site seems to represent a lithic workshop/retooling location that may have been used repeatedly by different people for several thousand years (Hoffecker and Elias, 2007; Rasic, 2003, 2011; Rasic and Gal, 2000).

3.3.2.4 Other sites

Several other sites contain compelling evidence for terminal Pleistocene occupations in eastern Beringia. Trail Creek Caves and Lime Hill Cave 1, located in western Alaska, at one time possessed dates of >10,000 cal. yr BP and could still possibly date to the terminal Pleistocene. The sites of Ground Hog Bay 2 and On Your Knees Cave are located in far southeastern Alaska and date to just after 10,000 cal. yr BP. Both contain compelling evidence of initial occupation of this maritime region of Beringia just after the Younger Dryas. Next is a brief review of these other Alaskan sites.

3.3.2.4.1 Trail Creek Caves

The Trail Creek Caves are located in the northeastern corner of the Seward Peninsula. These sites were first discovered by D. Hopkins in 1948 and later excavated by H. Larsen in 1949–1950. In a deeply buried context in Cave 2, microblades and slotted antler points were found associated with a caribou bone that provided a conventional ^{14}C date of 9070 ± 150 (K-980). For many years a couple of conventional ^{14}C dates, 13,070 ± 280 (K-1327) and 15,750 ± 350 (K-1210) cal. yr BP, on a butchered bison calcaneus and horse scapula from the lowest stratum in Cave 9 were thought to represent the oldest evidence of a human presence in eastern Beringia (West, 1981); however, recent taphonomic study of the bison bone indicates gnawing by carnivores, not butchering by humans (Vinson, 1993).

3.3.2.4.2 Lime Hills Cave 1

Lime Hills Cave 1, located in the Kuskokwim River valley in southwestern Alaska, was excavated by R. Ackerman in 1993. Cave sediments are relatively deep (about 120 cm), and Ackerman (1996) noted cultural materials from an upper component and possibly a lower component. In the upper component 58 microblade fragments, 8 debitage pieces, 4 slotted point fragments, and an unslotted point fragment were found loosely associated with nine ^{14}C dates, 8150 ± 80 (Beta-67668 [CAMS-9897]), 8480 ± 190 (WSU-4505), 8480 ± 260 (WSU-4504), 8740 ± 40 (SR-5036 [CAMS-55199]), 9080 ± 50 (NSRL-3800 [CAMS-39350]), 9170 ± 80 (SR-5041 [CAMS-56521]), 9530 ± 60 (Beta-67667 [CAMS-9896]), 9780 ± 40 (SR-5040 [CAMS-56520]), and 10,410 ± 40 (SR-5042 [CAMS-56519]) cal. yr BP. In a lower stratum, two caribou bones, thought to be modified by humans, provided dates of 13,130 ± 180 (Beta-67671) and 15,690 ± 140 (Beta-67669) cal. yr BP (Ackerman, 1996). Ackerman (2011) contended that this lower stratum does not contain evidence of human presence. Both Trail Creek Caves and Lime Hills Cave 1 provide a very similar story. Early on, some thought they both provided evidence for Late Glacial (13,000–16,000 cal. yr BP) occupation of western Alaska; however, both early occupations have been recently dismissed. In addition, both sites do provide evidence of initial use of caves in the region during the early Holocene.

3.3.2.4.3 Ground Hog Bay 2

Ground Hog Bay 2, located on a glacio-marine terrace along an embayment on southwest Chilkat Peninsula approximately 65 km west of Juneau in southeastern Alaska, was excavated in 1971 and 1973 (Ackerman, 1996). Sediments of the beach deposits contain several Holocene-aged cultural occupations beginning at about 9200 cal. yr BP. This lowest component contains wedge-shaped microblade cores and blade and flake cores, bifaces, scrapers, burins, gravers, and microblades.

3.3.2.4.4 On Your Knees Cave

On Your Knees Cave, located in a karst region on the northwest tip of Prince of Wales Island near Sumner Strait, was excavated by T. Heaton in 1994–1996 as part of a paleontological sampling project. Excavations produced several human remains, including the mandible of an adult male, four vertebra, a pelvis fragment, two rib fragments, and an incisor. All of the remains are believed to be from the same individual, isotopic analyses on these remains indicate a diet of almost exclusively marine resources, and two direct AMS ^{14}C dates of 9730 ± 60 (CAMS-29873) and 9880 ± 50 (CAMS-3238) indicate an earliest Holocene age of the individual (Dixon et al., 1997). Recent ancient DNA analyses indicate that the remains belong to the known Native American mtDNA haplogroup D and Y-Chromosome haplogroup Q-M3* (Kemp et al., 2007).

3.3.3 Summary of Paleolithic human dispersals into eastern Beringia at the end of the Pleistocene

Research into the peopling of eastern Beringia by Paleolithic peoples is ongoing. Though we currently have very limited data to paint a complete portrait, we can review the emerging patterns of the initial human dispersal into Alaska, a process that took over 2000 years to unfold between about 12,300 and 10,000 cal. yr BP.

3.3.3.1 First evidence of humans in eastern Beringia

The first unequivocal evidence of initial human settlement of eastern Beringia is represented by the lowest archaeological component (CZ4) at the Swan Point site in the middle Tanana River valley in interior Alaska, and possibly in Bluefish Caves 1 and 2; however, dates from Bluefish are not directly associated with lithic artifacts. Earliest inhabitants arrived by about 12,300–12,000 cal. yr BP during the Older Dryas chronozone. When these hunter-gatherers camped at the Swan Point location, they burned only animal fat and bone resources in their fires, likely because woody resources were not readily available to them at this time and place. They seem to have hunted birds and possibly horse, mammoth, and cervids. They carried wedge-shaped microblade core technology typical of the late Upper Paleolithic of Siberia and Russian Far East, indicating first human incursion in eastern Beringia was made by northeast Asians.

3.3.3.2 Settling interior Alaska during the Allerød interstade

The next evidence for human occupation occurred during the Allerød interstade between about 11,800–11,000 cal. yr BP and is represented at several sites in interior Alaska and the Canadian Yukon, including Broken Mammoth in the middle Tanana valley, Little John in the upper Tanana valley, and Dry Creek, Owl Ridge, Walker Road, Moose Creek, and possibly the Eroadaway site in the foothills of the Nenana valley. These people continued to inhabit lowland areas, but they also expanded into upland settings. Toolkits from these sites are different from the Siberian late Upper Paleolithic, in that they typically lack microblades and burin, and instead contain small bifacial points (Fig. 3.3.7). Interestingly, they are strikingly similar to lithic assemblages found at the western Beringian sites of Berelekh and Ushki Lake, which also date to the Allerød (Goebel et al., 2003, 2010; Pitul'ko, 2011, but see Dikov, 1977; Mochanov, 1977 for earlier dates). A comparative study of lithic artifacts from Allerød assemblages from Dry Creek and Ushki Lake indicate that both primary and secondary reduction activities were somewhat informal and a wide variety of lithic raw materials were used, indicating some knowledge of surrounding landscapes (Graf and Goebel, 2010). Similarly, faunal remains from the sites of Broken Mammoth and Dry Creek indicate that Allerød-aged Beringians hunted a wide variety of animals, both large and small game. It appears that between 11,800 and 11,000 cal. yr BP across Beringia, there was a shared set of behaviors and possible communication of ideas unique to Beringia.

3.3.3.3 Artifact assemblage variability during the Younger Dryas and settling new areas of Beringia

Between about 11,000 and 10,000 cal. yr BP during the Younger Dryas stade, there was diversification of lithic industries as people began to settle other areas of eastern Beringia. In the interior, where humans had previously inhabited Alaska, human populations were maintained throughout the Younger Dryas interval. At foothills and upland sites of the Alaska Range such as Dry Creek, Moose Creek, Owl Ridge, Panguingue Creek, Carlo Creek, and Bull River II in the Nenana valley and the Tangle Lakes in the upper Delta River valley, people were making microblade technologies (both wedge-shaped and *tortsovyi*) and other Denali complex diagnostics such as lanceolate bifaces and burins, as well as using concave-based bifacial points at Dry Creek and Owl Ridge (Fig. 3.3.7). In the lowlands of the Tanana valley lithic assemblage patterns are more complicated with both Little John and Swan Point containing microblade technology associated with the Denali complex and, in addition, Swan Point containing a combination of lanceolate, concave-based, teardrop-shaped and triangular bifacial points. At the same time, at Broken Mammoth, hunters made both triangular and concave-based bifacial points, but not the lanceolate bifaces and/or microblades typical of

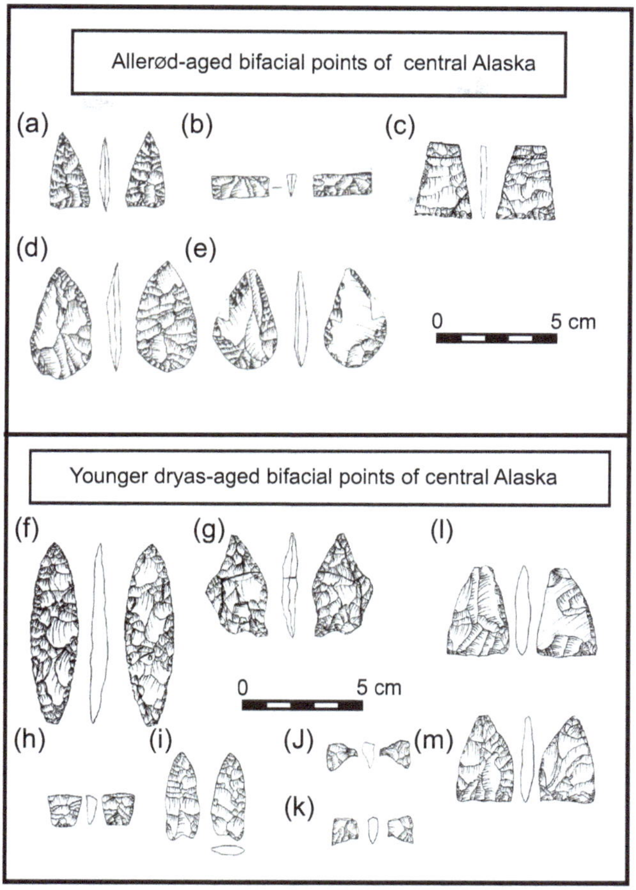

Figure 3.3.7 A sample of projectile points from central Alaska dating to the Allerød interstade and Younger Dryas stade. (a, b) Dry Creek, (c) Owl Ridge, (d, e) Walker Road, (f, g) Owl Ridge, (h, i) Dry Creek, (j–l) Swan Point, and (m) Broken Mammoth.

the Denali complex. Perhaps this technological diversity within central Alaska during the Younger Dryas can be tied to specific tasks employed at different sites.

Though the archaeological record in interior Alaska spans the Younger Dryas, some areas of eastern Beringia were initially inhabited much later. The first evidence of humans in northern and western Alaska comes from a suite of site occupations dating to the last few centuries of the Younger Dryas interval (c.10,400–10,000 cal. yr BP). These sites contain lithic assemblages with finely flaked lanceolate bifacial points similar to forms found in Paleoindian complexes south of the Canadian ice sheets. Some of these bifaces are fluted, while others are not. Assemblages from the sites of Mesa, Bedwell, Hilltop, Spein Mountain, Irwin Sluiceway, as well as some of the artifact concentrations at Tuluaq Hill, contain unfluted lanceolate points, while assemblages from Serpentine Hot Springs, Putu, and some artifact concentrations from Tuluaq Hill contain fluted points. Interestingly, microblade technologies are mostly missing from these Paleoindian sites except at Mesa and Tuluaq Hill, where discrete concentrations do contain microblade cores and microblades. Unfortunately, most of these assemblages are found in very shallow depositional contexts, making it difficult to assess whether artifact components resulted from palimpsests of activity by multiple groups of people or diverse toolkits left behind by one group. For instance, at Tuluaq Hill, lanceolate bifaces, fluted points, and microblades are all present, but we cannot be sure that this association reflects a single episode. Nevertheless, as a group these sites make up a northern Paleoindian complex that may have resulted from a late Younger Dryas migration of Paleoindians originating south of the ice sheets.

Southwestern and southeastern areas of Alaska seem to have been peopled rather late, after the Pleistocene. Archaeological sites emerge in these areas between about 10,000 and 8500 cal. yr BP at sites such as Lime Hills Cave, Ground Hog Bay, and On Your Knees Cave. In these instances, lithic assemblages are composed of microblade technologies and other artifacts resembling earlier Denali complex sites in the interior.

3.3.4 Conclusions

Eastern Beringia appears to have been colonized in steps. Initial settlement occurred during the Late Glacial by about 12,300 cal. yr BP when late Upper Paleolithic peoples migrated to interior Alaska from Siberia, possibly as the Late Pleistocene climate began to ameliorate and a shrub-tundra biome emerged (Hoffecker and Elias, 2007). During the Allerød interstade, humans seem to have become somewhat more familiar with interior landscapes in the Tanana and Nenana river valleys, and they adjusted successfully to the spread of shrub-tundra vegetation. They expanded their territories and modified technologies, possibly to suit the hunting of more varied fauna. During the Younger Dryas, the interior Alaskan climate may have become cooler and likely drier, so the composition of floral and faunal resources available to humans changed. Both the Nenana and Tanana valleys continued to be occupied, and lithic technologies became concentrated on the production of both biface projectile points and microblade composite tools, possibly for hunting large game (Graf and Bigelow, 2011). The archaeological record outside of interior Alaska suggests that near the end of the Younger Dryas, northern and western regions of Alaska were colonized by Paleoindian hunters, possibly predating on bison and other large herd animals. Whether these northern Paleoindians represent a direct migration from temperate North America or just the spread of technological knowledge is not known. The simultaneous presence of microblade technologies at some of these Paleoindian sites may suggest diffusion of Paleoindian technologies, but the associations of Paleoindian bifaces and microblades at a few northern Alaskan sites are equivocal. Based on current evidence, southern Alaska seems to have been colonized late, during the early Holocene, possibly by microblade-making hunter-gatherers from interior Alaska.

Section Four

Canadian North

Peopling of the Eastern Canadian Arctic

4.1

Julia M. Ross
Nunavut Government, Iqaluit, NU, Canada

Chapter Outline

4.1.1 Introduction

Human occupation of the Canadian Eastern Arctic was impossible prior to 5000 cal. yr BP because of ice sheets, which retreat from the region owing to the Holocene Thermal Maximum (HTM) during the Atlantic Phase. As deglaciation proceeded and formerly inundated land began to rebound, various animal species advanced into newly ice-free areas. Human hunters appear to have occupied the area not long after. The earliest archaeological sites associated with this human presence date to approximately 4500 cal. yr BP and comprise the traces of a people known as the Early Paleoeskimos. Part of the larger Arctic Small Tool tradition (ASTt), the Early Paleoeskimo period is conventionally dated between 4500 and 2800/2500 BP and is divided between three cultural groups known as Independence I, Pre-Dorset, and Saqqaq. Although Independence I and Pre-Dorset sites are widely known and reported throughout much of the region, Saqqaq is found primarily in Greenland and appears not to have contributed significantly to settlement of the Canadian Arctic Archipelago. For this reason, it is not considered further in this paper, which examines the first 1000 years of human expansion and colonization of the Canadian Eastern Arctic.

4.1.2 Glacial history

Details of the glacial history are complex and not agreed upon; even basic hypotheses such as whether ice was more extensive during the Early Wisconsin or Late

Human Colonization of the Arctic: The Interaction Between Early Migration and the Paleoenvironment
http://dx.doi.org/10.1016/B978-0-12-813532-7.00017-6

Wisconsin are debated (Vincent and Klassen, 1989). However, despite these ongoing debates, it is possible to provide a basic overview of these events. In Canada the Late Pleistocene or Wisconsin Stage is divided into three substages: Early (~80–65 ka), Middle (~65–23 ka), and Late (~23–10 ka[1]). During the Late Wisconsin, most of Canada was covered by the last glacial advance (Fulton, 1989), with the Laurentide and Innuitian ice sheets covering the Canadian Arctic. While the Innuitian Ice Sheet was primarily restricted to the Queen Elizabeth Islands, the Laurentide Sheet was more extensive and is divided between the Keewatin Sector (Arctic mainland and parts of the low western islands) and Baffin Sector (Baffin Island), each sector named for its area of likely inception and continuing outflow (Dyke, 2004). Ice coverage reached its southern and northeastern limits between 23 and 24 ^{14}C ka BP and attained its maximum southwestern and northern coverage between 20 and 21 ^{14}C ka BP (Dyke et al., 2002). While traditionally the greatest extent of ice was thought to correspond with the Late Glacial Maximum (LGM), "defined as occurring during a period of low global sea level as well as during a time of relative climate stability ~18 ^{14}C ka BP"(Dyke et al., 2002, p. 9), it now seems that the LGM occurred prior to and lasted beyond this period.

The retreat of the ice during the HTM was rapid, though temporally and spatially variable (Briner et al., 2007; Dyke, 2008). The Innuitian Ice Sheet rapidly disintegrated between 10 and 8 ka, while the Laurentide ice broke up in two stages: the first between 15 and 11.5 ka and second between 10 and 8 ka, although glaciers did readvance during the Younger Dryas event (~11–10 ^{14}C ka BP) and likely also during other less well dated periods associated with the 8.2 ka cold event (Bradley and England, 2008; Briner et al., 2006; Dyke, 2004). It is during the more recent period of deglaciation that ice retreated from much of the central Arctic (Dyke et al., 1989), with ice largely disappearing from Banks, Victoria, and Melville islands by 9.5 ka BP and from Lancaster Sound and most of the channels separating the islands of the Archipelago by 9.5–8.5 ka BP (Dyke, 2004). Most of the Canadian Arctic Archipelago was deglaciated by 6000 BP, although significant ice remained on western Baffin Island and portions of Melville Peninsula around Foxe Basin, as well as on parts of Ellesmere and Devon islands, until 5000 BP (Dyke, 2004, 2008).

However, while the ice was retreating and crustal rebound began, areas at the periphery of the ice sheets remained unavailable for human occupation because they were inundated by rising marine water levels. The limits of these postglacial seas in the Canadian Arctic range from lows of present sea level along the northwest coast of Banks Island to 250 m above sea level (ASL) near Committee Bay, at the bottom of the Gulf of Boothia (Andrews and Peltier, 1989). Most of the Canadian Arctic has followed a similar pattern of emergence resulting from isostatic rebound, although there are local variations dependent on glacial history and for bulge movements (Dyke and Peltier, 2000). In general, restrained or slow rebound occurs when the ice thins, then seas cover the once-glaciated areas (Andrews, 1970), from which there is rapid

[1] If a date is not associated with either ^{14}C or cal. yr BP, the original source was unclear if the dates presented were calibrated or uncalibrated, and dates are reported as found in the original text.

emergence, with a half-life of 1000–2000 years (the amount of time required for half of the remaining emergence to occur) (Dyke et al., 2005; Dyke and Peltier, 2000).

From a geological perspective, glacial emergence slowed in many part of the Canadian Arctic by approximately 5000 BP, although from a human standpoint the isostatic rebound that occurred between 5000 cal. yr BP and the present is significant. The location of archaeological sites in many parts of the Arctic corresponds with emergence rates, with Early Paleoeskimo (4500–2500 BP) sites typically being located 40–20 m ASL, Late Paleoeskimo (2500–1000/500 BP) sites between 20 and 10 m ASL, and Neoeskimo sites between 10 and 0 m ASL (Andrew et al., 1971; Dyke and Savelle, 2009). Currently, reconstructed average emergence rates for the Canadian Arctic are 2.4 m/1000 years, but these values vary by location as much as −4.4–7 m/1000 years (Andrews and Peltier, 1989).

4.1.3 Climatic history

The most common dates provided for the warm period known as the HTM are between 10 and 6 ka, which spans the Preboreal to Atlantic phases. Beyond the disappearance of the ice sheets, evidence indicating the warmth of this period is recorded in several proxy climate records including the location of the tree line (Bradley et al., 2003), pollen records (Gajewski et al., 2000), driftwood incursions (Barry et al., 1977), peat buildup (Broecker, 2001), borehole records (Pollack and Huang, 2000), and sea salt concentrations in ice cores (O'Brien et al., 1995). These sources indicate that the HTM was not consistently warm and was interrupted by an abrupt cold event at about 8.2 ka, identified in the ice core records (Barber et al., 1999; Grootes and Stuiver, 1997; Labeyrie et al., 2003). Some records indicate that this cold event lasted until 2000 BP, but the majority of information indicates a warm period both before and after the 8.2 ka event, as well as other fluctuations (Bradley, 1990; LeBlanc et al., 2004; Mudie et al., 2005; Short et al., 1994). The peak of this warm period, which was approximately 1–2°C warmer than present, may have occurred at about 6000 cal. yr BP in the Canadian Arctic (Barry et al., 1977; Dyke and Morris, 1990; Gajewski et al., 2000).

While 6000 BP may have been the HTM's warm peak, warm conditions lasted until approximately 4000 BP, just after the beginning of the Subboreal phase (Blake, 1989; Bradley, 1990; Koerner, 1989; LeBlanc et al., 2004; Stewart and England, 1983). However, evidence does suggest that cooling had already begun in some parts of the region by 6000 BP (Briner et al., 2006; Gajewski et al., 2000; Koerner and Fisher, 1990), and many climate records register 4500 BP as a time of almost consistently colder conditions. Ice cores from the Devon Ice Cap point to an overall cooling of between 2.7°C and 3.5°C in the High Arctic starting at 4500 BP (Koerner and Fisher, 1985), and there is an abrupt decrease in driftwood penetration at the same time, implying more severe ice conditions and the development of ice shelves (Lyons and Mielke, 1973).

The picture appears to be slightly more complex in the Western Canadian Low Arctic, consisting of Victoria Island and the Boothia and Melville peninsulas. Pollen records indicate that between 4500 and 3800 BP there was a short period of high temperature, as evidenced by an increase in Cyperaceae and *Salix* (Zabenskie and Gajewski, 2007). Contradicting this finding are two diatom records, where a lack of diatom diversity (as well as low overall concentrations) suggest cold periods dating to approximately 4600 and 4300 BP These cold periods may have lasted until approximately 1200 BP (Dyke and Savelle, 2009; LeBlanc et al., 2004).

Moving eastward, the majority of climate records from Baffin Island indicate cooling or cold climate signals that point to a cool period between 4800 and 2500 BP. There is evidence that lake water temperatures decreased gradually from a maximum of 8.2°C to about 6.5°C, and there is also an increased frequency of benthic diatom species (Lemmen et al., 1988), both signaling a response to progressive regional cooling during the late Holocene. This agrees with multiple episodes of glacial advance identified on Baffin Island (Joynt and Wolfe, 2001). At the same time, a dinoflagellate cyst record suggests two periods of warmer sea surface temperatures at 4275–4000 BP and 3750–3400 BP (Mudie et al., 2005), while pollen and foramina records suggest warmer conditions between 4500 and 3100 BP and approximately 4000 BP, respectively (Osterman and Nelson, 1989; Short et al., 1985). This discrepancy may be the result of local fluctuations, delays in the response time for different proxy indicators, or reflect differences between the terrestrial and marine environments.

Another cold climate threshold can be identified via several different proxy records in the High Arctic and on Baffin Island between 3500 BP and the end of the Subboreal phase at 2500 BP. Data from the Devon Ice Cap indicates a reduction in marine aerosols at 3500 BP, explainable as either a reduction in frequency of storms events or an increase in the extent of sea ice (Bradley, 1990). Either scenario has important ramifications for humans living in this region. Additional evidence obtained from pollen records in Baird Inlet on eastern Ellesmere Island marks 3500 BP as a period when vegetation stability came to an end and cooler conditions began, and glacial advances and formation of the Ward Hunt Ice Sheet further point to cold conditions (Bradley, 1990). Isotope values taken from ice cores also point to the past 3000 years as being cooler and drier than the preceding periods (Koerner, 1989). A change in pollen zones marking an increase in open ground and pioneer species on Baffin Island suggest that cold conditions were intensifying (Hyvärinen, 1985). Pollen records also suggest a pronounced cooling (>1°C over 1000 years) in the coastal regions of southern Baffin Island starting at approximately at 3000 BP (Kerwin et al., 2004), and most of these cold climate indicators persist well past 2500 BP (Bradley, 1990; Davis, 1985; Doner, 2001; Jacobs, 1985; Kerwin et al., 2004).

However, there are some proxy indicators (e.g., lake varves on Ellesmere Island and dinoflagellate cysts from High Arctic cores) that point to a possible reprieve from cold conditions in the High Arctic between 3400 and 3200 BP, contradicting findings suggestive of a renewed or intensified period of cold during this time (Lamoureux and Bradley, 1996; Mudie et al., 2005; Short et al., 1994). On Baffin Island, this reprieve is evident at 2800 BP, and while it is unclear how long it might have lasted, one record suggests as late as 2500 BP (Mudie et al., 2005; Williams, 1990).

Proxy data from the Western Low Arctic, including loss on ignition results and pollen concentrations, point to a clear cooling trend at 3800 BP (LeBlanc et al., 2004; Zabenskie and Gajewski, 2007), with a recovery to warmer conditions possibly occurring between 3100 and 2500 BP, after which pollen and diatom records suggest cooler conditions again prevailed (LeBlanc, 2002; Zabenskie and Gajewski, 2007).

To summarize, the climate records for the past 6000 years across the Eastern Canadian Arctic do not present a consistent or homogeneous picture. An overly simplified synopsis would be that conditions in the Eastern Arctic were warm from 10 kya BP to approximately 4500 BP, or from the Preboreal to Subboreal; however, by 4500 BP, many proxy indicators are sensitive enough to register that the climate was cooling, although conditions were still warmer than those found prior to the 21st century. This cooling trend primarily lasted throughout the Subboreal, with one intensification between approximately 3500–3000 BP and another possible at approximately 2500 BP. There are proxy records that suggest hiatuses in these cold periods; however, how widespread or intense these hiatuses were is not clear since these records are in the minority. At least one of these, produced by Mudie et al. (2005), focuses on sea surface temperatures using ocean cores, and it is not well established how proxy data from ocean cores articulate with the majority of terrestrial records.

In essence, during the critical 5000–4000 BP period when people were first inhabiting the Eastern Canadian Arctic, the climate was in a dynamic state of change, and while it was likely warmer than today, the climate was generally cooling, and the timing and intensity of the changes were not consistent across the region.

4.1.4 Site descriptions

Archaeological sites in the Eastern Canadian Arctic are predominately found on the surface with little to no overlying sediment. This means there is typically little or no stratigraphic separation within a site, with the possible exception of sites found on the Barrenlands and in some Low Arctic areas where postoccupation deposits are more likely to have developed. Raised beach ridge chronologies developed for discrete regions can be cautiously used to assign relative dates to sites without absolute dates; however, this form of dating relies on the assumption that people always settled on the lowest available beach and closest to the water. Complicating matters is the fact that, due to differing glacial histories, the elevation of archaeological sites from separate geographic locations cannot be directly compared. This means that sites at different elevations may be coeval, while other sites, located at the same height ASL, are not.

Owing to the fact the ASTt sites are often located on gravel ridges lacking any sort of sedimentary accumulations, very little paleoenvironmental work has been conducted on them, and broadly scaled regional or hemispheric climate records must of necessity be used to reconstruct a particular site's potential natural environment. In only a few cases has environmental data been obtained either by either using a site's zooarchaeological assemblage (Darwent, 2001; Knuth, 1967a,b; McGhee, 1979; Schledermann, 1990) or by conducting a multidisciplinary project dedicated to producing more locally specific (albeit not site-specific) environmental data (Alt and Garneau, 2000).

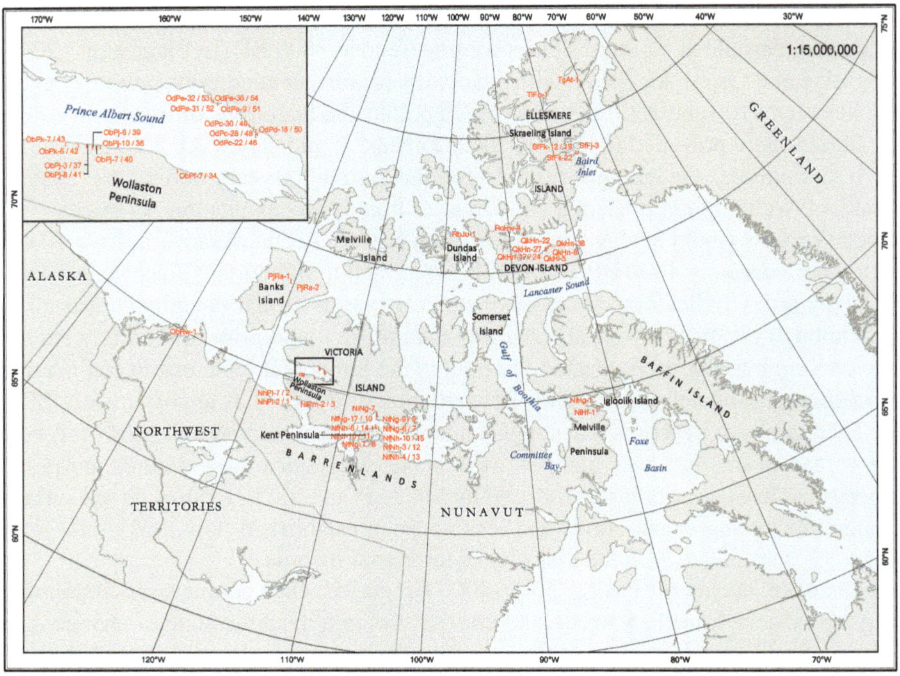

Figure 4.1.1 Main prehistoric sites relevant to the initial peopling of Northern Canada. The sites are indicated by numbers corresponding to the text and Table 4.1.1.

Unlike the situation in other areas, where investigations regarding the processes by which previously uninhabited lands are peopled emphasize the importance of one or two key sites (e.g., Blue Fish Caves or Monte Verde (Cing-Mars and Morlan, 1999; Meltzer et al., 1997)), there is no cornerstone site upon which reconstructions of the Early Paleoeskimo migration into the Canadian Eastern Arctic are based. There are, however, four study areas that provide the basis for most of our assumptions about Early Paleoeskimo lifeways and the nature of their migration. These include sites from Ellesmere Island (Schledermann and McCullough, 1988, 1990; Knuth, 1967a,b), two sequences of sites on Dundas Island at Port Refuge (McGhee, 1979, 1996) and the Truelove Lowlands of Devon Island (Helmer, 1982, 1984a,b, 1986, 1987, 1991), and Meldgaard's essentially unpublished work on Igloolik Island. Reference is also made to sites on Victoria and Banks islands (Archaeological Survey of Canada, 1972) (Figs. 4.1.1 and 4.1.2).

There are six well-described Independence I sites from Ellesmere Island. Cape Storm Beaches (RcHw-4) is at an elevation of between 14.94 and 20.12 m ASL and has produced a single radiocarbon assay, likely from driftwood charcoal, of 4360 ± 30 ^{14}C (GSC-2165) that calibrates to 4926–4966 (0.50)[2] cal. yr BP. There are 11 dwelling features; only one contains a midpassage, while the others represent vague scatters of

[2] Dates have been calibrated using Calib 4.42 to one sigma confidence intervals. Only one date range is provided, and this is the range with the greatest area under the probability curve; the probability factor is provided in brackets.

Figure 4.1.2 Site NfNh-10 demonstrates the typical lack of stratigraphy of Early Paleoeskimo sites in Canadian Arctic.
Photo credit Dr. James Savelle.

stone or patches of vegetation associated with burins, burin spalls, microblades, core, bifaces, and end blades. The second site, Lake View (SfFj-3), is at an elevation of between 21.00 and 27.00 m ASL. One date, on *Salix*, yielded an uncalibrated result of 3940 ± 70 ^{14}C (TO-993) or a calibrated range of 4283–4445 (0.83) cal. yr BP. Features include at least 19 central passage dwellings, as well as two scattered features; notable artifacts include bipointed projectiles, side blades, bifaces, flakes, microblades, burin, and burin spalls.

The third site is Kettle Lake (TiFo-1), which is at an elevation of 72–105 m ASL and consists of 23 habitation features, including some more substantial-appearing features with internal mid-passages. Four radiocarbon assays have been run (three on *Salix* sp. and the fourth on unidentified charcoal, K-5074): the oldest is 3930 ± 130 ^{14}C, which calibrates to 4216–4528 (0.86) cal. yr BP (K-1260); next is 3810 ± 130 ^{14}C, calibrating to 4080–4408 (0.91) cal. yr BP (K-1261); then 3760 ± 130 ^{14}C, calibrated between 3963 and 4297 (0.92) cal. yr BP (K-1262); and finally 3800 ± 80 ^{14}C, producing a calibrated range of 4084–4298 (0.89) cal. yr BP (K-5074). The fourth site, Burin Delta (TgAt-1), consists of six semisubterranean dwellings and is at an elevation of 20 m ASL. It has been dated by unidentified charcoal to 4200 ± 70 ^{14}C (Beta-36946, calibrated range of 4620–4764 (0.77) BP).

The fifth site is Camp View (SfFk-22), which is located 15 m ASL and dated to 4378–4532 (0.84) BP (TO-994, 3990 ± 70 ^{14}C uncalibrated). The site consists of six dwellings, some with midpassage hearth features, and the material culture includes microblades, burins, burin spalls, gravers, end triangular scrapers, concave side scarpers, flaked knives, biface choppers, biface fragments, leaf shape bipointed projectile points, asymmetrical scrapers, and serrated projectiles. The final site, Skraeling ASTt#7 (SfFk-12), is located on a small island off of the Ellesmere mainland at an elevation of between 18 and 21 m ASL. A single radiocarbon assay on driftwood was run, producing an uncalibrated date of 4400 ± 55 ^{14}C (GSC-3261), which calibrates

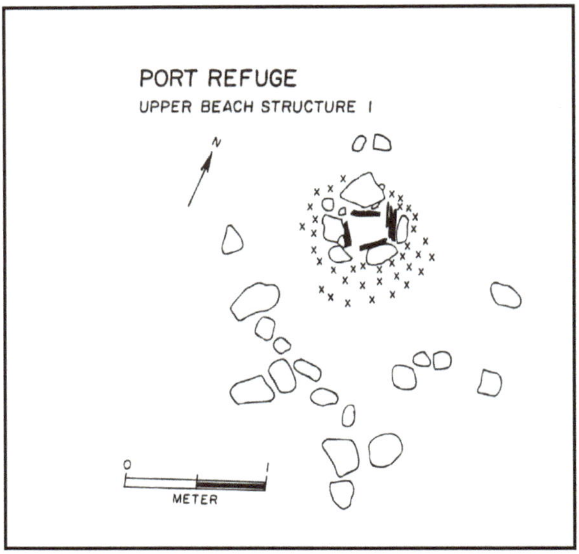

Figure 4.1.3 An example of a structure from Port Refuge (RbJu-1).

between 4869 and 5041 (1.00) BP. The site is small, with only one midpassage and a box hearth, but it produced microblades, burins, burin spalls, end scrapers, biface knives, and side blades.

The second study area is Port Refuge, located south of Ellesmere Island on Dundas Island. Port Refuge (RbJu-1) consists of 107 features distributed over four moderately distinct areas; these features include midpassages, hearths, caches, scattered tent rings, and isolated vegetation patches (Fig. 4.1.3). The Independence I features range from an elevation of 22–25 m ASL, and the recovered artifacts include microblades, microblade cores, burins, burin spalls, small and large stemmed bifaces, end blades, side blades, ovate bifaces, knives, large bifaces, concave side scrapers, flake knives, end scrapers, gravers, adze blades, retouched flakes, needles, harpoon heads, and barbed antler points. Two of the three radiocarbon estimates are on driftwood charcoal, and the third is on antler: the oldest date is GSC-1940, with an uncalibrated date of 4360 ± 45 ^{14}C (calibrated range of 4862–4968 (1.00) BP); next is GSC-1931, calibrated between 4567 and 4652 (0.43) BP (uncalibrated 4120 ± 60 ^{14}C); while the third date (CRNL-1230) is on antler and has an uncalibrated date of 4440 ± 1160 ^{14}C (calibrated range of 3554–6310 (1.00) BP).

The cultural affiliation of the Gneiss site (QkHn-8) site on Devon Island is uncertain, but based on edge serration, long tapered stems, and contracting stem bifaces (all traits associated with Independence I), the site is tentatively assigned to this affiliation. The site is between 8 and 9 m ASL and consists of two oval tent rings, one of which has a central hearth, while the other has a possible midpassage feature. The radiocarbon assay, Beta-12405, was run on *Salix* sp. and has an uncalibrated date of 4160 ± 180 ^{14}C, or a calibrated range of 4415–4872 (0.98) BP. The material culture includes triangular end blades with serrated edges, bifacially flaked burins, and burin spalls. Like

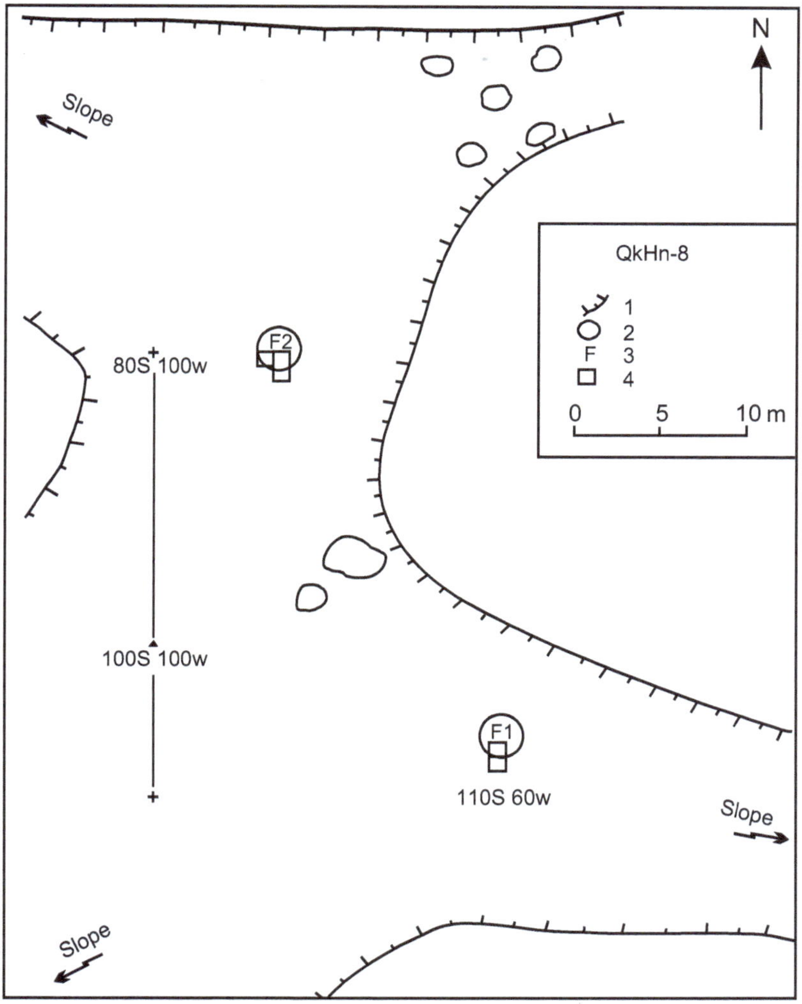

Figure 4.1.4 Sketch map of Gneiss site, QkHn-8. 1, approximate location of bedrock; 2, boulder; 3, feature; 4, excavation unit.

the Gneiss site, the cultural affiliation of the Far Site (QkHn-22) is also slightly uncertain, although artifact attributes point to Independence I. The site is at an elevation of 14 m ASL and there are two dates: the early assay, Beta-20783, is on charcoal and produced an uncalibrated date of 4110±90 ^{14}C (calibrated to 4522–4658 (0.54) BP). The other date, Beta-12406, is also on charcoal and returned an uncalibrated date of 4040±70 ^{14}C (calibrated to 4416–4586 (0.88) BP). This site consists of five oval tent rings, and artifacts include unifacial burins, a unifacial flake tool, microblades, serrated end blades, burin spalls, perforators, serrated and tapered bifaces, ovate side blades, concave scrapers, and unifacial and bifacial burins (Figs. 4.1.4 and 4.1.5).

There are seven Pre-Dorset sites with early dates and for which excavations have been conducted; however the information published on them varies. The Pre-Dorset

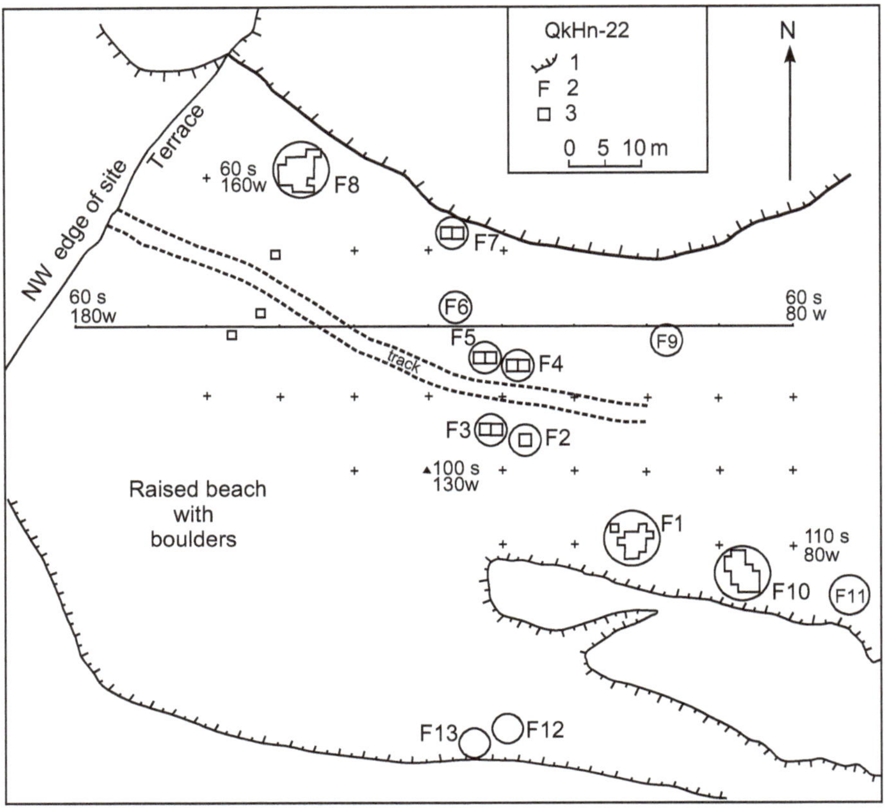

Figure 4.1.5 Sketch map of Far site, QkHn-22. 1, approximate location of bedrock; 2, feature; 3, excavation unit.

component, Gull Cliff, at Port Refuge is at an elevation of 18–25 m ASL. It consists of 30 features, 20 of which are thought to be habitation structures. Artifacts include microblades, microblade cores, burins, burin spalls, stemmed bifaces, triangular bifaces, side blades, knives, large bifaces, concave side scrapers, flake knives, end scrapers, drills, retouched flakes, needles, harpoon heads, barbed antler points, as well as more generalized worked bone objects. Unlike the neighboring Independence I sites, the burins found at Gull Cliff lack ground lateral edges and are generally smaller and more uniform. Three possible early Pre-Dorset sites are located in the Truelove Lowlands of northern Devon Island. The first site, Icy Bay (QkHl-5), is a Pre-Dorset site situated between 4 and 10 m ASL and has been dated via two driftwood charcoal radiocarbon assays. The older date, Beta-20780, is 4070 ± 80 ^{14}C uncalibrated, the calibrated date range is 4497–4646 (0.61) cal. yr BP; the younger assay is Beta-20781, 3770 ± 180 ^{14}C uncalibrated or calibrated to 3956–4409 (0.93) BP. The site consists of six tent rings, including two with central hearths, and the artifact assemblage includes burins, biface fragments, unifacial tools, and a microblade fragment.

The Hind site (QkHn-38) is situated between 7 and 8 m ASL and has been dated to 3700 ± 70 ^{14}C BP (Beta-25032, on caribou bone collagen), or calibrated between 3961 and 4098 (0.74) cal. yr BP (Fig. 4.1.6). Three features have been recorded including

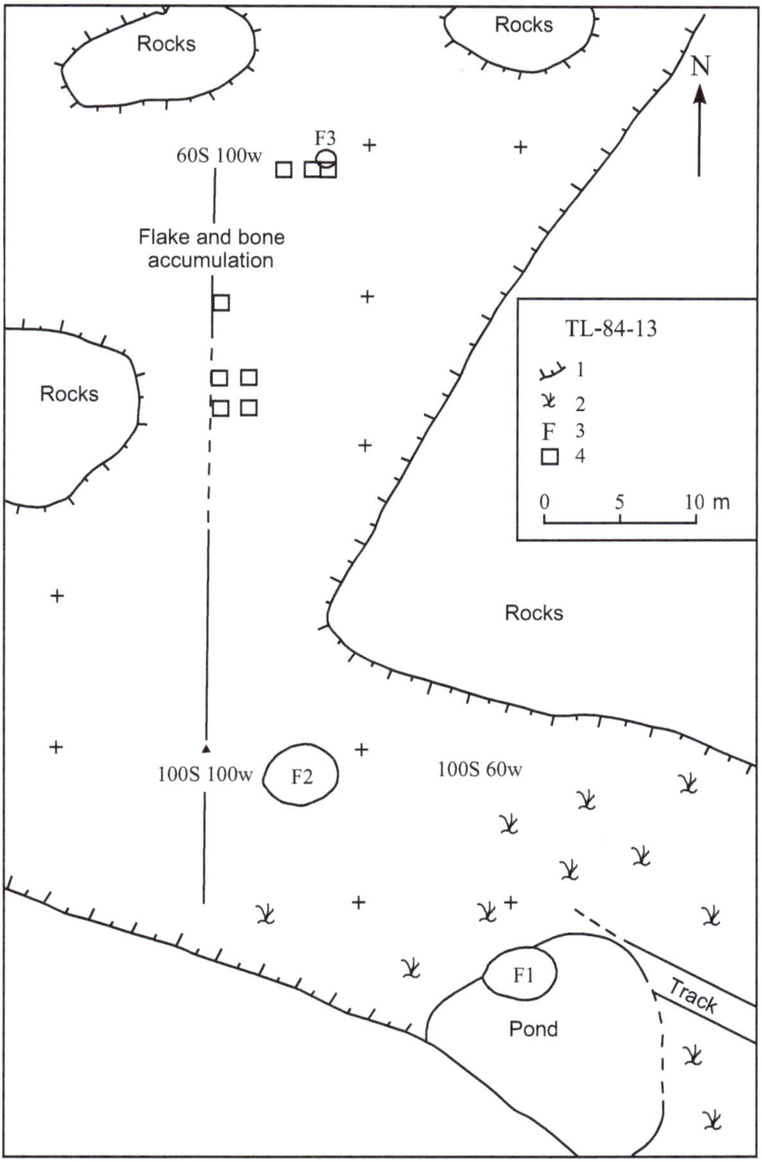

Figure 4.1.6 Sketch map of Hind site, QkHn-38. 1, approximate location of bedrock; 2, wet meadow; 3, feature; 4, excavation unit.

an oval tent ring with an axial midpassage arrangement, a partially buried oval tent ring, and an isolated hearth. The artifacts collected include triangular end blades, stemmed and asymmetric bifaces, unifacial and bifacial burins, a microblade core and microblades, and burin spalls. The third site included from this area is Rocky Point (QkHn-27), which has been dated, using possible *Salix* charcoal, to 4060±80 ^{14}C (Beta-16554, calibrated 4421–4626 (0.85) BP) and 3800±90 ^{14}C (Beta-15391, calibrated 4082–4300 (0.83) BP). Although an early affiliation is suggested by the dates, Helmer classifies this site as terminal Pre-Dorset (~2800–2500 BP) based upon the artifact assemblage, making its relevance to this discussion is uncertain. Rocky Point contains 15 oval tent rings, each approximately 3 m in diameter, and one exterior hearth feature. The artifacts include serrated edge end blades, side scrapers, burins (both bifacial and unifacial), parallel-sided and triangular bifaces, concave side scrapers, microblade cores, microblades, burin spalls, perforators, and needles.

The fourth study area, Igloolik Island, includes the Parry Hill or Kaleruserk site (NiHf-1), which, although described as an early site, is not well reported, and the site itself is now almost totally destroyed. The site is situated at about 50 m ASL, and Meldgaard reported at least 102 Pre-Dorset house ruins. The excavated artifact assemblage has not been described, although two dates (both on caribou bone) suggest an early use of the site (K-505 yielded an uncalibrated result of 3780 ± 300 ^{14}C, calibrated 3808–4531 (0.93) cal. yr BP, while P-208 is reported as 3640 ± 123 ^{14}C, calibrated range 3826–4099 (0.85) BP).

The Buchanan (NiNg-1) and Wellington Bay (NiNg-7) sites located near the banks of the Ekalluk River were thought to be the earliest Pre-Dorset sites located in the Western Canadian Low Arctic. However current research associated with these sites has led to the conclusion that they are more recent that first suspected (Friesen, 2000, 2002, 2003), and in addition the work of Savelle and Dyke (Dyke and Savelle, 2009; Savelle, 1999, 2004, 2006, 2007; Savelle and Dyke, 2002) have accumulated considerable data regarding the number of features, surface architecture, and chronological attribute for Paleoeskimo sites on the western part of Victoria Island. Radiocarbon assays obtained from a number of sites indicate that these sites, and not the Ekalluk River locations, represent the earliest Pre-Dorset presence in the Western Canadian Low Arctic. Although extensive excavations have not been conducted, limiting information regarding material culture and subsistence strategies, data critical for addressing the nature of the Early Paleoeskimo migration has accumulated and enables an overview of the general Paleoeskimo migration to be constructed. Savelle and Dyke's work has provided a large number of Early Paleoeskimo dates and indicates there were apparently several fluctuations in Pre-Dorset population in this area throughout the 4500–2800/2500 BP period.

4.1.5 Synthesis

The peopling of the Eastern Canadian Arctic was done by people referred to as the Arctic Small Tool traditions (ASTt). It has been suggested they originated from the Neolithic Siberian Bel'kachisk culture that dates to c.6000 BP (4000 BC) (McGhee, 1974, p. 844). Across the Bering Strait in Alaska, there are ASTt components known from Onion Portage and the Cape Denbigh area that appear suddenly and have no clear predecessor in the area, supporting the theory of a Siberian origin for ASTt populations (Ellis, 2008; Maxwell, 1985; McGhee, 1974, 1996). The Siberian origin seems to be taken as fact, and new analyses connecting the western and eastern expressions of ASTt have not been done. Current information indicates that people utilizing ASTt technologies moved eastward from Alaska into the Canadian Arctic; traditionally the date for this first move is placed between 4000 and 4500 BP (Maxwell, 1985; McGhee, 1974, 1979).

However, there are three points of controversy associated with the earliest migration into the Eastern Canadian Arctic by ASTt people: the speed at which this migration occurred; whether there were two groups (and, if so, did one migrate slightly after the other); and is there evidence that the quick migration might relate to the exhaustion of critical resources such as musk ox and fuel (Darwent, 2001; Dyke and Savelle, 2009; Knuth, 1967a,b; McGhee, 1974, 1979; Savelle and Dyke, 2002). While not necessarily connected to the peopling of the Eastern Arctic, two attributes of Early Paleoeskimo people are of note: one is the quality of their lithic assemblage, which is a diagnostic feature of their material culture, and the other is evidence indicative of several possible fluctuations in population levels.

The first research on early ASTt sites in Canada was by Eigil Knuth, who was tracing the origin of Independence I remains that he had originally worked on in Greenland (McGhee, 1979). Knuth (1967a,b), following Steensby (1917), suggested a rapid dispersal from Alaska to Ellesmere Island since radiocarbon dates indicated that occupation of these two distant areas were separated by only a few hundred years. Steensby and Knuth believed the quick dispersal of people from the low western Arctic to the high Eastern Arctic was related to following the musk ox, as the latter fled north during a period of climatic warming to escape the increasing temperatures. They further suggested that as wood and subsistence resources were exhausted, people had to move on, but it is unclear if this is forwarded to explain the original migration event or subsequent settlement patterns (Knuth, 1967a,b, p. 17).

Some suggest that slightly after the migration of Independence I was another expansion of ASTt culture-bearers, the Pre-Dorset. The Pre-Dorset are typically associated with the Low Arctic; however, there are Pre-Dorset sites in the High Arctic. In fact, it is at the Port Refuge sites on Dundas Island, where Pre-Dorset and Independence I occur on separate beach ridges, that evidence was first identified to support the idea of two migrations, with Pre-Dorset following Independence I (McGhee, 1979).

A lack of clear consensus among researcher regarding the significance, if indeed there is any, associated with the apparent differences between Independence I and Pre-Dorset material culture and an absence of justifications by researchers indicating why a specific cultural affiliations has been assigned limits the ability to address whether there was one migration or two and the directional nature of the migration. While the meaning associated with Independence I and Saqqaq have been fairly consistent, the meaning of Pre-Dorset has changed (Helmer, 1994). Several researchers have questioned whether a separation between Independence I and Pre-Dorset is valid since there are sites on Devon and Somerset Islands that have a mixture of attributes (Bielawski, 1988; Maxwell, 1980). In the literature, what an author intends when using the term Pre-Dorset is not always clear.

Some of the most recent literature, such as Savelle and Dyke (2002), who present new data on the earliest occupation of the Eastern Canadian Arctic by Paleoeskimo people, follow Helmer's (1994) classification system and use the term Pre-Dorset as inclusive of all evidence of Early Paleoeskimos. As a result of their approach, which is based on dates and cultural affiliation, it seems as if Independence I suddenly occurs on Ellesmere Island with no antecedent to the west. McGhee (1996), who first proposed that the two groups migrated separately, states that the different names affiliated with Early Paleoeskimo sites is more a product of archaeological history as all share an overall culture, differentiated only because of regional environments.

There are 11 dates from six sites affiliated with Independence I; there are 49 Pre-Dorset dates from 40 sites, and there are 20 dates from 14 sites with a more general "Paleoeskimo" affiliation, all having uncalibrated radiocarbon dates that run from approximately 4500 to 3500 BP (CARD, 1999). It is assumed that sites designated "Paleoeskimo" are not associated with diagnostic Early Paleoeskimo attributes, or they otherwise would have been attributed a more specific affiliation; this assumption in combination with the established fact that radiocarbon dates from the Arctic can be problematic means occupations and associated dates identified under a more general "Paleoeskimo" rubric will not be presented in the following discussion.

Uncalibrated Independence dates on terrestrial materials range from 4440 ± 1160 ^{14}C BP to 3760 ± 130 ^{14}C BP (Table 4.1.1), corresponding to the transition between the Atlantic and Subboreal phases. While only assays associated with terrestrial material are considered to avoid issues with marine reservoir effects, the materials on which the assays have been run can still be considered problematic. Radiocarbon assays on *Salix*, driftwood, and antler may provide more accurate results than assays on marine samples; however, these materials are associated with their own uncertainty that might result in dates being older than the cultural event with which they are assumed to be associated (Beschel and Webb, 1963; Bradley, 1990; Dyke et al., 1997; McGhee and Tuck, 1976; Morrison, 2001; Nagy, 1997; Nelson and McGhee, 2002; Nelson and Møhl, 2002; Savelle and Dyke, 2002; Tremblay et al., 1997; Whitridge, 1999). If CRNL-1230 on antler is removed owing to its large standard error and only the calibrations with the greatest probability are considered, then the date range for these earliest Independence I sites is 5040–3960 cal. yr BP. Nine of the dates have calibrations that are older than the 4500 BP marker taken as the accepted entry for ASTt into the Eastern Canadian Arctic. The sites from which these dates are from are predominately located on Ellesmere Island; however, one is on Devon Island.

There are 39 uncalibrated dates from 34 sites with a Pre-Dorset affiliation, and these dates range from 5245 ± 42 ^{14}C to 3640 ± 123 ^{14}C BP. Other than a few more assays conducted on terrestrial mammals, the type of material dated from Pre-Dorset sites is comparable with those used for Independence I sites. There are 13 dates that have calibrations (greatest probability) in the Atlantic Phase, just older than 4500 BP, and the calibrated range for the Pre-Dorset dates is from 6000 to 3800 cal. yr BP. Other than one date, Beta-20780, which is from the Icy Bay site on Devon Island, all these dates are from the Western Canadian Low Arctic (Victoria Island or the Boothia Peninsula). The interpretation of the sequence of migration rests on the acceptance or rejection of CRNL-1230. If the date is accepted, there is weak radiocarbon evidence that Independence I peoples were present on Devon Island before Pre-Dorset were present along the shores of Prince Albert Sound and the southern Wollaston Peninsula of Victoria Island; this requires the assumption that the whole calibrated range can be associated with an actual human occupation. If this date is rejected, as it likely should

Table 4.1.1 Early ASTt sites descriptions and dates, not directly discussed in text

Borden	Culture	Elevation above sea level (m)	Uncalibrated radiocarbon	Calibrated radiocarbon date with greatest probability	Material dated	Site features
QkHn-22 Far	Paleoeskimo Independence?	14–14	Beta-20783 4110 ± 90	4522–4658 (0.54)	Charcoal	
ObRw-1 Qugyuk	Paleoeskimo	3	AECV-1710C 3980 ± 80	4346–4532 (0.84)		
QkHn-17 Twin Pond	Paleoeskimo Late Pre-Dorset	8–10	Beta-15390 3680 ± 90	3888–4098 (0.87)	Charcoal (driftwood?)	At least 3 midpassage structures, and additional features
NhPl-2	Pre-Dorset	12.5	AA-41518 4172 ± 58	4620–4765 (0.81)	*Salix* charcoal	20 dwelling features, most have midpassages, there are some poorly defined tent rings as well
NfNg-6	Pre-Dorset	30	30356 3960 ± 15	4409–4444 (0.77)	*Salix* charcoal	2 midpassage features
NfNg-7	Pre-Dorset	29	30357 3775 ± 15	4089–4161 (0.75)	*Salix* charcoal	4 midpassage dwellings with slightly different from
NfNg-9	Pre-Dorset	37	4415 ± 20	4950–5047 (0.76)	*Salix* charcoal	2 midpassages and a circular tent ring
NfNg-17	Pre-Dorset	40	30359 4070 ± 15	4517–4583 (0.81)	*Salix* charcoal	Tent ring with midpassage
NfNf-10	Pre-Dorset	33	30363 3925 ± 15	4347–4422 (0.70)	*Salix* charcoal	1 midpassage and 5 other features
NfNh-3	Pre-Dorset	31	30365 3910 ± 15	4293–4418 (1.00)	*Salix* charcoal	4 features with midpassages
NfNh-4	Pre-Dorset	31	30366 3695 ± 15	3982–4086 (1.00)	*Salix* charcoal	Isolated box hearth

Continued

Table 4.1.1 **Continued**

Borden	Culture	Elevation above sea level (m)	Uncalibrated radiocarbon	Calibrated radiocarbon date with greatest probability	Material dated	Site features
NfNh-6	Pre-Dorset	29	30381 3955 ± 15	4404–4445 (0.83)	*Picea* charcoal	27 features recorded including midpassages, box hearths, sod-covered paved areas
NfNh-10	Pre-Dorset	43.	29241 3680 ± 20	3967–4086 (0.95)	Caribou or musk ox bone	8 features recorded, 3 at least seem to be rectangular tent rings with midpassage
NhPl-7	Pre-Dorset	12–13	TO-8531 4490 ± 90	5037–5295 (0.93)	*Picea* charcoal	3 midpassage dwellings on upper beaches; 25 on lower beaches
			TO-8530 3930 ± 80	4243–4444 (0.92)	*Picea* charcoal	
NiPm-2	Pre-Dorset	12.5–13	TO-8534 4410 ± 70	4866–5051 (0.84)	*Picea* charcoal	23+ features; consisting predominately of midpassages and hearths
			TO-8533 3950 ± 70	4345–4448 (0.55)	*Picea* charcoal	
ObPf-7	Pre-Dorset	14	AA-40578 3911 ± 42	4345–4415 (0.62)	*Salix* charcoal	9 Paleoeskimo features
ObPj-10	Pre-Dorset		AA-41499 3979 ± 47	4459–4524 (0.56)	*Salix* charcoal	22 midpassages occur in one group, and 11 hearths occur over several ridges
ObPj-3	Pre-Dorset		AA-40849 3826 ± 41	4149–4262 (0.85)	*Picea* charcoal	About 19 box hearths

ObPj-6	Pre-Dorset		AA-40853 5245 ± 42	5929–5997 (0.71)	*Picea* charcoal	7 midpassages and isolated hearths
ObPj-7	Pre-Dorset		AA-40577 4237 ± 43	4813–4851 (0.48)	*Salix* charcoal	2 amorphous stone scatters
ObPj-8	Pre-Dorset		AA-40854 3804 ± 41	4145–4245 (0.83)	*Salix* charcoal	1 central passage within loose tent ring
ObPk-6	Pre-Dorset		AA-40574 3828 ± 54	4147–4297 (0.88)	*Picea* charcoal	3 midpassages (2 in tent rings) and 2 hearths on single beach ridge
ObPk-7	Pre-Dorset		AA-40575 3945 ± 42	4350–4440 (0.74)	*Picea* charcoal	At least 4 large midpassages and 2 hearths
OdPc-22	Pre-Dorset		AA-40583 3830 ± 42	4150-4263 (0.81)	*Picea* charcoal	33+ Paleoeskimo occupation features
OdPc-28	Pre-Dorset		AA-40585 4154 ± 45	4612–4712 (0.57)	*Salix* charcoal	11 Paleoeskimo dwelling features
OdPc-30	Pre-Dorset		AA-40587 4216 ± 51	4691–4761 (0.53)	*Salix* charcoal	5 Paleoeskimo hearths and 8 midpassages
OdPd-18	Pre-Dorset		AA-40586 4133 ± 42	4758–4811 (0.30)	*Salix* charcoal	Over 13 Paleoeskimo dwelling and 11 hearths
OdPe-9	Pre-Dorset		AA-40591 4557 ± 45	5121–5177 (0.35)	*Salix* charcoal	58+ Paleoeskimo features on at least 6 different level
OePe-31	Pre-Dorset		AA-40863 4197 ± 41	4691–4761 (0.58)	*Salix* charcoal	8 Paleoeskimo areas included 1 hearth and 2 paved areas
OePe-32	Pre-Dorset		AA-41514 3786 ± 45	4090–4186 (0.69)	*Salix* charcoal	4 midpassages and 2 hearths
OePe-36	Pre-Dorset		AA-40590 4455 ± 52	5167–5278 (0.49)	*Picea* charcoal	45+ Paleoeskimo features on 4 beach levels

be owing to its large standard error, than Pre-Dorset were present about 1000 years prior to the first Independence I dates at approximately 5000 cal. yr BP.

If all dates prior to 4500 cal. yr BP, or the Subboreal phase, are rejected, then it would seem that Pre-Dorset (Beta-30359, 4583–4517 cal. yr BP) and Independence I (TO-994, 4532–4378 cal. BP) simultaneously migrated to the Canadian Arctic, albeit occupying geographically divergent locations (the Kent Peninsula and Skraeling Island). The same pattern is established if the next youngest dates are considered, even though the Pre-Dorset dates from Igloolik Island slightly reduce the geographic disparity between the oldest dates associated with these two cultural horizons. The use of Helmer's terminology by Savelle and Dyke precludes finding a definitive answer to whether or not Independence I and Pre-Dorset had different migration routes, as well as whether the timing of these migrations differed. It would seem, based on radiocarbon dates, that if Independence I and Pre-Dorset represent different AST traditions entering the Arctic, they did so at approximately the same time. The more likely interpretation is that Independence I and Pre-Dorset are slight regional variations representing the same culture, and the vast area of the Eastern Arctic was occupied simultaneously, suggesting the initial settlement of the area involved people who traveled great distances quickly and in a nondirectional manner. Early occupation may have been concentrated in a few areas, such as Ellesmere and Devon islands; however, it is more likely this pattern is the result of research activities that focused on these locations and not others.

The calibrated date ranges lead to the question of whether or not a date around 4500 cal. yr BP is still the most appropriate point for the first appearance of human groups in the Eastern Arctic, or if the date of initial settlement should be pushed back to approximately 5000 cal. yr BP. It is becoming clear that, to accommodate entry of ASTt culture-bearers into the Canadian Eastern Arctic prior to 4500 cal. yr BP, it is necessary to push back the date at which the Early Paleoeskimos first appeared along the Bering Strait coast. However, while evidence for earlier dates in Alaska exists, more collaborative evidence is needed before a conclusive statement can be made (Ackerman, 1998, p. 844; Harritt, 1998; McGhee, 1974). No matter the exact timing of the original migration, it is apparent that it coincided with the Atlantic–Subboreal transition or the Subboreal phase, depending on accepted dates.

The climate during this time was warmer than present; however, it is evident that it occurred as part of a cooling trend that essentially lasted, with regional fluctuations, until modern times. The state of "warmer" has been the focus of most research, as it is traditional in archaeological thought for "warmer" to mean better conditions for caribou and musk ox, which some ASTt sites, Far, Qugyuk, Shoran Lake (PjRa-2), and Umingmak (PjRa-1), seem to focus on. However, little consideration has been given to the likely dynamics of the environment during this more general cooling phase, and while the land was rebounding from the weight of glacial ice and while this time was warmer than present, the more generalized cooling trend apparent from various proxy records was almost certainly influencing and possibly stressing terrestrial species, potentially making them less predictable and reliable for human hunters. Environmental uncertainty and unpredictability may have been one cause of the apparently disperse nature of the earliest Paleoeskimo sites.

The availability and distribution of musk oxen, the original critical resource identified as influencing the speed and direction of the human migration eastward, would also be impacted by varying environmental conditions (Knuth, 1967a,b). The assumption is that these animals would have been plentiful owing to generations without human predation; however, their natural defense mechanism, developed in response to wolf predation, meant that they would be easy prey for humans armed with bows and arrow who may have overhunted the small herds (Savelle and Dyke, 2002). The hypothesis that Early Paleoeskimo could at least periodically overkill these animals is accepted by a number of researchers who point to Early Paleoeskimo sites such as the Shoran Lake and Umingmak sites on Banks Island, where musk ox compose between 60% and 80% of identified species, as evidence of overhunting (McGhee, 1996; Dyke and Savelle, 2009). However the calibrated and uncalibrated dates associated with the Umingmak site (GSC-651, 3420 ± 150 ^{14}C BP; 3800–3500 cal. yr BP; and GSC-669, 3440 ± 160 ^{14}C BP; 3800–3600 cal. yr BP) do not support the premise that it is one of the earliest sites. Thus, while the Banks Island sites support substantial reliance on musk ox, they cannot be used to support the theory that rapidly expanding ASTt peoples moved through the area in a quick and directional migration since the dates are not the earliest and there is no evidence for subsequent overreliance on musk ox in a west-to-east direction.

Darwent's (2001) reanalysis of faunal collections from the High Arctic indicates that while musk ox were a key resource, the Early Paleoeskimos did not overly rely on them; instead, her analysis points to an evenness in the use of a variety of species and does not provide evidence for overkilling. Darwent (2001) also found, as would be expected, resource use was reflective of the resources that were available in a given area, so marine sites had more seal than terrestrial animals, and vice versa. Smaller animals such as fox and hare also proved to be critical resources for early AST tradition people. Darwent, following Helmer (1994), does not differentiate Independence I from Pre-Dorset but instead groups all sites dating between 4200 and 3600 BP as early Pre-Dorset.

In another analysis of Early Paleoeskimo High Arctic subsistence patterns, McCartney and Helmer (1989) specifically analyzed sites dating between 3900 and 3400 ^{14}C BP (4450–3700 cal. yr BP) located in three physiographic zones and specifically examined lowland oases known to attract terrestrial resources. They did this in part to determine the dependence of ASTt people on terrestrial resources; they found that the marine component from the sites ranges from 55% to 80%, while the terrestrial component varied from 12% to 30%. They also discovered that caribou, and not musk oxen, comprised the major terrestrial contribution, a determination that casts further doubts on the idea that early ASTt peoples were following and targeting musk ox as they moved across the Canadian Arctic and toward Greenland.

More excavation and analysis in the Western Canadian Low Arctic needs to be done before a more conclusive statement can be made, but other than the Shoran Lake and Umingmak sites, it would seem that the Early Paleoeskimos maintained a dual economy that relied more or less evenly on marine and terrestrial animals. There is simply not enough evidence to support an overkill hypothesis. However, it could be that the demise of terrestrial resources, owing to environmental factors or human and nonhuman predation, might have contributed to the periodic abandonment or near-abandonment of localized areas given these animals provided critical raw materials resources (Savelle and Dyke, 2002; McCartney and Helmer, 1989).

Demographic estimates are difficult to make, although a total count of clearly identified habitation features indicates there are 1824 dwellings associated with Pre-Dorset and Independence I, all of which would have been constructed over a 2000-year period. This should be regarded as a minimal number, and based on recent survey results and the number of areas awaiting a comprehensive archaeological survey, this is clearly an underestimate of Early Paleoeskimo occupation. It does seem that overall population numbers throughout the early ASTt are small, and demographic changes (surmised from the density of sites in specific areas and during particular periods) are evident that indicate the occupation of many areas fluctuated across time and space. To date, reported early ASTt sites range from one habitation feature to 133 features, although the average number of dwellings at a site is about six. There are 17 sites with over 20 habitation features; however, it is unclear if all features were occupied simultaneously. At some of the larger sites, such as OdPe-9, the features are spread over a few beach ridges, implying that site habitation was sequential and not simultaneous; however, at other sites, including ObPj-10, the features all occur on one beach ridge, indicating that they may have been in use more or less simultaneously.

Savelle and Dyke (2002) have suggested population fluctuations for southwestern Victoria Island, with modest populations from 4500 to 4000 ^{14}C BP (5200–4400 cal. BP); a peak in population from 4000 to 3800 ^{14}C BP (4500–4000 cal. yr BP), followed by a population decline lasting for an unspecified duration. The decline is presumed to be short since they suggest populations were at moderate levels for about 600 years prior to 3200 ^{14}C BP (4000–3300 cal. yr BP) before another decline (based on the absence of features associated with radiocarbon dates) until after 3000 ^{14}C BP, at which point, there is again an increase of population between 3000 and 2800 ^{14}C BP (3400–2800 cal. yr BP).

The moderate population levels that are suggested to correspond with the earliest period of occupation are associated with the warmest conditions on record for the Canadian Arctic, while the period of peak population between 4500 and 4000 cal. yr BP, as well as the reduction to moderate population levels immediately following (4000–3300 cal. yr BP), is associated with a mixture of both warm and cold proxy climate indicators. Conversely, although cold conditions are found after 3800 cal. yr BP and the climate does not ameliorate again until between 3100 and 2500 BP, populations levels again rose by 3400 BP prior to climate recovery. This uncertainty indicates that more temporally and geographically specific climate records are needed to establish whether a link between overall population levels and climatic conditions can be made.

Dyke and Savelle (2009) have conducted similar work for the Low Central Arctic and have identified similar population fluctuations, although these occur at slightly different times. Their results are used to suggest that, after initial colonization, there is a peak in population from 4200 to 3600 ^{14}C yr BP (4600–3700 cal. yr BP), after which populations seem to have seriously declined before partially recovering between 3100 and 2500 ^{14}C yr BP (3400–2400 cal. yr BP). Again, for this area the peak population levels coincide with a time during which the environmental record shows a mix of warm and cold conditions. The reduction of population between 3700 and 3400 cal. yr BP might correspond with a cold period, but unfortunately, there are few climate records that specifically deal with this time period. The general impression is that populations recover at 3400 cal. yr BP, prior to climatic fluctuations veering to warmer at 3100 cal. yr BP (Zabenskie and Gajewski, 2007).

McGhee's (1996) Port Refuge data appear to imply a total abandonment of the area, as opposed to a population reduction; however, clear dates are not available, and he suggests occupations last no longer than 600 years. McGhee (1996, p. 111) later suggests a general Eastern Arctic population decline at 3500 BP. There are several climate records that either signal out 3500 cal. yr BP as a marker for clearly colder climate conditions, or indicate that the onset of cold conditions predates 3500 cal. yr BP but also last beyond this time (Bradley, 1990; Hyvärinen, 1985; LeBlanc et al., 2004; Zabenskie and Gajewski, 2007). Helmer (1991) suggests population declines occurred on north Devon Island in the late Pre-Dorset period, typologically dated to 3000–2800 BP, although the radiocarbon dates associated with the one site (Rocky Point), are 4060 ± 80 ^{14}C (Beta 16554) and 3800 ± 90 ^{14}C (Beta 15391), which have calibrated ranges from 4800 to 4400 cal. yr BP. and 4300 to 4000 cal. yr BP, respectively. If the radiocarbon dates are given priority, then the climate would be warm although unpredictable and trending toward overall cooler conditions, whereas if the typological dates are given preeminence, several climate records are involved that support a time that was colder the previous period (Bradley, 1990; Briner et al., 2006; Koerner, 1977, 1989; Williams and Bradley, 1985).

The lithic technology of the ASTt people is fairly standard across the Canadian Arctic and can be characterized as involving remarkably well-crafted flaked stone technology, with tools most often being made on fine-grained cherts. Beyond being a diagnostic trait for the identification of early ASTt sites, the analysis of ASTt stone tools has been used to suggest gender divisions, to identify individual knappers using apparent skill levels, and has also been used to support the division of Pre-Dorset from Independence I (Bielawski, 1988; McGhee, 1979; Milne, 2005). The finely crafted and standardized nature of ASTt stone may also relate to the fact they were a pioneering population, as has been suggested by Ellis (2008) in an examination of lithic technologies used by another pioneering population, the Paleo Indians.

Ellis (2008) has compared the Paleo Indian fluted point tradition with ASTt technologies, suggesting that a flaked technology that required very fine-grained raw materials such as chert is preferable to a ground stone technology during a migration period because flaked stone technology is more flexible and could be utilized for more diverse needs. Since people, ASTt or Paleo Indian, could be expected to neither know the landscape into which they were moving nor predict the location of resources, more generalized tools that people had experience making, could be quickly modified, and easily maintained would be preferred over ground stone technologies that were not as versatile or easily manufactured.

Although the lithics of the ASTt are diagnostic, whether the differences between the Independence I and Pre-Dorset took kits, both consisting today primarily of preserved lithic materials, mark different cultures or just different functional aspects of the same culture is not agreed upon (Bielawski, 1988). The collection of Independence I and Pre-Dorset assemblages from Port Refuge are the most complete from a comparative standpoint, and the attributes present on these assemblages suggest that diagnostic markers of Independence I include the presence of large unpolished burins, contracting stemmed end blades, triangular end blades with contracting stems and straight bases, and small bi-points. Independence I also shows a tendency to serration of bifacial edges and finer flaking patterns and thinner tool cross-sections than is seen in corresponding Pre-Dorset toolkits (Bielawski, 1988; McGhee, 1979).

4.1.6 Conclusion

If the Eastern Arctic is viewed at a broad scale, some evidence can be found to support the hypothesis that the migration of ASTt peoples into the region occurred during and was influenced by a period of warmer than present temperatures at approximately 4500cal. yr BP. There is also some suggestion that a subsequent cooling trend c.3500cal. yr BP corresponds with population declines in several parts of the Eastern Arctic. However, when population fluctuations and climatic records from more localized areas are examined in greater detail, it is evident that there is no clear connection between climate and population because, as best established by radiocarbon dating, the two do not coincide.

Perhaps the key environmental attribute that should be the focus during the critical period when larger scale migrations are believed to have occurred is the dynamic nature of the Eastern Arctic climate. Current evidence derived from radiocarbon sequences in the Low and High Arctic indicates overlapping date ranges in both regions and suggests that the ASTt expansion across the Eastern Arctic occurred quickly. Based solely on radiocarbon dates, there is no directional pattern to the migration, and while there are locations where apparent concentrations of early ASTt sites occur, this may an artifact of research design.

There is also no clear evidence for early overreliance on one or a few key species (e.g., musk ox); rather, renewed analyses of faunal assemblages indicate that subsistence strategies were diversified as hunters exploited many of the species, large and small, that were available to them. It is also evident that there were tendencies to rely somewhat more heavily on marine or terrestrial species, depending on site location, but specialization on one or a limited range of animals does not appear to have happened. In general the early ASTt sites are small, and preserved material culture, while limited, points to small and mobile populations. At the same time, evidence for periodic population aggregations indicates that while the migrating population was small, there was an overarching social organization linking these otherwise dispersed groups together.

Finally, the nonspecialized flaked stone tool industry produced by ASTt populations is a reflection of their highly mobile and pioneering nature considering this toolkit appears to have been of a generalized nature and geared toward unpredictable and/or unknown situations when the type of prey encountered was not clearly known. This is consistent with a population who were only beginning to explore and familiarize themselves with a largely unknown landscape still undergoing rapid and dynamic change.

Acknowledgments

Max Friesen for suggesting to the volume organizers that I write this paper, Karen Ryan and Douglas Hodgson for helpful comments on earlier drafts, and Steve Perry for assistance with databases and map.

Section Five

Greenland

First people in Greenland*

5.1

Jens F. Jensen[1], Ulla Odgaard[1], Svend Funder[2], Patrick Plumet[3]
[1]Arctic Center of Ethnographic Collections, the National Museum of Denmark, Copenhagen, Denmark; [2]University of Copenhagen, Copenhagen, Denmark; [3]Université de Québec, Montréal, Canada

Chapter Outline

5.1.1 Introduction

Greenland was first settled in the Holocene by the same Paleoeskimo peoples of the Arctic Small Tool tradition (ASTt) as are known from Eastern Canadian Arctic. The Paleoeskimos migrated into the Eastern Arctic (Northeastern Canada and Greenland)

* This article is dedicated to Patrick Plumet (1934–2010), our coauthor, whose opinions and arguments are included, although he did not get the time to approve of the final result, before he recently passed away. His remarkable knowledge about Arctic cultures, not only in the Eastern Arctic, but in the whole circumpolar area, will still continue to inspire researchers.

Human Colonization of the Arctic: The Interaction Between Early Migration and the Paleoenvironment
http://dx.doi.org/10.1016/B978-0-12-813532-7.00018-8

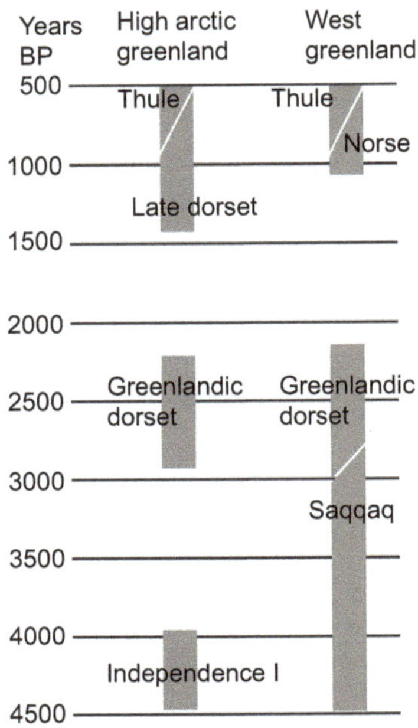

Figure 5.1.1 Prehistoric cultures in Greenland.

nearly 5000 years ago. They inhabited this vast area for around 3500 years, but they seem to have disappeared when the Thule culture (the ancestors of present Inuit) expanded to this region around AD 1200 (Friesen and Arnold, 2008; Gulløv, 2004) (Fig. 5.1.1). The archaeological remains, especially the stone tools, of the earliest Paleoeskimos have been typologically divided into three material cultures: "Independence I," "Saqqaq," and "Pre-dorset." This paper will give a picture of Saqqaq, who represent the first people in West Greenland, and Independence I, which is the pioneer culture in Northeast Greenland. In Greenland, Pre-Dorset is only known from a few localities along the Nares Strait, and for the moment, it seems as if this culture only settled the Nares Strait region and never dispersed to other parts of Greenland (Jensen, 2006a; Sørensen, 2011). Therefore we will only discuss the two pioneering cultures of Saqqaq and Independence I.

The Paleoeskimo immigrants entered Greenland over the Nares Strait (Fig. 5.1.2). With a width of c.30 km, this water body was not an obstacle, but the passage led into Greenland's most inhospitable areas and left the immigrants with the choice of turning south or north: both options were problematical. To the south, they would face the Melville Bugt where the ice sheet more or less forms the coast over a distance of c.300 km, presenting an insurmountable barrier to such animals as the musk ox, which never reached lush West Greenland. The northern route, chosen by the musk ox, leads into the world's northernmost land with permanent sea ice at the coast and 5 months' polar night.

The first immigrants at 5–4.5 ka BP chose both options. Throughout history, the geographic division of Greenland into High Arctic and Low Arctic regions have affected the life ways of peoples in High and Low Arctic Greenland differently, and the environmental differences also affect the preservation and appearance of the archaeological record distinctly different in the two regions. Next, we give a sketch

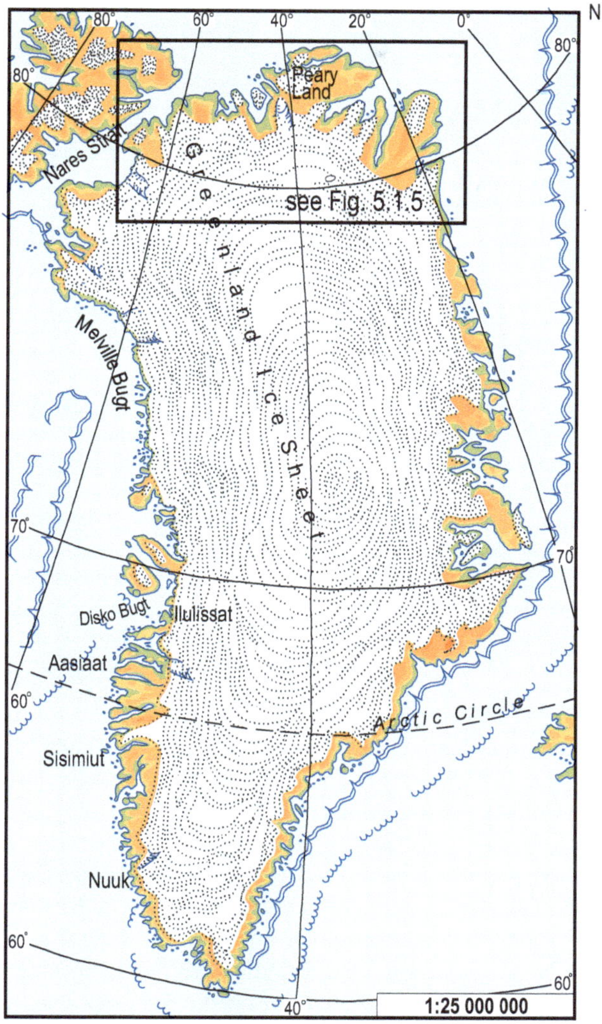

Figure 5.1.2 Map of Greenland (with Nares Strait, Melville Bugt, Disko Bugt, Sisimiut, and Nuuk).

of the land that these first immigrants met, based partly on a review of the ice sheet's history by Funder et al.

The ^{14}C ages have been calibrated with the CALIB version 6.0 conversion program (Reimer et al., 2013).

5.1.1.1 Setting the stage

During the last ice age the Greenland ice sheet merged with the Innuitian ice sheet, which covered the Canadian Arctic Archipelago, and a major ice stream filled the Nares Strait (Dyke, 2004). Melting of the ice sheets began at c.16 ka BP, and by

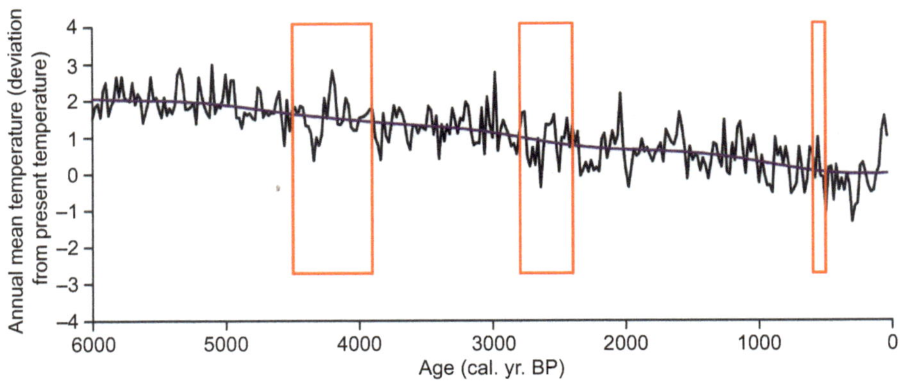

Figure 5.1.3 Annual mean temperatures over Greenland in the period since the first immigration, and duration of habitation in northernmost Greenland. Temperatures based on ice core measurements. *Black*, 20-year average, *blue*, smoothed with Gaussian filter. *Red boxes*, human habitation in North Greenland.
From Vinther, B.M., Buchardt, S.L., Clausen, H.B., Dahl-Jensen, D., Johnsen, S.J., Fisher, D.A., Koerner, R.M., Raynaud, D., Lipenkov, V., Andersen, K.K., Blunier, T., Rasmussen, S.O., Steffensen, J.P., Svensson, A.M., 2009. Holocene thinning of the Greenland ice sheet. Nature 461, 385–388; Grønnow, B., Jensen, J.F., 2003. The Northernmost Ruins of the Globe. Eigil Knuth's Archaeological Investigations in Peary Land and Adjacent Areas of High Arctic Greenland. Meddelelser om Grønland/Monographs on Greenland, vol. 29. Man & Society, p. 403.

c.10 ka BP the strait was cleared of ice. Two thousand years later, most of the presently ice-free land in Greenland was cleared, but up until c.6 ka, large ice caps still existed in eastern Canada (Dyke, 2004), and it was probably only after this time that the gateway was open to immigrants. This was also the time when gradual cooling began to set its marks on terrestrial and near shore marine biotopes. Ice core temperature records show that after a culmination shortly before 8 ka BP, annual mean temperatures over Greenland started a decline, which went on until preindustrial times (Vinther et al., 2009). The decline was caused mainly by decreasing insolation, and annual mean temperatures over this period dropped c.3°C. However, the decline was punctuated by short-term warm and cold fluctuations with almost the same amplitude (Fig. 5.1.3).

In the terrestrial and marine environments, different proxies with different sensitivities record the decline at different times. The proxies include such features as changes in terrestrial and limnic flora and fauna seen in pollen and microfauna in lake cores, changes in extent of sea ice, appearance of permanent lake ice in the north, build-up of local ice caps, and glacier advance. The earliest signs of temperature decline have been recorded at c.6.5 ka BP in a South Greenland lake (Bennike et al., 2010), but the most severe setback in terrestrial and nearshore marine environments in West Greenland occurred at c.3.5–3 ka BP (Böcher and Fredskild, 1993; Ren et al., 2009). In North and East Greenland the temperature decline is seen clearly at 5.5–5 ka BP, before the arrival of the first settlers. In all areas the cooling culminated during the Little Ice Age (AD c.1350–1920). In general, as in the ice cores, the proxies show that the decrease in temperatures since c.8 ka BP was 2–4°C, but the biological proxies generally reflect summer temperatures. Conditions similar to the present were attained at c.4–3 ka BP.

An important factor in Greenland's climate is advective heat brought from the south by air masses and ocean currents. Along the east coast the cold East Greenland Current carries polar water and sea ice from the Arctic Ocean to the south. Off southeast Greenland, it is combined with the warm Irminger Current, and together, they swing around the southern tip of Greenland and head northward into Baffin Bay as the warm West Greenland Current. This pattern of warm and cold currents dictates greatly different living conditions in east and west Greenland. The basic pattern of these currents can be traced back at least to the earliest Holocene, but during the Early Holocene the warm component—advected Atlantic water—reached a maximum, bringing new warmth-demanding mollusks into Greenland waters (Funder and Weidick, 1991), and probably extending the area for Atlantic fish, bird, and sea mammal species further north than ever seen since. This came to an end at 4–5 ka BP (Justwan et al., 2008; Lloyd et al., 2007; Ren et al., 2009), and distinctly Arctic species and the sea ice biome expanded toward the south again.

Among the main animals hunted by the early immigrants, most were present in the country, also in the north, some time before the arrival of humans: ringed and harp seal back to c.9 ka BP, reindeer to c.8 ka BP, polar bear to c.6 ka BP (Bennike, 1997), and walrus to c.7 ka BP (Bennike and Andreasen, 2007). The most significant omission from this list is musk ox, which might have arrived just shortly prior to the first immigrants, but only in north and east Greenland (Campos et al., 2010).

5.1.1.2 Greenland at 5–4 ka BP

When the first immigrants arrived in Greenland the ice sheet was somewhat smaller than now. Modeling from the pattern of isostatic uplift has indicated that the ice sheet reached its Holocene minimum around this time, but the reduction was mainly along the western margin (Simpson et al., 2009). The amount of ice margin retreat, up to 80 km behind the present, may be exaggerated, but for the first settlers in Disko Bugt the area for hunting of reindeer and other terrestrial mammals was undoubtedly larger than now, as suggested by Meldgaard (2004). In this area, which was a nucleus for the Saqqaq culture, temperatures were slightly higher and the vegetation richer until c.3 ka BP (Böcher and Fredskild, 1993). In the adjacent fjord, cooling had begun at 5 ka BP, but it only became prevalent from c.3.5 ka BP (Lloyd et al., 2007).

In the north, conditions were not very different from today's. The ice margin was close to the present and the extent of hunting grounds also similar to the present. In this area the cooling began to set its mark on the environment around the same time as the arrival of the Independence I people at c.4.5 ka BP when lakes became permanently ice covered and dwarf bush heath was transformed to polar desert (Blake et al., 1992).

Owing to the incomplete deglaciation of Greenland the pattern of isostatic uplift is complex (Funder and Hansen, 1996). Maximum uplift occurred in coastal areas, and when the first immigrants arrived, sea level was up to c.10 m above the present in the

Sisimiut area of Western Greenland and at the coast of Northeast Greenland (Long et al., 1999). In other areas the uplift was lower. The Saqqaq people in the interior Disko Bugt area saw a sea level close to the present (Long et al., 1999), and at the time of the Independence I people in interior Independence Fjord, sea level was c.5 m above the present (Funder and Abrahamsen, 1988).

5.1.1.3 The role of sea ice

Northern Greenland, which the immigrants had to pass through on their way to more pleasant places, ranks among the world's most hostile environments to humans. To explain how people were able to sustain a living here, resort has often been taken to the climate: was the climate better (warmer), when they came? From Fig. 5.1.3, it can be seen that although it was generally 1.5° warmer when the Independence I people lived here, then temperatures during the two following habitation periods were not significantly different from the present. As mentioned subsequently, the habitation in the north was short, much shorter than indicated by the boxes in Fig. 5.1.3, and the uncertainty in ^{14}C dating does not allow a more precise dating within the range given in Fig. 5.1.3. Through the three cultural phases the general temperature drop was overprinted by decadal to centennial cold–warm variations, as shown also by the proxies. This is similar to the variations known from historical time, and with their effect on hunting and sea ice, they probably presented a constant challenge to the inhabitants in all three phases. Therefore, warm temperatures do not appear to have been the incentive for the immigrants, and there is no correlation between a certain type of climate and habitation. The immigration of the "Thule people" indeed took place during the Little Ice Age when temperatures reached their minimum. The passage to Greenland had been free, probably at least a millennium before the Paleoeskimo immigrants used it, and ice sheet distribution and sea level at the gateway were not significantly different from today. Most of the game was already established in the area, so the immigration was not caused by any of these factors.

Among the game, only the musk ox apparently was not there. It apparently arrived together with the immigrants, or rather vice versa, as suggested already by Knuth (1967b). The Independence I peoples' dependency on musk ox in High Arctic Greenland is shown by the many musk ox bones at their camp sites, but the fauna material from Independence I sites in High Arctic Canada (Darwent, 2004) and the presence of coastal sites in Northeast Greenland show that they also mastered seal hunting (Andreasen, 2004). Since there is no marked change in climate or physiography at this time, the oxen's—and therefore the first immigrants'—passage into Greenland should probably be regarded as another step in an ongoing expansion. Musk oxen apparently were present in the central Arctic Archipelago already at c.8 ka BP (Harrington, 1980), and alternatively the passage into Greenland could be seen as a reaction to stress provoked by the effect of declining temperatures on the vegetation in their Canadian homeland.

However, both oxen and men had to come a long way across the Canadian Arctic Archipelago before reaching Greenland. From West Greenland Saqqaq sites, there is evidence of the use of kayak-like vessels (Grønnow, 1994), so the migration may to some extent have relied on sailing in open waters during the warm months. However, for the people, who migrated north around Greenland, an indispensable prerequisite for the trek could be the sea ice, which bridged the islands in the archipelago. Early descriptions of the life of the Polar Eskimos stress the importance of sea ice—especially landfast ice—as an indispensable means for transportation, social contact, and hunting (Boas, 1888; Rasmussen, 1908; Steensby, 1910). The immigrants to Greenland came trained to live a life where changing sea ice conditions were as important as changing sun and rain were to the European peasant.

Sea ice is an elusive agent in the geological record, but a clue to past sea ice conditions at the coasts can be read out of their record of driftwood and coastal landforms (England et al., 2008). Driftwood on the coasts of Greenland and the Canadian Arctic Archipelago is ferried across the ocean on sea ice, but absence or scarcity of driftwood in combination with lack of wave-generated landforms indicate long-lasting landfast ice along the coast, blocking the landing of wood. For northern Ellesmere Island and Greenland's north coast the driftwood record shows that periods with perennial or long-lasting landfast ice became more frequent from c.5.5 ka BP and especially after c.3 ka BP, culminating during the Little Ice Age. Similarly, in Lancaster Sound in the southeastern archipelago, an increase in yearly sea ice cover also began at 5.5 ka and proceeded to the present day, although the western part of the archipelago shows a reverse trend (Ledu et al., 2010). The coincidence of increasing duration of landfast sea ice with the paleoeskimoic expansion into the High Arctic and across the archipelago suggests that this development assisted the immigrants, providing longer seasons for hunting and travel.

5.1.2 Saqqaq culture: first people in West Greenland

The oldest Stone Age sites in Greenland are dated to just around 4500 cal. yr BP. In East Greenland the site of Røde Hytte in the vicinity of the town Ittoqqortoormiit has been dated to between 4030 and 4000 BP (Sandell and Sandell, 1996), and in West Greenland the site of Qivittup Nuua near the town of Sisimiut has been dated to 4010 BP (Kramer, 1996b). A few hundred kilometers further north is the little surface site of Ikerasak that has been radiocarbon dated to 3980 BP (Jensen, 2006b, p. 175). The aforementioned dates were conducted on charcoal of locally grown wood (Table 5.1.1). Almost similarly old dates are known from the site of Qeqertasussuk, where turf from the deepest layers has been dated to 3980 BP (Grønnow, 1994, p. 203). From around 3850 BP the suite of dates becomes more consistent, since from this time and onward there are numerous dates from different localities in West Greenland as well as in Northeast Greenland, and many dates are conducted on terrestrial mammal bone or antler giving more reliable dates than wood, which in particular in the high arctic environment can survive for many years on the surface. The suite of radiocarbon dates thus does not allow us to outline the details of the whereabouts of the early pioneering peoples in Greenland. Within a very short period around 4000 BP (4500 cal. yr BP) humans dispersed to Peary Land as well as West Greenland.

Table 5.1.1 **Table of Early Paleoeskimo ^{14}C dates from Greenland. The calibrated dates are given as published in referred publications**

Lab. no	Site name	Cultural significance	Material	Species	^{14}C age BP	±	δ ^{13}C	cal. BC ± standard deviation	Reference
Ua-2166	Qivittup Nuua	Saqqaq	Charcoal	*Betula nana*	4010	90		2840–2810, 2670–2400, 2380–2350	Kramer (1996a)
Ka-6990	Ikerasak	Saqqaq	Charcoal	*Betula nana*, *Salix* sp.	3980	70	−25.4	2580–2400, 2380–2350	Jensen (2006a,b)
K-4823	Qeqertasussuk	Saqqaq	Turf		3980	85	−25.5	2630–2340, 2320–2310	Grønnow (1994)
K-938	Pearylandville, 24	Independence I	Charcoal	*Salix* sp.	3950	120		2620–2200	Grønnow and Jensen (2003)
K-928	Portfjeldet, 1	Independence I	Charcoal	*Salix* sp.	3890	120		2560–2140	Grønnow and Jensen (2003)
K-4566	Qeqertasussuk	Saqqaq	Turf		3880	85	−22.5	2470–2270, 2260–2200	Meldgaard (2004)
K-3366	Solbakken, 2B	Independence I	Bone	*Ovibos moschatus*	3870	85	−18.8	2470–2200	Grønnow and Jensen (2003)
K-929	Portfjeldet, 2	Independence I	Charcoal	*Salix* sp.	3860	120		2480–2130	Grønnow and Jensen (2003)
K-4497	Deltaterrasserne, 14	Independence I	Bone	*Ovibos moschatus*	3850	90	−18.6	2460–2200	Grønnow and Jensen (2003)
K-939	Pearylandville, 10a	Independence I	Charcoal	*Salix* sp.	3840	120		2470–2130	Grønnow and Jensen (2003)
K-4498	Deltaterrasserne, 12	Independence I	Bone	*Ovibos moschatus*	3840	85	−19.6	2450–2150	Grønnow and Jensen (2003)

TUa-1689	Tupersui feature III	Saqqaq	Charcoal	*Betula*, *Salix*	3835	70	–26.1	2460–2420, 2410–2190, 2160–2140	Olsen (1998)
K-3364	Midternæs, 6	Independence I	Bone	*Ovibos moschatus*	3830	85	–19.3	2460–2140	Grønnow and Jensen (2003)
K-5125	Qeqertasussuk	Saqqaq	Bone	*Phoca* sp.	3820	60	–14.6	2390–2140	Meldgaard (2004)
TUa-1722	Tupersui feature III	Saqqaq	Charcoal	*Salix* sp., *Betula nana*	3815	65	–26.1	2400–2370, 2350–2190, 2180–2140	Olsen (1998)
Ka-6988	Kannala	Saqqaq	Charcoal	*Betula nana*	3810	40	–26.0	2310–2190, 2170–2140	Jensen (2006a,b)
K-1062	Vendenæs, 4	Independence I	Charcoal	*Salix* sp.	3800	120		2460–2040	Grønnow and Jensen (2003)
Ua-2164	Maniitsorsuaq	Saqqaq	Charcoal	*Salix* sp.	3800	90		2410–2370, 2360–2130, 2090–2040	Kramer (1996a)
K-3056	Tingit square B	Saqqaq	Charcoal	*Salix*, *Betula nana*	3800	85	–24.4	2410–2370, 2350–2130, 2080–2050	Kramer (1996a)
K-3362	Kap Peter Henrik, N-3	Independence I	Bone	*Ovibos moschatus*	3800	85	–18.5	2410–2050	Grønnow and Jensen (2003)
K-5788	Akia area A	Saqqaq	Charcoal	*Salix* sp., *Betula nana*	3790	85		2400–2380, 2350–2130, 2090–2040	Kramer (1996a)
K-4819	Qeqertasussuk	Saqqaq	Herbaceous turf		3780	85	–25.6	2400–2380, 2340–2120, 2100–2030	Meldgaard (2004)
K-930	Portfjeldet, 3	Independence I	Charcoal	*Salix* sp.	3790	120		2460–2030	Grønnow and Jensen (2003)
K-5076	Portfjeldet, 3	Independence I	Bone	*Ovibos moschatus*	3790	60	–19.8	2340–2060	Grønnow and Jensen (2003)

Continued

Table 5.1.1 Continued

Lab. no	Site name	Cultural significance	Material	Species	^{14}C age BP	±	δ ^{13}C	cal. BC ± standard deviation	Reference
K-932	Vandfaldsnæs, 15	Independence I	Charcoal	*Salix* sp.	3780	120		2410–2030	Grønnow and Jensen (2003)
K-4676	Adam C. Knuth Site, II, 1	Independence I	Antler	*Rangifer tarandus*	3780	80	−20.7	2340–2030	Grønnow and Jensen (2003)
T-12919	Tupersui feature III	Saqqaq	Charcoal	*Betula*, *Salix*	3770	75	−26.1	2300–2110, 2100–2030	Olsen (1998)
Ua-1687	Deltaterrasserne, 13	Independence I	Bone	*Ovibos moschatus*	3770	75		2300–2030	Grønnow and Jensen (2003)
K-1262	Kettle Lake, N, 4	Independence I	Charcoal	*Salix arctica*	3760	130		2400–1970	Grønnow and Jensen (2003)
K-1262	Kettle Lake, N, 4	Independence I	Charcoal	*Salix arctica*	3760	130		2400–1970	Grønnow and Jensen (2003)
K-1061	Vendenæs, 3	Independence I	Charcoal	*Salix* sp.	3760	120		2400–1970	Grønnow and Jensen (2003)
K-4821	Qeqertasussuk	Saqqaq	Heather		3760	80	−27.3	2300–2030	Meldgaard (2004)
K-4258	Kap Holbæk, I, A + b	Independence I	Bone	*Ovibos moschatus*	3760	70	−18.6	2290–2030 BC	Grønnow and Jensen (2003)
K-4565	Qeqertasussuk	Saqqaq	Peat		3750	80	−22.4	2290–2030	Meldgaard (2004)
T-12997	Tupersui feature II	Saqqaq	Charcoal	*Betula*, *Salix*	3730	110	−26.1	2300–1950	Olsen (1998)

K-5739	Bob's Site, Cache C,3	Independence I	Bone	*Ovibos moschatus*	3730	85	−20.5	2290–1970 BC	Grønnow and Jensen (2003)
K-4564	Qeqertasussuk	Saqqaq	Peat		3730	80	−24.3	2290–2250, 2240–2220, 2000–1970	Meldgaard (2004)
K-4563	Qeqertasussuk	Saqqaq	Peat		3720	80	−24.2	2280–2250, 2230–2220, 2210–2010, 2000–1970	Meldgaard (2004)
K-4562	Qeqertasussuk	Saqqaq	Peat		3690	80	−22.5	2200–2160, 2150–1950	Meldgaard (2004)
K-3768	Sermermiut section 13H layer 5	Saqqaq	Peat		3690	80	−23.3	2200–2160, 2150–1950	Jensen (2006)
Tua-1688	Niivertussannguaq feature III	Saqqaq	Charcoal	*Betula*, *Salix*	3685	75	−26.1	2200–2160, 2150–1950	Olsen (1998)
K-4817	Qeqertasussuk	Saqqaq	Herbaceous turf		3680	85	−26.6	2200–2160, 2150–1940	Meldgaard (2004)
K-6192	Nipisat	Saqqaq	Bone	*Rangifer tarandus*	3670	90		2200–2160, 2150–1920	Jensen (2006a,b)
K-3531	Adam C. Knuth Site, III, 1	Independence I	Bone	*Ovibos moschatus*	3670	80	−19.6	2200–1930 BC	Grønnow and Jensen (2003)
K-4058	Pearylandville, 7	Independence I	Bone	*Ovibos moschatus*	3670	75	−19.9	2190–1940 BC	Grønnow and Jensen (2003)
K-4818	Qeqertasussuk	Saqqaq	Heather		3650	85	−27.3	2140–1910, 1900–1890	Meldgaard (2004)
AAR-3630	Nipisat	Saqqaq	Charcoal	*Salix* sp.	3650	65		2130–1920	Gotfredsen and Møbjerg (2004)
K-4822	Qeqertasussuk	Saqqaq	Heather		3640	75	−26.7	2140–1910, 1900–1890	Meldgaard (2004)

Continued

Table 5.1.1 **Continued**

Lab. no	Site name	Cultural significance	Material	Species	^{14}C age BP	±	δ ^{13}C	cal. BC ± standard deviation	Reference
AAR-3027	Asummiut	Saqqaq	Charcoal	*Salix glauca*	3630	80		2140–2070, 2050–1880	Jensen (2006a,b)
K-3532	Adam C. Knuth Site, II, 1	Independence I	Bone	*Ovibos moschatus*	3630	80	−19.4	2140–1880 BC	Grønnow and Jensen (2003)
AAR-2957	Kuup Qalorsaa	Saqqaq	Bone	*Phoca* sp.	3630	60	−14	2130–2080, 2040–1880	Jensen (2006a,b)
K-1196	Gammel Strand Nord, 3	Independence I	Charcoal	*Salix* sp.	3620	110		2140–1770 BC	Grønnow and Jensen (2003)
K-1537	Tuapassuit	Saqqaq	Charcoal	*Salix* sp. *Betula* sp.	3620	100		2140–1870, 1840–1820, 1800–1780	Jensen (2006a,b)
Tua-1723	Tupersui feature 4	Saqqaq	Charcoal	*Salix* sp., *Betula nana*	3620	70	−26.1	2130–2080, 2040–1880	Olsen (1998)
T-12996	Niivertussannguaq feature II	Saqqaq	Charcoal		3615	85	−26.1	2140–2070, 2050–1870, 1840–1820	Olsen (1998)
AAR-3030	Asummiut	Saqqaq	Charcoal	*Betula nana*	3605	65		2120–2090, 2040–1870, 1840–1820	Jensen (2006a,b)
K-3365	Martins Site, A	Independence I	Bone	*Ovibos moschatus*	3590	60	−20.0	2040–1780 BC	Grønnow and Jensen (2003)
TUa-1725	Kuup Qalorsaa	Saqqaq	Charcoal	*Ericales*, *Betula nana*	3585	70	−26.1	2040–1870, 1850–1810, 1800–1770	Olsen (1998)
AAR-3029	Asummiut	Saqqaq	Charcoal	*Betula nana*	3575	65		2030–1870, 1850–1810, 1800–1770	Jensen (2006a,b)

K-144	Illorsuit I	Saqqaq	Charcoal		3570	150		2140–1730, 1710–1690	Larsen and Meldgaard (1958)
K-5740	Vandfaldsnæs, 15	Independence I	Bone	*Ovibos moschatus*	3570	80	–20.0	2030–1770 BC	Grønnow and Jensen (2003)
K-5127	Qeqertasussuk	Saqqaq	Bone	*Phoca* sp.	3570	80	–14.0	2020–1770	Meldgaard (2004)
AAR-3582	Nipisat	Saqqaq	Shell	*Mytilus edulis*	3570	60		1970–1780	Gotfredsen and Møbjerg (2004)
K-3904	Qajaa, C	Saqqaq	Twigs		3550	85	–26.7	2020–1990, 1980–1740	Jensen (2006a,b)
K-3899	Qajaa, E	Saqqaq	Twigs		3550	80	–25.6	2020–1990, 1980–1750	Jensen (2006a,b)
TUa-1686	Niivertussannguaq feature II	Saqqaq	Charcoal	*Betula, Ericales, Salix*	3545	75	–26.1	2010–2000, 1980–1740	Olsen (1998)
K-4104,	Qajaa, D	Saqqaq	Twig		3540	80	–23.6	2010–2000, 1960–1740	Jensen (2006a,b)
Ua-2167	Amitsuarsuk	Saqqaq	Charcoal	*Salix* sp.	3530	110		2030–1990, 1980–1730, 1720–1690	Kramer (1996a)
Ka-6079	Nuunguaq, structure 1	Saqqaq	Charcoal twigs		3530	60	–24.0	1940–1740	Appelt and Pind (1996)
T-12918	Tupersui feature I	Saqqaq	Charcoal	*Betula, Salix*	3525	85	–26.1	1960–1730	Olsen (1998)
AAR-3576	Nipisat I, phase 1	Saqqaq	Bone	*Rangifer tarandus*	3525	50		1910–1750	Gotfredsen and Møbjerg (2004)
AAR-3577	Nipisat I, phase 1	Saqqaq	Bone	*Rangifer tarandus*	3515	65		1910–1740	Gotfredsen and Møbjerg (2004)

Continued

Table 5.1.1 Continued

Lab. no	Site name	Cultural significance	Material	Species	^{14}C age BP	±	δ ^{13}C	cal. BC ± standard deviation	Reference
K-806	Sermermiut B, Section I	Saqqaq	Sand with humus		3510	120		2020–1990, 1980–1680	Jensen (2006a,b)
TUa-1690	Tupersui feature V	Saqqaq	Charcoal	*Betula*, *Salix*	3505	105	−26.1	1960–1680	Olsen (1998)
K-5126	Qeqertasussuk	Saqqaq	Bone	*Phoca* sp.	3500	80	−14.1	1920–1730, 1720–1690	Grønnow (1994)
K-3906	Qajaa, F	Saqqaq	Twig, modified		3490	140	−26.7	2020–1990, 1980–1620	Jensen (2006a,b)
K-6031	Nipisat I area B, layer 1	Saqqaq	Bone	*Rangifer tarandus*	3490	80		1920–1730, 1720–1690	Kramer (1996a)
K-5192	Angujaartorfik	Saqqaq	Bone	*Rangifer tarandus*	3490	80	−17.8	1920–1730, 1720–1690	Jensen (2006a,b)
AAR-3581	Nipisat I, phase 1	Saqqaq	Shell	*Hiatella arctica*	3490	50		1880–1740	Gotfredsen and Møbjerg (2004)
AAR-7376	Siorartalik	Saqqaq	Charcoal	Local	3486	38	−25.69	1880–1840, 1830–1740	Jensen (2006a,b)
AAR-3578	Nipisat I, phase 1	Saqqaq	Bone	*Rangifer tarandus*	3480	50		1880–1700	Gotfredsen and Møbjerg (2004)
K-3057	Tingit square A	Saqqaq	Charcoal	*Salix*, *Betula nana*	3470	125	−26.4	1950–1620	Kramer (1996a)
K-5124	Qeqertasussuk	Saqqaq	Bone	*Phoca* sp.	3460	80	−14.4	1880–1680	Meldgaard (2004)

TUa-1687	Niivertussannguaq feature I	Saqqaq	Charcoal	*Betula, Ericales, Salix*	3460	70	−26.1	1880–1680	Olsen (1998)
TUa-1724	Tupersui feature V	Saqqaq	Charcoal	*Salix, Betula nana*	3455	65	−26.1	1880–1680	Olsen (1998)
AAR-6699	Nipisat I, phase 1	Saqqaq	Bone	*Rangifer tarandus*	3450	45		1875–1690	Gotfredsen and Møbjerg (2004)
Ka-6944	Brinchip tupegarfikuutaa	Saqqaq	Charcoal	*Betula nana*	3450	40	−26.5	1880–1840, 1810–1790, 1780–1730, 1720–1680	Jensen (2006a,b)
K-3900	Qajaa, E	Saqqaq	Twigs		3440	80	−25.9	1880–1680, 1670–1630	Jensen (2006a,b)
AAR-3580	Nipisat I, phase 1	Saqqaq	Shell	*Littorina saxatilis*	3445	50		1870–1680	Gotfredsen and Møbjerg (2004)
K-5194	Angujaartorfik	Saqqaq	Bone	*Rangifer tarandus*	3440	80	−18.0	1880–1680, 1670–1630	Jensen (2006a,b)
AAR-6698	Nipisat I, phase 1	Saqqaq	Bone	*Rangifer tarandus*	3425	50		1855–1645	Gotfredsen and Møbjerg (2004)
K-5128	Qeqertasussuk	Saqqaq	Bone	*Phoca* sp.	3400	80	−14.9	1870–1840, 1780–1600	Grønnow (1994)
K-4561	Qeqertasussuk	Saqqaq	Peat		3390	80	−24.8	1660–1840, 1770–1600, 1570–1520	Meldgaard (2004)
AAR-3028	Asummiut	Saqqaq	Charcoal	*Salix* sp.	3370	65		1740–1600, 1570–1530	Jensen (2006a,b)
K-5191	Angujaartorfik	Saqqaq	Bone	*Rangifer tarandus*	3370	80	−17.6	1750–1520	Jensen (2006a,b)
K-807	Sermermiut B, Section I	Saqqaq	Peat		3360	120		1870–1840, 1810–1800, 1780–1510	Jensen (2006a,b)

Continued

Table 5.1.1 Continued

Lab. no	Site name	Cultural significance	Material	Species	^{14}C age BP	±	$\delta\ ^{13}C$	cal. BC ± standard deviation	Reference
K-855	Ikka	Saqqaq	Charcoal	*Picea* sp.	3360	120		1870–1840, 1810–1800, 1780–1510	Jensen (2006a,b)
AAR-8407	Kannala	Saqqaq	Charcoal	*Betula nana*	3360	45	25.44	1740–1710, 1690–1600, 1560–1530	Jensen (2006a,b)
K-146	Illorsuit I	Saqqaq	Charcoal		3340	140		1860–1840, 1780–1440	Larsen and Meldgaard (1958)
K-6459	Nipisat phase 3 area 1	Saqqaq	Bone	*Rangifer tarandus*	3010	95		1390–1080	Gotfredsen and Møbjerg (2004)
K-6195	Nipisat phase 2 area 2	Saqqaq	Bone	*Rangifer tarandus*	3310	85		1690–1490	Jensen (2006a,b)
K-4816	Qeqertasussuk	Saqqaq	Herbaceous turf		3310	80	−26.2	1690–1500	Grønnow (1994)
K-6196	Nipisat phase 2 area 2	Saqqaq	Bone	*Rangifer tarandus*	3300	85		1690–1490	Jensen (2006a,b)
AAR-6986	Nipisat I, phase 1	Saqqaq	Bone	*Rangifer tarandus*	3280	45		1620–1515	Gotfredsen and Møbjerg (2004)
K-5193	Angujaartorfik	Saqqaq	Bone	*Rangifer tarandus*	3190	75	−17.6	1600–1580, 1530–1380, 1340–1320	Jensen (2006a,b)
K-6198	Nipisat phase 2 area 2	Saqqaq	Charcoal	*Salix* sp., *Betula nana*	3180	125		1620–1310, 1280–1260	Jensen (2006a,b)
T-12917	Niivertussannguaq feature II	Saqqaq	Charcoal	*Salix*, *Ericales*	3160	85	−26.1	1530–1310	Olsen (1998)
K-4820	Qeqertasussuk	Saqqaq	Herbaceous turf		3150	80	−27.9	1520–1370, 1360–1310	Meldgaard (2004)
K-1193	Itinnera	Saqqaq	Charcoal	*Betula* sp. and *Salix* sp.	3140	120		1520–1370, 1360–1310	Jensen (2006a,b)

AAR-3570	Nipisat phase 2 area 2	Saqqaq	Bone	*Rangifer tarandus*	3085	45		1405–1265	Gotfredsen and Møbjerg (2004)
AAR-3572	Nipisat phase 3 area 3	Saqqaq	Bone	*Rangifer tarandus*	3065	40		1395–1265	Gotfredsen and Møbjerg (2004)
AAR-4158	Nipisat phase 3 area 1	Saqqaq	Bone	*Rangifer tarandus*	3060	45		1395–1230	Gotfredsen and Møbjerg (2004)
AAR-3571	Nipisat phase 3 area 3	Saqqaq	Bone	*Rangifer tarandus*	3040	50		1390–1210	Gotfredsen and Møbjerg (2004)
AAR-4160	Nipisat phase 3 area 3	Saqqaq	Bone	*Rangifer tarandus*	3025	40		1375–1210	Gotfredsen and Møbjerg (2004)
AAR-3583	Nipisat phase 3 area 1	Saqqaq	Shell	*Mytilus edulis*	3020	50		1380–1140	Gotfredsen and Møbjerg (2004)
K-6459	Nipisat, phase 3 area 1	Saqqaq	Bone	*Rangifer tarandus*	3010	95		1400–1120	Gotfredsen and Møbjerg (2004)
AAR-6987	Nipisat, phase 3 area 1	Saqqaq	Bone	*Rangifer tarandus*	2975	45		1290–1125	Gotfredsen and Møbjerg (2004)
K-588	Itinnera, house 1	Saqqaq	Wood	*Juniperus* sp.	2960	110		1370–1360, 1320–1010	Jensen (2006a,b)
AAR-3573	Nipisat phase 3 area 1	Saqqaq	Bone	*Rangifer tarandus*	2950	50		1260–1040	Gotfredsen and Møbjerg (2004)
K-5584	Nipisat I phase 3 area 3	Saqqaq	Bone	*Rangifer tarandus*	2940	80	−18,8	1260–1010	Kramer (1996a)
CAMS-82924 &25	Nipisat, phase 3 area 2	Saqqaq	Bone	*Rangifer tarandus*	2935	22			Gotfredsen and Møbjerg (2004)

Continued

Table 5.1.1 Continued

Lab. no	Site name	Cultural significance	Material	Species	^{14}C age BP	±	$\delta\ ^{13}C$	cal. BC ± standard deviation	Reference
K-6193	Nipisat phase 3 area 1	Saqqaq	Bone	*Rangifer tarandus*	2920	85		1260–990	Gotfredsen and Møbjerg (2004)
K-3905	Qajaa, C	Saqqaq	Peat		2910	75	−25.6	1260–1230, 1220–990	Jensen (2006a,b)
K-6460	Nipisat, phase 3 area 2	Saqqaq	Bone	*Rangifer tarandus*	2910	60		1220–1000	Gotfredsen and Møbjerg (2004)
K-5864	Nipisat phase 3 area 3	Saqqaq	Bone	*Rangifer tarandus*	2860	80		1190–1180, 1150–1140, 1130–910	Kramer (1996a)
AAR-4159	Nipisat phase 3 area 1	Saqqaq	Bone	*Rangifer tarandus*	2835	40		1015–920	Gotfredsen and Møbjerg (2004)
K-3901	Qajaa, E	Saqqaq	Peat		2830	80	−24.2	1130–890	Jensen (2006a,b)
AAR-3575	Nipisat phase 3 area 1	Saqqaq	Bone	*Rangifer tarandus*	2815	45		1005–905	Gotfredsen and Møbjerg (2004)
K-518	Saqqaq Niaqornaarsuk	Paleoeskimo	Charcoal	*Betula nana*	2760	100		1010–800	Jensen (2006a,b)
K-3894	Qajaa, A	Saqqaq	Peat		2700	75	−25.5	920–790	Jensen (2006a,b)
Ka-6992	Kangerlussorissunnguup Kangia C	Dorset	Charcoal	*Salix* sp.	2680	40	−25.4	900–875, 860–850, 840–800	Jensen (2006a,b)
K-6194	Nipisat, phase 3 area 1	Saqqaq	Bone	*Rangifer tarandus*	2670	85		970–950, 930–760	Gotfredsen and Møbjerg (2004)

In Peary Land the suite of reliable Independence I dates is focused in the centuries between 4450 and 3750 cal. yr BP. Whereas in West Greenland the Saqqaq culture last almost a millennium longer, since here the youngest Saqqaq culture sites are dated to around 2750 cal. yr BP. The age of the occupation of several sites in West Greenland indicate that the Saqqaq pioneers quickly after arrival dispersed to virtually all niches of the landscape, and they rapidly adapted to the local resources and raw materials. Many large Saqqaq sites have continuous suites of multiple radiocarbon dates beginning between 3850 and 3750 BP. The radiocarbon dates from West Greenland thus indicate that the initial colonizing population might have been relatively large compared to the Independence I culture to the north, and the first Saqqaq generations had a strong population growth throughout the first centuries of settlement in West Greenland. During the first centuries of the Saqqaq culture, many large settlements were established near key resources, and stratigraphy as well as radiocarbon dates indicate that many sites must have been populated simultaneously, presumably by different family clans. On a larger geographic scale the radiocarbon dates also show that populations established themselves in all of the major fjord systems and archipelagos of West Greenland.

Most Saqqaq settlements have been located in areas of Disko Bugt (Grønnow, 1994; Jensen, 2006b), district of Sisimiut (Kramer, 1996a; Gotfredsen and Møbjerg, 2004), and around the present Greenlandic capital of Nuuk (Fig. 5.1.2) (Hinnerson-Berglund, 2004; Appelt, 2006). Presumably the clusters of settlements in these areas reflect that Stone Age peoples preferred to settle near key resources, in particular in or near the outlet of large fjord systems, where hunting is generally good, and where a diverse range of ecological niches easily can be reached. However the relatively high number of prehistoric sites in these areas compared to other areas might also be the result of favorable conditions of preservation, in particular, postglacial isostatic uplift. Another influencing factor is that intensive reconnaissance has been conducted here during the last 20 years.

Central West Greenland thus holds the most thoroughly excavated and well-published Saqqaq sites, which might skew our knowledge of regional variation. Apart from the eponymous site of Saqqaq Niaqornaarsuk (Meldgaard, 1952) near the present settlement of Saqqaq, the most well-known sites are Sermermiut (Larsen and Meldgaard, 1958), Qajaa (Meldgaard, 1983), and Qeqertasussuk (Grønnow, 1994), all located in Disko Bugt, and the site of Nipisat near the town of Sisimiut (Gotfredsen and Møbjerg, 2004). The three Disko Bugt sites are characterized by massive stratified cultural accumulations. The two former with stratified Saqqaq layers covered by younger deposits from the Dorset and Thule cultures, and the latter with a well-preserved permafrozen midden accumulated over centuries of Saqqaq occupation from approximately 4450 to 3350 cal. yr BP. The stratigraphy of Sermermiut (Fig. 5.1.4) and Qajaa illustrate the chronological framework for this part of Greenland. At the bottom are deposits from the Saqqaq culture, which are covered by sterile peat. Then, there was a thin cultural layer of Dorset origin and, finally, a thick deposit of Thule culture and historical Inuit origin. A similar stratified sequence has been recorded on Qajaa, and the pattern with sterile horizons between the Saqqaq and Dorset layers also characterize stratigraphies on most other sites where Dorset layers have been found in association with Saqqaq deposits.

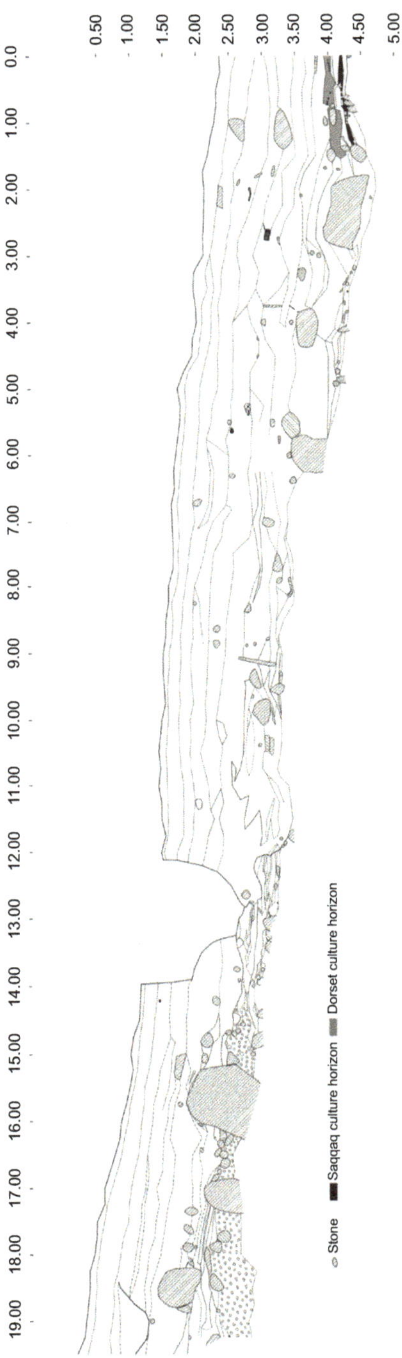

Figure 5.1.4 Stratigraphy from Sermermiut as it was recorded by Mathiassen's team in 1955. Only the section to the far right where the Paleoeskimo layers are seen below thick deposits from Thule culture and historic Inuit has been published hitherto.

5.1.2.1 Sermermiut

In the early 1900s, large-scale excavations were conducted at Sermermiut, but no well-defined Stone Age layers could be detected. New excavations were conducted in 1953 (Larsen and Meldgaard, 1958). This time, excavations focused in areas different from those investigated earlier (Jensen, 2006b, p. 33). Saqqaq layers were found preserved in the lowest part of the midden (Fig. 5.1.4), and in 1955, additional investigations were conducted, also including a pollen diagram. The excavations and cultural sequence at Sermermiut were thus multidisciplinary published, and the cultural sequence was interpreted in relation to climatic change (Mathiassen, 1958; Fredskild, 1967, 1996). The 1953 and 1955 excavations focused on sections along the eroding midden front in the southeastern part of the settlement, where preserved Saqqaq layers were found in three sections. In two of the sections the stratigraphy was well developed with Saqqaq layers at the bottom resting directly on sterile gravel and sand. Overlayering the Saqqaq deposits are sterile peat, followed by new cultural horizons of Dorset origin (Fig. 5.1.4).

The stratigraphic sections have been dated by a total of 14 radiocarbon dates. The Saqqaq horizons are dated by three dates (see Table 5.1.1), Dorset culture layers are dated by two dates: K-3769; 2260±70 BP and K-515; 1910±100 BP. Thule culture layers have been dated by two dates, K-556; 940±120 BP and K-557; 710±100 BP, and the remainder of the dates date the sterile peat layers separating the cultural horizons.

In contrast to the two stratified sites of Qajaa and Qeqertasussuk described subsequently, Sermermiut does not—or does only to a limited extent—have permafrozen midden layers, so only a little organic material such as strings of baleen have been found in the Paleoeskimo horizons at this locality. In the history of archaeology in Greenland, Sermermiut stands out as a large Stone Age settlement that already in the 1800s had become known among antiquarians as an important locality for the collection of artifacts, and later, it became the first site where the presence of Stone Age cultures in Greenland was documented in a stratigraphic section. The extent and thickness of the Saqqaq culture layers preserved at Sermermiut remain obscure because the entire area is covered by thick deposits from the Thule culture; however, the large museum collections that formerly have been collected from the place indicate that larger groups of Saqqaq people stayed at this settlement.

5.1.2.2 Qajaa

Similar to Sermermiut on the northern shore of Jakobshavn Isfjord, Qajaa on the southern side of the fjord was an intensively occupied locality with stratified deposits from the early Stone Age to historical time. However, due to the difficult access deep within the ice-blocked waters of the fjord, less research has been carried out on Qajaa than on Sermermiut. The first archaeological investigations were carried out here in 1870 (Nordenskiöld, 1871). The oldest known description of a stratigraphy is from Qajaa, but the first systematic excavations were conducted in 1981 and 1982 (Meldgaard, 1983, 1991). These excavations replicated the discontinuous stratified sequence noted by Fleischer (1871) and also known from the site of Sermermiut.

Further, it was discovered that Qajaa has permafrozen midden layers at the bottom of the stratified midden. In these layers, arrow shafts, lance shafts, harpoon heads, hafted knives, and numerous other organic implements, hitherto unknown from the Saqqaq culture, were found. In 2009, renewed partial excavations and monitoring of the preservation and erosion was initiated by the National Museum of Denmark (Hollesen et al., 2011). The stratigraphy has been recorded at several different exposed sections, and there are some variations in the stratigraphy between the different portions of the eroded midden. However the most complete section displays a series of layers comparable to the stratigraphy recorded at Sermermiut. At the bottom, Qajaa has thick permafrozen refuse layers from the Saqqaq culture resting directly on sterile gravel. On top of these layers is sterile turf; then follows a cultural horizon with Dorset culture artifacts, occasionally with preserved fauna; and on top of these layers is sterile turf in some places followed by accumulations from the Thule culture, and in other places capped by recent grass turf (Meldgaard, 1983).

The culture layers have been dated by a suite of 25 radiocarbon dates. In two different sections, the earliest human presence is dated to 3550 BP (Section C, K-3904, Section E, K-3899). This is several hundred years younger than the oldest West Greenland Saqqaq dates from Qivittup Nuua near Sisimiut and Ikerasaq in the vicinity of Aasiat in southern Disko Bugt. However the most recent investigations at Qajaa indicate that earlier excavations might not have penetrated to the deepest part of the midden (Grønnow personal communication). Furthermore the earliest traces of human occupations at Qajaa could easily have been obliterated by erosion or missed by the relatively limited archaeological investigations carried out so far. The youngest Saqqaq layers are dated to 2700 BP. On top of the Saqqaq layers are sterile turf, and then follows the culture layers of Dorset origin, which have been dated to 2210 BP.

In addition to a rich lithic material, more than 15,000 bones and bone fragments were identified from Qajaa (Møhl, 1986). Similar to the site of Qeqertasussuk, the fauna from Qajaa is dominated by harp seal (*Phoca groenlandica*) and Ringed seal (*Phoca vitulina*) which together make up 92% of the identified elements. When the fractions of unidentified harp or ringed seal bones are proportionally added to the harp and ringed seal fractions, and the unidentified bird bones are relatively divided into the identified bird species, the relative portions of the identified species are as follows: harp seal dominates with 68.1%, and ringed seal makes up 23.9%. Among the remainder 21 species identified, bones from gulls—glaucous gull (*Larus hyperboreaus*) and Iceland gull (*Larus glaucopides*)—make up the largest portion of 5.6%; also are fulmar (*Fulmarus glacialis*) (0.1%), eider (*Somateria* sp.) (0.6%), white-fronted goose (*Anser albifrons*) (+?), ptarmigan (*Lagopus mutus*) (0.2%), black-legged kittiwake (*Rissa tridactyla*) (+), dovekie (*Plotus alle*) (+), thick-billed murre (*Uria lomvia*) (0.6%), black guillemot (*Cepphus grylle*) (0.3%), and raven (*Corvus corax*) (0.1%).

5.1.2.3 Qeqertasussuk

The well-known Saqqaq site of Qeqertasussuk is situated on a point at the eastern side of the island of Qeqertasussuk in Disko Bugt. The site was located during survey in 1983 (Grønnow and Meldgaard, 1991; Grønnow, 1994), and subsequently, it was the focus of some of the most intensive interdisciplinary excavations in Greenland since the excavations on Sermermiut in the 1950s. Apart from numerous test pits, excavations focused on three principal areas, each documenting different aspects of the settlement (Grønnow, 1994).

On the top of the gravel ridge connecting the rocky point to the island, there are blowouts free of vegetation, where a couple of open-air hearths and tent ring sites are situated at an elevation of 11 m a.s.l.

Some meters downslope toward the shore and in the central part of the sloping terraces, there are thick peat layers with midden accumulations and dense occupation layers with several features located at different depths within the deposits. An excavated area of approximately 45 m^2 contains three midpassage dwellings, as well as several isolated hearths and activity areas (Grønnow and Meldgaard, 1991; Grønnow, 1994). Meticulous excavation and documentation revealed that a large midpassage feature, A8, was originally situated on the central axis of a dwelling. The 3-m-long structure extended from the entrance to the rear of the dwelling, thus dividing the interior into platform areas to the left and right, and a central hearth and activity area where two parallel rows of irregular head-sized stones framed the hearth filled with fire-cracked rocks. Toward the entrance, there was an area filled with beach pebbles, and in the hearth, there were several layers of fire-cracked rocks (Grønnow and Meldgaard, 1991). Analyses based on the type, size, degree of fire-cracking, and position and level of the rocks suggest that the structure was used during three episodes, of which the first included drying of fish and/or meat (Odgaard, 2003).

In the northwestern part of the site, there is a midden area facing the sea, where 10 m^2 have been excavated. The permafrozen conditions below the living vegetation caused the thousands of bones of game animals to be preserved in this part of the midden accumulations. The excavated material has been the subject to a comprehensive study of human ecology in the Saqqaq society of Sydostbugten (Meldgaard, 2004). The midden accumulations have been analyzed in three chronological units ranging from 4350 to 4050 cal. yr BP, 4050 to 3850 cal. yr BP, and 3850 to 3650 cal. yr BP. Based on the differences in the relative frequencies of game species and an array of detailed studies of age and season of death of the animals, Meldgaard (2004) suggested a chronological and demographic trend in the use of the site, developing from a year-round base camp type of settlement to a more specialized use of the Qeqertasussuk site as a summer and autumn hunting camp.

Qeqertassusuk was only a small settlement for a couple of families. However, this site has great importance for archaeological knowledge. In addition to the well-preserved hearths and stone set dwelling features, the Qeqertasussuk site revealed a multitude of organic tools and household utensils. Meticulously carved trays and scoops of wood, sperm whale tooth, and antler are among the well-preserved household utensils, and there are also numerous hafted lithic implements, enabling identification of the specific function of different end blade varieties. Similarly, there are numerous wooden shafts from harpoons, throwing darts, and arrows. The arrows were exquisitely made with fletchings and a nock in the proximal end and a blade bed for the projectile point in the distal end. Presumably, it was the first Paleoeskimo immigrants who introduced the bow and arrow technology to the New World, since the occurrence of arrow points in the Americas postdate the Paleoeskimo arrival (McGhee, 1996, p. 189). Comparative analysis of projectile points and shaft fragments from Qeqertasussuk show that arrow shafts with a diameter of 7–10 mm correspond to the diameter of the smallest points (arrow points), and thicker shaft fragments with a diameter of 12–16 mm fit the leisters, whereas the thickest shafts with a diameter between 12 and 27 mm were lances for the largest game (Grønnow, 1996).

5.1.2.3.1 Fauna

More than 220,000 bones and bone fragments have been recovered during the excavations of the Qeqertasussuk site. The well-preserved osteological material gives detailed information on the economy of the Saqqaq peoples in southern Disko Bugt more than 4000 years ago. The fauna from the midden excavated in area B was analyzed in three chronological units thus revealing temporal changes in the economy of the Qeqertasussuk settlement (Meldgaard, 2004).

The settlement periods from 4350 to 4050 cal. yr BP and 4050 to 3850 cal. yr BP are characterized as "base camp period I and II," whereas the settlement period from 3850 to 3650 cal. yr BP is characterized as the hunting camp period (Meldgaard, 2004, p. 169). The difficult task of documenting seasonality has been accomplished not only by evaluating the seasonality of migratory species, but also by the analysis of sections of teeth from fox, harp, and ringed seal. These studies indicate that during the base camp period I and II, the Qeqertasussuk site was occupied in all seasons. During the "base camp period II," ranging from 4050 to 3850 cal. yr BP, evidence of more specialized hunting with increased dependency on harp seal and birds and a relative decrease in ringed seal hunting during winter and spring has been documented. The decreased reliance on winter and spring ringed seal hunting may portend the development from year-round base camp toward a site used as a specialized hunting camp, most clearly seen in the bones from a period from 3850 to 3650 cal. yr BP.

A 40% decrease, from base camp period I to base camp period II, in the frequency of remains of harp seals born in the same year, i.e., "young-of-the-year," can be evidence of harp seal netting.

This decrease might result from the development of specialized hunting on flocks of migrating harp seals, which probably involved boats and nets. A more specialized hunting during the "hunting camp period" is also evidenced in a decrease in species diversity from 32 in the base camp period to 24 species documented in the latest portion of the midden. No young-of-the-year harp seals have been found from this latest period, and the bird bone assemblages indicate a shift toward more autumn fowling (Meldgaard, 2004).

The interpretation of the long-term trends seen in the osteological material are in line with some of the results of the palynological and entomological studies showing initial stages of occupation with a heavy pressure on the local vegetation and rapid build-up of midden accumulations (Böcher and Fredskild, 1993). In contrast the younger settlement phases are characterized by heavy growth of grasses and herbs in between more sporadic periods, when the site was occupied.

The results of the Qeqertasussuk excavations have been published in a large number of papers (Grønnow, 1990, 1994, 1996; Grønnow and Meldgaard, 1991; Meldgaard, 1995, 2004). Formerly, the sites of Sermermiut and Qajaa, with their ethnohistoric dimensions and Thule components, were most often used to illustrate the base camp type of settlement envisioned for the Saqqaq culture. The detailed studies of the Qeqertasussuk site have added hitherto unknown details concerning material culture, dwelling remains, and economy to this picture. The detailed studies of economy and documentation of seasonality of the Qeqertasussuk site are particularly important in this respect, because these studies enable future reconstructions of settlement patterns and site characteristics to be less dependent on analogy with ethnohistorical studies and Thule settlement patterns.

The stratigraphies have been dated by numerous radiocarbon dates (Table 5.1.1). Unfortunately, most of these are conventional radiocarbon dates conducted on peat, and only a few samples of heather from sleeping platforms are directly comparable to other dates for terrestrial material. However, the relative stratigraphic order of the layers in the deposits must be considered well dated. It is characteristic that most of the occupations occurred during the early and middle part of the Saqqaq period, from approximately 4400 to 3350 cal. yr BP. Human presence on Qeqertasussuk during the later Saqqaq sequence, extending from approximately 3350 to 2850 cal. yr BP, is poorly known since no organic materials are preserved in the uppermost layers and no well-preserved features were found in the most recent peat layers.

5.1.2.4 Nipisat

The Nipisat site, discovered in 1989, is situated on a small island approximately 15 km to the south of the modern town Sisimiut. Test pits showed that organic material was preserved, and from 1989 to 1994, more than 200 m^2 have been excavated (Gotfredsen and Møbjerg, 2004). The site is situated on level, raised beach deposits of sand and crushed shells, resulting in excellent conditions for the preservation of organic material. Below the recent vegetation the site has up to 40–50 cm stratified culture layers. The excavated material comprises more than 1000 artifacts, 20,000 flakes, and 70,000 bone fragments. Thirty-three radiocarbon dates reveal that Nipisat was visited by Saqqaq people throughout nearly 1500 years. Three different chronological phases were identified from a combination of stratigraphic observation and radiocarbon dating. Phase 1 lasted from 3970 to 3690 cal. yr BP, phase 2 partly overlaps, but is mainly younger than phase 1 and lasted from 3810 to 3275 cal. BP, and phase 3 is dated to 3260–2760 cal. yr BP. One date was very young (2470 cal. yr BP) and problematic due to the wiggling of the radiocarbon calibration curve in the period between 2750 and 2350 cal. yr BP. From phase 1, there is a midpassage structure with a box hearth. A ring of flagstones were seen around the structure. From phase 2, there is a well-defined box hearth. There was no clear outline of a tent ring surrounding the hearth. No dwelling structures were recognized from phase 3, but several shards of soapstone were recorded, indicating that the fire technology might have changed during the final Saqqaq phase from the characteristic use of hearths with "hot rocks" to one where the combustion of blubber in soap stone vessels was more common. From phase 1 and 2 the tool types are similar to those known from other Saqqaq sites in Greenland and Arctic Canada. From phase 3, however, previously unknown types have been recorded, reflecting the introduction of a new tool kit for sea mammal hunting, in the form of sturdy harpoon or lance heads made of antler. Also in the lithic inventory a new type of beveled and polished blades has been registered (Møbjerg, 2004).

The faunal assemblage of Nipisat yielded 28,823 identified bone fragments representing at least 42 animal species. The fish remains, comprising c.2% of the faunal material, consisted nearly entirely (98%) of fairly large-sized cod (*Gadus morhua*). The bird remains comprise c.47% of the material and derive from at least 24 bird species. Gulls are the dominant group (c.54% of the bird remains) followed by eider ducks (*Somateria* spp.) (24%) and *Branta* spp. (13%), presumably barnacle geese (*Branta leucopsis*), while auks (*Alcidae*) were found in lower frequencies. A total of 60 presumed whooper swan (*Cygnus cygnus*) remains constitute the hitherto largest, northernmost, and oldest occurrence of this species in Greenland, and skeletal remains of great auks (*Pinguinus impennis*) were found from the oldest phase, which indicate the slightly warmer (though cooling) climatic conditions that were present during the Saqqaq period. Fourteen mammalian species were identified, revealing a large proportion of caribou (*Rangifer tarandus*) (51% of the mammal remains), which is surprising for a coastal site situated on an island. Seals accounted for 45%, with the common seal (*P. vitulina*) as the most dominant component, which is quite different from the sites of Sermermiut, Qajaa, and Qeqertasussuk in Disko Bugt further north. Other marine mammals were walrus (*Odobenus rosmarus*) and harbor porpoise (*Phocoena phocoena*), which played an important but minor role. Polar bear (*Ursus maritimus*) hunting was documented by the presence of four fragments from the youngest phase. Saqqaq people were accompanied by fairly large and robust dogs (*Canis familiaris*).

The exploitation of fauna through the entire occupation period show only minimal chronological changes in the frequencies of the different game species and age groups. Eiders were more abundant in phase 1 (36%) than in phase 3 (17%), while gulls increased from 43% to 61% in the same time period. The same trend was found valid for geese, which increased over time, while the importance of auks decreased. Harbor porpoise seem to have decreased, while walrus increased in relative importance through time. Caribou seem to be of greater importance in phase 3 with 55% compared to 45% in phase 1. The slight shift in preferred resources may be explained by fluctuating abundance and availability of the game species combined with the development of new hunting tools. However, when the economy of Nipisat is compared to the previously mentioned Saqqaq sites in Disko Bugt, the most striking feature is the geographic variation, with the caribou and common seal being the highest ranked mammals at Nipisat in contrast to the dominance of harp seal and ringed seal on the sites from Disko Bugt (Gotfredsen, 2004, p. 190). The abundance of common seal at Nipisat clearly demonstrate the open water conditions of the Nipisat site in contrast to the localities in Disko Bugt, which even during the warmer conditions of the Saqqaq period must have experienced the reoccurring formation of landfast ice similar to the present formation of landfast ice during the winter time.

5.1.2.5 Niivertussannguaq

Niivertussannguaq is a Saqqaq site in the southern part of Disko Bugt. The site is situated where a rocky point is connected to a larger island by a saddle of raised beach terraces. On top of the saddle, there are large deflated areas, where features and artifacts have been exposed. In 1996, a "platform dwelling" and two open-air hearths and activity areas were excavated (Olsen, 1998). Features and artifacts are situated at the present surface with a few of the larger stones protruding the sparse vegetation. Artifacts and culture deposits are mixed with the top of the natural matrix of sand and crushed shells, but with some areas that are dominated by turf.

All together, 4263 flakes and 196 tools have been recovered from the excavated areas. The shell-gravel subsoil has assured rather good conditions for the preservation of a limited faunal material, which have been excavated from the features (Andreasen, 1998). Olsen (1998, p. 109) speculates that differences in the relative frequencies of fish and bird versus mammal bone fragments may result from differences in seasonality. The relatively high frequencies of bird and fish bones in the activity area and, in particular, in open hearth site, is thus believed to result from summer occupations, whereas the dominance of the stable food sources of harp and ringed seal bones in the dwelling feature is believed to result from winter occupation. This dwelling had an open floor area, which also was an activity and work area (Olsen, 1998) toward the entrance in the front.

Feature I has been dated to 3460±70 BP, the radiocarbon date for a sample of locally grown wood (TUa-1687); feature II is dated by two samples of locally grown wood: T-12996: 3615±85 BP and TUa-1686: 3545±75 BP; and feature III has been dated by one sample of locally grown wood: TUa-1688: 3685±75 BP.

5.1.2.6 Tupersui

The site of Tupersui is situated at the northwestern point of the island, bearing the same name, to the southeast of the town Aasiaat in southern Disko Bugt, West Greenland. Dwelling remains, open-air hearths, and artifacts are situated on an even gravel terrace 7–8 m a.s.l. (Olsen, 1998). Similar to the Niivertussannguaq site, Tupersui has well-preserved stone-built features with associated artifact scatters. The five excavated features are (1) one large midpassage structure filled with fire-cracked rocks, (2) one platform dwelling, (3) and (4) two box hearths, one of which may have been situated inside a dwelling, and (5) a possible cache built of large irregular stones and flagstones.

The midpassage structure has been dated by three radiocarbon dates that place the occupation in the earlier part of the Saqqaq sequence. The platform dwelling has been dated by two samples of locally grown wood, TUa-1690 dated to 3505±105 (3910–3630 cal. yr BP) and TUa-1724 dated to 3455±65 BP (3830–3630 cal. yr BP (Olsen, 1998)). A box hearth filled with fire-cracked rocks has been dated by way of a sample of charcoal from locally grown wood (T-12918). The sample was dated to 3525±85 BP (3830–3680 cal. yr BP). Tupersui has limited preservation of organic material, but the well-preserved dwelling features confirmed the formalized character of the large midpassage filled with fire-cracked rocks that also are known from the sites of Qeqertasussuk and Nipisat as well as of the "platform dwelling" also known from the site of Niivertussannguaq.

5.1.3 Northeast Greenland: Independence I

5.1.3.1 Landmarks of the Independence I pioneers in Peary Land

Independence I sites are known from all parts of North and Northeast Greenland (Grønnow and Jensen, 2003), but the settlements of the early Stone Age pioneers are remarkably clustered in the Wandel Dal corridor in central Peary Land (Fig. 5.1.5). To some extent, this concentration may have been reinforced by an overweight of archaeological research in the areas rich in sites, but recent surveys (Jensen and Pedersen, 2002) indicate that the distribution of sites is representative, and the area between Midsommer Søerne and Jørgen Brønlund Fjord indeed was a principal area of occupation of the Independence I.

5.1.3.2 Adam C. Knuth Site

Adam C. Knuth site is a large open-air site with dwelling ruins, caches, and lithic workshops situated on raised terraces overlooking a fossil fjord. Altogether the site holds at least 14 ruins, several being well-built midpassage ruins with rich associated find accumulations and 10 stone-built caches. In many ways, Adam C. Knuth site is comparable to Pearylandville (see later), but the lithic industries of Adam C. Knuth site are characterized by the proximity of the raw materials, which can be collected in the form of large boulders in the vicinity of the settlement. Therefore, Adam C. Knuth Site has many large flakes and debitage typical for primary reduction such as fragments of cores and nodules (Jensen and Pedersen, 2002). Among the ruins are some well-preserved midpassage ruins with boulder tent ring and midpassages of round boulders or flagstones set on edge. Some of these have the character of being winter dwellings (Grønnow and Jensen, 2003, p. 226), and the distribution of artifacts along the midpassage structure indicated that midpassage could have divided the tent space into a female seating and working area on one side (indicated by the presence of many needles), and a male seating area at the other side of the midpassage.

Fauna remains have been collected from the excavated ruins, and 767 out of a total of 1285 bones and bone fragments have been identified to species (Darwent, 2004, p. 344). Arctic Fox (*Alopex lagopus*) dominate with (45.1%), as well as musk ox (*Ovibos moschatus*) (31.6%), arctic hare (*Lepus arcticus*) (4.4%), caribou (*R. tarandus*) (0.1%), collared lemming (*Dicrostonyx torquatus*) (0.3%), which might be intrusive, naturally deposited bones, polar bear (*U. maritimus*) (0.4%), bearded seal (*Erignathus barbatus*) (1.8%), ringed seal (*Phoca hispida*) (1.3%), harp seal (*P. groenlandica*) (0.1%), Brent goose (*Branta bernicla*) (2.2%), rock ptarmigan (*L. mutus*) (7.7%), glaucous gull (*Larus hyperboreus*) (0.4%), red knot (*Calidris canutus*) (0.1%), and arctic char (*Salvelinus alpinus*) (4.4%). The relative high frequency of arctic fox is unusual, but apart from this evidence of fox hunting, which is likely to have been conducted in the cold months when the fur is at its best, the diet was dominated by musk ox, which is typical for Independence I sites in Peary Land. The occurrence of a harp seal ear bone is the northernmost known distribution of this species in Greenland and may well support other evidence for the slightly warmer conditions during the period of Independence I presence in Northern Greenland.

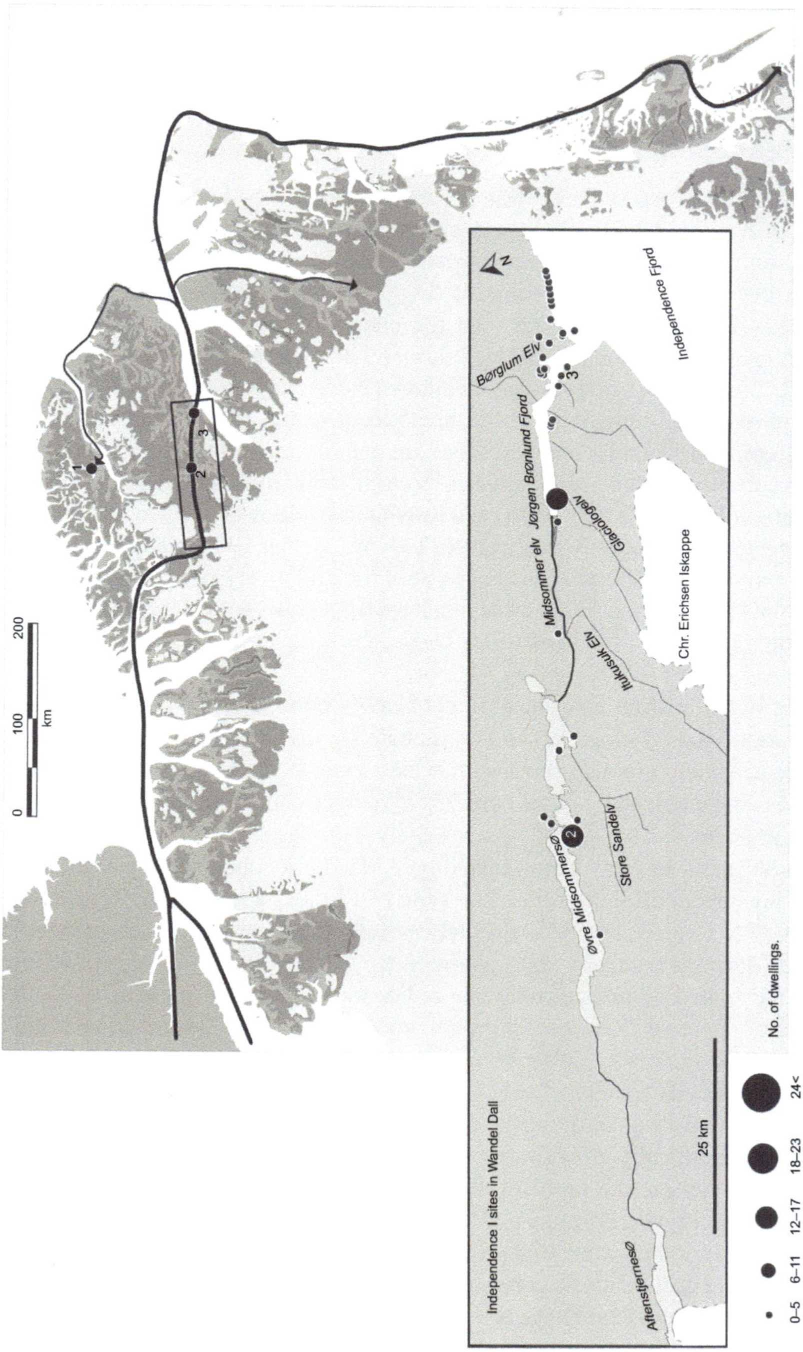

Figure 5.1.5 Map of Northern Greenland, with possible migration routes of Independence I pioneers marked. Inserted detailed map of Wandel Dal and Jørgen Brønlund Fjord show location of sites mentioned in the text. *1*, Adam C Knuth Site, *2*, Pearylandville, *3*, Vandfaldsnæs.

Occasionally the lithic and faunal material together with fire-cracked rocks have formed culture layers up to 30cm thick, but in general the features and artifacts are deposited on the surface, making the organic remains highly vulnerable to erosion and decomposition.

5.1.3.3 Pearylandville

Pearylandville is the largest Paleoeskimo camp site in Peary Land, and it plays a central role in Knuth's (1967a,b) reconstruction of the settlement pattern in the area of Jørgen Brønlund Fjord, Midsommersøerne. The site is located on a promontory at the eastern end of Nedre Midsommersø (Fig. 5.1.5). The substratum is sand and gravel partially mixed with boulders, and the undulating surface changes between plane gravel terraces and boulder fields. Pearylandville has a total of 35 reasonably well-described features: 11 midpassage dwellings, 3 tent rings, 1 undefined dwelling, 4 flagstone pavements, 7 open-air hearths, and 5 meat caches. Two features are interpreted as ceremonial features, and two areas are activity areas, revealed by open-air artifact concentrations. Some ruins are rich both in tools, lithic debitage, and fauna refuse, making it likely that they were used as winter dwellings during an occupation lasting several months. But only two features have more than 100 tools associated, so presumably most of the habitations had been used for a relatively short period of time. Altogether, the 820 tools and 5312 flakes that have been collected from Pearylandville only make up a fraction of the inventories known from the largest Stone Age sites in West Greenland.

The fauna material from Pearylandville is heavily dominated by musk ox, but in some ruins, arctic fox (*A. lagopus*) and arctic hare (*L. arcticus*) make up substantial components, and there are also ruins with bones from arctic char (*S. alpinus*). The avian fauna is characterized by some bones from rock ptarmigan (*L. mutus*) and a few geese and gull bones.

Pearylandville has recently been redated by AMS dates conducted on 13 samples of musk ox bone from different ruins. The suite of 13 dates all came out in the interval between 3725±45 BP and 3850±45 BP, indicating that probably Pearylandville was occupied during a relatively short phase of occupation. Previously, three of these ruins have been dated by one conventional radiocarbon dating of musk ox bone and two samples of charcoal from locally grown wood (*Salix*). The older date on a bulk sample of musk ox bone was 120 years younger than the recent AMS date, and in the case of the two older dates of charcoal of *Salix*, they are respectively 50 and 195 years older than the new AMS dates of musk ox bone from the same ruins. This tendency of dates on charcoal of locally grown wood to be older than dates on musk ox bone from the same ruins underline the problem of relying too heavily on the dating of wood from the High Arctic where trunks and branches from dead trees can be found and used many years after they were alive. None of the features were disturbed to such an extent that multiple occupations can be suggested, but one is rich in artifacts and fauna remains, indicating that it could have been occupied for several months, and of course, the dating of surface localities such as Pearylandville can be problematic due to the general difficulty of separating older remains from younger ones.

5.1.3.4 Vandfaldsnæs

Vandfaldsnæs ("waterfall point") is a large and complicated site with Independence I, Independence II, and Thule ruins scattered over a distance of several hundred meters on different terraces. The oldest and highest situated ruin is an Independence I ruin situated at a river delta terrace 11–12 m a.s.l. On a lower terrace 5–6.5 m a.s.l. are ruins from the younger Stone Age phase Greenland Dorset (which Knuth named Independence II), and at the lowest lying recent beach terraces situated 0.7–1.2 m a.s.l. are ruins from the Thule culture, the ancestors of the historical Inuit (see Fig. 5.1.6). Vandfaldsnæs is an open-air display of the entire culture history of Peary Land, since the isostatic uplift has resulted in the dispersal of the ruins and artifacts from different cultural epochs at different elevations. Near the present shore are the most recent ruins, and as one walks inland and up the gravel terraces, one moves back in time, reaching the settlements of the Independence I pioneers at an elevation of 11–12 m a.s.l. (Grønnow and Jensen, 2003, p. 141).

5.1.3.4.1 Discussion

In the previous sections, we have condensed the environmental history and shortly described the most spectacular archaeological sites left by the pioneering peoples of Greenland. From the presentation, it is clear that living conditions in the High Arctic (Northeast Greenland) were very different from those in the Low Arctic West Greenland. Similarly the qualities of the archaeological record differ greatly between the two regions. In the High Arctic regions the 4000-year-old Stone Age sites can be observed and recorded on the existing surface, which supplies a wealth of "horizontal" data on the archaeological remains. In West Greenland, most localities are obscured below the present vegetation, rendering difficult the observation of dwelling features and living floors. On the other hand the environment in West Greenland occasionally has resulted in the formation of stratified sites with well-preserved, sometimes even permafrozen, culture layers. These localities disclose a wealth of stratigraphic data, and the preservation of organic material has left an abundance of artifacts enabling detailed reconstructions of the material culture of the pioneer peoples of Greenland.

We do not know a single grave of Paleoeskimo origin in Greenland, so the peoples themselves, their physical anthropology, and mortuary practice remain unknown. The only skeletal remains of early Paleoeskimo origin is a newborn or premature child recovered from a dwelling floor at a site on Devon Island (Helmer, 1991) in Canada. Given the age of the individual, nothing can be said regarding biological affinities (ibid). In Western Greenland a few badly preserved human bones were found in the midden layers at the previously described Saqqaq site Qeqertasussuk (Grønnow and Meldgaard, 1991). However, the analysis of lumps of human hair with preserved DNA material from the Qeqertasussuk site demonstrate the scientific potential of the rare archaeological sites with permafrozen middens. The mtDNA extracted from the hair proved to be most closely related to present populations from the Aleuts and Siberian Sireniki Yuits (Asian Eskimos) (Gilbert et al., 2008), and an analysis of the ancient Greenlander's nuclear genome indicates that his father's ancestors are most closely related to the present populations in the Old World Arctic populations—Nganasans,

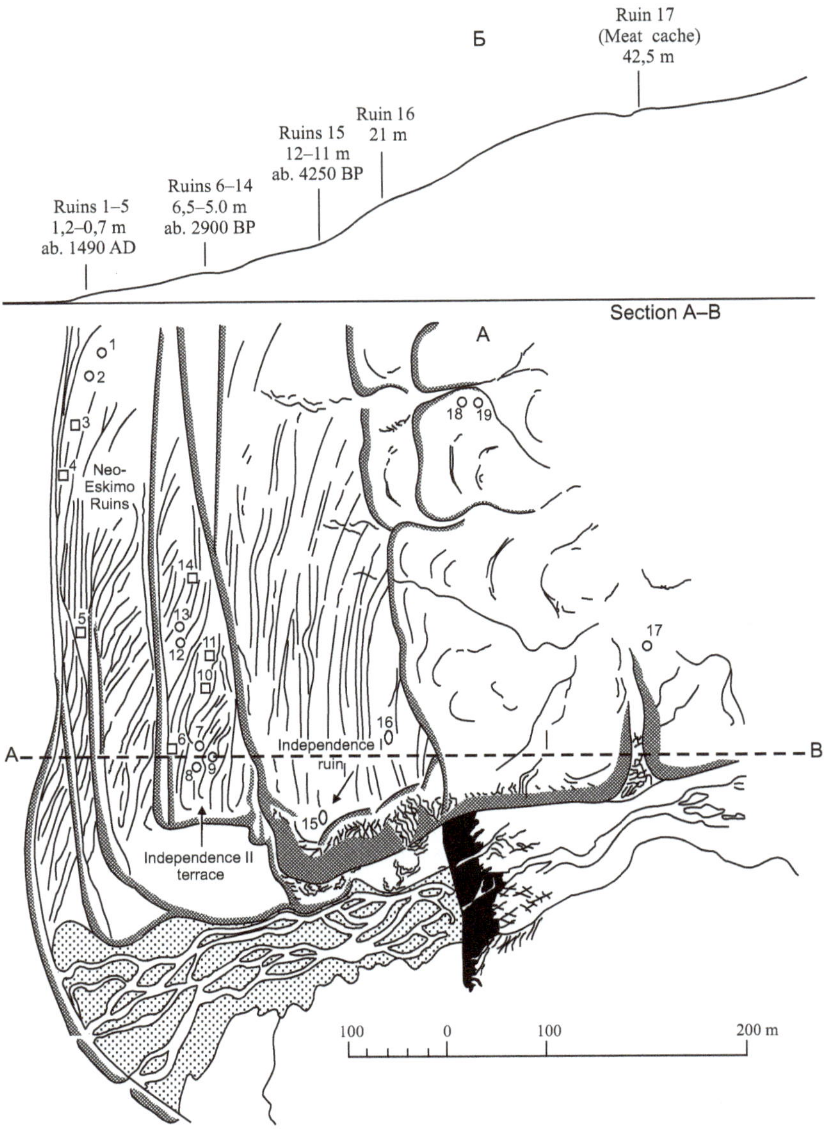

Figure 5.1.6 Map of the site Vandfaldsnæs, with ruins of different age situated at terraces with different elevations.
Map by E. Knuth from: Grønnow, B., Jensen, J.F., 2003. The Northernmost Ruins of the Globe. Eigil Knuth's Archaeological Investigations in Peary Land and Adjacent Areas of High Arctic Greenland. Meddelelser om Grønland/Monographs on Greenland, vol. 29. Man & Society, p. 403.

Koryaks and Chukchis—while being more distantly related to the New World groups (Amerinds, Na-Dene, and Greenland Inuit) (Rasmussen et al., 2010). These results of DNA analysis are in line with Dumond (1984), Wright's (1995), and McGhee's (1996) suggestions of a Siberian origin of the Paleoeskimo cultures. Inspired by similarities in a number of lithic tool types, Powers and Jordan (1990) suggested that the Arctic

Small Tool tradition might have been derived from the Bel'kachi, and the ultimate origins of this development lie even deeper in the preceding Neolithic (Syalakh) and pre-Neolithic (Sumnagin) cultures of the Siberian North. The DNA preserved in the frozen hair from Qeqertasussuk West Greenland appears to support this suggestion.

5.1.3.5 Settlement patterns

In Peary Land the lack of obscuring vegetation and high density of Independence I sites in the Wandel Dal-Jørgen Brønlund Fjord corridor enable inferences on the prehistoric settlement pattern and demography. The Independence I sites of Peary Land can be divided into "large" sites with multiple dwellings, some of which are rich in lithic tools, debitage from tool manufacturing, and fauna material, and "small" and "intermediate" sites where the ruins more often are poor in lithic artifacts and ecofacts (Grønnow and Jensen, 2003, p. 333). Also characteristic for the localities with many dwelling features is the fact that caches are more abundant on these sites than they are on the small sites with just one or a few dwellings. This pattern indicates that the largest Independence I sites are not palimpsests from consecutive temporal occupations. Indeed, they appear to be aggregation camps as was already suggested intuitively by Knuth (1967a,b). The large sites of Pearylandville and Deltaterrasserne in Wandel Dal and Brønlund Fjord both have several small sites with just a single or a few dwellings in their vicinity, making the suggestion reasonable that these localities were aggregation camps in a cyclic annual round, where people during the warmer months lived in smaller family units dispersed into the upland, where they could benefit from smaller predictable resources such as char runs, basking seals along ice edges, or migrating birds in polynyas. If this scenario is accepted, it becomes clear that probably this Independence I flourishing in Peary Land lasted for only a single or a few generations, simply because the total number of Independence I sites (n. 51) in Northernmost Greenland and their number of ruins (n. 244) and richness in artifacts is relatively limited. All of the Independence I sites known from northernmost Greenland thus have less artifacts and lithic debitage than retrieved from the excavated areas on just a single of the large Paleoeskimo sites in West Greenland. The short duration of the Independence I settlement in Peary Land is also indicated by the radiocarbon dates. All the most reliable dates conducted on musk ox bone thus gives dates between 3850 and 3570 BP, which in calibrated years give a maximum time span for the Independence I occupation of Peary Land to 4410–3810 cal. yr BP. However, when considering this time span in relation to the known campsites and their number of habitations, it seems likely that the period of Independence I settlement in Peary Land was much shorter. The limited number of Independence I dwellings in Peary Land also enables a qualified suggestion on the size of the Independence I population. Since the aggregation camps of Pearylandville, Adam C. Knuth Site, and Deltaterrasserne at the most could have 10–20 simultaneously occupied dwellings, the population at the most seems to have numbered a few hundred individuals. In Greenland, most Independence I sites have been investigated in the interior of Peary Land, where Knuth (Grønnow and Jensen, 2003) focused his lifelong effort at the discovery of Stone Age sites in northernmost Greenland. However the marine resources also played an important role to the earliest settlers of the High Arctic as known from contemporaneous sites in Arctic Canada (Darwent, 2004; Schledermann, 1990). In coastal Northeast Greenland the Independence I settlements often seem to be located in vicinity to polynyas (Gotfredsen, 2010), and recent surveys indicate that coastal Northeast Greenland might hold more Independence I sites than previously recognized.

In Western Greenland, there is more vegetation and peat covering the Saqqaq sites, which make it difficult to assess settlement patterns and the size of the population. However, based on the excavated material, it is obvious that the Saqqaq population was bigger than the Independence I. Several hundred Saqqaq sites are known from West Greenland, where Disko Bugt and the large fjord systems around the town of Sisimiut and Nuuk stand out as important population centers. Within these areas the annual round appears to have been a cyclic migration between temporarily occupied warm season camps and more permanently occupied winter camps. Predictable resources topographic bottlenecks during time of caribou migration or char runs may also have sustained gatherings of numerous families as it is known from historical sources and as indicated by large, warm season camps (Appelt, 2006).

5.1.3.6 Dwellings

The Paleoeskimo dwellings were oval or round tents with a diameter around 3–4 m, in some cases with a stone-built “midpassage” including the hearth, through the axis from the entrance to the rear (Fig. 5.1.7). Midpassages are found in about one-third of all Paleoeskimo dwellings in both Canada and Greenland from the earliest times until the end of the Paleoeskimo era more than 3000 years later. The midpassage has been suggested to be a symbolic marker of a cosmology, in analogy with ancient cosmological concepts of some people in Siberia (Odgaard, 2001, 2010). From the Saqqaq culture in West Greenland, we also know of the so-called platform dwellings, where the interior are characterized by a central stone set hearth, which to one side has a large platform occupying approximately one-third of the rear part of the interior. According to ^{14}C dates the platform dwelling is younger than the midpassage feature. Olsen (1998) suggests, accordingly, that the platform dwellings might be a later development resulting from changes in Saqqaq settlement patterns and residential mobility in the period from approximately 3950–3650 cal. yr BP.

5.1.3.7 Overwintering in the High Arctic

In the cold and icy Arctic areas with only a little vegetation, pyrotechnology is a subject of great importance, and this might be the reason why the early Paleoeskimos often payed meticulous attention to the construction of their characteristic hearths, the so-called midpassage hearth. The only source of heat in the tents of the earliest Paleoeskimos came from the hearth, since no oil or blubber lamps have been found from this early phase. Eigil Knuth made this observation, which has not been modified by later archaeological work: “The importance of the central fireplace in the building is thus disclosed to the archaeologist by the fact that a ruin on the 'Musk-ox way' is in the main simply equivalent to the ruin of a fireplace. The background for this is illustrated

Figure 5.1.7 Paleoeskimo midpassage dwelling from the Saqqaq site of Kannala, West Greenland. From Jensen, J.F., 2005. Independence culture. In: Nuttall, M. (Ed.), Encyclopedia of the Arctic. Routledge Publishing Co., New York, London, pp. 945–948.

by the result of the excavations of many years which – as mentioned before – establish that so far not a single fragment of a blubber lamp or a stone cooking pot has been found in the ruins of either the Independence cultures. In the dark period the hearth was a piece of technical inventory which possessed the triple function of warming oven, cooking stove and lighting installation" (Knuth, 1967b, p. 42).

Some of the 4500-year-old Independence I hearths are box hearths, built of slabs set on edge, and the size is most often around 40×40 cm. The contents of the box hearths can vary, but some of the better preserved hearths contain an interior layering of grease-saturated sand or gravel, topped with fist-sized stones, charred bone, and wood (Fig. 5.1.8) (Schledermann and McCullough, 1988; Knuth, 1967a). These hearths were probably used during the most extreme phase of the Arctic winter, consisting of 2 months total darkness and temperatures below −30°C.

There is little vegetation suitable for firewood in this region. Heating thus relied on the combustion of driftwood, bones, and grease from game animals. Another important issue is indoor climate. Many have speculated about how it was possible to survive during the High Arctic winter in a rather small tent with a fireplace, without either dying from too much smoke or from the extreme cold that sufficient draft would allow in. Schledermann and McCullough (1988) suggest that the tents were abandoned by midwinter in favor of

Figure 5.1.8 Box hearth from (possible winter) midpassage dwelling (Grønnow and Jensen, 2003).

temporary snow house camps on the sea ice or snow-banked tents on land. But the absence of lamps does not support this hypothesis. Knuth (1967a), McGhee (1979, 1996), and Maxwell (1985) suggest that winter was spent asleep in a kind of torpor in a frozen tent.

McGhee (1996, p. 64) writes: "We must imagine darkness broken for weeks on end only by the circling moon and the midday glow of dawn light in the southern sky. For months the cold so intense that an exposed hand freezes in minutes, and the cold increased by a light wind which swirls the snow around the dark tent of the winter camp. The heavy musk-ox skins of the tent move slightly, the driftwood poles creak in the wind, but this is the only sound or movement in the landscape. Within the tent it is totally dark, as cold as the outside world, and only the sounds of sleep emerge from the heap of musk-ox and bear skins beneath which two small families are living through the winter. Sleep is occasionally broken by the squall of a hungry baby, quickly silenced at his mother's breast" (McGhee, 1979).

However human beings are not able to go into hibernation like some animals can. Humans need food and sufficient warmth to survive. In addition to space heating for comfortable accommodation, heat is also of vital importance for giving access to drinking water. In the High Arctic, it is necessary to melt snow or ice for drinking water for most of the year, and on a diet that almost exclusively consists of meat and fish, it is necessary to drink much water. An ethnographic source reports about the Inuit: "The Eskimo drink great quantities of water. It is curious that with its world of congealed water the Arctic should be a veritable Sahara. Water is usually supplied by melted snow or ice and the lamp is brought into requisition for the purpose, though sometimes the warmth of the hut is sufficient, especially if the vessel containing snow is placed near the flame" (Hough, 1898, p. 1036).

On this background the winter box hearths of the Independence I culture attract attention, since they must contain part of the answer of how the overwintering took place. Archaeological analysis supplemented with experimental archaeology and heat-loss calculations have shown that this particular hearth type could actually provide a suitable temperature for a group of people in a small tent (Odgaard, 2001, 2003, 2005). The grease-saturated sand or gravel, topped with fist-sized stones, indicate that often the box hearth must have been fueled by the combustion of fat from game animals, which even on the interior sites in the High Arctic could have been obtained from seasonal exploitation of marine mammals caught when the Independence I people settled on coastal camps.

The hearth structure could have been heated with firewood and bones, but the continued combustion process was probably fed with seal-blubber or fat from other animals, as in a lamp. An experimental hearth thus showed the same traces as the archaeological hearths, and it is most likely that the Paleoeskimos did not go into a torpor, but with sufficient storage of fat, they spent a rather pleasant time of winter darkness in the skin tent (Odgaard, 2007a,b).

The people of the Saqqaq culture, who settled simultaneously with the Independence I culture, but in more southern regions of Greenland, faced a totally different fuel situation. Here, lots of driftwood was available on the beaches (Grønnow, 1996), enabling the Paleoeskimos to apply stone heating technology at a greater scale: a technology that requires lots of fuel. Therefore, especially in Western Greenland, the hearths in the dwellings are filled with fire-cracked rocks, i.e., rocks with signs of cracking after intensive heating. Rocks were used as heating elements throughout the Paleoeskimo tradition, but the method and intensity vary. Hot rocks are excellent for room heating, but another important aspect is the absence of fire-proof cooking vessels. Heating and boiling of liquid in containers made of organic material is only possible by the transference of heat through hot rocks. Studies of Saqqaq hearths, however, also show that drying of meat and/or fish over hot rocks has been a factor in the accumulation of fire-cracked rocks at the sites (Odgaard, 2003).

5.1.3.8 Saqqaq and Independence I lithic typology

It has been discussed whether the Saqqaq culture and Independence I are the remains of the same group of immigrants, or if they were the result of different immigrations of different groups of people.

There are many similarities in Saqqaq and Independence I technology and tool types testifying to the fact that both are local variants of the early ASTt and related to the Denbigh Flint Complex in Alaska (Giddings, 1964). However, there are also differences in technology and principles of tool manufacturing, showing, that Saqqaq and Independence I were local traditions with little or no interaction (Sørensen, 2011).

Initially, differences in lithic typology, in combination with dwelling styles, it was argued, underlined the separation of Independence Culture from West Greenland Saqqaq (Knuth, 1954, 1978, p. 23). More recently, Elling (1996) argued that the differences on Saqqaq and Independence I assemblages should be explained in the light of different economies and availability of raw materials and not as a result of cultural differences. Appelt (1997), however, found that there are more differences between the two than accepted by Elling. In the light of the radiocarbon dates, some of the differences, such as the lack of soap stone lamps in Independence I, should be seen as a parallel to the similar lack of soapstone in the older Saqqaq assemblages. Other differences such as the lack or very rare occurrence of triangular harpoon end blades in Independence I assemblages should probably be seen in the light of different economies, since in Greenland, the Independence I settlements are clearly concentrated in the interior where there is little use of harpoon blades. At the coastal Independence I sites of Port Refuge, McGhee (1979, p. 143) found harpoon heads similar to Saqqaq harpoon heads as well as straight-based harpoon end blades. The common use of polishing seen in Saqqaq assemblages is a major characteristic of the Saqqaq, but evidently, that the harder lithic varieties preferred by Independence I never allowed for polishing is also argued by Elling. Elling also believed that the restricted use of microblades could be ascribed to the differences in raw material availability. However, in West Greenland the Dorset I sites display a massive production of microblades of the harder lithic varieties, which are available from the same regions as where the Saqqaq acquired the raw material named killiaq, which they preferred for the bifacial tool production. So in this particular case, we would argue that something more than just regional differences in lithic resources separates Saqqaq from Independence I.

Apparently the West Greenland Saqqaq operated with only minimal use of microblades, and when they are present, they are often based on quartz crystal. To the contrary, Independence I had an abundant microblade production. This particular difference in lithic technologies is hard to explain by anything but differences in the lithic traditions of Saqqaq and Independence I (Jensen, 2006a; Grønnow and Sørensen, 2006).

5.1.4 Conclusion

The three major formations of the first people to enter Greenland and Eastern Arctic Canada: the Independence I, Saqqaq, and Predorset reflect three aspects of regional lifestyle and a material culture similar to—if not derived from—the Denbigh culture in Alaska and the Arctic Small Tool tradition.

The first Stone Age pioneers arrived to the Nares Strait region around 4500 cal. yr BP. The peculiar geography of Greenland, where the interior is capped by inhospitable inland ice and the ice-free land along the coasts , then resulted in the dispersal of one group into West Greenland toward the south where they archaeologically have become known as the Saqqaq culture, whereas another group of pioneers (archaeologically known as the Independence I culture) passed north around Greenland where they settled in Peary Land. Although temperatures were slightly higher, the environmental studies show that in the far north the physiography and plant cover in the land the immigrants arrived in were not significantly different from the present. The incentive for the immigration may have been the musk ox, which seems to have reached Greenland at the same time as the first immigrants. For both oxen and men the passage through the Canadian Arctic Archipelago and into Greenland may have been eased by increasing landfast ice along the coasts, serving as a road for transportation, a hunting ground, and a means of social contact.

In Greenland the Independence I culture lasted only a few centuries and numbered probably not more than 100–200 persons. Their way of life with regard to economy was much less diverse than in Saqqaq. The ecological system in High Arctic Greenland is unstable and the resources both vulnerable to overhunting and the deteriorating of environmental conditions.

Simultaneously with the High Arctic Independence I culture, the Saqqaq culture flourished in more southern regions in Greenland. In these areas, resources are much more diversified, allowing rapid response to ecological fluctuations by compensating for the deficiency of one resource by exploitation of another. This can explain why the Saqqaq population, unlike Independence I, continued for almost 2000 years before a substantial change in the economy took place, just like the Pre-Dorset culture in Canada.

Adaptability and versatility have probably been the main factors of an otherwise relatively weak cultural system during the Paleoeskimo culture. The Paleoeskimo period is characterized by long periods of seemingly stable cultures where no or only little change can be seen in the material culture. In between these periods are short, punctuated episodes of cultural change, which in Greenland always appear to be related to the arrival of new populations from Arctic Canada. The Paleoeskimo peoples must have already been well adapted in Alaska for the Arctic environment. Without doubt, the necessary flexibility for conducting an effective hunting life, despite extreme stress, has been assured by a high mobility of very small groups. The absence of predecessors in the vast area of Arctic Canada and Greenland also allowed these pioneers a maximum dispersion, while maintaining a minimal communication network with no risk of interruption by a foreign population.

Unlike the concurrent diverse populations of the Bering Strait and Alaska, who were in contact—and probably in conflict—with each other, the cultural and technological evolution of the eastern Paleoeskimos escaped any stimulation other than those generated by the environment, which explains its relative stability. Even on the southern fringes of the ecumene, the denuded interior of the Central Arctic and Labrador, where contact with Native Americans could occur, evidence of foreign influence is very weak (Plumet, 1980, 1985, 1986, 1994). This isolation and lack of interaction with foreigners continued throughout a large part of the late Paleoeskimo culture.

Section Six

Iceland

(photo by O.A. Markelova)

Iceland at the time of initial colonization

6.1

Olga A. Markelova
Moscow, Russia

Unlike most of the other northern regions, it was at a historical time that people arrived at Iceland, and the process of the settlement is described by the Old Norse as well as modern historians[1].

The official date of the settlement is considered to be AD 874; in modern Iceland, all the anniversaries of the settlement are connected to this date, but in reality, it could have occurred somewhat earlier or later. The first two centuries after the settlement are not documented at all. The earliest sources that tell about the settlement of this island are "The Book of Icelanders" (Íslendingabók) by Ari Thorgilsson Fróði (1067–1148), the Old Icelandic scholar who is regarded as the first "real" historian in Iceland, and "The Book of Settlements" (Landnámabók), compiled by Sturla Thordarsson in the first half of the 12th century. It is obvious that the texts describing events that have taken place 200 years ago cannot claim to be fully exact; nevertheless, they can give us a general picture of the process of the settlement[2].

Certain notions that there should be some land in the Atlantic Ocean northwest of the continent had existed long before Iceland was discovered. European texts on geography from the 8th century, among them works of the Irish monk Dicuil and the Anglo-Saxon scholar Bede the Venerable (674–735), tell about the mysterious land of Thule where the sun never sets in midsummer and never rises in winter. The Book of Settlement concludes from this description, that Thule is Iceland (Landnámabók, 1948, p. 1).

Before Iceland became populated by the outcomers from Northern Europe, it was practically uninhabited. According to the Old Icelandic sources, by the time when the first settlers arrived, there were Irish monks that had settled in Iceland and the nearby small islands; the sources call those monks *pápar* (Ari þorgilsson, 1930, p. 48; Landnámabók, 1948, p. 1). There is not much known about them; traditionally, they are never regarded as a "full-fledged" population of Iceland (i.e., as someone cultivating the land and having evident influence on the environment), as their goal had obviously been flight from the vanities of the world, not colonizing new lands. It

[1] Before 1970 the primal settlement of Iceland (and the sources that describe it) was the object of interest merely to historians and literary scholars doing research on Icelandic literature. Only quite recently has it come into the sphere of interest of paleogeographers and geologists (McGovern et al., 2007, p. 27).

[2] The Icelandic historians do not agree on how "The Book of the settlements" is valuable as a historical source (see, e.g., Jakobsson, 2000). But its symbolic status in Iceland's culture is high. The official date of the land's settlement was stated just according to what this book says, and common Icelanders tend to trace their descent from the settlers mentioned in this book.

Human Colonization of the Arctic: The Interaction Between Early Migration and the Paleoenvironment
http://dx.doi.org/10.1016/B978-0-12-813532-7.00019-X

is possible that they did not dwell there constantly, but visited the island from time to time, or they could have been living there for a short time when the Norwegian settlers arrived (Hjalmarsson, 2009, pp. 18–19). The "Book of Icelanders" says the monks left some cult items behind (Irish books, bells, the bishop's staff) (Ari þorgilsson, 1930, p. 48), but archaeologists have never found the traces of the monks' dwelling in Iceland (e.g., McGovern et al., 2007, p. 28). Some Icelandic toponyms keep the memories of those fabulous monks, for example, the name of the Pápey island.

According to Ari Fróði, the settlement process took 60 years, approximately from the 870s to 930s (Ari þorgilsson, 1930, p. 49).

The majority of the settlers came from Norway. The Old Norse sources tell (and in modern sources, including school textbooks, it has long become a topos) that the general reason for moving to the new lands was the discontentment with the political situation in Norway; by that time the increase of power of Harald Hárfagri, who united separate Norwegian kingdoms into one centralized state, had begun (Hjalmarsson, 2009, p. 26). In Old Norse texts, it is often mentioned that particular persons have left Norway for Iceland because of King Harald Hárfagri's abuse of power ("fyrir ofríki Haralds konungs hárfagra") (Landnámabók, 1948, pp. 118, 164, 177). But as the modern Icelandic historian Gunnar Karlsson points out, it is hardly correct to regard the first settlers of Iceland as the first dissidents in the history of Europe; as the archaeological excavations show, most of them were wealthy free farmers (in Icelandic sociological terminology, *bóndi*), so they were probably more attracted by new agricultural lands (Karlsson, 2000, p. 5). The first settlers were not monoethnic: in "The Book of Settlement," there were about 400 settlers listed by names, mostly Norwegians, but there are also a certain amount of inhabitants of the Scandinavian colonies from the British Isles, and Irish people, usually the settlers' Irish slaves (in the new land, they could quickly become free men and raise their own farms). But free Irish settlers are also mentioned there. The old sources pay special attention to the religious confessions of the settlers (the Christians from Britain and Ireland could settle apart from the pagan Norwegians) (Landnámabók, 1948, p. 70).

The settlers generally arrived to Iceland from Norway, so they got to the island from the east side (Ingólfur Árnarson landed at the East Fjords of Iceland). The settlers from the Orkney Islands (Suðreyjar) and from Ireland could land on the southwestern coast (Hjalmarsson, 2009, pp. 26–27). "The Book of Settlement" says that the east quarter of Iceland was settled first, and the northern quarter was the most populated (Landnámabók, 1948, pp. 12–13).

All the sources agree that the very first settler on the island was a man named Ingólfur Árnarson. He raised his house on a cape on the southwestern end of Iceland, the same place where Reykjavík was built many centuries later. However, the landing of the first settler and the raising of the capital of Iceland at the same place were not connected (Karlsson, 2000, p. 5).

At the time of the first settlement, there was very intensive sea traffic between Iceland and Norway. The same goes for traffic and connections within Iceland.

The island was settled along the coast. From the very beginning of the mass emigration from Norway, there had been rules set about how much land one person could take. "Some of them that had arrived first, settled closer to the mountains and marked their land possessions by how their cattle moved from the seaside to the mountains. But those that came later thought the settlers who had come first had taken too much land far and wide. So King Harald the Hair-fair ruled that no one may take a wider piece of land than he and his crew could surround by bonfires in one day..." (Landnámabók, 1948, p. 13).

The middle of the island, which is covered by high mountains, glaciers, and deserts, has for a long time been (and for the most still is) uninhabited[3]. The fact that the inner part of the land was uncultivated was reflected in the Icelandic folklore in a unique way. From the 17th and 18th centuries, legends of rich fertile valleys in the inner regions emerged, where the elves were said to live in happiness, while peasants who left their pauperized farms went to the mountains. However, already by the time of the first settlement, Iceland was said to be inhabited by various supernatural creatures; in "The Book of Settlements," there are mentions of water spirits, living in the sea (Landnámabók, 1948, p. 97), mountain trolls (Landnámabók, 1948, p. 295), and the land-keeping spirits (landvættir). Though this information is more related to folklore studies than paleogeography, it is possible that tales of some supernatural creatures could have influenced the settlement of people in some regions [In "The Book of settlements" one person chose the place for his farm based on a prophecy the water-spirit had made for him (Landnámabók, 1948, p. 97)].

The first inhabitants of Iceland were very mobile in spite of the long distances and difficulties of getting over mountains and through deserts. The first settlers were quick to move from one part of the island to the other as they explored the new land. Later, by the time the Old Icelandic society was formed (from AD 930), even the inhabitants of the most distant farms from the whole country went to the all Icelanders' popular assembly, the "althing" (Hjalmarsson, 2009, pp. 29–31).

The descriptions of what the land looked like in the eyes of the first settlers are rather poor in the Old Norse sources; one can easily call them lacking. The natural objects mentioned there are mountains, glaciers, and rivers with waterfalls (one of the explorers says, "The land that we found must be vast: the waterfalls are big here" (Landnámabók, 1948, p. 5; Landnámabók, 1948, p. 6 (AM 107 fol., 2v). There are also mentions of forests and marshes (Landnámabók, 1948, p. 78). It is also said there that the most suitable land for settlement was in the south of the island (Landnámabók, 1948, p. 13). As is well known, Old Icelandic literature used to be very short in descriptions of obvious and familiar things (Steblin-Kamenskiy, 1967, p. 168); one can conclude that the nature of the new lands did not strike the settlers because it reminded them of the familiar landscapes in the North of Norway (or that it was easy for them to get used to such landscapes and climate in the several generations that had passed from the settlement to the moment the Landnámabók was written). "The Book of settlements" claims that the settlers expected to find the same things on the new island as in their home land, e.g., the same animal species. For example, one of them sent his slaves to find a bear that should have killed his ox (Landnámabók, 1948, p. 9).

[3] This situation is well illustrated by the fact that there was no road around the coast of Iceland until the 20th century, but there was mainly sea traffic between the parts of the land (Hjalmarsson, 2009, pp. 138, 158).

There have never been big beasts of prey in Iceland. They also brought familiar methods of agriculture there.

It is curious to mention that the thermal activity of the earth was not described in the Old Icelandic texts on the first settlement; lava fields and steam from hot springs are only mentioned sometimes as toponyms. As far as glaciers are concerned, they have only undergone insignificant changes according to both the old texts and the modern geological studies. The glaciers the first settlers saw were maybe only 10%–20% smaller than nowadays. In the sagas of Icelanders, written in the 13th century, there are descriptions of prosperous farms at the places where nowadays there are preglacier deserts. The intensive growth of glaciers took place from the 14th century till the beginning of the 20th century (Sigurðsson, 2010).

The archaeological and paleogeographic data prove the information from the Old Norse texts.

The first settlers should have discovered colonies of sea birds on the island, nesting areas of migratory birds, and sea mammals (seals, walruses, whales). There are swans mentioned in the "Landnámabók" (Landnámabók, 1948, p. 78) and polar bears in the northern part of Iceland (Landnámabók, 1948, p. 197); it is mentioned that polar bears were not familiar to Norwegians.

The flora of Iceland of the 9th century was different from the modern. According to both the Norse sources and the newest archaeological researches, polar birch covered huge territories of the island (Figs. 6.1.1–6.1.3). According to the Icelandic Forest Research, it was up to 28,000 square kilometers, i.e., a quarter of the island. Some of these forests have survived till today (þór Kjartansson and Eggertsson, 2005; Sigurðsson, 1977).

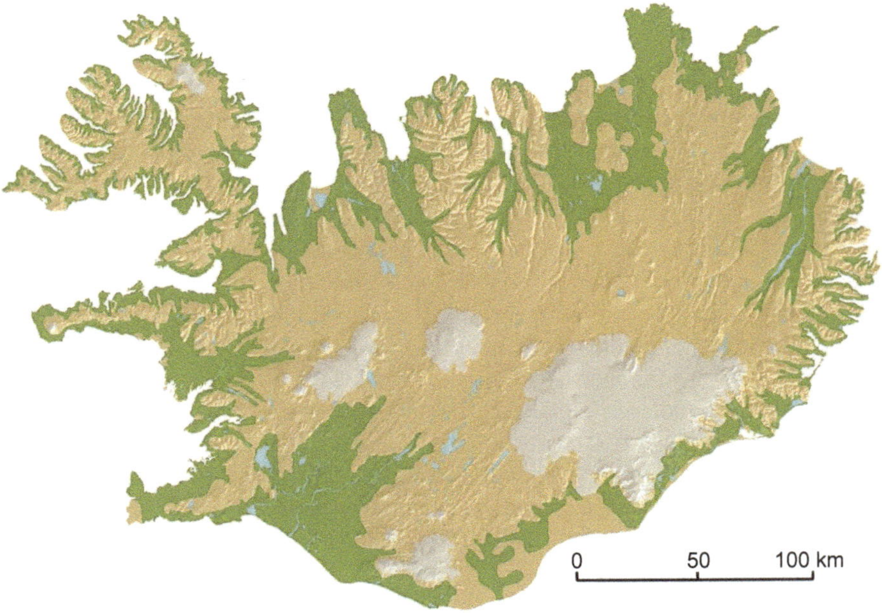

Figure 6.1.1 Birch forests in Iceland at the time of initial colonization. Forested areas are shown in green, ice caps in white.

The first Icelanders lived on fishing, either in rivers and creeks or at the sea (Landnámabók, 1948, p. 97). As archaeological excavations show, both sea fish and river fish bones can be found in the garbage from the same farm, even in the inner regions of Iceland (McGovern et al., 2007, p. 42). The first settlers also gathered eggs of wild sea birds and used whales thrown to the shore as food

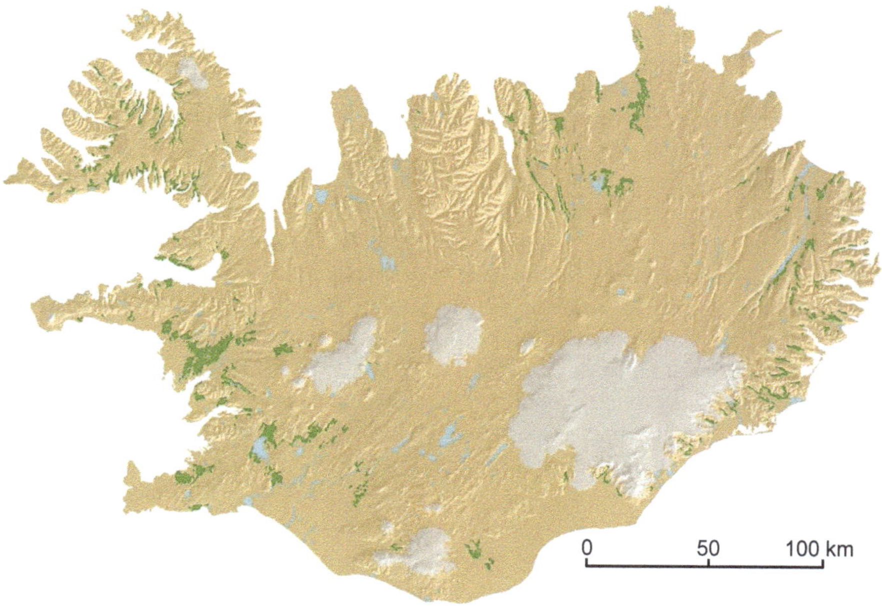

Figure 6.1.2 Birch forests in Iceland at present. Forested areas are shown in green, ice caps in white.

Figure 6.1.3 Dwelling of the epoch of the initial peopling (reconstruction).

(Landnámabók, 1948, p. 106). Agriculture developed quickly in the new land, especially animal husbandry. The settlers brought cattle with them, and the grass that grew in abundance on the island could provide enough food for it. But it was not possible to completely transfer the familiar agricultural methods to the new land with a new climate. Thus, it is widely known that there were goats and pigs in the "standart set" of domestic animals the settlers brought to the new land (Landnámabók, 1948, pp. 226, 295), but they already stopped raising them after the 1st century of their settlement. It was sheep farming that became the main area of Iceland agriculture for many centuries; the sheep were in the highland pastures all summer long without any shepherd, and gathered to the farms in autumn [such autumn sheep gathering in a special pen is mentioned in "The Book of settlements" (Landnámabók, 1948, p. 32)].

The houses of Icelanders in the first centuries after the settlements were of a similar type as in continental Scandinavia (Hjalmarsson, 2009, pp. 27–28). As centuries passed, the appearance of the buildings changed, but the main features remained the same: those were earth huts covered with turf, adjusted to the landscape. Such houses used to have sidewalls of stones and turf, wooden gables, and turf roofs. The roofs were usually low, their slopes stretched down to the ground (sometimes such buildings looked like steep hills with doors). If there were several rooms in the house, they were built side by side to each other in a row. Earth huts were used in the countryside until the 1950s, and then they were taken down everywhere and replaced by concrete houses, as Icelanders regarded these huts as unpleasant evidence of their century-long poverty.

The fact that even today the majority of lands in Iceland remain uninhabited and the natural powers are still active toward human beings (earthquakes, eruptions, sand drifts) has created an idea in the popular mind that the anthropogenic influence to the Icelandic nature must have been insufficient, and the nature remained "pristine" and "intact." Unfortunately, this hypothesis is not true: in the first centuries after the settlement the island underwent great human influence. First of all, the colonization greatly influenced the flora and the fauna of the island[4]: many species of farm animals and cultural plants have been imported, as well as some wild plants and insects from the continent, which all got to Iceland with the settlers (McGovern et al., 2007, p. 29). At that period, 90% of forests in Iceland were chopped down, and 40% of the soil underwent erosion, as the lands were often used as pastures. Some erosional processes that had started on sheep pastures of the 10th century continue in some places even now; the researcher Paul Bakland called landscapes formed in places where such pastures had been, "ovigenic landscape" (McGovern et al., 2007, p. 29). Tree felling is mentioned in "The Book of the settlement"; in particular, it is said that there used to be such a high forest somewhere in the west of

[4] It was also in the later times that the Icelandic fauna was enriched by "immigrant" species. The majority of mammals in Iceland are domestic animals, but some wild species have also been imported there by humans. Thus, in the 17th century, reindeer have been imported to the island for breeding, but now they live in the wild in the northeast of the country and are hunted; in the beginning of the 20th century, minks arrived and became well known as destroyers of birds' nests (see http://www.nat.is/Spendyr/villt_spendyr.htm).

the island that one of the local farmers built a ship of its wood suitable for sea sailing (Landnámabók, 1948, p. 69).

Nowadays, nearly all the tall trees in Iceland are on plantations. There is a natural high forest (also completed by plantations) in East Iceland at Hallormsstaðir near the glacier river of Lagafljót (Eysteinsson, 2005). It is a well-known fact that most of the natural forests in Iceland, at least near human settlements, were felled for firewood before the 17th century.

Part Two

Specific features of the initial colonization

(reconstruction by A.A. Velichko)

Expansion of ecumene northward and the diet of the Paleolithic population of Eurasia: Neanderthals and *Homo sapiens*

7

Mariya V. Dobrovolskaya
Institute of Archaeology, Russian Academy of Sciences, Moscow, Russia

Chapter Outline

A paleoecological approach allows us to update our views about mechanisms of the prehistoric human population dispersal. One of the important ecological characteristics is food resources. To a considerable extent, it defines the possibility for a certain group to survive on a given territory. It is well known that the choices of food resources define, to a large extent, the system of collective subsistence.

Thus, the questions of humans colonizing high latitudes hardly can be solved without analyzing and modeling food traditions of various representatives of the genus *Homo*. Modern methods, allowing one to investigate diets of Paleolithic men, provide many ways for getting ecological information about early man. One of the promising ways is the use of isotopic analysis. The distribution of stable isotopes of nitrogen ^{15}N and carbon ^{13}C follows ecological regularities that are well understood at present (see review by Tiunov, 2007). Accumulation of such isotopes in the basic structural protein of bone tissue (collagen) is controlled by these regularities. It allows us to put forward paleoenvironmental reconstructions, estimating average food rations for individuals. Characteristics obtained would describe the everyday food ration of adult humans.

As the ecology of a species' diet predetermines, to a considerable extent, its dispersal, nomadic or sedentary way of life, and other important characteristics, it seems important to discuss data about accumulation of these isotopes: markers of food potential as well as interpretation of obtained results.

Human Colonization of the Arctic: The Interaction Between Early Migration and the Paleoenvironment
http://dx.doi.org/10.1016/B978-0-12-813532-7.00020-6

In modern literature on the Paleolithic human ecology, the topic of diet adaptation has been analyzed in full length. Of the recent monographs on the subject, worth mentioning are these: "Meat-Eating and Human Evolution" (Stanford and Bunn, 2001), "Evolution of the Human Diet: known, unknown and unknowable" (Ungar, 2007), and "Human Diet: its origin and evolution" (Ungar and Teaford, 2002). The issue was considered also at the large international symposium, "The Evolution of Hominid Diets: Integrating approaches to the study of Paleolithic subsistence," organized and held by the Max Planck Institute for Evolutionary Anthropology (Leipzig, Germany) during May 17–21, 2006. Fundamental characteristics of *Homo sapiens* and his ancestor's diet are considered in two books published in Russia: "Man and its diet" (Dobrovolskaya, 2005) and "Human food" (Kozlov, 2005).

A vast array of scientific methods is introduced into investigation of the Paleolithic, thus enlarging informativeness of objects and creating more weighty system of evidence that our reconstructions are based on. As a result of active work on upgrading the methods for about 15 years or so, a sizeable volume of initial data on the Paleolithic diet in Europe has been accumulated; the materials under study refer mainly to the Mousterian and the Upper Paleolithic.

Obviously, the further dispersal of human groups over Eurasia in the Upper Pleistocene was largely controlled by the availability of food resources that could secure the population survival. It might be a good idea to compare data at hand about types of diet of Neanderthals and *Homo sapiens* to check up an earlier hypothesis about more diversified food resources that were mastered by the representatives of the now-existing species (Kozlovskaya, 2000).

7.1 Ecology of human diet at the Mousterian and Upper Paleolithic

Specific features of the Neanderthal's diet were first reconstructed by French researchers headed by Erve Bocherens (Institut des sciences de l'evolution, Université de Montpellier, France); the reconstructions were based on the isotope analysis. This publication (Bocherens et al., 1991) formulated main questions, answers to which are necessary for understanding specific characteristics of the Eurasian aboriginal diet at the Middle Paleolithic.

1. What was the position of Neanderthals and *Homo sapiens* in local trophic chains?
2. What was the space-time dynamics of their diet?
3. What carnivorous animals could Neanderthals and *Homo sapiens* compete with for food resources?
4. What species were typically hunted by the Late Pleistocene humans?
5. Can we judge about landscape choice of the Neanderthal communities, or had they occupied various kinds of biotopes within the limits of the colonized land?

All these questions are very important for an understanding of environmental preferences of a certain group as well as for reconstruction of some aspects of their behavior.

7.2 Adaptive process and food specialization in the Neanderthal communities

According to a traditional view in paleoanthropology, Neanderthals had an assortment of adaptive features related to a successful colonization of lands with a cold climate. At present, those customary concepts are to be altered to a certain extent.

Now, these familiar concepts underwent certain changes. Thus, T. Rae et al. developed, on the basis of computerized tomography, a three-dimensional model of paranasal sinuses of Neanderthals and Modern Man. The model showed that relative size of Neanderthal sinuses does not exceed that of modern humans (Rae et al., 2011). This data confirms the opinion that the efficiency of air warming in the process of breathing did not exceed that of an average modern European man.

T. Weaver (2009) reviewed critically the accepted views that typical features of cephalic and postcephalic morphology. The author emphasizes that it was genetic drift, and not the adaptation to low temperatures, that played the leading role in the formation of the Neanderthal specific morphology. The author supported his opinion by listing typical Neanderthaloid characteristics mostly lacking relation to adaptivity. Besides, from the author's point of view, the type of diet and specificity of the Neanderthal's physical activity can be also considered as the factors that influenced the body build. Nevertheless, such distinctive morphological characteristics as comparatively shortened distal segments of extremities in combination with considerable width of the shoulders and a massive thorax are in full accord with arctic adaptive type as described by Alekseyeva (1977). It might be better to state the reasons of these adaptations, with due consideration for the fact that Neanderthal groups lived mostly under conditions of a cold temperate climate incomparable with the arctic and subarctic climates of today.

Until recently the Neanderthal dispersal was thought to be bounded to the territories lying from the east of 68°E, as the easternmost site with traces of the Mousterian assemblages and human remains with an assortment of typical Neanderthaloid characteristics was the Teshik-Tash grotto.

One of the important achievements in Paleolithic archaeology is the discovery of the Mousterian in Northwestern Altay (Derevyanko and Markin, 1992; Derevyanko, 2007; Derevyanko et al., 2008). "About 50 thousand years ago a small group of Neanderthals with a lithic industry of Mousterian type penetrated into the Gorny Altai. In spite of the region having been reasonably well studied, now there are only two localities known at present: Okladnikov and Chagyrskaya caves, where culture-bearing horizons with Mousterian industry of the Sibiryachikha type were discovered" (Derevyanko, 2011: p. 49). Sequencing of a mitochondrial genome from a bone fragment found at the Okladnikov Cave and also morphological study of skeletal remains allowed the conclusion that this region was colonized by Neanderthals at the Paleolithic (Turner, 1990; Krause et al., 2007; Buzhilova, 2011; Mednikova, 2011). After these discoveries the boundary of the Neanderthal area shifted 2000 km eastward. Generally, the sites with archaeological evidence of the Mousterian culture and skeletal remains of Neanderthals do not occur north of 53°N in Eurasia.

7.3 Isotopic evidence for the diet of Neanderthals

Now, let us turn to isotope indexes of the collagen in bone tissue of Neanderthals recovered from various sites in Europe and Asia. By now, there have been analyzed samples of collagen derived from skeletal remains of 15 adults and 2 children (Richards and Trinkaus, 2009; Richards et al., 2008; Dobrovolskaya and Tiunov, 2011a). Summarized data point to a similarity of the environments chosen by various groups of European Neanderthals. The nitrogen indexes are stable, which indicates a distinct choice of hunted animals (Table 7.1). The exceptions are individuals from Feldhofer, which is the northernmost site. Those people seem to live in somewhat more humid boreal conditions, and the principal objects of their hunt were local herbivores of middle size, or animals feeding by tree branches and leaves of trees and bushes.

The majority of other groups preferred more open landscapes (forest-steppe, alternating forested and open areas). The European Neanderthals hunted mostly animals with $\delta^{15}N$ about 7%–8%. Extensive data on the $^{13}C/^{12}C$ and $^{15}N/^{14}N$ ratios in collagen

Table 7.1 Carbon and nitrogen isotopes from collagen of Eurasian Neanderthals

Site	$\delta^{13}C‰$	$\delta^{15}N‰$
Skladina, 4a-2	−19.9	10.9
Skladina -1в-4	−21.2	11.8
Spy 572a	−19.8	11
La Pradelle 10	−19.1	11.6
La Pradelle M300	−19.1	11.5
La Pradelle M400	−19.5	11.4
St.-Cesaire	−19.8	11.4
Le Rocher de Villeneuve	−19.0	11.6
Jonzak 1	−19.7	11.2
Vindiya 208	−20.2	10.3
Vindiya 207	−20.5	10.8
Feldhofer 1	−21.5	9
Feldhofer 2	−21.6	7.9
Okladnikov Cave, adult	−19.4	13.8
Okladnikov Cave, adult	−20.1	13.7
Okladnikov Cave, adult	−19.3	13.6
Okladnikov Cave, child	−19.2	13.9
Okladnikov Cave, child	−19.3	13.4

After Richards, M.P., Trinkaus, E., 2009. Isotopic evidence for the diets of European Neanderthals and early modern humans. Proceedings of the National Academy of Sciences of the USA (PNAS) 106, 16034–16039; Richards, M.P., Taylor, G., Steele, T., McPherron, S.P., Soressi, M., Jaubert, J., Orschiedt, J., Mallye, J.B., Rendu, W., Hublin, J.J., 2008. Isotopic dietary analysis of a Neanderthal and associated fauna from the site of Jonzac (Charente-Maritime), France. Journal of Human Evolution 55, 179–185; Dobrovolskaya, M.V., Tiunov A.V., 2013. Neanderthals from the Okladnikov Cave: environments and specific features of diet based on the data of isotope analysis. Archaeology, Ethnology and Anthropology of Eurasia 1, 78–88.

Table 7.2 The individual variability of δ^{13}C‰ and δ^{15}N‰ for the mammals from St.-Cesaire, La Berbie, and Camjac (Richards et al., 2008)

Species	δ^{13}C‰			δ^{15}N‰		
	Min	Max	SD	Min	Max	SD
Mammuthus primigenius	−21.8	−21.2	0.3	7.7	8.7	0.4
Coelodonta antiquitatis	−20.9	−19.6	0.5	5.4	8.4	1.0
Bison priscus	−20.9	−20.1	0.3	4.6	6.1	1.0
Rangifer tarandus	−19.4	−18.3	0.4	3.7	7.6	1.5
Equus cabalis	−21.7	−20.4	0.4	3.5	7.5	1.4
Crocuta crocuta	−19.9	−19.0	0.3	8.5	9.5	0.5

of various animals for the late Mousterian in Europe permits one to estimate the dynamics of the spatial and temporal changes in environments.

As follows from Table 7.2, individual variability of isotope indexes for different species is not equal. Intragroup variability of isotope indicators depends on a degree of mobility of the species' mode of life, the scale of migrations of herbivorous mammals, and variety of the animal's food ration. Based on summarized data, the highest individual variability is recorded for reindeer and bison.

Reindeer make regular long migrations feeding on grass, lichens, and small animals. An analysis of ^{87}Sr/^{86}Sr ratio in dental enamel of reindeer recovered from Jonzac (Richards et al., 2008) demonstrates that the animals procured by Neanderthals not far from their site had been migrating much farther than bisons procured by the same hunters. It should be pointed out, however, that in the Pleistocene, bisons could also migrate over long distances in steppes, and a high individual variability of δ^{15}N might be only the consequence of that. Another species that showed a significant individual variability of ^{15}N/^{14}N ratio was the wild horse, also known for long-range migrations. All the preceding shows that isotope indexes for some animals may vary within wide limits due to their behavior.

Studies of isotope composition have been carried out on bone tissue of Neanderthals from the Okladnikov Cave (Dobrovolskaya and Tiunov, 2013). As follows from the δ^{13}C values (Fig. 7.1), the bulk of the food chain was formed by the plants of a temperate climate. High values of δ^{15}N definitely testify to a high trophic level of these individuals (Dobrovolskaya and Tiunov, 2011a,b). Noteworthy also is a very low intragroup variability. Even taking into consideration the fact that some of the samples belong to the same skeleton, the minimum possible number of individuals is three. Such a similarity in the indexes suggests a common type of diet for adults and young men and women, as well as a common mode of life. A low individual variability may be considered an indicator of a sedentary way of life.

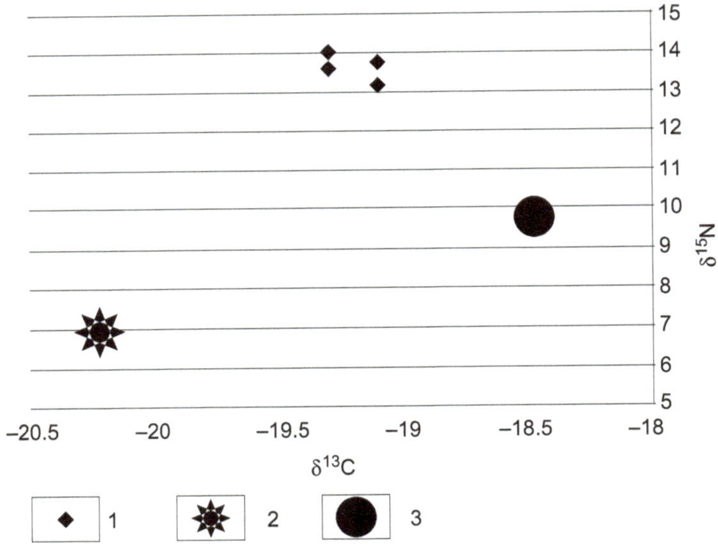

Figure 7.1 Isotope $\delta^{13}C$ and $\delta^{15}N$ indexes of the collagen extracted from the bone tissue of the Neanderthals from Okladnikov Cave (1), of bison (2), and wolf (3) from Chagyrskaya Cave, the northwest Altay Mountains.

As it was shown earlier (Ambrose and De Niro, 1986; Yakumin et al., 2000), the intensity of nitrogen isotope enrichment is higher under conditions of arid landscapes. So there is a good reason to believe that the higher value of $\delta^{15}N$, obtained for Neanderthals from the Okladnikov Cave, is attributable to specific features of the environments, as well as to the fact that the hunted animals migrated over long ranges in arid steppes. The archaeozoological materials recovered from the Pleistocene layers of cave sites from Northeastern Altay provided evidence of most mammals inhabiting open (steppe) landscape (Derevyanko et al., 2003a,b).

Another factor that could favor the rising of the isotope index values is a high proportion of young animals in the Neanderthal's prey. The isotope composition of collagen in bone tissue of herbivore young at a period of their being fed milk would be essentially different from that of adults. Milk is a protein-rich food and therefore serves to raise the $\delta^{15}N$ in the organism. Archaeozoological studies in the region also confirm a significant proportion of young animals (Derevyanko et al., 2003). However, the significance of this factor for the bone tissue composition of the Pleistocene hunters should not be overestimated, as they could provide it easily only for a short term in summer. Nevertheless, a combination of these two factors could account for a specific isotope composition of the Neanderthal bones recovered from the Okladnikov Cave.

The isotope characteristics of the principal species hunted by the Neanderthal group may be found by calculating the enrichment factor. Instead of using a precise value of the factor, the authors recommend to take its intervals, namely 0–2‰ for $\delta^{13}C$ and 3 up to 5‰ for $\delta^{15}N$ (Bocherens et al., 2005). Such a wide interval, however, makes our reconstructions less informative.

In accordance with recommendations, the isotope parameters of animal collagen derived from bones of animals hunted by Neanderthals from Okladnikov Cave were approximately as follows: $\delta^{13}C$ from −21.4 to −19.4; $\delta^{15}N$ from 8.7 up to 10.7. Unfortunately, the archaeozoological materials from Okladnikov Cave are not properly studied; we have, however, data on isotope composition of fauna samples from another cave located at the Northwestern Altay, Chagyrskaya. The cave yielded Mousterian lithics and remains of the Neanderthals (Viola et al., 2012). We carried out the collagen extraction and identified an isotope ratio $^{13}C/^{12}C$ and $^{15}N/^{14}N$ for the samples of herbivorous animal (bison) and beast of prey (wolf) (Tiunov and Dobrovolskaya, 2011). Isotope characteristics of the remains supposedly belonging to the hunter's prey are close to $\delta^{13}C$, which were determined in the bone tissue collagen of bison. The value of $\delta^{15}N$ (7.1‰), however, obtained for bison bone tissue is somewhat lower than the expected value. The higher meaning of $\delta^{15}N$ might be a result of regularly consumed river fish. In this case, however, the expected carbon indicators would be considerably lower. Therefore, the hypothesis of regular eating of river fish seems hardly convincing.

The main animal species hunted by the Eurasian Neanderthals were large herbivorous mammals. Where the accompanying fauna included mammoth and woolly rhinoceros, those species were dominant. In the absence of the latter, the most common hunted animals were aurochs and wild horse; less common were other herbivores, such as reindeer. It should be stressed that all the animals aforementioned inhabit open landscapes. That indicates that the area inhabited by the Neanderthals was bounded on the north by the cold steppe landscapes. There is reason to believe that the Neanderthal communities did not occupy all the open steppe regions, but only those adjoining foothills and piedmonts.

The question of importance is this: whether the Neanderthal hunters were hunters or scavengers. The well-known hypothesis of the use of animal carcasses being was repeatedly discussed in literature dealing with food strategy dynamics in the process of the human origins (cf. Foley, 1987).

Special calculations performed by a team under the direction of Bocherens et al. (2005) and based on a modified multisource mixing model developed by Phillips and Gregg (2001) and later analysis by Richards et al. (2008) demonstrated that the dietary behavior of Neanderthals differed from that of hyenas. The data obtained argued against the scavenging hypothesis. The conclusion seems to be important, as behavior patterns of hunters are essentially different from those of scavengers. Hunting calls for a high degree of coordination, so large mammal procurement could be done by groups of hunters.

Turning back to the question of a chronological dynamics of Neanderthal hunting, it would not be inconceivable to make a tentative suggestion that the choice of hunting objects persisted through tens of millennia: they always preferred large, herbivorous mammals. Such a choice forced them to dwell at a short distance from the animal grazing lands. Thus, we may conclude on a distinctly pronounced carnivorous dietary preference and hunting behavioral specialization.

7.4 Human diet of the Upper Paleolithic based on isotope analysis

Archaeological materials point to a more diversified choice of animal species used by anatomically modern humans. Now, we direct our attention to the evidence of collagen isotopes (nitrogen and carbon) in human bone tissues recovered from some Upper Paleolithic sites in Europe and Asia as an independent source.

As follows from the data given in Table 7.2 the environmental specificity of food resources actively used by the Upper Paleolithic people differs from those used by Neanderthals. The latter included not only herbivorous mammals, but probably water fauna (primarily freshwater species). Thus, in the opinion of the researchers, the individual from Kostenki 1 got about half of its animal protein from organisms related to water (fish, waterfowl). In the diet of humans from Dolní Věstonice 35 and Brno-Francuska-2 the proportion of protein is about 25% (Richards et al., 2001).

The earlier reconstruction of dietary pattern of individuals from the Upper Paleolithic burials in Sungir (Kozlovskaya, 2000) suggests a presence of water-origin food in a diet of at least one of the humans (C-2 boy). The suggestion is based on the bone tissue microelement analysis (Table 7.3). Such information on the usage of freshwater food resources by the inhabitants of one of the northernmost Upper Paleolithic site—Sungir—complements the general view (Dobrovolskaya and Tiunov, 2011a,b).

The environments of the Final Paleolithic time (12,000–11,000 BP) were controlled to a greater extent by the ice sheet size. Archaeological sites in Western Europe occur as far north as 54°N. Archaeological and archaeozoological materials recovered from the Kendrick's Cave site (northern coast of Wales, Great Britain) and attributed to the

Table 7.3 Stable ^{13}C and ^{15}N isotopes from bone tissue collagen of anatomically modern Eurasian *Homo sapiens*

Samples	$\delta^{13}C$	$\delta^{15}N$	Age (1000 yr BP)	References
Brno-Francuzska 2	−19.0	12.3	24	Richards et al. (2001)
Dolní Věstonice 35	−18.8	12.3	23	Richards et al. (2001)
Kostenki 1	−18.2	15.1	33	Richards et al. (2001)
Kostenki 8	−18.3	10.9	23	Vermeersch (2011) and Dobrovolskaya and Tiunov (2012)
Kostenki 18	−19.1	13.1	21	Richards et al. (2001)
Kostenki 14	−18.2	13.5	33	Marom et al. (2012) and Dobrovolskaya and Tiunov (2012)
Malta 1	−18.4	12.2	20	Richards et al. (2001)
Pavyland 1	−18.4	9.3	26	Richards et al. (2001)
Sungir 1	−19.2	11.3	30	Marom et al. (2012)
Sungir 3	−18.9	11.3	30	Marom et al. (2012)
Pokrovka 2	−18.4	10.4	28	Akimova et al. (2010)

Table 7.4 Stable ^{13}C and ^{15}N isotopes in bone tissue collagen of individuals found at Kendrick's Cave site (Richards et al., 2008)

Sample	$\delta^{13}C$	$\delta^{15}N$	Age
57	−17.9	13.8	11,880 ± 90 (OxA-7003)/13,700 ± 110 cal. yr BP/
59	−18.0	13.4	11,093 ± 90 (OxA-7003)/12,940 ± 100 cal. yr BP/
60	−17.7	13.9	12,090 ± 90 (OxA-7003)/13,950 ± 120 cal. yr BP/
69	−18.1	13.7	117,600 ± 90 (OxA-7003)/13,590 ± 90 cal. yr BP/

Late Paleolithic enabled the specialists to attempt at reconstructing the diet and subsistence pattern of the site inhabitants. Though the site had been excavated long ago (at the end of the 19th century), all the materials had been properly documented, the excavation process described in detail, and the bone remains kept safely in a museum. All the preceding made it possible to perform isotope studies of the materials. The data obtained are given in Table 7.4. Noteworthy is the absence of bones of marine animals (both vertebrates and invertebrates) among the remains discovered.

The investigators (Richards et al., 2005) emphasize that the $\delta^{13}C$ and $\delta^{15}N$ values indicate a considerable proportion of seafood in the daily diet of the early men. A comparison between the same isotope indexes obtained for large herbivores (bovids) and a gray seal permitted modeling the seafood proportion in the human's diet. The seafood share appeared to amount to about 30% of the protein. The authors suggest that the diet of the Kendrick's Cave inhabitants was primarily based on meat of large herbivores and that of seal.

The seafood resource use was confirmed by the studies of Italian archaeologists in the late 1980s. The bone tissue of 10 humans from Arene Candide (Liguria, Italy) was sampled in Epipaleolithic burials and studied using microelement analysis (Francalacci, 1989; Francalacci and Borgonini, 1988). As follows from the results obtained, the people of the Ligurian coast fed primarily on seafood. Noteworthy are the famous images of fish found at the Upper Paleolithic sites in France (Lespugue, Niaux, Lartet) attributed to the late Magdalenian.

Thus the tradition of the water resource usage existed and developed from the Gravettian to the late Magdalenian. The aquatic resources seem to have a great potential, as the majority of the early Mesolithic sites in the European north are confined to marine coasts or to the banks of large freshwater basins (Bailey and Spikins, 2010).

7.5 Conclusion

Summing up the results of comparison between dietary traditions of the Late Pleistocene *Homo* and taking into consideration the materials available, we can elucidate two different diet strategies typical for the northern parts of the areas of the Neanderthals and *Homo sapiens* in Eurasia. The former are distinct for a specialized hunt for herbivorous herd animals, while the latter are known for unspecialized hunting for terrestrial and aquatic animals. The choice of prey influenced migrations of hunter's communities. That is why the boundaries of northern limit of the area inhabited by the Neanderthals were delimited by the opportunity to hunt the large herd mammals on the open areas. The Upper Paleolithic hunters also hunted in the open steppes inhabited by herds of large herbivores; they also used, however, aquatic resources and were adapted to the coastal environments that essentially increased their potential for colonizing new territories lying to the north by way of sea coasts.

The Mesolithic population in the north of Eastern Europe (reconstruction of social activity according to archaeological and anthropological data)

8

Alexandra P. Buzhilova
Lomonosov Moscow State University, Moscow, Russia

Chapter Outline

According to the specialists, the way of life of hunter-gatherers and fishermen typical for the Holocene beginning was best adapted to the specific environments of the north of Eastern Europe. The studies of the cultural remains of that time yielded a number of artifacts that gave grounds for recognizing the Mesolithic as a specific phase in the human culture. A considerable diversification is noticeable in tools and hunting weaponry. This interval was marked by usage of the so-called distant weapons, such as bow and arrow. Fishing equipment (harpoons and hooks) occur far and wide, and skis come into general use. The tools are made from softer varieties of stones than before, with wood, bone, and antler being widely used. It opened possibilities for using a wider set of material processing techniques, including polishing and drilling (Oshibkina, 1989). As follows from the paleozoological data, prehistoric people hunted reindeer, elk, bear, boar, roe deer, beaver, ducks, swan, and loon. Fishing gains in importance at later stages of the cultural evolution (Gurina, 1956; Oshibkina, 2006; Buzhilova et al., 2008).

The north of Eastern Europe is the richest in archaeological finds belonging to the Mesolithic. The active excavations having been carried out in the region for the last decades provided information on various types of Mesolithic dwellings, such as lake dwellings, pit-houses, and frame aboveground houses. Another specific feature to be noted is related to the burial practice: a number of graveyards dated to the Mesolithic include graves different in funeral rites.

As a rule, a poor degree of preservation of skeletal material or its absence from an archaeological site prevents the degree of biological and cultural adaptation of the Mesolithic people from being fully estimated. Nevertheless, the years of excavation within the mentioned

Human Colonization of the Arctic: The Interaction Between Early Migration and the Paleoenvironment
http://dx.doi.org/10.1016/B978-0-12-813532-7.00021-8

area produced a considerable volume of collections that can be used (with certain reservations) as a basis for paleoenvironmental reconstructions. Such are human remains from burial grounds attributed to well-known archaeological cultures of Onega and Veretye.

S.V. Oshibkina (2006) argues that the early sites of the Veretye archaeological culture in the East Onega region are concentrated mainly near the lakes developed at the time of the last Ice Age. The environments of that time appeared to be favorable enough for life activities and cultural progress of hunter-gatherers. It is evident for archaeologists that the population exploited actively the natural resources of the region for a long time. The Onega culture area lies about 200 km northwest of that of the Veretye culture. An analysis of the artifact assemblages attributed to different cultures with due regard for physical anthropological data gave grounds to the suggestion that the people, bearers of the Veretye culture, migrated toward Lake Onega at the end of the Boreal. This migration not only accounts for a certain similarity in artifact assemblages of both cultures, but it can be also traced in the craniological complex morphology of the so-called Northern Caucasians. We'll try to assess this conjecture taking the known anthropological sources into the consideration.

8.1 Anthropological materials

The Oleni Ostrov graveyard is a huge necropolis dated to the late Stone Age. It is located on the Yuzhny Oleni Island in Lake Onega (Republic of Karelia, Russia). It has no equal in the number of discovered burials among archaeological sites dated to that period. The excavations were carried out about 70 years ago; their results have been published in several papers, including the detailed monograph written by Gurina (1956). The expeditions discovered 177 burials; their number, however, could be twice as great, as the central part of the burial ground had been destroyed by quarrying. The anthropological materials were studied in detail and published by leading Russian specialists (Zhirov, 1940; Yakimov, 1960; Alekseyev and Gokhman, 1984a,b; Benevolenskaya, 1984). The unique site attracted (and still attracts) a considerable interest of the specialists (see, for example a recent overview by Gerasimova and Pezhemsky, 2005).

There are several series of radiocarbon dates obtained on the Oleni Ostrov burials in different laboratories, the earliest one having been published by L.D. Sulerzhitsky (Mamonova and Sulerzhitsky, 1989). The results permitted attributing the burial ground to the later phases of the Mesolithic, except one seemingly older grave (burial no. 100) known as "the shaman's tomb" (9910 ± 80 yr BP GIN 4836/11,400 ± 140 cal BP/). The second series was analyzed in Oxford laboratory. As follows from the results, the dates varied over a wide range; the main body of data, however, was confined to the 6th millennium BC (Price and Jacobs, 1990).

The Popovo burial ground near the village of the same name (the Kargopolsky District, Arkhangelsk Region) was discovered and excavated by S.V. Oshibkina in 1979. The anthropological collection from the burial ground includes remains of only a few individuals. The skeletons are poorly preserved, and in a few cases are found in fragments (probably, due to specific burial rites). The materials have been thoroughly studied by Gokhman (1984).

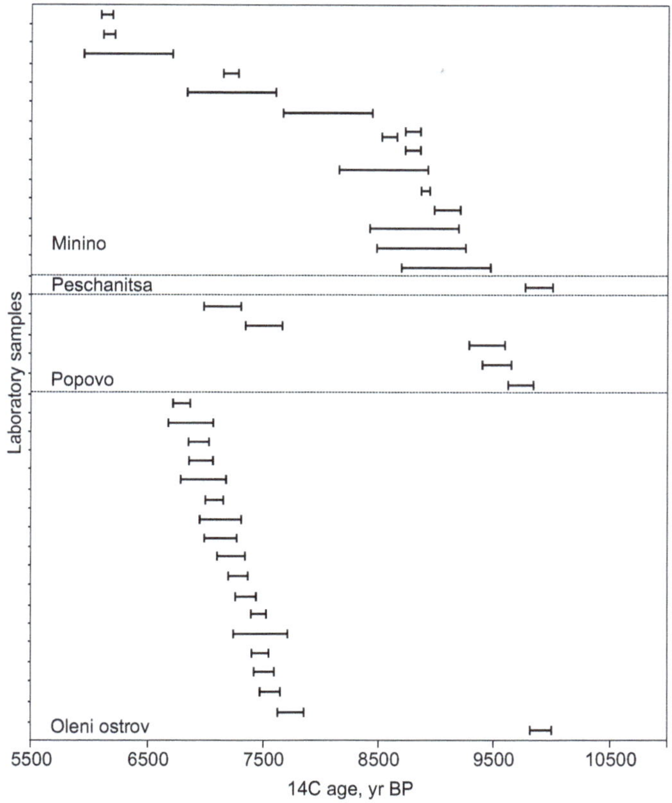

FIGURE 8.1 RADIOCARBON DATES FOR THE VERETYE CULTURE SITES (MAMONOVA AND SULERZHITSKIY, 1989; MAMONOVA, 1995; PRICE AND JACOBS, 1990; WOOD, 2006).

Carrying out works on chronologically different sites of the Minino archaeological cluster on Lake Kubenskoye (Minino I and Minino II, the Vologda Region), A.V. Suvorov identified and gave a systematic description of the Mesolithic and Neolithic materials (Suvorov, 2001). During the long-term excavations carried out by the Institute of Archaeology RAS 29, burials of the Stone Age have been studied, including 22 single, 5 double, and 2 triple ones[1]. A.V. Suvorov noticed that the Stone Age burials did not form a single burial site (graveyard), as individual graves within the limits of the site differed in age and in funeral rites (Suvorov, 1998). The radiocarbon dates obtained on samples from the Minino burials in laboratories of Geological Institute RAS (burials nos. 3 and 5), the Aarhus University, Denmark (burials nos. 4, 13, and 19), and at the Oxford University (burials nos. 3, 5, 6, 20, I, II, III, IV, V, and VII) proved that the age of individual graves within the burial ground varied within an interval of 3000 years or more (Suvorov and Buzhilova, 2004; Wood, 2006; Buzhilova, 2008) (Fig. 8.1).

[1] Besides the Stone Age burials, 75 graves of the Middle Ages were found.

An analysis of radiocarbon dates obtained on osteological materials from Minino and other archaeological sites in Northern Europe revealed a certain pattern in the burial complex position in the system of burials attributed to the Veretye and Onega culture. The oldest burials in Minino II complex (burials nos. VIII, I, IV, V, and II) and Minino I (burials nos. 20 and 19) correspond to the earlier stage of the Popovo burial ground (Oshibkina, 2006). According to A.V. Suvorov, the triple burial no. 22 from Minino I could be dated to the same time (Suvorov and Buzhilova, 2004). Most of the other dated objects at Minino I (burials nos. 6, 3, 5, 4, and 13) could be correlated with later stages of the Popovo complex and the main burial assemblages of Yuzhny Oleni Island (Zaitseva et al., 1997; Oshibkina, 2006); the rest of the burials from Minino represent the Early Neolithic (Table 8.1; Fig. 8.1).

The analysis of craniological characteristics of the male skull series from Minino showed some features similar to those of the population from the Veretye and Onega cultures. The skulls from Minino are poorly preserved. Only the brain case is suitable for studies, and in a few cases, dimensions of the facial skeleton have been measured from the male part of the sample. Table 8.2 gives some craniometric measurements.

The best preserved is the skull from burial no. V (Fig. 8.2). The maximum length of this skull is close to the upper limit of medium size, and its maximum breadth falls into the category of small sizes. The cranial index is 72.8, so it is attributable to the small-size category. The maximum height (basion-bregma), as well as the porion-bregma height, permits one to assign it to the category of hypsicranial (high-headed) forms. So it may be concluded that the skull from burial no. V is a dolichocranial form (hypsicranial by its cranial length-height index and acrocranial by its cranial breadth-height index). The minimum frontal breadth matches with the lower limit of the large-size category, and the maximum frontal breadth corresponds to the lower limit of the small sizes; according to the index, the front of that individual was very wide. The parietal bones are medium-sized, and parietal bosses are not developed. The occipital bone breadth falls into medium-size category; the superior Nuchal lines are distinct.

The facial bones are worse preserved, so we could not assess the bizigomatic breadth and medium face width. Visually it could be rated as medium-sized. The bizigomatic breadth seems to be greater than it actually is against the background of the narrow mandible, so the face appears triangular in outline (see Fig. 8.2). The upper face height, as well as the total face height, falls into the medium-size category, while the vertical craniofacial index falls into the small-size one.

The vertical profile could be estimated by the Flower Index only (No. 40; No. 5); it fits the medium-size category (96.2), so the skull may be defined as orthognathic. The lateral norm of the skull displays prognathism in the alveolar part, with the nose bones moderately sticking out. In estimating the horizontal profile the naso-molar angle may be used only; its value (143.5) corresponds to the upper limit of the medium-size category. Therefore, the face at the top level is slightly flattened.

When describing the frontal part of the face, the skull features to be noted are medium–high orbits and medium-wide nose. The absolute dimensions of the eye-socket are in the medium-size category, and the nose height and breadth are close to the upper limit of the small sizes. Simothic width is 6 mm, simotic height is 3 mm, and

Table 8.1 Results of radiocarbon dating of the samples from burial complex Minino

Laboratory index	Number of burial	Sample	Uncalibrated date (yrs BP)
Burial complex M-1			
OxA-16194	3	Human	7631±39
OxA-X-2178-30	3	Elk	7205±60
GIN-8837	3	Human	7240±160
AAR-5787	4	Human	6680±50
AAR-5788	4	Ungulate	6165±45
OxA-16195	5	Human	6669±38
OxA-X-2178-35	5	Elk	9070±110
GIN-8838	5	Human	6210±210
OxA-16196	6	Human	7568±38
OxA-16197	6	Elk	8975±55
AAR-5789	13	Human	6590±50
AAR-5790	13	Elk	6140±50
AAR-5791	19	Human (skeleton 2)	9320±55
AAR-5792	19	Elk (skeleton 2)	8960±55
AAR-5794	19	Animal (skeleton 2/3)	8790±55
AAR-5793	19	Human (skeleton 3)	9435±55
AAR-5794	19	Animal (skeleton 2/3)	8790±55
OxA-16198	20	Human	9385±40
OxA-16199	20	Elk	8895±40
Burial complex M-II			
OxA-X-2182-53	II	Human (skeleton 2)	9205±45
OxA-X-2182-51	II	Human	8885±45
OxA-X-2182-52	III	Human	8400±40
OxA-X-2182-50	IV	Human	9145±45
OxA-16200	V	Human	9435±40
OxA-X-2178-36	V	Elk	9090±110
OxA-X-2182-49	VIII (1)	Human	9430±45

After Wood, R., 2006. Chronometric and Paleodietary Studies at the Mesolithic and Neolithic Burial Ground of Minino, NW Russia (M.Sc. dissertation in Archaeological Science). Oxford University, Oxford.

simotic index belongs to the category of large sizes. Dacrial chord is 25 mm, dacrial height is 11 mm, and dacrial index falls into small-size category.

Hence, the individual from burial no. V presents a dolichohypsi-acrocranial form of the skull, with wide frontal bone, medium-broad and medium–high face close to triangle in outline, with medium–high orbits, and a nose of medium breadth and medium prognathism. Notable is the flattened horizontal profile of the face at the upper level and prognathic profile of the alveolar part.

Another skull identified as belonging to a young man and dated to the same age was recovered from burial no. 19. It is noted for a well-preserved braincase, while its facial

Table 8.2 Results of craniometrical data of male and female skulls from a series from Minino (indexes in the left column are given in accordance with the system developed by R. Martin (Alekseyev and Debets, 1965))

№	Measuring signs	Males					Females				
		N	Mean	SD	Min	Max	N	Mean	SD	Min	Max
1	Maximum length	7	183.3	2.56	180.0	186.0	1	177.0	–	177.0	177.0
8	Maximum breadth	6	135.0	2.53	131.0	138.0	1	135.0	–	135.0	135.0
17	Maximum height (basion-bregma)	2	142.0	0.00	142.0	142.0	–	–	–	–	–
5	Basio-nasion length	2	107.5	3.54	105.0	110.0	–	–	–	–	–
9	Minimum frontal breadth	6	97.3	3.72	92.0	103.0	1	96.0	–	96.0	96.0
45	Bizigomatic breadth	–	–	–	–	–	–	–	–	–	–
48	Upper face height (nasion-prosthion)	2	67.0	2.83	65.0	69.0	–	–	–	–	–
55	Nasal height	2	48.0	0.00	48.0	48.0	–	–	–	–	–
54	Nasal breadth	3	24.0	2.00	22.0	26.0	1	21.0	–	21.0	21.0
51	Breadth of orbit from *Maxillo-frontale*	3	42.0	0.00	42.0	42.0	1	39.0	–	39.0	39.0
51a	Orbit breadth from *Dacrion*	4	39.0	0.82	38.0	40.0	2	36.8	0.35	36.5	37.0
52	Orbit height	4	35.6	2.29	32.5	38.0	1	30.0	–	30.0	30.0
43 (1)	Bimolar chord	4	103.0	4.24	99.0	108.0	1	98.0	–	98.0	98.0
	Bimolar height	3	15.8	1.61	14.0	17.0	1	15.0	–	15.0	15.0
	Zigo-maxillar width	–	–	–	–	–	–	–	–	–	–
	Zigo-maxillar height	1	22.0	–	22.0	22.0	–	–	–	–	–
DC	Dacrial width	1	25.0	–	25.0	25.0	1	19.0	–	19.0	19.0

DS	Dacrial height	1	11.0	–	11.0	11.0	1	10.0	–	10.0	10.0
SC	Simothic width	1	6.0	–	6.0	6.0	1	8.0	–	8.0	8.0
SS	Simotic height	1	3.0	–	3.0	3.0	1	4.0	–	4.0	4.0
72	Total facial angle	–	–	–	–	–	–	–	–	–	–
75	Nasal angle	–	–	–	–	–	–	–	–	–	–
75(1)	Angle of outstanding nose	–	–	–	–	–	–	–	–	–	–
77	Horizontal profile (naso-molar angle)	3	145.6	2.52	143.5	148.4	1	146.0	–	146.0	146.0
‹zm	Zigo-maxillar angle	–	–	–	–	–	–	–	–	–	–
8:1	Cranial Index	6	73.0	3.35	67.2	76.7	1	76.3	–	76.3	76.3
17:1	Cranial length-height Index	2	77.0	0.88	76.3	77.6	–	–	–	–	–
17:8	Cranial breadth-height Index	2	77.0	0.88	76.3	77.6	–	–	–	–	–
9:8	Frontal-breadth Index	5	54.0	1.61	51.9	56.3	1	54.2	–	54.2	54.2
45:8	Cranial breadth – Face Index	–	–	–	–	–	–	–	–	–	–
48:17	Cranial height – Face Index	2	51.9	4.73	48.5	55.2	–	–	–	–	–
48:45	Face Index	–	–	–	–	–	–	–	–	–	–
54:55	Nose Index	2	52.1	2.95	50.0	54.2	–	–	–	–	–
52:51	Orbital Index	3	82.9	4.81	77.4	85.7	1	76.9	–	76.9	76.9
	Nasion-molar Index	3	15.5	1.20	14.1	16.5	1	15.3	–	15.3	15.3
	Zigo-maxillar Index	–	–	–	–	–	–	–	–	–	–
DS:DC	Dacrial Index	1	44.0	–	44.0	44.0	1	52.6	–	52.6	52.6
SS:SC	Simothic Index	1	50.0	–	50.0	50.0	1	50.0	–	50.0	50.0

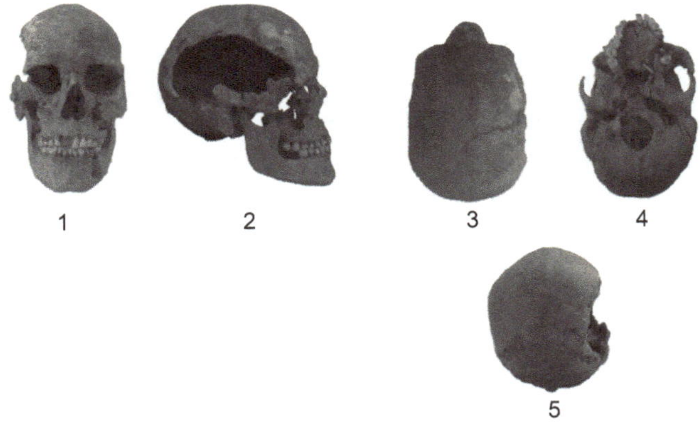

Figure 8.2 Graveyard of Minino II. Man's skull from burial V. Frontal (1), lateral (2), vertical (3), basal (4), and occipital (5) norms.

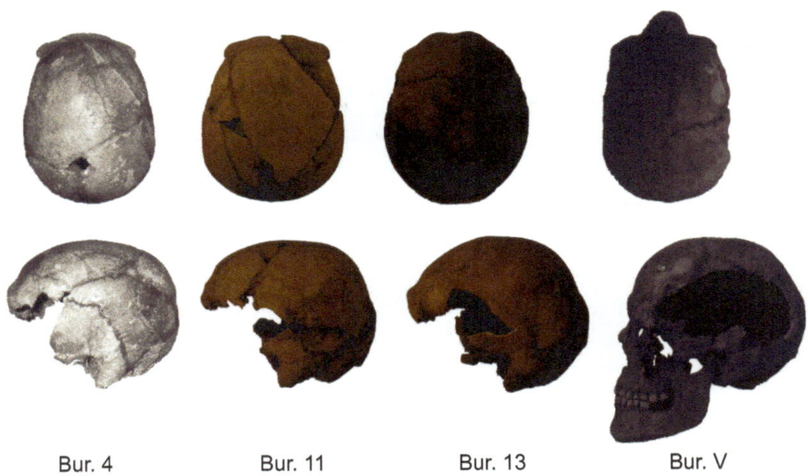

Figure 8.3 Graveyard of Minino I. Men skulls from burials nos. 4, 11, and 13; vertical and lateral norms of the skull from burial V (Minino II) are taken into consideration.

part is essentially destroyed. Morphological features permit one to place it into a category of dolichohypsi-acrocranial form with a wide frontal bone, a wide (at the top level) face, and with high and medium-broad orbits. Visible is some flattening at the top level of horizontal profiling of the face, as well as a mesognathism of the alveolar part of the profile. The assemblage of characteristics identified in this skull corresponds largely to those observed in the skull recovered from burial no. V. Even now, at the preliminary stage of the data processing, it is not inconceivable to suggest the existence of a specific anthropological variant that existed at the Mesolithic in the region of Lake Kubenoskoye.

Several skulls were dated by radiocarbon and attributed to a relatively later group. In the majority of the objects, only braincases are preserved, so any generalization cannot be but speculative (Fig. 8.3).

Along with the described specimen, the skulls in the latter group display dolichocranic forms only. The cranial index (74.8) permits one to place them into the small-size group, though the index is slightly higher than among the Mesolithic forms discussed earlier, the latter being noted typically for very low indexes. The individuals can be visually designated as medium- to high-headed variants. Part of them demonstrate for very wide and wide frontal bones (burials nos. 3 and 11), whereas others show a medium value of the maximum frontal breadth (nos. 4 and 13). Another group is identifiable by the configuration of eye sockets: there are skulls with high and medium-broad orbits (burials nos. 3 and 4). One of the skulls demonstrates a wide nose (no. 4), while another one (recovered from burial no. 3) is noted for a narrow nose. Nevertheless, the horizontal profile of both skulls shows an undoubted flatness of the face at the upper level.

The possibility of estimating anthropological options is mainly limited by the degree of the skull preservation. Though definitely dolichocranial, the latter skulls seem to be less narrow than the Mesolithic finds (see Fig. 8.3). It is not inconceivable, however, that the noted differences reflect only a degree of individual variability in the group under consideration. The latter is distinctly noticeable when estimating specific features of faces: there are faces with medium–high orbits and medium-broad nose, as well as those with high and medium-wide orbits and a broad or narrow nose. The unquestionably distinctive feature of the Minino skull series is the skull flatness at the upper level; the tendency being most obvious in younger groups.

The cranial dimensions measured from the preserved fragments of the male skulls from the Minino burial have been compared with series of the same age known from Northeastern Europe (Popovo, Peschanitsa, Yuzhny Oleni Ostrov, and Zvejnieki)[2]. The comparison permits outlining some trends.

First, the main part of the Minino sample (with exception of the Middle Mesolithic individuals from burials nos. 19/3 and V) displays a certain similarity to the skulls from the Yuzhny Oleni Island (burials nos. 139, 27, 131, 128 and 76). Judging from their basic parameters, those skulls correspond to Group I identified by Benevolenskaya (1984) on the basis of the braincase dimensions (including maximum cranial length, maximum cranial breadth, and basion-bregma height diameters), as well as the face height, face breadth, and horizontal profiling.

Secondly, the cranial index of the skull from Minino I, burial no. 5, shows a resemblance to a skull from Popovo, burial no. 3. Unfortunately, both skulls are very poorly preserved, which prevented the specialists from further searching for similarity.

In the third place, values of the braincase parameters measured in the skulls from Minino, burials nos. 19/3 and V, permit one to place them into the same cluster as the skull from the Peschanitsa site, though the latter differs from them in the characteristics of the face skeleton.

[2] In the analysis, there have been used the following parameters: individual skull dimensions measured on objects from Popovo (Gokhman, 1984); the same, measured on the Peschanitsa skull (Gerasimova and Pezhemskiy, 2005); individual and average dimensions of skulls from the Yuzhny Oleni Island, some dimensions having been corrected (Yakimov, 1960; Alekseyev and Gokhman, 1984b); and mean sizes of the Mesolithic sample from the Zvejnieki burial (Denisova, 1975).

As is obvious from the preceding, craniological data alone are not conclusive evidence of a similarity between the bearers of the Veretye and Onega cultures. The noted distinctions, however, do not rule out a possibility of the regional population to develop on a uniform anthropological substrate; that provides a circumstantial evidence in support of the Oshibkina's opinion on the existence of the so-called Northern Caucasians (Europoids).

8.2 Demographic features

The analysis of sex and age structure of the population in the studied archaeological complexes reveals some tendencies typical of the Mesolithic, namely, noticeable prevalence of the male remains over those of females, the children's burial being scarce or practically absent.

It can be easily seen that individual and group burials may vary considerably within a single archaeological complex: there are graves with two bodies of different sexes or of the same sex, graves in which adults are buried together with children, or those of children of various ages. A few archaeological sites are noteworthy for specific funeral rites: there were bodies positioned vertically in the grave (Yuzhny Oleni Island); graves with parts of one or several skeletons buried; and group burials with complete and partial skeletons in one grave (Suvorov and Buzhilova, 2004; Buzhilova, 2008). Typically, single male burials are dominant at all the archaeological complexes.

The calculated mean age of death in the Stone Age series varies in the range from 28.6 to 39.2 years in northern Europe (Buzhilova, 2005). The group of human remains studied in the Minino complex displays an age close to the upper limit of the interval, 36.6 years, and an insignificant difference between the male and female samples (36.6 and 37.2 years, respectively). In the series of the same age recovered from the Popovo graveyard the life span index comes up to the maximum values recorded in that region (Gokhman, 1984). Equally high values of the mean age at death, 39.2, were found in the group from Yuzhny Oleni Island.

By comparison with other materials (Mesolithic series from the Zvejnieki graveyard, Latvia, and from burials of the same age in Denmark) (Denisova, 1975; Bennike, 1985), we can see that the demographic indexes in the latter are essentially below those we discussed earlier. The average age at death in Zveniyeki series is about 29 years, and in the Danish groups, it is about 30–31 years. An unfavorable demographic situation is indicated by some other characteristics: the Zvejnieki group contains almost no elderly people, and the child mortality in Mesolithic samples from Denmark and Latvia is almost as high as 30%, while in the large series of the Yuzhny Oleni Island the child number is noticeably lower and comes up to 14% (Buzhilova, 2008).

So, taking into consideration the demographic indexes (indicative of a satisfactory situation for that period), the considered groups attributed to the Onega and Veretye cultures show evidence of a successful adaptation for the climate of Northeastern Europe.

8.3 Specific features of the funeral rites

A special interest has been attracted by the fact of several funeral traditions present in the same archaeological complex. Such a situation has been described in a few complexes. First to be noted are vertical, single (one-man) graves on the Yuzhny Oleni Island (4 cases), having no analogs in other graveyards of Europe. In two cases (individuals from burials nos. 100 and 125), some traumatic injuries of a bone tissue have been detected.

In the male skeleton from burial no. 100 (a vertical burial widely known as the shaman's grave), there have been recorded traces of a healed injury on the left tibia in lateral part of a diaphysis (MAE no. 7573-53). It may be concluded from the traces of healing that the wound was inflicted by a weapon with a sharp edge, the bone tissue having been partly damaged, as can be easily seen at the bone edge. Periostitis traces visible along the perimeter suggest an extensive inflammatory process related to the soft tissue infection. The man was buried in vertical position and leaned against the grave wall, the empty space around him having been filled with stones and sand. According to Gurina (1956), both the skeleton and associated tools were richly powdered with ocher; the tools recovered from the grave included two harpoons, four knives, a dagger, and several dozens of flint arrowheads. As follows from four vertebras found in anatomic order in the grave, a part of an animal carcass and birds were put into the grave.

Another male skeleton (from the burial no. 125) was positioned at an angle of 45° to the grave wall and lined with large stones on all sides. The skeleton was covered by ocher from the skull to pelvis. In the pelvis area, fragments of bone artifacts were found, and 83 elk incisors were discovered around both legs. A bear phalanx with ring cutting was found near the feet (Gurina, 1956). The male skull (MAE no. 5773-72) displays a perforation about 8 mm in diameter in the area of the left parietal boss. The perforation margins suggest it to be inflicted by a strong blow of a tool with pointed end. It seems that it was an arrow or a harpoon with a narrow point. D.G. Rokhlin and V.P. Yakimov (Yakimov, 1960) were inclined to believe that the wound could have been a cause of the man's death.

A particular attention has been drawn to the graves with two, three, or several individuals buried together. Burials of that kind had been described on Yuzhny Oleni Island, Minino, and Zvejnieki (Latvia). Analyzing the sex and age composition of the buried remains, it has been noted that both in Minino and Zvejnieki, as well as on Yuzhny Oleni Island, most of pair burials contain two adult individuals of different sex. We have tried to appraise the sample giving primary attention to causes of death, a particular emphasis being placed on injures and wounds.

Interesting results have been obtained from the analysis of the largest sample of Yuzhny Oleni Island. Special attention was given to the man and woman (nos. 65 and 66) buried together. Both skeletons were laid on the right side parallel to each other, the man behind the woman. The skeletons were densely powdered with ocher. As noted by Gurina (1956), the ocher powdering extends beyond the limits of the skeletons themselves. Around the neck of the man's skeleton, there were elk incisors (a necklace or a part of clothing decoration). Four plates from beaver incisors were found at the woman's skull. The woman's skeleton is preserved in fragments, so it was difficult to conclude on the possible cause of her death. The male skeleton shows a damaged area with traces of inflammation on the frontal bone above the left orbit (MAE no. 5773-17, burial no. 65). The injury analysis revealed distinct traces of an active inflammatory process. The inflammation could have brought about a sepsis that caused death. D.G. Rokhlin (Yakimov, 1960) considered the bone changes to be a result of fibrous dysplasia or a local hyperostosis of the partial leontiasis type. When describing damages of the facial skeleton, Yakimov (1960) noted an asymmetry of its right and left sides and, in particular, of the size of the eye sockets, left zygomatic thickened and nasal bones shifted to the right. In his opinion, these changes may suggest an extensive injury of the face that could result, in particular, in the loss of the left eye. According to Yakimov, the injury had happened at a young age, so the loss of eye resulted in a retarded growth of the left eye-socket. Unfortunately, the poor preservation of the skull gives no way of estimating the changes described by V.P. Yakimov. We have but fragments of the skull, so in the differential diagnostics, we should take into account at least two possible causes of the bone deformations complicated with an active inflammatory process, namely physical injury or a systemic disease.

There is a case of two women buried together (nos. 93 and 94) on the Yuzhny Oleni Island site. One of them (MAE no. 5773-41, no. 94) bears traces of a depressed fracture of the calvaria in the parietal area on the left, close to the coronal suture (Fig. 8.4). An active porosis marks a wide zone of inflammatory process both on the parietal and frontal bones. Taking into consideration the nature of the bone tissue changes, it seems that the head injury resulted from a blow with a heavy blunt object. Noteworthy is that there is one clear imprint of rectangular shape, 24×8 mm in size, easily recognizable among several impressions (Fig. 8.5). It may be supposed that the distinct imprint corresponds to the striking platform of the object the

Figure 8.4 Traces of depressed fracture on a skullcap, left parietal bone (see triangular point). A woman's burial from the Yuzhny Oleni Ostrov graveyard (MAE, No. 5773-41, burial 94).

Figure 8.5 Typical deformation on parietal bone (see the triangular point) indicative of the type of trauma. Woman's skull from the Yuzhny Oleni Ostrov graveyard (MAE, No. 5773-41, burial 94).

injury had been made with. It is possible that an extensive inflammatory process became the cause of the woman's death.

The skeletons in this burial were laid at full length, facing each other. No ritual objects or tools attributable to the remains of the woman with skull injury have been found. Near the other one (no. 93), there was an antler polisher lying close to the frontal bone and a flint knife between the jaws. That woman was older. Apart from the early tooth loss and a partial craniostenosis near the metopic suture, there is a genetically determined ankylosis of the second and third cervical vertebrae. The data obtained from the osseous system did not shed light on the cause of the second woman's death.

The cases described suggest a possibility that the body of a suddenly deceased individual, probably as a result of some tragic event, could be buried into a grave of another individual having died earlier and prepared already for funeral. If such was the case, it is still a debatable question: why had a separate grave not been prepared?

Another possible explanation consists in that the joint burials could be an element of complex funeral rites. To check this hypothesis, we tested the genetic diversity in the series taking into consideration funeral rites in cases of joint and single burials. As genetic markers, we used, in particular, signs of the tooth enamel defects (a distinct oval on the chewing surface of the first or second molars), probably controlled genetically. Another characteristic taken into account was a specific morphology of tibia (flattened diaphysial part). Of signs less commonly occurring, we analyzed an *intercondyloid fossa* on a humeral bone; a preservation of finger-shaped impressions on the endocranial surface of frontal and parietal bones; and perforation of the ear canal. It should be noted that the gender distribution of the tooth enamel defects strongly suggests that men and women buried together were not blood relatives but most probably were a man and his wife (Buzhilova, 2006).

In the series from the Yuzhny Oleni Island and Zvejnieki graveyards, besides graves of adults, there are also pair burials of an adult (man or woman) and a child. There have been recorded a few cases, both at the Yuzhny Oleni and Zvejnieki, when the child was placed between the adult person's femurs (burials nos. 2 and 3 in the Yuzhny Oleni graveyard, no. 42 in Zvejnieki). There were also burials of children of different age found on both burial grounds; the skeleton position does not differ from that of adult individual remains in joint burials.

Another feature worth mentioning is several (most often, three) individuals buried in one grave; such burials have been found in the Yuzhny Oleni Island and Zvejnieki graveyards (Suvorov and Buzhilova, 2004; Gurina, 1956; Buzhilova, 2008). In the Minino series the bones appeared to be sufficiently well preserved to permit the sex determination. In every case the bodies were arranged in the same order, namely, male–female–male (burials nos. 19 and 22), the woman's body being positioned between the two bodies of men. In burial no. 19 the human remains are vertically arranged in three conditional layers; the uppermost remains is only partial: it is a fragment of a male skull lying above the female skeleton. The archaeological evidence available strongly suggests all the three remains had been buried simultaneously. Or, at least, the lower (male) and middle (female) bodies were put into the grave at the same time, every one having been swathed heavily in a shroud (Suvorov and Buzhilova, 2004).

As to the triple burials on the Yuzhny Oleni Island, there is hardly seen any regularity. In one case, two women and a man between them were buried together (burials nos. 55-56-57). Another burial contains a man, a woman, and a third person of unidentifiable sex because of a poor preservation of the skeleton (burials nos. 26-27-28). In Zvejnieki, three buried persons (burials nos. 14-15-16), as well as children from burials nos. 32-33-34, are still of an undetermined sex for the same reason.

Further anthropological research of burials nos. 32-33-34 revealed the presence of female skeleton fragments among the child remains, so the grave may be considered a place of collective burial. Burials of this type are recorded only on Yuzhny Oleni Island and in Zvejnieki. There is only one such case on the first of the burial grounds mentioned, that is burials nos. 103-104-105; in common with the Zvejnieki case, it was described previously as a triple burial of children (Gurina, 1956). Our research revealed the presence of a woman's skeleton fragments together with remains of immature individuals (Buzhilova, 2008). A similar case is known from Zvejnieki, with an adult skeleton being surrounded by fragments of three children's remains (burials nos. 27-28-29-30-31). Thus, collective burials seem to be rather steady in their gender and age characteristics: it is usually one adult individual (woman?) buried together with several children.

In an attempt to specify social relations between individuals buried together in collective graves, we used the isotopic analysis. That analysis is commonly applied to dietary reconstructions. The most comprehensive data could be obtained by comparison of the Minino I and II series[3]. When comparing specific dietary features of individuals buried in single and multiple (double and triple) burials, the nitrogen isotope content is notably lower in human remains from joint burials, which might be

[3] The analysis was carried out in the Oxford laboratory (Wood, 2006).

attributed to a higher proportion of vegetable food in their diet. When considering the sample of remains from joint burials, there are two groups different in the carbon isotope content. Taking into account the reconstruction of the Neolithic human diet by Dobrovolskaya (2005) and with due consideration to the data from the Minino group, we came to the conclusion that those people were hunters using the animal meat for food. When compared directly with the sample from Popovo, it becomes clear that a part of the Minino population displays rather specific features of their diet (Kozlovskaya, 2003; Buzhilova, 2005).

In triple burials of Minino series the isotopes obtained from male skeletons (burial no. 19, skeleton 1; burial no. 22, skeleton 1) show a similar relation between the nitrogen and carbon isotopes. The other remains buried in the same graves with the males (burial no. 19, skeleton 3; burial no. 22, skeletons 2 and 3) differ in the isotope compositions, the carbon isotope proportion being distinctly increased. That tendency may be most clearly seen by the example of the woman from burial no. 19 (skeleton 2). The second group of individuals differs from the first one presumably by a more diversified diet (due to the increased proportion of vegetable food). In double burials the individuals differ in the proportion of the nitrogen isotope, which is relatively high in male remains mostly. On the whole, however, its level is lower as compared with males from single burials.

So, a comparison of human remains recovered from different kinds of graves different in number of buried individuals (single or joint), as well as in funeral rites, revealed certain morphological and physiological distinctions between humans from joint burials and those buried individually. The diet of those buried in the joint graves was seemingly more diversified and in all probability included vegetable products. Quite common are the cases of one adult buried together with several children. Noteworthy, a comparative analysis of genetically determined signs has shown that the adult humans of different sex buried together were not kin relatives; probably, they were related by marriage. It is not inconceivable that the burials of heterosexual adults could be attributed to some specific features of the funeral rites in the Mesolithic.

8.4 Reconstruction of social activity

The integrated analysis of the archaeological and paleozoological data provided information on the hunting practice typical for the North European population during the Mesolithic–Early Neolithic. As follows from studies of elements of clothes, ornaments, and tools (indicative of animal species commonly hunted), the way of life in the northern Europe was controlled by harsh environment, and first of all by a cold climate (Gurina, 1956; Oshibkina, 1983; Buzhilova et al., 2008). In the Mesolithic the principal hunted animals were the beaver, elk, and pine marten (*Martes martes*); less important were bear, fox, and water vole and fishing.

It should be noted that the dominance of beaver, elk, and pine marten bones in the Mesolithic and Neolithic sites is typical of the forest zone. Noteworthy also are the water vole remains found in Mesolithic layers of Minino (Buzhilova et al., 2008). This mammal inhabits most often coasts of drainless or poorly drained water basins. It swims well and leads a semiaquatic life feeding on plants. In summer, it makes nests above the water in heaps of a dry reed. Bones of this species are known from many Stone Age sites (Vereshchagin and Rusakov, 1979; Karkhu et al., 2003). The studies confirmed the suggestion that the water vole constituted a part of the early man die: the practice abandoned in later historical epochs (Karkhu et al., 2003).

Judging from the available fragmentary remains of canids recovered from various layers of Minino, it is difficult to come to a conclusion on whether the dog was domesticated at that time (Buzhilova et al., 2008). Other sources, however (see almost complete dog skeletons found at the Mesolithic site of Ivanovskoye VII), unambiguously indicate the dog presence (Karkhu et al., 2003).

An analysis of the physiological stress markers (according to paleopathological data) revealed a confusing picture of the environmental negative impact on the Mesolithic population in the European Northeast (Buzhilova, 1998). The analysis suggests harsh environments, with certain seasonality in forming stress factors.

That may be illustrated by the example of the so-called Harris lines, or the growth arrest lines, transverse lines on the tibia indicative of temporary arrested growth at the child age due to seasonal stresses (most probably winter cold). The X-ray analysis of the bones of adults from Minino permitted identifying the Harris lines in 80% of bone remains, and an average number of the lines varied between 7 and 11 per a metaphyseal area. When directly compared with another marker of growth rate slowdown, such as enamel hypoplasia on the tooth crown resulting from acute infections or other diseases the individual suffered in childhood, the latter marker appears to be practically lacking in the series. Such an inconsistency of the two markers in the series may indicate indirectly a seasonal distribution of the short-term stresses in the early age, probably related to a lack of food in a cold season, and not to diseases (Buzhilova, 2005).

Possible consequences of negative stresses, including those resulting from cold climate, have been estimated by means of the radiographic analysis applied to tubular bones of upper and lower extremities. On the basis of the obtained data calculations the index of compactness was calculated using a special formula for estimating the degree of cortical tissue development and the width of medullary canal. There exists an opinion that the human body exposure to low temperatures results in various morphophysiological changes, including the medullary canal widening and a noticeably narrower cortical tissue. Such tendency was noted, for example, studying long bones of Eskimos.

The comparative analysis of the human remains in the Minino series for the index of compactness revealed two types of human response to low temperatures of the environments. The response of the first type is closest to that recorded in the Upper Paleolithic individual 1 from Sungir studied by Mednikova (2000). It is noted for lower values of the index due to lower values of both width of medullary canal and that of cortical bone tissue. The second group of individuals from Minino displays a similarity to the Yuzhny Oleni Island and Chernaya Gora series. In that case the index of compactness is close to the first variant in the width of medullary cavity, though it differs in a much greater width of the cortical bone tissue. So, the sample of tubular bones from Minino series does not show a noticeable decrease in the cortical tissue volume; the latter is mostly typical for the modern inhabitants of Arctic. The index of compactness calculated for the Minino series is close to those of modern Russian and Mordva populations of the temperate zone (Buzhilova, 2008).

It is safe to assume therefore that a specific variant of the hemopoiesis function under conditions of temperate zone was developed in the north of Europe as early as the Mesolithic/Neolithic boundary.

As to the state of health, the human remains studied in the Minino and Yuzhny Oleni Island burials are noted for the absence of chronic and infectious diseases of teeth and bone systems. No case of caries has been documented, except for a case of dental pulp and alveolar inflammation (odontogenic osteomyelitis) in the area of the second premolar of an adult man (Yuzhny Oleni Island, burial no. 136) and an abscess in the first molar of an adult woman (burial no. 60). We interpret them as resulting from a mechanical damage of the tooth crown because of an excessively great load, presumably functional, on the dentition system at the early age.

D.G. Rokhlin who advised V.P. Yakimov about some paleopathological finds attributed the lifetime loss of 2M recorded in cases of adult women from burials nos. 93 and 142 to a functional load and odontogenic osteomyelitis (Yakimov, 1960). Noteworthy is that such a kind of damage is commonly found in fossil human remains in Arctic, where people often use teeth as a kind of "third hand" when processing hides of killed animals (Merbs, 1983). As additional evidence in support of that idea, we could be reminded of the cases of characteristic wear of frontal teeth with the chewing surface of back teeth seeming intact; such cases were recorded in series from the Yuzhny Oleni Island (burials nos. 31, 52, 60, 67, 75, 94, 111, 153, 158, etc.). A special analysis of the tooth crown wear has shown that men from Minino also used teeth as "the third hand." There have been recorded cases of specific (not age-related) wear on the molar chewing surface and characteristic breaks in crowns (Buzhilova, 2005).

In the studied series, there are practically no cases of chronic anemia, nor scurvy. A few individuals show traces of former rickets. Early (regardless of age) arthropathies—diseases of joints and of vertebral column—were recorded occasionally.

A few traces of wounds inflicted during lifetime were found in the Minino series, including healed fractures of forearm bones, clavicle, etc. Those might be interpreted as resulting from domestic accidents under conditions of strenuous physical activities. The Popovo series of the same age as the Minino one and geographically close to it is small in number, the remains being poorly preserved. The analysis of the finds has not revealed any injuries of the human remains in this group (Gokhman, 1984).

Our studies of the Yuzhny Oleni Island series recorded few cases of cranial traumas, some of them with traces of healing. There have been found also cases of healed fractures of tubular bones, as well as wounds inflicted with a sharp weapon from a distance. Fourteen cases have been recorded in the adult group solely (11.6%). The bone injuries are mostly typical of the male skeletons (males 19.3%, females 5.4%). Skull traumas make up about a half of the total number of the skeletal bone injuries, the lesser part of them being found in the female group.

One of the women's skulls (MAE no. 5773-73, burial no. 130) displays a severe injury in the occipital area. As no traces of active inflammatory process have been

Figure 8.6 Perforation on woman's skull from the Yuzhny Oleni Ostrov graveyard (MAE, No. 5773-73, burial 130).

found, it may be safely assumed that the blow on the head was inflicted only a short time before death or even at the moment of it. The trace of the blow has the appearance of a rounded perforation of 20×22mm in size in its maximal part and 17×19mm in minimal one. There is a characteristic platform showing the place of the blow, which broke the occipital bone (Fig. 8.6). It is most probable that the blow was struck with a great force onto the individual lying facedown on the ground. A depression formed at the point of contact resulted in a perforating fracture of greater size due to radial cracking of the skull's bones around the point of the blow striking. While making a differential diagnosis, it should be taken into consideration that the opening could be a result of an intentional operational perforation (trepanation) of the skull, or as a consequence of systemic disease (Gurina, 1956).

D.G. Rokhlin (see Yakimov, 1960) suggested that both the bigger opening and another hole of smaller size (6×4mm) positioned on the occipital bone at the same level might result from metastases of the malignant growth or could be a consequence of myelomatosis (malignant alteration). The diagnosis seems to be adequate at first sight, as myeloma is known to form apertures of different sizes in the skull bones, the main (primary) opening being usually the greatest in size.

However, we disagree with D.G. Rokhlin's conjecture as the characteristics of the aperture margins are in contradiction with the suggested diagnosis. In case of such disease the margins of the damaged area form a kind of "festoon" pattern, all the flat bones being damaged on their edge because of an active lysis of bone tissue (Buzhilova and Berezina, 2008).

In the case under consideration the described perforation had smooth margins with characteristic chips resulting from a mechanical impact on the upper plate of the skull (see Fig. 8.6). As to a possibility of the skull trepanation performed by the Stone Age men, the operations of that kind are known from Peru, where a special V-shaped, double-edged knife had been used for that purpose. The aperture outlines could help to recognize configuration of the tool used in the operation, though it gives no way

Figure 8.7 Typical parallel incisions on occipital bone made by a tool, woman's skull from the Yuzhny Oleni Ostrov graveyard (MAE, No. 5773-73, burial 130).

of deducing whether the perforation had been made intentionally. So, the issue of the perforation nature is still debatable.

Noteworthy are distinctive parallel hatches found on the calvaria above the openings on the left parietal bone and arranged along the sagittal suture (Fig. 8.7). At a first glance, the hatches seem to be postmortem. The suggestion, however, does not agree well with facts: judging from a distinctive form of the hatches and the preservation of the injury edges, the perforation was inflicted by a tool with a sharp cutting edge. The position of the defect on the calvaria does not give grounds for reconstructing a partial scalping process; it is more likely to be wounds from repeated blows with a sharp weapon. Taking all the evidence into consideration, the perforation as described may be considered to be a result of injury rather than of intentional trepanation.

Describing that burial, Gurina (1956) calls attention to the fact that the skeleton was laid on its back, being turned slightly to the left, the right arm placed slightly aside. At the right femur, not far from pelvic, a little antler human figurine was discovered. A fragment of a bone object was discovered close to the right shoulder. The skeleton was strewn with brown ocher powder. Initially the skeleton was considered to be male; later, however, Yakimov (1960) determined it as female. We share the opinion of the latter.

Another case of injury of the facial bones, and not of the braincase, is recorded as a scar with traces of healing above the right orbit of a mature man (MAE no. 5773-117, burial no. 151). The defect is no more than 22 mm long. Besides, there is another injury on the same skul: that is a scar about 42 mm long, bearing no signs of inflammation; it may be interpreted as a safely healed wound. Though the extremity bones have been preserved only fragmentary, one may note a pilaster developed on the preserved femur, a pronounced relief of the humeral bone, and enthesopathy on costoclavicular ligament, which prevented the man from making wide sweeping movement with the arm. All the characteristics of the man suggest his intensive everyday physical activity. Yakimov (1960) noted, in particular, a great length of the preserved collar bone that indicates big shoulders.

When describing the position of the skeleton in this single burial, Gurina (1956) specified that it was laid on its back, his head being turned slightly to the right. Besides dense powdering with ocher, there were found four elk incisors at the left forearm and the same at the left femur, and two bear canines on the chest, and one more bear canine at the left femur. There are remains of another individual discovered at the first individual's legs (fragments of a skull, pelvis, and ribs). The context of the second burial is not clear. N.N. Gurina suggested only that the bones belong to other burial. It is quite possible that the aforementioned elk incisors and the bear canine belong to the other individual remains, and not to the man from burial no. 151.

Depressed, partly healed fractures in the parietal-occipital area not exceeding 15–17 mm in diameter were recorded in two male remains (MAE no. 5773-40, burial (?), and MAE no. 5773-74, burial no. 142). The cranial vault perforation, undoubtedly inflicted after death, was noted in the remains of a young male (MAE no. 5773-69, burial no. 117). In spite of a poor preservation of those single burials, the skull bones appeared to be preserved well enough, so the injuries could be estimated.

As can be seen from the preceding, the injuries of male skulls in Oleni Island are more common than those documented in female skulls. Besides, in cases of the male skulls the blows were delivered onto the face (direct aggression), while women's skulls are injured in the occipital area, the blows being stricken from behind or on the side of skull.

Two cases in the Oleni Island series were diagnosed as healed fractures of tibial bones in the lateral parts of diaphysis (MAE no. 7573-53, burial no. 100 and MAE no. 7573-123, burial no. 158). In the case of the male from burial no. 100 (widely known vertical burial of the so-called shaman) the trace of a healed injury may be interpreted as resulting from a blow of a sharp-edged weapon, a fragment of the tibial bone having been cut off. Similar injuries on both tibial bones (almost at the same level) are found on the individual excavated from burial no. 158. He was buried at the traditional extended position with the head slightly turned to the right, two bone points being found on his chest (Gurina, 1956).

Another case from the same burial ground (MAE no. 5773-145, burial no. 42) is a male skeleton with numerous superficial bone injuries varying in outlines and dimensions found on the left femur in the middle of diaphysis. One of the injuries is rounded, no more than 9 mm in diameter, 2–3 mm deep. Other damages are irregular in outline and most probably resulted from a vast wound of the thigh soft tissues, with the bone surface being partly damaged. When describing the buried body position, N.N. Gurina noted that the body was laid on the back, his arms slightly bent at the elbows, the hands being placed behind the back (Gurina, 1956, p. 290). A bone point has been found at the chest of the buried man, and a flint flake near the right elbow.

Another kind of injuries is represented by an ossification of soft tissues due to muscle microtraumas resulting from hypothermia (myositis). They were found in remains of a middle-aged man at the ankle joint (MAE no. 5773-5, burial no. 59) and in another one on a femoral bone (MAE no. 7573-123, burial no. 158). As has been stated earlier, the man from burial no. 158 had injuries on both tibial bones. Burial no. 59, as the majority of others in this group, held many artifacts indicative of a rich funeral rite. The man's clothes were decorated with 113 elk teeth, a dog canine, and plates made from beaver incisors were found in the grave. A knife made of slate had been put into the grave near the head, and a flint point at the right shoulder (Gurina, 1956).

A comparative analysis of traumas and fractures found in the series from the Baltic region has shown that among the Mesolithic population of Denmark the cranial injuries make up 44% of the total number of traumas, while those of extremity bones are negligibly small (0.59%) (Bennike, 1985). The part of the Zvejnieki series (Baltic region) dated to the Mesolithic revealed two cases of spine injuries, most probably compression fractures in the lumbar area (burials nos. 58 and 157). V.Ya. Derums (1970) described a case of a skull injury (left side of the parietal bone). There are two defects in the middle part of the bone rounded in outline. Almost identical sizes of the defects (12–14×9–10 mm approximately) and a small depth (5 mm) gave grounds to a conclusion of the wound being inflicted with a harpoon. The same individual bears traces of a healed fracture on the skull, probably resulted from a blow with a blunt instrument. In that case the author calls the reader's attention to the obvious signs of the fracture having been healed.

As can be seen from the preceding, the greater part of the studied materials from the northeastern territories does not suggest a high level of traumatism. The groups of injuries, however, are in agreement with the archaeological data on the lifestyle of the Mesolithic population; the data suggest the male population was deeply involved in hazardous activities, including those related to a considerable physical exertion.

Heavy and long-lasting physical loads are also indicated by traces of a secondary venous stasis (mostly as a result of thrombophlebitis) found on the lower extremities and increased density of longitudinal striation on the tibial bone surface[4]. It should be noted that this characteristic is typically found in young men and cannot be ascribed to age-related changes.

A comparative analysis of synchronous groups revealed similar tendencies of the impact of heavy physical loads on arthropathies. V.Ya. Derums (1970), for example, described several cases of the deforming arthrosis in Mesolithic groups of the Baltic region. The cases could be related probably to silo loading in the lumbar area; hernia have been recorded (Zvejnieki, burials nos. 48, 58, 153, etc.). One of the individuals (burial no. 14) displays traces of considerable load on lower extremities that resulted in a specific restructuring of the bone tissue. It seems quite possible that in this environments hunters had to travel long distances, often carrying a considerable load.

The occurrence of the dental calculus (indicative of the type of human diet) and the absence of infectious caries suggest a general similarity in the food ingredients of men and women, the food being viscous, mostly of protein composition. In the male group, there are cases of a premature wear of tooth crowns (chewing surface).

[4] Recently, some Russian authors considered (without good reason) this feature as a consequence of hypothermia or inflammation resulting from a prolonged stay in cold water during hunting or other activities. In the author's opinion, such a definite interpretation of the secondary periostitis at its earliest stage cannot be conclusive without other facts in support of the idea about an adverse health effect of cold water. In our work, we use only cases of thrombophlebitis signs found together with vertical striation of tibial bone tissue (initial stage of the secondary periostitis) as indicative of hard physical work and an excessive load on the lower limbs.

It seems probable that the men happened to consume tough and fibrous food (such as half-cooked meat).

An analysis of trace elements in the bone tissue proved the protein diet of the Minino population (Dobrovolskaya, 2005). We performed a detailed comparative analysis of trace elements present in individual remains (materials obtained by Dobrovolskaya (2005) from the Minino and Popovo series); the copper, zinc, and strontium distribution have been analyzed using the method of principal components for estimating their closeness. Data on individual dietary elements known from the Sungir series were also taken into consideration. As a result, two typical variants of the diet have been identified. The first one is similar to that reconstructed for the male from Sungir (Sungir I). Having used various methodical approaches the specialists came to the conclusion that his diet was based mostly on the meat of terrestrial animals, the proportion of vegetable food being negligible (Kozlovskaya, 2000b). A similar variant of the diet may be reconstructed for individuals from Minino (burial no. 19, individual 3, and burial no. 15) and Popovo (burials nos. 6 and 9). Another man from Popovo shows a similar kind of the diet (burial no. 8). A kind of diet different from the aforementioned was characteristic for a girl from Sungir (Sungir 3) (Kozlovskaya, 2000a). It was also dominated by the meat of terrestrial animals (though the proportion of it was slightly less), but distinct for a somewhat increased share of vegetables. The individuals from Minino kept to this type of diet as follows from reconstructions for individuals from burials nos. 3, 5, 16, burial no. 19 (individual 2), and to a lesser extent, burials nos. 11 and 13. It should be noted that the identified groups do not show correlation with the chronological attribution of the buried persons, or with topography of the burial complexes. The author would like to stress that inhabitants of both Minino and Popovo lived mostly on animal meat.

In addition, the N and C isotope analysis performed on the practically entire Minino series contributed to the reconstruction of the human diet. The results suggested that fish was used actively for food from the beginning of the Neolithic (Wood, 2006). The individual data thus obtained on the N and C isotope distribution made possible estimating similarity in the individual diets. In common with the case discussed before, there are two variants of the diet distinguished in Minino; in addition, however, there is a recognizable trend toward consuming river fish. The latter tendency is distinct at the final chronological stages only (at the transition from the Mesolithic to Neolithic and during the Neolithic). The last trend was identified not only due to the isotope method, but mostly because practically all the individuals of the series have been sampled, while the trace element analysis covered less than two-thirds of the series (Buzhilova, 2008).

The gender-controlled differences in activities are confirmed by various biological signs. A significant difference exists, for example, in distribution of traumas on skull and skeleton: all traumas but two are recorded in the male sample. In most cases, those are damaged skull bones or healed wounds of extremity bones, which suggest aggressive environments or lifestyle of the men. Quite commonly, part of the males could stay in the open air for a long enough time, under conditions of a high humidity and low temperatures (the latter is suggested by the cold stress marks). Noteworthy is that the marks are absent from the female group. Symptoms of arthrosis and secondary venous stasis are also found in male groups only, which confirms the gender differentiation in the Mesolithic.

Summing up the preceding, we would call a special attention to the following: the archaeological and anthropological data considered together clearly point to the fact that the Mesolithic population in the north of Eastern Europe was primarily hunters, and the objects of their hunt were terrestrial mammals used for food and other needs. The studied anthropological groups represent a wide time interval and existed under certain climatic conditions. However, they have no specific biological features permitting one to distinguish between them. It is not inconceivable that the regional population developed on a single anthropological substrate; that is reflected in the similarity of craniological characteristics, as well as in resemblance of demographic parameters.

The regional population is distinct for almost complete absence of chronic diseases of tooth and bone systems, including infectious ones. No cases of chronic anemia or scurvy have been recorded. A small part of the studied individuals display markers of a physiological stress indicative of acute febrile diseases in childhood. In the general context, we consider these data as resulting from negative environmental factors, confined to particular seasons. This explanation is in agreement with ethnographic data and permits us to come closer to a more or less unbiased assessment of facts.

By the Mesolithic the studied groups of the population in the north of Eastern Europe had already acquired various skills and mastered the practice of making weaponry and hunting weapons, as well as complicated bone ornaments worn on the clothes. Eloquent are various traumas and injuries on human skulls inflicted by weapons of all kinds.

A complex reconstruction of specific features of the society's evolution revealed the division of social activities according to gender; the burial traditions appeared to be diversified with regard not only to the social status of the person, but also to kinship. Of special interest is the human diet dynamics, which reflect the complicacy of the social adaptation of the population at that time. Further investigation of the joint burials could enlarge our knowledge on the Mesolithic culture.

As follows from the preceding, the general anthropological analysis of the Mesolithic population adaptation to the environments (with due consideration for archaeological data) can give diversified information needed for reconstructing the lifeway of hunter-gatherers.

Way to North: Anthropological Evidence of Adaptivity of the First Inhabitants in the High Latitudes

9

Margarita M. Gerasimova
Institute of Ethnology and Anthropology, Russian Academy of Sciences, Moscow, Russia

Chapter Outline

The chapter presents some results of the works performed as a part of the project aimed at the elucidation of certain aspects of the people's initial penetration into the high latitudes and their adaptation to the extremely harsh environments. The correct solution of those problems is vital for an adequate sustenance of the Extreme North aborigines, for the preservation of the linguistic and cultural heritage of the Arctic people under present conditions, when the modern civilization affects enormously the environments of the native population and their way of life. When dealing with those problems, the anthropological aspects and those of physical (biological) anthropology, in particular, are of primary importance.

The indigenous people of the Arctic and Subarctic have been adapting themselves to the extreme environmental factors for thousands of years. The last decades, however, from the mid-20th century till the present days, introduced new factors resulting in drastic changes of the lifestyle and a necessity for an adaptation to new heterogeneous environments. That is distinctly seen in the results of the long-term studies of the health status and disease incidence among the northern people.

In the opinion of the specialists (Mayer et al., 1970; Barbashova and Likhnitskaya, 1972; Barbashova, 1976), the acclimatization process in the North is attended with adverse reactions. An analysis of certain forms of pathologies revealed the disease incidence among the newcomers (mostly construction workers) to be four times as great as that among native reindeer herders, fishers, and hunters; the recorded disease frequency of native people may be considered, however, as a dysadaptation to the changing environments.

Human Colonization of the Arctic: The Interaction Between Early Migration and the Paleoenvironment
http://dx.doi.org/10.1016/B978-0-12-813532-7.00022-X

As suggested by recent works, the aboriginal population experiences a considerable stress, including a physiological one, mostly conditioned by social and cultural factors (Gudkova, 1995). An analysis of physiological indexes, such as Hb level, serum protein, general cholesterin, and glucose content, revealed the males aged 40–49 years to be most sensitive to the acculturation stress in the Arctic zone. Changes in the sustenance system, nature management, and their assuming of the "European" type of diet markedly affected metabolism of the Chukchee and the Eskimo people whose diet had been dominated by proteins for many centuries. As a result, the proteins gave place to carbohydrates and lipids.

The changes having been observed in some members of a population induce a stress in physiological homeostasis, which interferes with the adaptation to the extreme environments at the population level. The stress may be considered a deconditioning factor of the Arctic native population at present.

9.1 Biological adaptation and the Arctic adaptive type: a synthesis

Before analyzing the specific materials, it seems pertinent to consider briefly the present notions of biological adaptation of human populations to the existence in various ecological niches, and, in particular, in the extremely severe environments of the Arctic zone.

As to the notion of "adaptation," there are a great many conceptual interpretations of the term. A comprehensive overview of various opinions on the subject is given by L.K. Gudkova (2003), who finally comes to an opinion we cannot but agree with: that an integral definition of the term is practically impossible to give and that it is actually a customary terminological image. She assumes, however, that the adaptation notion is centered on the concept of an "adaptive norm," the latter being "a historically formed assortment of genotypes having an optimum range of phenotype variability that provides the maximum adaptiveness to certain environments" (Altukhov and Kurbatova, 1990). In my opinion, such a definition is most adequate and promising as applied to the anthropological aspect of the morphofunctional variability. In spite of varying nuances in its interpretation, the term "adaptation" has been comprehensively defined in works on general biology; it is still used as a synonym of "acclimatization" in some papers on anthropology and medicine. By both those terms are meant any environment-induced changes (traits) in human organisms that may return to the previous state after the former conditions have been restored; those traits are not hereditary. Herewith, in the author's interpretation, the adaptation is *a micro-evolutionary process of the development of inheritable features serving the purpose of adaptation to the environments*.

Sometimes archaeologists and paleogeographers use the term "adaptation" in a broader sense. It was developed from long-term studies of the man–environment interaction (coevolution) and from integrated studies of the principal early man sites against the background of the landscape and climate dynamics. Three types of adaptation have been suggested to distinguish in the context of the "man–social-economic structure–environments" system, namely, migratory, autochthonous, and diffuse (Velichko et al., 2008a,b).

In the author's opinion, the suggested adaptation types, though presenting schematically an interrelation between the named objects, still give neither quantitative

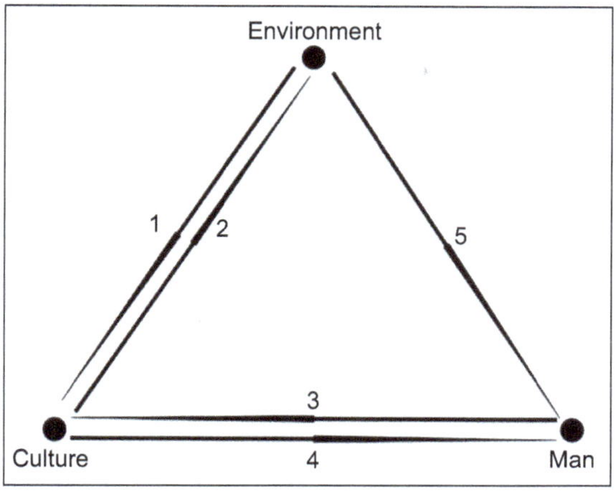

Figure 9.1 Interaction between the principal ecological constituents (Alekseyeva, 1998).

nor qualitative characteristics of the relations. As a physical anthropologist, I consider the scheme developed by V.P. Alekseyev (1993) to be more promising, as it takes into consideration the bilateral relationship between the culture and environments, as well as between the man and the culture. The impact of man (the latter being considered a biological entity) on nature is negligible as compared with man's influence on environments through culture. The action of all the environmental and cultural factors on human biology is unidirectional and has been quite considerable up to now (Fig. 9.1). The lack of space does not permit us to discuss in any detail the problems of human impact on culture. As to the channels of the cultural influence on human biology, the main ones are food, clothes, and dwelling.

A close relationship between the man and environment (both its biotic and abiotic constituents) provokes an intense adaptive process that had been exerting an influence on the development of the human population for many millennia. The human population dispersal, their entrance into new environments, overcoming the difficulties, and the adaptation to new, unfamiliar, and occasionally extremely harsh living conditions: all these promoted the genetic adaptivity and the growth of adaptive changeability potentially inherent in the entire species.

On the other hand, the biosphere zonality, the local (irregular) character in the distribution of chemical elements, the system organization of landscapes, specific features of the climate and weather, food potential, and a series of other factors are all the listed characteristics of the biosphere that result in humans often being confined to certain (limited) habitats (Alekseyev, 1993). That permits extending to humans the main "ecological rules," those formulated by C. Bergmann, J.A. Allen, C.L. Gloger, A. Thomson, and L.H. Buxton. It is pertinent to remind one that according to Bergmann's rule, there is a certain correlation between the temperatures and the body size: as Bergmann has initially formulated, larger animals have a lower surface area to volume ratio than smaller animals, so they radiate less body heat per unit of mass. As applied to humans, this ratio manifests itself in the body's length, weight, and surface. As Allen's rule states, the inhabitants of cold regions have shorter extremities and a heavier build than people living in warmer environments. The relationship is most distinct in the sitting height to full body length ratio, that is, the relative length of legs. Gloger's rule states that the intensity of pigmentation (skin, hair, eyes) in human beings depends on the latitude. The populations that evolved closer to the equator tend to be darker than those in middle latitudes. Finally, the rule established by A. Thomson and L.H. Buxton states that the nose width depends on the mean annual temperature, insolation, and the latitude of the region. All the cited rules are essentially statistical.

Studies performed on the present-day populations confined to various ecological niches made it possible to formulate a concept of adaptive types showing a subtle balance between the human populations and their environments (Alekseyeva, 1977, 1986, 1998). Five adaptive types have been identified and described: temperate, arctic, continental, equatorial, and alpine (high mountain) ones.

The arctic zone repeatedly attracted attention of anthropologists. It was mostly considered a test site for investigating adaptive processes typical of its population. The inhabitants of High Arctic, irrespectively of their ethnic background, are distinct for a specific morphofunctional complex. Though relatively short, their body is noted for a considerable weight, large chest, and buttock circumference, and well-developed bone and muscular tissue; also noteworthy is a high proportion of athletic constitution in the population. Besides, typical are a high basal metabolic rate and thermoregulation, and massive bones with a large medullary canal; the latter is favorable to hematosis under conditions of insufficient oxygen supply typical of the high latitudes (Alekseyeva et al., 2008). The morphofunctional complex includes also a high rate of bloodstream and a high temperature of fingers and toes at supercooling, which may be considered an adaptation to cold based on a genetically controlled reduced capacity for vessel narrowing. Such a morphofunctional complex of similar reactions fixed in inheritable traits and strictly confined to the high latitudes may be interpreted as an Arctic adaptive type developed in the process of adaptation to the environment and presenting the norm of reaction to it (Alekseyev, 1977, 1974a, 1979) (Fig. 9.2).

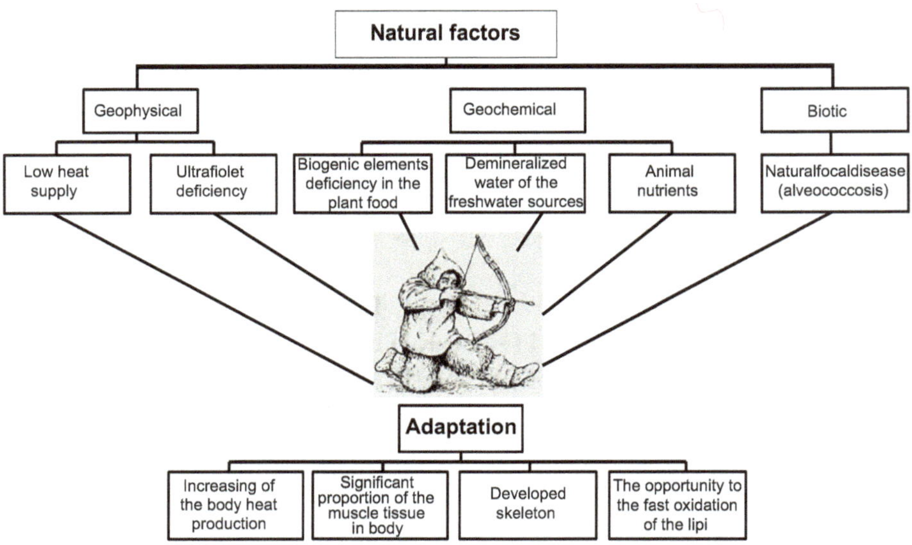

Figure 9.2 Action of natural factors on the human biology in the Arctic (Alekseyeva et al., 2008).

9.2 Sami, Nenets, Eskimo (Inuits): indigenous population of the North, representatives of the Arctic adaptive type

By way of example, we consider three nations living at present in the Extreme North and noted for a high degree of biological adaptation to the high latitude environments. Those are the Sami of the Kola Peninsula (known also as Lapps), forest Nenets, and the Eskimo of the Chukchi Peninsula; they have been extensively studied under a morpho-physiological program aimed at estimating the population adaptivity level (Alekseyeva et al., 1972, 1973, 2008; Chernilovskiy, 1983). Those people, though different in language families and anthropological types, as well as in their mode of economy, display, nevertheless, a considerable similarity in biological adaptation.

Sami, small people inhabiting northern Norway, Sweden, Finland, and the Kola Peninsula, form the indigenous population of the European North. The Sami are distinct from their North European neighbors in anthropological characteristics, originality of the language, and specific features of the material culture well adapted to the extreme conditions of the North. The Sami appearance in the European North may be dated to the 1st millennium AD. Judging from the written sources and toponymic data, at the 1st–2nd millennia boundary the Sami inhabited a large area, including the northern regions of Scandinavia, the Kola Peninsula, and the coasts of the Ladoga and Onega lakes. In the east, they occurred presumably in the lower reaches of Severnaya Dvina and Mezen rivers and in the Kanin tundra.

The modern Sami language belongs to the Fenno-Ugric family, where it occupies a rather isolated place. There is a substrate identified in it that may be traced to the Ugric-Samoyedic languages. Besides, the Sami language is divided into a number of dialects and local varieties, noticeably different in their phonetics, grammar, and dictionary, which essentially hinders mutual understanding. That may be a result of a small population density over a vast area and the contacts with neighboring peoples (Luk'yanchenko, 1971, 1980).

Variations in the natural conditions of the area inhabited by Sami (tundra and taiga zones and the Arctic Ocean coasts) seemingly contributed to the division of the Sami into two ethnic groups: forest and coastal (marine) Sami. The forest Sami were engaged in hunting reindeer and furred animals and fishing in lakes, as well as reindeer herding. The coastal Sami practiced fishing in the sea and hunting terrestrial and marine fauna (Luk'yanchenko, 1971, 1980) (Figs. 9.3 and 9.4).

The ancestral tribes of Sami formed presumably on the northern coasts of Norway, the extreme northwest of the Kola Peninsula, Eastern Finland, and Karelia. The first archaeological sites related to the Sami origin are attributed to the Komsa culture (Gurina, 1961; Luk'yanchenko, 1980). The sites are few in number and dated to the 9th to 7th millennia BC in Norway and to the 7th millennium BC on the Kola Peninsula. The lifestyle of the people was, most probably, semisedentary, the main occupation being sea mammal hunting. Supposedly, they came here migrating from Southern Scandinavia along the sea coast. By the middle of the 4th millennium, sites with specific ceramics of Sperrings type (of local origin) appeared in Karelia and a little later in Finland. It is with those sites that the specialists tend to associate some pre-Fenno-Ugric substrate in the language, some specific features of culture and

Figure 9.3 Sami: a hunter with a gun. The Kola Peninsula.
Photo from the archives of the Institute of Ethnology and Anthropology, Russian Academy of Sciences.

Figure 9.4 A Sami woman with kids and a cradle. The Kola Peninsula.
Photo from the archives of the Institute of Ethnology and Anthropology, Russian Academy of Sciences.

anthropological characteristics of the Sami people. The latter, however, is no more than a conjecture that has not been substantiated with data on the early penetration of the Sami ancestors into the high latitudes.

As to the anthropological data, the Sami show a specific assortment of features: they are short, with short legs, flatter face than that of their Europoid neighbors, with more prominent cheek-bones and more pronounced epicanthic fold, and scanty hair cover on the face; all those may be considered to be distinguishing features when differentiating between Europoid and Mongoloid populations. The Sami are typically noted for brachycrania, low and broad face, low concave nose bridge, and relatively dark skin color. The great majority of specialists identify the Laponoid type as a self-standing unit in the race systematics (Bunak, 1956, 1980; Gokhman et al., 1978; Khartanovich, 1978, 1980, 1991, 2004).

Of the morphophysiological characteristics, the most noticeable are these: large chest circumference indicative of heightened respiratory function; high content of hydroxyapatite (bone mineral) in the skeleton, that is, a high degree of mineralization that compensates at the population level for the minimum index of the skeleton massiveness typical of the Sami; and a high level of oxidizing processes (Alekseyeva et al., 1973). Among other morphofunctional characteristics of the modern Sami, the following should be mentioned: their high cold tolerance; an intensified blood supply of skin when exposed to cold; a greater proportion of fat-free body mass; and high content of total proteins and those of the globulin-rich fraction (Andersen et al., 1960; Alekseyeva et al., 1973).

The Sami origin is still debatable. To gain a better insight into this controversial problem, it is particularly important to get an adequate craniological material; the data thus obtained may be compared with those of paleoanthropology.

The Sami of Norway are represented in literature by several craniological series studied by K. Schreiner (1931, 1945). The series were recovered from burials in the southeast of Norway; they are dated at the 18th to early 19th centuries, except for one attributed to the 1st–2nd millennia boundary.

The Sami of Finland are known from the series studied by G.F. Debets and published by V.P. Alekseyev (1974b). They were excavated from burials in northern regions and also dated to the 18th–19th centuries. A comparison between the recent and medieval series showed that the latter is noted for a narrower braincase, a lower cephalic index, a wider face, and lower facial index. In the opinion of V.P. Alekseyev (1969, 1974b), the wide face has been typical for Sami since the ancient time, while the brachycrania is a recently acquired feature.

Five local samples recovered from old graveyards (the late 19th–20th centuries) of the Kola Peninsula have been minutely studied. The results proved a difference exists between the coastal groups and populations from the inner regions of the peninsula: morphological characteristics of the former suggest a late Europoid admixture, while groups from inner regions most fully preserved typical characteristics of the Sami anthropological complex: the skulls are very small, gracile, brachycranial, with large transversal diameters, and very low. The facial skeleton is also wide (as compared with European standard) and very low, and the entire face, nose bridge, and nasal bones are noticeably flattened. The listed features are most distinctly pronounced in the earliest series, at least 200 years old (Khartanovich, 1978, 1980, 1991, 2004).

The Sami display features similar to Mongoloid ones in almost all the systems of anthropological characteristics. That gave grounds to the miscegenation hypothesis advanced long ago by several specialists (Vitov et al., 1959; Debets, 1961, 1964; Alekseyev, 1969, 1974b) and still keeping its place, though somewhat modified. Its present-day supporters have rejected the facial skeleton profile as the main diagnostic character indicative of Asiatic impulse and a Mongoloid admixture; the recently obtained data proved a facial skeleton similar to Mongoloid in profile to occur in the regions where an appearance of Asian migrants is a priori hardly possible. The miscegenation hypothesis attributes the specific features of the Sami to mixing of Europoids with the representatives of the ancient formation known in the Russian anthropological literature as the Uralian race (Alekseyev and Gokhman, 1984a,b; Gokhman, 1986).

As for myself, I agree with V.I. Khartanovich in that the Sami preserved in their appearance some features of an ancient formation that differed from both racial trunks of today: European and Asian. It is confirmed by specific characteristics of their mitochondrial gene pool: the principal constituents of the latter indicate the Sami ancestral populations originate from Eastern Europe (Balanovskiy, 2008).

The Nenets belong to the peoples speaking Samoyedic languages. They are widely spread over West Siberia and the East European tundra almost as far west as the White Sea (Figs. 9.5 and 9.6). East of the Urals the Nenets live on the Yamal Peninsula, at the Ob' River mouth, along the Nadym River, and in the lower reaches of the Pur, Taz, and Yenisey rivers. A separate group is formed by "forest Nenets" living in the Pur region. The fact that Nenets are dispersed over a large territory and that there are distinguishable many ethnic groups, dialects, and subdialects suggests the people had been divided to a population level since the remote past.

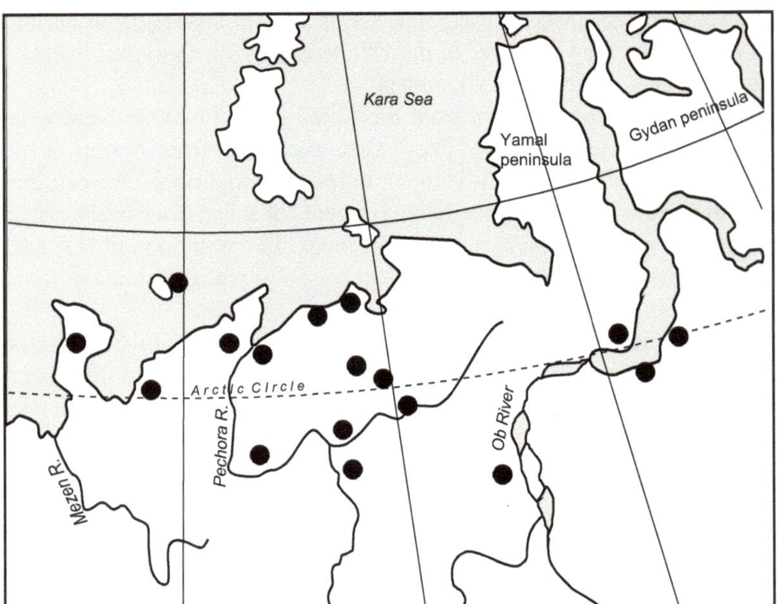

Figure 9.5 The studied populations of Nenets.

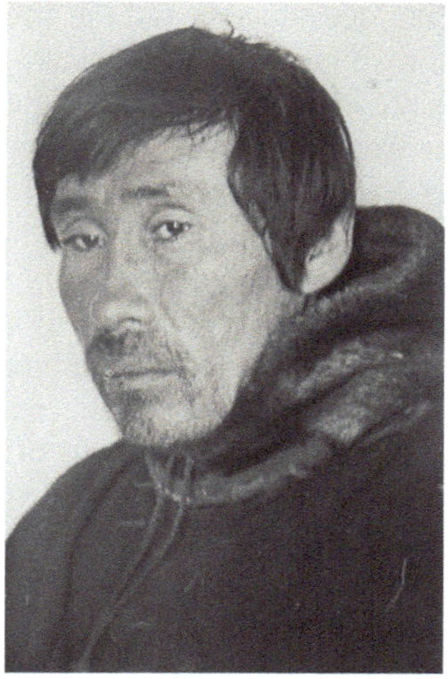

Figure 9.6 Kanin Nenets.
Photo from the personal archive of G.A. Aksyanova.

The ancient homeland of the northern Samoyeds is presumably the Altay-Sayan mountain steppes; that suggestion is supported by the fact that some elements of the Nenets culture (reindeer herding, in particular) are definitely traced to the Sayan region (Vasilevich and Levin, 1951; Vasil'ev, 1979, 1995). The fact that not so long ago the Samoyedic languages were widely spread among the Altay and Sayan peoples was thought to be another evidence of that. There is a reason to believe that the ancient Samoyeds had to leave their dwelling places in the south under the pressure of Huns and Turkic peoples; the northward migration occurred as a series of waves, each of them moving independently from others.

The process took about a millennium, and the ancestors of the present-day Nenets reached the tundra zone at the beginning of the 1st millennium (AD), that is later than many other tribes (Debets, 1941; Shluger, 1941; Zolotareva, 1974; Alekseyeva et al., 1972). It is noteworthy, however, that the anthropological characteristics of the modern population and rather scarce paleoanthropological materials might be interpreted otherwise: the Samoyedic peoples could penetrate to the Altay-Sayan uplands a little earlier than 2000 years ago. So that southern region could be in fact the periphery of the later settled area and not the area of the initial formation of the population (Alekseyev and Gokhman, 1984a,b).

Most of the authors agree in that the modern Samoyedic population of the Extreme North descended not only from the incoming tribes, but also from the aboriginal people. As the tundra was being colonized and the indigenous people were assimilated by the newcomers, the Nenets culture developed from two constituents. Deer herding and some relics of the open outer clothes of men may be considered as elements of the southern origin. The elements taken from the indigenous people include slipover clothes without clasps or fastening, and the technique of sea animal hunting (Khomich, 1966, 1976). Judging from the archaeological evidence, this ancient aboriginal substrate ascended to the culture of reindeer and sea mammal hunters, the ethnicity of the latter being questionable and the anthropological characteristics unknown (Chernetsov, 1973; Vasil'ev, 1979; Khlobystin, 1969, 1975).

Those aboriginal people are known in folklore literature under the name of "Siirtya" or "Sihirtya." It was from the ancient inhabitants that the Nenets' ancestors borrowed the ways of hunting sea animals and wild reindeer. They, in turn, introduced to the tundra people the reindeer herding skill. Since the middle of the 1st millennium the Nenets' ancestors had been increasing steadily the livestock, the aboriginal inhabitants having been crowded, destroyed, and partly assimilated (Vasil'ev, 1995; Golovnev et al., 1994; Golovnev, 1996). No paleoanthropological material has been preserved from the time of the ancient substrate existence. Among the early carriers of the Samoyedic, there were probably the inhabitants of the Irtysh drainage basin in the Omsk region; they are known from the Sargat culture artifacts dated to the late 1st millennium BC to the the 1st millennium AD.

That Samoyedic accounts for the northward expansion of the bearers of ethnic names that played an important part in the development of the Siberian tundra and forest Nenets. Some of the components of the same ethnic group took part in the origin of the European forest Nenets. The studies of some paleoanthropological data revealed a long-term infiltration of the taiga low-skull Mongoloid elements into the population bearing the Sargat culture (Bagashev, 2000).

It should be mentioned that the paleoanthropological data on the immediate predecessors of Nenets are extremely scarce, which may be attributed to environmental characteristics or to specific features of the burial rituals. However, the available craniological materials suggest a complicated process of the race formation in the region. There is a composite series collected from the vast territory of the Nenets settlement. It includes materials from Lower Ob' region as well as those from the European North and the lower reaches of the Yenisey. That series fits best of all to the general Samoyedic type (Dremov, 1984). Besides, there are two moderate-size series collected in the late 19th century on the Yamal Peninsula (lower Shchuchya River, Yar-Sale and Khamanel points) (Bagashev, 2003). Those series are distinct for subbrachycrania, small height of skull, high and wide flat face, and high and slightly protruding nose. On the basis of the craniological data the Nenets may be considered to be representatives of the Yenisey anthropological type of the North Asian Mongoloid race (Bagashev, 1998).

The forest Asiatic Nenets have been most comprehensively studied. The Siberian forest Nenets differ noticeably from other Nenets' groups (inhabiting Bolshezemelskaya and Malozemelskaya tundra, Taz River basin, Northern and Southern Yamal). Their specificity is expressed in stronger manifestation of Mongoloid traits, brachycephaly, high face, and relatively narrow nose together with relatively light-colored eyes. The latter was occasionally attributed to the Europoid admixture; that hardly may be taken as a plausible explanation, however, as the degree of the Mongoloid characteristic abatement does not correspond to the degree of depigmentation. It resulted most probably from spontaneous genetic processes taking place in isolated populations, as may be seen in separate groups in Scandinavia, high mountain regions of the Central Caucasus, and Western Pamir (Alekseyev, 1971).

The forest Nenets of Siberia have typically a small body with a large chest circumference, the thoracic cage being of cylindrical configuration and considerably exceeding a half of the body length; also typical is a gracile constitution, weakly developed muscles, and a low degree of fat deposition, though present are a high content of structural and immune proteins in the blood serum, a high rate of the bloodstream, and a high level of oxidative processes. Neither athletic nor asthenic types of the body build have been found among the Nenets. It follows from all the preceding that the forest Nenets form a group of a high morphofunctional status and are biologically adapted to the life in the Extreme North. The specialists studying the forest Nenets consider them to be well adapted to cold and, somewhat worse, to vitamin deficiency (Alekseyeva et al., 1972). Their lifestyle and economy are fully aimed at the adaptation to cold, while the modern Nenets' diet is vitamin deficient.

Regrettably, those investigations have been performed only for that group of Nenets population, though at present there are data on somatology, odontology, and craniology available from all the regions of the Nenets settlement (Aksyanova, 2003; Bagashev, 2003). All the data point to the originality of the people's physical type and its specific position in the racial system of both West Siberia and entire Northern Eurasia. A comparison between different populations shows their considerable diversity, which manifests itself in several local variants coexisting within a "certain anthropological unity" (Bagashev, 2003).

The European Nenets are noted for maximum originality (Fig. 9.7). The varieties inhabiting Yamal (Fig. 9.8) and the Asiatic forest Nenets are more homogeneous and closer to the averaged Samoyedic complex in the somatological characteristics. In their anthropological characteristics the Nenets have no direct analogs in the Finno-Ugric group and display well-pronounced West Siberian features as distinct from Mongoloids in Siberia; in their own region, however, they have definitely pronounced Mongoloid traits. Somatological data, as well as odontological, suggest the Nenets' kinship not with their closest neighbors (Enets, Nganasans), but with southern Samoyedic people (Selkups of Narym). This is somewhat inconsistent with close

Figure 9.7 Asian tundra (Yamal) Nenets. A young man. Photo from the personal archive of G.A. Aksyanova.

Figure 9.8 Asian tundra (Yamal) Nenets. A young woman.
Photo from the personal archive of G.A. Aksyanova.

racial and genetic relations between the Asian Nenets and the South Siberian populations inferred from the craniological characteristics.

There is no doubt as to the Nenets anthropological composition being complex and multicomponent. According to A.I. Bagashev (2003), the Nenets racial specifics developed as follows: the 1st layer was related to the ancient population from northeast Europe and areas east of the Urals, who were the carriers of the Uralian complex of anthropological characteristics; the second layer was formed by the carriers of the northern Paleoasiatic anthropological complex coming from the Lower Yenisey and the Northeast of West Siberia; and the 3rd—and the latest—corresponds to the carriers of the South Siberian anthropological complex typical for Southern Siberia and the Altay-Sayan region.

The Eskimo inhabited traditionally Greenland, the north of Canada, Alaska, and Chukchi Peninsula. In spite of living at present on different continents, the Eskimo and Aleut were formerly a single people known as Eskimo-Aleut, or Eskaleut. There are many theories about the Eskimo origin differing in the opinions on the time and place of that culture's development; most of the specialists agree, however, that the Eskimo culture has roots in Asia, and the area of the Arctic race formation was far south of its present range. The Eskimo culture is distinct from those of neighboring ethnic groups in its orientation toward marine hunting (as suggested by usage of kayak, toggling harpoon, oil lamp), but most of all in its mobility and in the ability of the people to adapt themselves to changing environments, to pass from mostly marine hunting to that on land and vice versa. Its versatility and adaptability in the process of colonizing vast areas in the Arctic is shown in many variations in hunting ways: they hunt marine animals on open water, from the ice edge, through ice holes, etc. It was due to their enhanced cultural adaptivity that when the whale catching became unproductive (in the 17th–18th centuries), part of the Eskimo people switched from it to caribou hunting and fishing in inner lakes and rivers (Fainberg, 1964, 1967, 1980) (Figs. 9.9–9.11).

Figure 9.9 Eskimo. A fishing scene (Fainberg, 1964).

Figure 9.10 An Eskimo hunter with a bow (Fainberg, 1964).

The Eskimo are attributed to Paleoasian people. The Eskimo and Aleuts speak Eskimo-Aleut languages, the subdivision of those into two seems to take place no later than 4000 yr BP. The further differentiation of the separated Eskimo resulted in the subdivision of the language into two branches—Yupik and Inupik—in the 1st millennium AD. In the

Figure 9.11 An Eskimo women with a child. Sireniki settlement, Chukotka National district. Photo from the archives of the Institute of Ethnology and Anthropology, Russian Academy of Sciences.

Figure 9.12 An Eskimo.
Photo from the archives of the Department of Anthropology, Moscow State University.

subsequent centuries the Yupik-speaking tribes lived at a considerable distance from each other, which accounts for a development of several Yupik dialects. The Inupik-speaking tribes inhabited Arctic America and remained closer to each other in language and culture. In the course of time, when the populations migrated northward and southward for a distance of thousands of kilometers over vast coastal territories, more than 20 Eskimo dialects developed. Noteworthy is that some dialects spoken in neighboring settlements differ no less than in those placed at a great distance. Examples are provided by the Eskimo dialects in Chaplino, Sireniki, and Naukan settlements.

Anthropologically, the Eskimo are typical representatives of the Arctic race (Figs. 9.12 and 9.13). They are noted for large face, a certain tendency toward prognathism,

Figure 9.13 An Eskimo woman.
Photo from the archives of the Department of Anthropology, Moscow State University.

a dark skin, very dark and coarse hair, and dark eyes; specific Mongoloid features in the upper eyelid are poorly pronounced, while the tertiary hair cover is rather better developed in comparison with continental Mongoloids. All those traits suggest East Asian, rather than Siberian, origin. As compared with the continental Mongoloids, the Eskimo on Chukotka display a less typical assortment of the Mongoloid traits: a slightly higher nose bridge, a less flattened face, and a smaller epicanthic fold. They have, however, some characteristics—straight, extremely coarse, dark hair, dark skin color, full lips, and a pronounced procheilon—in common with eastern and southern Mongoloids; all the three together forming a Pacific branch of the Mongoloid race (Cheboksarov, 1947).

That the Asian and American Eskimo belong to the same race is proved by the materials on their blood groups: distribution of the latter in various widely separated communities displays a considerable similarity. Craniological and somatological characteristics revealed some differences within the Eskimo physical type. At least two variants may be recognized: eastern (or Greenland) type noted for a pronounced dolichocrania and a very narrow nose and western (or Bering Sea) type, less dolichocranic and with a slightly wider nose. The Eskimo differ from other Mongoloid groups by their narrow noses, which is their distinctive feature.

As to the interrelations between the two variants—eastern (Greenland) and western (Bering Sea)—there is no consensus of opinions among the specialists. Some authors consider the specific features of the Greenland variant to be a result of environmental impact and a relative isolation. The less specialized Bering Sea variant is thought to be older (Debets, 1951). In the opinion of others, it is the dolichocephalic narrow-nosed type of Eastern Eskimo that should be considered the most ancient, persisting in Greenland (Levin, 1958; Stewart, 1959). A comparison of the skulls from ancient burials with recent series shows the dolichocranial type to be more widespread in the ancient times; that seemingly points to its greater age than that of the mesocephalic type (Levin, 1958, 1964; Pestryakov and Grigoryeva, 2003, 2011).

When estimating differences between separate local groups of Eskimo, one should take into account the influence of prolonged isolation on one hand and that of mixing on the other. In the case of mixing formerly isolated parts of the same population, quite possible are changes in the head form and a shift toward brachycephaly due to changes in the conditions of individual trait development (Bunak, 1951). It is among the western Eskimo groups that the mixing processes were most active. In the Russian literature the dominant point of view is that the Eskimo culture development and the formation of the Eskimo anthropological type were a result of long-term complicated processes on a vast territory adjacent to the Bering Sea.

Morphophysiological characteristics of Eskimo are taken as the standard of the adaptive Arctic type. Although their body is rather short, it is distinct for a considerable weight, a large circumference of the chest, waist, and buttocks, and well-developed bone and muscle constituent. The proportion of adults with an athletic constitution among Eskimo is up to 56.1% (in comparison with 32.1% recorded in Nenets populations). Typically, they have a large cylindrical thoracic cage, which ensures a high vital lung capacity facilitating the gas exchange processes under conditions of hypoxia.

Among morphological characteristics of the human body the most important are those of the skeleton, the latter being less susceptible to exogenic factor influence than the fat or muscle mass; besides, it is a carrier of valuable genetic information. A few skeletons of the Asian Eskimo were recovered from an abandoned cemetery dated to the early 20th century (the Naukan graveyard). The studied skeletons are distinguished for a relatively small size of long bones (bearing a well-developed relief, big epiphyses and a large diameter of diaphyses, relatively and absolutely thin walls of diaphyses, and a large medullary cavity).

The listed specific features are related, on one hand, to a necessarily greater density of the skeleton under conditions of a high physical activity, and on the other to enhanced blood formation needed to compensate for the oxygen deficiency in the North (Alekseyeva and Kovalenko, 1980).

The solidity of the Eskimo skeleton combined with a rather small length of individual elements is in a reasonable agreement with a general body build of the modern Arctic populations (Klevtsova and Smirnova, 1974; Klevtsova, 1976). There are interesting data on the skeleton massiveness in different and widely dispersed Eskimo groups (Fig. 9.14; Table 9.1). Using those data and having calculated the average index of massiveness of the long bones, we can see that the index tends to grow as the environments become less favorable. The greatest values of the index have been found in the groups from Naukan (the Chukchi Peninsula), northeast Greenland, and Point Barrow (Jorgensen, 1953, Alekseyeva and Kovalenko, 1980).

In common with most Arctic groups, the Eskimo are typically noted for a high basal metabolic rate, a considerable content of proteins and lipids in the blood serum controlling the energy balance in the organism, and a relatively high content of oxyapatite in bones. The latter may be considered an adaptation to insufficient insolation, particularly in the processes of growth and development (Alekseyeva et al., 2008).

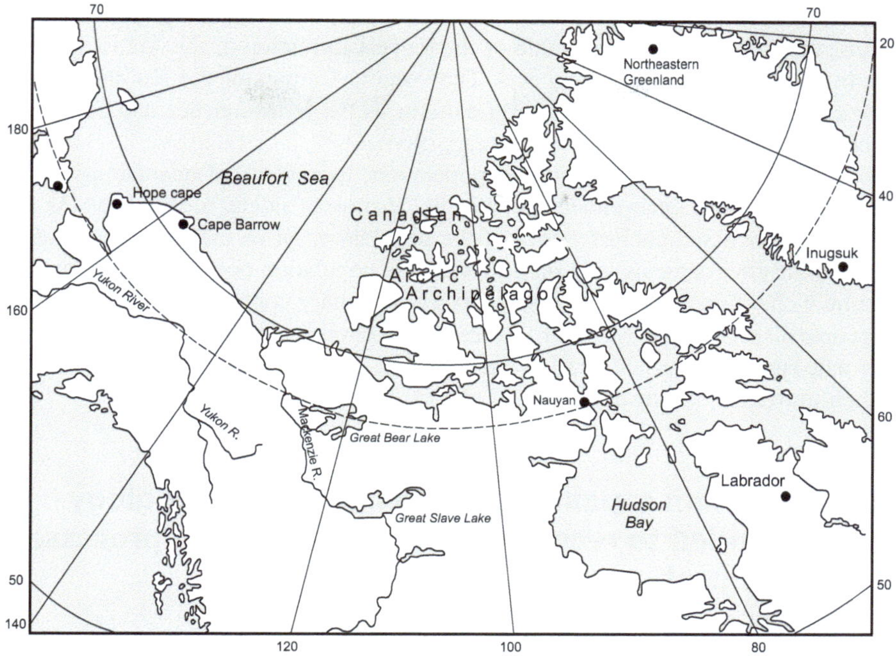

Figure 9.14 Position of the Eskimo burial grounds of the 17th–middle 20th centuries. After Alekseyeva, T.I., Kovalenko, V.Yu., 1980. Morphofunctional characteristics of the postcranial skeleton of Asian Eskimo. In: Okladnikov, A.P., Alekseyev, V.P. (Eds.), Paleoanthropology of Siberia. Nauka, Moscow, pp. 131–153 (in Russian).

Table 9.1 Bone massiveness index of the Eskimo skeletons, 17th–20th centuries

Series	Index	References
Inutsuk, southwest Greenland	1.17	Jorgensen (1953)
Labrador, Canada	2.33	Jorgensen (1953)
Point Hope, Alaska	3.50	Jorgensen (1953)
Nauyan, Canada	4.33	Jorgensen (1953)
Point Barrow, Old Igloo, Alaska	5.00	Jorgensen (1953)
Northeastern Greenland.	5.50	Jorgensen (1953)
Cape Dezhnev, Naukan, Chukotka	5.67	Alekseyeva and Kovalenko (1980)

The preceding materials show specific features in the adaptive processes of the Sami, Nenets, and Eskimo. The morphofunctional status of the Sami and that of Nenets have a great many features in common. That suggests the same adaptive morphofunctional complexes to form on a different genetic basis in similar, though not identical, geographic (environmental) conditions; the Nenets, however, came into the

Arctic relatively later than the Sami (Alekseyeva et al., 1972). In comparison with the Nenets, the Eskimo give an example of the longest formation of the Arctic adaptive type in the most severe environments. The specific features of the Eskimo anthropological type are traceable in as early cultures as the Independence and Saqqaq in Greenland and Dorset in Canada (Fainberg, 1964).

We can see, therefore, that in any population, its reserve of adaptability when meeting unfavorable environments, that is, its "margin of safety," depends on the initial morphofunctional characteristics of the population, on its experience in dealing with various environments, and on how long the population occupies that particular extreme niche. The adaptability reserve exerts a primary control over migrations, or the people dispersal, and over their final results. From that viewpoint, we'll try to consider paleoanthropological materials showing the adaptation processes in populations on their movement northward.

9.3 Resolution capabilities of the paleoanthropology as applied to reconstruction of adaptation processes in the past

Here, we try to discuss in more detail complexes of morphological variations developed in the course of humans' adaptation to environments and forming a certain adaptive type by way of selection. It should be noted that the racial variation is another adaptive component; in common with adaptive types, it shows a certain zonation, which obeys to some extent the ecological rules formulated by C. Bergmann, J.A. Allen, A. Thomson, and L.H. Buxton. From that point of view, it is particularly interesting to consider, apart from craniological and odontological data, the paleoanthropological materials on the human skeleton giving ideas on the body height and constitution, proportions, and general habitus of ancient people. The materials include morphometric characteristics of skeletal bones, proportions, inner structure of the long bones, and their histology, that is, characteristics that preserve "memory" of their origin (Fedosova, 1989, 1991).

We have only limited possibilities of studying the human population adaptability and factors controlling it on the paleoanthropological materials. The materials available do not always permit identifying significant biological characteristics of the adaptive type in the ancient population (because of incomplete finds, poorly preserved bones, a necessity of too expensive procedures, etc.). The result is that the principal method of the adaptation process investigation became studies of the paleodemographic characteristics of prehistoric populations and an analysis of the biological stress markers on the skull, skeleton, and teeth.

Widely used in paleoenvironmental reconstructions is the physiological stress model suggested by American specialists (Goudman et al., 1984, 1989). Studies along this line were performed by the group of physical anthropologists of the Institute of Archaeology RAS (Buzhilova et al., 1998; Alekseyev and Fedosova, 1992). Some skeletal pathology, such as Harris lines, stenosis of the spinal canal, hypoplasia of tooth enamel, porotic hyperostosis, osteoporosis, etc., are considered to be indicators of the physiological stress and form the basis for modeling its causes and results.

Another interesting and promising source of information on the indigenous people's lifestyle is an analysis of physical load on the skeletal bones or on teeth and jaws (also known as "physical activity analysis"). It includes recording indicators of the bone relief development, specific features of articular surfaces in large joints and the tooth chewing surface, a registration of physical injuries, and pathologies related to mechanical loads. Of special interest is the reconstruction of typical movements often performed by the man and indicative of a certain kind of activities (such as "horse riding" or "usage of teeth as a third hand"). The indexes are by no means to be considered adaptive; they are, in fact, phenotypic changes, not controlled genetically and not inheritable.

In my opinion (which I share with T.A. Alekseyeva (1998)), those studies are mostly centered on the acclimatization process. Environmental stresses, and in particular those due to low temperatures and food deficiency, as well as chronic diseases and pathologies not predetermined genetically may be considered evidence of environments unsuitable for a man or population, which may result from acclimatization usually taking place shortly after the group's migration into a new territory or some drastic changes in living conditions.

Data on people's injuries, on morphological characteristics resulting from customary movements (known as "rider's complex," "squatting," "using teeth as the third hand," etc.), and physical activity indexes contribute essentially to archaeological reconstructions of the living environments of individuals or human populations; they cannot, however, be used as characteristics of biological adaptation contributing to the homeostasis preservation. The data on the acclimatization are indisputably important as they provide a useful insight into the initial stages of the humans' adaptation. It should be remembered, however, that the phenotypical variations that appeared in the process of acclimatization are hardly comparable in importance with adaptive changes developed in the long-term adaptation of the populations to environments of various ecological niches.

9.4 Specific paleoanthropological materials on the humans' dispersal over the high latitudes

The population of the polar regions of Eurasia developed as a result of a complicated and prolonged process. Formation of ethnic groups inhabiting the Extreme North, with their specific culture and physical characteristics, proceeded in the course of long-term infiltration and northward shifts of various groups that originated in the south. In the course of time the displaced groups acquired specific features ("divergent evolution") related to the necessity to adapt themselves to new environments by developing new modes of life. Whereas archaeological data on the initial penetration of humans into the high latitudes are numerous and conclusive enough, the anthropological materials are extremely scarce (Table 9.2).

When analyzing the Paleolithic colonization of the East European North, the *site of Sungir* is usually taken as the starting point. Until recently, it was the northernmost locality of the Upper Paleolithic man bones found in Eastern Europe. Lately, however, there has been published information on the finding of a mandible belonging to a

Table 9.2 Paleoanthropological finds demonstrating the initial penetration of humans into the high latitudes

Region	Late Paleolithic	Mesolithic	Neolithic	Historical time	Present days
Northeast of Eastern Europe and Fennoscandia	–	South Oleni Island, Popovo, Peschanitsa, Zveynieki, Minino		Lävanluhta	Sami
East European Plain	Sungir	Mayak (Sedelkino)			
Northeast of Europe	Krutaya Gora		Bolshoy Oleni		Nenets
West Siberia				Sargat culture	
East Siberia (Yakutia)			Diring-Yuryakh, Bugachan, Tuoy-Khaya, Rodinka II		
Northeast of Asia, Beringia			Ust'-Belskiy, Kamenka II	Uelen, Ekven, Ippiutak, Tigara	Eskimo

child about 10 years old. It was discovered in the Upper Paleolithic cultural layer of the *Krutaya Gora* site on the Pechora River (Timofeyev et al., 2005). A short and not very informative preliminary description suggests the form attributable to the modern man. The enamel hypoplasia discernible on the permanent teeth—incisor and canine crowns—indicated, in the authors' opinion, periods of acute food deficiency. On the lingual side the teeth are noted for side crest development on the crown indicative of the so-called shovel-shaped incisors. A conjecture made tentatively by the authors about the child belonging to the Eastern odontological stem is beneath criticism, as the diagnostic property is the shovel-like shape of the upper, and not lower, incisors.

The skeleton materials from Sungir were thoroughly studied by a number of programs, and the results were published, so we need not describe them here (Alekseyev, 1984a; Alekseyeva and Bader, 2000). They belong to several humans, male and female, different in age; they have been described by several authors under the same numbers (indexes), namely, Sungir 1 – a skeleton of an elderly man (grave 1); Sungir 2 – boy's skeleton; Sungir 3 – girl's skeleton from the twin burial (grave 2); Sungir 4 – femur without epiphysis in the twin tomb of children; Sungir 5 – female skull; and Sungir 6 – occasional find of mandible.

Figure 9.15 Sculptural portrait of a man reconstructed from the Sungir' 1 skull by M.M. Gerasimov.

In the context of our discussion, it is most interesting that the Sungir materials show an extreme degree of the craniological diversity, its range being of the same order as the variability of all the Upper Paleolithic human fossils in Europe. The racial traits of the skulls, as they are understood today, are not very distinct: in the male skull the facial bones and nose are somewhat flattened; one of the children's skulls—the boy's—is noted for a narrow nose opening and a pronounced prognathism; two other skulls—those of a little girl and of a woman—also belong to two different craniotypes. A skull similar to Sungir 1 (except that the latter is distinct for a very large face) may be found in any European population. As has been stated by G.F. Debets, it is a representative of the Cro-Magnon type in a broad sense of the term that is applicable to any of the Upper Paleolithic men of Europe.

The combination of all the characteristics present on the Sungir 1 skull is exceedingly rare in the present-day racial variants. As to fossil skulls, the nearest analogs are found, on one hand, with the Predmosti III skull and, on the other with Choukoutien 101 skull (Bunak, 1973; Bunak and Gerasimova, 1984; Gerasimova, 2000) (Fig. 9.15). The Sungir 5 skull differs from that of Sungir 1 in a noticeable dolichocrania and a broad, low face. A disharmony in facial and cranial indexes is no less pronounced than in the Cro-Magnon type described by French anthropologists. There is a distinct similarity between this female skull and that from the Kostenki 2 site (Gerasimova, 1984). Skeletons and skulls of children from twin burial 2 are morphologically different, though the genetic analysis seemingly indicates their kinship: "The coincidence of nucleotide sequences in Sungir 2 and Sungir 3 suggests their maternal kinship (in theory, they may be children of one mother). The degree of possibility cannot be estimated without information on the identified metatype frequency in the Sungir paleopopulation" (Poltaraus et al., 2000, p. 354) (Figs. 9.16 and 9.17).

Figure 9.16 Sculptural portrait of a girl reconstructed from the Sungir' 3 skull recovered from the twin burial by G.V. Lebedinskaya.

Figure 9.17 Sculptural portrait of a boy reconstructed from the Sungir' 2 skull from the twin burial by T.S. Surnina.

Therefore, if we consider all the buried on the Sungir site to be members of a single population, we will have to acknowledge that its anthropological composition is exceedingly diversified. There is a considerable difference between the age of the man's burial and that of children, and the radiocarbon dates on bone material and charcoal from the cultural layer vary from 28 to 25 ka BP; it is hardly possible that the settlement and the burial ground were functioning for such a long time interval. In all probability, the locality was a suitable place for killing reindeer at the river crossing during seasonal migrations of the herds; various populations of Paleolithic people could visit it repeatedly over 2 or 3 millennia.

A similar situation has been reconstructed at the Kostenki region on the Don River. A high concentration of the Upper Paleolithic sites on a rather limited area seemingly indicates a kind of uniformity in the lifestyle of various populations, so they usually chose dwelling places near the river, on the floodplain, or low terrace near scarps of higher terraces or interfluves where they hunted mammoths (Velichko and Zelikson, 2006). Probably, a most typical social unity was a small family group; now and then, such groups were getting together for common hunting or feasts. A prolonged contact between the groups necessary for the anthropological type consolidating (at a small rate of the population growth) took place relatively seldom.

The Sungir 1 postcranial skeleton was meticulously studied by E.N. Khrisanfova; the completeness of the skeleton permitted specifying the paleoanthropological characteristics of the fossil hominid constitution and reconstructing their habitus (Khrisanfova, 1979, 1980, 1984, 2000). The author gave a reasonable evidence of osteological polymorphism in the postcranial skeleton of the modern man predecessors manifested in variations of the body size and proportions. She also suggested that the polymorphism is essentially an adaptive reaction of the early man populations.

Typically the classical Neanderthal has a short body, big shoulders, brachymorphic upper body, and on the whole, a heavyset, massive constitution. The fossil humans found in Western Asia are morphologically different from the aforementioned: they are noted for a greater height, dolichomorphism, narrower shoulders, and longer legs, showing therefore a tendency toward more elongated, gracile constitution. The typological variations of habitus found in late Neanderthals (classical European and West Asian ones) are interpreted in the context of morphological adaptation to various living conditions and considered to be adaptive.

Many characteristics of the Sungir man morphotype are similar to those of the recent Arctic populations and to a certain extent to the Neanderthals (Table 9.3). Among them should be noted an exceedingly solid body build, a definite brachymorphia of the upper body, well-developed mesomorphic constituent, and a very massive skeleton. The body weight to surface area ratio corresponds to a group maximum of the modern humans and close to that value of the classical Neanderthals (Khrisanfova, 2000). On the other hand, that author indicated the presence of specific, archaic characteristics of the postcranial skeleton that make the Sungir man closer to the "sapientoid Mousterians" in Eastern Europe (Romankovo, Samara, Skurlat) and "proto-Cromagnons" of Western Asia (the Skhoul group). Among those characteristics are a considerable height and absolutely and relatively elongated lower arm (forearm) and tibia, with the trunk relatively shortened; all those features are present in the original population and are most

Table 9.3 Characteristics of the "Arctic" adaptive type of the Middle and Upper Paleolithic humans

Characteristics	Neanderthal	Sungir 1
CISV	7.85	9.44
W	70?	80.0
L	161.2	179.5
W/L^3	1.67	1.40
W/S	41.1	40.2

Abbreviations used in Tables 9.3 and 9.4: *CISV*, conventional index of skeleton volume; *CLL*, conventional leg length, cm; *L*, body length, cm; *S*, body surface; *W*, weight, kg; W/L^3, height-weight index (Rohrer index); *W/S*, weight to body surface ratio.
After Khrisanfova, E.N., 1979. Paleoanthropological aspect of the constitution. Voprosy Antropologii 62, 3–13 (in Russian); Khrisanfova, E.N., 1984. Post-cranial skeleton of an adult man Sungir 1. Femoral bone Sungir 4. In: Alekseyev, V.P. (Ed.), Sungir. Anthropological Analysis. Nauka, Moscow, pp. 100–140 (in Russian).

typical to the groups of southern origin. The materials recovered from the Sungir site lead to the conclusion that the modern man expansion northward set going the process of biological adaptation to the harsh environments of periglacial steppes.

It might be suggested that the complex of characteristics adaptive to low temperatures and oxygen deficiency typical to ice ages began to develop at the time of classical European Neanderthals. It turns out, however, that the morphofunctional adaptation of *Homo* could proceed in two ways (Khrisanfova, 2000; Dobrovolskaya et al., 2008). Those different ways are reflected in the inner massiveness of long bones of the postcranial skeleton. The adaptivity of the modern man (as has been shown by studies of the present-day indigenous people of the high latitudes) to hypoxia and low temperatures consists usually in increased bone marrow volume in long bones and, therefore, in intensified blood formation. The late European Neanderthals had typically thick-walled, tubular (long) bones, a small bone marrow volume, and, therefore, a reduced blood formation function. Having colonized various habitats (which enhanced their adaptability), the Neanderthals did not penetrate nevertheless into the high latitudes, as no sites attributable to them are known north of 55°N (Svendsen et al., 2008). The Neanderthals' confinement to warmer biotopes suggests that the modern man was more adaptive and, unlike the Neanderthals, could live in periglacial steppe (Dolukhanov, 2008).

As archaeological materials from the Mamontova Kurya show, the Upper Paleolithic *Homo sapiens* first penetrated into East European North about 34–37 ka BP and then repeatedly visited it; he developed a migratory type of economic and cultural adaptation (according to A.A. Velichko). The archaeological data indicate that humans did not leave those lands even at the time of the maximum Late Pleistocene cooling (LGM). The success of humans in colonizing the periglacial steppe was ensured, first of all, by the genetic adaptation of the organism to the extremely severe environments.

A dramatic example of the lack of such an adaptive complex is provided by a unique burial of a young man discovered in the Kostenki 14; the body was laid in a tomb below the floor of the cultural layer III (Rogachev, 1955; Rogachev and

Table 9.4 Characteristics of the body of the Upper Paleolithic humans on the East European Plain

Finds	CLL	CISV	W	L	W/L³	S	W/S
Sungir 1	50.71	9.44	80.0	179.5	1.40	1.99	40.2
Kostenki 14	47.98	5.88	54.3	160.8	1.32	1.56	34.80

For explanation see Table 9.3.
After Debets, G.F., 1955. Paleoanthropolical finds in Kostenki sites. Sovetskaya Etnografiya 1, 39–46 (in Russian).

Sinitsyn, 1982), the industry of the latter has not been yet attributed. The found remains are known as the "Negroid from Markina Gora" studied by G.F. Debets (1955). The burial is dated to the early Upper Paleolithic. That may account for the fact that the author who first described it stressed the absence of Neanderthaloid traits on the skull, though made notes on the presence of some archaic features in the extremities.

There was no difficulty in the species determination: it was certainly *H. sapiens fossilis*. The skull, however, presented an unusual combination of traits that permitted defining it as "Negroid" or "Papuan-like." In all the indexes describing the linear proportions the skeleton from Kostenki 14 stays within the range of variations typical of the European man. Calculated indexes (conventional index of skeleton volume, weight, height, Rohrer height-weight index, body surface, weight to body surface ratio) permit one to characterize the constitutional habitus of the Kostenki 14 man as noted for small weight, low height, gracility, and low density of the body (Table 9.4).

Taking the data on the habitus of the man buried in the Kostenki 14 site and extrapolating them to the population level, it may be supposed that they are indicative of the environments more hospitable that those of the Sungir man (Gerasimova, 1982, 2010). The recent investigations, however, discovered an artifact-bearing strata below cultural layer III; a series of ^{14}C dates obtained on layer III fall within 30–32 ka BP interval. The closeness between ages of the Sungir and Markina Gora human remains does not give grounds to consider their living conditions to be diametrically opposite. Quite possibly, the human remains discovered in the Markina Gora locality give evidence of an occasional early penetration onto the East European Plain of an individual representative of a population accustomed to a much milder climate (even the interstadial conditions appeared too severe for him); in all probability, it was an occasional episode of far-distance migrations.

So, it may be supposed on the basis of the archaeological data that the Upper Paleolithic humans migrated from southern regions of Europe onto the East European Plain and followed along the valleys of large rivers to the central regions of the plain and farther north. One wave of migrants separated from the main stream, went far eastward, crossed the steppe zone, and settled in the mountains and river valleys in the Central Siberia and the Baikal region (Dolukhanov, 2008). The presence of the European impulse in the Upper Paleolithic settlement of those territories is suggested by the western odontological complex of infants from Malta site (Zubov, 2004). Unfortunately, we have practically no paleoanthropological materials to illustrate that migration; occasional finds of teeth and mandible bone fragments in Novoselovo, Listvenka, and Solovyinaya Luka are inadequate from that point of view (Gerasimova et al., 2007).

A kind of milestone marking the Pleistocene–Holocene boundary is a recent find of three burials on the Bolshoy Cheremshan River, east of the Volga River. The burials (Sedelkino, Mayak Mt.), partly destroyed by quarrying, yielded some paleoanthropological materials.

The human bones have been ^{14}C dated at 10,030 ± 50 yr BP, GIN-11,528 (uncal.), or 11 550 cal. yr BP (Kuznetsova and Ponomarenko, 2003; Stashenkov, 2003). The skeletons do not show any signs of the Arctic adaptive type; they are rather dolichomorphic, with narrow shoulders (Borutskaya and Vasil'ev, 2003). The time of those people's life corresponds to the Allerød (11.8–11 ka BP) marked by a considerable warming and a wide distribution of birch and pine forests in the temperate zone of Eastern Europe (Dolukhanov, 2008). The preserved skull of an adult man has an assortment of specific traits that, though slightly modified, have been typical of certain populations in steppes and forest-steppes of the Volga-Urals region throughout the subsequent historical epochs. The characteristic traits include dolichocrania, broad and low face flattened in its upper part, a narrow forehead, nasal bone concave in profile, and moderate prognathism (Khokhlov and Yablonskiy, 2003). No analogs of that assortment of characteristics have been found either among the known craniological complexes of the Western European populations attributed to the Upper Paleolithic, or on the East European Plain in the Upper Paleolithic and Mesolithic populations.

It is quite possible that the specific features of the skull recovered from the Mayak Mt. (Gora Mayak, Sedelkino) burial are indicative of the origin of the ancient formation considered by V.V. Bunak as belonging to the very beginning of the Uralian race. The skull morphology argues against the population migrating from east to west "following the retreating ice sheet" and taking part in formation of the Mesolithic population in the Northwest of Eastern Europe. The earliest known Mesolithic finds in the Baltic region and in the Russian North are quite different in the anthropological appearance. Besides, that does not agree with data on the changing environments since 12 ka BP, when the degrading permafrost destabilized the surface and prevented the man migrations westward (Gribchenko and Kurenkova, 2007). Most probably, the man went to the north; such a hypothesis is in better agreement with paleogeographic information and with modern ideas about the dispersal of reindeer hunters (Khlobystin, 1969, 1975).

The Mesolithic era is often considered an intermediate stage between the Paleolithic and Neolithic (Zubov, 2004). But it is quite clear at present that during the Mesolithic the human race lived through a serious crisis. A gradual amelioration of the climate and environments took place along with the Scandinavian ice sheet retreating, shrinking, and decaying. Those global changes resulted in a northward shift of all the natural zones in Europe, of tundra and forest-tundra in particular; some new landscape zones came into being, and a redistribution of floristic and faunistic assemblages took place. At the beginning of the Holocene, man had to face the fact of disruption of food chains and changing relationships between species; at the same time, some species of megafauna (wooly rhinoceros, mammoth) became extinct, along with some large herd animals that served as the main source of food for the Upper Paleolithic man.

The disintegration of a single hyperzone and development of various communities of plants and animals, more or less similar to those of today, resulted in the destruction of the Upper Paleolithic ecosocial subsystems adapted to periglacial landscapes and keeping their stability under those conditions (Khotinskiy, 1977). The search for new ways of food procuring and the occupation of new ecological niches resulted in numerous local movements and displacements of the population. With increasing human dispersal over the north, northwest, and northeast of Europe, the population density would necessarily lessen. The Mesolithic population movement in pursuit of migrating reindeer herds left its imprint on the cultural evolution; it is to be expected that human biology would be also influenced. The reduced population density and the isolation of some mobile groups contributed to the anthropological differentiation.

Along with the isolation, a few other processes took place in the course of the Mesolithic people dispersal and his physical characteristic evolution; those processes included a disintegration of isolates; an extinction of some separate groups (and therefore elimination of some parts of the gene pool); fluctuations of local population size; a genetic drift (spontaneous genetic processes); and finally, characteristics of the initial group gene pool and a degree of its ecological adaptability.

This important stage in the high latitude colonization is best presented in the anthropological materials from the Mesolithic sites in the Northwest of Eastern Europe. The oldest site is Peschanitsa (9890 ± 120 yr BP, GIN-4558/11,410 ± 210 cal. yr BP/); the next is Popovo site as indicated by several dates (9730 ± 110 yr BP, GIN-4556/11,070 ± 180 cal. yr BP/; 9352 ± 130 yr BP, GIN-4442/10,600 ± 210 cal. yr BP/; 9430 ± 150 yr BP, GIN-4447/10,730 ± 230 cal. yr BP/). The Mesolithic burials from the Zvejnieki cemetery are dated to the end of 7th to the first half of the 5th millennium BC. The earliest date is 8240 ± 70, Ua-3634 (Eriksson et al., 2003; Zagorska, 2006). The dates on burials of the Yuzhny Oleni Ostrov vary between 7470 ± 240 and 6790 ± 80 yr BP (Mamonova and Sulerzhitskiy, 1989). The burial complex Minino at Kubenskoye Lake was formed for more than 3000 years. The earliest dates for that complex are 9320 ± 55 yr BP (Minino I) and 9385 ± 40 yr BP (Minino II) (Suvorov, 1998; Suvorov and Buzhilova, 2004). The problem of the northwest of Eastern Europe and Eastern Baltic regions was discussed by Russian anthropologists more than once (Zhirov, 1940; Mark, 1956, 1975; Yakimov, 1956, 1960, 1961; Debets, 1961; Denisova, 1975; Alekseyev, 1984b; Gokhman, 1966, 1984, 1986; Mamonova, 1995; Gerasimova and Pezhemsky, 2005; Gerasimova, 2008).

When considering the colonization of the European North, the central place traditionally belongs to the materials from the famous *Oleni Ostrov graveyard*, South Oleni island in Lake Onega. The anthropological materials recovered from it were studied more than half a century ago; however, the morphology of skulls from that burial is still questionable in their racial attribution (Gokhman, 1984; Benevolenskaya, 1984) (Fig. 9.18). All the authors who studied the series identified, besides the skulls with high face and distinct profile, another type with low and somewhat flattened face (Zhirov, 1940; Yakimov, 1960; Denisova, 1975; Gokhman, 1984; Benevolenskaya, 1984). The second variant was, however, interpreted in different ways by different authors. Some of them discerned in the skulls' morphology some traces of miscegenation with ancient Mongoloids, coming, in all probability, from beyond the Urals (Gerasimov, 1955; Debets, 1948; Denisova, 1975, 1998) (Fig. 9.19). Others, though

Figure 9.18 Sculptural portrait of a man reconstructed from the skull found on Yuzhny Oleni Island graveyard by M.M. Gerasimov.

Figure 9.19 Sculptural portrait of a man reconstructed from the skull found on Yuzhny Oleni Island graveyard by M.M. Gerasimov.

agreed on the importance of the characteristics and considered that the flattened face developed independently from the Mongoloid admixture (Yakimov, 1953; Gokhman, 1984, 1986; Alekseyev, 1984b).

The studies of the skeletons recovered from the Oleni Ostrov burial led the specialists to the conclusion that the people had no distinct adaptive characteristics typical of the forest zone inhabitants. At the same time, specific features of the bone inner structure characteristic of the studied specimens, namely their relatively thin walls of long bones and large volume of medullary cavity, suggest an adaptation to low-temperature environments; in the opinion of Mednikova (2006, 2009), that makes them similar to the Sungir man and to Khanty and Eskimo of today. Unfortunately, corticomedullary indexes, though thoroughly investigated in various groups of modern Europeans (where the index value varies within 46.1–51.2), are insufficiently studied in other groups representative of various ecological niches. For example, it does not show much difference between ecologically different populations, such as Khanty and Eskimo. Mednikova (2006) believes, however, that certain characteristics of the postcranial skeleton, some linear proportions, and narrow shoulders suggest certain neutrality in their morphofunctional status; there are not features indicative of a special adaptation to cold stress. Longer forearm and tibia typical of the Oleni Ostrov inhabitants suggest the local population kept some traces of the southern origin in their morphology.

Somewhat younger paleoanthropological series are known from the eastern Baltic region (Latvia), *Zvejnieki* burial (Zagorska and Zagorskis, 1977; Zagorska, 2006). Mesolithic burials from that cemetery are dated to the end of the 7th to the first half of the 5th millennia BC. The series had been studied by R.Ya. Denisova (1975, 1998), who distinguished two variants of skulls. One of them is noted for a pronounced dolichocrania, a distinct face profile, and a narrow, protruding nose; that type is traceable far beyond the Baltic region, from Brittany in the west to the Dnieper River in the east. The second variant, with somewhat flattened upper part of the face and less protruding nose, is considered by that author to be an indication of the Baltic region settlement proceeding from the east.

There is, however, a wealth of data obtained by now about the described complex of characteristics present in various local groups of Mesolithic and Neolithic people not related genetically to Mongoloids (Gokhman, 1986; Alekseyev, 1979, 1984b); such groups were found on the Balkan Peninsula, in the north of Africa, and in the middle reaches of the Dnieper. So we can dismiss the hypothesis of Mongoloid admixture as insufficiently substantiated and look for other sources of that population. As to morphological characteristics of the people who left the Zvejnieki and Oleni Ostrov burials, they cannot give an unambiguous answer to the question about the genetic relations of the early inhabitants of the northeast of Eastern Europe. Some features of the Kunda culture seemingly traced in those two burials are indicative of the post-Swiderian people migrations. Unfortunately, we have not material enough to characterize that people. We do have, however, some materials dated to an earlier time than those from South Oleni island and Zvejnieki: the findings from Popovo and Peschanitsa sites of the Veretye culture belonging to the same group (Kol'tsov, 2003; Oshibkina, 1982, 2006; Shumkin, 1991, 1993, 2007). In spite of their paucity, the paleoanthropological materials of the Veretye culture are important in solving problems of the northern East Europe colonization and getting a better insight into anthropological composition of the population of Northern and Eastern Europe.

Though displaying clearly archaic features (an assortment of traits typical of *H. sapiens fossilis*), the skulls from *Popovo* (Gokhman, 1984) and the skull from *Peschanitsa* (Mamonova, 1995; Gerasimova and Pezhemsky, 2005) bear distinct Europoid characteristics. The most specific feature of them is a very high braincase, its height exceeding the skull width. Both variants identified in the Popovo, Oleni Ostrov, and Zvejnieki are traceable in a few finds in Western Europe, while skulls bearing features typical of those in Peschanitsa are found in burials in southern Sweden (Gerasimova and Pezhemsky, 2005). That enables us to conclude that the population arrived there in the Mesolithic from the west, and not from the east, and even groups east of Onega Lake showed the western assortment of characteristics. The recorded morphological diversity of the Mesolithic people in the Northwest of Eastern Europe might be attributed to repeated mixing with populations of western and southern origin.

That hypothesis suggested long ago by I.I. Gokhman (1984, 1986) seems to agree best with the data. As to the manifestations of the Arctic adaptive type that could be discerned in a postcranial skeleton, they are seemingly absent. And there is hardly a reason to look for them. Both archaeological and paleogeographic data indicate that the early Mesolithic man's dispersal toward the northwest was a natural result of the enlargement of the ecological niche traditionally occupied by the boreal forest hunters. Their dispersal was by no means related to the colonization of new lands with radically new environments (Lisistsyn and Gerasimov, 2008). The Veretye culture bearers were tall dolichomorphous people. Their osteological characteristics (large epiphyses of the bones, well developed macro- and micro-relief, etc.) suggest an adaptation to considerable physical load and a specialization in walking and running (Gokhman, 1984; Gerasimova and Pezhemsky, 2005).

Another block of interesting data on the northwest of the East European Plain was obtained from the *Minino* graveyard on Lake Kubenskoye (the Vologda Region). The assemblages contain burials of different—Mesolithic and Neolithic—age (Suvorov, 2001; Suvorov and Buzhilova, 2004). Single Mesolithic burials were considered typical of the known burial assemblages of the Onega Lake region (Oleni Ostrov, Popovo, Peschanitsa). Craniological characteristics of the Minino series bear a definite resemblance to those of the synchronous populations of the adjacent regions of the northwest. In the opinion of A.P. Buzhilova, the inner bones of skeletons from Minino (unlike those from the burial on the Yuzhny Oleni Island) display a trend toward development of two possible reactions to the cold stress. There are also noted two variants of the cortical tissue development and the medullary cavity volume: one is closest to the Sungir variant, and the other is distinct for thicker walls of the medullary canal, the width of the latter being the same. The characteristics of the second variant fall within the range known for the present-day Mordovians and Russians. "It might be suggested that a certain variant of the blood formation in the temperate climatic zone had formed on the Russian Plain as early as the Mesolithic/Neolithic boundary" (Buzhilova, 2008, p. 123). Unfortunately, this author did not give the statistics that proved the variants to really exist. The indexes of physical activity and physiological stresses, as well as the absence of chronic and infectious diseases of the tooth and bone systems provide evidence of a successful enough biological adaptation of the people to their environments (Buzhilova, 2008, 2009).

There are convincing archaeological data indicative of an early penetration of humans beyond the Arctic Circle; the fact is also confirmed by a body of linguistic evidence, which suggest a single linguistic (probably, Uralian) substrate having been in existence in the high latitudes. However, we still have no anthropological materials attributable to the early period of human habitation in the Arctic. Ethnographers are fairly confident that the economy of the tundra inhabitants was dominated by reindeer hunting all the way from Scandinavia to Chukotka (Simchenko, 1976, 1980). As a mass killing of the animals (spearing) could be performed only twice a year, the people's survival under unfavorable conditions required other sources of food to be found.

The graveyard on *Bolshoy Oleni* Island in the Kol'skiy Bay of the Barents Sea reveals traces of the search for such sources. Though dated to the 2nd millennium BC, it is really a cemetery left by the population of the so-called "survived Neolithic" culture. As may be inferred from the archaeological materials, the population hunted sea animals in summer and moved southward, in the inner areas of the Kola Peninsula, in winter. The anthropological characteristics of the buried humans suggest some Mongoloid admixture (Yakimov, 1953), which was in reasonable agreement with then accepted concept of the northern way of the Mongoloid group penetration into Eastern Europe. The origin of the Mongoloid type was then undetermined. Recently performed studies of the complemented craniological collection make plausible a suggestion of their similarity to Nenets or Nganasan ancestors (V.I. Khartanovich, personal communication). Osteological and osteometric studies of postcranial skeletons yielded somewhat controversial results (Vasil'ev and Borutskaya, 2008).

On one hand, the authors stated that the proportions of extremities were analogous to those of the modern Arctic people, though the calculated intravital body length was above the average. The ancient humans were noted for their narrow shoulders and wide lower part of torso (which does not comply very well with somatic characteristics of the Arctic adaptive type); at the same time, they show a brachymorphous constitution and shortened forearm and tibia. Most of the identified pathologies in the skeletons are related to attempts at acclimatization; the population seemingly existed in unfriendly environments and was insufficiently adapted to them. The pathologies were due to the cold stress, vitamin deficiency, and an insufficient income of solar radiation. The periostitis symptoms observed in leg bones could be attributed to staying long in cold water, and a pronounced deformity of femoral bones to rickets the men had suffered in their childhood (Borutskaya, 2005).

On the other hand, it is well known that at present the children of the Chukotka natives never have rickets, unlike the newcomers' children. The Chukchi and Eskimo skeletons are typically rich in oxyapatites, which may be considered an adaptation to insufficient insolation (Alekseyeva et al., 2008). So, the population that left the burial on Bolshoy Oleni Island displays definite signs of changes aimed at acclimatization; that suggests a rather short duration of the population under conditions of the Extreme North.

Most of the materials derived from vast spaces of Siberia and its northern territories in particular are dated to later periods. Many specialists, when discussing the genesis of the northern ethnic groups, admit implicitly that there were some aboriginal people having colonized the circumpolar zone of Europe and Siberia as early as the Mesolithic time, long before the Samoyedic or the northeastern Paleoasiatic peoples appeared there. The spatial distribution of the Neolithic sites shows, however, that at that time, man still inhabited mostly forest-steppe and southern taiga regions. It is there that most of Neolithic cultures are concentrated, and sources of paleoanthropological materials are localized (Chikisheva, 2012). No anthropological evidence of the human presence in the higher latitudes has been found. It has been postulated that the hypothetical ancient population mixed with groups of the Neolithic hunters that came to northern Yakutia from Southern Siberia, the Amur drainage basin, and the Maritime (Primorye) regions; the hunters inherited partly the culture of the older inhabitants and formed the basis of the northeastern Paleoasiatic people.

Such reasoning seems to be rather speculative. There is, however, a small series of skulls recovered from Neolithic burials that presents a considerable interest from that point of view. The burials known as the *Diring-Yuryakh* are in a valley of a small tributary of the Lena River. They were explored by Yu.A. Mochanov in 1982. The results definitely confirmed the Mongoloid appearance of the ancient population of the East Siberian taiga. In the craniological characteristics of those, quite unique, skulls, there are discernible features of large racial groups present in the modern population of Siberia: Arctic, Baikalian, and Central Asian races in their initial, weakly differentiated form. The brachycranial braincase with a wide forehead is combined with a great height of skull. The wide orthognathic, very high face with a very narrow nose and slightly protruding nasal bones is equally flattened on the higher and middle levels.

Among the modern people, we can find a close analogy with Yakutian skulls and with those of the Arctic race representatives (Gokhman and Tomtosova, 1983; Alekseyev and Gokhman, 1984a). A brachycranial skull from the *Tuoy-Khaya* burial on the Chon River, a tributary of the Vilyui River, tentatively dated to the Neolithic, also presents some Mongoloid features, such as horizontally flattened face, slightly protruding nasal bones, and a flat nose bridge; the face is very wide (152 mm), though rather low for a Mongoloid (71 mm) (Debets, 1956).

A somewhat younger skull was recovered from an Early Bronze burial on the bank of the *Bugachan* River entering the Lena River at 68°N. It presents an assortment of features typical of northern continental Mongoloids: strongly flattened facial skeleton; weakly pronounced canine fossa; a wide, though not very high (for a Mongoloid) face; a low degree of the nose protrusion; extraordinary wide ramus of mandible, etc. (Yakimov, 1950).

For understanding the initial stages of the Arctic race formation, a single—up to now—paleoanthropological find appeared to be extremely important. It was recovered from the *Ust'-Belskiy* graveyard in the extreme northeast, on the Anadyr River, and dated to the 2nd millennium BC. The burials were left by ancient reindeer hunters and fishers when their culture had already spread all over the Chukchi Peninsula and far to the west and east of it (Dikov, 1961). The skull was studied by I.I. Gokhman (Gokhman, 1961; Alekseyev and Gokhman, 1984a,b). G.V. Lebedinskaya reconstructed the appearance (sculptural portrait) of the ancient Chukotka settler from that skull (Fig. 9.20). The morphological type of the man presents a combination of features typical of the

Figure 9.20 Sculptural portrait of a man reconstructed from the skull recovered from the Ust'-Belsky graveyard by G.V. Lebedinskaya.

Arctic and Siberian continental Mongoloids. The face is rather large, which is typical of the continental Mongoloids; it is noted for weakly protruding nasal bones, mesognathic facial skeleton, and a high dolichocranic braincase with a roof-like cranial vault (which is typical of the arctic Mongoloids). The question arises whether it is a demonstration of the Mongoloid race being yet undifferentiated or a result of miscegenation.

There are other facts indicative of an early formation of the Arctic race complex in the northeast of Asia; among them are morphological characteristics of a skull recovered from Neolithic layers of the *Rodinka II* at the mouth of the Kolyma River (Gokhman and Tomtosova, 1983); the characteristics are analogous to those of the Ust'-Belskiy skull. And finally, in the middle part of the Kolyma drainage basin, at the mouth of the Kamenka River (66°N), there was discovered the *Kamenka II* graveyard attributed to the Ymyyakhtakh culture and dated to the 2nd millennium BC. The burial contained remains of three children. An analysis of craniological data led the specialists to the conclusion of a combination of different racial characteristics, namely those of the Baikal variant of the North Asian continental Mongoloid race and of the Arctic race. The closest analog of the findings is the Ust'-Belskiy skull. The recorded mosaic pattern of the craniological and odontological characteristics is most probably due to miscegenation processes (Alekseyev and Gokhman, 1984a).

To continue a review of paleoanthropological findings related to the initial human penetration to the high latitudes, we would turn our attention to the historical time and concentrate on the materials on the ancestry of the aforementioned representatives of the Arctic adaptive type. The materials to be discussed refer to the immediate ancestors of Sami and Eskimo only.

Let us return to Europe. There are skeletons of several women and children found in Finland, near *Lävanluhta* (~63°N), buried in peat. The place is not an ordinary burial ground; more likely, it was a place for sacrifice or a result of some tragic event. The finds were archaeologically dated to the 6th–7th centuries AD. The paleoanthropological material from the site, scarce as it was, was studied by G.F. Debets (1964). The first to attract attention was a dolichocranial braincase. Judging from that, the earlier authors considered the remains as belonging to representatives of a German tribe of Scandinavian origin. Some other characteristics—small size and gracility of long bones (with proper consideration to the sex dimorphism), the facial skeleton profile, specific feature of the nose bridge, and a degree of the nose protrusion—showed that the studied series is much closer anthropologically to Sami than to the Swedes. G.F. Debets noted the similarity between that series and the wide assortment of forms combined under the name of the Uralian protomorphous race. He attributed, however, a flattened facial skeleton to the contacts with Mongoloid populations. "A thousand years before the women and children bodies were thrown into the lake, – Debets wrote with reference to the paper by Yakimov (1953), – the Kola Peninsula was inhabited by people noted for even flatter and wider faces." In his opinion, there is little doubt in the Siberian Mongoloid participation in the development of that type.

Unfortunately, no traces of Proto-Eskimo culture had been found in Chukotka that could be comparable in age with Independence or Saqqaq cultures in Greenland. The traces of sea mammal hunters are dated to a much later time, mostly younger than the 2nd century BC (excavations on the Uelen and Ekven cemeteries). The most extensive excavations were performed at the ancient settlement Ekven and adjacent burial ground on the coast of the Bering Strait and in the Paipelgak settlement at the Chegitun River mouth on the Arctic Ocean coast (Dneprovskiy, 2007). The ancient Eskimo—sea mammal hunters—inhabited practically the entire coast of the Chukchi Peninsula, from the Kolyma mouth to the Anadyr River, the remains of settlements being found in abundance at present. The settlements are usually a few kilometers apart, which facilitated the Eskimo traveling long distances along the coastline.

The *Uelen and Ekven graveyards* dated to the 2nd century BC to the 14th century AD yielded rich and diversified materials shedding light on the earlier stages of the Old Eskimo culture, namely Old Bering Sea, Birnirk, and Punuk (Arutyunov and Sergeyev, 1969, 1975) (Fig. 9.21(a)–(e)). The Punuk stage is close to the recent Eskimo culture (the second half of the 2nd millennium AD).

The paleoanthropological materials were studied by M.G. Levin (1964), and later by G.F. Debets (1975). In the latter's opinion, the skulls characteristics suggest a combination of traits typical to Eskimo and other Arctic people ascending to a high antiquity. Typically, the face is wider than the braincase, and the latter is narrow and high, with roof-like outlines from the back. This type is best represented in the Greenland Eskimo populations. There is a reason to believe that the aforementioned characteristic of the skull structure reflects an adaptive variability. Such outlines of the skull are typical not only of the Arctic race, but also of other races living in the high polar latitudes (Alekseyev, 1985).

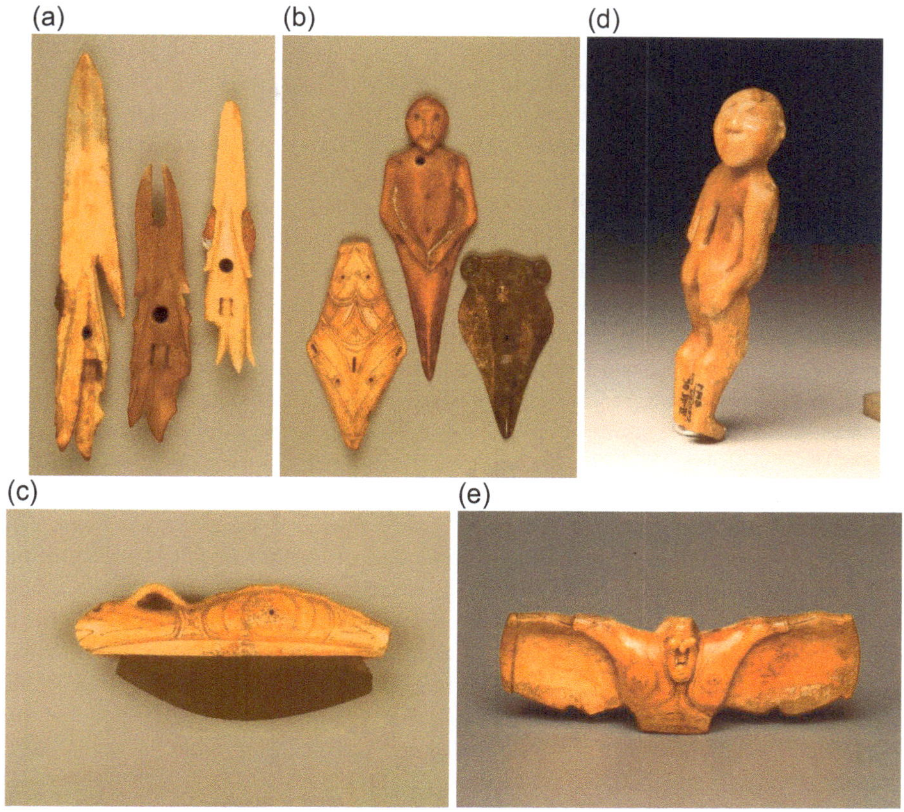

Figure 9.21 Artifacts of the ancient cultures of the Chukchi Peninsula. (a) Toggling harpoons made from walrus tusk and reindeer antler. Birnirk–early Punuk culture. (b) Handles of ritual ladles. The two lower ones are from the Ekven cemetery, burials 233 and 302, the Old Bering Sea culture. The item in the middle of the upper row is an anthropomorphous handle of a ladle from the Paipelgak settlement, Birnirk–early Punuk culture. (c) A traditional woman's knife with a zoomorphic handle. Ekven, burial 313. The Old Bering Sea culture. (d) An anthropomorphic figurine. Ekven. The Old Bering Sea culture. (e) A winged object. Ekven. The Old Bering Sea culture. Photos by E.I. Zheltov from the personal archive of K. Dneprovsky.

Throughout the time of the Uelen and Ekven burials functioning the anthropological type was held constant. In the opinion of T.I. Alekseyeva, the lack of significant morphological differences in the considered series may be assigned to the natural selection that ensured the preservation of that type as the best fit to the arctic environments (Alekseyeva et al., 2008). Unfortunately, it is not always so evident. That different groups of the Eskimo belong to the same race is undisputable fact, supported not only by the morphological analysis, but also by blood type studies (Levin, 1958). There are, however, considerable variations in craniological characteristics between individual groups isolated from each other and different in age. First of all, the skulls from ancient Eskimo burials on Chukotka differ both from modern Eskimo skulls from Naukan (Chukchi Peninsula) and from those from Chaplino.

The ancient population of Alaska is represented by two series recovered from burials on *Point Hope*, which is the westernmost point on Alaska north of the Bering Strait. One series is attributed to the Ipiutak culture, the age of which is estimated at 1500 years. That culture developed synchronously with the line beginning from the Old Bering Sea one, though independently from the latter (Larsen and Rainey, 1948). The second series originates from the same locality, the *Tigara* cemetery, which is no more than five centuries old and belongs undoubtedly to the direct ancestors of the modern Eskimo. As stated by G.F. Debets (1986), there is no essential difference between the series in the most important characteristics: flattened face and protruded nose. The Ipiutak skulls, however, are devoid of typical Eskimo traits, such as a high skull and the face breadth exceeding the breadth of skull. In those characteristics, they differ from Tigara people and from modern Eskimo, particularly from the central and eastern groups; their difference from the modern western (Asian) ones is less pronounced. As has been noted by G.F. Debets (1986), the facts give the impression that three to four tens of generations that separate the two population groups were insufficient for development of an essential dissemblance between them.

As is seen from the preceding, the proper Eskimo type existed already in Chukotka (Uelen and Ekven burials) at the Ipiutak time and on Alaska (Point Barrow) at the Birnirk time; they are separated by an interval no more than a ten of generations, which was hardly sufficient for the development of a serious difference. It should be noted, however, that in Chukotka the so-called "Greenland" variant (Uelen and Ekven burials) preceded the western variant (the *Naukan* burial). On the Point Barrow (Birnirk burial ground) the differences are similar to those recorded in Chukotka, while on the Point Hope (Alaska) the western anthropological variant (the *Ipiutak* burial) is more ancient than the Greenland one (the Tigara burial) (Debets, 1975).

Quantitative analysis once performed by V.P. Alekseyev (Alekseyev and Baluyeva, 1976) confirmed the validity of distinguishing Western (or the Bering Sea) and Eastern (Greenland) variants in the Eskimo population. The Western Eskimo of Chukotka appeared to be almost twice as differentiated as the Eskimo of Canada or Greenland. The Eskimo are in general more differentiated morphologically than the Finno-Ugric or Caucasian peoples. The latter, though, include several local races, while all the Eskimo belong to the same local race.

Table 9.5 is based on materials from works by G.F. Debets (1951, 1975, 1986) and V.P. Alekseyev (Alekseyev and Baluyeva, 1976). It gives distinctive characteristics of individual local series within a single Eskimo type, namely height of skull, cephalic index, width of skull, jugal face width, and width of nose. The Greenland type is noted for the braincase width smaller than its height and width of face; the Alaska Eskimo have a wider and shorter skull. The ancient series from Chukotka are nearer on the whole to those of modern Eskimo from Greenland and Labrador than to the series of the modern Chukotka Eskimo. In the past the Greenland type occurred both in Alaska and in Chukotka.

As to interrelations between Greenland and Asian variants, there is a difference in opinions. According to A. Hrdlicka (1945), the oldest type is mesocephalic. It is best preserved in the Alaska Eskimo and underwent considerable changes in the Eastern (Greenland) Eskimo in the course of adaptation to Arctic environments. This

Table 9.5 Variability of the skull characteristics (braincase size, face width, nose width) in various ancient and modern groups of Eskimos

Series	1	8:1	17	18	45	54
Naukan, Chukotka	183.3	143.1	136.8	78.1	142.3	25.2
Chaplino, Chukotka	181.8	140.7	135.0	77.6	137.5	24.4
Uelen, Chukotka	189.3	134.3	139.3	70.9	140.8	24.3
Ekven, Chukotka	192.2	134.2	139.1	70.0	140.9	24.0
Laurentian peninsula	182.6	140.5	136.4	76.9	140.5	24.7
Southwest Alaska	182.9	140.3	135.7	76.7	140.6	24.7
Seward Peninsula	186.2	136.7	137.9	73.4	140.5	23.6
Point Barrow, Alaska	187.6	137.8	136.6	73.5	141.4	23.6
Birnirk, Point Barrow	189.7	133.4	141.3	70.3	140.9	23.7
Ippiutak, Point Hope	182.7	139.1	133.0	76.2	137.6	24.3
Tigara. Point Hope	186.1	137.4	139.9	73.9	142.2	24.0
Labrador Peninsula, Canada	187.2	134.6	135.5	71.9	135.8	22.6
Western Greenland	188.6	135.0	138.3	71.5	139.8	23.0
Eastern Greenland	187.8	133.3	138.9	71.1	137.9	22.9

1 – maximum length, 8 – maximum breadth, 17 – maximum height (bision-bregma), 8:1 – cranial index, 45 – bizigomatic breadth, 48 – upper face height.

opinion had been shared by G.F. Debets, who considered the Bering Sea type to be less specialized (Debets, 1951); later, taking into consideration the data on the Ipiutak and Tigara burials, he came to the conclusion that the problem was still to be solved (Debets, 1986).

Comparison of the skulls from ancient burials with the recent series suggests the dolichocranial type to be more widely spread in the past; that may be considered as an argument in support of its being older than the mesocephalic (western) one. From the point of view of Levin (1958), the recorded differences between local groups might have resulted, on one hand, from a prolonged isolation of individual groups, and on the other from mixing of formerly isolated parts of a single population. The mixing may change the growth conditions, and changes in the rate of growth may lead to the development of brachycephalic characteristics (Bunak, 1951). To gain an insight into the cephalic index variations, it should be noted that the archaeological data indicate the most intensive mixing to occur in western groups of the Eskimo populations, and not in the eastern groups.

We have no reason to agree with T.I. Alekseyeva (2003) in that the lack of significant morphological difference between those series may be attributed to the natural stabilizing selection, which was instrumental in preservation of the type found in the Uelen and Ekven burials as the most adequate. Contrary to that, the morphological differentiation of the Eskimo was promoted by the dispersive selection; it was especially effective under conditions of a vast area of settlement under extreme conditions and kindred populations widely spaced (Alekseyev, 1974a).

It should be noted that the "skeletal constitution" of people who left the Uelen and Ekven burials is typical of the Arctic adaptive type (Fedosova, 1991). The characteristics include a considerable volume of medullary cavities in the long bones; relatively shortened distal segments of the limbs; and brachymorphism. In comparison with the modern series, those from Uelen and Ekven burials are noted for a smaller size of the skeleton long bones and, therefore, for lesser total length of the body.

The same tendency is traceable in the skeletons from the burials on Point Hope (Dobrovolskaya et al., 2008). The authors, however, do not agree with the hypothesis that the Eskimo population tends to grow in the body dimensions throughout the last 2 millennia; they are inclined to assign the observed difference to the local population variability. In fact, as is seen from comparative data on the long bone size from Asian and North American burials, modern and ancient ones, the differences in the skeleton massiveness (solidity) and in the bone dimensions of Eskimo (Tables 9.1 and 9.6) reflects mostly local variability (which is also confirmed by the craniological data), and not the chronological differences.

The aforementioned authors admit, however, that the local variability of the long bones may be controlled by the environmental characteristics. From their viewpoint, that is confirmed by the fact of a significant improvement in general characteristics of physical development in case of even slightly mitigated negative environmental factors. Such regularity has been confirmed by the studies of the present-day aboriginal populations on Chukotka and Kamchatka peninsulas (Dobrovolskaya et al., 2008).

Unfortunately, no matter how eager we were for it, we do not have any single convincing evidence of the direct adaptation to environments becoming inheritable. It is the dispersive selection that may be considered as the main mechanism of microevolutionary changes and formation of inheritable adaptive traits (Alekseyev, 1974a,b).

Table 9.6 Comparative data on the dimensions of long bones recovered from different Eskimo burials

	Femoral bone		Tibia		Humerus	
Series	F_1	F_8	T_1	T_{10b}	H_1	H_7
Uelen, Chukotka	423.7	91.0	334.7	71.7	298.8	62.6
Ekven, Chukotka	421.9	90.3	338.1	73.9	300.6	64.7
Naukan, Chukotka	439.0	97.0	355.7	79.4	316.0	72.8
Ippiutak, Alaska	414.4	86.2	326.0	74.5	302.5	60.9
Tigara, Alaska	438.4	90.7	357.0	74.8	310.2	63.8
Old Igloo, Alaska	438.6		356.0		311.7	
Point Hope, Alaska	440.6		364.0		310.7	
Inugsuk, Southwest Greenland	421.7	87.8	330.6	70.7	302.9	59.5
Northeast Greenland	441.3	93.8	363.7	78.3	310.7	68.8
Labrador	424.2		343.8		293.9	
Nauyan, Canada	425.4	95.9	359.0	79.7	309.6	68.8

There are observations made by ethnologists and archaeologists (Arutyunov and Sergeyev, 1975; Bronshtein, 2007) concerning a mosaic structure of the Old Eskimo ethnocultural tradition. It is supposedly related to the existence of stable communities of the sea mammal hunters, which were fully conscious of their identity and difference from other groups; that could appreciably raise their survivability in the extremely harsh environments. It is not inconceivable that the high level of the morphological specificity in various local Eskimo groups acted in the same direction.

Unfortunately, we have to leave aside many questions and highly diversified answers to them as to relative antiquity of the Ipiutak and Greenland anthropological types, to anthropological essence of difference between those types, and to relative kinship of the individual Eskimo groups. The solving of those problems strongly depends on the data on the substrate phenomena in culture, language, and specific characteristics of the Eskimo physical type.

Some of the questions—in particular, those of the scale and character of human migrations in the Arctic and Subarctic—may be answered on the basis of paleoanthropological materials. In the case of the American Eskimo, the role of the substrate could be played by the Paleoeskimo Ipiutak culture; the ethnicity of the latter is not quite clear yet (Rainey, 1958). Craniological characteristics of the Ipiutak people may be considered as not fully specialized; however, the idea of them being the original form does not agree with the relationship between types observed in other regions. In the northeast of Asia the pre-Eskimo tradition may be traced as far back as the 9th millennium BC, judging from the lower layers of the Ushki site (Dikov, 1971); unfortunately, there are no paleoanthropological materials available, except for the earlier described skulls from the Anadyr (Ust'-Belskiy burial) and Kolyma (Rodinka II site) rivers.

9.5 Conclusion

Though we know of the human penetration to the high latitudes as early as the Paleolithic, there is no anthropological evidence of the fact, or data on the man's adaptivity. The anthropological data found occasionally are not older than the Mesolithic; their occurrence depends on the state of our knowledge of individual regions in Northern Eurasia, on the degree of the paleoanthropological material preservation.

We have a rather limited capability of identifying biological traits of adaptation from the paleoanthropological materials dated to the initial human colonization of the high latitudes. The adaptive characteristics may be recognized if the "ecological rules" established for animals are extended to humans. Among the rules are those by C. Bergmann, J.A. Allen, and C.L. Gloger, as well as the rules by A. Thomson and L.H. Buxton; the latter states that the nose width depends on the mean annual temperature, insolation, and the latitude of the locality. Numerous and thorough investigations proved that the adaptive factor was and still is of great importance in human morphogenesis; the adaptive processes and adaptive abilities (adaptivity) are specific to various racial and constitutional types.

The Arctic zone is the region of the extremely harsh conditions for human living. The specialists succeeded in identifying a complex of similar reactions and morphophysiological variability developed in the process of adaptation to the Arctic environment irrespective of racial characteristics of populations; the complex is known as the *Arctic adaptive complex*. In the context of the paleoanthropological studies, the most significant are craniological characteristics indicative of potential adaptivity; apart from them, the following features may be of interest: the skeleton volume; distal to proximal segment ratio in the limb bones; structure of long bone diaphyses; bone tissue histology; mineral saturation of bones; and some other data.

The extrapolation of the adaptive morphophysiological features of three indigenous northern peoples to the paleoanthropological material gave us an insight into the process of biological adaptation that made possible colonization of the North by the ancient populations.

Genetic data on peopling high latitude areas

10

Oleg P. Balanovsky
Vavilov Institute of General Genetics, Russian Academy of Sciences, Moscow, Russia

Chapter Outline

10.1 Introduction

Gene geography studies genetic variation in geographic space and discovers both common trends of gene pool structures and partial passages of their history. The main approach is retrospective: from present to the past. Namely, gene geography studies the gene pool of modern populations and—knowing the laws of changes in gene pools—it could suppose the genetic composition of ancestral populations and estimate which population events happened from ancient times to modern day. The alternative approach is analysis of ancient DNA from paleoanthropological samples. In this chapter, we first consider the modern gene pools data, hypothesize which migrations in high latitudes could form this current picture, and finally consider the data of ancient gene pools in high latitudes.

Modern researches of gene pools use four main systems of genetic markers: mitochondrial DNA (mtDNA), Y-chromosome, genome-wide panels including up to millions of autosomal markers, and sequencing the full genome. Though full-genome studies recently started specifying the picture of global genetic variation (Pagani et al., 2016; Mallick et al., 2016), the populations in high latitudes are poorly sampled, and the current amount of full genomes in these populations is not enough for the reliable conclusions. For this reason, in this chapter, we use generally mtDNA and Y-chromosome data when considering modern

Human Colonization of the Arctic: The Interaction Between Early Migration and the Paleoenvironment
http://dx.doi.org/10.1016/B978-0-12-813532-7.00023-1

gene pools. When analyzing Sami origin, we also used genome-wide panels. With regard to ancient DNA, all kinds of genetic data were used, though most conclusions were based on the mtDNA data, as mtDNA is better preserved in paleoanthropological specimens.

10.2 Cartographic analysis of gene pool structure in Eurasia

10.2.1 Two atlases

To understand the position of the high latitude populations in the system of Eurasian gene pools, we need to describe this Eurasian system first. For this purpose, we used the cartographic approach.

Many papers appear annually describing the diversity of mtDNA and Y-chromosome in various populations. The data accumulated allow tracing in detail the geographic distribution of each variant of mtDNA (or Y-chromosome). These variants are called *haplogroups*, and each includes a group of related *haplotypes*. Haplotype is one of many mtDNA variants (as one leaf on the tree), so a haplogroup is a branch carrying many leaves, and all variants of mtDNA in humanity form the entire human mtDNA tree. For many haplogroups, we can presumably describe not only their modern distribution but also the time and place of their origin, in the other words, the time of their separation from the common stem.

A frequency distribution map of a haplogroup is a necessary tool in phylogeographic studies (Cruciani et al., 2007, 2010; Semino et al., 2004; Myres et al., 2011; Rootsi et al., 2007; Tofanelli et al., 2009; Underhill et al., 2010). But despite the significance of the map of a single haplogroup being widely recognized, there is lack of systematic map collections or cartographic atlases. This gap is caused by several reasons, including the lack of systematic global databases, which should be the background for a cartographic study. To the best of my knowledge, the only successful previous attempt to chart wide spectra of haplogroups (for Y-chromosome) was made in the D.Sci thesis of V.A. Stepanov (Stepanov et al., 2001). During the more than a decade since that study a huge amount of data has been collected, and new haplogroups have been discovered. Moreover, there are no elaborate cartographic atlases for mtDNA at all. So we have performed the broad-scale cartographic research: we produced an atlas of Y-chromosome variation and atlas of mtDNA variation in Eurasia. Atlases are based on the *Y-base* (for Y-chromosome) and MURKA (for mtDNA) databases, which were developed by the research team headed by the author.

10.2.2 MtDNA atlas

The MURKA database gathered most published data on mtDNA variation in the world populations. It allowed us to map this dataset and produce the mtDNA Atlas of Eurasia (Balanovskiy, 2012; Balanovskaya and Balanovskiy, 2007; www.genofond.ru). This atlas shows the distribution of mtDNA haplogroups in indigenous populations of Eurasia. The atlas maps reveal not just the patterns of mtDNA haplogroups distribution (most of them were known previously), but they also specify details of their spread, and importantly, they allow data generalization by revealing the main continuums in the mitochondrial pool.

Although the database includes information for hundreds of haplogroups, we have selected haplogroups with frequencies more than 0.1% in Eurasian populations, and the atlas consists of 82 maps, each showing the frequency distribution of 82 haplogroups. Instead of characterizing all the separate maps, we describe here their typology. Distinguishing East Eurasian and West Eurasian types became a tradition in human population genetics, though attributing a certain haplogroup to a certain type often differs depending on the researcher. Hence, we have developed the method for objective haplogroup classification into geographic continuums. At the first stage, haplogroup frequencies in populations were analyzed by principal components, and variables (haplogroups) were plotted on the resulting graph instead of plotting observations (populations). Associations of haplogroups into clusters (continuums) on the plot means their correlation with each other and correspondence to the common geographic pattern. So, at the second stage, we produced the frequency maps for continuums. Each continuum map presents the sum of frequencies of haplogroups that belong to this continuum. Finally, the continuums were named by the geographic area where they are most frequent. Figs. 10.1–10.3 demonstrate this approach applied for the mitochondrial pool of Eurasian populations.

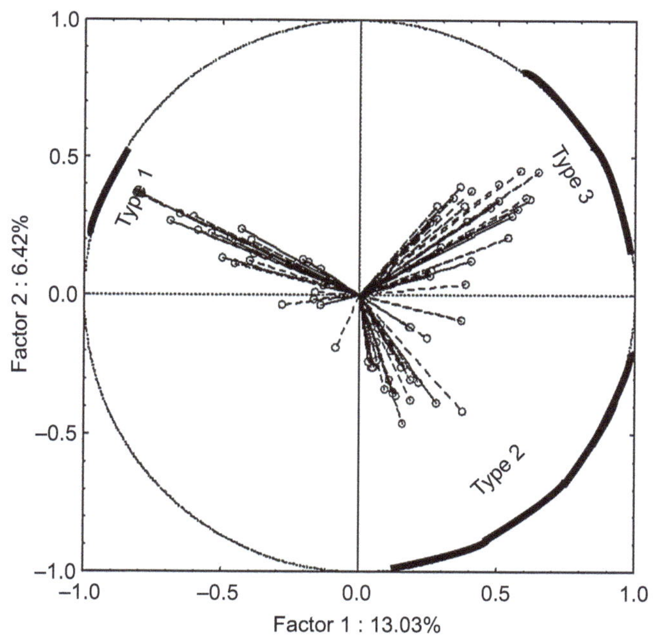

Figure 10.1 Geographic classification of the mitochondrial DNA haplogroups. The plot presents principal component analysis (PCA) of the mtDNA haplogroup frequencies in Eurasian populations. The PCA results contain information on both observations (populations) and variables (haplogroups). Usually population geneticists plot the populations, but here we plotted the haplogroups. *Circles* designate all 82 studied haplogroups, and *dashed lines* link them with the origin of the coordinate system. The analysis revealed three clear types (continuums) of haplogroups. *Thick arcs* indicate intervals of values of the first and second principal components that fall into each continuum.

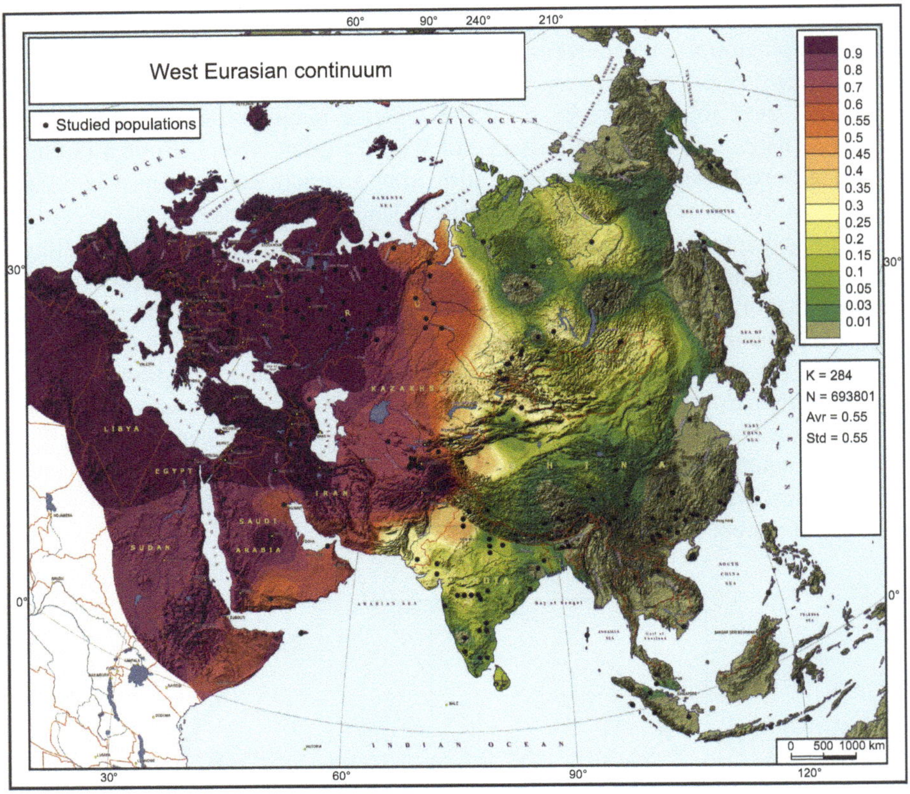

Figure 10.2 The map of the total frequency of mtDNA haplogroups of the West Eurasian continuum of the Eurasian gene pool (Type 1 on Fig. 10.1). Different colors designate the corresponding frequency (as shown in the inset scale); *K* – number of studied populations; *N* – number of grid nodes (nodes in sea not included); *min* – minimum frequency; *max* – maximum frequency; *avr* – average frequency; *Rlat* – correlation with the latitude; *Rlon* – correlation with the longitude.

The principal component analysis graph (Fig. 10.1) revealed three continuums. The first one (№ 1, Fig. 10.1) corresponds to the West Eurasian haplogroups. It became obvious when looking on the map of this continuum (Fig. 10.2). It occupies the western parts of Eurasia, and the east boundary of the area of this continuum follows the 80th meridian through the Trans-Ural region. Similarly, continuum 3 (Fig. 10.1) corresponds to the East Eurasian haplogroups (Fig. 10.3). The newly revealed continuum 3 follows the specific geographic pattern. It may be called South-Chinese because its maximal frequencies (in Southeast Asia, South China, Indonesia, and Philippines) are located around South China Sea. We do not present the map of this continuum here because its area is far from the high latitudes.

10.2.3 Y-chromosomal atlas

We have produced a similar atlas for the Y-chromosome variation in Eurasia. It includes frequency distribution maps for 79 haplogroups (this list includes all haplogroups

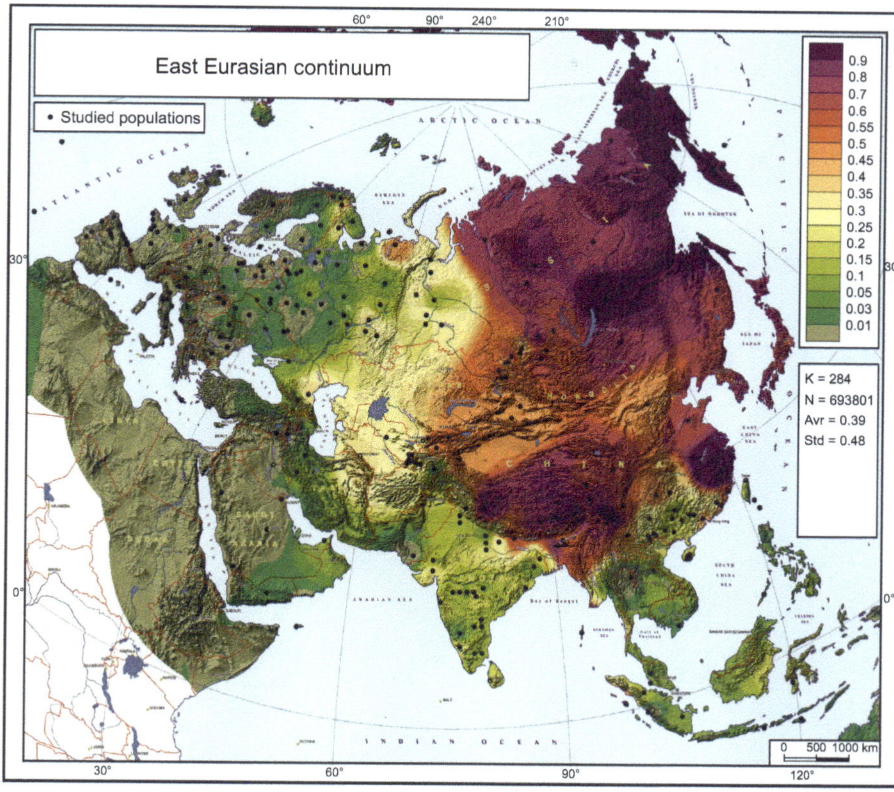

Figure 10.3 The map of the total frequency of mtDNA haplogroups of the East Eurasian continuum (Type 3 on Fig. 10.1). See the legend for Fig. 10.2.

revealed in Eurasia with frequency more than 1% for at least one to two populations). The joint analysis of these maps resulted in the map of genetic barriers (Fig. 10.4). This map shows the main "break zones" of the Eurasian gene pool. These zones appeared not in the Ural mountains, the geographic boundary between Europe and Asia, but they go obliquely across the continent. This "break zone" consists of several cuts: Caucasian, South Uralian, Northern Kazakhstan, and Middle Siberian (along the Yenisey River). So, the main genetic boundary separates the gene pool of Eurasia into west and east parts.

10.2.4 West–East

The maps of both atlases confirmed the existence of the West Eurasian and East Eurasian gene pools. The first occupy nearly all the west half of Eurasia (Europe, Mediterranean, Southwest Asia), while another conversely resides in the east, among the indigenous populations of East Asia, Southeast Asia, and Siberia; some of these haplogroups have been also found in the Americas, demonstration of the Siberian roots of native Americans. There are several "boundary" haplogroups (for example, those spread in the Indian subcontinent), for which attribution to a western or eastern cluster

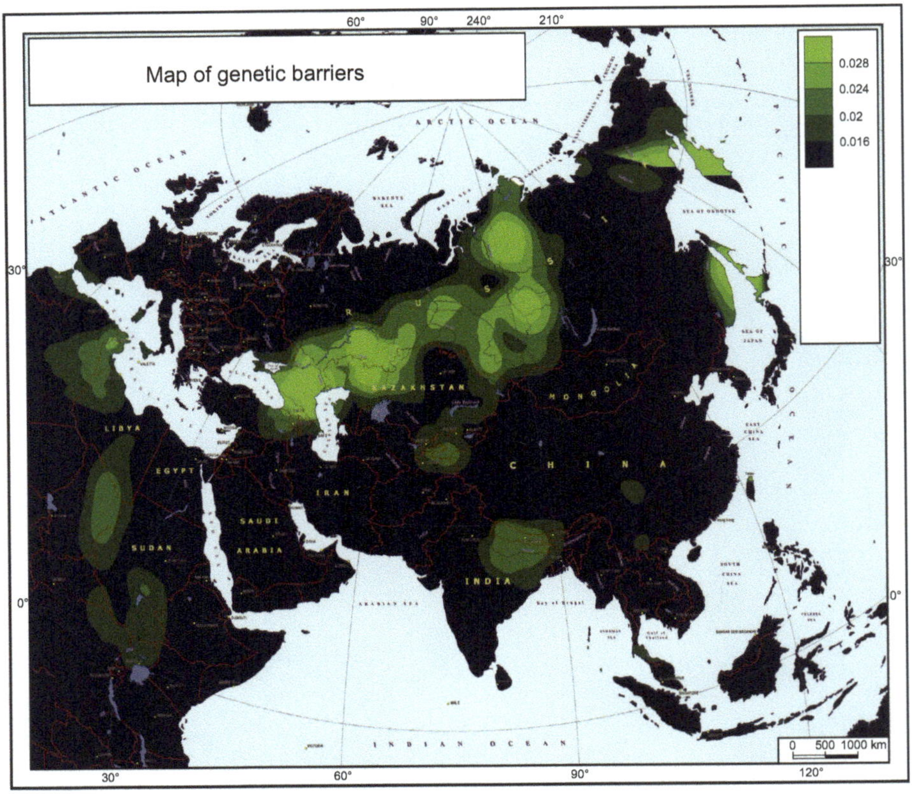

Figure 10.4 The structure of the Eurasian gene pool: map of genetic barriers as revealed by the Y-chromosome. Each grid node contains the value of interpopulation genetic variation in the 1000 × 1000 km window surrounding the node. The different values are shown by the corresponding colors, as the inset scale designates.

is uncertain; however, some of them have been found also in the West but not in the East. In addition, as by the general haplogroup spectrum, South Asia overlays sooner West than East Eurasia, so it would be preferable to consider Indian haplogroups as West Eurasian ones.

Besides the West Eurasian and East Eurasian gene pool, there is the wide transient zone. Its center is located beyond the Urals, and for this reason the population there appears genetically intermediate/mixed. This observation, revealed by both atlases, is employed in our further discourses about origin of the gene pools in high latitudes.

10.2.5 North

The general look at the mtDNA variability maps (Figs. 10.2 and 10.3) makes it clear that the division of the gene pool into west, east, and transient zones persists even in the high latitudes. Isolines—marking gradual replacement of one gene pool by another—go mostly meridionally across Eurasia. But there is a local exception on the map of East Eurasian influence: a transient zone goes far to the West along the Barents Sea and White Sea.

The Y-chromosome variation (Fig. 10.4) also confirms that the division of the gene pool into the western, eastern, and transient zones persists at the high latitudes. The transient gene pool zone (the genetic boundary area) is located in West Siberia as well as in the case of mtDNA.

10.2.6 Two migrations

What could this picture tell us about origin of high latitude populations? It indicates that the North has been populated by two parallel, independent waves of migration: the first wave passed from the middle latitudes of western Eurasia to the north of Europe; the second one is expected to pass from Central Asia to north of Siberia and then to the Americas.

We can sketch these ancient migrations rather schematically, but the west–east dichotomy kept in the northern gene pool suggests that the two parts of the northern gene pool derived from two (western and eastern) genetic provinces of Eurasia. The exact dating of these migrations using genetic approaches is ahead of time now. Though it is, in principle, possible to obtain the time of each haplogroup arrival by calculating intrahaplogroup variation among northern populations employing the molecular clock method, it is hard to separate the variation that populations accumulated after their arrival into high latitudes from that inherited from their homeland. Ignoring this limitation might result in overestimating the time depth. An alternative approach for estimating the upper time limit of colonization is the usage of the whole mtDNA sequences for the dating the haplogroups that emerged after the colonization. Several such haplogroups (A4b, C5, Z1, N2) emerged around 20–25 kyr BP (1000 years ago), but most of the intrahaplogroup variation in high latitude populations was accumulated 3–10 kyr BP.

Of course, the two colonization waves of high latitudes interacted with each other. It is obvious from the gradual pattern of gene pool changes from west to east (Figs. 10.2 and 10.3). Based on the mtDNA data the interpenetration of two gene pools was mutual but not equal in intensity. We suppose that the gene flow from the east to the west predominated over the reverse migrations, which has led to the occurrence of the transient gene pool zone in northern Europe. This interpenetration of gene pools is supported by the Y-chromosome data. In this case, the genetic barrier (interpenetration zone) in the north is even wider than in other parts of Eurasia and includes all West Siberia and also part of Middle Siberia. It may indicate the high intensity of interaction of western and eastern gene pools in high latitudes.

10.3 The origin of the Sami gene pool

The presented sketch of forming the gene pool in high latitudes is too general. Thus, we also consider the origin of the gene pool of one ethnic group as a case study. Sami is the best genetically studied population in high latitudes. For this reason, Sami is the best example. The following description is based mainly on the results obtained by a large international team including the author and published in several articles (Tambets et al., 2004, 2011).

The peculiarities of Sami anthropological type lead one to question their possible Trans-Ural origin. Not only physical anthropology but also data on classical genetic markers supported the peculiarity of the Sami gene pool that was considered a "genetic outlier" in the analysis of European populations (Cavalli-Sforza et al., 1994). Even using different genetic markers (Evseeva et al., 2001), specialists could not determine the reason of their genetic peculiarity for decades: whether their origin was beyond Europe or peculiarities emerged due to the long isolation from other Europeans. Such genetic changes in the isolated populations are well known in population genetics as gene drift.

Using mtDNA markers—because of their clear geographic distribution—allows one to distinguish two scenarios. It turned out (Tambets et al., 2004) that the mitochondrial pool of Sami consists of the usual European haplogroups at unusual frequencies, which are quite different from the frequencies in all populations of western (and more so, eastern) Eurasia. The conclusion has been that Sami derived not from the East Eurasian (Siberian) populations but from a European population of the Mesolithic epoch. Among the wide spectra of European haplogroups, only a few survived in Sami, while all others were eliminated due to genetic drift during their long isolation. As a result, the frequencies of these haplogroups became quite unusual.

Let us describe these results in more detail. Five mtDNA haplogroups with the following frequencies have been found in three populations of Sami (from Sweden, Norway, and Finland): **U5b** (48%), **V** (42%), **H** (3%), **D5** (3%), and **Z** (1%); other haplogroups were found in single persons only. Although the three studied Sami populations were genetically different, haplogroup frequencies in all the populations fluctuated around these average values. It is notable that the Sami population from Russia (Lovosero) studied later also had the same profile of haplogroup frequencies (Tonks et al., 2006).

So, 90% of the Sami gene pool are represented by two haplogroups only: **V** and **U5b**. Both are West Eurasian haplogroups, common in Europe (particularly in East Europe) and rare elsewhere (Balanovsky and Balanovskiy, 2007). Nowhere in the world do these haplogroups reach such high frequencies as in Sami. One may suppose that these haplogroups were brought to the north of Scandinavia (from where the ice sheet was just retreating) by the first colonizers, probably from East Europe. Likely, that initial gene pool of this proto-Sami population consisted of more haplogroups with frequencies similar to other Europeans. But during the long habitation on the periphery of Europe and limited contacts with other Europeans the frequencies of some haplogroups decreased by chance till most of them were lost, but the two that remained increased up to 90% of population's gene pool.

The extreme power of genetic drift in Sami populations explains nicely why earlier researches revealed their genetic uniqueness but could not reveal their origin. These classical researches have been based primarily on the gene frequencies, which changed in Sami radically and unpredictably because of the strong drift. So, it is no wonder that Sami appeared far from both European and Siberian populations, demonstrating occasionally a small and accidental similarity with others. Most of the genetic markers are distributed all over the world; thus Europeans and Siberians differ not in spectra but only in frequencies of these alleles. MtDNA (and Y-chromosomal) markers, however, have high geographic specificity, and they are distributed very differently between Europe and Siberia.

The question emerges about the remaining 10% of the Sami gene pool. As indicated earlier, this share consists of haplogroups **H**, **D5**, and **Z**. Haplogroup **H** is a typical West Eurasian haplogroup; it takes more than one-third of the mitochondrial pool in every European population, except Sami. In East Eurasian populations, this haplogroup is rare or absent. So, presence of this haplogroup in Sami also indicates the West Eurasian rather than East Eurasian origin of Sami. Its frequency in Sami, unusually low for Europe, could be also explained by genetic drift. It is even possible that the initial drift had eliminated this haplogroup in Sami completely, but later contacts with Swedes and Norwegians brought this—most frequent among them—haplogroup into Sami gene pool.

Considering the two remaining haplogroups, **D5** and **Z**, we finally meet in Sami the clear signal of eastern influence: haplogroup **Z** is frequent in Siberia and in the single enclaves in Central Asia, but it is extremely rare in Europe. In other words, haplogroup **Z** was likely brought into the Sami gene pool by migrations from Siberia. But the proportion of this East impact is tiny, as the frequency of haplogroup **Z** in Sami is less than 1%.

The last Sami haplogroup **D** is the most typical one for East Eurasia (as haplogroup **H** is most typical for West Eurasia), but the specific Sami variant within haplogroup **D5** was found in Altai, the Urals, and at higher frequency in northeast Europe, so European origin is more likely. Anyway, its frequency in Sami is only 3%, so even the less probable Siberian version of its origin could not shake the main conclusion about the mainly European roots of the Sami gene pool.

The chart (Fig. 10.5) summarizes our considerations on the structure of the mitochondrial gene pool in Sami and on the origin of its components. The figure was modified from (Tambets et al., 2004) with some improvement in the light of new knowledge of mtDNA geography added. In particular, our atlas indicates clearly that East (rather than West) Europe is a zone of maximal frequencies of haplogroup **V**, and the diversity of this haplogroup is also higher in East Europe. So we considered East Europe as the more likely source of the haplogroup **V** in Sami. As we see (Fig. 10.5), 99% of the Sami gene pool could be traced back to East Europe.

These evidences came from mtDNA results, and analysis of the Y-chromosome leads to similar conclusions. Y-chromosome haplogroups predominating in Sami indicate their European, rather than Siberian, origin (Tambets et al., 2004). Recent research of the same team including the author (Tambets et al., 2011) analyzed the Sami gene pool by the third genetic system: a panel of 610,000 autosomal genetic markers. The same panel was also studied in other populations of the Uralic language family and neighboring populations from East Europe and Siberia. The pattern of variability for this system is the same as for mtDNA and Y-chromosome: West Eurasian and East Eurasian zones are clear at high latitudes, and the transient gene pool zone is located in the Urals and West Siberia. The gene pools of the most studied populations are similar with gene pools of their geographic neighbors. But Sami is an exclusion: the proportion of Siberian component in their gene pool is higher than in surrounding

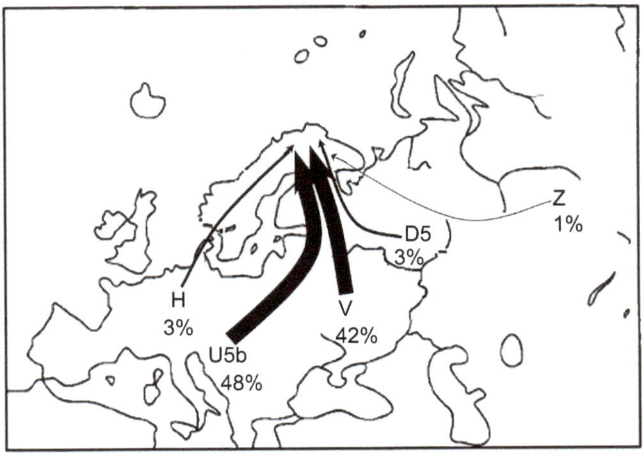

Figure 10.5 The schematic map showing the sources of the Sami mitochondrial pool. *Letters* designate mitochondrial DNA haplogroups found in Sami. *Arrows* indicate the possible migration ways of the haplogroup carriers that entered northern Scandinavia and became ancestors of the Sami population. The thickness of each *arrow* is proportional to the frequency of the corresponding haplogroup in the Sami gene pool.
Tambets, K., Rootsi, S., Kivisild, T., Help, H., Serk, P., Loogvali, E.-L., Tolk, H.-V., Reidla, M., Metspalu, E., Pliss, L., Balanovsky, O., Pshenichnov, A., Balanovska, E., Gubina, M., Zhadanov, S., Osipova, L., Damba, L., Voevoda, M., Kutuev, I., Bermisheva, M., Khusnutdinova, E., Gusar, V., Grechanina, E., Parik, J., Pennarun, E., Chaventre, A., Moisan, J.-P., Barac, L., Pericic, M., Rudan, P., Terzic, R., Mikarezi, I., Krumina, A., Baumanis, V., Beckman, L., Villems, R., 2004. The western and eastern roots of the extreme European genetic outliers – the origin of mtDNAs and Y-chromosomes of the Sami. American Journal of Human Genetics 74 (4), 661–682.

populations (Finns, Karelians, and Vepses), similar to the proportion of this component in populations on the eastern edge of Europe (Komi, Mari, and Udmurts), but two to three times lower than the proportion of this component in West Siberian populations. So this genetic system also shows the mainly European roots of Sami, while their partial eastern affinities are limited to Northeast Europe rather than to Siberia.

10.4 Ancient DNA

10.4.1 Concept

The concept of ancient DNA analysis is simple: instead of reconstructing the ancient gene pools from their present-day descendants, it concentrates on directly studying the ancient gene pool using DNA from the ancient specimens. So, ancient DNA analysis is genetics, but at the same time, it belongs to the area of paleoanthropology. It adds to main paleoanthropological systems of traits (odontology, craniometry, and others) a new tool: "DNA-metry." However, approaches for statistical analysis and reasoning of the results in the case of ancient DNA are the same as in classical paleoanthropology. Both paleoanthropologists and ancient DNA researchers often explore the same samples, compare them with modern populations, and make conclusions about migrations and other events of the demographic history.

10.4.2 Methodology

Analysis of paleoDNA differs from analysis of modern DNA by strict demands to the purity of the experiment. Even a few molecules of human DNA is enough to obtain the results (as a rule, only a few molecules are preserved in an ancient sample). But the molecules of human DNA in the same concentrations fly in the air in any room and appear on any object touched by a researcher. Hence, there is a serious danger that a result was obtained not from the ancient DNA of the specimen but from a modern DNA that contaminated this specimen. For this reason, ancient DNA labs are designed to minimize the risk of contamination of ancient samples by the modern DNA, and the workflow includes many precautions. The results described next were obtained in cooperation with one of the best designed labs: Australian Center for Ancient DNA.

Another problem is the limited number of paleoanthropological material usable for ancient DNA analysis. So most ancient DNA studies are based on the low samples sizes, while dozens and hundreds of samples are normally needed for the accurate analysis of a gene pool. There are two ways for solving this problem. In the first, once the ancient DNA technology is improved and new laboratories are established, the number of studies and thus the cumulative sample size increase. Second, the lack of data on ancient samples could be partly compensated by large databases on the present-day populations used for the comparative analysis.

So, the advantages and limitations of ancient DNA and modern DNA analysis are opposite. In the case of modern DNA the plentiful data of gene pools are available, but it is problematic to reconstruct the genetic structure of the previous epochs. In the case of ancient DNA the immediate data of ancient gene pools are available, but the reconstruction is limited by the number of samples. Here describe the results of combined analysis of ancient and modern DNA data centered on the two populations from the north of the East European plain.

10.4.3 Yuzhny Oleni Ostrov

One example of ancient DNA analysis in comparison with the modern gene pools is the study of Mesolithic populations of Yuzhny Oleni Ostrov (Southern Reindeer Island). This study, we performed in collaboration with Australian Center for Ancient DNA, Museum of Ethnography and Anthropology (Kunstkamera), and Institute of Archaeology, Russian Academy of Sciences and published in 2013 (Der Sarkissian et al., 2013).

Yuzhny Oleni Ostrov burial site in South Karelia (Russia) is an important Mesolithic site in the forest zone of East Europe. The anthropological traits of this population (their partly Mongoloid phenotype) induced the unceasing discussion of the intensity, direction, and geographic scale of the migrations between western and eastern parts of North Eurasia in that time. The recent anthropological research (Khartanovich and Moiseyev, 2012) analyzed Yuzhny Oleni Ostrov individuals in the context of many ancient and modern populations and suggested that this population was an early member of the so-called "Urals race," which was intermediate but not necessarily admixed with Caucasoid and Mongoloid ones. Thus, the genetic makeup on this population becomes of great interest.

We managed to get the reproducible and full results on mtDNA haplotype and haplogroup for 9 out of 42 studied samples. We also analyzed two samples from Popovo burial, which is anthropologically and archaeologically close to Yuzhny Oleni Ostrov, thus increasing the sample size up to 11. Four of these samples belonged to haplogroup **U4**, three samples represented haplogroup **C**, two samples were of haplogroup **U2e**, and haplogroups **U5a** and **H** were represented by one sample each.

The most frequent haplogroup, **U4**, emerged in the southern regions of Eurasia, and it is now spread across West Eurasia with the maximum frequencies in the Ural region (both on the European side of the Ural Mountains and in West Siberia). The presence of haplogroup U4 in ancient Yuzhny Oleni samples indicated that it might have had a similar geographic pattern 7000 years ago. This also highlights the genetic relation of the Yuzhny Oleni Ostrov population with modern Ural populations.

The second most frequent in Yuzhny Oleni Ostrov, haplogroup **C**, is typical for contemporary Siberian populations. The rare occurrence in some northeast European populations reported in a few studies was always considered as evidence of gene flow from Siberia. The same interpretation could be had for occurrence of haplogroup **C** in the Mesolithic population.

The remaining three haplogroups found in Yuzhny Oleni Ostrov burial are typical for West Eurasia and dominate in populations of both East and West Europe.

So the haplogroup spectrum in the Mesolithic population of Yuzhny Oleni Ostrov is similar to that in modern Europeans, but haplogroup frequencies differ pronouncedly. Haplogroup **H**, the most frequent in modern Europeans, was found in a single individual, and two remaining typical European haplogroups, **T** and **J**, were not found at all. The haplogroup frequencies in the studied Mesolithic population are similar to the Mesolithic populations from Central and West Europe (Bramanti et al., 2009), particularly by the high frequency of haplogroup **U** variants (**U4** and **U5a**). This similarity of Mesolithic populations from East and West Europe was revealed not only by haplogroup frequencies ("branches" in phylogenetic tree) but also in haplotype spectra ("leaves" of this tree). So about one-third of the haplotypes found in Yuzhny Oleni Ostrov were found also in other Mesolithic Europeans (Der Sarkissian et al., 2013).

Thus, ancient DNA analysis revealed the intermediate genetic makeup of the population leaved Yuzhny Oleni Ostrov burial. In general, it belongs to the large group of Mesolithic European populations. However, the Yuzhny Oleni Ostrov gene pool contained also traces of a migration from the Urals or Siberia. The map (Fig. 10.6) shows to which degree different modern populations are similar to the Yuzhny Oleni one (genetic distances from mitochondrial gene pool of Yuzhny Oleni Ostrov to modern populations are plotted on the map). The high genetic distances (brown color on the corresponding territories) designate larger differences of modern populations from the ancient gene pool of Yuzhny Oleni Ostrov. Low genetic distances (light color) show modern populations genetically similar to the ancient gene pools of Yuzhny Oleni Ostrov. The map reveals these similar populations (light color) in West Siberia.

This result could be interpreted in a more decisive or in a more cautious way. The first interpretation suggests West Siberia was the homeland of this Mesolithic population in East Europe. The more cautious interpretation suggests that the proportion of

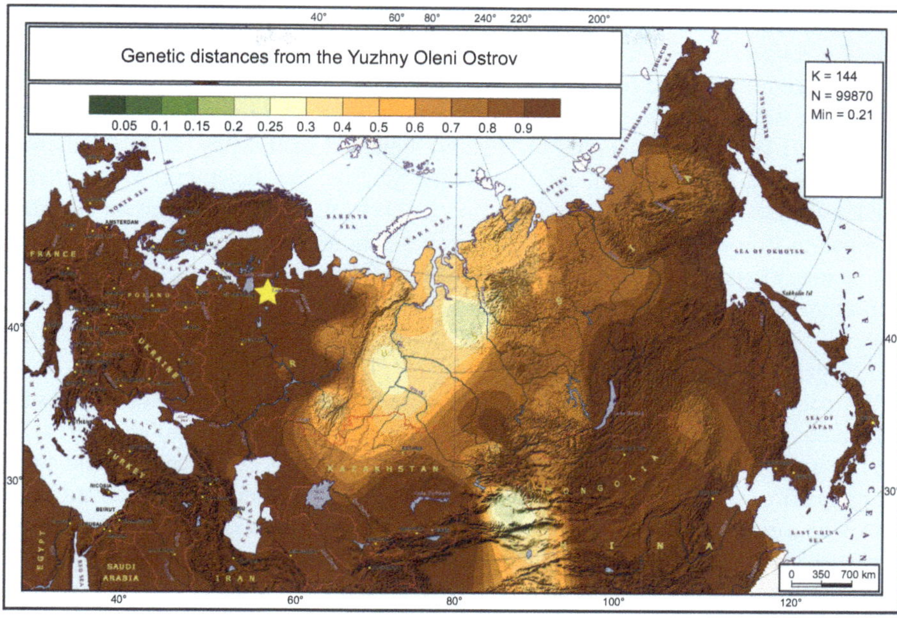

Figure 10.6 The map of genetic distances from the Yuzhny Oleni Ostrov population. The change in colors (from light green to brown) indicates the change of genetic distances between the present-day population of the given area and the studied ancient population. *K* – the number of the studied present-day populations; *N* – number of the grid nodes (nodes in sea not included); the *yellow asterisk* indicates geographic location of the ancient populations studied (aUzPo, the united Mesolithic population of the Bolshoy Oleni and Popovo graveyards).

western and eastern gene pools in the Yuzhny Oleni population was the same as it is in the modern West Siberians. In comparing with modern East Europeans, this proportion was shifted to the eastern component. This pattern might originate if Mesolithic peoples on both sides of the Ural mountains (East Europe and West Siberia) had a common source, which might be located somewhere close to the Urals.

10.4.4 Bolshoy Oleni Ostrov

This burial field located in the north of Kola Peninsula is dated about 3500 years ago. It belonged to the Early Metal epoch. Anthropological analysis demonstrated its intermediate position between Caucasoid and Mongoloid populations (Moiseyev and Khartanovich, 2012). The first canonical variable marked the west-to-east axis in phenotypic diversity, and Bolshoy Oleni Ostrov by this variable is even more shifted toward the typical Mongoloid populations from Siberia than Yuzhny Oleni Ostrov. The authors of that study emphasized also a second canonical variable, which differentiated the Ural populations from all other Eurasians. Both Oleni Ostrov populations are clearly Ural-like by this variable.

The ancient DNA analysis reproduced the first part of these anthropology conclusions. Due to nice DNA preservation in human remains from Bolshoy Oleni Ostrov, we managed to study 23 samples, which is two times more than from the Yuzhny one

(Der Sarkissian et al., 2013). We identified the following haplogroups in the Bolshoy Oleni Ostrov (ordered by decreasing frequency): **C**, **U5**, **D**, **Z**, **U4**, and **T**. Similar to Yuzhny Oleni Ostrov the list contains both European and Siberian haplogroups. But in contrast to Yuzhny Oleni, the Siberian haplogroups predominated. The Siberian haplogroups **D** and **Z** (not found in Yuzhny) appeared in Bolshoy, and the Siberian haplogroup **C** is the most frequent one. So, the Siberian genetic component predominated in the Bolshoy Oleni Ostrov mitochondrial pool. It implies a migration not from the close-to-Europe Trans-Ural areas but from the more distant inner parts of Siberia.

The map of genetic distances from the Bolshoy Oleni Ostrov population (Fig. 10.7) supported these suggestions. The map shows that modern populations from Siberia, especially in the Yenisey River basin in Central Siberia are genetically close to the ancient gene pool of Bolshoy Oleni Ostrov.

We concluded that the Bolshoy Oleni Ostrov population on the Kola Peninsula originated by a migration from the center of Siberia. But later the population decayed and has not left any traces in the modern gene pool of that area. Indeed, no one modern population of Europe (with the exception of Tundra Nenets, which arrived from Siberia historically recently) carry so high a frequency of Siberian haplogroups. Even more, no one haplogroup shows genetic continuity between the Bolshoy Oleni Ostrov population and modern Sami living in the Kola Peninsula now. Thus, the ancient DNA analysis revealed that west-to-east interactions in high latitudes in the past were even more intensive than one may suggest analyzing the present-day populations only.

This study has been supported by the Russian Foundation for Basic Research, project 16-06-00303.

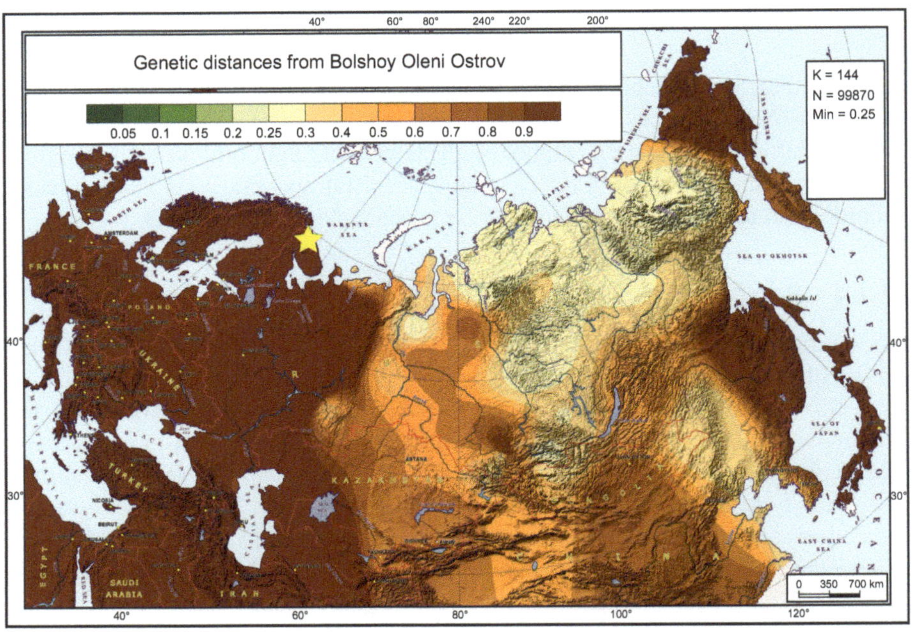

Figure 10.7 The map of genetic distances from the Bolshoy Oleni Ostrov population. **aBOO**: Early Metal Age population from the Bolshoy Oleni graveyard (see the *yellow asterisk* on the map).

Stages of the initial human colonization of Arctic and Subarctic

11

Andrey A. Velichko[1], *Sergey A. Vasil'ev*[2], *Yuriy N. Gribchenko*[1], *Elena I. Kurenkova*[1]

[1]Institute of Geography, Russian Academy of Sciences, Moscow, Russia; [2]Institute for the Material Culture History, Russian Academy of Sciences, St. Petersburg, Russia

Chapter Outline

11.1 Introduction

The initial penetration of the early man into the High latitudes proceeded against the background of an active dispersal of human groups over the plains of Northern Eurasia. It was a complex and multiple-stage process, and the chronology of successive stages could be noticeably different in different regions on the plains. The dispersal process left its distinct traces and may be reconstructed from the early man sites and signs of human economic activities.

The stages of human dispersal may be correlated with the main natural events of the Late Pleistocene; such attempts were repeatedly made and reflected in various stratigraphic schemes (Fig. 11.1). The vital importance of dating the Upper Paleolithic sites is quite understandable, as it is the dating on which the chronostratigraphy and periodicity of the initial colonization are based. In that way, cultural layers may be correlated with regional stratigraphic horizons, thus providing the base for revealing general patterns and local features of environmental changes. For example, an integrated analysis of the Late Pleistocene formations in the northeast of Europe gives an

Human Colonization of the Arctic: The Interaction Between Early Migration and the Paleoenvironment
http://dx.doi.org/10.1016/B978-0-12-813532-7.00024-3

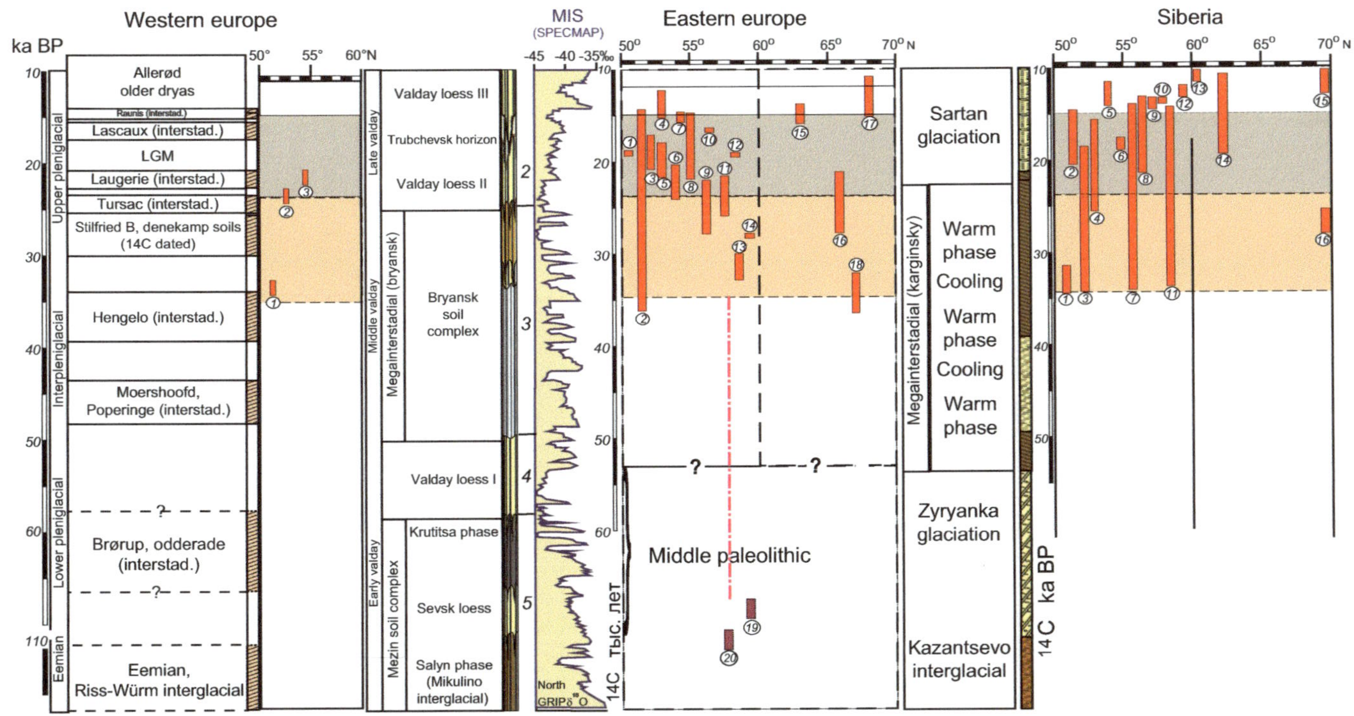

Figure 11.1 The Late Pleistocene chronostratigraphy and Paleolithic sites in Northern Eurasia north of 50°N. Western Europe: 1, Dolní Věstonice, 2, Gönnersdorf, 3, Gronau; Eastern Europe: 1, Kirillovskaya; 2, Kostenki; 3, Avdeyevo; 4, Yudinovo; 5, Pogon, Puchkari I; 6, Khotylevo II; 7, Shatrishchi; 8, Zaraisk; 9, Sungir; 10, Karacharovo; 11, Gornova; 12, Talitskogo; 13, Zaozerye; 14, Garchi I; 15, Medvezhya Cave; 16 –Byzovaya; 17, Pymva-Shor; 18, Mamontova Kurya; 19, Garchi I; 20, Peshcherny Log. Siberia: 1, Kara-Bom; 2, Mal'ta; 3, Varvarina Gora; 4, Sokhatino; 5, Tashtyk; 6, Tarachikha; 7, Kurtak; 8, Novoselovo; 9, Kokorevo; 10, Volchya Griva; 11, Ust'-Mil'; 12, Dyuktay; 13, Gari; 14, Lugovskoye; 15, Berelekh; 16, Yanskaya.

insight into the character of natural processes; the latter, in turn, determined the choice of ways the early human groups penetrated into high latitudes.

It is obvious that when dealing with geoarchaeological problems the dating results can be used only in combination with lithological, geochemical, paleontological, paleopedological, and other data. That integrated approach makes possible a mutual control of results obtained by different methods and an estimate of possible discrepancies. Such an approach is especially important in studies of complex multilayered sites and those of the Early Upper Paleolithic with partly redeposited material. In those cases the radiocarbon date series from one site may show a considerable chronological scatter. It is no less important to take into consideration that the Paleolithic people could use bones of mammals taken from the older death sites ("mammoth cemeteries"). In such a situation, the ^{14}C dates may differ considerably from the real time of the site functioning.

The Paleolithic man chose ways for his migrations and initial settling in the northern regions of Eastern Europe and Siberia against the background of complicated climatic fluctuations and recurrent transformations of environments in the Late Pleistocene. Having performed an integrated analysis of the available materials on chronology, archaeology, and environmental changes the authors were able to reconstruct the main stages of the Stone Age people dispersal and to get some idea of how the early man interacted with the environments.

The stages distinguished in the initial migration into and the subsequent colonization of the northern plains have no sharp chronological boundaries, and their identification is based both on archaeological criteria and on the pronounced changes of climate and environments recorded on a global scale.

As a first approximation, five stages may be recognized in the human settlement of the high latitudes.

Stage I—Middle Paleolithic—included the Mikulino (Eemian) Interglacial and the Early Valday interstadials; its chronological boundaries are tentatively defined as 115–55 (60) ka BP (until new information becomes available).

Stage II—Early Upper Paleolithic—might correspond to various phases of the Middle Valday mega-interstadial, 55 (60)–25 ka BP.

Stage III—Middle Upper Paleolithic—25 (23)–16 ka BP, covers the Late Valday (Late Würm, Weichselian), that is, the maximum Pleistocene cooling (Late Glacial Maximum, LGM).

Stage IV—final stage of the Upper Paleolithic—16 (15)–11 (10) ka BP, Late Glacial.

Stage V—Mesolithic and Neolithic, the first half of the Holocene—approximately 11–6 ka BP.

Before proceeding to characteristics of these stages, we should discuss briefly the first appearance of humans in the middle latitudes of Eurasia. At present, we have reliable evidence for the human appearance in the extratropical latitudes of Europe at the beginning of the Eopleistocene (Dmanisi in Georgia). There are Early Paleolithic sites attributable to Eopleistocene and Lower Pleistocene in the North Caucasus and on the southern coasts of the Sea of Azov (Aynikab I and Muhkay I in Daghestan, Kermek, Bogatyri/Sinyaya Balka, the Krasnodar Territory) (Amirkhanov, 2008; Shchelinskiy, 2011). For a long time the early humans stayed within the mountains and in the foothills where they were better protected from the northern cold winds. It was, however, as early as the Middle Pleistocene (400–300 ka BP) that the people expanded beyond the mountains onto the plain. That is witnessed by the sites Khryashchi and Mikhaylovskoye in the lower reaches of the Don and Severskiy Donets rivers (Praslov, 1969). At the beginning of the Valday epoch the people penetrated farther to the north along the Volga and Dnieper valleys (Sukhaya Mechetka and Khotylevo I).

The first appearance of man in Siberia was recorded at its southern margin, on the Altay-Sayan highlands, where man seemingly came from Central Asia. The human presence is evidenced by Acheulian industry with typical hand axes such as those discovered in Tuva by Astakhov (2008).

Karama on the Anuy River is considered to be the most ancient Early Paleolithic site of the Altay and of entire Northern Asia (Derevyanko et al., 2002a,b). The multidisciplinary studies suggested the earliest episode of the human occupation dated to the Early Pleistocene (800–600 ka BP), when the climate in the region was suitable for human life (Zykin et al., 2005; Bolikhovskaya et al., 2006; Derevyanko, 2009a; Derevyanko and Shunkov, 2009).

The Middle Pleistocene age is attributed to the complexes of tools recovered from caves (Denisova) and open sites (Ust'-Karakol-1) of the Altay Mountains. Samples from the basal part of the lower horizon in the Denisova Cave are dated at 282 ± 56 ka BP (RTL-548). Somewhat younger dates were obtained on artifacts recovered from the fluvial sediments at the base of the sequence at the Ust'-Karakol-1: 210 ± 42 ka BP (RTL-640) and 207 ± 41 ka BP (RTL-662) (Derevyanko, 2009a).

Another locality supposedly of the Middle Pleistocene age (most probably, corresponding to the time of the Tobolsk glaciation) is MK-1 in the Kuznetsk Depression that yielded a few artifacts (Derevyanko and Markin, 2007). A number of localities presumably also attributable to the Middle Pleistocene are known from the middle reaches of the Yenisey River, in the Kurtak region (Razlog, Kamenny Log, Ust'Izhul, etc.) and on the Angara River (Igetei Mountain). Among the most debatable sites of the Stone Age in Siberia is Diring-Yuryakh (middle reaches of the Lena River). Mochanov (1992) considers it to be of Eopleistocene age, though there are other opinions on the subject among the specialists (Alekseyev et al., 1990; Ranov and Tseitlin, 1991). Another site in Eastern Siberia also attributed by this author to the Early Paleolithic is Mungkharyma on the Viluy River (Mochanov, 2007; Mochanov and Fedoseyeva, 2001). At present, however, the data available are insufficient for the site chronological position to be precisely determined (see Section 11.2 in the monograph).

11.2 Middle Paleolithic

The sites and localities belonging to Stage I are dated to the interval from c.115 to 55 (up to 35) ka BP (Fig. 11.2). The epoch formerly known as Mousterian is now denoted by a broader term "Middle Paleolithic." The term is applied to the industries with diversified flake tools and widely (though not universally) spread Levallois techniques. It spans a broader time interval than the Mousterian and includes some sites formerly attributed to Late and Final Acheulian. The industries considered to be

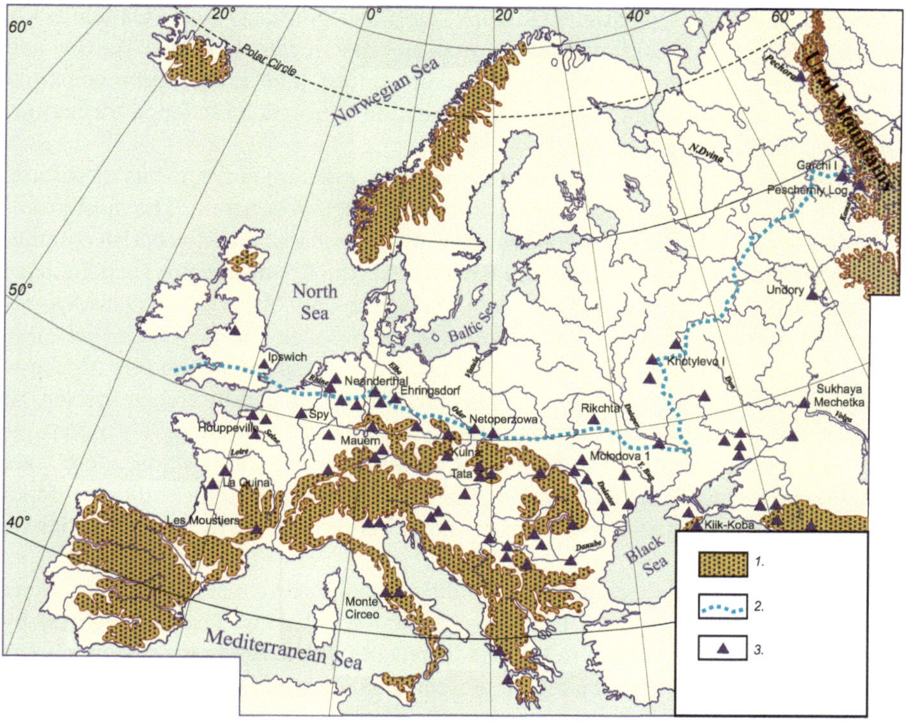

Figure 11.2 Middle Paleolithic sites in Europe. 1, mountains; 2, the maximal glaciation boundary (Saale – Dnieper); 3, sites.
After Mongait, A.L., 1973. Archaeology of Western Europe. Stone Age. Nauka, Moscow, 356 pp. (in Russian); Gerasimov, M.M., 1964. Stone Age People. Nauka, Moscow, 201 pp. (in Russian); Praslov, N.D., 1982. Man's dispersal over Europe in the Late Pleistocene. In: Gerasimov, I.P., Velichko, A.A. (Eds.), Paleogeography of Europe during the Last 100,000 Years. Nauka, Moscow, pp. 113–115 (in Russian).

Middle Paleolithic are diversified and include the Mousterian *sensu stricto*, Micoquian, Seclinian, Epi-Acheulian, etc.

Traces of human penetration onto the plains of temperate zones in extratropical Eurasia are known from Western Europe and the southern margins of Central and Eastern Europe, as well as from Siberia (Velichko, 1997; Derevyanko and Markin, 1992; Ivanova, 1965; Markov and Velichko, 1967; Praslov, 1969). There is no evidence of long-term settlements, and individual sites are separated in time by considerable intervals. That may be interpreted as a result of short-term occupation or individual tentative episodes of the human invasion (raids "there and back again").

The data available on most of open sites of that time (including fossil soils, pollen and spore assemblages, and faunal assemblages) suggest the site functioning fell mostly on the interstadials of the first half of the Early Valday (Ivanova, 1977, 1982; Velichko, 1988). The majority of the sites occurs in middle latitudes on plains and forelands of Europe (Le Moustier, La Ferrassie, Neanderthal, Steinheim,

Molodova 1, Korman' 4, Stinka, Rozhok, etc.) and Northern Asia (Okladnikova, Denisova, Strashnaya, and other caves sites at Altay; localities on the Yenisey and Angara rivers; and those in the Transbaikal region). The landscape reconstructions based on pollen data suggest a combination of forest and steppe formations (Grichuk, 1982, 1989, 2002).

Most of the sites of that stage are represented by disturbed fragments of cultural layers, so it is hardly possible to reconstruct their original structure. The most promising sites for the purposes of reconstructing environments of the Mousterian communities in the central Eastern Europe are Khotylevo I (the Bryansk region) and Sukhaya Mechetka (the Volgograd region), the latter being the only one with stone inventories in abundance, faunal remains, and hearth structures (Moskvitin, 1962; Grishchenko, 1965). Pollen assemblages suggest the prevalence of steppes with patches of forests (pine) in river valleys. The faunal remains are also dominated by steppe species. At Khotylevo I located farther north than Sukhaya Mechetka, the cultural layer contains stone tools and animal bones. As follows from the fauna composition, there were mostly open, and not forested, spaces in the region at the time of the settlement functioning. The site is tentatively dated at the Early Valday (Zavernyaev, 1972; Velichko, 1988).

There are also distinct signs of the Middle Paleolithic human penetration farther north in Eastern Europe at the beginning of the Early Valday. That is suggested by localities (not very numerous) in the middle reaches of the Kama River (Peshcherny Log, Garchi I). At that time the region presented an alternation of open treeless landscapes and open forests of spruce and birch. At Garchi I, stone implements of the Mousterian appearance have been collected from the beach at the base of the exposure. The original geological context of the findings and their initial position in the loess-soil sequence is still not clear (Pavlov and Makarov, 1998; Svendsen et al., 2008). Pavlov et al. (2004) identified the position of a few implements within a soil complex. The latter bears traces of cryogenic deformations probably related to the beginning of the Early Valday cooling.

Judging from the results of recent investigations, the range of the Middle Paleolithic localities in Siberia was close to that of the Early Paleolithic. It included the Altay, Kuznetsk Depression, valleys belonging to the drainage basins of the upper and partly middle Yenisey, Angara, the upper reaches of the Lena, and the Transbaikal region. Most of the Paleolithic localities are confined to rugged areas in mountains of South Siberia, in the upper reaches of large rivers. The region is noted for mosaic structure of landscapes, which apparently was favorable for a wide dispersal of the primeval human groups.

11.3 The Early Upper Paleolithic

At the second stage the process of Eurasia colonization from 55 (45) to 25 (23) ka BP was associated with early and middle Upper Paleolithic cultures: Chatelperronian, Szeletian, Aurignacian, and Gravettian (Anikovich et al., 2007; Renault-Miskovsky, 1985). In spite of the characteristics of the sites varying over a wide range, all of them belong to the prolonged and complicated Middle Valday mega-interstadial, an interval of milder climate separating the early and late stages of the Late Pleistocene glacial time. The mega-interstadial was identified on the basis of palynological and chronostratigraphic data (Arslanov et al., 1981; Velichko et al., 2008a,b; Spiridonova, 1983). Within this mega-interstadial, and in its first half in particular, there have been identified phases of rather high values of heat supply typical of an interstadial. That gave grounds to define the interval as mega-interstadial MIS 3.

As follows from the paleobotanic data, the landscape characteristics indicative of a higher heat supply appear from 35–32 kyr BP. In the northwest of European Russia, that time was characterized by subtaiga forests with considerable proportion of broadleaf species, the landscapes not unlike an interglacial (Spiridonova, 1983).

Similar landscapes were reconstructed in central European Russia, at the sites of Kostenki XVII (Fedorova, 1963) and Kostenki XII (Levkovskaya et al., 2005). Broadleaf species are found in pollen spectra from Sungir and Zaozerye (northeast of the East European Plain) dated to the Bryansk interval. The second half of the MIS3 stage, therefore, featured conditions close to interglacial in the heat supply level (Velichko, 2009a). Taking all these into consideration, it has been suggested to replace the term "mega-interstadial" in reference to MIS 3 stage with the more general term "mega-interval" (or "Bryansk mega-interval" if relating to Eastern Europe).

Sites attributed to this stage of the initial colonization are found in many regions of Northern Eurasia. They are dated mostly by radiocarbon, and multidisciplinary studies have been carried out on many of them. Some of the sites have distinct cultural layers that contain lithic and bone tools, mammal bones, and remains of bone and ground constructions and hearths; the findings suggest either continuous long-term habitation on the site, or repeated episodes of colonization.

In Eastern Europe, large settlements of that time are found not only in the south, but also in the central part of the plain (Kostenki I, XII, Sungir, etc.). Several northernmost sites of the Upper Paleolithic—Zaozerye, Garchi I, Byzovaya, and Mamontova Kurya—are also dated to that chronological interval. The most noticeable feature in the spatial distribution of the sites is that they are concentrated within certain areas in drainage basins of the Dniester, Dnieper, Desna, Don, Oka, Kama, and Pechora rivers. The sites are diversified in age and structure. By the radiocarbon dates available at present, they may be classified into several groups (Table 11.1).

It should be noted that the territories colonized underwent complicated transformations of landscapes at the previous stage of their history. Both landscapes and topography of the north of Eastern Europe repeatedly experienced an impact of ice sheets at the time of Dnieper glaciation (the later part of the Middle Pleistocene) and of the last—Valday—glacial epoch. The main center of ice sheet formation was Scandinavia, though the ice sheets centered on Novaya Zemlya and Polar Urals were also of importance (Andreicheva, 1992; Gribchenko, 1994, 2005; Lavrov, 1974; Lavrov and Potapenko, 2005; Yakovlev, 1956).

An important factor in the relief transformations that predetermined its principal features by the time of the initial penetration of Paleolithic humans to the north was erosion and deposition by glacial meltwater. The relief of the river drainage basins the main Paleolithic sites are concentrated in (the Kama, Oka, Desna, middle reaches

Table 11.1 **Paleolithic sites of the second stage of colonization in Eastern Europe north of 50°N according to the radiocarbon age of the cultural layers**

Intervals of ^{14}C dates, ka BP	Drainage basins of rivers				
	Dnieper	Oka	Don	Kama	Pechora
32–25	Yurovichi	Sungir' Rusanikha	Kostenki XII Kostenki XIV	Zaozerye, Garchi I	Byzovaya
40–33	–	–	Kostenki XII Kostenki XIV	–	Mamontova Kurya

of the Dnieper) developed mostly on glaciofluvial and glacio-lacustrine sediments (Butakov, 1986; Vereshchagina, 1965; Goretskiy, 1964; Gribchenko, 2005, 2008; Lider, 1976; Obedientova, 1977). Those sediments are overlain with subaerial formations including a series of loess and eolian-deluvial loams with fossil soil horizons between them. It is those formations that the Upper Paleolithic cultural layers are associated with in the Kama drainage basin (Garchi I, Zaozerye).

The Late Pleistocene archaeological localities known from the north of Eastern Europe are less numerous as compared with its central and southern regions. The Paleolithic sites discovered as yet are concentrated in the eastern part of the plain and in the Ural forelands (the Kama and Pechora drainage basins). Only a few finds of the Late Paleolithic lithics are known from the Severnaya Dvina basin (Devyatova, 1982). There are, however, numerous localities of the Late Pleistocene mammoth fauna in the region, which suggests environments suitable for large mammals. If such is the case, groups of the Upper Paleolithic hunters could occasionally visit the region.

Farther north, in the Pechora drainage basin, the most important sites (Byzovaya, Mamontova Kurya) are related to small valley (*balka*) deposits and usually bear traces of redeposition (Guslitser and Liyva, 1972; Kanivets, 1976; Pavlov, 1997).

The Pleistocene formations have been studied in the regions of Byzovaya, Mamontova Kurya, Garchi I, and Zaozerye sites as a part of the PECHORA and ICEHUS Research Projects (Svendsen et al., 2008; Pavlov et al., 2004; Svendsen and Pavlov, 2003). In the same years, integrated paleogeographic studies were carried out on the sites of Zaozerye and Garchi I in the middle reaches of the Kama River, their lithology and stratigraphy, as well as paleosols and paleogeomorphic characteristics, being compared with sequences in the adjacent regions (Gribchenko, 2005, 2008; Pavlov et al., 2005; Timireva and Velichko, 2008). The results permit a comparison with similar data from other regions of Northern Europe.

Mamontova Kurya is in the Usa River valley at the latitude of the Arctic Circle; it is the earliest Upper Paleolithic locality in the region. Eight radiocarbon dates obtained on mammoth bones are in the range from 31 880±390, T-15727 /35 790±430 cal. yr

BP/ to 37 360±630, LU-4001 /41 790±490 cal. yr BP/ (Gribchenko, 2008; Pavlov, 2002; Svendsen and Pavlov, 2003). The pebble and gravel balka deposits contain abundant bones of mammals (mostly mammoths). Several stone tools were found nearby, as well as a mammoth tusk with incisions (Pavlov, 1997). Various and rather complicated interpretations given as yet to the locality genesis and age call for further multidisciplinary investigation.

Similar is the situation with Byzovaya (vicinities of Pechora town) (Guslitser and Liyva, 1972; Kanivets, 1976; Pavlov, 2002). The initial position of the site and its structure are closely related to fluvial gravels bearing abundant mammoth bones; it is not inconceivable that the bones have been redeposited. ^{14}C dates obtained on the bones (12 dates) vary from 25 450±380, TA-121b /29 640±470 cal. yr BP/ to 29 170±340, ЛУ-3983 /33 300±390 cal. yr BP/ and 33 180±2020, ЛУ-4007 /38 200±2410 cal. yr BP/ (Gribchenko, 2008; Pavlov, 2002; Svendsen and Pavlov, 2003). As is Mamontova Kurya, Byzovaya is associated with large mammoth death sites.

Zaozerye and Romakhino (the Chusovaya River valley), as well as Garchi I in the Kama valley, are related to the third terrace. The latter forms several levels in the region, each of them presumably corresponding to a certain stage of the maximal Middle Pleistocene glaciation (Gribchenko, 2008). The structure and stratigraphy of those geomorphic elements are traceable distinctly in river valleys of the Kama and its tributaries (Generalov, 1965; Goretskiy, 1964; Gromov, 1948; Obedientova, 1977).

The sequence of sediments includes a pedocomplex (Mezin) that records several stages of environmental and climatic changes: there are soil horizons related to warmings and a series of cryogenic deformations indicative of episodes of cooling. Above the Mezin pedocomplex, there is another soil, Bryansk, with the Zaozerye cultural layer in its upper part. The layer contains bones of large mammals, mostly of horse, along with bones of woolly rhinoceros, reindeer, hare, and mammoth. L.D. Sulerzhitsky obtained several ^{14}C dates on the bones, three of them around 31 ka BP and one at 31.5 ka BP. Besides, there are four ^{14}C AMS dates in the range of 30–31 ka BP and one AMS date on bone at about 35.1 ka BP, the latter being taken as "most reliable" out of a large series. The calibrated dates give an older age of the site: up to 38–39 cal. ka BP (Svendsen et al., 2008). If such is the case, the site may be considered Early Upper Paleolithic or transitional from the late Mousterian to the Initial Upper Paleolithic.

The position of another Early Upper Paleolithic site—Garchi—in the Kama valley on the third terrace is analogous to Zaozerye on the Chusovaya River. The cultural layer sampled in excavation no. II is dated by ^{14}C AMS on charcoal at 28 750±795, TUa-941 /32 800±830 cal. yr BP/ and several OSL dates (33±2, 38±5, 45±2, 46±7 ka BP) were obtained for samples from paleosols and loess (Svendsen et al., 2008). The correlation between OSL and radiocarbon dates should be a subject of special consideration. At present, we may only tell with confidence that the episodes of Upper Paleolithic occupation at Zaozerye and Garchi were undoubtedly close to each other in time and similar in environments.

To estimate paleoenvironmental characteristics of the Upper Paleolithic sites in the Kama drainage basin in comparison with other sites in Eastern Europe, it seems advisable to compare them with a large site of Sungir in the Oka drainage basin (vicinities of the city of Vladimir), the northernmost site in a group of Early Upper

Paleolithic settlements in the center of the East European Plain. Considering the problems of initial human colonization of the European North, Sungir is of special interest as its age (about 28 ka) and characteristics of its stone industry (known as belonging to the Streletskaya–Sungir) are similar to those described at Garchi I and Byzovaya (Pavlov, 1997; Pavlov and Makarov, 1998; Svendsen et al., 2008). The Sungir cultural layer occurrence in the loess-paleosol series has much in common with Garchi I and Zaozerye (Bader, 1978; Gribchenko, 2008; Lavrushin and Spiridonova, 1998).

Detailed chronostratigraphic studies of paleosol and the cultural layer at Sungir revealed a cyclicity in environmental changes in the region of the settlement (Lavrushin et al., 2000). The soil horizon displays a complicated structure related to deluvial (sheet and rill wash) and cryogenic processes. The observed cyclicity of natural processes at Sungir (indicative of the multilayered character of the site) provided new insight into similar data obtained on northern sites, such as Gornova, Zaozerye, Garchi I, and Byzovaya.

Such complex structure is most distinctly seen in the Mezin pedocomplex. The time of its development spanned two warm epochs: the Mikulino Interglacial (Salyn phase) and Brørup and Amersfoort interstadials (Krutitsa phase) (Velichko and Morozova, 2005); the two warm epochs are separated by the cooling phase that marked the onset of the Valday cold epoch (Sevsk loess horizon).

Similar regularities are noted in the Bryansk buried soil, its humus horizon being dated by radiocarbon in the range of 31–23 ka BP. When occurring in the loess sequences, the soil includes various phases of the longer Middle Valday mega-interstadial from 55 (60) to 23 (25) ka BP. In case the deposition took place under specific conditions of gentle slopes and hollows, several individual phases are often distinguishable in pedocomplexes. A similar pattern has been observed in the loess-paleosol sequences at Willendorf and Dolni Vestonice Paleolithic sites, as well as at multilayered sites in Eastern Europe on the Dniester, Dnieper, and Don rivers. All the listed sites have been most thoroughly studied and repeatedly dated by radiocarbon. The data available on the archaeology and paleogeography of the open sites in different regions of Northern Europe are easily correlatable, which suggests a similarity of environments and natural processes all over the plain territory colonized by the Upper Paleolithic people in the Late Valday. The earliest presence of man (before 35 ka BP) is recorded mostly in southern and central regions of Europe, on the middle Don in particular: Kostenki XII (Anikovich et al., 2007), Kostenki XIV (Sinitsyn, 2004). It may be safely supposed that at the Early Upper Paleolithic the modern man communities still adhered to Middle Paleolithic traditions when choosing migration ways or places for settlements.

Specific features recognizable in the chronostratigraphy of sites and in the initial colonization of the East European North are well comparable with some sites of Central and Western Europe. The most interesting in this regard are sites of Willendorf (Austria), Dolni Vestonice, Pavlov, Predmosti (Czech Republic), Krakow, and Spadzista (Poland). The cultural layers of the sites are related to fossil soil horizons and are substantiated with a series of ^{14}C dates. Those sites are attributed to a group of the so-called "Loess Paleolithic" and are confined to the areas of loess (eolian) formations of the Late Pleistocene (Damblon et al., 1996; Escutenaire et al., 1999; Klima et al., 1961; Svoboda, 2000).

The central European sites have been compared with the Upper Paleolithic sites of the East European Plain using various archaeological and paleogeographic criteria (assuming they belonging to a single—Willendorf–Pavlov–Kostenki–Avdeevo—culture) (Markov and Velichko, 1967; Grigoryev, 1968; Klima et al., 1961).

More complicated was the situation at the early stage of the Upper Paleolithic in Northern Asia. Most of archaeological sites of that time are concentrated in Southern Siberia, from Altay to Transbaikal (Abramova, 1989; Velichko et al., 2002a,b). The early dates obtained on wood samples from the Aldan sites (Mochanov, 1977; Tseitlin, 1979) are criticized by an increasing number of specialists. It is difficult, therefore, to come to a definite conclusion on the real penetration of humans into Northeastern Asia at that stage. The northernmost site complex at the lower Yana River (70°43′N) belonging to the same time interval is dated at about 27–28.5 ka BP and related to the warm phase of the Karginsky (Pitul'ko and Pavlova, 2010). Integrated studies of that complex and reconstructions of the paleoenvironments suggest the human habitation under conditions of the climate warmer and dryer than at present. The dominant landscape was tundra-steppe dominated by xerophytes (*Artemisia*), with shrub birch on the river floodplains. The fauna was also dominated by tundra and steppe species: mammoth, bison, horse, musk ox, reindeer, polar fox, etc. The archaeological finds permit the site attribution to a transitional type from Early to Middle Upper Paleolithic (Pitul'ko et al., 2007).

11.4 Middle Upper Paleolithic

Northern Eurasia

Pleniglacial

The time frame of the third stage of human colonization is defined approximately as 25 (23)–16 ka BP (Fig. 11.4). Many archaeological sites and localities all over Northern Eurasia are related to this stage. Judging from the radiocarbon dates, most of the sites were functioning at the Pleniglacial time, which is the main phase of the Late Valday glaciation.

In spite of the Scandinavian ice sheet expanding over a considerable territory of the East European Plain, the Upper Paleolithic groups did not leave the plain even at the time of maximum cooling. Though most of the sites concentrated in its central and southern parts, the human settlements existed in the northeast of the plain, in the Kama River basin (Talitskogo), and even in the upper reaches of the Pechora River (the Medvezhya Cave). The northernmost localities are dated at 18–17 ka BP. They are insufficiently studied, but the data available at present suggest that the prehistoric humans were well adapted to the harsh environments and climate.

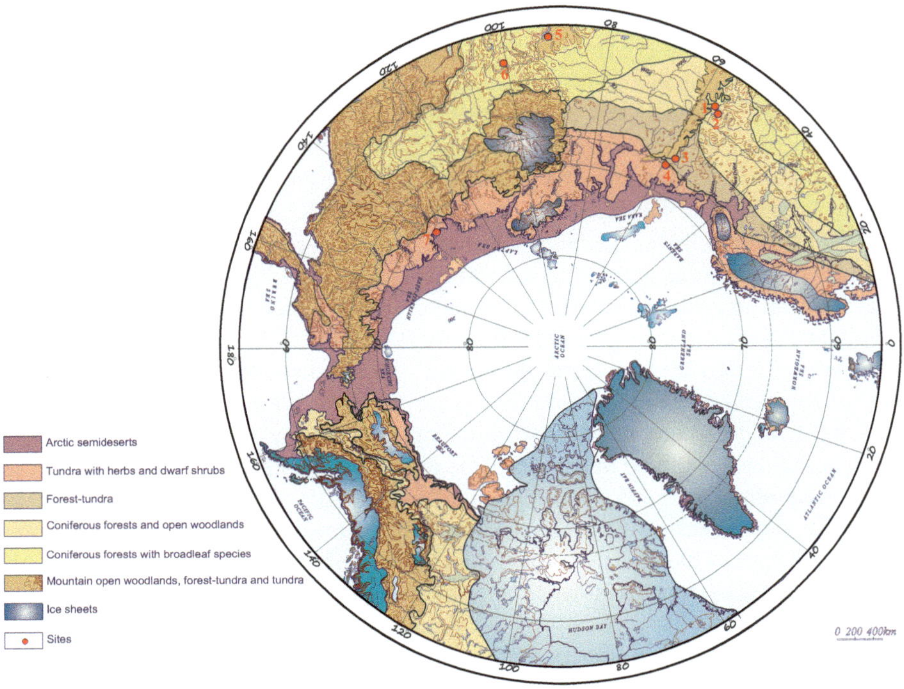

Figure 11.3 Reconstructed environments of the human settlement at the time of MIS 3 warm mega-interstadial (Paleolithic sites dated by 40–25 ka BP). Sites marked by numbers on the map: 1, Zaozerye; 2, Garchi I; 3, Byzovaya; 4, Mamontova Kurya; 5, Sabanikha, Kurtak-4, Kashtanka-1; 6, Ust'-Kova; 7, Yanskaya.

It may be supposed that some specific natural processes hindered the human dispersal during the last glacial epoch. Among them, one might name an active subaerial sedimentation of loess material (silt) and widely spread ice-dammed lake basins, as well as restructuring of river valleys. That imposed limitations on the choice of place for living at that time more than at any other time of the human colonization. Similar natural conditions existed in the most of northern regions of the Eurasian plains. Considering the severity of environments, it is quite possible that some part of the population could migrate from the high latitudes to the south at the time of extremely drastic changes in the relief-forming processes and the loss of paleo-surface stability. Due to such migrations a quantity of sites could increase in the central and southern regions and decrease in the north.

A characteristic feature of the considered stage of initial colonization is the occurrence of the Upper Paleolithic settlements in various regions of Eastern Europe. Most landforms in river valleys (the Dniester, Dnieper, Desna, Seim, Oka, and Kama rivers) and river terraces in particular were stable enough for the early man to perform long-distance migrations and even to form stable settlements. Many sites were positioned in

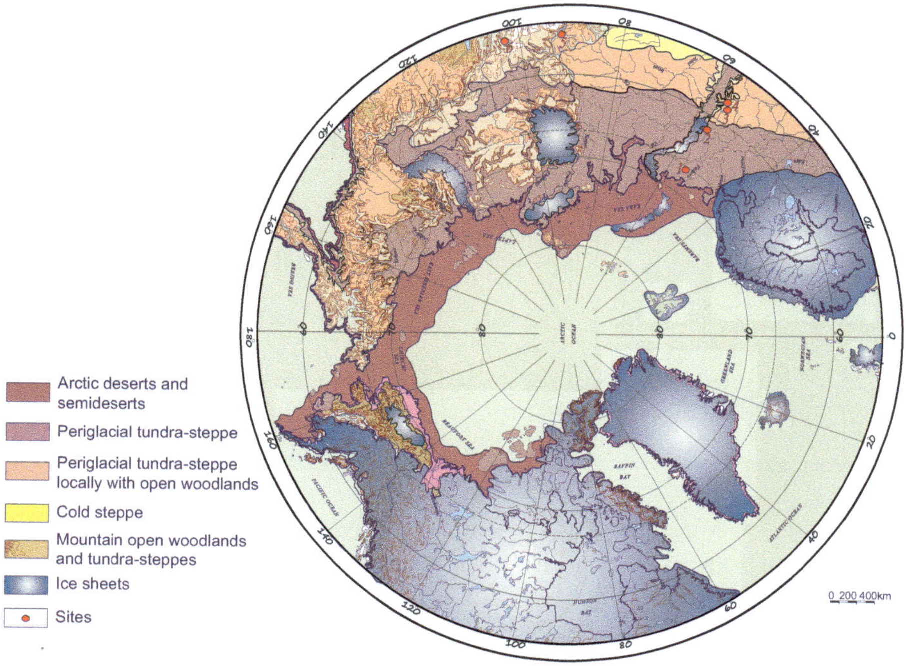

Figure 11.4 Reconstructed environments of the human settlement at the time of maximum cooling of the Pleistocene (MIS 2) (sites dated by 20–18 ka BP). Sites marked by numerals on the map: 1, Pymva-Shor; 2, Medvezhya Cave; 3, Shirovanovo II; 4, Talitskogo; 5, Shlenka, Tarachikha; 6, Krasny Yar I, Cheremushnik.

the upper part of the valley slopes or on low (most often on the second) terraces. Such are many sites described in the Dnieper drainage basin: Radomyshl (19 ka BP), Kirillovskaya (19.2 ka BP), Berdyzh (22.5 ka BP), Khotylevo II (23.6 ka BP), Pushkari I and Pogon (22 ka BP), Avdeevo (22.4 ka BP), Byki (17 ka BP), etc. A group of sites (Zaraisk) existed in the Oka drainage basin on a high interfluve between 21.6 and 16.7 ka BP (Table 11.2).

At the same time, there existed the Upper Paleolithic settlements in the Kama (Talitskogo, Gornova) and the Pechora basins (Medvezhya Cave, Stolbovoy rock shelter) indicative of human penetration far to the north along the eastern forelands of the Ural Mountains (Gromov, 1948; Kanivets, 1976; Pavlov, 2008a).

In Siberia the considered stage of the Upper Paleolithic human dispersal fell on the last (Sartan) glacial time. No Upper Paleolithic sites of that time have been found north of 60°N. The known sites mostly concentrate in Southern Siberia: the area traditionally inhabited by the Paleolithic groups. A series of sites dated to that interval are known in the south of Western and Eastern Siberia (Tomskaya, Shestakovo, Sabanikha, Krasny Yar 1, Sanny Mys, etc.) (Abramova, 1989; Lisitsyn and Svezhentsev, 1997). There are, however, signs of human penetration farther northeast along the Lena and

Table 11.2 Paleolithic sites of the third stage of colonization in Eastern Europe north of 50°N according to the radiocarbon age of the cultural layers

Intervals of ^{14}C dates, ka BP	Drainage basins of the rivers				
	Dnieper	Desna	Oka	Kama	Pechora
18–16	–	Eliseyevichi Byki	–	Talitskogo	Medvezhya
20–18	Kirillovskaya Radomyshl'	Novgorod-Seversky	Zaraisk	Bezymyanny	–
24–20	Berdyzh	Khotylevo II Avdeevo Pushkari I	–	–	Stolbovoy

Aldan valleys (the Khaiyrgas Cave dated by radiocarbon at 21–16 ka BP; Verkhne-Troitskaya site dated at about 18 ka BP on charcoal).

11.5 Late Upper Paleolithic

Northern Eurasia

Late Glacial

The fourth stage dated at 15–11 (10) ka BP (Fig. 11.5) has been studied much better compared with the Pleniglacial. The sites attributed to that stage are found on different geomorphic levels, mostly on higher fragments of the first terraces and on interfluves. The archaeological sites of that age attributed to the Final Paleolithic are widely spread over Northern Eurasia and Alaska. The cultural layers of the sites often contain remnants of large constructions of bones and earth dwellings of complex structure, semisubterranean structures. They mostly occur in the central and southern regions, that is, outside the high latitudes of Eurasia.

There is evidence of human penetration into northern regions at that time found in the middle reaches of the Kama (Gornaya Talitsa, Kapova Cave, Smelovskaya, Bobylek) and Pechora (Pymva-Shor, etc.). Sites in the northern Urals are usually located in caves or rock shelters.

The choice of locations suitable for long-term settlements seems to have been limited in the Kama drainage basin; that may be inferred, on one hand, from the absence of large dwelling places (at least, known at present), and on the other, from numerous localities bearing traces of the Paleolithic hunter visits. The latter suggests a high mobility of the Late Paleolithic groups oriented on short-term stay in seasonal hunting camps.

In Northern Siberia the fourth stage of human colonization continued at least to 10 kyr BP; the ^{14}C dates of the Late and Final Upper Paleolithic sites vary in the range

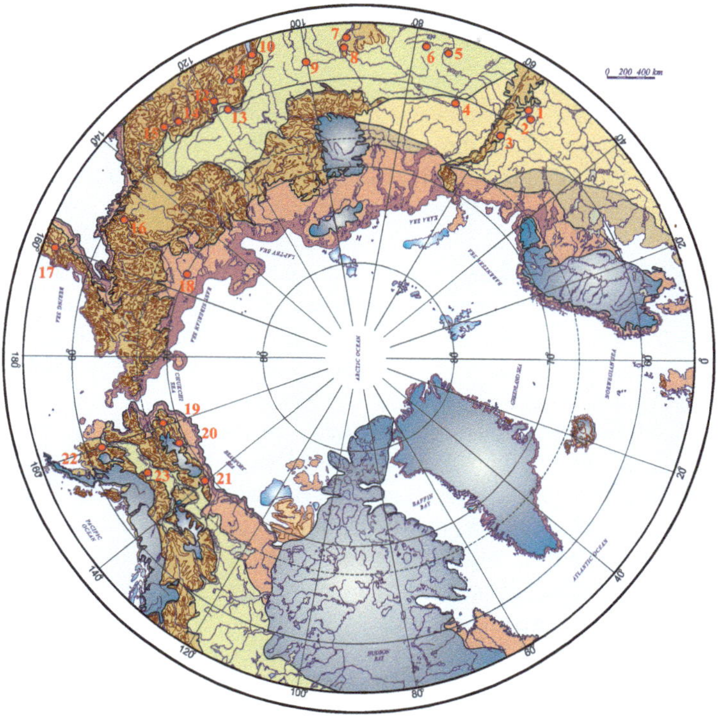

Figure 11.5 Reconstructed environments of the human settlements at the Late Glacial time (at the end of MIS 2 interval). For explanations, see Fig. 11.4. Sites marked by numerals on the map: 1, Kapova Cave; 2, Talitsa; 3, Gari; 4, Lugovskoye; 5, Chernoozerye-2; 6, Volch'ya Griva; 7, Afontova Gora, Listvenka, Biryusa-1, etc.; 8, Druzhinikha; 9, Ust'-Kova; 10, Kurla; 11, Bolshoy Yakor', Avdeikha; 12, Novy Leten-1; 13, Khaiyrgas; 14, Ust'-Timpton; 15, Dyuktay; 16, Siberdik; 17, Ushki; 18, Berelekh; 19, Tuluaq; 20, Mesa; 21, Bedwell, Hilltop; 22, Spein Mountain; 23, Broken Mammoth, Dry Creek, Owl Ridge, etc.

14–10 kyr BP. The earliest sites are concentrated in Southern Siberia (Velichko et al., 2002a,b). The northernmost sites are found in the Lena drainage basin at about 60° and farther north. Among them are the Dyuktay cave dated at 14–12 ka BP (Mochanov, 1977) and at least some of the materials from the Berelekh site on the Indigirka River.

In north of West Siberia, there are only a few sites known, including Gari at 59°N (eastern forelands of the Urals) and Lugovskoye at 61°N (center of West Siberia, near the city of Khanty-Mansiysk).

The latter locality yielded stone tools, mostly redeposited, and abundant bones of large mammals dominated by mammoth (Mashchenko et al., 2006). The faunal remains included also bones of woolly rhinoceros, horse, bison, reindeer, musk ox, etc. The radiocarbon dates vary from 20 to 12 ka BP. So the site may be considered evidence of the northernmost settlement of the Late Upper Paleolithic human groups in West Siberia.

North America

Late Glacial

A few last millennia of the Pleistocene were marked by an extremely important event in human history: the colonization of the New World. As the process was of prime importance, we will dwell on the archaeological evidence of the first settlers in North America [for more details, see the reviews (Haynes, 2002; Meltzer, 2009; Graf et al., 2013)].

The origin of the Native American ancestors is one of the most challenging problems; it calls for a synthesis of information on paleogeography, physical anthropology, ethnography, folkloristics, linguistics, and archaeology. Here we will try to estimate the available archaeological data and to present a review of the ideas on the time and ways of the Paleolithic man's penetration into North America.

The recent findings have changed our views on the process of colonization; the latter is seen now not as a single unidirectional migration wave directed from the Northeast of Asia through the Beringian Land Bridge to Alaska (that is how the colonization of America is described up to now in most textbooks and monographs), but as a complicated system of migrations having taken place at different times and in various directions, including a seemingly reversed (with reference to the main trend) movement from south to north.

We will leave aside the fantastic version of man's coming to America from Europe over ice of the frozen Atlantic Ocean or skirting the Laurentide Ice Sheet from the north, crossing the Canadian Arctic Archipelago on their way to the Atlantic zone of North America. On the exclusion of those versions, there is, in fact, only two possible ways of the first men's penetration into the New World.

According to the first scenario, groups of hunters could migrate over Beringia eastward, that is, from Chukchi Peninsula to the central part of Alaska. If such was the case, they could only follow valleys between the Brooks Range with alpine glaciers in the north and more heavily glaciated Alaska Range in the south. Their further penetration into the continent would necessarily follow along the so-called ice-free corridor along the Mackenzie River, between the Laurentide and Cordilleran ice sheets. Such a scenario of the America colonization prevailed in the special literature until the 1990s, when it was replaced by the coastal migration variant, now generally recognized.

When the specialists tried to substantiate the hypothesis of the interior route of the migration, they met with considerable difficulties in finding traces of the early man and reconstructing paleoenvironments within the ice-free corridor. We will return to the subject subsequently.

According the second scenario—"coastal"—the first settlers of America moved along the Pacific coast using boats to pass by the glacier tongues that came occasionally into the sea. Unfortunately, the postglacial rise of the sea level concealed the supposed migrant traces. At present, considerable efforts are being made by underwater archaeologists intending to prove or to disprove the concept.

We now turn our attention to the oldest archaeological materials from Alaska. Judging from the recent results of geneticists, the settlement of Beringia occurred even before the LGM. Needless to say, it is theoretical reasoning only, not confirmed by archaeological evidence yet. Besides, various groups of geneticists differ considerably in their chronological estimates. The only (and not inarguable) evidence of the very early appearance of man in the region was obtained from the Bluefish Caves (the Yukon Territory, Canada). The stone tools and bones recovered from the deposits are dated to the interval 12–25 ka BP (Fig. 11.6).

There is still a controversy as to whether the stratigraphy of cave deposits can be ascertained with precision or objects of different age may be mixed together. If we agree with the previously cited age of the Bluefish findings, they may be regarded as evidence of the earliest human appearance in Eastern Beringia, that is, in perfect agreement with estimates made by geneticists.

The oldest credibly dated sites in Alaska belong to the final Pleistocene, beginning from the interval between Allerød and Bølling, but the main body of sites is dated to the very end of the Pleistocene, the Younger Dryas. Even such scarce chronological

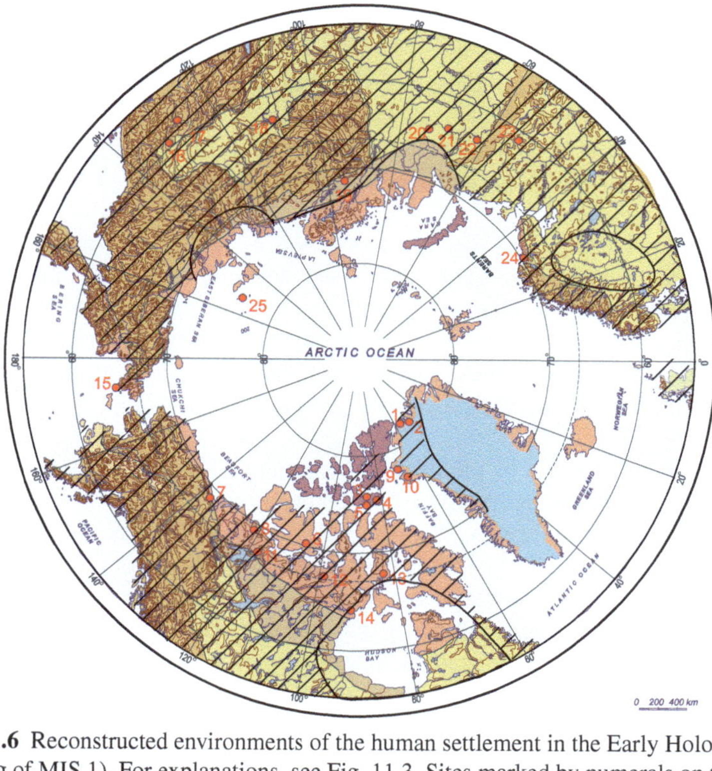

Figure 11.6 Reconstructed environments of the human settlement in the Early Holocene (beginning of MIS 1). For explanations, see Fig. 11.3. Sites marked by numerals on the map: 1, Kettle Lake; 2, Sojourn; 3, Cape Storm; 4, Port Refuge; 5, West Wind; 6, Roky Point; 7, Engigstciak; 8, Umingmak; 9, Digarmulen; 10, Scraeling ASTt 7; 11, Bloody Falls; 12, Wellington Bay; 13, Mittimatalik; 14, Kaleruserk (Parry Hill); 15, Nayvan; 16, Belkachi-1; 17, Ust'-Timpton; 18, Ust'-Chirkuo; 19, Tagenar-6; 20, Et-To-1; 21, Amnya-1; 22, Ches-Tyy-Yag; 23, Parch-2, Cherdyb-1, 2; 24, Zarubikha-2, Orlovka-6, 7; 25, Zhokhov Island.

data testify against the interior route (the first variant), as the oldest sites in the central part of North America (south of the ice sheets) are much older than those in Alaska. Moreover, the earliest sites concentrate in the south and southeast of North America. When considering the geographic distribution of the sites, one has the impression that after the first human arrival to the continent the further colonization proceeded from the south to north and northwest.

Turning back to the Alaskan sites, the known sites of the Pleistocene age are concentrated in the center of the region, in the Tanana and Nenana valleys. According to conventional ^{14}C dates, the age of the sites is about 10–12 ka BP (14.7 ka BP and less on the calibrated scale). The stone industries of the oldest sites belong to two traditions (in all probability, they coexisted); the main difference between them consists in the presence or absence of microblade technology. The first tradition known under the name of Nenana is noted for the presence of specific teardrop-shaped points. Another tradition known as Denali features an advanced microblade technology based on wedge-shaped cores and diversified leaf-shaped bifaces. The latter variant of industry has close analogs in Late Upper Paleolithic cultures in the east of Asia (Dyuktay sites on the Aldan River, some components of the Ushki sites in Kamchatka, Selemdzha culture in the Far East of Russia, etc.).

The aforementioned variation has been long discussed in American archaeology. In common with other Paleolithic regions, the characteristics of stone industry were probably influenced both by the culture traditions of different groups and by many other factors (various functions and seasonality of sites, type of economic activities, intra-site variability, etc.).

Besides, there has been recorded one more culture tradition on Alaska dated to the end of the Pleistocene and different from the aforementioned Nenana and Denali traditions. The complexes attributed to it occur mostly on the periphery of Alaska (in the north and west of the region) and include various forms of projectile points. Because of a noticeable similarity of these industries to the cultures of the North American continent south of the ice sheets, they were given the name of Northern Paleoindian traditions.

There are three variants of the described complexes known as Mesa, Sluiceway (after the names of sites on Alaska), and complexes with fluted points. The sites with those industries are dated to 13.2–10 ka BP (calibrated). In the opinion of some researchers (Smith et al., 2013), those assemblages may be result of a backward wave of migration at the end of the Pleistocene from the main part of North America to the northwest by the ice-free corridor, which enlarged considerably at that time.

That such migration could occur is partly confirmed by the genetically proven origin of the Alaskan bison from their ancestors formerly inhabiting the Great Plains. Paleoenvironmental reconstructions show much more favorable natural conditions in the Mackenzie Corridor than was thought before. Unfortunately, all the traces of Paleoindians known in the region are dated to the Younger Dryas and concentrated in the southern part of the region. Among other things, the Clovis points have been found there. That is another evidence of the cultural tradition being propagated northward.

Now, we consider the material from the Pacific coastal zone. In the opinion of a considerable group of scientists, it is along that zone that the routes of the initial migration into the New World ran. Unfortunately, the reliably documented sites in the region are dated either to the very end of the Pleistocene, or (mostly) to the Early Holocene, and therefore bear no relation to the discussed problem of the initial colonization of the American continent. The most ancient traces of humans on the Pacific coasts are known from regions farther south: from the present-day Washington State (Manis) to sites on islands near the California coasts. In the latter case the human presence is dated at 13 ka BP (calibrated). The coastal route version is corroborated by recently developed paleoenvironmental reconstructions; according to the scenarios, there existed ice-free and covered with vegetation lands on the northwestern coast of America. The sites known in the Haida Gwai archipelago, however, are dated at 10.7 ka BP (calibrated), so they cannot be related to the initial period of colonization.

On the main territory of North America south of the Pleistocene ice sheets the initial colonization traces are marked by Pre-Clovis and Clovis sites.

The Pre-Clovis sites are few in number and scattered over the area from the Pacific to Atlantic, though most of sites known are discovered in the south and east of the continent. Among the Pre-Clovis sites, some evidence of human hunting are found, such as a mastodon bone pierced with a point (Manis), or a mastodon tusk with human treatment from Page-Ladson in Florida. Noteworthy is a series of mammoth remains found together with stone tools at Hebior and Schaefer in Wisconsin and attributed to Pre-Clovis. There are other traces of human presence before the Clovis, such as human coprolites and fragments of stemmed points found in Paisley Caves and dated at 16–13 ka BP (calibrated). Needless to say, we do not take into consideration numerous fragmented remains of the Pleistocene megafauna if no stone implements have not been found nearby (in such cases, the presence of humans was inferred by investigators from some characteristic cut and breakage marks on the bones). The actual data show that practically all kinds of bone breakage may result from the bone collision with stone fragments, and therefore, there is no reliable criteria for distinguishing a man-made split from a natural one.

More reliable and archaeologically significant evidence is presented by the blade and biface industry discovered below (stratigraphically) Clovis tools at Cactus Hill (Virginia). Noticeably older dates have been obtained there, 16–20 ka BP. No less significant stratified finds were recovered from Topper (South Carolina), Gault (Texas), and Meadowcroft rock shelter (Pennsylvania). In those cases the cultural layer was dated at approximately 15–16 ka BP.

Clovis is the most ancient and traceable distinctly over the entire area of North America cultural tradition; it occurred from the Atlantic to Pacific and penetrated southward, through Central America into the most northern part of South America. It should be noted that the area inhabited by the Clovis hunters was really much greater as it included a broad, flooded at present, zone along the Atlantic coast. When put on the map, the finds of typical Clovis fluted points appear to concentrate in the south and southeast of the continent, where this cultural tradition developed on the basis of the Pre-Clovis industries. It should be stressed that the fluted point technology was invented by the early man colonizing the New World and has no analogs in the Old World Paleolithic (except for a few implements with channel flakes obtained accidentally when making thin bifaces). The Clovis site dates fall within a narrow chronological interval between 13.4 and 12.7 ka BP (calibrated) or 11.6–10.8 ka BP (conventional).

The Clovis technology was based on using high-quality raw material (various siliceous rocks, jasper, obsidian, etc.) that was transported for long distances. Various bifaces served as preforms. Such a specific technological system has no analogs in the Paleolithic of the Old World.

A uniform appearance of the culture all over the vast continent (taking into consideration a rather short period of its existence) suggests mobile groups of hunters spreading quickly from a single center. The main object of hunting was bison, though horse and deer remains were also found on some sites, and those of reindeer in the northeast of the continent. There are a series of localities with mammoth remains found together with Clovis points. Quite probably, men killed wounded or ill animals on squashy ground, or were butchering carcasses there of animals already dead from natural causes. As distinct from later Paleoindian cultures, Clovis hunting was not specialized.

As to the role of humans in extermination of the Late Pleistocene megafauna on the American continent, it is still hotly debated. Some data suggest the importance of human factors in this process have been overestimated. Many species began dying out before man appeared in the New World, and it is not likely that the process could be essentially influenced by isolated groups of the Paleolithic hunters.

In the western part of the North American continent the Clovis culture coexisted, in all probability, with the Stemmed Point tradition.

The Clovis sites are most often located on the coasts of former (dried up at present) ponds and along small rivers and streams. It should be taken into consideration that it was the time of ice sheet melting, so the area was wetter than at present.

Apart from settlements and hunter camps for butchering carcasses of bison, mammoths, and mastodons, there are caches; part of the latter was functional and contained biface preforms and points kept for later use, while another part served seemingly for sacral rituals. The oldest burial at Anzick (Montana) is also attributed to the Clovis.

So the available archaeological data on the initial peopling of the New World are extremely random. They can give but a rough idea of the complex migration processes having taken place over a vast area in changeable environments of the Late Pleistocene. It can be stated with confidence that man came to the New World after the LGM. The possibility of some groups to have come there earlier should not be ruled out completely, but their penetration would be hardly traceable, considering the low density of the early population and a small size of the hunter groups. It should be taken into account that the initial dispersal of humans was not like a purposeful advance: it was rather a very slow, chaotic movement of hunter groups in pursuit of animals: objects of hunt. Intervals of the habitable territory expansion alternated with episodes of its reduction as a result of epidemics, drastic changes in the animal number, and conflicts with neighboring groups, etc.

On the whole, when comparing data on the initial dispersal of man in Northern Eurasia and North America, one may see two different types, kinds of different geographic models of colonization on a global scale: the process oriented from south to north in Northern Eurasia and the oppositely directed process of colonization—from north to south—in America.

11.6 Mesolithic and Neolithic. Holocene

The last stage in the human dispersal falls on the Holocene and is dated at 11 (10)–6 ka BP. The regions of active human dispersal and colonization in Northern Eurasia by the early hunting groups were noticeably enlarged as early as the Pleistocene–Holocene boundary (Fig. 11.6).

About 10 ka BP the people of the Final Paleolithic and Mesolithic cultures moved to the European North and Scandinavia following the receding ice sheet. Numerous Mesolithic sites and settlements dated to the Early Holocene (9.5–7.5 ka BP) are scattered over Northern Europe; Mesolithic man successfully colonized the area that had been unsuitable or inhospitable for Paleolithic man at the time of glaciation (Volokitin, 1997; Dolukhanov and Khotinskiy, 2002; Kol'tsov, 1989).

As a result of the initial dispersal of the Mesolithic communities, the Mesolithic sites occur not only in the regions immediately west of the Urals (the Kama and Pechora drainage basins), but are found also farther west, in the drainage basins of the Vychegda, Severnaya Dvina, and Onega rivers. As the Scandinavian ice sheet melted, humans gradually colonized the Kola Peninsula, Karelia, and Finland.

At the boundary between the Pleistocene and the Holocene, the global changes in environments and climates brought about a change in economic activity of the prehistoric hunters. The hunt in the Middle and Upper Paleolithic was oriented mostly to large mammal procurement (reindeer, horse, bison); the Mesolithic hunters, along with hunting on forest animals (red deer, elk, roe deer), hunted birds, and they also practiced fishing and sea mammal procurement (Fig. 11.7). As the hunting abilities of man improved, he became able to colonize various regions of Northern Eurasia and to penetrate deep into the High Arctic.

In Siberia, Mesolithic sites have been mostly found in the regions of the concentration of the Paleolithic sites. There are known, however, a few sites in Arctic Asia (Korchagi 1 in the north of West Siberia dated at about 7.2 ka BP, Tagenar 4 and Pyasina on Taymyr, etc.). It may be concluded that the colonization of the high latitudes in Eurasia gained in intensity in the Early Holocene, after the changes in environments had made possible long-distance migrations of the Mesolithic communities.

A wide range of potentially suitable locations for settlements included terraces, and slopes of river valleys, coasts of lakes and seas, and even wetlands ("peat sites"), the choice of locality depending on the type of the community specialization. Harsh climates and environments were no longer an obstacle to the active colonization. Prehistoric sites have been studied in Finland, Karelia, on the Kola Peninsula, in the lower reaches of the Severnaya Dvina, Pechora, Ob' rivers, on the Taymyr Peninsula, on the Lower Lena, Yana, Indigirka, Kolyma rivers, and on the Zhokhov Island in the East Siberian Sea (Dikov, 1977; Timofeyev et al., 2004; Pitul'ko and Pavlova, 2010).

The recent stage of the High Arctic colonization by man (which is still in process) included Paleoeskimo migrations into the Eastern Arctic (Northeastern Canada and Greenland). The Paleoeskimo pioneering communities inhabited vast areas of Chukchi Peninsula, Alaska, and Canadian Arctic for about 1000 years (between 5000 and 4000 ka BP). The data available on the earliest Paleoeskimo lifeways permit one to understand principal features of their economy. The latter was based mostly on the

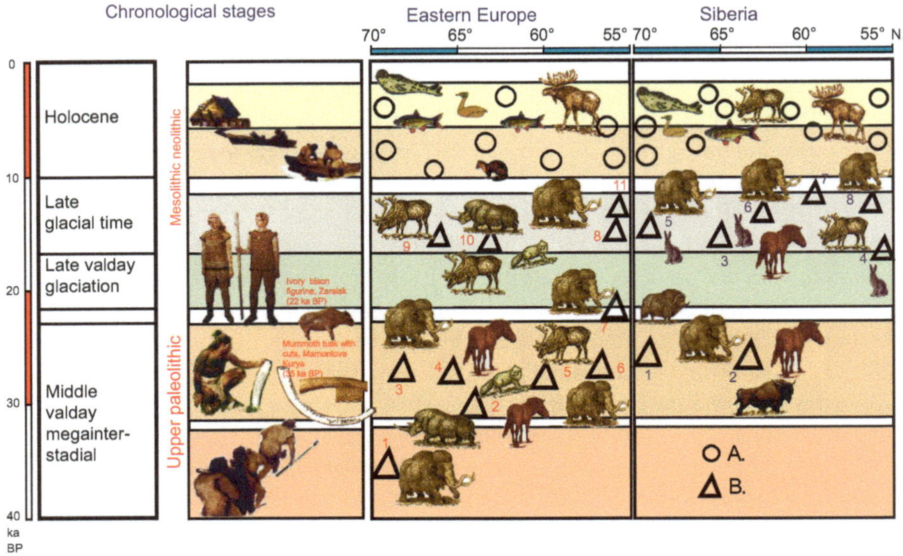

Figure 11.7 Chronostratigraphy and position of the main Late Paleolithic, Mesolithic and Neolithic sites. The principal objects of hunt. 1, Mesolithic, Neolithic; 2, Upper Paleolithic. The main Upper Paleolithic sites: Eastern Europe: 1, Mamontova Kurya; 2, Zaozerye; 3, Byzovaya; 4, Garchi I; 5, Gornova; 6, Sungir'; 7, Zaraisk; 8, Shatrishchi; 9, Medvezhya Cave; 10, Talitskogo; 11, Karacharovo. Siberia: 1, Yanskaya; 2, Ikhine; 3, Dyuktai; 4, Kokorevo; 5, Berelekh; 6, Lugovskoye; 7, Gari; 8, Volchya Griva.

sea mammal hunt (seals, walruses, whales), and to a lesser extent on the hunt of musk ox, caribou, and birds.

It is not inconceivable that the climate deterioration at the end of the Holocene optimum imposed certain limitations on the long-distance migrations of the animals, and therefore reduced opportunities to migrate for humans. The process was completed with colonization of Greenland 4.5–4 ka BP.

The latest event in man's dispersal in the Arctic was Viking colonization of Iceland in the 9th century. Russian settlements appeared on the Spitsbergen archipelago at the end of the 16th century. The first signs of the Russian settlers on the Novaya Zemlya are also of that age.

* * *

The chronostratigraphic, archaeological and paleogeographic data available have elucidated the principal features of the initial colonization of the high latitudes in the Northern Hemisphere. As follows from reliable data, the Late Paleolithic man came to Northern Eurasia about 35–33 ka BP, that is, at the time of warming between two glacial epochs of the Late Pleistocene. There is some evidence of the Middle Paleolithic man appearance here earlier, during the Early Valday ice age, though it calls for additional verification.

As man was certainly present in high latitudes at the Late Valday (LGM) time, he seems to have been partly adapted to the extreme cryoarid environments, though part of the population moved to southern regions.

During the subsequent mitigation of climate (Late Glacial, 16–11 ka BP) the ecumene was greatly enlarged. At that time, man passed over the Beringia Land Bridge and began the colonization of the high latitudes in the Western Hemisphere, having penetrated into Alaska and the Yukon Territory. The recent investigations have shown, however, that the part of the New World south of the ice sheets was colonized still earlier by the man who used the coastal migration route along the Pacific coast.

When considering the initial colonization process, some essential irregularities become noticeable (Fig. 11.8). Two principal directions in the high latitudes peopling are easily identified between 35 and 12 ka BP: the Kama and Pechora drainage

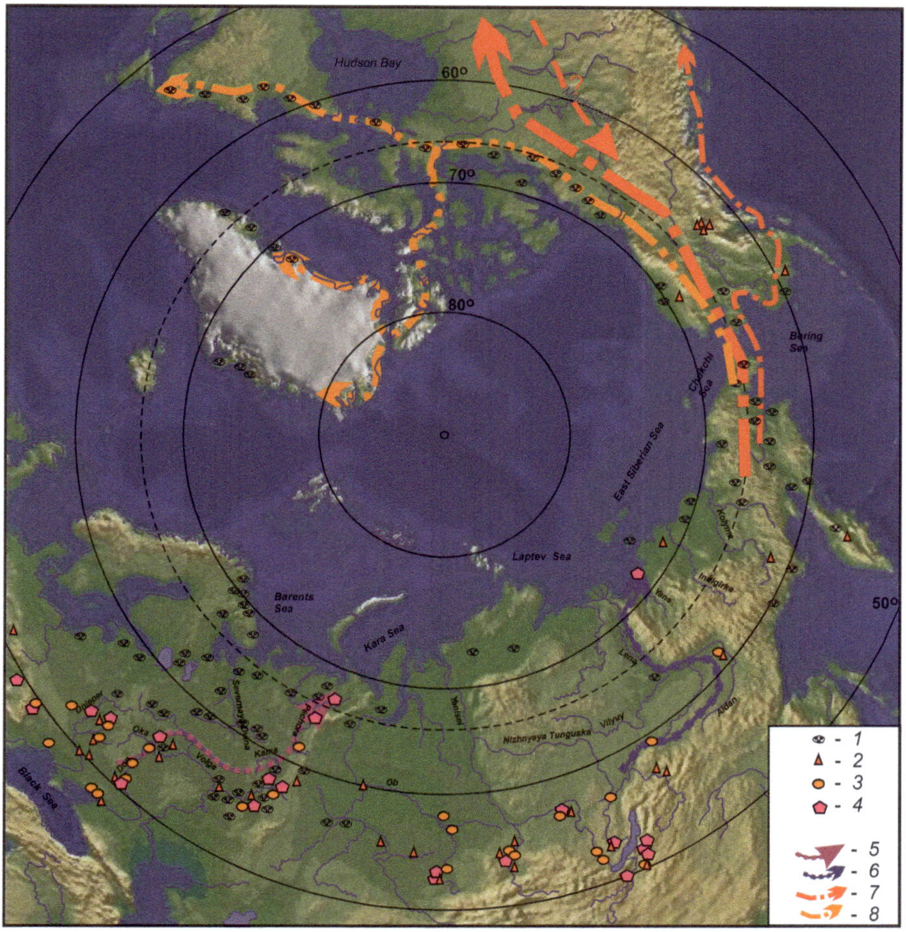

Figure 11.8 Principal routes of the early man dispersal and distribution of the main sites at different stages of the initial colonization of Arctic, Subarctic, and adjacent territories. *1–4, main sites attributed to*: *1*, Mesolithic, Neolithic; *2–4*, Upper Paleolithic (*2* – 15–12 ka BP; *3* – 23–16 ka BP; *4* – 35–24 ka BP); *5–8, probable routes of the initial migration of the earliest settlers in Arctic and Subarctic*: 5, Eastern Europe; *6*, East Siberia; *7*, Northeastern Asia – Beringia – North America; *8, Chukchi Peninsula – north of Canada – Greenland.*

basins in Eastern Europe and the Lena drainage basin in Northeastern Asia. In the Late Glacial an active colonization of West Siberia took place, and at the same time the initial penetration of man occurs from the Eastern to Western Hemisphere, from Chukchi Peninsula to Alaska by way of the Bering Land Bridge. The next stage of the active high latitude colonization occurred at the Mesolithic and Neolithic (9–4 ka BP), when man reached the northern coasts of Greenland. The Arctic archipelagoes were populated later.

Conclusion

A.A. Velichko

Summing up the results of the work presented in this volume, the process of the initial penetration of humans into the high latitudes and the colonization of arctic regions have been described and interpreted on a global scale for the first time.

The data at hand demonstrate the complicated nature of concepts aimed at explaining the colonization process. An analysis of the process based on the data presented in the monograph calls for a number of factors of human life support to be taken into consideration. Among them, there are the level of culture development, type of economy, physical characteristics of human groups, and an advance in adaptive capabilities. The factors aforementioned provided prerequisites for human migration onto new territories and the initial colonization of the latter. The colonization itself might proceed in two ways, conventionally termed "mobile" and "domestic," with some transitional forms. The significance of individual factors, as well as their characteristics, may be essentially different at different stages of colonization and varies depending on the level of the society development.

During the earliest stages of human penetration into the high latitudes at the beginning of the Upper Paleolithic (and probably at the end of Middle Paleolithic) the linear mobile type of migration was dominant. Groups of prehistoric hunters were moving along river valleys following their main resource, herds of large mammals (mammoths, reindeers, horses), which were the principal objects for hunt in the Paleolithic. The animals were mostly grazing in valleys where they were adequately supplied with forage and drinkable water. Forest (taiga) ecosystems of interfluvial areas were less favorable for herbivores. Due to the specialized hunting activities and somewhat limited adaptivity, prehistoric hunters mostly visited the northern areas during warm seasons and retreated to southern regions with the advent of cold weather ("forth and back" tactics). Such a system is suggested by the character of habitations located in drainage basins of the Kama River (Northeastern Europe) and the Lena River (Eastern Siberia).

A different type of the initial colonization was typical of the intermediate epoch, the transition from the Upper Paleolithic to Mesolithic, when the environments were changing from the Late Glacial to the Holocene. At that time, early man exploited not only river valley ecosystems but also lake coasts (see, for example, the Veretye sites). Advanced technology of toolmaking ensured successful mammal hunting, along with fishing and bird netting. With development of improved adaptation systems, the "domestic" type of colonization became prevalent. Sites dated to that interval in the European North, Siberia, Kamchatka, and Alaska provided evidence of permanent human habitation throughout the year.

The colonization of the North became even more intensive during the Neolithic (the Holocene), when in addition to the previously acquired skills, man began to actively hunt sea animals near the shores. Considering its spatial characteristics, such a type of initial colonization could be defined as diffusional.

The human penetration into the extreme North was a result of a complicated combination of factors involved. That is especially true for northern North America, the Canadian Arctic Islands, and the northern part of Greenland. In the Paleoeskimo movement to the north of the continent and onto the islands the two types of colonization—"mobile" and "domestic"—were used in combination. One of the important culture elements playing a crucial part in this process was the use of storage pits or storehouses for keeping frozen meat. In Greenland, the colonization proceeded in two ways. On the western coast under conditions of harsh coastal ecosystems, the domestic type was dominant. Another, mobile, type of colonization was typical for the human settlement in the northernmost region of Greenland, the Peary Land. Meticulous studies of the traces left there by the Paleoeskimo enabled specialists to suggest that the early settlers could occasionally spend a polar night in this region and then return to some more hospitable or familiar environments. Such a system may be tentatively termed "back and forth" tactics.

As follows from the materials presented in the monograph, different types of the initial colonization were related to climate changes. The first, and highly dynamic, stage of colonization coincided with the Middle Valdai (MIS3, Bryansk, Karginsky) mega-interstadial that was close to an interglacial in temperature characteristics of its maximum. The second interval of active human dispersals occurred at the time of warming at the end of Late Glacial and in the Holocene. Contrary to that, at the coldest time of the Sartan (Late Valdai) glaciation (MIS2) the sites were reduced in number, indicative of a reduced colonization process, or even of migration of some hunter communities southward. The data suggest the colonization process intensity was heavily influenced by the climatic and environmental dynamics. Thus, large-scale restructuring of environments might account for a wave-like pattern of the colonization, as well as for occasional interruptions in the process. On the contrary, the waves of warmth opened possibilities for human advance northward.

New results of geoarchaeological studies give grounds for revision of the views on the initial colonization of the northern latitudes as a linear one-way process (from south to north).

Of particular interest in the light of the new data is the problem of early man's arrival to the New World ("the first discovery of America"). For a long time, it was accepted that the initial human penetration from the Eastern Hemisphere into the Western Hemisphere occurred at the very beginning of the Holocene via the Bering Land Bridge from the Chukchi Peninsula to Alaska. Even leaving aside the debatable dating of the Bluefish caves (Yukon, Canada) to MIS3 mega-interstadial, we have to admit that the Bering Land Bridge was first used at the beginning of the Late Glacial, 15 ka BP at the latest, as indicated by the age of the Pre-Clovis sites. The occurrence of the sites in the Great Lakes region and farther south suggests that man crossed the Bering Land Bridge moving to the south. Individual groups, however, could later move northward and meet other communities that came to Alaska by way of the Beringia later, at the beginning of the Holocene. The third stage of human penetration into the Western Hemisphere occurred in the Middle Holocene, when Paleoeskimo from Providence Bay (Chukchi Peninsula) arrived at Greenland, confirmed by the results of the DNA analyses of samples from burials in Greenland, as well as by radiocarbon dates.

The results summarized in the monograph give grounds to the statement that the investigations performed on global scale have elucidated the principal landmarks in the initial colonization of the arctic regions by the prehistoric man. At the same time, there are still a lot of problems that cannot be solved due to deficiency of geoarchaeological data on the arctic and subarctic regions. The main problem still consists in an unequal study of the giant territory. The Late Paleolithic sites, for example, are discovered in greater number in the north of Eastern Europe compared with the northern regions of Western and Central Siberia. It still remains unclear whether the difference is due to the diversified colonization process or to inadequate exploration of the regions. It seems that the second version is more plausible. As an example, we refer to the Yana site (located at the lower reaches of the Yana River, Eastern Siberia), exceedingly rich in archaeological finds, while no comparable sites have been found for many hundreds of kilometers around, which is difficult to explain. Still far from being clear is the tracing of the ways of human penetration into the high latitudes of North America.

All the cited examples show beyond a doubt that integrated geoarchaeological investigations in the future would make essential corrections in our understanding of the initial human penetration into the arctic regions. The colonization of the high latitudes became the concluding stage of the human dispersal all over the planet, the process that started 2.5 million years ago in equatorial Africa and came to its end about 4000 years ago in the polar regions of Greenland.

References

Abbott, M.B., Edwards, M.E., Finney, B.P., 2010. A 40,000-year record of environmental change from Burial Lake in northwest Alaska. Quaternary Research 74, 156–165.

Abramova, Z.A., 1979. On the problem of the age of the Paleolithic on the Aldan River. Sovetskaya Arkheologiya 4, 5–14 (in Russian).

Abramova, Z.A., 1983. The Paleolithic site of Tarachikha on the Yenisey River. Kratkiye Soobshcheniya Instituta Arkheologiyi USSR Academy of Sciences 173, 43–50 (in Russian).

Abramova, Z.A., 1989. Paleolithic of Northern Asia. In: Boriskovskiy, P.I. (Ed.), Paleolithic of the World: Paleolithic of the Caucasus and Northern Asia. Nauka, Moscow, pp. 145–243 (in Russian).

Abramova, Z.A., Matyushchenko, V.I., 1973. New data on the Tomsk Paleolithic site. In: On the History of Siberia, Iss. 5. Tomsk Univ. Press, Tomsk, pp. 16–23 (in Russian).

Ackerman, R.E., 1996. Cave I, Lime Hills. In: West, F.H. (Ed.), American Beginnings: The Prehistory and Palaeoecology of Beringia. Chicago Univ. Press, Chicago, pp. 470–477.

Ackerman, R.E., 1998. Early maritime traditions in the Bering, Chukchi, and East Siberian seas. Arctic Anthropology 35, 247–262.

Ackerman, R.E., 2001. Spein Mountain: a Mesa-complex site in southwestern Alaska. Arctic Anthropology 38 (2), 81–97.

Ackerman, R.E., 2011. Microblade assemblages in Southwestern Alaska: an early Holocene adaptation. In: Goebel, T., Buvit, I. (Eds.), From the Yenisei to the Yukon: Interpreting Lithic Variability in Late Pleistocene/Early Holocene Beringia. Peopling of the Americas Publications Series. Texas A&M Univ. Press, College Station, pp. 255–269.

Agadjanian, A.K., 2001. Spatial pattern of the Late Pleistocene mammal fauna in Northern Eurasia. Archaeology, Ethnography and Anthropology of Eurasia 2 (6), 2–18.

Ager, T.A., 1983. Holocene vegetational history of Alaska. In: Wright, H.E. (Ed.), Late Quaternary Environments of the United States. The Holocene. Univ. of Minnesota Press, Minneapolis, pp. 128–141.

Akimova, E., Higham, T., Stasyuk, I., Buzhilova, A., Dobrovolskaya, M., Mednikova, M., 2010. A new direct radiocarbon AMS date for an Upper Palaeolithic human bone from Siberia. Archaeometry 52, 1122–1130.

Aksyanova, G.A., 2003. Somatology. Odontology. In: Bagashev, A.N. (Ed.), Ethnography and Anthropology of Yamal. Nauka, Novosibirsk, pp. 200–343 (in Russian).

Alekseyev, V.P., 1969. Origin of the East European Peoples (Craniological Study). Nauka, Moscow. 324 pp. (in Russian).

Alekseyev, V.P., 1971. Forest Nenets (somatological observations). Voprosy Antropologii 39, 61–75 (in Russian).

Alekseyev, V.P., 1974a. Geography of Human Races. Mysl', Moscow. 351 pp. (in Russian).

Alekseyev, V.P., 1974b. Craniological characteristic of the Eastern Fennoscandia population (based on the data of G.F. Debets and the author). In: Zolotareva, I.M. (Ed.), Race Genesis in the Ethnos History. Nauka, Moscow, pp. 85–105 (in Russian).

Alekseyev, V.P., 1979. Adaptation and heredity. In: Ecological Environment and Heredity. Medgiz, Moscow, pp. 69–79 (in Russian).

Alekseyev, V.P. (Ed.), 1984a. Sungir: Anthropological Investigation. Nauka, Moscow. 215 pp. (in Russian).

Alekseyev, V.P., 1984b. Physical characteristics of the Mesolithic and Early Neolithic people in Eastern Europe in connection with the problem of ancient peopling of the area. In: Gokhman, I.I. (Ed.), Anthropological Problems of the Ancient and Present-Day Population of the Eurasian North. Nauka, Leningrad, pp. 28–36 (in Russian).

Alekseyev, V.P., 1985. The Man: Taxonomy and Evolution (Some Theoretical Problems). Nauka, Moscow. 285 pp. (in Russian).

Alekseyev, V.P., 1993. Essays on Human Ecology. Nauka, Moscow. 191 pp. (in Russian).

Alekseyev, V.P., Baluyeva, T.S., 1976. Materials on the Naukan Eskimo craniology (to the differentiation of the Arctic race). Sovetskaya Etnografiya 1, 84–100 (in Russian).

Alekseyev, V.P., Debets, G.F., 1965. Craniometry: Methods of Anthropological Studies. Nauka, Moscow. 128 pp. (in Russian).

Alekseyev, V.P., Fedosova, V.N. (Eds.), 1992. Ecological Aspects of Paleoanthropological and Archaeological Reconstructions. Institute of Archaeology Russian Academy of Sciences, Moscow. 204 pp. (in Russian).

Alekseyev, V.P., Gokhman, I.I., 1984a. Anthropology of the Asian Part of the USSR. Nauka, Moscow. 208 pp. (in Russian).

Alekseyev, V.P., Gokhman, I.I., 1984b. Expert appraisement of the craniometric index reliability in studies of anthropological materials recovered from the burials on Yuzhny Oleni island, Onega Lake (with reference to their preservation and specificity of restoration). In: Gokhman, I.I. (Ed.), Anthropological Problems of the Ancient and Present-Day Population of the Eurasian North. Nauka, Leningrad, pp. 6–27 (in Russian).

Alekseyev, M.N., Kamaletdinov, V.A., Ziegert, C., Grinenko, O.V., Gnibidenko, Z.N., Gravis, G.F., Shamshina, E.A., 1990. Problems of Geology of the Diring-Yuryakh Paleolithic Site. Academy of Sciences, USSR, Siberian Branch, Yakutsk. 48 pp. (in Russian).

Alekseyeva, T.I., 1977. Geographical Environment and Human Biology. Mysl', Moscow. 302 pp. (in Russian).

Alekseyeva, T.I., 1986. Adaptive Processes in Human Populations. Moscow State Univ. Press, Moscow. 215 pp. (in Russian).

Alekseyeva, T.I., 1998. Adaptation of Man in Various Ecological Niches of the Earth (Biological Aspects). MNEPU, Moscow. 280 pp. (in Russian).

Alekseyeva, T.I., 2003. Ancient and present-day population of Beringia. In: Horizons of Anthropology. Nauka, Moscow, pp. 367–370. Materials of International Conf. to the Memory of V.P. Alekseyev (in Russian).

Alekseyeva, T.I., Bader, N.O. (Eds.), 2000. Homo Sungirensis. The Upper Paleolithic Man: Ecological and Evolutionary Aspects. Nauchny Mir, Moscow. 468 pp. (in Russian).

Alekseyeva, T.I., Kovalenko, V.Yu, 1980. Morphofunctional characteristics of the postcranial skeleton of Asian Eskimo. In: Okladnikov, A.P., Alekseyev, V.P. (Eds.), Paleoanthropology of Siberia. Nauka, Moscow, pp. 131–153 (in Russian).

Alekseyeva, T.I., Volkov-Dubrovin, V.P., Golubchikova, Z.A., Pavlovskiy, O.M., Smirnova, N.S., Spitsyn, V.A., 1972. Anthropological studies of forest Nenets (morphology, physiology, population genetics). Part I. Voprosy Antropologii 41, 19–35 Part II. Voprosy Antropologii 42, 31–54 (in Russian).

Alekseyeva, T.I., Volkov-Dubrovin, V.P., Golubchikova, Z.A., Pavlovskiy, O.M., Smirnova, N.S., Shchekochikhina, L.K., 1973. Sami. Morphofunctional essay. Voprosy Antropologii 43, 52–67 (in Russian).

Alekseyeva, T.I., Buzhilova, A.P., Mednikova, M.B., Dobrovolskaya, M.V. (Eds.), 2008. Anthropoecology of the Northeastern Asia (Chukotka, Kamchatka, Komandorskie Islands). Taus, Moscow. 368 pp. (in Russian).

Alexander, H.L., 1987. Putu: A Fluted Point Site in Alaska. Publ. 19. Simon Fraser University, Burnaby. 127 pp.

Alfimov, A.V., Berman, D.I., 2004. Distribution of thermophyte tundra-steppes and climate at the Last Glacial maximum in the Northeast of Asia. Kriosfera Zemli 8 (4), 78–87 (in Russian).

Allen, J.R.M., Long, A.J., Ottley, C.J., Pearson, D.G., Huntley, B., 2007. Holocene climate variability in northernmost Europe. Quaternary Science Reviews 26, 1432–1453.

Alt, B.T., Garneau, M., 2000. Summary of results and recommendations. In: Garneau, M., Alt, B.T. (Eds.), Environmental Response to Climate Change in the Canadian High Arctic. Geological Survey of Canada, Ottawa, pp. 391–400.

Altukhov, Iu.P., Kurbatova, O.L., 1990. Problem of the adaptive norm in human populations. Genetika 26 (4), 583–598 (in Russian).

Ambrose, S.H., DeNiro, M.J., 1986. Reconstruction of African human diet using bone collagen carbon and nitrogen isotope ratios. Nature 319, 321–324.

Amirkhanov, Kh.A., 2008. Comparative typological-statistical characteristic of the inventory from the Muhkay I in Central Daghestan (based of materials of excavations in 2007). In: Early Paleolithic of Europe: New Discoveries. Materials of International Conference, Krasnodar-Temryuk, September 1–6, 2008. Rostov-on-Don, pp. 28–31 (in Russian).

Andersen, K.L., Lyning, J., Nelms, J.D., Wilson, O., Fok, R.H., Bolstad, A., 1960. Metabolic and thermal response to a moderate cold exposure in nomadic Lapps. Journal of Applied Physiology 15 (4), 649–653.

Anderson, P.M., Brubaker, L.B., 1994. Vegetation history of north-central Alaska: a mapped summary of late Quaternary pollen data. Quaternary Science Reviews 13, 71–92.

Anderson, P.M., Edwards, M.E., Brubaker, L.B., 2004. Results and paleoclimate implications of 35 years of paleoecological research in Alaska. In: Gillespie, A.R., Porter, S.C., Atwater, B.F. (Eds.), The Quaternary Period in the United States. Elsevier, Amsterdam, pp. 427–440.

Anderson, P.M., Lozhkin, A.V., 2001. The stage 3 interstadial complex (Karginsky/middle Wisconsinan interval) of Beringia: variations in paleoenvironments and implications for paleoclimatic interpretations. Quaternary Science Reviews 20 (1–3), 93–125.

Anderson, P.M., Lozhkin, A.V. (Eds.), 2002. Late Quaternary Vegetation and Climates of Siberia and the Russian Far East (Palynological and Radiocarbon Database). Northeastern Interdisciplinary Research Institute, Far Eastern Branch RAS, Magadan. 370 pp. (in Russian).

Anderson, P.M., Lozhkin, A.V., 2004. Foreword to the pollen records of lacustrine sediments in Beringia. In: Onopriyenko, Yu.I., Lozhkin, A.V. (Eds.), Climatic Records in Quaternary Deposits in Beringia. Northeastern Interdisciplinary Research Institute, Far Eastern Branch, RAS, Magadan, pp. 4–26 (in Russian).

Anderson, P.M., Lozhkin, A.V., 2011. Climatic reconstruction based on analogs of the recent pollen spectra with reference to Lake Elgygytgyn, Chukotka. In: Problems of the Modern Palynology. Materials of XIII Russian Palynological Conference, vol. 1. Komi Research Center, Uralian Branch, Russian Academy of Sciences, Syktyvkar, pp. 77–79 (in Russian).

Andreasen, C., 2004. Independence i kulturen. In: Gulløv, H.C. (Ed.), Grønlands forhistorie. Gyldendal, Copenhagen, pp. 35–64.

Andreasen, T.N., 1998. Nivertussannguaq – a survey of the faunal remains from a Saqqaq settlement in the Disko Bay area of western Greenland. Acta Borealia 15 (2), 129–137.

Andreev, A.A., Forman, S.L., Ingylfsson, Y., Manley, W.F., 2006. Middle Weichselian environments on western Yamal Peninsula, Kara Sea, based on pollen records. Quaternary Research 65 (2), 275–281.

Andreev, A.A., Siegert, C., Klimanov, V.A., Derevyagin, A.Y., Shilova, G.N., Melles, M., 2002. Late Pleistocene and Holocene vegetation and climate on the Taymyr lowland, northern Siberia. Quaternary Research 57 (1), 138–150.

Andreev, A.A., Tarasov, P.E., Siegert, C., Ebel, T., Klimanov, V.A., Melles, M., Bobrov, A.A., Derevyagin, A.Y., Lubinski, D.J., Hubberten, H.-W., 2003. Late Pleistocene and Holocene vegetation and climate on the northern Taymyr Peninsula, Arctic Russia. Boreas 32 (3), 484–505.

Andreev, A.A., Tarasov, P.E., Klimanov, V.A., Melles, M., Lisitsyna, O.M., Hubberten, H.-W., 2004a. Vegetation and climate changes around the Lama Lake, Taymyr Peninsula, Russia, during the Late Pleistocene and Holocene. Quaternary International 122, 69–84.

Andreev, A., Tarasov, P., Schwamborn, G., Ilyashuk, B., Ilyashuk, E., Bobrov, A., Klimanov, V., Rachold, V., Hubberten, H.-W., 2004b. Holocene paleoenvironmental records from Nikolay Lake, Lena River Delta, Arctic Russia. Palaeogeography, Palaeoclimatology, Palaeoecology 209 (1–4), 197–217.

Andreicheva, L.N., 1992. Ground Tills in the European Northeast of Russia: Their Lithostratigraphic Importance. Nauka, St.-Petersburg. 125 pp. (in Russian).

Andreicheva, L.N., 2002. Pleistocene of the European Northeast. Uralian Branch, Russian Academy of Sciences, Ekaterinburg. 322 pp. (in Russian).

Andreicheva, L.N., Konovalenko, L.A., 2008. The Pleistocene deposit sequence and environments of deposition in the Southwestern Timan region. In: Cermnykh, V.A., Loseva, E.I. (Eds.), Biostratigraphy of Phanerozoic in the Timan-Pechora Province. Institute of Geology, Komi Research Center, Syktyvkar, pp. 75–83 (in Russian).

Andreicheva, L.N., Duryagina, D.A., Marchenko, T.I., 1999. Stages of the Cenozoic sedimentation in the Timan-Pechora-Vychegda region. In: Geology and Mineral Resources of the European Northeast: New Results and New Prospects. Materials of the 13th Geological Conference of the Komi Republic, vol. II, pp. 80–83 (in Russian).

Andreicheva, L.N., Marchenko-Vagapova, N.I., 2003. Evolution of Environments and Climate in the Northeast of Europe in the Anthropogene. Institute of Geology, Komi Research Center, Syktyvkar. 22 pp. (in Russian).

Andreicheva, L.N., Bratushchak, Yu.V., Marchenko-Vagapova, N.I., 2006. The Environment and Climate Evolution in the Pleistocene and Holocene in the North of European Russia. Institute of Geology, Komi Research Center, Syktyvkar. 23 pp. (in Russian).

Andreicheva, L.N., Marchenko-Vagapova, N.I., Golubeva, YuV., 2008. Evolution of Environments and Climate in the Late Pleistocene and Holocene in the European North of Russia. Institute of Geology, Komi Research Center, Syktyvkar. 52 pp. (in Russian).

Andrews, J.T., 1970. A Geomorphological Study of Post-glacial Uplift with Particular Reference to Arctic Canada. Alden Press, Oxford. 156 pp.

Andrew, J.T., McGhee, R., McKenzie-Pollock, L., 1971. Comparison of elevations of archaeological sites and calculated sea levels in Arctic Canada. Arctic 24, 210–228.

Andrews, J.T., Peltier, W.R., 1989. Quaternary geodynamics in Canada. In: Fulton, R.J. (Ed.), Quaternary Geology of Canada and Greenland. Geological Survey of Canada, Ottawa, pp. 541–572.

Andreyeva, S.M., 1980. North Siberian Lowland at the Karginsky time: paleogeography, radiocarbon chronology. In: Ivanova, I.K., Kind, N.V. (Eds.), Geochronology of the Quaternary. Nauka, Moscow, pp. 183–191 (in Russian).

Andrianova, L.S., Vasil'eva, N.B., 2000. Mesolithic site Poboishchnoye 1 on the lower Sukhona. In: Tver' Collection of Archaeological Papers, Iss. 4. Triada Publ, Tver', pp. 99–110 (in Russian).

Anikovich, M.V., 1977. Stone inventory from the lower layers of the Volkovskaya site. In: Praslov, N.D. (Ed.), Problems of Paleolithic of Eastern and Central Europe. Nauka, Leningrad, pp. 94–112 (in Russian).

Anikovich, M.V., 1998. Dnieper-Don historical-cultural region of the mammoth hunters: from "Eastern Gravettian" to "Eastern Epigravettian". In: Amirkhanov, Kh.A. (Ed.), Eastern Gravettian. Nauchny Mir, Moscow, pp. 35–66 (in Russian).

Anikovich, M.V., Anisyutkin, N.K., 1995. Man and mammoth in the Upper Paleolithic. In: The First International Mammoth Symposium. Abstract Volume. St.-Petersburg, p. 597 (in Russian).

Anikovich, M.V., Rogachev, A.N., 1984. Late Paleolithic of the Russian Plain and Crimea. In: Rybakov, B.A. (Ed.), Archaeology of the USSR. V. I: Paleolithic of the USSR. Nauka, Moscow, pp. 162–271 (in Russian).

Anikovich, M.V., Anisyutkin, N.K., Vishnyatsky, L.B., 2007. Key Problems of the Transition to the Upper Paleolithic in Eurasia. Nestor-Istoriya Publ., St.-Petersburg. 335 pp. (in Russian).

Antipina, T.G., Panova, N.K., 2009. Botanical and palynological characteristics of a peat bog near the Bolskaya Umytya-69 settlement. In: The Khanty-Mansi Autonomous Area in the Mirror of the Past, Iss. 7. Tomsk Univ. Press, Tomsk, pp. 190–197 (in Russian).

Antoine, P., Rousseau, D.D., Zoller, L., Lang, A., Munaut, A.V., Hatte, C., Fontugne, M., 2001. High resolution record of the Last Interglacial – Glacial cycle in the loess palaeosol sequences of Nussloch (Rhine Valley – Germany). Quaternary International 76/77, 211–229.

Anundsen, K., 1996. The physical conditions for earliest settlement during the last deglaciation in Norway. In: Larsson, L. (Ed.), The Earliest Settlement of Scandinavia and Its Relationship with Neighbouring Areas (Acta Archaeologica Lundensia. No. 24), pp. 207–217.

Appelt, M., 1997. The construction of an archaeological "Culture". Similarities and differences in early Paleo-Eskimo cultures of Greenland. In: Gilberg, R., Gulløv, H.C. (Eds.), Fifty Years of Arctic Research: Anthropological Studies from Greenland to Siberia. The National Museum of Denmark, Copenhagen, pp. 33–40.

Appelt, M., 2006. Gathering sites as focal places in prehistoric Greenland. In: Arneborg, J., Grønnow, B. (Eds.), Dynamics of Northern Societies. Proceedings of the SILA/NABO Conference on Arctic and North Atlantic Archaeology, Copenhagen, May 10th–14th, 2004, pp. 215–224.

Appelt, M., Pind, J., 1996. Nunnguaq – a Saqqaq Site from the Godthâbsfiord. In: Grønnow, B. (Ed.), The Paleo-Eskimo Cultures of Greenland – New Perspectives in Greenlandic Archaeology. Danish Polar Center, Copenhagen.

Archaeological Survey of Canada, 1972. CMC Archaeological Sites Database. Canadian Museum of Civilization, Ottawa.

Argunov, V.G., 1990. Stone Age in Northwestern Yakutia. Nauka, Novosibirsk. 213 pp. (in Russian).

Arkhipov, S.A., Volkova, V.S., 1994. Geological History, Landscapes and Climates of the Pleistocene in West Siberia. Scientific Publisher Center, United Institute of Geology, Geophysics and Mineralogy, Siberian Branch RAS, Novosibirsk. 106 pp. (in Russian).

Arkhipov, S.A., Astakhov, V.I., Volkov, I.A., Panychev, V.A., 1980. Paleogeography of the West Siberian Plain at the Zyryanka Glaciation Maximum. Nauka, Novosibirsk. 107 pp. (in Russian).

Armand, A.D., 1961. New data on the last glaciation on the Kola Peninsula. Doklady. Academy of Sciences, USSR 138 (4).

Armand, A.D., Armand, N.N., 1965. Marginal zones of the Valday glaciation on the Kola Peninsula. In: Gerasimov, I.P. (Ed.), The Last Ice Sheet in Europe. Nauka, Moscow, p. 44 (in Russian).

Armand, A.D., Armand, N.N., 1969. Murmansk Region. In: Gerasimov, I.P. (Ed.), The Last Ice Sheet in the North-West on the European Part of the USSR. Nauka, Moscow, pp. 225–267 (in Russian).

Armand, A.D., Armand, N.N., Nikonov, A.A., 1963. Specific features of the last glaciation in the northeast of Fennoscandia. Izvestiya. Academy of Sciences, USSR Series. Geogr. 2, 55–60 (in Russian).

Arslanov, Kh.A., Koshechkin, B.I., Chernov, B.S., 1974. Absolute chronology of the late glacial and post-glacial basin sediments on the Kola Peninsula. Vestnik Leningradskogo Universiteta Ser. Geol. 12, 132–138 (in Russian).

Arslanov, Kh.A., Lavrov, A.S., Nikiforova, L.D., 1981. On stratigraphy, geochronology, and climate changes in the Middle and Late Pleistocene and Holocene in the Northeast of the Russian Plain. In: Velichko, A.A., Faustova, M.A. (Eds.), Pleistocene Glaciations on the Russian Plain. Nauka, Moscow, pp. 37–52 (in Russian).

Arslanov, Kh.A., Vereshchagin, N.K., Lyadov, V.V., Ukraintseva, V.V., 1980. On the chronology of the Karginsky Interglacial and reconstruction of the Siberian landscapes based on studies of mammoths' carcasses and accompanying faunal remains. In: Ivanova, I.K., Kind, N.V. (Eds.), Geochronology of the Quaternary. Nauka, Moscow, pp. 208–213 (in Russian).

Arutyunov, S.A., Sergeyev, D.A., 1969. Prehistoric Cultures of the Asian Eskimo. Uelen Burial Ground. Nauka, Moscow. 206 pp. (in Russian).

Arutyunov, S.A., Sergeyev, D.A., 1975. Problems of the Ethnic History in the Bering Sea Region: Ekven Burial Ground. Nauka, Moscow. 240 pp. (in Russian).

Aseyev, A.A., Bronguleyev, V.V., Makkaveyev, A.N., 1982. Reconstruction of the ice sheet in Europe. In: Gerasimov, I.P., Velichko, A.A. (Eds.), Paleogeography of Europe During the Last 100,000 Years. Nauka, Moscow, pp. 27–31 (in Russian).

Astakhov, S.N., 2008. Paleolithic Sites of Tuva. Nestor-Istoriya Publ., St.-Petersburg. 180 pp. (in Russian).

Astakhov, V.I., Svendsen, J.I., 2008. Environments during the time of initial human settlement of the northern part of the Ural region. In: Velichko, A.A., Vasil'ev, S.A. (Eds.), Way to North: Paleoenvironment and Earliest Inhabitants of Arctic and Subarctic. Institute of Geography RAS, Moscow, pp. 98–106 (in Russian).

Atlas of the Murmansk Region, 1971. GUGiK SSSR, Moscow. 33 pp.

Averin, V.A., Zhilin, M.G., 2001. Stone inventory of the Mesolithic layers of Stanovoye settlement. In: Nizhegorodskiye issledovaniya po krayevedeniyu i arkheologii (Studies of Regional Ethnography and Archaeology in Nizhny-Novgorod), Iss. 5, pp. 8–16 (in Russian).

Äyräpää (Europaeus), A., 1925. Stenålderskeramik från kustboplaster i Finland. Suomen Muinaismuistoyhdistyksen Aikakauskirja 34 (1), 45–77.

Bader, O.N., 1947. Initial colonization of the Urals and the Volga-Kama drainage basins by man. Uchenye Zapiski, Perm' State University 5 (2), 93–134 (in Russian).

Bader, O.N., 1950. Problems of paleogeography of the Urals and Northeastern Europe. In: Materialy po chetvertichnomy periodu SSSR, Iss. 2. Academy of Sciences of the USSR Publ., Moscow-Leningrad, pp. 140–149 (in Russian).

Bader, O.N., 1974. Man, his culture, and environments at the northern margin of the European oecumene. In: Gerasimov, I.P., Velichko, A.A. (Eds.), Early Man, His Material Culture and Environments in the Pleistocene and Holocene. Institute of Geography, Academy of Sciences of the USSR, Moscow, pp. 117–125 (in Russian).

Bader, O.N., 1978. Sungir: Upper Paleolithic Site. Nauka, Moscow. 271 pp. (in Russian).

Bader, O.N., Flint, V.E., 1977. Engraving on mammoth tusk from Berelekh. In: Mammoth Fauna of the Russian Plain and East Siberia, vol. 72. Transactions Zoological Institute, pp. 68–72 (in Russian).

Bader, O.N., Serikov, Yu.B., 1975. The northernmost Paleolithic site in the Trans-Ural region. In: Archaeological Discoveries of 1974. Nauka, Moscow, p. 138 (in Russian).
Bader, O.N., Serikov, Yu.B., 1981. Gari Paleolithic locality on the Sos'va River. Sovetskaya Arkheologiya 3, 242–248 (in Russian).
Bagashev, A.N., 1998. Population of the Lower Ob' region. In: Bagashev, A.N. (Ed.), Essays on Culturogenesis of the West Siberian Peoples. V. 4. Race Genesis of the Aboriginal Population. Tomsk Univ. Press, Tomsk, pp. 124–140 (in Russian).
Bagashev, A.N., 2000. Paleoanthropology of West Siberia. Forest-Steppe at the Early Iron Age. Nauka, Novosibirsk. 374 pp. (in Russian).
Bagashev, A.N., 2003. Craniology. In: Bagashev, A.N. (Ed.), Ethnography and Anthropology of the Yamal Peninsula. Nauka, Novosibirsk, pp. 344–356 (in Russian).
Bailey, G., Spikins, P., 2010. Mesolithic Europe. Cambridge Univ. Press, Cambridge. 479 pp.
Bakhmutov, V.G., 2006. Paleo-Secular Geomagnetic Variations. Naukova Dumka Publ., Kiev. 295 pp. (in Russian).
Balanovskaya, E.V., Balanovskiy, O.P., 2007. Russian Genofond on the Russian Plain. Luch Publ., Moscow. 416 pp. (in Russian).
Balanovskiy, O.P., 2008. Gene pool of the High Latitudes of Eurasia, or where the Sami came from? In: Velichko, A.A., Vasil'ev, S.A. (Eds.), Way to North: Paleoenvironment and Earliest Inhabitants of Arctic and Subarctic. Institute of Geography RAS, Moscow, pp. 277–282 (in Russian).
Balanovskiy, O.P., 2012. Variation of the Gene Pool in Space and Time: Synthesis of Data on Gene Geography of Mitochondrial DNA and Y-Chromosome (Thesis. Dr.Sc in Biology). Moscow, 45 pp. (in Russian).
Ballin, T.B., 1999. The middle Mesolithic in Southern Norway. In: The Mesolithic of Central Scandinavia. Universitets Oldsaksamlings Skrifer. No. 22, Oslo, pp. 203–216.
Baltrunas, V., 2006. Distribution of raw material for prehistoric flint artifacts in South Lithuania. Geografija (Vilnius) 42 (2), 41–47.
Bang-Anderson, S., 2003. Southwest Norway at the Pleistocene/Holocene transition: Landscape development, colonization, site types, settlement patterns. Norwegian Archaeological Review 36 (1), 5–25.
Barbashova, Z.I., 1976. Adaptation to continental polar regions of newly arrived migrants and of local inhabitants of aboriginal and and non-aboriginal nationalities. In: Bauer, O.N., Smirnov, N.N. (Eds.), Resources of Biosphere. Iss. 3. Human Adaptation. Nauka, Leningrad, pp. 99–119 (in Russian).
Barbashova, Z.I., Likhnitskaya, I.I. (Eds.), 1972. Adaptation of Man. Nauka, Leningrad. 267 pp. (in Russian).
Barber, D.C., Dyke, A., Hillaire-Marcel, C., Jennings, A.E., Andrews, J.T., Kerwin, M.W., Bilodeau, G., McNeely, R., Southon, J., Morehead, M.D., Gagnon, J.-M., 1999. Forcing the cold event of 8200 years ago by catastrophic drainage of Laurentide Lakes. Nature 400, 344–348.
Barry, R.G., Arundale, W.H., Andrews, J.T., Bradley, R.S., Nichols, H., 1977. Environmental change and cultural change in the eastern Canadian Arctic during the last 5000 years. Arctic and Alpine Research 9, 193–210.
Basilyan, A.E., Anisimov, M.A., Pavlova, E.Yu., Pitulko, V.V., Nikolskiy, P.A., 2009. Key section of the Quaternary on the Yana-Indigirka lowland, lower reaches of the Yana River. In: Fundamental Problems of Quaternary: Results of the Studies and the Main Trends in Further Investigations. Materials of VI All-Russia Conference on the Quaternary Studies, Novosibirsk, November 19–23, 2009. Siberian Branch RAS Publ., Novosibirsk, pp. 63–65 (in Russian).

Beebe, B.F., 1983. Evidence of carnivore activity in a Late Pleistocene/Early Holocene archaeological site (Blue Fish cave 1), Yukon Territory, Canada. In: LeMoine, G.M., MacEachern, A.S. (Eds.), A Question of Bone Technology. The Archaeological Association of the Univ. of Calgary, Calgary, pp. 1–14.

Belov, N.P., Baranovskaya, A.V., 1969. Soils of the Murmansk Region. Nauka, Leningrad. 99 pp. (in Russian).

Benevolenskaya, Yu.D., 1984. On the morphological nonuniformity of the craniological series from the burial on Yuzhny Oleni Island. In: Gokhman, I.I. (Ed.), Problems of Anthropology and Ancient and Present-Day Population of Northern Eurasia. Nauka, Leningrad, pp. 37–54 (in Russian).

Bennike, O., 1997. Quaternary vertebrates from Greenland: a review. Quaternary Science Reviews 16, 899–909.

Bennike, O., Andreasen, C., 2007. Radiocarbon dating of walrus (*Odobenus rosmarus*) remains from Greenland. Polar Record 43, 361–365.

Bennike, O., Anderson, N.J., McGowan, S., 2010. Holocene palaeoecology of southwest Greenland inferred from macrofossils in sediments of an Oligosaline Lake. Journal of Paleolimnology 43, 787–798.

Bennike, P., 1985. Paleopathology of Danish Skeletons: A Comparative Study of Demography, Disease and Injury. Akademisk Forlag, Copenhagen, Denmark.

Beschel, R.E., Webb, D., 1963. Growth Ring Studies on Arctic Willows, Jacobsen-McGill Arctic Research Expedition 1959–1962: Preliminary Report 1961–1962. McGill University, Montreal, pp. 189–198.

Besprozvanny, E.M., 1997. Mesolithic of the taiga zone of West Siberia (preliminary results). In: Salvage Archaeological Studies on the Middle Urals. Iss. 1. Ekaterinburg, pp. 26–38 (in Russian).

Besprozvanny, E.M., Pogodin, A.A., 1998. On the cultic notions of the Mesolithic people inhabiting the Konda River basin. Voprosy Arkheologii Urala 23, 48–62 (in Russian).

Besprozvanny, E.M., Pogodin, A.A., 2006. Mesolithic sites in the lower reaches of the Bolshaya Uchinya River. In: Yakovlev, Ya.A. (Ed.), Khanty-Mansi Autonomous Area as Reflected in the Mirror of the Past. Iss. 3. Tomsk-Khanty-Mansiysk, pp. 4–19 (in Russian).

Bever, M.R., 2001. Stone tool technology and the Mesa complex: developing a framework of Alaskan Paleoindian prehistory. Arctic Anthropology 38 (2), 98–118.

Bever, M.R., 2006. Too little, too late? The radiocarbon chronology of Alaska and the peopling of the new world. American Antiquity 71 (4), 595–620.

Bezus'ko, L.G., Bezus'ko, A.G., Mosyakin, S.L., Gubin, S.V., Dumko, I.O., 2008. On the history of vegetation on the Kolyma Lowland in the Late Pleistocene (based on palynological data). In: Velichko, A.A., Vasil'ev, S.A. (Eds.), Way to North: Paleoenvironment and Earliest Inhabitants of Arctic and Subarctic. Institute of Geography RAS, Moscow, pp. 243–248 (in Russian).

Bielawski, E., 1988. Paleoeskimo variability: the early Arctic Small Tool Tradition in the central Canadian Arctic. American Antiquity 53, 52–74.

Bigelow, N.H., Edwards, M.E., 2001. A 14,000 yr paleoenvironmental record from Windmill Lake, central Alaska: Late Glacial and Holocene vegetation in the Alaska Range. Quaternary Science Reviews 20, 203–215.

Bigelow, N.H., Powers, W.R., 1994. New AMS dates from the Dry Creek Paleoindian Site, central Alaska. Current Research in the Pleistocene 11, 114–116.

Bjerck, H.B., 1989. Forskningsstyrt kulturminneforvaltning på Vega, Nordland. En studie av steinaldermenenskenes boplassmøønstre og arkeologiske letemetoder. Gunneria, vol.61, 212 pp.

Bjerck, H.B., 1990. Mesolithic site types and settlement patterns at Vega, Northern Norway. Acta Archaeologica 16, 11–32.

Bjerck, H.B., 1994. Nordsjøfastlandet og pionerbosetningen i Norge. Viking LVII, 25–58.

Bjerck, H.B., 1995. The North Sea Continent and the pioneer settlement of Norway. In: Fischer, A. (Ed.), Man and Sea in the Mesolithic: Coastal Settlement above and below Present Sea Level. Proceedings of the International Symposium, Kalundborg, Denmark, 1993. Oxbow Books, Oxford, pp. 131–144.

Bjerck, H.B., 2008. Norwegian Mesolithic trends. In: Bailey, G., Spikins, P. (Eds.), Mesolithic Europe. Cambridge Univ. Press, Cambridge, pp. 60–106.

Björck, S., 1995. Late Weichselian to early Holocene development of the Baltic Sea – its implications for coastal settlements in the southern Baltic region. In: Fischer, A. (Ed.), Man and Sea in the Mesolithic: Coastal Settlements Above and Below Present Sea Level. Proceedings of the International Symposium, Kalundborg, Denmark, 1993. Oxbow, Oxford, pp. 23–34.

Björck, S., Berglund, B.E., Digerfeldt, G., 1988. New aspects on the deglaciation chronology of South Sweden. Geographia Polonia 55, 37–49.

Blake Jr., W., 1989. Inferences concerning climate change from a deeply frozen lake on Rundfjeld, Ellesmere Island, Arctic Canada. Journal of Paleolimnology 2, 41–54.

Blake Jr., W., Boucherle, M.M., Fredskild, B., Janssens, J.A., Smol, J.P., 1992. The geomorphological setting, glacial history and Holocene development of 'Kap Inglefield Sø', Inglefield Land, North-West Greenland. Meddelelser om Grønland, Geoscience 27, 42.

Blankholm, H.P., 2004. Earliest Mesolithic site in Northern Norway? A reassessment of Sarnes B4. Arctic Anthropology 41 (1), 41–57.

Blankholm, H.P., 2008. Melsnes 1. An Early Post-glacial Coastal Site in Northern Norway. Oxbow Books, Oxford. 108 pp.

Boas, F., 1888. The Central Eskimo. Bureau of American Ethnology, Sixth Annual Report. The Smithsonian Institution, Washington, DC, pp. 409–669.

Böcher, J., Fredskild, B., 1993. Plant and arthropod remains from the Palaeo-Eskimo site at Qeqertasussuk, West Greenland. Meddelelser om Grønland, Geoscience 30, 35 pp.

Bocherens, H., Drucker, D.G., Billiou, D., Patou-Mathis, M., Vandermeersch, B., 2005. Isotopic evidence and subsistence pattern of the Saint-Cesare I Neanderthal: review and use of multi-source mixing model. Journal of Human Evolution 49, 71–87.

Bocherens, H., Fizet, M., Mariotti, A., Lange-Badre, B., Vandermeersch, B., Borel, J.-P., Bellon, G., 1991. Isotopic biogeochemistry (13C, 15N) of fossil vertebrate collagen: implications for the study of fossil food web including Neanderthal man. Journal of Human Evolution 20, 481–492.

Bøe, J., Nummedal, A., 1936. Le Finnmarkien. Les origines de la civilisation dans l'extrême nord de l'Europe. Instituttet for Sammenlignende Kulturforskning, Oslo. Serie B: Skrifter 32.

Boeskorov, G.G., 2003. Composition of the Late Pleistocene and Holocene theriofauna in Yakutia (based on archaeological data). In: Alekseyev, A.N. (Ed.), Ancient Cultures in North-Eastern Asia: Astroarchaeology Paleoinformatics. Nauka, Novosibirsk, pp. 27–43 (in Russian).

Bolikhovskaya, N.S., Derevianko, A.P., Shun'kov, M.V., 2006. The fossil palynoflora, geological age and climatostratigraphy of the earliest deposits of the Karama site (Early Paleolithic, Altai Mountains). Paleontological Journal 40, 558–566.

Bølviken, E., Helskog, E.K., Holm, O.L., 1982. Correspondence analysis: an alternative to principal components. World Archaeology 14 (1), 41–60.

Borodin, A.V., Strukova, T.V., Ulitko, A.I., Chairkin, S.E., Bachura, O.P., 2000. Cheremukhovo-1 – a new historical-environmental and archaeological object in the Northern Urals (location and stratigraphy). In: Kosintsev, P.A. (Ed.), Pleistocene and Holocene Faunas of the Urals. Rifey Publ., Chelyabinsk, pp. 36–58 (in Russian).

Borutskaya, S.B., 2005. Paleopathology of the late Neolithic burial ground on the Bolshoy Oleni Island, the Barents Sea. Vestnik Arkheologii, Institute of Ethnography and Archaeology RAS 12, 98–105 (in Russian).

Borutskaya, S.B., Vasil'ev, S.V., 2003. Preliminary analysis of postcranial skeletons of two adult humans of the Mayak site. In: Zones of Contact in Eurasia at the Boundary of Centuries. Center "Integratsiya", Samara, pp. 28–32 (in Russian).

Bowers, P., Reuther, J.D., 2008. AMS re-dating of the Carlo Creek site, Nenana valley, central Alaska. Current Research in the Pleistocene 25, 58–61.

Boyarskiy, H.V. (Ed.), 1994. Novaya Zemlya. Transactions of the Marine Arctic Complex Expedition. vol. 3(4) Russian Research Institute of Cultural and Natural Heritage. 239 pp. (in Russian).

Bradley, R.S., 1990. Holocene paleoclimateology of the Queen Elizabeth Islands, Canadian high Arctic. Quaternary Science Reviews 9, 365–384.

Bradley, R.S., England, J.H., 2008. The Younger Dryas and the sea of ancient ice. Quaternary Research 70, 1–10.

Bradley, R.S., Briffa, K.R., Cole, J.E., Hughes, M.K., Osborn, T.J., 2003. The climate of the last millennium. In: Alverson, K., Bradley, R.S., Pedersen, T.F. (Eds.), Paleoclimate, Global Change, and the Future. Springer Verlag, Berlin, pp. 105–141.

Bramanti, B., Thomas, M., Haak, W., Unterlaender, M., Jores, P., Tambets, K., Antanaitis-Jacobs, I., Haidle, M.N., Jankauskas, R., Kind, C.J., Lueth, F., Terberger, T., Hiller, J., Matsumura, S., Forster, P., Burger, J., 2009. Genetic discontinuity between local hunter-gatherers and central Europe's first farmers. Science 326, 137–140.

Brigham-Grette, J., Lozhkin, A.V., Anderson, P.M., Glushkova, O.Yu, 2004. Paleoenvironmental conditions in Western Beringia before and during the Last Glacial maximum. In: Madsen, D.B. (Ed.), Entering America: Northeast Asia and Beringia Before the Last Glacial Maximum. Univ. of Utah Press, Salt Lake City, Utah, pp. 29–61.

Briner, J.P., Michelutti, N., Francis, D.R., Miller, G.H., Axford, Y., Wooller, M.J., Wolfe, A.P., 2006. A multi-proxy lacustrine record of Holocene climate change on northeastern Baffin Island, Arctic Canada. Quaternary Research 65, 431–442.

Briner, J.P., Overeem, I., Miller, G., Finkel, R., 2007. The deglaciation of Clyde Inlet, northeastern Baffin Island, Arctic Canada. Journal of Quaternary Science 22, 223–232.

Broadbent, N., 1979. Coastal resources and settlement stability: a critical study of a Mesolithic site complex in northern Sweden. In: AUN 3. Institute of North European Archaeology, Uppsala. 268 pp.

Broecker, W.S., 2000. Abrupt climate change: causal constraints provided by the paleoclimate record. Earth Science Reviews 51, 137–154.

Broecker, W.S., 2001. Was the Medieval warm period global? Science 291, 1497–1499.

Bronshtein, M.M., 2007. Ekven and Paipelgak seen by ethnologists. In: World of the Arctic Hunters: Steps into Unknown. Catalogue of Exposition. Museum of East, Moscow, Anadyr, pp. 26–47. Museum Center "Chukotka Heritage" (in Russian).

Brubaker, L.B., Anderson, P.M., Hu, F.S., 2001. Vegetation ecotone dynamics in southwest Alaska during the late Quaternary. Quaternary Science Reviews 20, 175–188.

Brubaker, L.B., Garfinkel, H.L., Edwards, M.E., 1983. A late-Wisconsin and Holocene vegetation history form the central Brooks Range: implications for Alaskan paleoecology. Quaternary Research 20, 194–214.

Brusnitsyna, A.G., Oshchepkov, K.A., 2000. Archaeological sites on Middle Yamal. In: Golovnev, A.V. (Ed.), Antiquities of Yamal. Iss. 1. Uralian Branch RAS, Ekaterinburg-Salekhard, p. 82 (in Russian).
Bunak, V.V., 1951. Craniological Types in the West European Paleolithic in Comparison with Older Ones. Short Communications of the Institute of Ethnography, Academy of Sciences USSR, Iss. XIII, pp. 38–46 (in Russian).
Bunak, V.V., 1956. Human races and ways of their formation. Sovetskaya Etnografiya 1, 86–104 (in Russian).
Bunak, V.V., 1973. Fossil man from the Sungir site: its place among other late Paleolithic fossils. In: Reports of Soviet Specialists, IX IUAES Congress (Chicago, September 1973). Nauka, Moscow (in Russian).
Bunak, V.V., 1980. Genus Homo, Its Origin and Subsequent Evolution. Nauka, Moscow. 329 pp. (in Russian).
Bunak, V.V., Gerasimova, M.M., 1984. The Upper Paleolithic skull Sungir 1 and its place among other Upper Paleolithic skulls. In: Alekseyev, V.P. (Ed.), Sungir: Anthropological Study. Nauka, Moscow, pp. 14–99 (in Russian).
Burke, A., Cinq-Mars, J., 1996. Dental characteristics of *Equus lambei* from the Bluefish Caves, Yukon Territory. Geographie Physique et Quaternaire 50 (1), 81–93.
Burke, A., Cinq-Mars, J., 1998. Paleoethological reconstruction and taphonomy of *Equus lambei* from the Bluefish Caves, Yukon Territory, Canada. Arctic 51 (2), 105–115.
Burov, G.M., 1967a. Archaeological Sites of the Vychegda Valley. Komi Publ., Syktyvkar. 96 pp. (in Russian).
Burov, G.M., 1967b. Ancient Sindor: On the History of Tribes of the European Northeast from the 7th Millennium BC to the 1st Millennium AD. Nauka, Moscow. 220 pp. (in Russian).
Burov, G.M., 1974a. Archaeological Cultures of the North of European Part of the USSR (Northern Dvina Territory). Manual for History Students, Ulyanovsk. 120 pp. (in Russian).
Burov, G.M., 1974b. Permanent sedentary life and fishery of Neolithic tribes in North-Eastern Europe. In: Gerasimov, I.P., Velichko, A.A. (Eds.), Early Man, His Material Culture and Environments in the Pleistocene and Holocene. Institute of Geography, Academy of Sciences, USSR, Moscow, pp. 283–287 (in Russian).
Burov, G.M., 1986. The Extreme North-East of Europe at the Mesolithic, Neolithic, and early Metal epochs (Thesis Dr.Sc.). Novosibirsk, 37 pp. (in Russian).
Burov, G.M., 1993. Neolithic and Eneolithic pit-houses in the extreme Northeast of Europe with view to social organization of its early population. Voprosy Arkheologii Urala 21, 23–41 (in Russian).
Burov, G.M., 2009. Mesolithic wooden work pieces of new categories and types from the Vis I settlement in the Vychegda drainage basin. Rossiyskaya Arkheologiya 2, 17–29 (in Russian).
Burov, G.M., 2011. Fishery in the Mesolithic in the European North of Russia. Rossiyskaya Arkheologiya 2, 5–11 (in Russian).
Burov, G.M., Romanova, E.N., Sementsov, A.D., 1972. Chronology of the wooden constructions and objects found in the Severnaya Dvina drainage basin. In: Kolchin, B.A. (Ed.), Problems of Absolute Dating in Archaeology. Nauka, Moscow, pp. 76–79 (in Russian).
Butakov, G.P., 1986. The Pleistocene Periglacial Morphogenesis in the East of the Russian Plain. Kazan' Univ. Press, Kazan'. 144 pp. (in Russian).
Buzhilova, A.P., 1998. Paleo-pathology in bioarchaeological reconstrctions. In: Buzhilova, A.P., Kozlovskaya, M.V., Mednikova, M.B. (Eds.), Historical Ecology of Man. Methods of Biological Investigation. Staryi Sad Publ., Moscow, pp. 87–146 (in Russian).
Buzhilova, A.P., 2005. Homo sapiens: Medical History. Yazyki Slavyanskoy Kul'tury Publ., Moscow. 320 pp. (in Russian).

Buzhilova, A.P., 2006. An analysis of the Mesolithic population social structure studied on materials from the Oleni Ostrov burial: anthropological reconstruction. In: Martynov, A. Ya (Ed.), Prehistoric and Medieval History and Culture of the European North. Problem of Investigations and Scientific Reconstruction. Solti Publ., Solovki Settl, Arkhangelsk Region, pp. 382–390 (in Russian).

Buzhilova, A.P., 2008. Anthropological materials dated to Mesolithic and Neolithic epochs from the Minino archaeological complex on Lake Kubenskoye. In: Sorokin, A.N. (Ed.), Man, Adaptation, Culture. GRIFiK Publ., Moscow, pp. 113–126 (in Russian).

Buzhilova, A.P., 2009. Mesolithic population in the European North. In: Gromov, A.V., Khartanovich, V.I. (Eds.), Micro-evolutionary Processes in Human Populations. MAE (Museum of Anthropology and Ethnography), St.-Petersburg, pp. 57–74 (in Russian).

Buzhilova, A.P., 2011. Odontometry of homo deciduous teeth from Late Pleistocene layers of Altai caves, Siberia. In: Derevianko, A.P., Shun'kov, M.V. (Eds.), Characteristic Features of the Middle to Upper Paleolithic Transition in Eurasia. Publishing Department of the Institute of Archaeology and Ethnography Siberian Branch RAS, Novosibirsk, pp. 24–39.

Buzhilova, A.P., Berezina, N.Ya, 2008. A probable case of multiple myeloma (based of anthropological materials dated to the 5th century AD, the Northern Caucasus). In: Buzhilova, A.P. (Ed.), OPUS: Interdisciplinary Investigations in Archaeology, 6. Paralleli Publ., Moscow, pp. 343–351 (in Russian).

Buzhilova, A.P., Kozlovskaya, M.V., Mednikova, M.B., 1998. Historical Ecology of Man. Methods of Biological Investigation. Staryi Sad Publ., Moscow. 260 pp. (in Russian).

Buzhilova, A.P., Suvorov, A.V., Krylovich, O.A., 2008. On the Reconstruction of the Population Way of Life during the Late Stages of the Stone Age (Based of Archeological Materials from the Minino Complex on Lake Kubenskoye). Short Communications of Institute of Archeology, vol. 222, pp. 1–18 (in Russian).

Callahan, E., 1987. An Evaluation of Lithic Technology in Middle Sweden During the Mesolithic and Neolithic. Aun 8. Societas Archaeologica Uppsaliensis, Uppsala.

Campos, P.F., Willerslev, E., Sher, A., Orlando, L., Axelsson, E., Tikhonov, A., Aaris-Sørensen, K., Greenwood, A.D., Kahlke, R.-D., Kosintsev, P., Krakhmalnaya, T., Kuznetsova, T., Lemey, P., MacPhee, R., Norris, C.A., Shepherd, K., Suchard, M.A., Zazula, G.D., Shapiro, B., Thomas, M., Gilbert, M.T., 2010. Ancient DNA analyses exclude humans as the driving force behind late Pleistocene musk ox (*Ovibos moschatus*) population dynamics. Proceedings of the National Academy of Sciences of the United States of America 107, 5675–5680.

CARD, 1999. Canadian Archaeological Radiocarbon Database (CARD).

Carpelan, Ch, 1999. On the postglacial colonization of Eastern Fennoscandia. In: Dig it All. Papers Dedicated to Prof. Ari Siiriainen, pp. 151–172 Jyvaskyla, Helsinki.

Carpelan, Ch, 2008. On the History and Recent Studies of the "Antrea Net Find". Iskos 16. Karelian Isthmus. Stone Age Studies in 1998–2003, pp. 88–127.

Cavalli-Sforza, L.L., Menozzi, P., Piazza, A., 1994. History and Geography of Human Genes. Princeton Univ. Press, Princeton. 1059 pp.

Chabay, V.P., 2004. Middle Paleolithic of the Crimea. Shlyakh Publ., Simferopol. 304 pp. (in Russian).

Cheboksarov, N.N., 1947. Main Directions in Race Differentiation in Eastern Asia, vol. 2. Transactions of Institute of Ethnography (New Series), Moscow–Leningrad, pp. 24–83 (in Russian).

Chebotareva, N.S., Makarycheva, I.A., 1982. Geochronology of natural changes in the glaciated regions of Eastern Europe at the Valday Ice Age. In: Gerasimov, I.P., Velichko, A.A. (Eds.), Paleogeography of Europe during the Last 100,000 Years. Nauka, Moscow, pp. 16–27 (in Russian).

Cherdyntsev, V.V., Zavelskiy, F.S., Kind, N.S., Sulerzhitskiy, L.D., Forova, V.S., 1969. Radiocarbon dates, GIN, Academy of Sciences, USSR. Bulletin of Commission for Study of the Quaternary 36, 172–193 (in Russian).

Chernilovskiy, Z.M., 1983. At the Chukotka and Alaska Contact. Nauka, Moscow. 231 pp. (in Russian).

Chernov, G.A., 1974. On the Quaternary deposits and geomorphology of the Vangyr region, the Subpolar Urals. Bulletin of Commission for Study of the Quaternary 42, 66–81 (in Russian).

Chernov, G.A., 1985. Atlas of Archaeological Sites in the Bolshezemel'skaya Tundra. Nauka, Moscow, (in Russian).

Chikisheva, T.A., 2012. Dynamics of the Anthropological Differentiation of the Southern West Siberia Population at the Neolithic and Early Iron Ages. Institute of Archaeology and Ethnography, Siberian Branch RAS, Novosibirsk. 468 pp. (in Russian).

Cinq-Mars, J., 1979. Bluefish Cave I: a Late Pleistocene Eastern Beringian Cave deposits in the Northern Yukon. Canadian Journal of Archaeology 3, 1–32.

Cinq-Mars, J., 1990. La place de Grottes du Poisson Bleu dans la préhistoire béringienne. Revista de Arqueologia Americana 1, 9–32.

Cinq-Mars, J., 2001. On the significance of modified mammoth bones from eastern Beringia. In: Cavaretta, G., Giolo, P., Mussi, M., Palombo, M.R. (Eds.), The World of Elephants. International Congress, Proceedings of the 1st International Congress, Rome, October 16–20, 2001. Consiglio Nazionale della Ricerche, Rome, pp. 424–428.

Cinq-Mars, J., Morlan, R., 1999. Bluefish Caves and Old Crow Basin: a new rapport. In: Bonnichsen, R., Turnmire, K.L. (Eds.), Ice Age Peoples of North America: Environments, Origins, and Adaptations of the First Americans. Oregon State Univ. Press, Corvallis, pp. 200–212.

Clark, J.G.D., 1975. The Earlier Stone Age Settlement of Scandinavia. Cambridge Univ. Press, Cambridge, pp. 221–227.

Cook, J.P., 1969. Early Prehistory of Healy Lake, Alaska Unpublished (Ph.D. dissertation). Department of Anthropology, Univ.of Wisconsin.

Cook, J.P., 1996. Healy Lake. In: West, F.H. (Ed.), American Beginnings: The Prehistory and Palaeoecology of Beringia. Univ. of Chicago Press, Chicago, pp. 323–327.

Corner, G.D., Yevzerov, V.Ya., Kolka, V.V., Møller, J.J., 1999. Isolation basin stratigraphy and Holocene relative sea-level change at the Norwegian-Russian border north of Nikel, north-west Russia. Boreas 28 (1), 146–166.

Corner, G.D., Kolka, V.V., Yevzerov, V.Ya., Møller, J.J., 2001. Postglacial relative sea-level change and stratigraphy of raised coastal basins on Kola Peninsula, northwest Russia. Global and Planetary Change 31, 155–177.

Cruciani, F., La Fratta, R., Trombetta, B., Santolamazza, P., Sellitto, D., Colomb, E.B., Dugoujon, J.-M., Crivellaro, F., Benincasa, T., Pascone, R., Moral, P., Watson, E., Melegh, B., Barbujani, G., Fuselli, S., Vona, G., Zagradisnik, B., Assum, G., Brdicka, R., Kozlov, A.I., Efremov, G.D., Coppa, A., Novelletto, A., Scozzari, R., 2007. Tracing past human male movements in Northern/Eastern Africa and Western Eurasia: new clues from Y-chromosomal haplogroups E-M78 and J-M12. Molecular Biology and Evolution 24 (6), 1300–1311.

Cruciani, F., Trombetta, B., Sellitto, D., Massaia, A., Destro-Bisol, G., Watson, E., Colomb, E.B., Dugoujon, J.-M., Moral, P., Scozzari, R., 2010. Human Y chromosome haplogroup R-V88: a paternal genetic record of early mid Holocene trans-Saharan connections and the spread of Chadic languages. European Journal of Human Genetics 18 (7), 800–807.

Damblon, F., Haesaerts, P., Van der Plicht, J., 1996. New datings and considerations on the chronology of Upper Palaeolithic sites in the Great Eurasiatic plain. Préhistoire Européenne 9, 177–231.

Darwent, C.M., 2001. High Arctic Paleoeskimo Fauna: Temporal Changes and Regional Differences (Ph.D. dissertation). Univ. of Missouri, Columbia.

Darwent, C.M., 2004. The highs and lows of high Arctic mammals: temporal change and regional variability in Paleo-Eskimo subsistence. In: Mondini, M., Mucoz, S., Wickler, S. (Eds.), Colonization, Migration, and Marginal Areas: A Zooarchaeological Approach. Oxbow Books, Oxford, UK, pp. 62–73.

Davis, P.T., 1985. Neoglacial moraines on Baffin Island. In: Andrews, J.T. (Ed.), Holocene Pollen Records from West Greenland. Allen & Unwin, Boston, pp. 682–718.

Davis, L.G., Sisson, D.A., 1998. An early stemmed point cache from the Lower Salmon River Canyon of West-Central Idaho. Current Research in the Pleistocene 15, 12–14.

Davydova, N., Servant-Vildary, S., 1996. Late Pleistocene and Holocene history of the lakes in the Kola Peninsula, Karelia and the northwestern part of the East European plain. Quaternary Science Reviews 15, 997–1012.

Debets, G.F., 1941. Problem of North-West Siberia Colonization on the Paleoanthropological Evidence. Short Communications of the Institute for the Material Culture History, Academy of Sciences, USSR, vol. 9, pp. 14–18 (in Russian).

Debets, G.F., 1948. Paleoanthropology of the USSR. 392 pp. (Transactions of Institute of Ethnography, V. 4). USSR Academy of Sciences Publ., Moscow–Leningrad, (in Russian).

Debets, G.F., 1951. Anthropolical Investigations in the Kamchatka Region. 263 pp. Transactions of Institute of Ethnography, vol. 17. USSR Academy of Sciences Publ., Moscow, (in Russian).

Debets, G.F., 1955. Paleoanthropolical finds in Kostenki sites. Sovetskaya Etnografiya 1, 39–46 (in Russian).

Debets, G.F., 1956. Ancient Skull from Yakutia. Short Communications of the Institute of Ethnography, USSR Academy of Sciences, vol. 25, pp. 60–63 (in Russian).

Debets, G.F., 1961. On the routes of colonization of the northern Russian Plain and Eastern Baltic region. Sovetskaya Etnografiya 6, 51–69 (in Russian).

Debets, G.F., 1964. On the anthropological type of the ancient population of Finland. Modern Anthropology. Transactions of the Moscow Society of Naturalists 14, 233–239 (in Russian).

Debets, G.F., 1975. Paleoanthropolical materials from the Old Bering Sea burials Uelen and Ekven. In: Arutyunov, S.A., Sergeyev, D.A. (Eds.), Problems of Ethnic History of the Bering Sea Region. Nauka, Moscow, pp. 198–237 (in Russian).

Debets, G.F., 1986. Paleoanthropology of ancient Eskimo (Ipiutak, Tigara). In: Velikanova, M.S., Zolotareva, I.M. (Eds.), Ethnical Relationships between Peoples of Asian North and North America Based on Anthropological Data. Nauka, Moscow, pp. 6–149 (in Russian).

Denisov, V.P., 1976. New Paleolithic localities and sites at the Kama reservoir. In: Archaeological Discoveries – 1975. Nauka, Moscow, p. 174 (in Russian).

Denisova, R.Ya, 1975. Anthropology of the Ancient Balts. Zinatne, Riga. 404 pp. (in Russian).

Denisova, R.Ya, 1998. Zvejnieki – a Stone Age burial ground: inner structure, chronology, anthropological composition. In: Abstracts. The 1st International Conference 'Race: Myth or Reality?' Moscow, October 7–9, 1998. Staryi Sad Publ., Moscow, pp. 45–46 (in Russian).

Der Sarkissian, C., Balanovsky, O., Brandt, G., Khartanovich, V., Buzhilova, A., Koshel, S., Zaporozhchenko, V., Gronenborn, D., Moiseyev, V., Kolpakov, E., Shumkin, V., Alt, K.W., Balanovska, E., Cooper, A., Haak, W., Adhikarla, S., Adler, C.J., Bertranpetit, J., Clarke, A.C., Comas, D., Dulik, M.C., Gaieski, J.B., Haber, M., Jin, L., Kaplan, M.E., Li, S., MartHnez-Cruz, B., Matisoo-Smith, E.A., Mitchell, R., Owings, A.C., Parida, L., Pitchappan, R., Platt, D.E., Quintana-Murci, L., Renfrew, C., Lacerda, D.R., Royyuru, A.K., Santos, F.R., Schurr, T.G., Soodyall, H., Hernanz, D.F., Swamikrishnan, P., Tyler-Smith, C., Santhakumari, A.V., Vieira, P.P., Vilar, M.G., Wells, R., Zalloua, P.A., Ziegle, J.S., Ganesh Prasad, A., Merchant, N.C., 2013. Ancient DNA reveals prehistoric gene-flow from Siberia in the complex human population history of North East Europe. PLoS Genetics 9 (2), e1003296.

Derevyanko, A.P., 2001. Transition from Middle to Upper Paleolithic on the Altai. Archaeology, Ethnography and Anthropology of Eurasia 3, 70–103.

Derevyanko, A.P., 2005. The most ancient migrations of humans into Eurasia and the problem of the upper Paleolithic formation. Archaeology,. Ethnography and Anthropology of Eurasia 2, 22–36.

Derevyanko, A.P., 2007. On the Problem of the Neanderthals Presence in Central Asia and Siberia. Institute of Archaeology and Ethnography, Siberian Branch RAS Publ., Novosibirsk. 76 pp. (in Russian).

Derevyanko, A.P., 2009a. The Most Ancient Migrations of Humans in Eurasia in the Early Paleolithic. Institute of Archaeology and Ethnography, Siberian Branch RAS Publ., Novosibirsk. 231 pp. (in Russian).

Derevyanko, A.P., 2009b. Transition from Middle to Upper Paleolithic and the Problem of *Homo sapiens sapiens* Development in Eastern, Central and Northern Asia. Institute of Archaeology and Ethnography, Siberian Branch RAS Publ., Novosibirsk. 328 pp. (in Russian).

Derevyanko, A.P., 2011. The Upper Paleolithic in Africa and Eurasia and Development of Man of the Modern Anatomical Type. Institute of Archaeology and Ethnography, Siberian Branch RAS Publ., Novosibirsk. 559 pp. (in Russian).

Derevyanko, A.P., Markin, S.V., 1992. The Mountain Altai Mousterian (Based on Materials from Okladnikov Cave). Nauka, Novosibirsk. 224 pp. (in Russian).

Derevyanko, A.P., Markin, S.V., 1998. Paleolithic of the northwestern Altai–Sayany. Rossiyskaya Arkheologiya 4, 17–34 (in Russian).

Derevyanko, A.P., Markin, S.V., 2007. Lower Paleolithic site MK-1 in the southeast of West Siberia. In: Amirkhanov, Kh (Ed.), The Caucasus and Initial Colonization of the Old World by Humans. Peterburgskoye Vostokovedeniye (St.-Petersburg Centre for Oriental Studies), St.-Petersburg, pp. 149–155 (in Russian).

Derevyanko, A.P., Rybin, E.P., 2003. The most ancient manifestation of the Paleolithic man symbolic activities in the Mountain Altai. Archaeology, Ethnography and Anthropology of Eurasia 3, 27–50.

Derevyanko, A.P., Shun'kov, M.V., 2009. Development of early human culture in northern Asia. Paleontological Journal 43 (8), 31–39.

Derevyanko, A.P., Shun'kov, M.V., 2004. Formation of the Upper Paleolithic traditions in the Altai. Archaeology, Ethnography and Anthropology of Eurasia 3, 12–40.

Derevyanko, A.P., Markin, S.V., Vasil'ev, S.V., 1994. Paleolithic Studies. Nauka, Novosibirsk. 288 pp. (in Russian).

Derevyanko, A.P., Volkov, P.V., Hon-John, L., 1998. Selemdzha Late Paleolithic Culture. Institute of Archaeology and Ethnography, Siberian Branch RAS Publ., Novosibirsk. 336 pp. (in Russian).

Derevyanko, A.P., Petrin, V.T., Rybin, E.P., 2000a. Nature of Mousterian – Late Paleolithic transition on the Altai (based on materials from the Kara-Bom site). Archaeology, Ethnography and Anthropology of Eurasia 2, 33–52.

Derevyanko, A.P., Zenin, V.N., Leshchinskiy, S.V., Mashchenko, E.N., 2000b. Specific features of the mammoth bone accumulation near the Shestakovo site, West Siberia. Archaeology, Ethnography and Anthropology of Eurasia 3, 42–55.

Derevyanko, A.P., Shun'kov, M.V., Zykin, V.S., Markin, M.M., 2002a. A new Early Paleolithic complex in the Gorny Altay. In: Problems of Archaeology, Ethnography, and Anthropology of Siberia and Adjacent Territories. Proceedings of the Yearly Session of IAET Siberian Branch RAS, vol. 8. Institute of Archaeology and Ethnography, Siberian Branch RAS, Novosibirsk, pp. 84–89 (in Russian).

Derevyanko, A.P., Volkov, P.V., Petrin, V.T., 2002b. Origin of the Microblade Knapping Technology. Institute of Archaeology and Ethnography, Siberian Branch RAS Publ., Novosibirsk. 170 pp. (in Russian).

Derevyanko, A.P., Molodin, V.I., Zenin, V.N., Leshchinskiy, S.V., Mashchenko, E.N., 2003a. Shestakovo – Late Paleolithic Locality. Institute of Archaeology and Ethnography, Siberian Branch RAS Publ., Novosibirsk. 168 pp. (in Russian).

Derevyanko, A.P., Shun'kov, M.V., Agadjanyan, A.K., Baryshnikov, G.F., Malayeva, E.M., Ulyanov, V.A., Kulik, N.A., Anoykin, A.A., 2003b. Environments and Man in the Paleolithic in the Mountain Altai. Institute of Archaeology and Ethnography, Siberian Branch RAS Publ., Novosibirsk. 447 pp. (in Russian).

Derevyanko, A.P., Shun'kov, M.V., Bolikhovskaya, N.S., Zykin, V.S., Zykina, V.S., Kulik, N.A., Ulyanov, V.A., Chirkin, K.V., 2005. The Early Paleolithic Site Karama on the Altai Mts. Institute of Archaeology and Ethnography, Siberian Branch RAS Publ., Novosibirsk. 88 pp. (in Russian).

Derevyanko, A.P., Zenin, V.N., Shewkomud, I.Y., 2006. Paleolithic of the Priamurye (Amur river basin). In: Archaeology of the Russian Far East: Essays in Stone Age Prehistory. BAR International Series 1540. Oxford, pp. 55–73.

Derevyanko, A.P., Markin, S.V., Zykin, V.S., 2008. Chagyrskaya cave – a new middle Paleolithic site on the Altai Mts. In: Problems of Archaeology, Ethnography, Anthropology of Siberia and Adjacent Territories, vol. 14. Institute of Archaeology and Ethnography, Siberian Branch RAS Publ., Novosibirsk, pp. 52–55 (in Russian).

Derums, V.Ya, 1970. Diseases and Medical Treatment in Ancient Baltic Region. Zinatne, Riga. 200 pp. (in Russian).

Devyatova, E.I., 1976. Geology and Palynology of the Holocene and Chronology of the Prehistoric Epoch Sites in the Region Southwest of the White Sea. Nauka, Leningrad. 121 pp. (in Russian).

Devyatova, E.I., 1982. The Late Pleistocene Environments and Their Impact on the Human Dispersal over the Severnaya Dvina Drainage Basin and in Karelia. Karelia Publ., Petrozavodsk. 155 pp. (in Russian).

Dikov, N.N., 1961. On excavations of the Ust'Belskiy burial ground in 1958. In: Proc. Chukchi Regional Museum, Iss. 2. Magadan, pp. 4–14 (in Russian).

Dikov, N.N., 1971. Ancient Cultures of Kamchatka and Chukotka. Novosibirsk State Univ. Press, Novosibirsk. 48 pp. (in Russian).

Dikov, N.N., 1977. Archaeological Sites of Kamchatka, Chukotka, and the Upper Kolyma. Nauka, Moscow. 391 pp. (in Russian).

Dikov, N.N., 1979. Ancient Cultures of Northeastern Asia. Nauka, Moscow. 352 pp. (in Russian).

Dikov, N.N., 1990. The Paleolithic of Kamchatka and Chukotka in connection with the problems of peopling of America via Beringia. In: Han, S.-B., Kim, K.-O. (Eds.), Traditional Cultures of the Pacific Societies: Continuity and Change. Seoul National Univ. Press, Seoul, pp. 55–66.

Dikov, N.N., 1993a. Asia at the Interface of America in Ancient Times. Nauka, St.-Petersburg. 304 pp. (in Russian).

Dikov, N.N., 1993b. Paleolithic of Kamchatka and Chukotka in Relation to the Problem of Initial Colonization of America. Northeastern Interdisciplinary Research Institute, Far East Branch RAS, Magadan. 68 pp. (in Russian).

Dikov, N.N., 1996. Archaeological past of the inhabitants of North-East of Asia. In: Dikov, N.N. (Ed.), Northeast of Russia Since Ancient Times to the Present Day: New Excursions into History. Northeastern Interdisciplinary Research Institute, Far East Branch RAS, Magadan, pp. 13–31 (in Russian).

Dixon, J.E., Heaton, T.H., Fifield, T.E., Hamilton, T.D., Putnam, D.E., Grady, F., 1997. Late Quaternary regional geoarchaeology of Southeast Alaska Karst: a progress report. Geoarchaeology 12 (6), 689–712.

Djindjian, F., Kozlowski, J., Bazile, F., 2003. Europe during the Early Upper Palaeolithic (40,000 – 30,000 BP): a synthesis. In: Zilhвo, J., D'Errico, F. (Eds.), The Chronology of the Aurignacian and the Transitional Technocomplexes – Dating, Stratigraphies, Cultural Implications. Trabalhos de Archaeologia, 33, Lisboa, pp. 29–48.

Dneprovskiy, K.A., 2007. Ancient Eskimo sites studied by the Chukotka archaeological expedition. In: Brusnitsyna, A.G., Dneprovskiy, K.A., Sukhorukova, E.S. (Eds.), World of Arctic Hunters. Steps into Unknown. Catalogue of Exposition. Moscow State Museum of the East, Moscow, Anadyr, pp. 15–25 (in Russian).

Dobretsov, N.L., Zykin, V.S., Zykina, V.S., 2006. Desertification of mid-latitude Northern Asia and global change periodicity in the Quaternary. In: Environmental Security and Sustainable Land Use of Mountain and Steppe Territories of Mongolia and Altai. NATO Science Series 2, Environmental Security. Springer, Dordrecht/Boston/London, pp. 3–18.

Dobrovolskaya, M.V., 2005. Man and His Diet. Food Specializations and Problems of Human Origins. Nauchny Mir, Moscow. 367 pp. (in Russian).

Dobrovolskaya, M.V., Tiunov, A.B., 2011a. Stable isotope (13C/12C and 15N/14N) evidence for Late Pleistocene hominides' paleodiets in Gorny Altai. In: Derevianko, A.P., Shun'kov, M.V. (Eds.), Characteristic Features of the Middle to Upper Paleolithic Transition in Eurasia. Publishing Department of the Institute of Archaeology and Ethnography Siberian Branch RAS, Novosibirsk, pp. 81–89.

Dobrovolskaya, M.V., Tiunov, A.V., 2011b. Stable isotope evidence for Upper Paleolithic human diets from Kostenki. In: Stevens, R., Joes, M. (Eds.), Early Anatomically Modern Humans in Eurasia: Coping with Climatic Complexity, vol. 32. The Kavli Royal Society International Centre.

Dobrovolskaya, M.V., Tiunov, A.V., 2013. Neanderthals from the Okladnikov Cave: environments and specific features of diet based on the data of isotope analysis. Archaeology, Ethnography and Anthropology of Eurasia 1, 78–88.

Dobrovolskaya, M.V., Buzhilova, A.P., Mednikova, M.B., 2008. Adaptive process study in the groups of ancient and modern Eskimo population. In: Buzhilova, A.P., Dobrovolskaya, M.V., Mednikova, M.B. (Eds.), Actual Trends in Anthropology. Institute of Archelogy RAS, Moscow, pp. 101–107 (in Russian).

Dolukhanov, P.M., 1969. History of the Baltic Region. Nauka, Moscow. 118 pp. (in Russian).

Dolukhanov, P.M., 1971. Geological Age of Sites on the Rybachiy Peninsula. Short Communications of Institute of Archeology RAS, vol. 126, pp. 100–102 (in Russian).

Dolukhanov, P.M., 1986. Natural environment and the Holocene settlement pattern in the north-western part of the USSR. Fennoscandia Archaeological 3, 3–16.

Dolukhanov, P.M., 2000. Origin of Ethnos. Evropeiskiy Dom Publ., St.-Petersburg. 221 pp. (in Russian).

Dolukhanov, P.M., 2008. Evolution of environment and early human dispersal in Northern Eurasia. In: Velichko, A.A., Vasil'ev, S.A. (Eds.), Way to North: Paleoenvironment and Earliest Inhabitants of Arctic and Subarctic. Institute of Geography RAS, Moscow, pp. 33–47 (in Russian).

Dolukhanov, P.M., Khotinskiy, N.A., 2002. Prehistoric society and environments. Neolithic. In: Velichko, A.A. (Ed.), Dynamics of Terrestrial Landscape Components and Marginal Seas of Northern Eurasia During the Last 130 000 Years. GEOS Publ., Moscow, pp. 146–156 (in Russian).

Dolukhanov, P.M., Shukurov, A., Tarasov, P., Zaitseva, G., 2002. Colonisation of Northern Eurasia by modern humans: radiocarbon chronology and environment. Journal of Archaeological Science 29, 593–606.

Dolukhanov, P.M., Kosheleva, E.A., Subetto, D.A., 2009. Spatial and temporal reconstruction of lanscapes and initial colonization of Fennoscandia. In: Fundamental Problems of Quaternary: Results of the Studies and the Main Trends in Further Investigations. Materials of VI All-Russia Conference on the Quaternary Studies, Novosibirsk, November 19–23, 2009. Siberian Branch RAS Publ., Novosibirsk, pp. 189–192 (in Russian).

Doner, L.A., 2001. Late Holocene Palaeolimnology and Paleoclimatology from Sub-arctic Lakes in Nunavut, Canada and Iceland (Ph.D. dissertation). Univ. of Colorado, Colorado.

Dremov, V.A., 1984. Race differentiation of Ugric and Samoyedic groups in West Siberia on craniological evidence. In: Gokhman, I.I. (Ed.), Problems of Anthropology of the Ancient and Modern Population in the North of Eurasia. Nauka, Leningrad, pp. 106–132 (in Russian).

Dumond, D., 1984. Prehistory of the Bering Sea Region. In: Damas, D. (Ed.), Handbook of North American Indians. vol. 5. Arctic. Smithsonian Institution, Washington, DC, pp. 94–105.

Dumond, D.E., 2001. The archaeology of eastern Beringia: some contrasts and connections. Arctic Anthropology 38 (2), 196–205.

Duryagina, D.A., Konovalenko, L.A., 1993. Palynology of Pleistocene of the Northwestern European Russia. Nauka, St.-Petersburg. 124 pp. (in Russian).

Dyakov, V.I., 2000. Primorye in Early Holocene (Ustinovka IV Mesolithic Site). Dalnauka Publ., Vladivostok. 228 pp. (in Russian).

Dyke, A.S., 2004. An outline of North American deglaciation with emphasis on central and northern Canada. In: Ehlers, J., Gibbard, P.L. (Eds.), Quaternary Glaciations – Extent and Chronology. Elsevier, Amsterdam, The Netherlands, pp. 373–425.

Dyke, A.S., 2008. The Steensby Inlet Ice Stream in the context of the deglaciation of Northern Baffin Island, Eastern Arctic Canada. Earth Surface Process and Landforms 33, 573–592.

Dyke, A.S., Morris, T.F., 1990. Postglacial History of the Bowhead Whale and of Driftwood Penetration; Implications for Paleoclimate, Central Canadian Arctic, vol. 89(24). Geological Survey of Canada, pp. 1–17.

Dyke, A.S., Peltier, W.R., 2000. Forms, response times and variability of relative sea-level curves, glaciated North America. Geomorphology 32, 315–333.

Dyke, A.S., Savelle, J.M., 2009. Paleoeskimo demography and sea-level history, Kent Peninsula and King William Island, central Northwest Passage, Arctic Canada. Arctic 62, 371–392.

Dyke, A.S., Vincent, J.S., Andrews, J.T., Dredge, L.A., Cowan, W.R., 1989. The Laurentide ice sheet and an introduction to the Quaternary geology of the Canadian Shield. In: Fulton, R.J. (Ed.), Quaternary Geology of Canada and Greenland. Geological Survey of Canada, Ottawa, pp. 178–189.

Dyke, A.S., England, J., Reimnits, E., Jettй, H., 1997. Changes in driftwood delivery to the Canadian Arctic archipelago: the hypothesis of postglacial oscillations of transpolar drift. Arctic 50, 1–16.

Dyke, A.S., Andrews, J.T., Clark, P.U., England, J.H., Miller, G.H., Shaw, J., Veillette, J.J., 2002. The Laurentide and Innuitian ice sheets during the Last Glacial maximum. Quaternary Science Reviews 21, 9–31.

Dyke, A.S., Dredge, L.A., Hodgson, D.A., 2005. North American deglacial marine-and lake-limit surfaces. Gйographie Physique et Quaternaire 59, 155–185.

Easton, N.A., MacKay, G.R., Schnurr, P., Young, P.B., Baker, C., 2007. The Little John site (KdVo-6), a multi-component (Nenana-Denali complex) site in the far Southwest of Yukon Territory, Canada. Current Research in the Pleistocene 24, 82–84.

Easton, N.A., Yesner, D.R., Hutchinson, V., Schnurr, P., Baker, C., 2009. Wisconsin interstadial, Terminal-Pleistocene, and Early-Holocene radiocarbon dates from the Little John site, Southwest Yukon Territory, Canada. Current Research in the Pleistocene 26, 47–50.

Easton, N.A., MacKay, G.R., Young, P.B., Schnurr, P., Yesner, D.R., 2011. Chindadn in Canada? Emergent evidence of the Pleistocene transition in Southeast Beringia as revealed by the Little John site, Yukon. In: Goebel, T., Buvit, I. (Eds.), From the Yenisei to the Yukon: Interpreting Lithic Assemblage Variability in Late Pleistocene/Early Holocene Beringia. Peopling of the Americas Publications Series. Texas A&M Univ. Press, College Station, pp. 289–322.

Edovin, E.G., 2009. A new group of Mesolithic sites in the Arkhangelsk North. In: Tver' Collection of Archaeological Papers, Iss. 7. Triada Publ., Tver', pp. 98–107 (in Russian).

Edovin, E.G., Belichenko, A.E., 2009. Local groups of Mesolithic population in the Arkhangelsk North. In: Tver' Collection of Archaeological Papers. Iss. 7. Triada Publ., Tver', pp. 108–122 (in Russian).

Edwards, M.E., Brubaker, L.B., Lozhkin, A.V., Anderson, P.M., 2005. Structurally novel biomes: a response to past warming in Beringia. Ecology 86, 1696–1703.

Edwards, M.E., Mock, C.J., Finney, B.P., Barber, V.A., Bartlein, P.J., 2001. Potential analogues for paleoclimatic variations in eastern interior Alaska during the past 14,000 yr: atmospheric-circulation controls of region temperature and moisture responses. Quaternary Science Reviews 20, 189–202.

Elias, S.A., Shor, S.K., Nelson, C.H., Birk, H.H., 1996. Life and times of the Bering Land Bridge. Nature 382, 60–63.

Elina, G.A., 1981. Principles and Methods of the Holocene Vegetation Reconstructing and Mapping. Nauka, Leningrad. 159 pp. (in Russian).

Elina, G.A., Lebedeva, R.M., 1982. The Holocene dynamics of landscape zones in the North-West of the European part of the USSR. In: Velichko, A.A., Spasskaya, I.I., Khotinskiy, N.A. (Eds.), Evolution of Environments of the USSR in the Late Pleistocene and Holocene. Nauka, Moscow, pp. 148–154 (in Russian).

Elina, G.A., Lukashov, A.D., Yurkovskaya, T.K., 2000. Late Glacial and Holocene of Eastern Fennoscandia (Paleo-Vegetation and Paleogeography). Karelian Research Center RAS, Petrozavodsk. 242 pp. (in Russian).

Elling, H., 1996. The Independence I culture and the Old Nuulliit culture in relation to the Saqqaq culture. In: The Paleo-Eskimo Cultures of Greenland. New Perspectives in Greenlandic Archaeology. Danish Polar Center, Copenhagen, pp. 191–198 Papers from a symposium at the Institute of Archaeology and Ethnography, Univ.of Copenhagen, May 21–24, 1992.

Ellis, C., 2008. The fluted point tradition and the Arctic Small Tool Tradition: what's the connection? Journal of Anthropological Archaelogy 27, 298–314.

Engelstad, E., 1989. Mesolithic house sites in Arctic Norway. In: Bonsall, C. (Ed.), The Mesolithic in Europe. Proceedings of the 3rd International Symposium. John Donald Publishers, Edinburgh, pp. 331–337.

Engelstad, E., 1991. The Mesolithic on the Outer Coast of Eastern. Finnmark, Tromsö.

England, J.H., Lakeman, T.R., Lemmen, D.S., Bednarski, J.M., Stewart, T.G., Evans, D.J.A., 2008. A millennial-scale record of Arctic Ocean sea ice variability and the demise of the Ellesmere Island ice shelves. Geophysical Research Letters 35, L19502. http://dx.doi.org/10.1029/2008GL034470.

Eriksson, G., Lougas, L., Zagorska, I., 2003. Stone Age hunter–fisher–gatherers at Zvejnieki, northern Latvia: radiocarbon, stable isotope and archaeozoology data. Before Farming: The Archaeology and Anthropology of Hunter–Gatherers 1 (2), 1–25.

Escutenaire, C., Kozlowski, J., Sitlivy, V., Sobczyk, K., 1999. Les chasseurs de mammouths de la vallée de la Vistule. Musées Royaux d'art et d'Histoire, Bruxelles. 99 pp.

Evseyeva, I.V., Wales, C.P., Balanovskiy, O.P., Spitsyn, V.A., 2001. Genetic structure of three northern populations – Sami, Nenets, and Pomors. In: Gerasimova, M.M., Dubova, N.A. (Eds.), Theory of Anthropology and Its Methods: Origination and Development. Stary Sad Publ., Moscow, pp. 65–67. To 110th anniversary of V.V. Bunak (in Russian).

Evzerov, V.Ya, 1979. Evolution of sedimentation in the coastal areas of the Barents and White sea in the late-glacial and post-glacial time. In: Gershanovich, D.E. (Ed.), Late Quaternary History and Sedimentogenesis in Marginal and Inner Seas. Nauka, Moscow, pp. 29–33 (in Russian).

Evzerov, V.Ya., Kagan, L.Ya., Koshechkin, B.I., Lebedeva, R.M., 1976. Aquatic sediments of the White Sea in the context of the evolution of environments in the Holocene. Izvestiya. All-Union Geografical Society 108 (5), 421–429 (in Russian).

Evzerov, V.Ya., Møller, J.J., Kolka, V.V., Corner, G.D., 2007. History of deglaciation of Rybachiy and Sredniy peninsulas (Kola Peninsula) in the Late Pleistocene. Izvestiya RAS Ser. Geogr 5, 58–63 (in Russian).

Eysteinsson þröstur, 2005. Hver er þriðji stærsti skógur Íslands? Vísindavefurinn. http://visindavefur.is/svar.php?id=5227.

Fainberg, L.A., 1964. Social Structure of Eskimo and Aleutians. Nauka, Moscow. 267 pp. (in Russian).

Fainberg, L.A., 1967. Population of Greenland in the 2nd millennium BC to 1 millenium AD. In: Grigulevich, I.R. (Ed.), From Alaska to Tierra del Fuego. Nauka, Moscow, pp. 202–211 (in Russian).

Fainberg, L.A., 1980. Origin of Eskimo and Aleutians. Ethnogenesis of paleo-Asiatic people of North-Eastern Asia. In: Gurvich, I.S. (Ed.), Ethnogenesis of Northern Peoples. Nauka, Moscow, pp. 227–240 (in Russian).

Fedorova, R.V., 1963. Environments of the Upper Paleolithic man habitation at the Kostenki vicinities (Voronezh Region) inferred from the pollen analysis of the Spitsyn site deposits, (Kostenki XVII). Materialy I Issledovaniya po Arkheolohii SSSR 121, 220–229 (in Russian).

Fedosova, V.N., 1989. Morphofunctional Variability of the Human Tubular Bones (with Reference to Problems of Paleoecology) (Ph.D. thesis), Moscow. (in Russian).

Fedosova, V.N., 1991. Ecological interpretations of the temporal variations of Asian Eskimo limb bones. Homo 42 (1), 244–264.

Filatova, V.F., 2006. Problems of origin and ethno-cultural attribution of the Mesolithic population. In: Kochkurkina, S.I., Kosmenko, M.G. (Eds.), Problems of Ethno-Cultural History of the Karelian Population (Mesolithic–Middle Ages). Karelian Research Center RAS, Petrozavodsk, pp. 14–72 (in Russian).

Finstad, I., Grydeland, S.E., 2009. Fv 53 Kroken-Tonsnes, Tromso kommune: Rapport fra arkeologisk undersokelser 2008//Tromura, Kulturvitenskap. Rapport No. 38. Universitetet i Tromso, Tromso. http://hdl.handle.net/10037/2936.

Firsov, L.V., Orlova, L.A., 1971. Radiocarbon dating of the mammoth bone from the Volchya Griva site. In: Materials of Field Studies of the Far East Archaeological Expedition, Iss. 2. Institute of History, Philology, and Philosophy, Siberian Branch, Academy of Sciences, USSR, Novosibirsk, pp. 132–134 (in Russian).

Fix, A.G., 2005. Rapid deployment of the five founding Amerind mtDNA haplogroups via coastal and riverine colonization. American Journal of Physical Anthropology 128, 430–436.

Fleischer, K.C., 1871. 23/8. Letter from C. Fleischer to Japetus Steenstrup with Description of Qajaa Stratigraphy – NKS 3460, 4. The Letter Base, The Royal Library, The National Library and Copenhagen Univ. Library.

Foley, R., 1987. Another Unique Species: Patterns in Human Evolutionary Ecology. Longman Scientific and Technical, and Wiley, Harlow, Essex, England, New York. 302 pp.

Forman, S.L., Ingylfsson, Y., Gataullin, V., Manley, W., Lockrantz, H., 2002. Late Quaternary stratigraphy, glacial limits, and paleoenvironments of the Marresale area, western Yamal Peninsula, Russia. Quaternary Research 57 (3), 355–370.

Fornander, E., 2006. The Wild Side of the Neolithic: A Study of Pitted Ware Diet and Ideology through the Analysis of Stable Carbon and Nitrogen Isotopes from Korsinдs, Grцdinge Parish, Sцdermanland. Arheologiska Forsknings Laboratoriet Stockholms Universitet, Stockholm. 40 pp.

Forsberg, O., 2006. Janisjoen reitin varhaisin asutus – inventointituloksia Laatokan pohjaispuolelta. Muinastutkija 1, 2–15.

Francalacci, P., 1989. Dietary reconstruction in Arene Candide Cave (Liguria, Italy) by means of trace elements analysis. Journal of Archaeological Science 16, 109–124.

Francalacci, P., Borgonini, T.S., 1988. Multi-elementary analysis of trace elements and preliminary results on stable isotopes in two Italian prehistoric sites. Methodological aspects. In: Grupe, G., Hermann, B. (Eds.), Trace Elements in Environmental History. Springer Verlag, Berlin, pp. 41–52.

Frechen, M., Zander, A., Zykina, V., Boenigk, W., 2005. The loess record from the section at Kurtak in Middle Siberia. Palaeogeography, Palaeoclimatology, Palaeoecology 228, 228–244.

Fredskild, B., 1967. Paleobotanical investigations at Sermermiut, Jakobshavn, West Greenland. Meddelelser om Grønland 178 (4), 54.

Fredskild, B., 1981. The natural environment of the norse settlers in Greenland. In: van Holk, A.G.F., Jacob, H.K., Temmingh, A.A.H.J. (Eds.), Proceedings of the International Symposium on Early European Exploitation of the Northern Atlantic 800–1700. University of Gröningen Arctic Centre, Groningen, pp. 27–42.

Fredskild, B., 1996. Holocene climatic changes in Greenland. In: Grønnow, B. (Ed.), The Paleo-Eskimo Cultures of Greenland. New Perspectives in Greenlandic Archaeology. Danish Polar Center, Copenhagen, pp. 243–251.

Freundt, E., 1948. Komsa – Fosna – Sandarna. Problems of the Scandinavian Mesolithicum. Acta Archaeologica 19, 1–68.

Friesen, T.M., 2000. The Iqaluktuuq Project: Archaeology on the Ekalluk River Near Cambridge Bay, Nunavut. Government of Nunavut, Igloolik.

Friesen, T.M., 2002. The Iqaluktuuq Project: 2002 Field Season. Archaeology on the Ekalluk River Near Cambridge Bay, Nunavut. Government of Nunavut, Igloolik.

Friesen, T.M., 2003. The Iqaluktuuq Project: 2003 Field Season: Archaeology on the Ekalluk River Near Cambridge Bay, Nunavut. Government of Nunavut, Igloolik.

Friesen, T.M., Arnold, C.D., 2008. The timing of the Thule migration: new dates from the Western Canadian Arctic. American Antiquity 73 (3), 527–538.

Fu, Q., Li, H., Moorjani, P., Jay, F., Slepchenko, S.M., Bondarev, A.A., Johnson, P.L.F., Petri, A.A., Prufer, K., de Filippo, C., Meyer, M., Zwyns, N., Salazar-Garcia, D.C., Kuzmin, Y.V., Keates, S.G., Kosintsev, P.A., Razhev, D.I., Richards, M.P., Peristov, N.V., Lachmann, M., Douka, K., Higham, T.F.G., Slatkin, M., Hublin, J.-J., Reich, D., Kelso, J., Viola, T.B., Paabo, S., 2014. The genome sequence of a 45,000-year-old modern human from western Siberia. Nature 514 (7523), 445–450.

Fuglestvedt, I., 2005. Pionerbosetningens fenomenologi. Sørvest-Norge og Nord-Europa 10 200/10 000–9500 BP. AmS-Nett nr. 6. Arkeologisk Museum, Stavanger.

Fulton, R.J., 1989. Forward-perspectives on Canadian Ice Sheet to the Quaternary geology of Canada and Greenland. In: Fulton, R.J. (Ed.), Geology of Canada and Greenland. Geological Survey of Canada, Ottawa, pp. 1–11.

Funder, S., Abrahamsen, N., 1988. Palynology in a Polar desert, eastern North Greenland. Boreas 17, 195–207.

Funder, S., Hansen, L., 1996. The Greenland ice sheet – a model for its culmination and decay during and after the last glacial maximum. Bulletin of the Geological Society of Denmark 42, 137–152.

Funder, S., Weidick, A., 1991. Holocene boreal molluscs in Greenland – palaeoceanographic implications. Palaeogeography, Palaeoclimatology, Palaeoecology 85, 123–135.

Gajewski, K., Vance, R., Sawada, M., Fung, I., Gignac, L.D., Halsey, L., John, J., Philippe, M., Mandell, P., Mudie, P.J., Richard, P.J.H., Sherin, A.G., Soroko, J., Vitt, D.H., 2000. The climate of North America and adjacent ocean waters ca. 6 ka. Canadian Journal of Earth Science 37, 661–681.

Galibin, V.A., Timofeyev, V.I., 1993. A new approach to the problem of search for the siliceous raw material for the Stone Age cultures in East Baltic regions. Arkheologicheskiye Vesti (St.-Petersburg) 2, 13–19 (in Russian).

Galimova, M.Sh, 2001. Late Paleolithic and Mesolithic Sites at the Kama River. Mouth. Yanus-K Publ., Kazan'. 272 pp. (in Russian).

Gamble, G., Davies, W., Pettitt, P., Hazlewood, L., Richards, M., 2006. The Late Glacial ancestry of Europeans, combining genetic and archaeological evidence. Documenta Praehistorica 32, 1–10.

Ganeshin, G.S. (Ed.), 1976. Map of the Quaternary Deposits of the USSR, Scale 1:2 500 000. Aerogeologiya Publ., Moscow. 16 sheets. (in Russian).

Gaustad, F., 1969. Stone Age investigations in northern Norway. Norwegian Archaeological Review 2, 86–93.

Generalov, P.P., 1965. Quaternary deposits of the western slope of the Urals. In: Yakhimovich, V.L., Lider, V.A. (Eds.), Stratigraphy of the Quaternary (Anthropogenous) Deposits of the Urals. Nedra Publ., Moscow, pp. 37–62 (in Russian).

Gening, V.F., Petrin, V.T., 1985. Late Paleolithic Epoch in the South of West Siberia. Nauka, Novosibirsk. 89 pp. (in Russian).

Gerasimov, I.P., Velichko, A.A. (Eds.), 1982. Paleogeography of Europe during the Last 100,000 Years. Nauka, Moscow. 156 pp. (in Russian).

Gerasimov, I.P., Komar, I.V., Chikishev, A.G. (Eds.), 1968. Environments and Natural Resources of the USSR. The Urals and Adjoining Regions. Nauka, Moscow. 461 pp. (in Russian).

Gerasimov, M.M., 1955. Reconstruction of Human Appearance from Skulls. USSR Academy of Sciences Publ., Moscow. 585 pp. (in Russian).

Gerasimov, M.M., 1964. Stone Age People. Nauka, Moscow. 201 pp. (in Russian).

Gerasimova, M.M., 1982. Paleoanthropological finds in Kostenki site: 1879–1979. In: Praslov, N.D., Rogachev, A.N. (Eds.), Paleolithic of the Kostenki–Borshchevo Region on the Don River. Nauka, Leningrad, pp. 245–257 (in Russian).

Gerasimova, M.M., 1984. A short description of the Sungir 5 skull. In: Alekseyev, V.P. (Ed.), Sungir: Anthropological Study. Nauka, Moscow, pp. 140–144 (in Russian).

Gerasimova, M.M., 2000. Upper Paleolithic skull Sungir 1 and its place in the series of other Upper Paleolithic skulls. In: Alekseyeva, T.I., Bader, N.O. (Eds.), Homo Sungirensis. The Upper Paleolithic Man: Ecological and Evolutionary Aspects of the Studies. Nauchny Mir, Moscow, pp. 158–174 (in Russian).

Gerasimova, M.M., 2008. Paleoanthropological data on the continuity, dispersal and prehistoric migrations of the Mesolithic inhabitants of Western and Eastern Europe and Mediterranean. In: Velichko, A.A., Vasil'ev, S.A. (Eds.), Way to North: Paleoenvironment and Earliest Inhabitants of Arctic and Subarctic. Institute of Geography RAS, Moscow, pp. 267–276 (in Russian).

Gerasimova, M.M., 2010. Kostenki findings revisited. Etnograficheskoye Obozreniye 2, 26–36 (in Russian).

Gerasimova, M.M., Pezhemsky, D.V., 2005. Mesolithic Man from Peschanitsa. Integrated Anthropological Analysis. Institute of Ethnography and Anthropology RAS, Moscow. 126 pp. (in Russian).

Gerasimova, M.M., Astakhov, S.N., Velichko, A.A., 2007. Paleolithic Man, Its Material Culture and Environments. Nestor-Istoriya Publ., St.-Petersburg. 240 pp. (in Russian).
Gerlach, C.S., Hall Jr., E.S., 1996. Two sites on Red Dog Creek, De Long Mountains. In: West, F.H. (Ed.), American Beginnings: The Prehistory and Palaeoecology of Beringia. Univ. of Chicago Press, Chicago, pp. 490–497.
German, K.E., 2002. Sperrings culture: chronology and periodization. In: Tverskoy Arkheologicheskiy Sbornik, Iss. 5. Triada Publ., Tver', pp. 261–273 (in Russian).
German, K.E., 2006. Sperrings Ceramics and Sдгдisniemi Ceramics in Russian Karelia. http://kizhi.karelia.ru/specialist/pub/library/index.htm.
Gervais, B.R., MacDonald, G.M., Snyder, J.A., Kremenetski, C.V., 2002. *Pinus sylvestris* treeline development and movement on the Kola Peninsula of Russia: pollen and stomate evidence. Journal of Ecology 90, 627–638.
Gey, V.P., Kotlukova, I.V., 1982. Middle Pleistocene sequence within the limits of the Moscow Ice Sheet. The Upper Sukhona basin. In: Goretskiy, G.I., Cheboitareva, N.S., Shik, S.M. (Eds.), The Moscow Ice Sheet in Eastern Europe. Nauka, Moscow, pp. 95–97 (in Russian).
Giddings, J.L., 1964. The Archaeology of Cape Denbigh. Brown Univ. Press, Providence, Rhode Island, p. 331.
Gilbert, M.T., Kivisild, T., Gronnow, B., Andersen, P.K., Metspalu, E., Reidla, M., Tamm, E., Axelsson, E., Gotherstrom, A., Campos, P.F., Rasmussen, M., Metspalu, M., Higham, T.F., Schwenninger, J.L., Nathan, R., De Hoog, C.J., Koch, A., Moller, L.N., Andreasen, C., Meldgaard, M., Villems, R., Bendixen, C., Willerslev, E., 2008. Paleo-Eskimo MtDNA genome reveals matrilineal discontinuity in Greenland. Science 320, 1787–1789.
Giterman, R.E., 1953. Data on the vegetation history in the lower reaches of the Chusovaya River in the Quaternary. Bulletin of Commission for Study of the Quaternary 17, 91–100 (in Russian).
Giterman, R.E., 1985. History of Vegetation of the North-East of the USSR in Pliocene and Pleistocene. Nauka, Moscow. 96 pp. (in Russian).
Gjessing, G., 1945. Norges Steinalder. Norsk Arkeologisk Selskap, Oslo.
Gjessing, G., 1948. Some problems in Northeastern archaeology. American Antiquity 13, 298–302.
Glushkova, O.Yu, 1984. Morphology and paleogeography of the Late Pleistocene glaciations in the North-East of Asia. In: Pleistocene Glaciations in the East of Asia. Northeastern Interdisciplinary Research Institute, Far Eastern Branch, RAS, Magadan, pp. 28–42 (in Russian).
Glushkova, O.Yu, 2001. Geomorphological correlation of Late Pleistocene glacial complexes of Western and Eastern Beringia. Quaternary Science Reviews 20, 405–417.
Glushkova, O.Yu, 2009. Neopleistocene glaciations in the mountain of western Chukotka. In: Proc. Conf. To the Memory of Acad. K.V. Simakov, Magadan, November 25–27, 2009. Northeastern Interdisciplinary Research Institute, Far Eastern Branch, RAS, Magadan, pp. 55–56.
Glushkova, O.Yu., Sedov, R.V., 1984. Late Quaternary and recent glaciation of the Pekulney Ridge glaciation. In: Pleistocene Glaciations in the East of Asia. Northeastern Interdisciplinary Research Institute, Far Eastern Branch, RAS, Magadan, pp. 131–140 (in Russian).
Glushkova, O.Yu., Smirnov, V.N., 2008. A new archaeological site on Lake Elgygytgyn (Chukotka). In: Materials of Dikov's Conference. V. Kordis Publ., Magadan, pp. 71–73 (in Russian).
Goebel, T., 2004. The search for a Clovis progenitor in sub-Arctic Siberia. In: Madsen, D.B. (Ed.), Entering America: Northeast Asia and Beringia Before the Last Glacial Maximum. Univ. of Utah Press, Salt Lake City, pp. 311–356.

Goebel, T., Bigelow, N.H., 1996. Panguingue Creek. In: West, F.H. (Ed.), American Beginnings: The Prehistory and Palaeoecology of Beringia. Univ. of Chicago Press, Chicago, pp. 366–371.

Goebel, T.E., Slobodin, S.B., 1999. The colonization of Western Beringia: technology, ecology, and adaptations. In: Bonnichsen, R., Turnmire, K.L. (Eds.), Ice Age Peoples of North America: Environments, Origins and Adaptations of the First Americans. Oregon State Univ. Press, Corvallis, Oregon, pp. 104–155.

Goebel, T., Powers, W.R., Bigelow, N.H., 1991. The Nenana complex of Alaska and Clovis origins. In: Bonnichsen, R., Turnmire, K.L. (Eds.), Clovis: Origins and Adaptations. Center for the Study of the First Americans, Corvallis, Oregon, pp. 49–79.

Goebel, T., Powers, W.R., Bigelow, N.H., Higgs, A.S., 1996. Walker Road. In: West, F.H. (Ed.), American Beginnings: The Prehistory and Palaeoecology of Beringia. Univ. of Chicago Press, Chicago, pp. 356–363.

Goebel, T., Waters, M.R., Dikova, M., 2003. The archaeology of Ushki Lake, Kamchatka, and the Pleistocene peopling of the Americas. Science 301, 501–505.

Goebel, T., Slobodin, S.B., Waters, M.R., 2010. New dates from Ushki-1, Kamchatka, confirm 13,000 cal BP age for earliest Paleolithic occupation. Journal of Archaeological Science 37, 2640–2649.

Gokhman, I.I., 1961. Old skull from Chukotka. In: Proc. Chukotka Regional Museum (Magadan), Iss. 2, pp. 14–18 (in Russian).

Gokhman, I.I., 1966. Population of the Ukraine in the Mesolithic and Neolithic. Anthropological Essay. Nauka, Moscow. 224 pp. (in Russian).

Gokhman, I.I., 1984. New anthropological finds of Mesolithic age in the Kargopol region. In: Gokhman, I.I. (Ed.), Anthropological Problems of Ancient and Modern Population of the Eurasian North. Nauka, Leningrad, pp. 6–27 (in Russian).

Gokhman, I.I., 1986. Anthropological specific features of the ancient population of the north of the European USSR and ways of their development. In: Gokhman, I.I., Kozintsev, A.G. (Eds.), Anthropology of Modern and Ancient Population of the European USSR. Nauka, Leningrad, pp. 216–222 (in Russian).

Gokhman, I., Tomtosova, L., 1983a. The anthropological composition of the population of the eastern Siberia in the light of the latest paleoanthropological finds in Yakutia. In: Studies in Ethnography and Anthropology. Papers Presented by Soviet Participants to XI ICAES. Part. 1. Moscow, pp. 57–67.

Gokhman, I.I., Tomtosova, L.D., 1983b. On the time of the Arctic race formation. In: Summary of Communications Presented at the Session on the Scientific Results of the Work on the 10th Five-Year Plan. Nauka, Leningrad, pp. 7–9 (in Russian).

Gokhman, I.I., Luk'yanchenko, T.V., Khartanovich, V.I., 1978. On the burial rites and craniology of Lapp. In: Field Investigations of the Institute of Ethnography in 1976. Institute of Ethnography, Moscow, pp. 51–67 (in Russian).

Golovanova, L.V., Hoffecker, J.F., 2000. Micoquian in the Northern Caucasus. Arkheologicheskiy almanac (Donetsk) 9, 35–64 (in Russian).

Golovnev, A.V., 1996. Sihirtya: traces of extinct people. In: Tomilov, N.A. (Ed.), Peoples of Siberia and Adjacent Territories. Tomsk State Univ. Press, Tomsk, pp. 37–43 (in Russian).

Golovnev, A.V., Zaitsev, G.S., Pribilskiy, Yu.P., 1994. History of Yamal. Yar-Sale Publ., Tobolsk. 144 pp. (in Russian).

Golubeva, Yu.V., 2008. The Holocene climate and vegetation on the Komi Republic territory. Litosfera (Ekaterinburg) 2, 124–132 (in Russian).

Gordon, B., 2003. The enigma of the far Northeast European Mesolithic: Reindeer herd followers or semi-sedentary hunters? In: Larsson, L., Kindgren, H., Knutssom, K., Leffler, D., Akerlund, A. (Eds.), Mesolithic on the Move. Oxbow books, Oxford, pp. 115–118.

Goretskiy, G.I., 1937. Some data on the Neolithic sites of the Kola Isthmus. In: Proc. Soviet Section of INQUA, Iss. 3, pp. 13–27 (in Russian).

Goretskiy, G.I., 1964. Alluvial Deposits of the Great Pre-rivers of the Russian Plain. Pre-rivers of the Kama Drainage Basin. Nauka, Moscow. 415 pp. (in Russian).

Gotfredsen, A.B., 2004. Mammals. In: Gotfredsen, A.B., Møbjerg, T. (Eds.), Nipisat – A Saqqaq Culture Site in Sisimiut, Central West GreenlandMeddelelser om Grønland, Man and Society, vol. 31, pp. 142–191.

Gotfredsen, A.B., 2010. Arctic Polynya and ice-edge habitat. Geografisk Tidsskrift, Danish Journal of Geography 110 (2), 142–191.

Gotfredsen, A.B., Møbjerg, T., 2004. Nipisat – A Saqqaq Culture Site in Sisimiut, Central West Greenland//Meddelelser om Grønland, vol. 331. Man and Society, vol. 31.

Goudman, A.H., Lallo, J., Armelagos, C.J., Rose, J.S., 1984. Health changes at Dickson mounds, Illinois (A.D. 950–1300). In: Cohen, M.N., Armelagos, G.J. (Eds.), Paleopathology at the Origins of Agriculture. Academic Press, Orlando, pp. 271–305.

Goudman, A.H., Brook, R.T., Smediund, A.C., Armelagos, G.J., 1989. Biocultural perspectives on stress in prehistoric, historical and contemporary population research. Yearbook of Physical Anthropology 31, 271–305.

Graf, K.E., Bigelow, N.H., 2011. Human response to climate during the Younger Dryas chronozone in central Alaska. Quaternary International 242, 434–451.

Graf, K.E., Blong, J., 2009. Reinvestigations of the Owl Ridge Site in Central Alaska: A First Look. Paper presented at the 17th Arctic Conference, Boulder.

Graf, K.E., Goebel, T., 2010. New Excavations at the Owl Ridge Site: An Update. Paper presented at the 37th Annual Meeting of the Alaska Anthropological Association, Anchorage, Alaska.

Graf, K.E., Ketron, C.V., Waters, M.R. (Eds.), 2013. Paleoamerican Odyssey. Center for the Study of the First Americans, College Station. 573 pp.

Gribchenko, Yu.N., 1994. Dynamics of Middle Pleistocene ice sheets on the Russian Plain (by lithological data). Izvestiya RAS Ser. Geogr. 4, 91–99 (in Russian).

Gribchenko, Yu.N., 2005. Specific features of the glacial lithogenesis in the Centre and North of the Russian Plain. In: Quarter-2005. Materials of the 4th All-Russian Conf. on Quaternary. Geoprint Publ., Syktyvkar, pp. 96–98 (in Russian).

Gribchenko, Yu.N., 2006. Lithology and stratigraphy of loess-soil series and cultural layers of Late Palaeolithic campsites in Eastern Europe. Quaternary International 152, 153–163.

Gribchenko, Yu.N., 2008. Initial colonization of the East European North by the Paleolithic and Neolithic man. In: Velichko, A.A., Vasil'ev, S.A. (Eds.), Way to North: Paleoenvironment and Earliest Inhabitants of Arctic and Subarctic. Institute of Geography RAS, Moscow, pp. 107–133 (in Russian).

Gribchenko, Yu.N., Kurenkova, E.I., 2006. Lithology and facial characteristics of cultural layers in some long-term settlements of the Late Paleolithic in Eastern Europe. In: Uzyanov, A.A. (Ed.), Cultural Layers of Archaeological Sites. Theory, Methods, Practice. National Information Agency Priroda, Moscow, pp. 160–180 (in Russian).

Gribchenko, Yu.N., Kurenkova, E.I., 2007. Significance of the relief formation cycles in the late Paleolithic communities over the East European Plain. Vestnik Antropologii (Institute of Ethnography and Anthropology RAS) 15, 120–127 (in Russian).

Grichuk, V.P., 1982. Vegetation of Europe in the Late Pleistocene. In: Gerasimov, I.P., Velichko, A.A. (Eds.), Paleogeography of Europe during the Last 100,000 Years. Nauka, Moscow, pp. 92–109 (in Russian).

Grichuk, V.P., 1989. History of Flora and Vegetation on the Russian Plain in the Pleistocene. Nauka, Moscow. 183 pp. (in Russian).

Grichuk, V.P., 2002. Vegetation in the Late Pleistocene. In: Velichko, A.A. (Ed.), Dynamics of Terrestrial Landscape Components and Marginal Seas of Northern Eurasia during the Last 130 000 Years. GEOS Publ., Moscow, pp. 64–88 (in Russian).

Grichuk, V.P., Borisova, O.K., 2009. Vegetation. Late Pleistocene. In: Velichko, A.A. (Ed.), Paleoclimates and Paleoenvironments of Extra-Tropical Area of the Northern Hemisphere. Late Pleistocene – Holocene. GEOS Publ., Moscow, pp. 70–74 (in Russian).

Grigoryev, G.P., 1968. Beginning of the Upper Paleolithic and the *Homo sapiens* Origin. Nauka, Leningrad. 225 pp. (in Russian).

Grishchenko, M.N., 1965. Geology of the Sukhaya Mechetka (Volgograd) site on the Volga and Rozhok 1 site in the Azov region. In: Bader, O.N. (Ed.), Stratigraphy and Periodization of Paleolithic of the Eastern and Central Europe. Nauka, Moscow, pp. 141–156 (in Russian).

Gromov, V.I., 1948. Paleontological and Archaeological Substantiation of the Quaternary Continental Deposit Stratigraphy on the USSR Territory. Transaction of Institute of Geological Sciences USSR Academy of Sciences, Iss. 64, 524 pp. (in Russian).

Grönlund, T., Kauppila, T., 2002. Holocene history of Lake Soldatskoje (Kola Peninsula, Russia) inferred from sedimentary diatom assemblages. Boreas 31, 273–284.

Grønnow, B., 1990. Prehistory in permafrost. Investigations at the Saqqaq site, Qeqertasussuk, Disco Bay, West Greenland. Journal of Danish Archaeology 7, 103–144 Odense Univ. Press, Odense.

Grønnow, B., 1994. Qeqertasussuk – the archaeology of a frozen Saqqaq site in Disko Bugt, West Greenland. In: Morrison, D., Pilon, J.-L. (Eds.), Threads of Arctic Prehistory: Papers in Honour of William E. Taylor, Jr. Canadian Museum of Civilization, vol. 149. Archaeological Survey of Canada, Mercury Series, Hull, pp. 197–238.

Grønnow, B., 1996. Driftwood and Saqqaq culture woodworking in West Greenland. In: Cultural and Social Research in Greenland 95/96. Essays in Honour of Robert Petersen. Nuuk, Greenland: Ilisimatusarfik/Atuakkiorfik, pp. 73–89.

Grønnow, B., Jensen, J.F., 2003. The northernmost ruins of the globe. Eigil Knuth's archaeological investigations in Peary Land and adjacent areas of high Arctic Greenland. Meddelelser om Grønland/Monographs on Greenland, Man and Society, vol. 29, p. 403.

Grønnow, B., Meldgaard, M., 1991. De første Vestgrønlændere. Resultaterne af 8 års undersøgelser på Qeqqertasussuk-bopladsen i Disko Bugt. Tidsskriftet Grønland 4–7, 103–144.

Grønnow, B., Sørensen, M., 2006. Palaeo-Eskimo migrations into Greenland: the Canadian connection. In: Arneborg, J., Grønnow, B. (Eds.), Dynamics of Northern Societies. Proceedings of the SILA/NABO Conference on Arctic and North Atlantic Archaeology, Copenhagen, May 10th–14th, 2004. The National Museum of Denmark, Copenhagen, pp. 59–74.

Grootes, P.M., Stuiver, M., 1997. Oxygen 18/16 variability in Greenland snow and ice with 10^{-3} to 10^{5}-year time resolution. Journal of Geophysical Research 102, 26455–26470.

Grydeland, S.E., 2000. Nye perspektiver på eldre steinalder i Finnmark. En studie fra indre Varanger. Viking, pp. 10–50.

Grydeland, S.E., 2005. The pioneers of Finnmark. From the earliest coastal settlement to the encounter with the inland people of Northern Finland. In: Knutsson, H. (Ed.), Pioneer Settlements and Colonization Processes in the Barents RegionVuollerim Papers on Hunter-Gatherer Archaeology, vol. 1, pp. 27–49.

Grydeland, S.E., 2006. Nytt lys pe elder steinalder i Finnmark (Unpublished manuscript).

Gubin, S.W., Zanina, O.G., Maksimovich, S.V., 2008. Pleistocene vegetation and soils on the plains of the North-East of Eurasia. In: Velichko, A.A., Vasil'ev, S.A. (Eds.), Way to North: Paleoenvironment and Earliest Inhabitants of Arctic and Subarctic. Institute of Geography RAS, Moscow, pp. 238–242 (in Russian).

Gudkova, L.K., 1995. Physiological indexes of blood and the adaptation of the North-Eastern Asia peoples. Voporsy Antropologii 88, 26–37 (in Russian).

Gudkova, L.K., 2003. Ecological aspects of the population physiology. In: Aleksyseva, T.I. (Ed.), Horizons of Anthropology. Nauka, Moscow, pp. 474–477 (in Russian).

Gudkova, L.K., 2008. Age-related dynamics of the man physiological status in ecologically contrasting populations. In: Buzhilova, A.P., Dobrovolskaya, M.V., Mednikova, M.B. (Eds.), Actual Trends in Anthropology. Institute of Archaeology RAS, Moscow, pp. 101–107 (in Russian).

Gulløv, H.C. (Ed.), 2004. Grønlands forhistorie. Gyldendal, Copenhagen. 434 pp.

Gurina, N.N., 1956. Oleni Ostrov Burial Ground. Materials and Studies on the Archaeology of the USSR, No. 47. USSR Academy of Sciences. Publ., Moscow–Leningrad. 429 pp. (in Russian).

Gurina, N.N., 1961. Ancient History of the North-West of the European Part of the USSR. USSR Academy of Sciences Publ., Moscow–Leningrad. 534 pp. (in Russian).

Gurina, N.N., 1971. New Investigations in the Northwestern Part of the Kola Peninsula. Short Communications of the Institute of Archaeology, Iss. 126, pp. 94–99 (in Russian).

Gurina, N.N., 1973. Ancient sites of the Kola Peninsula. In: Gurina, N.N. (Ed.), Ethnic-Cultural Communities of the Forest and Forest-Steppe Zone of the European Part of the USSR in the Neolithic Epoch (Materials and Studies on the Archaeology of the USSR, No. 172). Nauka, Leningrad, pp. 45–53 (in Russian).

Gurina, N.N., 1989. Mesolithic of Latvia and Estonia. In: Rybakov, B.A. (Ed.), Mesolithic of the USSR. Archaeology of the USSR. Nauka, Moscow, pp. 46–54 (in Russian).

Gurina, N.N., 1997. Cultural History of the Ancient Inhabitants of the Kola Peninsula. Peterburgskoye Vostokovedeniye Publ., St.-Petersburg. 230 pp. (in Russian).

Gurina, N.N., Koshechkin, B.I., Strelkov, S.A., 1974. Prehistoric cultures and evolution of environments in the Late Pleistocene and Holocene on the European Arctic coasts. In: Gerasimov, I.P., Velichko, A.A. (Eds.), Early Man, His Material Culture and Environments in the Pleistocene and Holocene. Institute of Geography, the USSR Academy of Sciences. Publ., Moscow, pp. 215–234 (in Russian).

Gusev, S.V., 2002. The early Holocene Naivan site in the Bering Strait (Chukchi Peninsula). In: Materials of Dikov's Conference II. Northeastern Interdisciplinary Research Institute, Far Eastern Division, RAS, Magadan, pp. 356–363 (in Russian).

Gusev, S.V., 2006. Study of the Early Holocene fireplaces at Naivan site, East Chukotka. In: Lebedintsev, A.I. (Ed.), Neolithic and Paleo-Metal Age in the North of the Far East. Northeastern Interdisciplinary Research Institute, Far Eastern Division, RAS, Magadan, pp. 7–21 (in Russian).

Guslitser, B.I., 1960. The upper Pechora valley: structure and history of evolution. Transactions of the Moscow Society of Naturalists 35 (3), 73–81 (in Russian).

Guslitser, B.I., 1976. On unreliability of some localities of Paleolithic archaeological and fossil fauna localities in the Komi Republic. Bulletin of Commission for Study of the Quaternary 45, 140–151 (in Russian).

Guslitser, B.I., Kanivets, V.I., 1962a. The first Paleolithic site on the Pechora Urals. Bulletin of Commission for Study of the Quaternary 27, 21–27 (in Russian).

Guslitser, B.I., Kanivets, V.I., 1962b. Caves of the Pechora region as a source of data on the Quaternary. Izvestiya. All-Union Geographic Society, Komi Branch 7, 45–59 (in Russian).

Guslitser, B.I., Kanivets, V.I., 1965a. Caves in the Pechora Urals. Nauka, Moscow–Leningrad. 134 pp. (in Russian).

Guslitser, B.I., Kanivets, V.I., 1965b. Paleolithic sites on the Pechora River. In: Bader, O.N. (Ed.), Stratigraphy and Periodization of Paleolithic of the Eastern and Central Europe. Nauka, Moscow, pp. 86–103 (in Russian).

Guslitser, B.I., Liyva, A., 1972. On the age of the fossil remains of Pleistocene mammals and Byzovaya Paleolithic site on the Middle Pechora. Izvestiya. Estonian Academy of Sciences. Ser. Biology 21 (3), 250–254 (in Russian).

Guslitser, B.I., Pavlov, P.Yu, 1987. On the Initial Colonization of the North-East of Europe (New Data). Academy of Sciences, USSR, Komi Branch, Syktyvkar. Preprints "Scientific Reports", Iss. 172, 24 pp. (in Russian).

Guslitser, B.I., Pavlov, P.Yu, 1988. The Upper Paleolithic site Medvezhya Cave (new data). In: Sites of the Stone and Metal Ages in the Northern Urals. Materials on Archaeology of European North-East. Iss. 11. Syktyvkar, pp. 5–15 (in Russian).

Guslitser, B.I., Pavlov, P.Yu, 1989. Initial colonization of Northeastern Europe. In: Cultural Adaptation at the Upper Paleolithic Time (Soviet-American Field Seminar), pp. 39–41 Abstracts. Leningrad (in Russian).

Guslitser, B.I., Pavlov, P.Yu, 1990. Paleo-Microtheriological Method as Applied to the Studies of the Upper Paleolithic Medvezhya Cave Site (The Upper Pechora). Short Communications of the Institute of Archeology, USSR Academy of Sciences, vol. 202. Nauka, Moscow, pp. 110–114 (in Russian).

Guslitser, B.I., Pavlov, P.Yu, 1993. Man and nature in Northeastern Europe in the Middle and Late Pleistocene. In: Soffer, O., Praslov, N.D. (Eds.), From Kostenki to Clovis. Upper Paleolithic – Paleoindian Adaptations. Plenum Press, New York and London, pp. 175–188.

Guslitser, B.I., Kanivets, V.I., Timofeyev, E.M., 1965. Byzovaya site – a Paleolithic site at the Arctic Circle. Sovetskaya Arkheologiya 2, 135–141 (in Russian).

Guslitser, B.I., Duryagina, D.A., Kochev, V.A., 1985. The age of relief-forming till in the lower Pechora drainage basin and the last ice sheet limit. In: Molin, V.A., Guslitser, B.I. (Eds.), Subdivision and Correlation of Phanerozoic Sediments in the North of European Part of USSR. Transaction of Institute of Geology, Komi Branch of the USSR Academy of Sciences, Iss. 54, Syktyvkar, pp. 97–107 (in Russian).

Guthrie, R.D., 1982. Mammals of the mammoth steppe as paleoenvironmental indicators. In: Hopkins, D.M., Matthews, J.V., Schweger, C.E., Young, S.B. (Eds.), Paleoecology of Beringia. Academic Press, New York, pp. 307–326.

Guthrie, R.D., 1990. Frozen Fauna of the Mammoth Steppe. Univ. of Chicago Press, Chicago. 323 pp.

Hacquebord, L., 1983. De walvistvaart: geschiedenes van een bedrijfstag. In: Smeerenburg: Holland opz'n koudst. Arktisch Centrum of Groningen, Netherlands, pp. 13–28.

Halinen, P., 2005. Prehistoric hunters of northernmost Lapland. Settlement patterns and subsistence strategies. Iskos (Helsinki) 14, 1–222.

Hamilton, T.D., Goebel, T., 1999. Late Pleistocene peopling of Alaska. In: Bonnichsen, R., Turnmire, K.L. (Eds.), Ice Age Peoples of North America: Environments, Origins and Adaptations of the First Americans. Oregon State Univ. Press, Corvallis, pp. 104–155.

Hammarlund, D., Barnekow, L., Birks, H.J.B., Buchardt, B., Edwards, T.W.D., 2002. Holocene changes in atmospheric circulation recorded in the oxygen-isotope stratigraphy of lacustrine carbonates from northern Sweden. The Holocene 12, 339–351.

Harrington, C.R., 1980. Radiocarbon dates on some Quaternary mammals and artifacts from Northern North America. Arctic 33 (4), 815–832.

Harrington, C.R., Cinq-Mars, J., 1995. Radiocarbon dates on saiga antelope (*Saiga tatarica*) fossils from Yukon and the Northwest Territories. Arctic 48 (1), 1–7.

Harritt, R.K., 1998. Paleo-Eskimo beginnings in North America: a new discovery at Kuzitrin Lake, Alaska. Inuit Studies 22, 61–81.

Hattestrand, C., Kolka, V., Stroeven, A.P., 2007. The Keiva ice marginal zone on the Kola Peninsula, northwest Russia: a key component for reconstructing the palaeoglaciology of the northeastern Fennoscandian Ice Sheet. Boreas 36, 352–370.

Hauglid, M.A., 1993. Mellom Fosna og Komsa. En Preboreal "avslagskultur" i Salten, Nordland (Magister thesis). Department of Archaeology, Univ. of Tromsø.

Haynes, G., 2002. The Early Settlement of North America: The Clovis Era. Cambridge Univ. Press, Cambridge. 345 pp.

Helmer, J.W., 1982. The Devon Island Archaeology Project. Prince of Wales Northern Heritage Centre, Yellowknife.

Helmer, J.W., 1984a. Final Report on the 1984 Devon Island Archaeology Project. Prince of Wales Northern Heritage Centre, Yellowknife.

Helmer, J.W., 1984b. Report on the Devon Island Archaeology Project (1982–1984) High Arctic, Canada. Prince of Wales Northern Heritage Centre, Yellowknife.

Helmer, J.W., 1986. Report on the 1986 Devon Island Archaeology Project. Prince of Wales Northern Heritage Centre, Yellowknife.

Helmer, J.W., 1987. The 1985–1987 Devon Island Archaeology Project: A Progress Report to the Social Sciences and Humanities Research Council on the 1985 Season. Prince of Wales Northern Heritage Centre, Yellowknife.

Helmer, J.W., 1991. The Palaeo-Eskimo prehistory of the North Devon lowlands. Arctic 44 (4), 301–317.

Helmer, J.W., 1994. Resurrecting the spirit(s) of Taylor's "Carlsberg Culture": cultural traditions and cultural horizons in Eastern prehistory. In: Morrison, D., Pilon, J.-L. (Eds.), Threads of Arctic Prehistory: Papers in Honour of William E. Taylor Jr. Canadian Museum of Civilization. Archaeological Survey of Canada, Mercury Series. No 149, Ottawa, pp. 15–34.

Helskog, K., 1974. Stone Age settlement patterns in the interior North Norway. Arctic Anthropology 9, 266–271.

Hesjedal, A., Niemi, A.R., 2003. Melkøya: Dokumentasjon av mennesker og miljø gjennom 10 000 er. Ottar 248. Tromsø Museum, Tromsø.

Hesjedal, A., Damm, C., Olsen, B., Storli, I., 1996. Arkeologi pe Slettnes. Dokumentasjon av 11 000 ers bosetning. Tromsø Museums Skrifter XXVI, Tromsø.

Hicks, S., Hyvдrinen, H., 1997. The vegetation history of Northern Finland. In: Schulz, E.L., Carpelan, C. (Eds.), Varhain Pohjoisessa. Early in the North. Helsinki Papers in Archaeology, 10, pp. 25–34.

Higuera, P.E., Brubaker, L.B., Anderson, P.M., Brown, T.A., Kennedy, A.T., Hu, F.S., 2008. Frequent fires in ancient shrub tundra: implications of paleorecords for arctic environmental change. PLoS One 3 (3), e0001744. http://dx.doi.org/10.1371/journal.pone.0001744. 7 pp.

Hinnerson-Berglund, M., 2004. Mobilitet och Estetik: Nuukfjorden pe Grцnlands vдskust som mдnniskornas livsvдrld for 4000 er sedan. Gцteborgs Univ., Arkeologiska Institutionen, Gцteborg, Sweden.

Hjálmarsson, J.R., 2009. History of Iceland: From the Settlement to the Present Day. Forlagip Publishing, Reykjavik.

Hoffecker, J.F., 1985. The Moose Creek site. National Geographic Society Research Reports 19, 33–48.

Hoffecker, J.F., 1996. Moose Creek. In: West, F.H. (Ed.), American Beginnings: The Prehistory and Palaeoecology of Beringia. Univ. of Chicago Press, Chicago, pp. 363–366.

Hoffecker, J.F., Elias, S.A., 2007. Human Ecology of Beringia. Columbia Univ. Press, New York, p. 290.

Hoffecker, J.F., Powers, R.W., 1996. Little Panguingue Creek. In: West, F.H. (Ed.), American Beginnings: The Prehistory and Palaeoecology of Beringia. Univ. of Chicago Press, Chicago, pp. 371–374.

Hoffecker, J.F., Powers, W.R., Goebel, T., 1993. The colonization of Beringia and the peopling of the new world. Science 259, 46–53.

Hoffecker, J.F., Powers, W.R., Phippen, P.G., 1996. Owl Ridge. In: West, F.H. (Ed.), American Beginnings: The Prehistory and Palaeoecology of Beringia. Univ. of Chicago Press, Chicago, pp. 353–356.

Hollesen, J., Matthiesen, H., Jensen, J.B., 2011. Qajaa, en frossen arkæologisk perle. In: Gulløv, H.C., Paulsen, C., Rønne, B. (Eds.), Nordlige Verdener – ændringer og udfordringer. Rapport fra workshop 1 på Nationalmuseet, September 29, 2010, pp. 90–95.

Holmes, C.E., 1988. An Early Post-Paleo-Arctic Site in the Alaska Range. Paper Presented at the 15th Annual Meeting of the Alaska Anthropological Association, Anchorage, Alaska.

Holmes, C.E., 1996. Broken Mammoth. In: West, F.H. (Ed.), American Beginnings: The Prehistory and Palaeoecology of Beringia. Univ. of Chicago Press, Chicago, pp. 312–318.

Holmes, C.E., 2001. Tanana River valley archaeology circa 14,000 to 9,000 BP. Arctic Archaeology 38 (2), 154–170.

Holmes, C.E., 2011. The Beringian and Transitional periods in Alaska: technology of the East Beringian tradition as viewed from Swan Point. In: Goebel, T., Buvit, I. (Eds.), From the Yenisei to the Yukon: Interpreting Lithic Assemblage Variability in Late Pleistocene/Early Holocene Beringia. Peopling of the Americas Publications Series. College Station Texas A&M Univ. Press. Center for the Study of the First Americans, pp. 179–191.

Holmes, C.E., VanderHoek, R., Dilley, T.E., 1996. Swan Point. In: West, F.H. (Ed.), American Beginnings: The Prehistory and Palaeoecology of Beringia. Univ. of Chicago Press, Chicago, pp. 319–323.

Holmes, C.E., Potter, B.A., Reuther, J.D., Mason, O.K., Thorson, R.M., Bowers, P.M., 2008. Geological and cultural context of the Nogahabara I site. American Antiquity 73 (4), 781–790.

Holmes, C.E., Reuther, J.D., Bowers, P., 2010. The Eroadaway Site: Early Holocene Lithic Technological Variability in the Central Alaska Range. Paper presented at the 37th Annual Meeting of the Alaska Anthropological Association, Anchorage, Alaska.

Holtedal, O., 1960. Geology of Norway, vol. 208. Norwegian Geological Survey (NGU). 540 pp.

Hood, B.C., 1991. The Prehistoric Foragers of the North Atlantic: Perspectives on Lithic Procurement and Social Complexity in the North Norwegian Stone Age and the Labrador Maritime Archaic Massachusets (Electronic Doctoral dissertations for UMass Amherst). Paper AAI9219445.

Hopkins, D.M., 1976. Sea level history in Beringia during the past 250 000 years. In: Kontrimavichus, V.L. (Ed.), Beringia in Cenozoic. Reports of All-Union Symposium "The Bering Land Bridge and and Its Role in the History of Holarctic Floras and Faunas in the Late Cenozoic. Far Eastern Scientific Centre, USSR Academy of Sciences, Vladivostok, pp. 9–27 (in Russian).

Hopkins, D.M., 1985. Comment to A.B. Dolitsky "Siberian Paleolithic archaeology: approaches and analytic methods". Current Anthropology 26 (3), 371–372.

Hornberg, G., Bohlin, E., Hellberg, E., Bergman, I., Zackrisson, O., Olofsson, A., 2006. Effects of Mesolithic hunter-gatherers on local vegetation in a non-uniform glacioisostatic land uplift area, northern Sweden. Vegetation History and Archaeobotany 15, 13–26.

Hough, W., 1898. The Lamp of the Eskimo (Annual Report), U. S. National Museum, 1896 [1898], pp. 1025–1057.

Housley, R.A., Gamble, C.S., Street, M., Pettitt, P., 1997. Radiocarbon evidence for the late-glacial human recolonisation of northern Europe. Proceedings of the Prehistoric Society 63, 25–54.

Housley, R.A., Higham, T.F.G., Anikovich, M.V., 2006. New AMS radiocarbon dates from Kostenki 12. In: Anikovich, M.V. (Ed.), The Early Upper Paleolithic of Eurasia: General and Local. Materials of International Conference Devoted to the 125-Year Anniversary of the Paleolithic Discovery in Kostenki (August 23–26, 2006). Nestor-Istoriya Publ., St.-Petersburg, pp. 152–155.

Hrdlicka, A., 1945. The Aleutian and Commander Islands and Their Inhabitants. The Wistar Institute of Anatomy and Biology, Philadelphia. 630 pp.

Hu, F.S., Brubaker, L.B., Anderson, P.M., 1993. A 12,000 year record of vegetation change and soil development from Wien Lake, central Alaska. Canadian Journal of Botany 71, 1133–1142.

Hu, F.S., Brubaker, L.B., Anderson, P.M., 1995. Postglacial vegetation and climate change in the Northern Bristol Bay region, Southwestern Alaska. Quaternary Research 43, 382–392.

Hu, F.S., Lee, B.Y., Kaufman, D.S., Yoneji, S., Nelson, D.M., Henne, P.D., 2002. Responses of tundra ecosystems in southwestern Alaska to Younger Dryas climatic oscillations. Global Change Biology 8, 1156–1163.

Huure, M., 2003. Viipurin Iaan kivikausi. In: Karjalan synty: Viipurin laanin historia 1. Jyvaskyla, pp. 151–244.

Hyvärinen, H., 1985. Holocene pollen stratigraphy Baird Inlet, east-central Ellesemere Island, Arctic Canada. Boreas 14, 19–32.

Hyyppä, E., 1937. Post-glacial changes of shorelines in south Finland. Bulletin de la Commission Géologique de Finlande 120 225 pp.

Iacumin, P., Nikolaev, V., Ramigni, M., 2000. C and N stable isotope measurements on Eurasian fossil mammals, 40,000 to 10,000 years BP: herbivore physiologies and palaeoenvironmental reconstruction. Palaeogeography, Palaeoclimatology, Palaeoecology 163 (1–2), 33–47.

Ilyashuk, E.A., Ilyashuk, B.P., Hammarlund, D., Larocque, I., 2005. Holocene climatic and environmental changes inferred from midge records (*Diptera: Chironomidae, Chaoboridae, Ceratopogonidae*) at Lake Berkut, southern Kola Peninsula, Russia. Holocene 15 (6), 897–914.

Indrelid, S., 1976. The site Hein 33. Typological and chronological problems of the New Stone Age of Southern Norway. Norwegian Archaeological Review 9 (1), 7–44.

Indrelid, S., 1978. Mesolithic economy and settlement patterns in Norway. In: Mellars, P.A. (Ed.), The Early Post-glacial Settlement of Northern Europe: An Ecological Perspective. Duckworth, London, pp. 147–176.

Isayeva, L.L., Andreyeva, S.I., Bardeyeva, M.A., Kolpakov, V.V., Laukhin, S.A., Reverdatto, M.V., Rindzyunskaya, N.M., Shofman, I.L., 1993. Middle Siberia. In: Velichko, A.A. (Ed.), Evolution of Landscapes and Climates of Northern Eurasia. Late Pleistocene – Holocene – Elements of Prognosis. I. Regional Paleogeography. Nauka, Moscow, pp. 46–59 (in Russian).

Ivanishcheva, M.V., 2002. Stone inventory of the Mesolithic site Berezovaya Slobodka X, on the lower Sukhona. In: Tverskoy Arkheologicheskiy Sbornik, Iss. 5. Triada Publ., Tver', pp. 159–170 (in Russian).

Ivanov, G.V., 1993. Stone Tool Findings on the European Arctic Islands (1989–1992). AD POLUS. Archaeological Survey, Iss. 10. Farn Publ., St.-Petersburg, pp. 47–54 (in Russian).

Ivanova, I.K., 1965. Geological Age of the Fossil Man. Nauka, Moscow. 192 pp. (in Russian).

Ivanova, I.K., 1977. Geology and paleogeography of the Korman' IV site on the background of the Stone Age in the middle reaches of the Dniester River basin. In: Goretskiy, G.I., Tseitlin, S.M. (Eds.), Multilayered Paleolithic Site Korman IV on the Middle Dniester. Nauka, Moscow, pp. 126–181 (in Russian).

Ivanova, I.K., 1982. Geology and paleogeography of the Mousterian Molodova 1 site. In: Goretskiy, G.I., Ivanova, I.K. (Eds.), Molodova 1: A Unique Mousterian Settlement on the Middle Dniester. Nauka, Moscow, pp. 188–235 (in Russian).

Jaanits, L., 1965. Über die Ergebnisse der Steinzeitforschung in Sowjetestland, vol. 72. Finskt Museum, Helsingfors, pp. 5–46.

Jaanits, L., Jaanits, K., 1975. Frühmesolithische Siedlung von Pulli. Izvestija AN Estonskoi SSR 24, 64–70.

Jaanits, L., Jaanits, K., 1978. Ausgrabungen der frühmesolithischen Siedlung von Pulli. Eesti NSV Teaduste Akadeemia Toimetised. Uhiskonnateadused. No. 1, Tallinn, pp. 56–63.
Jacobs, J.D., 1985. Environment and prehistory, Baffin Island. In: Andrews, J. (Ed.), Holocene Pollen Records from West Greenland. Allen & Unwin, Boston, pp. 719–739.
Jakobsson, S., August 19, 2000. Hvert er heimildargildi Landnámu? Hvenær er talið að hún hafi verið notuð? Vísindavefurinn. http://visindavefur.is/svar.php?id=837.
Jensen, J.F., 2005. Independence culture. In: Nuttall, M. (Ed.), Encyclopedia of the Arctic. Routledge Publishing Co., New York, London, pp. 945–948.
Jensen, J.F., 2006a. Different areas, different cultures? In: Arneborg, J., Grønnow, B. (Eds.), Dynamics of Northern Societies. Proceedings of the SILA/NABO Conference on Arctic and North Atlantic Archaeology, Copenhagen, May 10th–14th, 2004. The National Museum of Denmark, Copenhagen, pp. 75–85.
Jensen, J.F., 2006b. The Stone Age of Qeqertarsuup Tunua (Disko Bugt): a regional analysis of the Saqqaq and Dorset cultures of Central and West Greenland. In: Coll. "Man and Society", No. 32. Copenhague: Meddelelser om Grønland, p. 272.
Jensen, J.F., Pedersen, K.B., 2002. Jordens nordligste stenalder. Nye undersøgelser af pionerbebyggelsen i Grønland. In: Nationalmuseets Arbejdsmark, pp. 71–83.
Jorgensen, J.B., 1953. The Eskimo skeleton. Contribution to the physical anthropology of the aboriginal Greenlanders. Meddelelser an Greenland 146 (2), 1–154 Kobenhavn.
Joynt III, E.H., Wolfe, A.P., 2001. Paleoenvironmental inference models from sediment diatom assemblages in Baffin Island lakes (Nunavut, Canada) and reconstruction of summer water temperature. Canadian Journal of Fisheries and Aquatic Sciences 58, 1222–1243.
Jungner, H., 1979. Radiocarbon Dates I. Radiocarbon Dating Laboratory, vol. 1. Univ. of Helsinki, Rep. 131 pp.
Jussila, T., Kriiska, A., 2004. Shore displacement chronology of the Estonian Stone Age. Eesti Arheoloogiaajakiri 8, 3–32.
Jussila, T., Matiskainen, H., 2003. Mesolithic settlement during the Preboreal period in Finland. In: Mesolithic on the Move. Papers Presented at the Sixth International Conference on the Mesolithic in Europe, Stockholm, 2000. Oxbow Books, Oxford, pp. 664–670.
Jussila, T., Kriiska, A., Rostedt, T., 2007. The Mesolithic settlement in NE Savo, Finland, and the earliest settlement in the eastern Baltic Sea. Acta Archaeologica 78 (2), 143–162.
Jussila, T., Kriiska, A., Rostedt, T., 2012. Saarenoja 2: an Early Mesolithic site in South-Eastern Finland: preliminary results and interpretations of studies conducted in 2000 and 2008–10. Fennoscandia Archaeology 29, 3–28.
Justwan, A., Koc, N., Jennings, A.E., 2008. Evolution of the Irminger and East Icelandic Current systems through the Holocene, revealed by diatom-based sea surface temperature reconstructions. Quaternary Science Reviews 27, 1571–1582.
Kagan, L.Ya., Koshechkin, B.I., Lebedeva, R.M., 1992. The Kola Peninsula. In: Davydova, N.N. (Ed.), History of Lakes on the East European Plain. Nauka, St.-Petersburg, pp. 20–35 (in Russian).
Kaletskaya, M.S., 1965. Quaternary deposits of the Pechora basin and western slope of the Polar Urals. In: Yakhimovich, V.L., Lider, V.A. (Eds.), Stratigraphy of the Quaternary (Anthropogenous) Deposits of the Urals. Nedra Publ., Moscow, pp. 12–24 (in Russian).
Kanivets, V.I., 1969. Paleolithic man on the Pechora River. In: Gerasimov, I.P. (Ed.), Nature and the Early Human Society Evolution on the Territory of the European USSR. Nauka, Moscow, pp. 136–142 (in Russian).
Kanivets, V.I., 1973. Mesolithic sites on the Middle Pechora and USA rivers. In: Sites of the Stone and Copper-Bronze Ages on the Pechora and USA Rivers. Materials on Archaeology of European North-East. Iss. 4. Syktyvkar, pp. 3–23 (in Russian).

Kanivets, V.I., 1976. Paleolithic of the Extreme North-East of Europe. Pechora Drainage Basin. Nauka, Moscow. 95 pp. (in Russian).

Kaplina, T.N., 1979. Pollen spectra of the "Ice Complex" on the coastal lowlands of Yakutia. Izvestiya. Academy of Sciences, USSR Ser. Geogr. 2, 85–93 (in Russian).

Kaplina, T.N., Chekhovkiy, A.L., 1987. Reconstructions of paleoenvironments of the Holocene climatic optimum on the coastal lowlands of Yakutia. In: Pohialainen, V.P. (Ed.), Quaternary Period in the North-East of Asia. Northeastern Interdisciplinary Research Institute, Far Eastern Division, RAS, Magadan, pp. 145–151 (in Russian).

Kaplina, T.N., Giterman, R.E., 1983. Molotkovskiy Kamen' – key section of the Late Pleistocene deposits on the Kolyma Lowland. Izvestiya. Academy of Sciences, USSR Ser. Geol. 6, 79–83 (in Russian).

Kaplina, T.N., Lozhkin, A.V., 1982. The age of the "Ice Complex" on the coastal lowlands of Yakutia. Izvestiya. Academy of Sciences, USSR Ser. Geogr. 2, 84–95 (in Russian).

Karkhu, A.A., Kirillova, I.V., Zhilin, M.G., 2003. Hunting of the early inhabitants of the Ivanovskoye VII site. In: Antipina, E.E., Chernykh, E.N. (Eds.), The Recent Archaeo-Zoological Investigations in Russia. Yazyki Slavyanskoy Kultury Publ., Moscow, pp. 137–156 (in Russian).

Karlsson, G., 2000. Íslandssaga í stuttu máli: Mál og menning. Rvík 4–9.

Kashin, V.A., 1983. The Yubileinyi site and its place in the Yakutian Stone Age culture. In: Vasilyevskiy, P.S. (Ed.), The Late Pleistocene and Early Holocene Cultural Relations Between Asia and America. Nauka, Novosibirsk, pp. 93–102 (in Russian).

Kashin, V.A., 2003. Paleolithic of the Northeastern Asia: History and Results of the Studies, 1940–1980. Nauka, Novosibirsk. 235 pp. (in Russian).

Kasparov, A.K., 1998. Mammals of the Late Paleolithic sites in Southern Yakutia. In: Labutin, Yu.B. (Ed.), The Anthropogene Mammals of Yakutia. Yakutian Research Center, Siberian Branch RAS, Yakutsk, pp. 138–150 (in Russian).

Katasonov, E.M., Ivanov, M.S., Pudov, G.G., Ziegert, C., 1979. Structure and Absolute Age of the Alases in Central Yakutia. Nauka, Moscow. 95 pp. (in Russian).

Kedrowski, B.L., Crass, B.A., Behm, J.A., Luetke, J.C., Nichols, A.L., Moreck, A.M., Holmes, C.E., 2009. GC/MS analysis of fatty acids from ancient hearth residues at the Swan Point archaeological sites. Archaeometry 51 (1), 110–122.

Kemp, B.M., Malhi, R.S., McDonough, J., Bolnick, D.A., Eshelman, J.A., Rickards, O., Martinez-Labarga, C., Johnson, J.R., Lorenz, J.G., Dixon, E.J., Fifield, T.E., Heaton, T.H., Worl, R., Smith, D.G., 2007. Genetic analysis of early Holocene skeletal remains from Alaska and its implications for the settlement of the Americas. American Journal of Physical Anthropology 132, 605–621.

Kerwin, M.W., Overpeck, J.T., Webb, R.S., Anderson, K.H., 2004. Pollen-based summer temperature reconstructions for the eastern Canadian boreal forest, subarctic, and Arctic. Quaternary Science Reviews 23, 1901–1924.

Khartanovich, V.I., 1978. Results of the studies of a new Sami craniological series. In: Field Studies of the Institute of Ethnography – 1976. Nauka, Moscow, pp. 181–188 (in Russian).

Khartanovich, V.I., 1980. New Data on the Craniology of the Sami on the Kola Peninsula. Investigations into Paleoanthropology and Craniology of the USSR, vol. 36. Transaction of Museum of Anthropology and Ethnography (MAE), St.-Petersburg, pp. 35–47 (in Russian).

Khartanovich, V.I., 1991. On the interrelations between the Sami and Karelian anthropological types by craniological data. In: Aksyanova, G.A. (Ed.), Origin of the Sami. Nauka, Moscow, pp. 19–33 (in Russian).

Khartanovich, V.I., 2004. New craniological materials on the Sami on the Kola Peninsula. In: Kozintsev, A.G. (Ed.), Paleoanthropology, Ethnic Anthropology, Ethnogenesis. Museum of Anthropology and Ethnography (MAE) RAS, St.-Petersburg, pp. 108–125 (in Russian).

Khartanovich, V.I., Moiseyev, V.G., 2012. Craniological materials on the origin of the early inhabitants of the Extreme North of Europe. In: Terebikhin, N.M. (Ed.), Geohistorical and Geo-Ethnocultural Images and Symbols of the Arctic Area Colonization. (Pomors' Conference on the Culture Semiotics. Iss. 6). Solti Publ., Solovki Settl., Arkhangelsk Region, pp. 128–145 (in Russian).

Khersonskiy, S., 1898. On the animal hunting in the Okhotsk district of the Prymorsky region. In: Proceedings of the Imperial Russian Geographical Society, Amur Dept. vol. 4, Iss. 1. Khabarovsk, pp. 1–26 (in Russian).

Khlobystin, L.P., 1969. On the dispersal of the Samoedic people predecessors in the Bronze Age (the 2nd millennium BC). In: Materials of Conf. on Ethnogenesis of People in Northern Asia, Iss. 1. Institute of History, Philology and Philosophy, Academy of Sciences, USSR, Siberian Branch, Novosibirsk, pp. 113–135 (in Russian).

Khlobystin, L.P., 1975. Sites in Siberian polar regions and their relationship to the taiga zone cultures. In: Interrelation between the Ancient Siberian Cultures and Cultures of Adjacent Territories. Nauka, Novosibirsk, pp. 100–110 (in Russian).

Khlobystin, L.P., 1987. Salekhard Findings. Short Communications of the Institute of Archaeology Academy of Sciences, USSR, Iss. 189. Nauka, Moscow, pp. 108–111.

Khlobystin, L.P., 1998. Ancient History of the Polar Taymyr Region and the Problems of Formation of the Northern Eurasian Cultures. Dmitriy Bulanin Publ., St.-Petersburg. 341 pp. (in Russian).

Khlobystin, L.P., 2005. Taymyr: the archaeology of Northernmost Eurasia. In: Fitzhugh, W.W., Pitulko, V.V. (Eds.), Contributions to Circumpolar Anthropology, vol. 5. Arctic Studies Center, National Museum of Natural History, Smithsonian Institution, Washington, DC. 235 pp.

Khokhlov, A.A., Yablonskiy, L.T., 2003. The oldest anthropological materials from the Middle Volga regions (Preliminary communication). In: Zones of Contact in Eurasia at the Boundary of Centuries. Center "Integratsiya", Samara, pp. 25–28 (in Russian).

Khomich, L.V., 1966. Nenets. Historical-Ethnographical Study. Nauka, Moscow–Leningrad. 329 pp. (in Russian).

Khomich, L.V., 1976. Problems of the Ethnogenesis and the Nenets Ethnic History. Nauka, Leningrad. 190 pp. (in Russian).

Khomutova, V.I., Sokolova, V.B., Buslovich, A.L., Garkusha, V.I., Kotlukova, I.V., 1998. History of lakes in the regions of the Kubenskoye-Sukhona Depression and Severnye Uvaly ridge. In: Khomutova, V.I. (Ed.), History of the Pleistocene Lakes of the East European Plain. Nauka, St.-Petersburg, pp. 75–99 (in Russian).

Khotinskiy, N.A., 1977. The Holocene of Northern Eurasia. Nauka, Moscow. 198 pp. (in Russian).

Khotinskiy, N.A., 1982. Paleoecological reconstructions of the Holocene environments (a model of the present-day interglacial). In: Gerasimov, I.P., Velichko, A.A. (Eds.), Paleogeography of Europe during the Last 100,000 Years. Nauka, Moscow, pp. 123–127 (in Russian).

Khotinskiy, N.A., Klimanov, V.A., 2002. The Holocene vegetation. In: Velichko, A.A. (Ed.), Dynamics of Terrestrial Landscape Components and Marginal Seas of Northern Eurasia during the Last 130 000 Years. GEOS Publ., Moscow, pp. 89–105 (in Russian).

Khrisanfova, E.N., 1979. Paleoanthropological aspect of the constitution. Voprosy Antropologii 62, 3–13 (in Russian).

Khrisanfova, E.N., 1980. Skeleton of the Upper Paleolithic man from Sungir. Voprosy Antropologii 64, 40–68 (in Russian).

Khrisanfova, E.N., 1984. Post-cranial skeleton of an adult man Sungir 1. Femoral bone Sungir 4. In: Alekseyev, V.P. (Ed.), Sungir. Anthropological Analysis. Nauka, Moscow, pp. 100–140 (in Russian).

Khrisanfova, E.N., 2000. Post-cranial skeleton of an adult man Sungir 1. In: Alekseyeva, T.I., Bader, N.O. (Eds.), Homo Sungirensis. The Upper Paleolithic Man: Ecological and Evolutionary Aspects. Nauchny Mir, Moscow, pp. 193–217 (in Russian).

Kienast, F., Siegert, C., Derevyagin, A., Mai, D.H., 2001. Climatic implications of Late Quaternary plant macrofossil assemblages from the Taymyr Peninsula, Siberia. Global and Planetary Change 31 (1–4), 265–281.

Kind, N.V., 1974. The Late Anthropogene Geochronology Based on Isotope Data. Nauka, Moscow. 256 pp. (in Russian).

Kiryak, M.A., 1993. Archaeology of Western Chukotka. Nauka, Moscow. 224 pp. (in Russian).

Kiryak, M.A., 1996. Bolshoi Elgakhchan 1 and 2, Omolon River basin, Magadan district. In: West, F.H. (Ed.), American Beginnings: The Prehistory and Palaeoecology of Beringia. Univ. of Chicago Press, Chicago, pp. 228–236.

Kiryak, M.A., 2005. Late Pleistocene assemblages of the Ushki V site: history of discovery and prospects for investigation. In: Goryachev, N.A., Kokorev, E.M., Mumortsev, Yu.I. (Eds.), Northern Pacific: Cultural Adaptations at the End of Pleistocene –Holocene. Materials of International Conference, August 29–September 08, 2005. Northern International Univ. Publ., Magadan, pp. 62–66 (in Russian).

Kiryak, M.A., Makarov, I.V., 2006. New archaeological finds in the Lake Elgygytgyn region. In: Lebedintsev, A.I. (Ed.), Neolithic and Paleo-Metal Age in the North of the Far East. Northeastern Interdisciplinary Research Institute, Far Eastern Division, RAS, Magadan, pp. 8–16 (in Russian).

Kiryak, M.A., Glushkova, O.Yu., Brown, T.A., 2003. The Upper Paleolithic assemblages of the Tytylvaam valley (transpolar Chukchi Peninsula). Archaeology, Ethnography and Anthropology of Eurasia 3, 2–15.

Kistenev, S.P., 1980. New archaeological sites in the Kolyma drainage basin. In: New Data on the Yakutian Archaeology, Yakutsk Branch, USSR Academy of Sciences, Yakutsk, pp. 74–87 (in Russian).

þór Kjartansson, B., Eggertsson, Ó., January 13, 2005. Er til einhver skógur frá landnámsöld á Íslandi? Vísindavefurinn. http://visindavefur.is/?id=4712.

Klevtsova, N.I., 1976. Somatological specificity of Siberian Mongoloids: comparative data. Voprosy Antropologii 58, 151–163 (in Russian).

Klevtsova, N.I., Smirnova, N.S., 1974. Morphological characterisics of Chukchi and Eskimo bodies. Voprosy Antropologii 48, 18–33 (in Russian).

Klima, B., Kukla, J., Ložek, V., De Vries, H., 1961. Stratigraphie des Pleistozans und Alter des palaolithischen Rastplatzes in der Ziegelei von Dolni Věstonice (Unter–Wisternitz). Anthropozoikum 11, 59–78.

Klimanov, V.A., Shofman, I.L., 1982. On climatic fluctuations in the Upper Pleistocene and Holocene in Western Yakutia. In: Pletnev, S.P., Pushkar', V.S. (Eds.), Recent Sedimentation and Quaternary Morphogenesis in the Far East. Far Eastern Research Center. USSR Academy of Sciences, Vladivostok, pp. 38–47 (in Russian).

Knuth, E., 1954. The Paleo-Eskimo culture of Northeast Greenland elucidated by three new sites. American Antiquity 19 (4), 367–381.

Knuth, E., 1967a. Archaeology of the Musk-Ox Way. Contributions du Centre d'Études Arctiques et Finno–Scandinaves. No 5. École Pratique des Hautes Études Sorbonne, Paris.

Knuth, E., 1967b. The ruins of the musk ox way. Folk 8–9, 191–219.

Knuth, E., 1978. The "Old Nugdlik Culture" site at Nugdlit Peninsula, Thule district and the "Meso-Eskimo" site below it. Folk 19/20, 15–47.

Knutsson, K., 1999. The Mesolithic in Eastern Central Sweden. In: The Mesolithic of Central Scandinavia, vol. 22. Universitets Oldsaksamlings Skrifer, Oslo, pp. 87–124.

Kochev, V.A., 1984. Dating of Quaternary Deposits by Fossil Remains of Collared Lemmings. Ser. New Research Methods, Iss. 12. Komi Branch of the USSR Academy of Sciences, Syktyvkar. 16 pp. (in Russian).

Kochev, V.A., 1991. Evolution of the Pleistocene Rodent Fauna in the North-East of the European Part of the USSR. Geonauka Publ., Syktyvkar. 82 pp. (in Russian).

Koerner, R.M., 1977. Devon Island ice cap: core stratigraphy and paleoclimate. Science 196, 15–18.

Koerner, R.M., 1989. Ice core evidence for extensive melting of the Greenland ice sheet during the last interglacial. Science 244, 964–968.

Koerner, R.M., Fisher, D.A., 1985. The Devon island ice core and the glacial record. In: Andrews, J. (Ed.), Holocene Pollen Records from West Greenland. Allen & Unwin, Boston, pp. 309–327.

Koerner, R.M., Fisher, D.A., 1990. A record of Holocene summer climate from a Canadian High-Arctic ice core. Nature 343, 630–631.

Kokorowski, H.D., Anderson, P.M., Mock, C.J., Lozhkin, A.V., 2008. A re-evaluation and spatial analysis of evidence for a Younger Dryas climatic reversal in Beringia. Quaternary Science Reviews 27, 1710–1722.

Koksharov, S.F., Pogodin, A.A., 2000. Mesolithic settlement on Zatumannaya Kodna. Rossiyskaya Arkheologiya 4, 109–127 (in Russian).

Kolesnik, A.V., 2003. Middle Paleolithic of Donbass. Lebed' Publ., Donetsk, (Arkheologicheskiy Almanakh, No. 12). 292 pp. (in Russian).

Kol'tsov, L.V. (Ed.), 1989. Mesolithic of the USSR. Nauka, Moscow. 352 pp. (in Russian).

Kol'tsov, L.V., 2003. On the problem of the initial colonization of Northern Europe. In: Alekseyeva, N.I. (Ed.), Horizons of Anthropology. Nauka Publ., Moscow, pp. 2002–2005 (in Russian).

Kononenko, N.A., 2001. Environments and archaeological cultures dynamics in the Zerkal'naya River valley in the end of Pleistocene – beginning of Holocene. Archaeology, Ethnography and Anthropology of Eurasia 1, 40–59.

Konovalenko, L.A., 2006. Conclusion based on the pollen analysis results of the Parch-2 section, dwelling 2. In: Volokitin, A.V. (Ed.), Mesolithic Sites Parch 1 and Parch 2 of the Vychegda, Suppl. 3. Komi Research Center, Uralian Branch, RAS, Syktyvkar, pp. 111–113 (in Russian).

Konstantinov, M.V., 1994. Stone Age of the Eastern Region of the Baikalian Asia. Buryat Research Center Siberian Branch RAS, Ulan-Ude. 256 pp. (in Russian).

Korenyuk, S.N., Melnichuk, A.F., 1985. Investigations of the Stone Age Sites. In: A rchaeological Discoveries – 1983. Nauka, Moscow, pp. 154–155 (in Russian).

Koryakin, V.S., 1991. Materials on the History of the Arctic Colonization. Novaya Zemlya. Research Institute of Culture, Moscow. 57 pp. (in Russian).

Korzhuev, S.S., Fedorova, R.V., 1962. Chekurovo mammoth and his environments. Doklady. USSR Academy of Sciences 145 (1), 181–183 (in Russian).

Koshechkin, B.I., Kagan, L.Ya., Kudlayeva, A.L., Malyasova, E.S., Pervuninskaya, N.A., 1973. Coastal landforms of the late- and post-glacial marine basins in the South of the Kola Peninsula. In: Strelkov, S.A., Koshechkin, B.I. (Eds.), Paleogeography and Morphostructures of the Kola Peninsula. Nauka, Leningrad, pp. 87–133 (in Russian).

Kosinskaya, L.L., 2007. Mesolithic sites in the lower Vychegda drainage basin. In: Volokitin, A.V. (Ed.), Stone Age of the European North. Institute of Linguistic, Literature, and History. Komi Research Centre, Syktyvkar, pp. 125–147 (in Russian).

Kosinskaya, L.L., Volokitin, A.V., 1993. Typology of the Mesolithic sites of the European North-East. In: Interaction of Cultures of the North Ural Region in the Antiquity and Middle Ages (Materials on Archaeology of European North-East. Iss. 12). Komi Research Centre, Syktyvkar, pp. 18–29 (in Russian).

Kosorukova, N.V., 1996. Mesolithic sites in the Kolpi drainage basin. In: Drevnosti Russkogo Severa (The Russian North Antiquities), Iss. 1. Vologda, pp. 35–42 (in Russian).

Kosorukova, N.V., 2000. Mesolithic sites in the Sheksna drainage basin. In: Tverskoy Arkheologicheskiy Sbornik. Iss. 4. Triada Publ., Tver', pp. 91–98 (in Russian).

Kosorukova, N.V., 2005. Mesolithic complex of the Kurevanikha V-VI site. In: Arkheologiya Severa. Iss. 1. Cherepovets, pp. 4–14 (in Russian).

Kosorukova-Kondakova, N.V., 1991. The Mesolithic site on the Lotova Gora in the Sheksna drainage basin. In: Kashkin, A.V. (Ed.), Archaeology of the Upper Volga Region. Nizhniy Novgorod, pp. 30–41 (in Russian).

Kostyukevich, V.V., Ivanov, I.E., Nesterenko, S.A., 1980. List of radiocarbon dates, Laboratory of Geochemistry, Institute for Permafrost studies. Bulletin of Commission for Study of the Quaternary 50, 193–196 (in Russian).

Kotlyakov, V.M., Troitskiy, L.S., 1985. New data on the glaciation of Spitsbergen. Vestnik, USSR Academy of Sciences 2, 128–136 (in Russian).

Kovnurko, G.M., 1963. Some data on the siliceous rock outcrops in the North-West of the USSR. In: Materialy i Issledovaniya po arkheologii SSSR (Materials and Studies on the USSR Archaeology), No. 172, pp. 39–44 (in Russian).

Kozlov, A.V., 2005. Human's Food. Vek 2 Publ., Moscow. 271 pp. (in Russian).

Kozlovskaya, M.V., 2000a. Results of the chemical analysis of the osseous tissue taken from adolescents Sungir 2 and Sungir 3. In: Alekseyeva, T.I., Bader, N.O. (Eds.), Homo Sungirensis. The Upper Paleolithic Man: Ecological and Evolutionary Aspects. Nauchny Mir, Moscow, pp. 299–301 (in Russian).

Kozlovskaya, M.V., 2000b. Nutritional system of the Upper Paleolithic communities. In: Alekseyeva, T.I., Bader, N.O. (Eds.), Homo Sungirensis. The Upper Paleolithic Man: Ecological and Evolutionary Aspects. Nauchny Mir, Moscow, pp. 411–420 (in Russian).

Kozlovskaya, M.V., 2003. The Mesolithic people diet in the North of the European Russia: natural and cultural traditions. In: Zones of Contact in Eurasia at the Boundary of Centuries. Center "Integratsiya", Samara, pp. 51–52 (in Russian).

Kozlowski, J.K., 1980. Technological and typological differentiation of lithic assemblages in the Upper Palaeolithic: an interpretation attempt. In: Schild, R. (Ed.), Unconventional Archaeology, Wydawnictwo Polskiej Akademii Nauk, Wroclaw-Warszawa, pp. 33–56.

Kramer, F.E., 1996a. The Paleo-Eskimo cultures in Sisimiut district, West Greenland. Aspects of chronology. In: Grønnow, B., Pind, J. (Eds.), The Paleo-Eskimo Cultures of Greenland. New Perspectives in Greenland Archaeology. Papers from a symposium at the Institute of Archaeology and Ethnography, University of Copenhagen, May 21–24, 1992. Danish Polar Center, Copenhague, pp. 39–64.

Kramer, F.E., 1996b. Akia and Nipisat I: two Saqqaq sites in Sisimiut district, West Greenland. In: Grønnow, B., Pind, J. (Eds.), The Paleo-Eskimo Cultures of Greenland. New Perspectives in Greenland Archaeology. Papers from a symposium at the Institute of Archaeology and Ethnography, University of Copenhagen, May 21–24, 1992. Danish Polar Center, Copenhague, pp. 65–96.

Krasinski, K.E., 2005. Intrasite Spatial Analysis of Late Pleistocene/Early Holocene Archaeological Material from the Broken Mammoth Site (Unpublished M.A. thesis). Department of Anthropology, Univ. of Alaska, Anchorage.

Krasinski, K., Yesner, D.R., 2008. Late Pleistocene/Early Holocene site structure in Beringia: a case study from the Broken Mammoth site, interior Alaska. Alaska Journal of Anthropology 6 (1/2), 27–42.

Krasnov, I.I. (Ed.), 1983. The All-Union Conference on Stratigraphy of Pre-Cambrian, Paleozoic, and Quaternary system of Central Siberia (Novosibirsk, 1979). Part III. Quaternary system. Explanatory notes to regional stratigraphic schemes of the Quaternary. VSEGEI, Leningrad. 84 pp. (in Russian).

Krause, J., Orlando, L., Serre, D., Viola, B., Pufer, K., Richards, M.P., Hublin, J.-J., Hanni, C., Derevianko, A.P., Pääbo, S., 2007. Neanderthals in central Asia and Siberia. Nature 449, 902–904.

Kremenetski, C.V., Patyk-Kara, N.G., 1997. Holocene vegetation dynamics of the southeast Kola Peninsula, Russia. Holocene 7, 473–479.

Kremenetski, C., Vashchalova, T., Goriachkin, S., Cherkinsky, A., Sulerzhitsky, L., 1997. Holocene pollen stratigraphy and bog development in the western part of the Kola Peninsula, Russia. Boreas 26, 91–102.

Kremenetski, C., Vaschalova, T., Sulerzhitsky, L., 1999. The Holocene vegetation history of the Khibiny Mountains: implications for the post-glacial expansion of spruce and alder on the Kola Peninsula, northwestern Russia. Journal of Quaternary Science 14 (1), 29–43.

Kremenetski, C.V., MacDonald, G.M., Gervais, B.R., Borisova, O.K., Snyder, J.A., 2004. Holocene vegetation history and climate change on the northern Kola Peninsula, Russia: a case study from a small tundra lake. Quaternary International 122, 57–68.

Krestinin, V.V., 1805. Geographical news about Novaya Zemlya islands. In: Voyage of Acad. Ivan Lepekhin. Part 4. St.-Petersburg, pp. 4–6 (in Russian).

Kriiska, A., Lougas, L., 1999. Late Mesolithic and Early Neolithic Seasonal Settlement at Kopu, Hiiumaa Island, Estonia. PACT 57. Rixensart 157–172.

Kriiska, A., Hertell, E., Manninen, M.A., 2011. Stone Age flint technology in South-Western Estonia: results from the Parnu Bay area. In: Mesolithic Interfaces: Variability in Lithic Technologies in Eastern Fennoscandia: Monographs of the Archaeological Society of Finland, vol. 1, pp. 64–93.

Krupyanko, A.A., Tabarev, A.V., 2001. Archaeological Assemblages of the Stone Age in the Eastern Prymorye. Siberian Univ. Press., Novosibirsk. 104 pp. (in Russian).

Kundyshev, A.S., 1992. Cenozoic Palynostratigraphy and Landscapes of the Vankarem Depression (Chukchi Peninsula). Far Eastern Branch, Academy of Sciences, USSR, Vladivostok. 131 pp. (in Russian).

Kunz, M.L., Mann, D.H., 1997. The Mesa project: interactions between early prehistoric humans and environmental change in Alaska. Arctic Research of the United States 11, 55–62.

Kunz, M.L., Reanier, R., 1994. Paleoindians in Beringia: evidence from Arctic Alaska. Science 263, 660–662.

Kunz, M.L., Reanier, R., 1995. The Mesa site: a Paleoindian hunting lookout in Arctic Alaska. Arctic Anthropology 32, 5–30.

Kunz, M., Bever, M., Adkins, C., 2003. The Mesa Site: Paleoindians above the Arctic Circle U.S. Department of the Interior, Bureau of Land Management-Alaska Open File Report 86, Anchorage.

Kuratov, A.A., 1978. Archaeological Sites of the Arkhangelsk Region. Catalogue. Severo-Zapadnoe Izdatelstvo (Northwestern Publ. House), Arkhangelsk, (in Russian).

Kurenkova, E.I., 2008. Prehistoric community interaction with environments at different stages of the initial colonization with the north of Eastern Europe. In: Abstracts. Scientific Conf. "Russia Contribution to IPY", Sochi, pp. 70–71 (in Russian).

Kutzbach, J., Gallimore, R., Harrison, S., Behling, P., Selin, R., Laarif, T., 1998. Climate and biome simulations for the past 21 000 years. Quaternary Science Reviews 17, 473–506.

Kuzmin, Ya.V., 1994. Prehistoric colonization of Northeastern Siberia and migration to America: radiocarbon evidence. Radiocarbon 36 (3), 367–376.

Kuzmin, Ya.V., 2008. Siberia at the Last Glacial maximum: environment and archaeology. Journal of Archaeological Research 16 (3), 163–221.

Kuzmin, Ya.V., Tankersley, K.B., 1996. The colonization of Eastern Siberia: an evaluation of the Paleolithic age radiocarbon dates. Journal of Archaeological Science 23 (4), 577–585.

Kuzmin, Ya.V., Krivonogov, S.K., 1999. More about Diring-Yuriakh: unsolved geoarchaeological problems at a "Lower" Paleolithic site in Central Siberia. Geoarchaeology 14 (4), 351–359.

Kuzmin, Ya.V., Orlova, L.A., 1998. Radiocarbon chronology of the Siberian Paleolithic. Journal of World Prehistory 12 (4), 1–53.

Kuzmin, Ya.V., Zolnikov, I.D., Orlova, L.A., Zenin, V.N., 2004. Paleogeography of the West Siberian plain at LGM (with reference to findings of mammoths and Paleolithic sites). Doklady RAS 398 (4), 542–544 (in Russian).

Kuzmin, Ya.V., Speakman, R.J., Glascock, M.D., Popov, V.K., Grebennikov, A.V., Dikova, M.A., Ptashinsky, A.V., 2008. Obsidian use at the Ushki Lake complex, Kamchatka Peninsula (Northeastern Siberia): implications for terminal Pleistocene and early Holocene human migrations in Beringia. Journal of Archaeological Science 35, 2179–2187.

Kuzmin, Ya.V., Kosintsev, P.A., Razhev, D.I., Hodgins, G.W.L., 2009. The oldest directly-dated human remains in Siberia: AMS 14C age of talus bone from the Baigara locality, West Siberian Plain. Journal of Human Evolution 57, 91–95.

Kuzmina, I.E., 1971. Formation of mammal fauna on the Northern Urals in the Late Anthropogene. In: Bykhovskiy, B.E. (Ed.), Materials on the Anthropogene Faunas of the USSR. Nauka, Leningrad (Transaction of Zoological Institute, USSR Academy of Sciences, vol. 49), pp. 44–122 (in Russian).

Kuznetsova, L.V., Ponomarenko, E.V., 2003. On the age of the burial place "Gora Mayak". In: Zones of Contact in Eurasia at the Boundary of Centuries. Center "Integratsiya", Samara, pp. 16–24 (in Russian).

Kvasov, D.D., 1975. Late Quaternary History of Large Lakes and Inner Seas of Eastern Europe. Nauka, Leningrad. 278 pp. (in Russian).

Labeyrie, L., Cole, J., Alverson, K., Stocker, T., 2003. The History of Climate Dynamics in the Late Quaternary, Paleoclimate, Global Change and the Future. Springer Verlag, Berlin. 220 pp.

Lamoureux, S.F., Bradley, R.S., 1996. A late Holocene varved sediment record of environmental change from northern Ellesmere Island, Canada. Journal of Paleolimnology 16, 239–255.

Lapteva, E.G., 2007. Climatic stratigraphy and stages of the vegetation development on the eastern slope of the Urals in the second half of the Late Neopleistocene and Holocene (based on data of pollen analysis of loose sediments infilling karst caverns). In: Materials of the V All-Russia Conference on Quaternary (Moscow, November 7–9, 2007). GEOS Publ., Moscow, pp. 218–219 (in Russian).

Larsen, H., Meldgaard, J., 1958. Paleo-Eskimo cultures in Disko Bugt, West Greenland. In: Meddelelser om Grønland, vol. 161(2), pp. 1–75.

Larsen, H., Rainey, F., 1948. Ipiutak and the Arctic whale hunting culture. Anthropological Papers of the American Museum of Natural History 42 New York.

Larsson, L., 1996. The colonization of South Sweden during the deglaciation. In: Larsson, L. (Ed.), The Earliest Settlement of Scandinavia and Its Relationship with Neighbouring Areas (Acta Archaeologica Lundensia. No. 24). Stockholm, pp. 141–155.

Laukhin, S.A., 1997. Specific features of the Northern Asia colonization by the Paleolithic man. In: Velichko, A.A., Soffer, O. (Eds.), The Man Colonizes the Earth. Global Dispersal of Gominids. Institute of Geography RAS, Moscow, pp. 156–172 (in Russian).

Laukhin, S.A., Drozdov, N.I., 1989. The first finding of artifacts of the early Upper Paleolithic in the north of the Chukchi Peninsula. In: Conference on the Problems of Regional Studies, March 27–30, 1989. Abstracts of Papers. Ussuriysk, pp. 37–41 (in Russian).

Laukhin, S.A., Shilova, G.N., Velichkevich, F.Yu, 2006. Paleobotanical characteristics and paleoclimates of the Karginsky time on the West Siberian Plain. Vestnik Archeologii, Antropologii i Etnografii (Tyumen) 7, 203–225 (in Russian).

Lavrov, A.S., 1974. Late Pleistocene ice sheets in the northeast of the European part of the USSR. Bulletin of Commission for Study of the Quaternary 41, 48–55 (in Russian).

Lavrov, A.S., 1977. Kola-Mezen ice flow. Barents-Pechora ice flow. In: Chebotareva, N.S. (Ed.), Structure and Dynamics of the Ice Sheet in Europe. Nauka, Moscow, pp. 83–89 (in Russian).

Lavrov, A.S., Potapenko, L.M., 2005. Neopleistocene of the Northeast of the Russian Plain. Aerogeologiya Publ., Moscow. 222 pp. (in Russian).

Lavrov, A.S., Nikiforova, L.D., Potapenko, L.M., 1986. Dynamics of the Pleistocene ice sheets, vegetation and climate in the northeast of the European part of the USSR. In: Sidnev, A.V., Nemkova, V.K. (Eds.), New Materials on the Pleistocene Paleogeography and Stratigraphy. Bashkirian Branch of the USSR Academy of Sciences, Ufa, pp. 69–78 (in Russian).

Lavrova, M.A., 1960. Quaternary Geology of the Kola Peninsula. USSR Academy of Sciences Publ., Moscow–Leningrad. 234 pp. (in Russian).

Lavrova, M.A., 1969. Late glacial history of the White Sea. In: Gerasimov, I.P. (Ed.), Last Ice Sheet in the North-West of the European Part of the USSR. Nauka, Moscow, pp. 267–276 (in Russian).

Lavrushin, Y.A., Spiridonova, E.A., 1998. Geological and paleoecological events and environments at the Sungir Paleolithic settlement. In: Bader, N.O., Lavrushin, Yu.A. (Eds.), Late Paleolithic Settlement Sungir (Burials and Environments). Nauchnyi Mir, Moscow, pp. 189–218 (in Russian).

Lavrushin, Y.A., Sulerzhitskiy, L.D., Spiridonova, E.A., 2000. The age of the Sungir archaeological site and specific features of the environments at the time of the early man presence. In: Alekseyeva, T.I., Bader, N.O. (Eds.), Homo Sungirensis. The Upper Paleolithic Man: Ecological and Evolutionary Aspects. Nauchny Mir, Moscow, pp. 35–42 (in Russian).

LeBlanc, M., 2002. A Diatom-Based Holocene Palaeoenvironment Record from a Mid-artic Lake on Boothia Peninsula, Central Mid-arctic Nunavut, Canada (Master of Science thesis). Carlton, Ottawa.

LeBlanc, M., Gajewski, K., Hamilton, P.B., 2004. A diatom-based Holocene palaeoenvironmental record from a mid-Arctic lake on Boothia Peninsula, Nunavut, Canada. The Holocene 14, 417–425.

Ledu, D., Rochon, A., de Vernal, A., St-Onge, G., 2010. Holocene paleoceanography of the Northwest passage, Canadian Arctic Archipelago. Quaternary Science Reviews 29, 3468–3488.

Lemmen, D.S., Robert, G., Smol, J.P., Hall, R.I., 1988. Holocene sedimentation in glacial Tasikutaaq Lake, Baffin Island. Canadian Journal of Earth Sciences 25, 810–823.

Leshchinskiy, S.V., 2006. Paleoecological studies, taphonomy, and genesis of the Lugovskoye site. Archaeology, Ethnography and Anthropology of Eurasia 1, 33–40.

Leshchinskiy, S.V., Mashchenko, E.N., Ponomareva, E.A., Orlova, L.A., Burkanova, E.N., Konovalova, V.A., Teterina, I.I., Gevlya, K.M., 2006. Multidisciplinary paleontological-stratigraphic studies on the Lugovskoye locality (2002–2004). Archaeology, Ethnography and Anthropology of Eurasia 1, 54–69.

Levin, M.G., 1958. Ethnic anthropology and problems of ethnogenesis of the Far Eastern peoples. In: Transanction of North-Eastern Expedition, vol. 2. USSR Academy of Sciences Publ., Moscow. Transaction of Institute of Ethnography, vol. 36. (in Russian).

Levin, M.G., 1964. On the anthropological type of ancient Eskimo. Proceedings of Moscow Society of Naturalists 14, 262–269 (in Russian).

Levkovskaya, G.M., Khlobystin, L.P., Sementsov, A.A., Romanova, E.N., 1972. Absolute and relative age of the Tagenar VI site. In: Kolchin, B.A. (Ed.), Problems of Absolute Dating in Archaeology. Nauka, Moscow, pp. 130–133 (in Russian).

Levkovskaya, G.M., Hoffeker, J.F., Anikovich, M.V., Forman, S.L., Holliday, V.T., Popov, V.V., Kartseva, L.A., Stegantseva, V.Y., San'ko, A.F., 2005. Climatostratigraphy of the oldest Paleolithic layers of the Kostenki-12 site (Volkovskaya). In: Anikovich, M.V. (Ed.), Problems of the Early Upper Paleolithic in the Kostenki-Borshchevo Region and Adjacent Areas. Institute for the Material Culture History RAS, St.-Petersburg, pp. 93–130 (Transaction of Kostenki-Borshchevo Archaeological Expedition, Iss. 3). (in Russian).

Lider, V.A., 1976. Quaternary Deposits of the Urals. Nedra Publ., Moscow. 114 pp. (in Russian).

Lisitsyn, N.F., 2000. Late Paleolithic of the Chulym-Enisey Interfluve. Peterburgskoye Vostokovedeniye Publ., St.-Petersburg. 232 pp. (in Russian).

Lisitsyn, N.F., Svezhentsev, Yu.S., 1997. Radiocarbon chronology of the upper Paleolithic of Northern Asia. In: Sinitsyn, A.A., Praslov, N.D. (Eds.), Radiocarbon Chronology of the Paleolithic of Eastern Europe and Northern Asia. Problems and Prospects. Institute for the Material Culture History RAS, St.-Petersburg, pp. 67–108 (in Russian).

Lisitsyn, S.N., 2006a. Prehistoric settlements on the Litorina Sea coasts in the eastern part of the Gulf of Finland. An attempt of geomorphological periodization. In: Martynov, A.Y. (Ed.), Prehistoric and Medieval History and Culture of the European North. Problem of Investigations and Scientific Reconstruction. Solti Publ., Solovki Settl., Arkhangelsk Region, pp. 112–121 (in Russian).

Lisitsyn, S.N., 2006b. Modeling of the initial colonization of the formerly glaciated regions with reference to the North-West. In: In Situ; to the 85th Anniversary of Professor Stolyar A.D. St.-Petersburg Univ. Press, St.-Petersburg, pp. 87–109 (in Russian).

Lisitsyn, S.N., Gerasimov, D.V., 2008. Man and environment of Southwestern Fennoscandia in the Early Holocene. In: Velichko, A.A., Vasil'ev, S.A. (Eds.), Way to North: Paleoenvironment and Earliest Inhabitants of Arctic and Subarctic. Institute of Geography RAS, Moscow, pp. 134–151 (in Russian).

Lloyd, J.M., Kuijpers, A., Long, A., Moros, M., Park, L.A., 2007. Foraminiferal reconstruction of mid- to late-Holocene ocean circulation and climate variability in Disko Bugt, West Greenland. The Holocene 17, 1079–1091.

Long, A.J., Roberts, D.H., Wright, M.R., 1999. Isolation basin stratigraphy and Holocene relative sea-level change on Arveprinsen Ejland, Disko Bugt, West Greenland. Journal of Quaternary Science 14, 323–345.

Lopatina, D.A., Zanina, O.G., 2006. Paleobotanic analysis of material recovered from fossil souslik burrows and enclosing Upper Pleistocene deposits in the Lower Kolyma region. Stratigraphy and Geological Correlation 14 (5), 94–107.

Lougas, L., 1997. Subfossil seals finds from archaeological coastal sites in Estonia: East part of the Baltic Sea. Anthropozoologica 25/26, 699–706 Paris.

Lozhkin, A.V., 1977. Environments of the Berelekh population of mammoths. In: Svetovidov, A.N. (Ed.), Mammoth Fauna of the Russian Plain and East Siberia. Transaction of Zoological Institute, USSR Academy of Sciences, vol. 72. Nauka, Leningrad, pp. 67–68 (in Russian).

Lozhkin, A.V., 1997. Evolution of the Beringian environments in the Late Pleistocene and Holocene: some results of joint Russian–American investigations. In: Gagiev, M.K. (Ed.), Late Pleistocene and Holocene of Beringia. Northeastern Interdisciplinary Research Institute, Far Eastern Division, RAS, Magadan, pp. 5–22 (in Russian).

Lozhkin, V.A., 1998. On the radiocarbon dating and palynological characteristic of the mammoth 'cemetery' on the Berelekh River, in the lower reaches of the Indigirka River. In: Simakov, K.V. (Ed.), Changes in the Beringia Environments in the Quaternary. Northeastern Interdisciplinary Research Institute, Far Eastern Division, RAS, Magadan, pp. 45–62 (in Russian).

Lozhkin, A.V., 2002. Boundaries of Beringia in the Late Pleistocene and Holocene. In: Simakov, K.V. (Ed.), Quaternary Paleogeography of Beringia. Northeastern Interdisciplinary Research Institute, Far Eastern Division, RAS, Magadan, pp. 4–12 (in Russian).

Lozhkin, A.V., 2008. Tentative reconstruction of environments in the studies of multilayered archaeological sites in the North of the Far East. In: Velichko, A.A., Vasil'ev, S.A. (Eds.), Way to North: Paleoenvironment and Earliest Inhabitants of Arctic and Subarctic. Institute of Geography RAS, Moscow, pp. 253–259 (in Russian).

Lozhkin, A.V., Vazhenina, L.N., 1987. Evolution of the vegetation in the Kolyma Lowland in the Early Holocene. In: Pohialainen, V.P. (Ed.), Quaternary Period in the North-East of Asia. Northeastern Interdisciplinary Research Institute, Far Eastern Division, RAS, Magadan, pp. 135–144 (in Russian).

Lozhkin, A.V., Kotov, A.N., Ryabchun, V.K., 2000. Specific features of the paleobotanical characteristics and radiocarbon dating of the sediments exposed in the Ice Scarp. In: Simakov, K.V. (Ed.), Beringia in the Quaternary. Northeastern Interdisciplinary Research Institute, Far Eastern Division, RAS, Magadan, pp. 118–131 (in Russian).

Lozhkin, A.V., Anderson, P.M., Matrosova, T.V., Minyuk, P.S., 2007. The pollen record from El'gygytgyn Lake: implications for vegetation and climate histories of northern Chukotka since the late Middle Pleistocene. Journal of Paleoliminology 37, 135–153.

Lozhkin, A.V., Anderson, P.M., Matrosova, T.V., Minyuk, P.S., Pakhomov, A.V., Solomatkina, T.V., 2008a. An uninterrupted record of climate and vegetation changes in the North Okhotsk region spanning the last 70 ka. In: Velichko, A.A., Vasil'ev, S.A. (Eds.), Way to North: Paleoenvironment and Earliest Inhabitants of Arctic and Subarctic. Institute of Geography RAS, Moscow, pp. 249–252 (in Russian).

Lozhkin, A.V., Anderson, P.M., Matrosova, T.V., Solomatkina, T.V., 2008b. An attempt at the studies of lacustrine pollen record with a view to reconstructing Quaternary environments in Beringia. Vestnik. Far Eastern Division RAS 1, 24–32 (in Russian).

Luho, V., 1956. Die Askola-Kultur. Suomen muinaismuisto yhdistyksen aikakauskirja, (SMYA) 57 (1), 167 Finnish Antiquarian Society, Helsinki.

Luho, V., 1967. Die Suomujarvi-Kultur: Die mittel- und spät mesolitische zeit in Finnland. SMYA 66.

Luk'yanchenko, T.V., 1971. Material Culture of Sami (Lapps) of the Kola Peninsula in the Late 19th–20th Centuries. Nauka, Moscow. 167 pp. (in Russian).

Luk'yanchenko, T.V., 1980. Sami ethnogenesis. In: Gurvich, I.S. (Ed.), Ethnogenesis of the Northern Peoples. Nauka, Moscow, pp. 28–40 (in Russian).

Luzgin, V.E., 1973. Neolithic sites in the Central Timan. In: Settlements of the Stone and Copper-Bronze Age on the Pechora and USA Rivers. Materials on the Archaeology of the European North-East. Iss. 4. Syktyvkar, pp. 24–46 (in Russian).

Lyons, J.B., Mielke, J.E., 1973. Holocene history of a portion of northernmost Ellesmere Island. Arctic 26, 314–323.

MacDonald, G.M., Gervais, B.R., Snyder, J.A., Tarasov, G.A., Borisova, O.K., 2000a. Radiocarbon dated *Pinus sylvestris* L. wood from beyond treeline on the Kola Peninsula, Russia. The Holocene 10, 143–147.

MacDonald, G.M., Velichko, A.A., Kremenetski, C.V., Borisova, O.K., Goleva, A.A., Andreev, A.A., Cwynar, L.C., Riding, R.T., Forman, S.L., Edwards, T.W.D., Aravena, R., Hammarlund, D., Szeicz, J.M., Gattaulin, V.N., 2000b. Holocene treeline history and climate change across northern Eurasia. Quaternary Research 53 (3), 302–311.

Makarov, E.Y., Pavlov, P.Y., 2007. Shirovanovo II site – a new site of the Late Paleolithic in the upper Kama basin. In: Volokitin, A.V. (Ed.), Stone Age of the European North. Institute of Linguistic, Literature, and History, Komi Research Centre, the Urals Division, RAS, Syktyvkar, pp. 7–21 (in Russian).

Makarov, S.S., 2008. Peopling and colonization of West Siberian Plain in the Late Pleistocene (chronology and periodization). In: Velichko, A.A., Vasil'ev, S.A. (Eds.), Way to North: Paleoenvironment and Earliest Inhabitants of Arctic and Subarctic. Institute of Geography RAS, Moscow, pp. 188–199 (in Russian).

Mamonova, N.N., 1995. Paleoanthropological materials on the Mesolithic of the Northern regions of Russia (preliminary communication). Voprosy Antropologii 88, 117–125 (in Russian).

Mamonova, N.N., Sulerzhitskiy, L.D., 1989. 14C dating of the Holocene burial grounds in the Baikal region. Sovetskaya Arkheologiya 1, 19–32 (in Russian).

Mangerud, J., Svendsen, J.I., Astakhov, V.I., 1999. Age and extent of the Barents and Kara ice sheets in Northern Russia. Boreas 28, 46–80.

Mangerud, J., Jakobsson, M., Alexanderson, H., Astakhov, V.I., Clarke, M., Henriksen, M., Hjort, C., Krinner, G., Lunkka, J.P., Mцller, P., Murray, A., Nikolskaya, O., Saarnisto, M., Svendsen, J.I., 2004. Ice-dammed lakes and rerouting of the drainage of northern Eurasia during the Last Glaciation. Quaternary Science Reviews 23, 1313–1332.

Mann, D.H., Reanier, R.E., Peteet, D.M., Kunz, M.L., Johnson, M., 2001. Environmental change and Arctic Paleoindians. Arctic Anthropology 38 (2), 119–138.

Mark, K.Yu, 1956. Problems of ethnic history of the Estonian people in the light of the facts of anthropology. In: Moor, H.A. (Ed.), Problems of Ethnic History of the Estonian People. Insitute of History, Estonian Academy of Sciences Publ., Tallinn, pp. 219–242 (in Russian).

Mark, K.Yu, 1975. Antropology of Baltic-Finnish Peoples. Valgus Publ., Tallinn. 136 pp. (in Russian).

Markov, K.K., Velichko, A.A., 1967. Quaternary Period (Ice Age – Anthropogene). vol. 3. Continents and Oceans. Nedra Publ., Moscow. 440 pp. (in Russian).

Markova, A.K., 2005. Reconstructions of faunal assemblages of the Late Pleistocene. In: Quarter-2005. Materials of the 4th All-Russia Conference on the Quaternary. Geoprint Publ., Syktyvkar, pp. 253–255 (in Russian).

Marom, A., McCullagh, J., Higham, T., Sinitsyn, A., Hedges, R., 2012. Single amino acid radiocarbon dating of Upper Palaeolithic modern humans. Proceedings of the National Academy of Sciences. http://dx.doi.org/10.1073/pnas.1116328109.

Mashchenko, E.N., Shubina, Yu.V., Telegina, S.N., 2006. Lugovskoye Site: Landscape Against the Glacier Background. State Museum of Nature and Man, Khanty-Mansiysk; Basko Publ., Ekaterinburg, pp. 35–38 (in Russian).

Mathiassen, T., 1958. The Sermermiut excavations, 1955. Meddelelser om Grønland 161 (3) 52 pp. (København).

Matiskainen, H., 1989. Studies in chronology, material culture and subsistence of the Finnish Mesolithic, 10,000 – 6,000 BP. Iksos (Hesinki) (8).

Matiskainen, H., 1996. Discrepances in deglaciation chronology and the appearance of man in Finland. In: Larsson, L. (Ed.), The Earliest Settlement of Scandinavia and Its Relationships with Neighbouring Area (Acta Archaeologica Lundensia. No. 24). Stockholm, pp. 251–262.

Maxwell, M.S., 1980. Archaeology of the Arctic and subarctic zones. Annual Review of Anthropology 9, 161–185.

Maxwell, M.S., 1985. Prehistory of the Eastern Arctic. Academic Press, Orlando. 327 pp.

Mayer, G.I., Merkachev, I.T., Moskovkin, T.P., Prokhorov, B.B., 1970. Materials of the symposium on climatic and physiological problems of Siberia and Far East. In: Matyukhin, V.A. (Ed.), Problems of Bioclimatology and Climatophysiology. Institute of Physiology, Academy of Sciences, USSR, Novosibirsk, pp. 123–124 (in Russian).

McCartney, P.H., Helmer, J.W., 1989. Marine and terrestrial mammals in High Arctic Paleoeskimo economy. Archaeozoologia 3 (1/2), 143–160.

McGhee, R., 1974. The peopling of Arctic North America. In: Ives, J.D., Barry, R.G. (Eds.), Arctic and Apline Environments. William Clowes and Sons Ltd., London, pp. 831–855.

McGhee, R., 1979. The Palaeoeskimo Occupation at Port Refuge, High Arctic Canada. Archaeological Survey of Canada Mercury Series. Paper No. 92 National Museums of Canada, Ottawa. 176 pp.

McGhee, R., 1996. Ancient People of the Arctic. Canadian Museum of Civilization, Hull. 157 pp.

McGhee, R., Tuck, J.A., 1976. Un-dating the Canadian Arctic. In: Maxwell, M.S. (Ed.), Eastern Arctic Prehistory: Paleoeskimo Problems. Memoirs of the Society for American Archaeology, vol. 31, pp. 6–14.

McGovern, T.H., Vésteinsson, O., Fripriksson, A., Church, M.J., Lawson, I.T., Simpson, I.A., Einarsson, B., Dugmore, A.J., Cook, G.T., Perdikaris, S., Edwards, K.J., Thomson, A.M., Adderle, W.P., Newton, A.J., Lucas, G., Aldred, O., 2007. Landscapes of settlement in Northern Iceland: historical ecology of human impact and climate fluctuation on the millennial scale, invited paper in special issue on the archaeology of global change. American Anthropologist 109, 27–51.

Mednikova, M.B., 2000. Roentgen morphology of children from burial 2. In: Alekseyeva, T.I., Bader, N.O. (Eds.), Homo Sungirensis. The Upper Paleolithic Man: Ecological and Evolutionary Aspect of Studies. Nauchny Mir, Moscow, pp. 286–298 (in Russian).

Mednikova, M.B., 2006. The Russian North population in the Late Paleolithic by the data of the skeletal morphology. In: Martynov, A.Y. (Ed.), Prehistoric and Medieval History and Culture of the European North. Problem of Investigations and Scientific Reconstruction. Solti Publ., Solovki Settl., Arkhangelsk Region, pp. 128–133 (in Russian).

Mednikova, M.B., 2009. An attempt of bioarchaeological reconstruction based on data of the postcranial morphology (Burial ground of the Yuzhny Oleni Island). In: Gromov, A.V., Khartanovich, V.I. (Eds.), Micro-evolutionary Processes in Human Populations. Museum of Anthropology and Ethnography, St.-Petersburg, pp. 131–142 (in Russian).

Mednikova, M.B., 2011. Postcranial Morphology and Taxonomy of the Homo Genus Remains from Okladnikov Cave, the Altai Mts. Institute of Archaeology and Ethnography, Siberian Branch RAS Publ., Novosibirsk. 127 pp. (in Russian).

Meinander, K., 1982. Finns as a part of the Northeastern Europe population. In: Zubov, A.A., Shlygina, N.V. (Eds.), Finno-Ugric Collection: Anthropology, Archaeology, Ethnography. Nauka, Moscow, pp. 10–32 (in Russian).

Meldgaard, J., 1952. A Palaeo-Eskimo culture in West Greenland. American Antiquity 17 (3), 222–230.

Meldgaard, J., 1983. Qajaa, en køkkenmodding i dybfrost Feltrapport fra arbejdsmarken i Grønland. Nationalmuseets Arbejdsmark, Copenhagen, pp. 83–96.

Meldgaard, J., 1991. Bopladsen Qajaa i Jakobshavn Isfjord. In: Qeqertassuk. De første mennesker i Vestgrønland. Tidsskriftet Grønland, vol. 4, pp. 191–205.

Meldgaard, M., 1995. Resource pulses in a marine environment: a case study from Disko Bugt, West Greenland. In: Fisher, A. (Ed.), Man and Sea in the Mesolithic, vol. 53. Oxbow Books, Oxford, pp. 361–368.

Meldgaard, M., 2004. Ancient Harp Seal Hunters of Disko Bay. Subsistence and Settlement at the Saqqaq Culture Site Qeqertasussuk (2400–1400 BC), West Greenland. Meddelelser om Grønland, Man and Society, vol. 30, pp. 196–200.

Melekestsev, I.V., Glushkova, O.Y., Kiryanov, V.Y., Lozhkin, A.V., Sulerzhitskiy, L.D., 1991. Origin and age of the Magadan volcanic ash. Doklady, Academy of Sciences 317 (5), 1188–1192 (in Russian).

Melekestsev, I.V., Braitseva, O.A., Bazanova, L.I., Ponomareva, V.V., Sulerzhitskiy, L.D., 1996. A specific type of catastrophic explosive eruptions: the Holocene sub-caldera eruptions of Hangar, Khodutkinskiy 2 maar, Baraniy amphitheater (Kamchatka). Vulcanologiya I Seismologiya 2, 3–24 (in Russian).

Mellars, P., 2006. A new radiocarbon revolution and the dispersal of modern humans in Eurasia. Nature 429, 931–935.

Melnichuk, A.F., Pavlov, P.Yu, 1987. The Gornaya Talitsa site of the Chusovaya River and the Early Mesolithic problems in the Kama Basin. In: Nagovitsin, L.A. (Ed.), Problems of the Early History of Udmurtiya. Research Institute at the USSR Council of Ministers, Izhevsk, pp. 5–18 (in Russian).

Meltzer, D.J., 2009. First Peoples in a New World: Colonizing Ice Age America. Univ. California Press, Berkeley. 446 pp.

Meltzer, D.J., Grayson, D.K., Ardila, G., Barker, A.W., Dincauze, D.F., Haynes, C.V., Mena, F., Nъcez, L., Stanford, D.J., 1997. On the Pleistocene antiquity of Monte Verde, Southern Chile. American Antiquity 62, 659–663.

Merbs, C.F., 1983. Patterns of Activity Induced Pathology in Canadian Inuit Population. National Museum of Man (Mercury Series). Archaeological Survey of Canada. No. 119. 199 pp.

Miettinen, A., 2002. Relative Sea Level Changes in the Eastern Part of the Gulf of Finland during the Last 8000 Years. Annales Academiae Scientiarum Fennicae, Geologica, Geographica, 162. Suomalainen Tiedeakatemia, Helsinki. 100 pp.

Miettinen, A., Savelieva, L., Subetto, D., Dzhinorudze, R., Arslanov, K., Hyvarinen, H., 2007. Palaeoenvironment on the Karelian Isthmus, the easternmost part of the Gulf of Finland, during the Litorina Sea stage of the Baltic Sea history. Boreas 36 (4), 441–458.

Mikkelsen, E., 1975. Mesolithic in south-eastern Norway. Norwegian Archaeological Review 8 (1), 19–35.

Milne, B.S., 2005. Palaeo-Eskimo novice flintknapping in the eastern Canadian Arctic. Journal of Field Archaeology 30, 329–345.

Minyuk, P.S., Nowaczyk, N.P., Glushkova, O.Y., Smirnov, V.N., Brigham-Grette, J., Melles, M., Cherepanova, M.V., Lozhkin, A.V., Anderson, P., Matrosova, T.V., Hubberten, H.W., Belaya, B.V., Borkhodoeyv, V.Y., Forman, S.L., Asikainen, S., Leyer, P., Nolan, M., Prokein, P., Liston, G., Huntzinger, R., Sharpton, V.L., Nissen, F., 2003. Paleoclimatic data on Lake Elgygytgyn (multi-disciplinary investigations). In: Malakhov, M.I. (Ed.), The Process of Post-depositional Magnetization and Characteristic Changes of the Earth's Magnetic Field and Climate in the Past. Northeastern Interdisciplinary Research Institute, Far Eastern Division, RAS, Magadan, pp. 91–135 (in Russian).

Møbjerg, T., 2004. Cultural position of the Nipisat site. In: Gotfredsen, A.B., Møbjerg, T. (Eds.), Nipisat – a Saqqaq Culture Site in Sisimiut, Central West Greenland Meddelelser om Grønland. Man and Society, vol. 31, pp. 192–206.

Mochanov, Y.A., 1977. The Most Ancient Stages of the Northeastern Asia Colonization by Humans. Nauka, Novosibirsk. 263 pp. (in Russian).

Mochanov, Y.A., 1988. The oldest Paleolithic of the Diring site and the problem of an extra-tropical place of human origin. In: Alekseyev, A.N. (Ed.), Archaeology of Yakutia. Yakutsk Univ. Press, Yakutsk, pp. 15–54 (in Russian).

Mochanov, Y.A., 1992. The Oldest Paleolithic of the Diring Site and the Problem of an Extratropical Place of Human Origin. Nauka, Novosibirsk. 254 pp. (in Russian).

Mochanov, Y.A., 2007. The Dyuktay Bifacial Tradition of the Paleolithic of Northern Asia (History of Its Identifiation and Studies). Center of Arctic Archaeology and Paleoecology of the Man, Yakutsk. 200 pp. (in Russian).

Mochanov, Y.A., Fedoseeva, S.A., 1996. Berelekh, Allachovsk Region. In: West, F.H. (Ed.), American Beginnings: The Prehistory and Palaeoecology of Beringia. Chicago Univ. Press, Chicago, pp. 218–222.

Mochanov, Y.A., Fedoseyeva, S.A., 2001. Noosphera and archaeology. Nauka I Tekhnika v Yakutii 1, 28–33 (in Russian).

Mochanov, Y.A., Savvinova, G.M., 1980. The environment of the man at the Stone Age and the Early Metal Ages (based on materials of archaeological sites). In: Mochanov, Yu.A. (Ed.), New Development in the Archaeology of Yakutia. Yakutian Branch of the USSR Academy of Sciences, Yakutsk, pp. 14–27 (in Russian).

Mochanov, Y.A., Bagynanov, N.G., Kistenev, S.P., 1978. The first archaeological survey in the upper reaches of the Maly Anyui River. (Central Chukotka). In: Archaeology and Ethnography of Eastern Siberia. Abstracts of Conference, April 5–7, 1978, Irkusk. Irkutsk State Univ. Press, Irkutsk, pp. 64–65 (in Russian).

Mochanov, Y.A., Fedoseyeva, S.A., Alekseyev, A.N., Kozlov, V.I., Kochmar, N.N., Shcherbakova, N.M., 1983. Archaeological Sites of Yakutia. The Aldan and Olekma Rivers. Drainage Basins. Nauka, Novosibirsk. 391 pp. (in Russian).

Mochanov, Y.A., Fedoseyeva, S.A., Konstantinov, I.V., Antipina, N.V., Argunov, V.G., 1991. Archaeological Sites of Yakutia. The Vilyui, Anabar and Olenek Drainage Basins. Nauka, Novosibirsk. 224 pp. (in Russian).

Møhl, J., 1986. Dog remains from a Paleoeskimo settlement in West Greenland. Arctic Anthropology 23 (1/2), 81–89.

Moiseyev, V.G., Khartanovich, V.I., 2012. Craniological materials recovered from a burial of the Early Metal Age on Bolshoy Oleni Island, the Barents Sea. Archaeology, Ethnography and Anthropology of Eurasia 1 (49), 145–154.

Mökkönen, T., Nordqvist, K., Bel'skij, S., 2007. The Rupunkangas 1a site in the archipelago of ancient Lake Ladoga: a housepit with several rebuilding phases. Fennoscandia Archaeologica (Helsinki) 24, 3–28.

Møller, J.J., Yevzerov, V.Y., Kolka, V.V., Corner, G.D., 2002. Holocene raised-beach ridges and sea-ice-pushed boulders on the Kola Peninsula, northwest Russia: indicators of climatic change. The Holocene 12 (2), 169–176.

Mongait, A.L., 1973. Archaeology of Western Europe. Stone Age. Nauka, Moscow. 356 pp. (in Russian).

Morlan, R.E., 1988. Pre-Clovis people: early discoveries of America? In: Carlisle, R.C. (Ed.), Americans before Columbus: Ice Age Origins. (Ethnology Monographs No. 12). Univ. of Pittsburgh, Department of Anthropology, Pittsburgh, pp. 31–43.

Morlan, R.E., 2003. Current perspectives on the Pleistocene archaeology of eastern Beringia. Quaternary Research 60, 123–132.

Morlan, R.E., Cinq-Mars, J., 1982. Ancient Beringians: human occupation in the Late Pleistocen of Alaska and the Yukon Territory. In: Hopkins, D.H., Matthews, J.V., Schweger, C.E., Young, S.B. (Eds.), Paleoecology of Beringia. Academic Press, New York, pp. 353–398.

Morrison, D., 2001. Radiocarbon dating the Birnirk-Thule transition. Papers of the Univ. of Alaska, New Series 1 (1), 73–85.
Moskvitin, A.I., 1962. On the Wьrm subdivisions and relative position of Middle and Upper Paleolithic in Europe. Izvestiya. Akademii Nauk SSSR Ser. Geol. 7, 35–44 (in Russian).
Moskvitin, A.I., 1976. The Pleistocene Stratigraphy of the European Part of the USSR. Nauka, Moscow, (Transaction of GIN, Iss. 156), 236 pp. (in Russian).
Mudie, P., Rochon, A., Levac, E., 2005. Decadeal scale sea ice changes in the Canadian Arctic and their impacts on humans during the past 4,000 years. Environmental archaeology. Journal of Human Palaecology 10, 113–126.
Murygin, A.M., Karmanov, V.N., Klenov, M.V., 2012. New Archaeological Survey in Tundra of the Northeast of Europe. Institute of Linguistic, Literature, and History, Komi Research Centre, the Urals Division, RAS, Syktyvkar. 68 pp. (in Russian).
Myres, N.M., Rootsi, S., Lin, A.A., Jдrve, M., King, R.J., Kutuev, I., Cabrera, V.M., Khusnutdinova, E.K., Pshenichnov, A., Yunusbayev, B., Balanovsky, O., Balanovska, E., Rudan, P., Baldovic, M., Herrera, R.J., Chiaroni, J., Di Cristofaro, J.D., Villems, R., Kivisild, T., Underhill, P.A., 2011. A major Y-chromosome haplogroup R1b Holocene era founder effect in Central and Western Europe. European Journal of Human Genetics 19 (1), 95–101.
Nagy, M., 1997. Palaeoeskimo cultural transition: a case study from Ivujivik, Eastern Arctic. In: Nunavik Archaeology Monograph Series, No. 1. Avataq Cultural Institute, Inukjuak, Nunavik.
Nedomolkina, N.G., 2006. The Mesolithic site Nikolayevskaya ("Belltower"). In: Martynov, A. Ya (Ed.), Prehistoric and Medieval History and Culture of the European North. Problems of Investigations and Scientific Reconstruction. Solti Publ., Solovki Settl., Arkhangelsk Region, pp. 59–67 (in Russian).
Nelson, D.E., McGhee, R., 2002. Aberrant radiocarbon dates on an Inuit arrowhead. Arctic 55 (4), 345–347.
Nelson, D.E., Møhl, J., 2002. Radiocarbon dating caribou antler and bone: are they different? Arctic 56, 262–265.
Nemkova, V.K., 1976. Pollen spectra from a peats near Dutovo settlement, the Upper Pechora River. In: Yakhimovich, V.L. (Ed.), Stratigraphy and Correlation of the Pliocene and Pleistocene in West Uralian Forelands (Preduralye). Bashkir Branch of the USSR Academy of Sciences Publ., Ufa, pp. 68–72 (in Russian).
Nemkova, V.K., 1978. Stratigraphy of the late glacial and postglacial deposits in the piedmonts of the Urals. In: Yakhimovich, V.L. (Ed.), On the History of the Late Pleistocene and Holocene of the Southern Urals and Preduralye. Bashkir Branch of the USSR Academy of Sciences Publ., Ufa, pp. 4–44 (in Russian).
Neustadt, M.I., Khotinskiy, N.A., Kraynov, D.A., 1982. Vegetation and areas of early Neolithic cultures in the European part of the USSR at the Holocene climatic optimum. In: Gerasimov, I.P., Velichko, A.A. (Eds.), Paleogeography of Europe During the Last 100,000 Years. Nauka, Moscow, pp. 121–123 (in Russian).
Nikiforova, L.D., 1982. Dynamics of the landscape zones in the Northeast of the European part of the USSR in the Holocene. In: Velichko, A.A., Spasskaya, I.I., Khotinskiy, N.A. (Eds.), Evolution of the Environments on the USSR Territory During the Late Pleistocene and Holocene. Nauka, Moscow, pp. 154–162 (in Russian).
Nikolskaya, M.V., 1980. Paleobotanical characteristics of the Upper Pleistocene and Holocene deposits of Taymyr. In: Saks, V.N. (Ed.), Paleopalynology of Siberia. Nauka, Moscow, pp. 97–111 (in Russian).
Nikolskiy, P.A., Basilyan, A.E., Sulerzhitsky, L.D., Pitulko, V.V., 2010. Prelude to the extinction: revision of the Achchagyi-Allaikha and Berelyokh mass accumulations of mammoth. Quaternary International 219, 16–25.

Nikonov, A.A., 1966. Stratigraphy and paleogeography of the Anthropogene of the Kola Peninsula. In: Grichuk, V.P. (Ed.), Upper Pleistocene. Stratigraphy and Absolute Geochronology. Nauka, Moscow, pp. 92–105 (in Russian).

Nordenskiuld, A.E., 1871. Redogorelse for en Expedition til Gronland Er 1870. Цfvers. afkungl. Vet. Akad. Forh. No. 10, Stockholm.

Nummedal, A., 1929. Stone Age Finds in Finnmark. Instituttet for Sammenlignende Kulturforskning, Oslo.

Nucez, M., 1987. A model for the early settlement of Finland. Fennoscandia Archaeologica 4, 3–18.

Obedientova, G.V., 1977. Erosional Cycles and the Volga Valley Development. Nauka, Moscow. 240 pp. (in Russian).

O'Brien, S.R., Mayewski, P.A., Meeker, L.D., Meese, D.A., Twickler, M.S., Whitlow, S.I., 1995. Complexity of Holocene climate as reconstructed from a Greenland ice core. Science 270, 1962–1964.

Odess, D., Rasic, J.T., 2007. Toolkit composition and assemblage variability: the implications of Nogahabara I, northern Alaska. American Antiquity 72 (4), 691–718.

Odgaard, U., 2001. Palaeo-Eskimoic Shamanism. In: Vestergaard, T.A. (Ed.), Shamanism and Traditional Beliefs. (North Atlantic Studies. vol. 4). Aarhus Univ. Press, Aarhus, pp. 25–30.

Odgaard, U., 2003. Hearth and home of the Palaeo-Eskimos. Etudes/Inuit/Studies 27 (1–2), 349–374.

Odgaard, U., 2005. The most extreme situation. Contextual experiment with an Arctic hearth performed at Lejre Experimental Center. In: Experimental Pyrotechnology Group Newsletter. No. 2. National Univ. of Arts, Bucharest.

Odgaard, U., 2007a. Hearth, heat and meat. In: Gheorgiu, D. (Ed.), Fire as an Instrument: The Archaeology of Pyrotechnologies. British Archaeological Reports. International Series 1619, Oxford, pp. 7–18.

Odgaard, U., 2007b. The fireplace as centre of life. In: Gheorgiu, D., Nash, G. (Eds.), The Archaeology of Fire. Understanding Fire as Material Culture. Archaeolingua, Budapest, pp. 61–84.

Odgaard, U., 2010. Feeding the fire in the circumpolar hearths. In: Westerdahl, C. (Ed.), A Circumpolar Reappraisal: The Legacy of Gutorm Gjessing (1906–1979), British Archaeological Reports International Series, pp. 225–239.

Odner, K., 1966. Komsa-kulturen i Nesseby og Sor-Varanger, vol. XII. Tromsø Museums Skrifter, Tromsø. Universitetet i Tromsø.

Odner, K., 1970. The relations between the Komsa-Culture and the more eastern and western Stone Age cultures. In: Proc. VII International Congress on Anthropology and Ethnography, Moscow.

Okladnikov, A.P., 1950. Antiquities of the Lena, Iss. 3. USSR Academy of Sciences Publ., Moscow – Leningrad. 195 pp. (in Russian).

Okladnikov, A.P., Grigorenko, B.G., Alekseyeva, E.B., Volkov, I.A., 1971. The Upper Paleolithic site Volch'ya Griva. In: Field Materials of the Far East Archaeological Expedition, Iss. 2. Nauka, Novosibirsk, pp. 23–34 (in Russian).

Olsen, B., 1994. Bosetning og samfunn i Finnmarks forhistorie. Universitetsforlaget, Olso.

Olsen, B., 1998. Saqqaq housing and settlement in southern Disko Bay, West Greenland. Acta Borealia 15 (2), 81–128.

Olyunina, O.S., Karevskaya, I.A., Tsekina, M.V., 2005. Diatom and pollen analysis of peat on the western coast of Lake Umbozero (Kola Peninsula). In: Modern Problems of Paleofloristics, Paleophytogeography, and Phytostratigraphy. International Paleobotanical Conf. Abstracts. GEOS Publ., Moscow, pp. 92–93 (in Russian).

Olyunina, O.S., Polyakova, E.I., Romanenko, F.A., 2008. Diatom communities in the Holocene deposits on the Kola Peninsula. Doklady RAS 423 (3), 370–374 (in Russian).

Orlova, L.A., 1998. Radiocarbon dating of archaeological sites in Siberia and Far East of Russia. In: Derevyanko, A.P., Kholyushkin, Y.P., Rostovtsev, P.S., Voronin, V.T. (Eds.), Statistical Analysis of Late Paleolithic Sites in Northern Asia. Appendix 2. Novosibirsk Univ. Press, Novosibirsk, pp. 74–88 (in Russian).

Orlova, L.A., Zenin, V.N., Stuart, A.J., Higham, F.G., Grootes, P.M., Leshcinsky, S.V., Kuzmin, Ya.V., Pavlov, A.F., Maschenko, E.N., 2004. Lugovskoe, Western Siberia: a possible extra-arctic mammoth refugium at the end of the Late Glacial. Radiocarbon 46 (1), 363–368.

O'Shea, J., Zvelebil, M., 1984. Oleneostrovskii Mogilnik: reconstructing the social and economic organisation of prehistoric foragers in northern Russia. Journal of Anthropological Archaeology 3, 1–40.

Oshibkina, S.V., 1982. Mesolithic burial ground Popovo on the Kineshma River. Sovetskaya Arkheologiya 3, 122–138 (in Russian).

Oshibkina, S.V., 1983. Mesolithic in the Sukhona Drainage Basin and the Region East of Onega Lake. Nauka, Moscow. 295 pp. (in Russian).

Oshibkina, S.V., 1989. Mesolithic of the central and northeastern regions of the European part of the USSR. In: Kol'tsov, L.V. (Ed.), Mesolithic of the USSR. Institute of Archaeology, USSR Academy of Sciences, Moscow, pp. 32–45 (in Russian).

Oshibkina, S.V., 1996. North of Eastern Europe. In: Oshibkina, S.V. (Ed.), Neolithic of Northern Eurasia. Nauka, Moscow, pp. 212–242 (in Russian).

Oshibkina, S.V., 1997. Veretye 1 – Mesolithic Settlement in the North of Eastern Europe. Nauka, Moscow. 204 pp. (in Russian).

Oshibkina, S.V., 2006. Mesolithic of the Eastern Onezhskoye Region. Veretye Culture. Institute of Archaeology RAS. 322 pp. (in Russian).

Osterman, L.E., Nelson, A.R., 1989. Latest Quaternary and Holocene paleoceanography of the eastern Baffin Island continental shelf, Canada: benthic foraminiferal evidence. Canadian Journal of Earth Sciences 26, 2236–2248.

Oswald, W.W., Brubaker, L.B., Anderson, P.M., 1999. Late Quaternary vegetational history of the Howard pass area, northwestern Alaska. Canadian Journal of Botany 77, 570–581.

Overpeck, J.T., Prentice, I.C., Webb III, T., 1985. Quantitative interpretation of fossil pollen spectra: dissimilarity coefficients and the method of modern analog. Quaternary Research 23, 87–108.

Pagani, L., Lawson, D.J., Jagoda, E., Mörseburg, A., Eriksson, A., Mitt, M., Clemente, F., Hudjashov, G., DeGiorgio, M., Saag, L., Wall, J.D., Cardona, A., Mägi, R., Sayres, M.A., Kaewert, S., Inchley, C., Scheib, C.L., Järve, M., Karmin, M., Jacobs, G.S., Antao, T., Iliescu, F.M., Kushniarevich, A., Ayub, Q., Tyler-Smith, C., Xue, Y., Yunusbayev, B., Tambets, K., Mallick, C.B., Saag, L., Pocheshkhova, E., Andriadze, G., Muller, C., Westaway, M.C., Lambert, D.M., Zoraqi, G., Turdikulova, S., Dalimova, D., Sabitov, Z., Sultana, G.N., Lachance, J., Tishkoff, S., Momynaliev, K., Isakova, J., Damba, L.D., Gubina, M., Nymadawa, P., Evseeva, I., Atramentova, L., Utevska, O., Ricaut, F.X., Brucato, N., Sudoyo, H., Letellier, T., Cox, M.P., Barashkov, N.A., Škaro, V., Mulahasanovic, L., Primorac, D., Sahakyan, H., Mormina, M., Eichstaedt, C.A., Lichman, D.V., Abdullah, S., Chaubey, G., Wee, J.T., Mihailov, E., Karunas, A., Litvinov, S., Khusainova, R., Ekomasova, N., Akhmetova, V., Khidiyatova, I., Marjanović, D., Yepiskoposyan, L., Behar, D.M., Balanovska, E., Metspalu, A., Derenko, M., Malyarchuk, B., Voevoda, M., Fedorova, S.A., Osipova, L.P., Lahr, M.M., Gerbault, P., Leavesley, M., Migliano, A.B., Petraglia, M., Balanovsky, O., Khusnutdinova, E.K., Metspalu, E., Thomas, M.G., Manica, A., Nielsen, R., Villems, R., Willerslev, E., Kivisild, T., Metspalu, M., 2016. Genomic analyses inform on migration events during the peopling of Eurasia. Nature. http://dx.doi.org/10.1038/nature19792.

Palsi, S., 1920. Ein Steinzeitlichen Moorfund bei Korpilahti Kirchspeil Antrea, Lan Viborg. Suomen muinaismuistoyhdistyksen aikakauskirja (SMYA) 28 (2), 3–19.

Pankrushev, G.A., 1978. Mesolithic and Neolithic in Karelia. Part 1. Institute of Linguistic, History, and Literature, Karelian Research Centre, USSR Academy of Sciences, Perozavodsk. 138 pp. (in Russian).

Pankrushev, G.A., 1980. Origin of the Karelians (from archaeological data). In: New Archaeological Sites of Karelia and Kola Peninsula. Karelian Research Centre, USSR Academy of Sciences, Petrozavodsk, pp. 48–159 (in Russian).

Pavlov, A.F., Mashchenko, E.N., Zenin, V.N., Leshchinskiy, S.V., Orlova, L.A., 2002. Preliminary results of the multi-disciplinary studies of the Lugovskoye locality (Khanty-Mansi Autonomous area). In: Derevyanko, A.P., Molodin, V.I. (Eds.), Problems of Archaeology, Ethnography, Anthropology of Siberia and Adjacent Territories. (Materials of Annual Session of Institute of Archaeology and Ethnography Siberian Branch RAS, 2002), vol. VIII. Institute of Archaeology and Ethnography Siberian Branch RAS Publ., Novosibirsk, pp. 165–172 (in Russian).

Pavlov, P.Yu, 1988. Paleolithic of the North-East of the European Part of the USSR (Thesis Ph.D.). Leningrad. 20 pp. (in Russian).

Pavlov, P.Yu, 1994. Upper Paleolithic sites and "mammoth cemeteries" in the North-East of Europe. In: European North: Interaction of Cultures in the Antiquity and Middle Ages. Abstracts of International Conference. Komi Research Centre, Syktyvkar, pp. 67–68 (in Russian).

Pavlov, P.Yu, 1996. Paleolithic Sites in the Northeast of European Russia. Komi Research Centre, Uralian Branch of RAS. 212 pp. (in Russian).

Pavlov, P.Yu, 1997. Paleolithic. In: Saveliyeva, E.A., Pavlov, P.Yu., Korolev, K.S., Murygin, A.M. (Eds.), Archaeology of the Komi Republic. DiK Publ., Moscow, pp. 44–90 (in Russian).

Pavlov, P.Yu, 2002. The oldest stages in the Northern Eurasia colonization: Northeast of Europe in the Paleolithic. In: Northern Archaeological Congress, Khanty-Mansiysk, September 9–14, 2002. Papers. Akademkniga Publ., Ekaterinburg – Khanty-Mansiysk, pp. 192–209 (in Russian).

Pavlov, P.Yu, 2004. The Early Upper Paleolithic in the Northeast of Europe (Based of Materials from the Zaozeye Site) Nauchnye Doklady, Iss. 467. Komi Research Centre, Uralian Branch of RAS, Syktyvkar. 36 pp. (in Russian).

Pavlov, P.Yu, 2007. Environments and life support systems of the Upper Paleolithic population in the northeast of Europe. In: Matveyeva, N.P. (Ed.), Ecology of the Ancient and Traditional Communities, Iss. 3. Vector Book Publ., Tyumen, pp. 43–47 (in Russian).

Pavlov, P.Yu, 2008a. Main phases of the Northeastern Europe colonization by Paleolithic humans. In: Velichko, A.A., Vasil'ev, S.A. (Eds.), Way to North: Paleoenvironment and Earliest Inhabitants of Arctic and Subarctic. Institute of Geography RAS, Moscow, pp. 69–78 (in Russian).

Pavlov, P.Yu, 2008b. Paleolithic of northeastern Europe: new data. Archaeology. Ethnography and Anthropology of Eurasia 1, 33–45.

Pavlov, P.Yu, 2009. Zaozerye site – the site of the Early Upper Paleolithic in the northeast of Europe. Rossiyskaya Arkheologiya 1, 5–17 (in Russian).

Pavlov, P., Indrelid, S., 2000. Human occupation in Northeastern Europe during the period 35,000–18,000 BP. In: Roebroeks, W., Mussi, M. (Eds.), Hunters of the Golden Age. The Mid Upper Palaeolithic of Eurasia 30,000–20,000 BP. Leiden University, Leiden, pp. 165–172.

Pavlov, P.Yu., Makarov, E.Y., 1998. Garchi 1 – a site in the Northeast of Europe attributed to the Kostenki-Streletskaya culture. In: The Northern Urals at the Stone and Metal Ages. (Materials on Archaeology of European North-East. Iss. 15). Syktyvkar, pp. 4–17 (in Russian).

Pavlov, P.Yu., Indrelid, C., Svendsen, J.I., Hufthummer, A.K., Smirnov, N.G., Andreicheva, L.N., 1996. Investigations of the Paleolithic sites in the Northeast of Europe. In: Archaeological Discoveries – 1995. Nauka, Moscow, pp. 72–73 (in Russian).

Pavlov, P.Yu., Svendsen, J.-I., Indrelid, S., 2001. Human presence in the European Arctic nearly 40,000 years ago. Nature 413, 64–67.

Pavlov, P.Yu., Roebroeks, W., Svendsen, J.I., 2004. The Pleistocene colonization of Northeastern Europe: a report on recent research. Journal of Human Evolution 47 (1–2), 3–17.

Pavlov, P.Yu., Gribchenko, Yu.N., Kurenkova, E.I., 2005. Problems of Paleolithic stratigraphy and paleogeography of the Northeastern regions of the Russian Plain. In: Quarter – 2005. Materials of the 4th All-Russia Conference on the Quaternary. Geoprint Publ., Syktyvkar, pp. 310–312 (in Russian).

Pavlov, P.Yu., Volokitin, A.V., Karmanov, V.N., 2010. Main stages in the Northeastern Europe colonization by humans in the Stone Age. In: Derevyanko, A.P., Kudelin, A.B., Tishkov, V.A. (Eds.), Adaptation of Peoples and Cultures to Changes of Environments, Social and Technogenic Transformations. Russian Political Encyclopedia Publ., Moscow, pp. 31–37 (in Russian).

Pearson, G.A., 1999. Early occupations and cultural sequence at Moose Creek: a Late Pleistocene site in Central Alaska. Arctic 52 (4), 332–345.

Pesonen, P.E., 1978. Mesolithic sites on the Kandalaksha coast. In: Savvateyev, Y.A. (Ed.), Mesolithic Sites of Karelia. State Publishing House of Karelian Autonomous SSR, pp. 94–160 (in Russian).

Pesonen, P., 2005. Sarvingin salaisuus – Enon Rahakankaan varhaismesoliittinen ajoitus. Muinaistutkija 2, 2–13.

Pestryakov, A.P., Grigoryeva, O.M., 2003. Variability of the Eskimo brain case in the process of their historical dispersal. In: Belyaev, D.D., Ershova, G.G. (Eds.), Ancient Civilizations of the Old and New World: Cultural Distinctiveness and a Dialogue of Interpretations. Ippolitov Publ., Moscow, pp. 159–165 (in Russian).

Pestryakov, A.P., Grigoryeva, O.M., 2011. Anthropological variability in Eskimo (craniology and ABO blood groups) in the process of their geographic dispersal. Vestnik Moskovskogo Universiteta (Moscow Univ. Bulletin.) Ser. III. Anthropology 4, 55–68 (in Russian).

Petrin, V.T., 1986. Late Paleolithic Sites on the West Siberian Plain. Nauka, Novosibirsk. 144 pp. (in Russian).

Pettitt, P., 2000. Chronology of the Mid Upper Paleolithic: the radiocarbon evidence. In: Roebroeks, W., Mussi, M., Svodoba, J., Fennema, K. (Eds.), Hunters of the Golden Age: The Mid Upper Paleolithic of Eurasia 30,000–20,000 BP. Leiden Univ. Press, Leiden, pp. 21–30.

Péwé, T., 1965. INQUA Guidebook for Field Conference F, Central and South-Central Alaska. Division of Geological and Geophysical Surveys, Alaska Department of National Resources, Fairbanks.

Péwé, T., Reger, R.D., 1983. Guide to Permafrost and Quaternary Geology, Richardson and Glenn Highways, Alaska. Division of Geological and Geophysical Surveys, Alaska Department of Natural Resources, Fairbanks.

Phillips, D.L., Gregg, J.W., 2001. Uncertainty in source partitioning using stable isotopes. Oecologia 127, 171–179.

Phippen, P.G., 1988. Archaeology at Owl Ridge: A Pleistocene-Holocene Boundary Age Site in Central Alaska (Unpublished M.A. thesis). Department of Anthropology, Univ. of Alaska, Fairbanks.

Pisaric, M.F.J., MacDonald, G.M., Velichko, A.A., Cwynar, L.C., 2001. The Lateglacial and Postglacial vegetation history of the northwestern limits of Beringia, based on pollen, stomata and tree stump evidence. Quaternary Science Reviews 20 (1–3), 235–245.

Pitul'ko, V.V., 1998. The Zhokhov Site. D. Bulanin Publ., St.-Petersburg. 185 pp. (in Russian).

Pitul'ko, V.V., 2003. The Holocene Stone Age in Northeastern Asia. In: Nikolskiy, P.A., Pitul'ko, V.V. (Eds.), The Natural History of the Russian Eastern Arctic in the Pleistocene and Holocene. GEOS Publ., Moscow, pp. 99–141 (in Russian).

Pitul'ko, V.V., 2006a. Chronology of the Stone Age cultures in the Northeast of Asia. In: The 2nd Northern Archaeological Congress, September 24–30, 2006. Khanty-Mansiysk. Proceedings. Charoid Publ., Ekaterinburg, pp. 306–321 (in Russian).

Pitul'ko, V.V., 2006b. The Paleolithic Yanskaya site. In: Derevyanko, A.P., Molodin, V.I. (Eds.), The Current Problems of Archaeology of Russia, vol. 1. Institute of Archaeology and Ethnography, Siberian Branch RAS Publ., Novosibirsk, pp. 134–136 (in Russian).

Pitul'ko, V.V., 2008a. Secrets of Berelekh. Transaction of Institute for the Material Culture History RAS, vol. 3, pp. 98–117 (in Russian).

Pitul'ko, V.V., 2008b. The main scenarios of excavations under conditions of permafrost (an experience of works on Zhokhov and Yanskaya sites), Northern Yakutia. Archaeology, Ethnography and Anthropology of Eurasia 2 (34), 26–33.

Pitul'ko, V.V., 2010. The Late Neopleistocene dispersal of humans in the Northeast of Asia and adaptation. In: Derevyanko, A.P., Kudelin, A.B., Tishkov, V.A. (Eds.), Adaptation of Peoples and Cultures to Changes of Environments, Social and Technogenic Transformations. Russian Political Encyclopedia, Moscow, pp. 38–46 (in Russian).

Pitul'ko, V.V., 2011. The Berelekh quest: a review of forty years of research in the mammoth graveyard in northeast Siberia. Geoarchaeology 26 (1), 5–32.

Pitul'ko, V.V., Pavlova, E.Yu, 2010. Geoarchaeology and Radiocarbon Chronology of the Stone Age in Northeastern Asia. Nauka, St.-Petersburg. 264 pp. (in Russian).

Pitul'ko, V.V., Pavlova, E.Y., 2007. The age of Paleolithic sites of the Yana-Indigirka Lowland and radiocarbon dating of the Ice Complex deposits. In: Zaitseva, G.I., Kulkova, M.A. (Eds.), Radiocarbon in Archaeological and Paleoenvironmental Studies. Institute for Material Culture History RAS, St.-Petersburg, pp. 155–164 (in Russian).

Pitul'ko, V.V., Kasparov, A.K., Anisimov, M.A., 2003. Oleni Ruchei site, central Taymyr. In: Nikolskiy, P.A., Pitul'ko, V.V. (Eds.), The Natural History of the Russian Eastern Arctic in the Pleistocene and Holocene. GEOS Publ., Moscow, pp. 50–73 (in Russian).

Pitulko, V.V., Nikolsky, P.A., Girya, E.Y., Basilyan, A.E., Tumskoy, V.E., Koulakov, S.A., Astakhov, S.N., Pavlova, E.Y., Anisimov, M.A., 2004. The Yana RHS site: humans in the Arctic before the Last Glacial maximum. Science 303 (5654), 52–56.

Pitul'ko, V.V., Pavlova, E.Yu., Kuzmina, S.A., Nikolskiy, P.A., Basilyan, A.E., Tumskoy, V.E., Anisimov, M.A., 2007. Environmental and climatic changes on the Yana-Indigirka Lowland at the end of Karginsky time and the living conditions of the Upper Paleolithic people in the north of East Siberia. Doklady RAS 417 (1), 103–108 (in Russian).

Pitul'ko, V.V., Nikolskiy, P.A., Basilyan, A.E., 2009. Radiocarbon age of the Berelekh archaeological complex. In: Fundamental Problems of Quaternary: Results of the Studies and the Main Trends in Further Investigations. Materials of VI All-Russia Conference on the Quaternary Studies, Novosibirsk, November 19–23, 2009. Siberian Branch RAS Publ., Novosibirsk, pp. 474–478 (in Russian).

Plumet, P., 1980. Essai d'analyse descriptive: Les témoins façonnés pré-dorsétiens de Poste-de-la-Baleine, Québec. Montréal, Université du Québec à Montréal, Laboratoire d'archéologie. Paléo-Québec. 257 pp.

Plumet, P., 1985. Archéologie de l'Ungava: Le site de la Pointe aux Bélougas (Qilalugarsiuvik) et les maisons longues dorsétiennes. Montréal, Université du Québec à Montréal, Laboratoire d'archéologie, Paléo-Québec. 472 pp.

Plumet, P., 1986. Questions et réflexions concernant la préhistoire de l'Ungava. In: Palaeo-Eskimo Cultures in Newfoundland, Labrador and Ungava. Reports in Archaeology, No. 1. Memorial Univ. of Newfoundland, pp. 151–160.

Plumet, P., 1994. Le Paléoesquimau dans la baie du Diana (Arctique québécois). In: Morrison, D., Pilon, J.-L. (Eds.), Threads of Arctic Prehistory: Papers in Honour of William E. Taylor, Jr. Hull, Canadian Museum of Civilization, vol. 149. Archaeological Survey of Canada, Mercury Series, pp. 145–164.

Pogodin, A.A., 1998. On the time of the peopling of the Northeast of West Siberia. In: Alekseyev, V.V. (Ed.), The Urals in the Past and at Present. Proc. of Scientific Conference. P.I. Institute of History and Archaeology Publ., Uralian Branch RAS, Ekaterinburg, pp. 99–103 (in Russian).

Pogodin, A.A., Besprozvanny, E.M., 1999. The new studies of the Stone Age in the taiga zone of the West Siberian North. In: Kovaleva, V.T. (Ed.), 120 Years of the Archaeology of the Eastern Slope of the Urals. Materials of Scientific Conference. Part 2. Recent Discoveries of the Uralian Archaeologists. Uralian State Univ. Press, Ekaterinburg, pp. 8–12 (in Russian).

Pollack, H.N., Huang, S., 2000. Climate reconstructions from subsurface temperatures. Annual Review of Earth and Planetary Sciences 28, 339–365.

Poltaraus, A.B., Kulikov, E.E., Lebedeva, I.A., 2000. Molecular analysis of DNA from remains of three skeletons from the Sungir site (preliminary results). In: Alekseyeva, T.I., Bader, N.O. (Eds.), Homo Sungirensis. The Upper Paleolithic Man: Ecological and Evolutionary Aspects. Nauchny Mir, Moscow, pp. 351–358 (in Russian).

Ponkratova, I.Yu, 2007. The Ushki-5 site (Kamchatka): recent investigations, problems and prospects. In: Shirokov, A.I. (Ed.), Humanitarian Almanac of Kolyma, Iss. 2. Kordis Publ., Magadan, pp. 13–20 (in Russian).

Ponomarenko, A.K., 2000. The Ancient Culture of the Itelmens on Kamchatka. Operativnaya Poligrafiya Publ., Petropavlovsk-Kamchatskiy. 312 pp. (in Russian).

Ponomarev, D.V., 2001. Large Mammals of the European Northeast in the Late Pleistocene and Holocene. Komi Research Centre Scientific Reports 434 46 pp. (in Russian).

Potter, B.A., Reuther, J.D., Bowers, P.M., Gelvin-Reymiller, C., 2008. Little Delta Dune site: a Late-Pleistocene multicomponent site in Central Alaska. Current Research in the Pleistocene 25, 132–135.

Potter, B.A., Irish, J.D., Reuther, J.D., Gelvin-Reymiller, C., Holliday, V.T., 2011. A terminal Pleistocene child cremation and residential structure from eastern Beringia. Science 331, 1058–1062.

Poutiainen, H., 2002. Muinaista asutusta jaljittamassa. Sukupolvien maisema, Jyvaskyla 68–69.

Powers, W.R., Hoffecker, J.F., 1989. Late Pleistocene settlement in the Nenana valley, Central Alaska. American Antiquity 54 (2), 263–287.

Powers, W.R., Jordan, R.H., 1990. Human biogeography and climate change in Siberia and Arctic North America in the fourth and fifth millennia BP. Philosophical Transactions of Royal Society of London. Series A 330 (1615), 665–670.

Powers, W.R., Guthrie, R.D., Hoffecker, J.F., 1983. Dry Creek: The Archaeology and Paleoecology of a Late Pleistocene Alaskan Hunting Camp. Report submitted to U.S. National Park Service.

Powers, W.R., Goebel, F.E., Bigelow, N.H., 1990. Late Pleistocene occupation at Walker Road: new data on the Central Alaskan Nenana complex. Current Research in the Pleistocene 7, 40–43.

Praslov, N.D., 1969. Pre-Mousterian and Mousterian sites in the south of the Russian Plain. In: Gerasimov, I.P. (Ed.), Nature and the Early Human Society Evolution on the Territory of the European USSR. Nauka, Moscow, pp. 119–127 (in Russian).

Praslov, N.D., 1982. Man's dispersal over Europe in the Late Pleistocene. In: Gerasimov, I.P., Velichko, A.A. (Eds.), Paleogeography of Europe during the Last 100,000 Years. Nauka, Moscow, pp. 113–115 (in Russian).

Praslov, N.D., Rogachev, A.N. (Eds.), 1982. Paleolithic of the Kostenki-Borshchevo Region on the Don River. Some Results of Field Studies. Nauka, Leningrad, pp. 1879–1979. 283 pp. (in Russian).

Praslov, N.D., Nedomolkina, N.G., Zheltova, M.N., 2011. The oldest peopling of the Vologda region. In: Tverskoy Arkheologicheskiy Sbornik, Iss. 8. Triada Publ., Tver', pp. 48–56 (in Russian).

Price, T.D., Jacobs, K., 1990. Oleni' Ostrov: first radiocarbon dates from a major Mesolithic cemetery in Karelia, USSR. Antiquity 64, 849–853.

Ptashinskiy, A.V., 2010. Discovery of a new site dated to the end of Late Paleolithic in Kamchatka. In: Materials of Dikov's Conference VI, March 23–25, 2010. North East Interdisciplinary Research Institute, Far Eastern Branch RAS, Magadan, pp. 62–65 (in Russian).

Rae, T.C., Koppe, T., Stringer, C.B., 2011. Neanderthal face is not cold adapted. Journal of Human Evolution 60 (3), 234–239.

Rainey, F., 1958. Archaeology of American Arctic. Sovetskaya Etnografiya 2, 55–62 (in Russian).

Ramsay, W., 1898. Über die geologische Entwicklung der Halbinsel Kola in der Quartärzeit. Fennia 16 (1).

Ramsay, W., 1902. Om ett sannolikt fund of kambrisk lera i Viborgs län. Fennia 19 (3).

Ramsden, P., Murray, M., 1995. Identifying seasonality in Pre-Dorset structures in Back Bay, Prince of Wales Island, NWT. Arctic Anthropology 32 (2), 106–117.

Ramstad, M., Hesjedal, A., Niemi, A., 2005. The Melkøya project: maritime hunter-fisher island settlements and the use of space through 11,000 years on Melkøya, Arctic Norway. Antiquity. 70 (304). http://antiquity.ac.uk/ProjGall/304.html.

Rankama, T., Kankaanpää, J., 2007. The earliest postglacial inland settlement of Lapland. In: Volokitin, A.V. (Ed.), Stone Age of the European North. Institute of Linguistic, Literature, and History. Komi Research Centre RAS, Syktyvkar, pp. 44–60.

Rankama, T., Kankaanpää, J., 2008. Eastern arrivals in post-glacial Lapland: the Sujala site 10,000 cal BP. Antiquity 82, 884–899.

Ranov, V.A., Tseitlin, S.M., 1991. Diring Paleolithic site seen by geologist and archaeologist. Bulletin of Commission for Study of the Quaternary 60, 79–87 (in Russian).

Rasic, J., 2003. Ancient hunters of the western Brooks Range: integrating research and cultural resource management. Alaska Park Science 2 (2), 20–25.

Rasic, J.T., 2011. Functional variability in the Late Pleistocene archaeological record of eastern Beringia: a model of late Pleistocene land use and technology from Northwest Alaska. In: Goebel, T., Buvit, I. (Eds.), From the Yenisei to the Yukon: Interpreting Lithic Assemblage Variability in Late Pleistocene/Early Holocene Beringia. (Peopling of the Americas Publications Series). Texas A&M Univ. Press, College Station, pp. 128–164.

Rasic, J.T., Gal, R., 2000. An early lithic assemblage from the Tuluaq Site, Northwest Alaska. Current Research in the Pleistocene 17, 66–68.

Rasmussen, K., 1908. The People of the Polar North. Kegan, Trench, Trьbner, London. 358 pp.

Rasmussen, M., Li, Y., Lindgreen, S., Pedersen, J.S., Albrechtsen, A., Moltke, I., Metspalu, M., Metspalu, E., Kivisild, T., Gupta, R., Bertalan, M., Nielsen, K., Gilbert, M.T.P., Wang, Y., Raghavan, M., Campos, P.F., Kamp, H.M., Wilson, A.S., Gledhill, A., Tridico, S., Bunce, M., Lorenzen, E.D., Binladen, J., Guo, X., Zhao, J., Zhang, X., Zhang, H., Li, Z., Chen, M., Orlando, L., Kristiansen, K., Bak, M., Tommerup, N., Bendixen, C., Pierre, T.L., Grønnow,

B., Meldgaard, M., Andreasen, C., Fedorova, S.A., Osipova, L.P., Higham, T.F.G., Bronk Ramsey, C., Hansen, T.O., Nielsen, F.C., Crawford, M.H., Brunak, S., Sicheritz-Pontňn, T., Villems, R., Nielsen, R., Krogh, A., Wang, J., Willerslev, E., 2010. Ancient human genome sequence of an extinct Palaeo-Eskimo. Nature 463 (7282), 757–762.

Reanier, R.E., 1995. The antiquity of Paleoindian materials in northern Alaska. Arctic Anthropology 32 (1), 31–50.

Reimer, P.J., Bard, E., Bayliss, A., Beck, J.W., Blackwell, P.G., Ramsey, C.B., Buck, C.E., Cheng, H., Edwards, R.L., Friedrich, M., Grootes, P.M., Guilderson, T.P., Haflidason, H., Hajdas, I., Hatte, C., Heaton, T.J., Hoffmann, D.L., Hogg, A.G., Hughen, K.A., Kaiser, K.F., Kromer, B., Manning, S., Niu, M., Reimer, R.W., Richards, D.A., Scott, E.M., Southon, J.R., Staff, R.A., Turney, C.S.M., van der Plich, T.J., 2013. IntCal13 and marine13 radiocarbon age calibration curves, 0–50000 years cal BP. Radiocarbon 55 (4), 1869–1887.

Ren, J., Jiang, H., Seidenkrantz, M.S., Kuijpers, A., 2009. A diatom-based reconstruction of Early Holocene hydrographic and climatic change in a southwest Greenland fjord. Marine Micropaleontology 70, 166–176.

Renault-Miskovsky, J., 1985. L'environnement au temps de la préhistoire: Méthodes et modèles. Masson, Paris. 183 pp.

Richards, M.P., Trinkaus, E., 2009. Isotopic evidence for the diets of European Neanderthals and early modern humans. Proceedings of the National Academy of Sciences of the USA (PNAS) 106, 16034–16039.

Richards, M.P., Pettitt, P.B., Trinkaus, E., Smith, F.H., Paunovic, M., Karavanic, I., 2000. Neanderthal diet at Vindija and Neanderthal predation: the evidence from stable isotopes. Proceedings of the National Academy of Sciences of the USA (PNAS) 97 (13), 7663–7666.

Richards, M.P., Pettitt, P.B., Stiner, M.C., Trinkaus, E., 2001. Stable isotope evidence for increasing dietary breadth in the European Mid-Upper Paleolithic. Proceedings of the National Academy of Sciences of the United States of America 98 (11), 6528–6532.

Richards, M.P., Taylor, G., Steele, T., McPherron, S.P., Soressi, M., Jaubert, J., Orschiedt, J., Mallye, J.B., Rendu, W., Hublin, J.J., 2008. Isotopic dietary analysis of a Neanderthal and associated fauna from the site of Jonzac (Charente-Maritime), France. Journal of Human Evolution 55, 179–185.

Rikhter, G.D., 1936. Physiographic Regions of the Kola Peninsula. Transaction of Institute of Physical Geography, Iss. 19. USSR Academy of Science Publ., Moscow–Leningrad, (in Russian).

Rogachev, A.N., 1955. Paleolithic burial at Kostenki XIV site (Markina Gora). Sovetskaya Etnografiya 1, 29–38 (in Russian).

Rogachev, A.N., 1957. Multilayered Sites of the the Kostenki-Borshchevo Region on the Don River and Problems of the Culture Evolution at the Late Paleolithic Epoch on the Russian Plain. Materials and Studies of Archaeology in the USSR, No. 59. The USSR Academy of Science Publ., Moscow, pp. 9–134 (in Russian).

Rogachev, A.N., Sinitsyn, A.A., 1982. Kostenki XIV site (Markina Gora). In: Praslov, N.D., Rogachev, A.N. (Eds.), Paleolithic of the Kostenki-Borshchevo Region on the Don River. Some Results of Field Studies, 1879–1979. Nauka, Leningrad, pp. 145–162 (in Russian).

Rootsi, S., Zhivotovsky, L.A., Baldovic, M., Kayser, M., Kutuev, I.A., Khusainova, R., Bermisheva, M.A., Gubina, M., Fedorova, S.A., Ilumäe, A.M., Khusnutdinova, E.K., Voevoda, M.I., Osipova, L.P., Stoneking, M., Lin, A.A., Ferak, V., Parik, J., Kivisild, T., Underhill, P.A., Villems, R., 2007. A counterclockwise northern route of the Y-chromosome haplogroup N from Southeast Asia towards Europe. European Journal of Human Genetics 15 (2), 204–211.

Rozanov, L.L., 1977. Methods of Structural-Geomorphological Studies of River Valleys (with Special Reference to Northwest of the Russian Plain). Mysl' Publ., Moscow. 136 pp. (in Russian).

Arctic, Russian, 2001. Reference Book for Government Officials. Drofa Publ., Moscow. 344 pp. (in Russian).

Ryabkov, N.V., 1976. Old ice-dammed basins at the Kama, Pechora, and Vychegda interfluves and their relicts. Bulletin of Commission for Study of the Quaternary 45, 94–105 (in Russian).

Sandell, H., Sandell, B., 1996. Paleo-Eskimo sites and finds in the Scoresby Sund area. In: Grønnow, B. (Ed.), The Paleo-Eskimo Cultures of Greenland – New Perspectives in Greenlandic Archaeology. Danish Polar Center Publications, No. 1, Copenhagen, pp. 161–176.

Sandmo, A., 1986. Restoff og redskap – mer enn teknologisk hjelpemiddel. Om Symbolfunksjonen som et aspekt ved materiell kultur. Skisse av etableringsforløpet i en nordvesteuropeisk kystsone 10,000–9000 BP (Magister thesis). Department of Archaeology, Univ. of Tromsø.

Sandmo, A., 1989. Stilistisk variasjon og tidlig postglacial bosetning I Troms, vol. 22. Tromsø Museums Skrifter.

Sapelko, T.V., Ludnikova, A.V., Shumkin, V.Ya, 2009. Changes in environments in the southern part of the Kola Peninsula at the late glacial and post-glacial time. In: Fundamental Problems of Quaternary: Results of the Studies and the Main Trends in Further Investigations. Materials of VI All-Russia Conference on the Quaternary Studies, Novosibirsk, November 19–23, 2009. Siberian Branch RAS Publ., Novosibirsk, pp. 534–537 (in Russian).

Sattler, R.A., 1991. Paleoecology of a Late Quaternary Cave Deposit in Northeast Alaska (Unpublished M.A. thesis). Department of Anthropology, Univ. of Alaska, Fairbanks.

Savelle, J.M., 1999. Preliminary Report of Archaeological Investigations on Wollaston Peninsula, Western Victoria Island, Nunavut. July–August 1999. Dept. of Culture, Language, Elders, and Youth, Igloolik.

Savelle, J.M., 2004. Preliminary Report of Archaeological Investigations on Boothia Peninsula, Nunavut. July–August 2004. Dept. of Culture, Language, Elders, and Youth, Igloolik.

Savelle, J.M., 2006. Preliminary Report of Archaeological Investigations on Kent Peninsula and King William Islands, Nuanvut. Dept. of Culture, Language, Elders, and Youth, Igloolik.

Savelle, J.M., 2007. Preliminary Report of Archaeological Investigations on Somerset Island and Boothia Peninsula, Nunavut. Dept. of Culture, Language, Elders, and Youth, Igloolik.

Savelle, J.M., Dyke, A.S., 2002. Variability in Palaeoeskimo occupation on south-western Victoria Island, Arctic archaeology. World Archaeology 33, 508–522.

Savvinova, V.V., Fedoseeva, S.A., Mochanov, Y.A., 1996. Pollen records from archaeological sites in the Aldansky Region, Sakha Republic. In: West, F.H. (Ed.), American Beginnings: The Prehistory and Palaeoecology of Beringia. Univ. of Chicago Press, Chicago, pp. 97–107.

Schanche, A., Olsen, B., 1985. Var de alle nordmenn? En etnopolitisk kritikk av norsk arkeologi. In: Nжss, J.-R. (Ed.), Arkeologi og etnisitet. Arkeologisk museum i Stavanger, Stavanger, pp. 87–99. AmS-Varia 15.

Schanche, K., 1988. Mortensnes – en boplass i Varanger. En studie av samfunn og materiell kultur gjennom 10.000 er (M.A. thesis). Univ. of Tromsø, Tromsø.

Schirrmeister, L., Siegert, C., Kuznetsova, T., Kuzmina, S., Andreev, A., Kienast, F., Meyer, H., Bobrov, A., 2002. Paleoenvironmental and paleoclimatic records from permafrost deposits in the Arctic region of Northern Siberia. Quaternary International 89, 97–118.

Schledermann, P., 1990. Crossroads to Greenland: 3000 Years of Prehistory in the Eastern High Arctic. Komatic Series, 2. Calgary: The Arctic Institute of North America, Univ. of Calgary, Calgary. 221 pp.

Schledermann, P., McCullough, K., 1988. Hearth of darkness: structural variability in ASTt dwellings in the Canadian High Arctic. In: Paper Presented at the 21st Chacmool Conference, Calgary.
Schreiner, K., 1931. Zur Osteologie der Lappen. Oslo.
Schreiner, K., 1945. Further Note on the Craniology of the Lapps. Oslo.
Schultz, H.P., 1990. On the Mesolithic quartz industry in Finland. Iskos (Helsinki) 9, 7–23.
Schultz, H.P., 1996. Pioneerit pohjoisessa: Suomen varhaismesoliittinen asutus arkeologisen aineston valossa. Suomen Museo 103, 5–45.
Schweger, C.E., 1981. Chronology of late glacial events from the Tangle Lakes, Alaska Range. Arctic Anthropology 18, 97–101.
Semin, M.I., Shelkovaya, S.O., 1991. Paleolithic localities in the upper reaches of the Nizhnyaya Tunguska. In: Drozdov, N.B., Akimova, E.V., Novykh, L.B. (Eds.), Problems of Archaeology and Ethnography of Siberia and Far East, Iss. 1. Krasnoyarsk State Pedagogical Univ. (KGPI), Krasnoyarsk, pp. 148–150 (in Russian).
Semino, O., Magri, Ch., Benuzzi, G., Lin, A.A., Al-Zahery, N., Battaglia, V., MacCioni, L., Triantaphyllidis, C., Shen, P., Oefner, P.J., Zhivotovsky, L.A., King, R., Torroni, A., Cavalli-Sforza, L.L., Underhill, P.A., Santachiara-Benerecetti, A.S., 2004. Origin, diffusion, and differentiation of Y-chromosome haplogroups E and J: inferences on the Neolithization of Europe and later migratory events in the Mediterranean area. American Journal of Human Genetics 74 (5), 1023–1034.
Seppä, H., 1996. Post-glacial dynamics of vegetation and tree-lines in the far North of Fennoscandia. Fennia 174 (1), 1–96.
Seppä, H., Birks, H.J.B., 2001. July mean temperature and annual precipitation trends during the Holocene in the Fennoscandian tree-line area: pollen-based climate reconstructions. The Holocene 11, 527–539.
Seppä, H., Birks, H.J.B., 2002. Holocene climate reconstructions from the Fennoscandian treeline area based on pollen data from Toskaljavri. Quaternary Research 57, 191–199.
Serikov, Y.B., 1998. Mesolithic sites on the Konda River. In: Grebenyukov, V.I. (Ed.), Archaeology of West Siberia. Nizhnevartovsk Pedagogical Institute Publ., Nizhnevartovsk, pp. 3–23 (in Russian).
Serikov, Y.B., 2000. Paleolithic and Mesolithic of the Middle Transuralian Region. Poligrafist Publ., Nizhniy Tagil. 430 pp. (in Russian).
Serikov, Y.B., 2007. The Gari Paleolithic Site and Certain Problems of the Paleolithic Studies in the Urals. Nizhnietagilskaya State Social Teaching Academy, Nizhniy Tagil. 138 pp. (in Russian).
Shakhnovich, M.M., 2007a. Mesolithic of Northern and Western Karelia (Ph.D. thesis). St.-Petersburg, 31 pp. (in Russian).
Shakhnovich, M.M., 2007b. Kirkkolahti 1 Mesolithic settlement in the northern Ladoga region. In: Zhilin, M.G. (Ed.), Originality and Specific Adaptivity of the Forest Zone Cultures of Northern Eurasia in the Final Pleistocene – Early Holocene. Institute of Archaeology RAS, Moscow, pp. 163–181 (in Russian).
Shchelinskiy, V.E., 2011. A new Early Paleolithic site on the Taman' Peninsula (Southern Azov region). In: Gavrilov, K.N. (Ed.), Paleolithic and Mesolithic of Eastern Europe. Taus Publ., Institute of Archaeology RAS, Moscow, pp. 37–58 (in Russian).
Shcherbakova, T.I., 1994. Materials of the Upper Paleolithic Talitskogo Site (Ostrovsakaya). Uralian Branch RAS Publ., Ekaterinburg. 95 pp. (in Russian).
Sher, A.V., 1971. Mammals and Pleistocene Stratigraphy of the Extreme Northeast of the USSR and North America. Nauka, Moscow. 310 pp. (in Russian).
Sher, A.V., Kuzmina, S.A., Kuznetsova, T.V., Sulerzhitsky, L.D., 2005. New insights into the Weichselian environment and climate of the East Siberian Arctic, derived from fossil insects, plants, and mammals. Quaternary Science Reviews 24 (5–6), 533–569.

Shilo, N.A., Lozhkin, A.V., Anderson, P.M., Vazhenina, L.N., Glushkova, O.Yu., Matrosova, T.V., 2008. The first data on the larch *Larix gmelinii* (Rupr.) expansion into the Arctic regions of Beringia in the Early Holocene. Doklady RAS 422 (5), 1–3 (in Russian).

Shluger, S.A., 1941. Anthropological studies of Nenets. In: Short Communications on the Works of the Anthropological Institute and Museum of Moscow State University, in 1938–1940. Moscow Univ. Press, Moscow (in Russian).

Shofman, I.L., Kind, N.V., Pakhomov, V.V., Prokopchuk, B.I., Vinogradova, S.N., Sulerzhitskiy, L.D., Forova, V.S., 1977. New data on the age of the low terrace deposits in the Viluy drainage basin. Bulletin of Commission for Study of the Quaternary 47, 100–107 (in Russian).

Short, S.K., Mode, W.N., Davis, P.T., 1985. The Holocene record from Baffin Island: modern and fossil pollen studies. In: Andrews, J.T. (Ed.), Holocene Pollen Records from West Greenland. Allen & Unwin, Boston, pp. 608–642.

Short, S.K., Andrews, J.T., Williams, K.M., Weiner, N.J., Elias, S.A., 1994. Late Quaternary marine and terrestrial environments of Northwestern Baffin Island, Northwest territories. Géographie physique et Quaternaire 48, 85–95.

Shumkin, V.Y., 1984. Mesolithic of the Kola Peninsula. Sovetyskaya Arkheologiya 2, 15–33 (in Russian).

Shumkin, V., 1990. On the ethnogenesis of the Sams. Acta Borealia 7 (2), 320.

Shumkin, V.Ya, 1991. Ethnogenesis of Sami (archaeological aspect). In: Aksyanova, G.A. (Ed.), Origin of Sami. Nauka, Moscow, pp. 129–149 (in Russian).

Shumkin, V.Ya, 1993. Early Stone Age of the western European Arctic (Mesolithic of the Northern Scandinavia). In: Masson, V.M., Nosov, E.N., Ryabinin, E.A. (Eds.), Antiquities of the Northeast of Russia. Peterburgskoye Vostokovedeniye Publ., St.-Petersburg, pp. 34–59 (in Russian).

Shumkin, V.Ya, 1996. Neolithic of the Russian North. In: Antiquities of the Russian North. Iss. 1. (Vologda), pp. 67–74 (in Russian).

Shumkin, V.Ya, 2007. Stone Age of the eastern Lapland and initial colonization of Arctic. In: Volokitin, A.V. (Ed.), Stone Age of the European North. Institute of Linguistic, Literature, and History. Komi Research Centre, Syktyvkar, pp. 36–43 (in Russian).

Sigurðsson, S., 1977. Birki á Íslandi, útbreiðsla og ástand.Skógarmál bls. In: Skógarmál. Edda hf, Reykjavík, pp. 146–172.

Sigurðsson, O., June 24, 2010. Hver er munur á stærð íslenskra jökla í dag og á landnámsöld? Vísindavefurinn. http://visindavefur.is/?id=54186.

Siiriäinen, A., 1970. Archaeological background of ancient Lake Päijänne and geological dating of the Meso-Neolithic boundary in Finland. Bulletin of Geological Society of Finland 42, 119–127.

Siiriäinen, A., 1971. Shoreline dating of the Saraisniemi-1 ceramics in Finland. Suomen Museo 78, 9–19.

Siiriäinen, A., 1972. A gradient time curve for dating Stone Age shorelines in Finland. Suomen Museo 79, 5–18.

Siiriäinen, A., 1973. Studies relating to shore displacement and Stone Age chronology in Finland. Finskt Museum 80, 5–22.

Siiriäinen, A., 1974. Quartz, chert and obsidian – a comparison of raw materials in a Late Stone Age. Finskt Museum 81, 15–29.

Siiriäinen, A., 1978. Archaeological shore displacement chronology in Northern Ostrobotnia. Finland. Iskos 2, 5–23.

Siiriäinen, A., 1981. Problems of the Eastern Fennoscandian Mesolithic. Finskt Museum 84 (12), 5–31.

Siiriäinen, A., 1982. Shore displacement and archaeology in Finland. In: Aartolahti, T., Eronen, M. (Eds.), Studies in Baltic Shorelines and Sediments Indicating Sea Level ChangesAnnales Academiae Scientiarum Fennicae, Ser. A III. Geol. Geograph, vol. 134, pp. 173–184.
Simchenko, Yu.B., 1976. Culture of Reindeer Hunters in Northern Eurasia. Nauka, Moscow. 311 pp. (in Russian).
Simchenko, Yu.B., 1980. The early stages of ethnogenesis of the peoples attributed to the Uralian language family. In: Gurvich, I.S. (Ed.), Ethnogenesis of Northern Peoples. Nauka, Moscow, pp. 11–18 (in Russian).
Simonsen, P., 1961. Varangerfunnene II: fund og udgravninger på fjordens sydkyst. Tromsø Museums Skrifter 7 (2) Tromsø Museum, Tromsø. 502 pp.
Simonsen, P., 1975. When and why did occupational specialization begin at the Scandinavian north coast? In: Fitzhugh, W. (Ed.), Prehistoric Maritime Adaptations of the Circumpolar Zone. Mouton Publishers, The Hague, pp. 75–86.
Simonsen, P., 1986. Fortsatte undersøkelser ved Virdnejavri. In: Engelstad, E., Holm-Olsen, I.M. (Eds.), Arkeologisk feltarbeid i nord-Norge 1985. Tromura kulturhistorie, vol. 6. Tromsø Museum, pp. 1–11.
Simonsen, P., 1988. Trekk av Mesolithikum-forskningens historie. Tromsø.
Simonsen, P., 2001. Alta-kraftverkene. Kulturhistoriske Registreringer Og Utgravninger 1984-1987. In: Del A: Vir'dnejav'ri nord, Tromura, kulturhistorie. No. 34. Tromsø Museum.
Simpson, M.J.R., Milne, G.A., Huybrechts, P., Long, A.J., 2009. Calibrating a glaciological model of the Greenland ice sheet from the Last Glacial Maximum to present-day using field observations of relative sea level and ice extent. Quaternary Science Reviews 28, 1631–1657.
Sinitsyn, A.A., 2004. Earliest Upper Paleolithic Layers at Kostenki 14 (Markina Gora). Preliminary Results of the 1998–2001 Excavations: Acts of the XIV UISPP Congress, Liege, 2001. Section 6. The Upper Paleolithic. Oxford. BAR Intern, Ser. No. 1240, pp. 181–190.
Sinitsyn, A.A., Praslov, N.D., Svezhentsev, Yu.S., Sulerzhitskiy, L.D., 1998. Radiocarbon chronology of the Upper Paleolithic in Eastern Europe. In: Sinitsyn, A.A., Praslov, N.D. (Eds.), Radiocarbon Chronology of the Paleolithic of Eastern Europe and Northern Asia. Problems and Prospects. Institute for the Material Culture History RAS, St.-Petersburg, pp. 21–66 (in Russian).
Skandfer, M., Henriksen, S., Stensrud, G., Valen, C., 2009. Tønsnes havn og næringsområde. Rappport fra arkeologiske undersøkelser. Tromura, kulturvitenskap.
Slobodin, S., 1999a. Northeast Asia in the Late Pleistocene and Early Holocene. World Archaeology 30 (2), 251–266.
Slobodin, S.B., 1999b. Archaeology of the Kolyma and Okhotsk Continental Regions in the Late Pleistocene – Early Holocene. Northeastern Interdisciplinary Research Institute, Far Eastern Branch, RAS, Magadan. 234 pp. (in Russian).
Slobodin, S.B., 2000. Archaeological prospects of the early complexes in the Northeast of Asia. Archaeology, Ethnography and Anthropology of Eurasia 4 (4), 49–60.
Slobodin, S., 2001a. Western Beringia at the end of the Ice Age. Arctic Anthropology 38 (2), 31–47.
Slobodin, S.B., 2001b. The Upper Kolyma and Okhotsk Continental Regions in the Neolithic and Early Metal Ages. Northeastern Interdisciplinary Research Institute, Far Eastern Branch, RAS, Magadan. 202 pp. (in Russian).
Slobodin, S.B., 2002. Stemmed points of the Ushki type in Beringia. In: Materials of Dikov's Conference II. Northeastern Interdisciplinary Research Institute, Far Eastern Branch, RAS, Magadan, pp. 308–314 (in Russian).

Slobodin, S.B., 2005. Stages in the Okhotsk–Kolyma mountain land colonization by humans at the Pleistocene-Holocene boundary. In: Kharinskiy, A.V. (Ed.), Sociogenesis in Northern Asia. Irkutsk Technical Univ. Press, Irkutsk, pp. 158–163 (in Russian).

Slobodin, S.B., 2007. Obsidian raw material occurrence in the archaeological objects of the Kolyma, Kamchatka, and Chukotka (The Northeast of Russia). In: Matveyeva, N.P. (Ed.), Ecology of Ancient and Traditional Societies, Iss. 3. Vector Book Publ., Tyumen, pp. 136–140 (in Russian).

Slobodin, S.B., 2010a. The first Ushki type stemmed point from Upper Kolyma (Western Beringia). Current Research in the Pleistocene 27, 36–38.

Slobodin, S.B., 2010b. New data on the age of the Paleolithic layer VII of the Ushki site (based on datings of the coal samples from Dikov's excavations). In: Materials of Dikov's Conference VI. Northeastern Interdisciplinary Research Institute, Far East Branch RAS, Magadan, pp. 66–77 (in Russian).

Slobodin, S.B., 2011. Late Pleistocene and Early Holocene cultures of Beringia: the general and the specific. In: From the Yenisei to the Yukon: Interpreting Lithic Assemblage Variability in Late Pleistocene/Early Holocene Beringia. Center for the Study of First Americans, College Station. Texas A&M University, Texas, pp. 91–116.

Slobodin, S.B., Glushkova, O.Yu., Smirnov, V.N., 2008. Archaeology and paleogeography of the Paleolithic sites in the Upper Kolyma valley. In: Velichko, A.A., Vasil'ev, S.A. (Eds.), Way to North: Paleoenvironment and Earliest Inhabitants of Arctic and Subarctic. Institute of Geography RAS, Moscow, pp. 222–237 (in Russian).

Smirnov, N.G., Andreicheva, L.N., Korona, O.M., Zinovyev, E.V., Golovachev, I.B., Pavlov, P. Yu., Hufthammer, A.K., 1999. On the biota characteristics in the Uralian Subarctic at the Holocene optimum. In: Smirnov, N.G. (Ed.), Biota of the Uralian Subarctic in the Late Pleistocene and Holocene. "Ekaterinburg" Publ., Ekaterinburg, pp. 23–60 (in Russian).

Smirnov, V.I., 1937. Finds of large Quaternary mammal bones in the Northern region. Transaction of Commission for Study of Quaternary 5 (1), 47–68 (in Russian).

Smith, H.L., Rasic, J.T., Goebel, T., 2013. Biface traditions of Northern Alaska and their role in the peopling of Americas. In: Graf, K.E., Ketron, C.V., Waters, M.R. (Eds.), Paleoamerican Odyssey. Center for the Study of the First Americans, College Station, pp. 105–123.

Smith, R.A., 1912. On the date of Grime's Graves and Cissbury flint mines. Archaeologica 63.

Snyder, J.A., Forman, S.L., Mode, W.N., Tarasov, G.A., 1997. Postglacial relative sea-level history: sediment and diatom records of emerged coastal lakes, north-central Kola Peninsula, Russia. Boreas 26, 329–346.

Snyder, J.A., MacDonald, G.M., Forman, S.L., Tarasov, G.A., Mode, W.N., 2000. Postglacial climate and vegetation history, north-central Kola Peninsula, Russia: pollen and diatom records from Lake Yarnyshnoe-3. Boreas 29, 261–271.

Soffer, O., 1985. The Upper Paleolithic of the Central Russian Plain. Academic Press, Orlando, p. 539.

Solovieva, N., Jones, V.J., 2002. A multiproxy record of Holocene environmental changes in the central Kola Peninsula, northwest Russia. Journal of Quaternary Science 17 (4), 303–331.

Solovieva, N., Tarasov, P., MacDonald, G., 2005. Quantitative reconstruction of Holocene climate from Chuna Lake pollen record, Kola Peninsula, northwest Russia. The Holocene 15, 141–148.

Sørensen, M., 2011. Technology and tradition in the Eastern Arctic, 2500 BC – AD 1200: A dynamic technological investigation of lithic inventories in the Palaeo-Eskimo Traditions. Meddelelser om Grønland 250 Man and Society 40.

Sorokin, A.N., Oshibkina, S.V., Trusov, A.V., 2009. At the Turn of Epochs. Grif and C. Publ., Moscow. 388 pp. (in Russian).

Spiridonova, E.A., 1983. Palynological characteristics of the megainterstadial: its significance for reconstructing flora and vegetation evolution on the Russian Plain. Bulletin of Commission for Study of the Quaternary 32, 29–42 (in Russian).

Spiridonova, E.A., 1989. The main stages in the vegetation evolution in the Late Pleistocene in the unglaciated zone of Eastern Europe. In: Cultural Adaptation at the Upper Paleolithic Time (Soviet-American Field Seminar). Abstracts. Leningrad, pp. 40–43 (in Russian).

Spiridonova, E.A., 1991. Evolution of Vegetation in the Don Drainage Basin in the Upper Pleistocene – Holocene. Nauka, Moscow. 221 pp. (in Russian).

Spiridonova, E.A., 1997. Conclusions from the pollen analysis of the Veretye 1 site, Arkhangelsk Region. In: Oshibkina, S.V. (Ed.), Veretye 1 – Mesolithic Settlement in the North of Eastern Europe. Nauka, Moscow, pp. 189–190 (in Russian).

Spiridonova, E.A., Aleshinskaya, A.S., 1999. An attempt of applying pollen analysis to the periodization of Mesolithic on the Oka-Volga Interfluve. Transaction State Historical Museum 103, 127–141 (in Russian).

Stanford, C.B., Bunn, H.T., 2001. Meat-Eating and Human Evolution. Oxford Univ. Press, Oxford. 384 pp.

Stanford, D., Bonnichsen, R., Meggers, B., Steele, D.G., 2005. Paleoamerican origins, models, evidence, and future directions. In: Bonnichsen, R., Lepper, B.T., Stanford, D., Waters, M.R. (Eds.), Paleoamerican Origins: Beyond Clovis. Center for the Study of the First Americans. Texas A&M Univ.Press, College Station, pp. 313–353.

Starkov, V.F., 2001. An Outline of the Arctic Colonization History. V. II. Russia and the Northeastern Passage. Nauchnyi Mir, Moscow. 116 pp. (in Russian).

Stashenkov, D.A., 2003. Complex of sites near the Sidelkino village. In: Zones of Contact in Eurasia at the Boundary of Centuries. Center "Integratsiya", Samara, pp. 8–14 (in Russian).

Steblin-Kamenskiy, M.I., 1967. Culture of Iceland. Nauka, Leningrad. 185 pp. (in Russian).

Steensby, H.P., 1910. Contributions to the Ethnography and anthropogeography of the Polar Eskimos. Meddelelser om Grønland 34, 254–406.

Steensby, H.P., 1917. An anthropogeographical study of the origins of Eskimo culture. Meddelelser om Grønland 53, 39–288.

Stepanov, A.D., Kirillin, A.S., Vorobyev, S.A., Solovyeva, E.N., Efimov, N.N., 2003. The Khaiyrgas cave site in the middle reaches of the Lena River. (results of the survey in 1998–1999). In: Alekseyev, A.N. (Ed.), Ancient Cultures of the Northeastern Asia. Astro-Archaeology, Paleo-Informatics. Nauka, Novosibirsk, pp. 98–113 (in Russian).

Stepanov, V.A., Kharkov, V.N., Puzyrev, V.P., Soltobayeva, Zh.O., Stegniy, V.N., 2001. Y-chromosomes in the populations of Central Asia. Genetics 37 (2), 256–259.

Stewart, T.D., 1959. Skeletal Remains from the Vicinity of Point Barrow, Alaska. Anthropological Papers 47 (1), 245–255 American Museum of Natural History, New York.

Stewart, T.G., England, J., 1983. Holocene sea-ice variations and paleoenvironmental change, northernmost Ellesmere Island N.W.T., Canada. Arctic and Alpine Research 15, 1–17.

Stokolos, V.S., Korolev, K.S., 1984. Archaeological Map of the Komi ASSR. Nauka, Moscow. 126 pp. (in Russian).

Strelkov, S.A., 1971. The coastline migrations in the western Soviet Arctic over the last 12 000 years. In: Strelkov, S.A. (Ed.), Problems of the Sediments Formation on the Kola Peninsula. Nauka, Leningrad, pp. 5–17 (in Russian).

Strukova, T.V., Bachura, O.P., Borodin, A.V., Stefanovskiy, V.V., 2006. The first findings of mammal fauna in alluvial-speleogenic formations of the Late Neopleistocene and Holocene of the Northern Urals (Cheremukhovo-1). Stratigraphy and Geological Correlation 14 (1), 98–108.

Subetto, D.A., Sevastyanov, D.V., Savelyeva, L.A., Arslanov, Kh.A., 2002. Lake bottom sediments in the Leningrad Region as a record of the Baltic transgressions and regressions. Vestnik. St.-Petersburg State University, Ser. 7 (4), 75–85 (in Russian).

Sudoyo, H., Letellier, T., Cox, M.P., Barashkov, N.A., Škaro, V., Mulahasanovic, L., Primorac, D., Sahakyan, H., Mormina, M., Eichstaedt, C.A., Lichman, D.V., Abdullah, S., Chaubey, G., Wee, J.T., Mihailov, E., Karunas, A., Litvinov, S., Khusainova, R., Ekomasova, N., Akhmetova, V., Khidiyatova, I., Marjanović, D., Yepiskoposyan, L., Behar, D.M., Balanovska, E., Metspalu, A., Derenko, M., Malyarchuk, B., Voevoda, M., Fedorova, S.A., Osipova, L.P., Lahr, M.M., Gerbault, P., Leavesley, M., Migliano, A.B., Petraglia, M., Balanovsky, O., Khusnutdinova, E.K., Metspalu, E., Thomas, M.G., Manica, A., Nielsen, R., Villems, R., Willerslev, E., Kivisild, T., Metspalu, M., September 21, 2016. Genomic analyses inform on migration events during the peopling of Eurasia. Nature. http://dx.doi.org/10.1038/nature19792.

Sulerzhitskiy, L.D., 1997. Radiocarbon chronology of mammoths in Siberia and in the north of Eastern Europe. In: Velichko, A.A., Soffer, O.A. (Eds.), Man Colonizes the Earth. Global Dispersals of Hominids. Institute of Geography RAS, Moscow, pp. 184–200 (in Russian).

Suvorov, A.V., 1998. Minino 1 burial ground on Lake Kubenskoye: results of excavations of 1993, 1996. In: Tverskoy Arkheologicheskiy Sbornik, Iss. 3. Triada Publ., Tver', pp. 193–202 (in Russian).

Suvorov, A.V., 2001. From the Mesolithic to the Early Iron Age. In: Makarov, N.A. (Ed.), Lake Kubenskoye. View through Millennia. (6 Years of the Minino Archaeological Complex Studies). Antiquities of the North Publ., Vologda, pp. 7–14 (in Russian).

Suvorov, A.V., Buzhilova, A.P., 2004. Uncommon burial complexes of the Stone Age at the Minino Village on Lake Kubenskoye. In: OPUS: Interdisciplinary Studies in Archaeology, Iss. 3. Institute of Archaeology Publ., Moscow, pp. 41–54 (in Russian).

Svendsen, J.-I., Pavlov, P., 2003. Mamontova Kurya: an enigmatic nearly 40 000 years old Palaeolithic site in the Russian Arctic. In: Zilhвo, J., D'Errico, F. (Eds.), The Chronology of the Aurignacian and the Transitional Technocomplexes – Dating, Stratigraphies, Cultural Implications. Trabalhos de Archaeologia. No 33, Lisboa, pp. 109–120.

Svendsen, J.I., Alexanderson, H., Astakhov, V.I., Demidov, I., Dowdeswell, J.A., Funder, S., Gataulin, V., Henriksen, M., Hjort, Ch, Houmark-Nielsen, M., Hubberten, H.W., Ingylfsson, Ó., Jakobsson, M., Kjær, K.H., Larsen, E., Lokrantz, H., Lunkka, J.P., Lyså, A., Mangerud, J., Matiouchkov, A., Murray, A., Möller, P., Niessen, F., Nikolskaya, O., Polyak, L., Saarnisto, M., Siegert, Ch., Siegert, M., Spielhagen, R., Stein, R., 2004. Late Quaternary ice sheet history of northern Eurasia. Quaternary Science Reviews 23 (11–13), 1229–1271.

Svendsen, J.I., Pavlov, P.Yu., Heggen, H., Mangerud, Y., Hufthammer, A.K., Roebroeks, W., 2008. Pleistocene environments and Paleolithic sites at the western slope of the northern Urals. In: Velichko, A.A., Vasil'ev, S.A. (Eds.), Way to North: Paleoenvironment and Earliest Inhabitants of Arctic and Subarctic. Institute of Geography RAS, Moscow, pp. 79–97 (in Russian).

Svezhentsev, Yu.S., 1993. Radiocarbon chronology for the Upper Paleolithic sites on the East European Plain. In: Soffer, O., Praslov, N.D. (Eds.), From Kostenki to Clovis: Upper Paleolithic – Paleo-Indian Adaptation. Plenum Publ., NewYork – London, pp. 23–30.

Svitoch, A.A., 2002. Quaternary Geology, Paleogeography, Marine Pleistocene, Salt Tectonics. Russian Academy of Agriculture Publ., Moscow. 650 pp. (in Russian).

Svitoch, A.A., 2003. Marine Pleistocene of the Coasts of Russia. GEOS Publ., Moscow. 362 pp. (in Russian).

Svoboda, J., 2000. Dolni Věstonice – Pavlov: Historie a současnost archeologickeho fenomenu. Jižni Morava 36, 21–44.

Takala, H., 2004. The Ristola Site in Lahti and the Earliest Postglacial Settlement of South Finland. Lahti City Museum, Lahti, Jyvaskyla.

Talitskiy, M.V., 1939. Paleolithic Site Peshchernyy Log. Short Communications of the Institute for the Material Culture History, Iss. 12. USSR Academy of Sciences, Leningrad, pp. 9–14 (in Russian).

Talitskiy, M.V., 1940a. Ostrov Paleolithic Site Near the City of Perm'. Short Communications of the Institute for the Material Culture History, Iss. 4. USSR Academy of Sciences, Leningrad, pp. 41–42 (in Russian).

Talitskiy, M.V., 1940b. Paleolithic site on the Chusovaya River. Bulletin of Commission for Study of the Quaternary 6–7, 136–140 (in Russian).

Tambets, K., Rootsi, S., Kivisild, T., Help, H., Serk, P., Loogvali, E.-L., Tolk, H.-V., Reidla, M., Metspalu, E., Pliss, L., Balanovsky, O., Pshenichnov, A., Balanovska, E., Gubina, M., Zhadanov, S., Osipova, L., Damba, L., Voevoda, M., Kutuev, I., Bermisheva, M., Khusnutdinova, E., Gusar, V., Grechanina, E., Parik, J., Pennarun, E., Chaventre, A., Moisan, J.-P., Barac, L., Pericic, M., Rudan, P., Terzic, R., Mikarezi, I., Krumina, A., Baumanis, V., Beckman, L., Villems, R., 2004. The western and eastern roots of the extreme European genetic outliers – the origin of mtDNAs and Y-chromosomes of the Saami. American Journal of Human Genetics 74 (4), 661–682.

Tambets, K., Rootsi, S., Metspalu, M., Yunusbayev, B., Metspalu, E., Ilumäe, A.M., Reidla, M., Dibirova, K., Balanovsky, O., Evseeva, I., Osipova, L., Khusnutdinova, E., Villems, R., 2011. Synthesis of Autosomal and Gender-Specific Genetic Structures of the Uralic-Speaking Populations. The Annual Meeting of American Society for Human Genetics. Abstract Book http://www.ichg2011.org/cgi-bin/showdetail.pl?absno=21055.

Tarasov, A.Yu, 2007. Technological analysis of the quartz knapping products on the Kirkkolahti 1 Settlement north of the Ladoga Lake (based on materials of the 2005 excavations). In: Zhilin, M.G. (Ed.), Originality and Specific Adaptivity of the Forest Zone Cultures of Northern Eurasia in the Final Pleistocene – Early Holocene. Institute of Archaeology RAS, Moscow, pp. 182–189 (in Russian).

Tarasov, P.E., Volkova, V.S., Webb III, T., Guiot, J., Andreev, A.A., Bezusko, L.G., Bezusko, T.V., Bykova, G.V., Dorofeyuk, N.I., Kvavadze, E.V., Osipova, I.M., Panova, N.K., Sevastyanov, D.V., 2000. Last Glacial Maximum biomes reconstructed from pollen and plant macrofossil data from northern Eurasia. Journal of Biogeography 27 (3), 609–620.

Tarberger, T., Street, M., 2002. Hiatus or continuity? New results for the question of pleniglacial settlement of Central Europe. Antiquity 76, 691–698.

Ten Brink, N.W., Waythomas, C.F., 1985. Late Wisconsin glacial chronology of the north-central Alaska Range: a regional synthesis and its implications for early human settlements. National Geographic Society Research Reports 19, 15–32.

Thommesen, T., 1996. The early settlement of Northern Norway. In: Larson, L. (Ed.), The Earliest Settlement of Scandinavia and Its Relationship with Neighbouring Areas. (Acta Archaeologica Lundensia. No. 24). Almquist and Wiksell International, Stockholm, pp. 235–240.

Thorson, R.M., Hamilton, T., 1977. Geology of the Dry Creek site; a stratified early man site in interior Alaska. Quaternary Research 7, 149–176.

Timireva, S.N., Velichko, A.A., 2008. Subaerial processes at the time of human habitation on the Zaozerye site on the Chusovaya River. In: Conference on the Contribution of Russia to IPY. Abstracts. Sochi, p. 74 (in Russian).

Timofeyev, E.M., Bader, N.O., Buzhilova, A.P., 2005. New data on the Upper Paleolithic site Krutaya Gora on the Pechora River. In: Quarter – 2005. Materials of the 4th All-Russia Conference on the Quaternary. Geoprint Publ., Syktyvkar, pp. 421–422 (in Russian).

Timofeyev, V.I., 1993. Mesolithic and Neolithic sites of the Peterburg vicinities and their place in the system of the Stone Age cultures on the Baltic region. In: Masson, V.M., Nosov, E.N., Ryabinin, E.A. (Eds.), Antiquities of the Northeast of Russia. Peterburgskoye Vostokovedeniye Publ., St.-Petersburg, pp. 8–34 (in Russian).

Timofeyev, V.I., Zaitseva, G.I., Dolukhanov, P.M., Shukurov, A.M., 2004. Radiocarbon Chronology of Neolithic in Northern Eurasia. TEZA Publ., St.-Petersburg. 157 pp. (in Russian).

Tiunov, A.V., 2007. Stable Isotopes of Carbon and Nitrogen as Applied to Soil Ecology Studies. Izvestiya. Russian Academy of Sciences Ser. Biol. 4, 475–489 (in Russian).

Tiunov, A.V., Dobrovolskaya, M.V., 2011. Stable Isotope (13C/12C and 15N/14N) Evidence for Mid-Upper Paleolithic Hominines' Palaeodiets in Gorny Altai. Abstracts of Meeting of European Society of the Study of Human Evolution, Leipzig, September 23–24, 2011, p. 37.

Tofanelli, S., Ferri, G., Bulayeva, K., Caciagl, L., Onofri, V., Taglioli, L., Bulayev, O., Boschi, I., Alù, M., Berti, A., Rapone, C., Beduschi, G., Luiselli, D., Cadenas, A., Dafaallah, A.K., Mariani-Costantini, R., Eldin, E.N., Verginelli, F., Pilli, E., Herrera, R., Gusmão, L., Paoli, G., Capelli, C., 2009. J1-M267 Y lineage marks climate-driven pre-historical human displacements. European Journal of Human Genetics 17 (11), 1520–1524.

Tomskaya, A.I., 1981. Palynology of the Cenozoic in Yakutia. Nauka, Novosibirsk. 221 pp. (in Russian).

Tomskaya, A.I., Savvinova, G.M., 1971. Pollen spectra of the Pleistocene deposits in the middle reaches of the Yana River. In: Tomskaya, A.I., Vozin, V.F. (Eds.), Palynological Characteristics of the Paleozoic, Mesozoic, and Cenozoic Sediments in Yakutia. Yakutian Publ. House, Yakutsk, pp. 131–146 (in Russian).

Tomskaya, A.I., Savvinova, G.M., 1975. Pollen spectra from the deposits of the Aldan River valley enclosing Paleolithic sites. In: Mochanov, Yu.A. (Ed.), Yakutia and Its Neighbors in the Past. Yakutian Branch, Siberian Division, the USSR Academy of Sciences, Yakutsk, pp. 31–37 (in Russian).

Tonks, S., Winney, B.J., Evseeva, I., 2006. Comparison of Sex-Linked and Autosomal Markers in Orkney and Other North European Populations. Data Deposited in Genbank. 2006. www.ncbi.nlm.nih.gov.

Torvinen, M., 1997. Sar 1, comb ware of the Säräisniemi style. In: Kirpichnikov, A.N. (Ed.), Slavs and Finno-Ugric Peoples. Nauka, St.Petersburg, pp. 21–31.

Torvinen, M., 1998. Sar 1 keramiikkaa kayttanyt vaesto-etnisia kysymyksia. Muinaistutkija 4, 38–45.

Torvinen, M., 2000. Säräisniemi 1 Ware. Fennoscandia Archaeologica 16, 3–36.

Tremblay, L.-B., Mysak, L.A., Dyke, A.S., 1997. Evidence from driftwood records for century-to-millennial scale variations of the High Latitude atmospheric circulation during the Holocene. Geophysical Research Letters 24, 2027–2030.

Tseitlin, S.M., 1979. Geology of the Paleolithic of Northern Asia. Nauka, Moscow. 284 pp. (in Russian).

Turner, C.G., 1990. Paleolithic teeth of the Central Siberian Altai Mountains. In: Derevyanko, A.P. (Ed.), Chronostratigraphy of the Paleolithic in North, Central, East Asia and America. Proceedings of International Symposium. Institute of History, Philology and Philosophy, USSR Academy of Sciences, Siberian Branch, Novosibirsk, pp. 239–243.

Ukkonen, P., 2001. Shaped by the Ice Age: Reconstructing History of Mammals in Finland in the Late Pleistocene and Early Holocene. Yliopistopaino, Helsinki. 41 pp.
Ukkonen, P., 2002. The early history of seals in the Northern Baltic. Annales Zoologici Fennici 39, 187–207.
Ukraintseva, V.V., 2002. Vegetation and Climate of Siberia in the Epoch of Mammoths. International Forest Institute, RAS, East Siberian Branch, Krasnoyarsk. 192 pp. (in Russian).
Underhill, P.A., Myres, N.M., Rootsi, S., Metspalu, M., Zhivotovsky, L.A., King, R., Lin, A.A., Chow, C.E., Semino, O., Battaglia, V., Kutuev, I., Järve, M., Chaubey, G., Ayub, Q., Mohyuddin, A., Mehdi, Q., Sengupta, S., Rogaev, E.I., Khusnutdinova, E.K., Pshenichnov, A., Balanovsky, O., Balanovska, E., Jeran, N., Augustin, D.H., Baldovic, M., Herrera, R.J., Thangaraj, K., Singh, V., Singh, L., Majumder, P., Rudan, P., Primorac, D., Villems, R., Kivisild, T., 2010. Separating the post-Glacial coancestry of European and Asian Y chromosomes within haplogroup R1a. European Journal of Human Genetics 18 (4), 479–484.
Ungar, P.S., 2007. Evolution of the Human Diet: The Known, the Unknown, and the Unknowable. Oxford Univ. Press, Oxford. 432 pp.
Ungar, P.S., Teaford, M.F., 2002. Human Diet: Its Origin and Evolution. Bergim & Garvey Publ., Westport, Connecticut. 108 pp.
Vaikmae, R.A., Punning, J.-M., 1982. Isotope geochemical studies on the Vavilov ice dome, Severnaya Zemlya archipelago. In: Data of Glaciological Studies. Chronicles. Discussions 44, 145–149 (in Russian).
van Linschoten, I.N., 1611. Voyage ofte schipvaert. Franken.
Vartanyan, S.L., 2007. Wrangel Island at the End of Quaternary: Geology and Paleogeography. Ivan Limbakh Publ., St.-Petersburg. 144 p. (in Russian).
Vashchanova, T.V., Klimanov, V.A., 1987. Quantitative Climatic Reconstructions in Khibines as Scenarios of Future Climates. Vestnik, Moscow State Univ., Ser.5 1, 84–88 (in Russian).
Vasilevich, G.M., Levin, M.G., 1951. Types of reindeer breeding and their origin. Sovetskaya Etnografiya 1, 63–87 (in Russian).
Vasil'ev, S.A., 2004. The Earliest Cultures of North America. Peterburgskoye Vostokovedeniye (St.-Petersburg Centre for Oriental Studies) St.-Petersburg. 144 pp. (in Russian).
Vasil'ev, S.A., Kuzmin, Y.V., Orlova, L.A., Dementiev, V.N., 2002. Radiocarbon-based chronology of the Upper Paleolithic of Siberia and its relevance to the peopling of the New World. Radiocarbon 44 (2), 503–530.
Vasil'ev, S.V., Borutskaya, S.B., 2008. Bolshoy Oleni island (the Barents Sea): physical load and stress markers of the ancient Arctic population. In: Velichko, A.A., Vasil'ev, S.A. (Eds.), Way to North: Paleoenvironment and Earliest Inhabitants of Arctic and Subarctic. Institute of Geography RAS, Moscow, pp. 152–160 (in Russian).
Vasil'ev, V.I., 1979. Problem of the North Samoedic Peoples. Nauka, Moscow. 244 pp. (in Russian).
Vasil'ev, V.I., 1995. Siirtya go under ground. In: Tomilov, N.A. (Ed.), From the Urals to the Yenisey. Peoples of West and Central Siberia. Tomsk Univ. Publ., Tomsk, pp. 5–22 (in Russian).
Vasilyevskiy, R.S., 1973. Ancient Cultures of the Pacific North. Nauka, Novosibirsk. 267 pp. (in Russian).
Vasilyevskiy, R.S., Gladyshev, S.A., 1989. The Upper Paleolithic in Southern Primorye. Nauka, Novosibirsk. 184 pp. (in Russian).
Velichko, A.A., 1965. Cryogenic relief in the periglacial zone (cryolithozone) of Eastern Europe. In: Gromov, V.I., Ivanova, I.K., Markov, K.K., Fedorov, P.V., Shantser, E.V. (Eds.), Quaternary Period and Its History. Nauka, Moscow, pp. 104–120 (in Russian).

Velichko, A.A., 1968. The main climatic boundary of the Pleistocene. Izvestiya. USSR Academy of Sciences Ser Geogr. 3, 37–51 (in Russian).
Velichko, A.A., 1973. Natural Process in the Pleistocene. Nauka, Moscow. 256 pp. (in Russian).
Velichko, A.A., 1982. Periodization of the Late Pleistocene events in the periglacial region. In: Gerasimov, I.P., Velichko, A.A. (Eds.), Paleogeography of Europe during the Last 100,000 Years. Nauka, Moscow, pp. 67–70 (in Russian).
Velichko, A.A., 1988. Geology of the Mousterian in East Europe and the adjacent areas. In: Otte, M. (Ed.), L'Homme de Nиandertal, vol. 2. Eraul, Liege, pp. 181–206.
Velichko, A.A. (Ed.), 1993. Evolution of Landscapes and Climates of Northern Eurasia: Late Pleistocene – Holocene – Elements of Prognosis. I. Regional Paleogeography. Nauka, Moscow. 107 pp. (in Russian).
Velichko, A.A., 1997. Global initial colonization as a part of problem of the man and environment coevolution. In: Velichko, A.A., Soffer, O. (Eds.), Man Colonizes the Earth. Global Dispersal of Hominids. Institute of Geography RAS, Moscow, pp. 255–275 (in Russian).
Velichko, A.A. (Ed.), 2002a. Dynamics of Terrestrial Landscape Components and Inland and Marginal Seas of Northern Eurasia during the Last 130 000 Years (Atlas-Monograph. Evolution of Landscapes and Climates of Northern Eurasia. Late Plesitocene – Holocene – Elements of Prognosis. Issue II. General Paleogeography). GEOS Publ., Moscow (in Russian).
Velichko, A.A., 2002b. Specific features of environmental changes in Northern Eurasia in the Late Pleistocene and Holocene. In: Velichko, A.A. (Ed.), Dynamics of Terrestrial Landscape Components and Inland and Marginal Seas of Northern Eurasia during the Last 130,000 Years (Atlas-Monograph). GEOS Publ., Moscow, pp. 156–164 (in Russian).
Velichko, A.A., 2009a. Middle Valday Zyryanka-Sartan megainterstadial and the climatic rank of its optimum. In: Fundamental Problems of Quaternary: Results of the Studies and the Main Trends in Further Investigations. Materials of VI All-Russia Conference on the Quaternary Studies, Novosibirsk, November 19–23, 2009. Siberian Branch RAS Publ., Novosibirsk, pp. 107–109 (in Russian).
Velichko, A.A., 2009b. The landscape sphere structure at the time of the maximum glaciation. In: Velichko, A.A. (Ed.), Paleoclimate and Paleoenvironments of Extra-Tropical Area of the Northern Hemisphere. Late Pleistocene – Holocene. GEOS Publ., Moscow, pp. 95–98 (in Russian).
Velichko, A.A., Faustova, M.A., 2009. Glaciations during the Late Pleistocene. In: Velichko, A.A. (Ed.), Paleoclimate and Paleoenvironments of Extra-tropical Area of the Northern Hemisphere. Late Pleistocene – Holocene. GEOS Publ., Moscow, pp. 32–41 (in Russian).
Velichko, A.A., Morozova, T.D., 2005. Evolution of soil formation in the Pleistocene. In: Glazovskiy, N.F. (Ed.), Multifaceted Geography. Further Development of Ideas Put Forward by I.P. Gerasimov. KMK Publ., Moscow, pp. 65–75 (in Russian).
Velichko, A.A., Nechaev, V.P., 2005. Climate and cryolithozone dynamics in the Pleistocene. In: Quarter – 2005. Materials of the 4th All-Russia Conference on the Quaternary. Geoprint Publ., Syktyvkar, pp. 62–64 (in Russian).
Velichko, A.A., Timireva, S.N., 2005. West Siberia as a great Late Glacial desert. Priroda 5, 54–62 (in Russian).
Velichko, A.A., Vasil'ev, S.A. (Eds.), 2008. Way to North: Paleoenvironment and Earliest Inhabitants of Arctic and Subarctic. Materials of International Conference. Institute of Geography RAS, Moscow. 310 pp. (in Russian).
Velichko, A.A., Zelikson, E.M., 2006. Periglacial environments as a resource base of the late mammoth existence at the Upper Paleolithic time on the East European Plain. In: Anikovich, M.V. (Ed.), The Early Upper Paleolithic of Eurasia. Materials of International

Conference to the 125-Year Anniversary of the Paleolithic Discovery in Kostenki (August 23–26, 2006). Nestor-Istoriya Publ., St.-Petersburg, pp. 9–25 (in Russian).
Velichko, A.A., Berdnikov, V.V., Nechaev, V.P., 1982. Reconstructions of the permafrost zone and stages of its development (Maps 4, 5, 6). In: Gerasimov, I.P., Velichko, A.A. (Eds.), Paleogeography of Europe During the Last 100,000 Years. Nauka, Moscow, pp. 74–81 (in Russian).
Velichko, A.A., Gribchenko, Yu.N., Kurenkova, E.I., 1997. Geoarchaeology of the Paleolithic in the East European Plain. Anthropologie 35 (2), 215–232.
Velichko, A.A., Gribchenko, Yu.N., Kurenkova, E.I., Novenko, E.Yu, 1999. Geochronology of the Paleolithic on the East European Plain. In: Gribchenko, Yu.N., Nikolayev, V.I. (Eds.), Changes in Environments and Climates, Wildlife, and Humans in the Late Pleistocene and Holocene. Institute of Geography RAS, Moscow, pp. 19–50 (in Russian).
Velichko, A.A., Faustova, M.A., Kononov, Yu.M., 2002a. Glaciation. In: Velichko, A.A. (Ed.), Dynamics of Terrestrial Landscape Components and Inland and Marginal Seas of Northern Eurasia during the Last 130,000 Years (Atlas-Monograph). GEOS Publ., Moscow, pp. 13–23 (in Russian).
Velichko, A.A., Gribchenko, Yu.N., Abramova, Z.A., Kurenkova, E.I., Praslov, N.D., 2002b. Prehistoric society and environments. Late Paleolithic. In: Velichko, A.A. (Ed.), Dynamics of Terrestrial Landscape Components and Marginal Seas of Northern Eurasia during the Last 130,000 Years. GEOS Publ., Moscow, pp. 138–146 (in Russian).
Velichko, A.A., Gribchenko, Yu.N., Kurenkova, E.I., 2005. Environmental background of the Northern Eurasia colonization by humans in the Late Pleistocene. In: Glazovskiy, N.F. (Ed.), Multifaceted Geography. Further Development of Ideas Put Forward by I.P. Gerasimov. KMK Publ., Moscow, pp. 101–113 (in Russian).
Velichko, A.A., Timireva, S.N., Kremenetski, K.V., Macdonald, G.M., Smith, L., 2007. West Siberian Plain as a late glacial desert. Izvestiya RAS Ser. Geogr. 4 (16–28) (in Russian).
Velichko, A.A., Dolukhanov, P.M., Kurenkova, E.I., 2008a. Adaptation systems (man – social economic structures – environments) in the Late Paleolithic, Mesolithic and Neolithic of Eastern Europe. In: Velichko, A.A., Vasil'ev, S.A. (Eds.), Way to North: Paleoenvironment and Earliest Inhabitants of Arctic and Subarctic. Institute of Geography RAS, Moscow, pp. 14–32 (in Russian).
Velichko, A.A., Vasil'ev, S.A., Gribchenko, Yu.N., Kurenkova, E.I., Timireva, S.N., 2008b. The main characteristics of the initial human dispersal in High Latitudes under conditions of changing environments. In: Kotlyakov, V.M. (Ed.), Climate and Environment Changes, Natural and Related Technogenous Catastrophes. vol. 3. Part 2. Natural Processes in Polar Regions of the Earth. Institute of Geography RAS, Institute of Physics of the Earth RAS, Moscow, pp. 309–321 (in Russian).
Vereshchagin, N.K., 1977. Mammoths 'cemetery' in Berelekh. In: Svetovidov, A.N. (Ed.), Mammoth Fauna of the Russian Plain and East Siberia. Nauka, Leningrad, pp. 5–50 (Transactions of Zoological Institute, v. 72) (in Russian).
Vereshchagin, N.K., 1979. Mammal remains from the Paleolithic layer of Ushki I site. In: Dikov, N.N. (Ed.), New Archaeological Sites in the North of the Far East. Magadan Publ. House, Magadan, pp. 12–17 (in Russian).
Vereshchagin, N.K., Mochanov, Yu.A., 1972. The northernmost traces of the Upper Paleolithic. Sovetskaya Arkheologiya 3, 332–336 (in Russian).
Vereshchagin, N.K., Rusakov, O.S., 1979. Ungulates of the Northwest of the USSR (History, Way of Life, Practical Use by Humans). Nauka, Leningrad. 309 pp. (in Russian).
Vereshchagina, I.V., 2010. Mesolithic and Neolithic of the Extreme Northeast of Europe. Peterburgskoye Vostokovedeniye Publ., St.-Petersburg. 232 pp. (in Russian).

Vereshchagina, V.S., 1965. Stratigraphy of the Quaternary deposits of the western slope of Middle Urals and its forelands. In: Yakhimovich, V.L., Lider, V.A. (Eds.), Stratigraphy of the Quaternary (Anthropogene) Deposits of the Urals. Nedra Publ., Moscow, pp. 106–129 (in Russian).

Vermeersch, P.M., 2011. Radiocarbon Palaeolithic Europe Database V14. Dept. of Earth and Environmental Sciences, Katholieke Universiteit, Leuven. http://ees.kuleuven.be/geography/projects/14c-palaeolithic/index.html.

Vincent, J.-S., Klassen, R.W., 1989. Introduction. Quaternary geology of the Canadian Interior Plains. In: Fulton, R.J. (Ed.), Quaternary Geology of Canada and Greenland. Geological Survey of Canada, Ottawa, pp. 99–100.

Vinson, D.M., 1993. Taphonomic Analysis of Faunal Remains from Trail Creek Caves, Seward Peninsula, Alaska (Unpublished M.A. thesis, Department of Anthropology). University of Alaska, Fairbanks.

Vinther, B.M., Buchardt, S.L., Clausen, H.B., Dahl-Jensen, D., Johnsen, S.J., Fisher, D.A., Koerner, R.M., Raynaud, D., Lipenkov, V., Andersen, K.K., Blunier, T., Rasmussen, S.O., Steffensen, J.P., Svensson, A.M., 2009. Holocene thinning of the Greenland ice sheet. Nature 461, 385–388.

Viola, B.Th., Markin, S.V., Buzhilova, A.P., Mednikova, M.B., Dobrovolskaya, M.V., Le Cabec, A., Shun'kov, M.V., Derevianko, A.P., Hublin, J.-J., 2012. New Neanderthal remains from Chagyrskaya Cave (Altai Mountains, Russian Federation). American Journal of Physical Anthropology 147 (Suppl. 54), 293–294.

Vitov, M.V., Mark, K.Yu., Cheboksarov, N.N., 1959. Ethnic Anthropology of Eastern Baltic Region. In: Transaction of the Baltic Joint Integrated Expedition, vol. 2. USSR Academy of Sciences Publishing House, Moscow. 238 pp. (in Russian).

Vize, V.Yu, 1949. Old Russian names on the Novaya Zemlya Islands. In: Letopis Severa (Northern Annals), Iss. 1, pp. 107–122 Moscow–Leningrad (in Russian).

Volkov, I.A., 1971. Late Quaternary Subaerial Formation. Nauka, Moscow. 254 pp. (in Russian).

Volkov, I.A., 1973. Paleogeographic significance of some radiocarbon dates in the south of West Siberia. Geologiya i geofizika 2, 3–8 (in Russian).

Volkova, V.S., 1991. Climate fluctuations in West Siberia in the Late Pliocene and Quaternary. In: Zakharov, V.A. (Ed.), Evolution of Climate, Biota and Human Environments in the Late Cenozoic of Siberia. United Institute of Geology, Geophysics and Mineralogy, Novosibirsk, pp. 30–51 (in Russian).

Volkova, V.S. (Ed.), 2000. Unified Regional Stratigraphic Scheme of the Quaternary of West Siberian Plain. Explanatory Notes. Siberian Research Institute Geology. Geophysicist and Minerals (SNIIGGiMS) Publ., Novosibirsk. 64 pp. (in Russian).

Volkova, V.S., 2001. Paleogeography of the Karginsky Interglacial (Interstadial) in West Siberia 50(55)–23 ka BP. Bullettin of the Commission. for the Study of the Quaternary Period 64, 89–93 (in Russian).

Volkova, V.S., Zykina, V.S., 2000. Stratigraphy of Neopleistocene and environments of the Tobolsk region of the Irtysh drainage basin. In: Vaganov, E.A., Derevyanko, A.P., Zykin, V.S., Markin, S.V. (Eds.), Problems of Reconstruction of Climate and Environments of the Holocene and Pleistocene in Siberia, Iss. 2. Institute of Archaeology and Etnography (Siberian Branch RAS), Novosibirsk, pp. 75–83 (in Russian).

Volkova, V.S., Arkhipov, S.A., Babushkin, A.E., Kulkova, I.A., Guskov, S.A., Kuzmina, O.B., Levchuk, L.K., Mikhailova, I.V., Sukhorukova, S.S., 2003. Cenozoic of West Siberia. GEO Publ. (Siberian Branch RAS), Novosibirsk. 247 pp. (in Russian).

Volokitin, A.V., 1997. Stone Age. Mesolithic. In: Savelyeva, E.A. (Ed.), Archaeology of the Komi Republic. DiK Publ., Moscow, pp. 91–145 (in Russian).

Volokitin, A.V., 2002. Studies of the Mesolithic Dwellings of the Parch 2 Site, on the Vychegda River. Tverskoy Arkheologicheskiy Sbornik, Iss. 5. Triada Publ., Tver', pp. 183–191 (in Russian).
Volokitin, A.V., 2005. The Mesolithic Lek-Lesa site on the Izhma River. In: Zhilin, M.G. (Ed.), Stone Age of the Forest Zone in Eastern Europe and Transuralia. Academia Publ., Moscow, pp. 198–205 (in Russian).
Volokitin, A.V., 2006. Mesolithic Sites Parch 1 and Parch 2 on Vychegda. Komi Research Centre, the Urals Division, RAS, Syktyvkar. 126 pp. (in Russian).
Volokitin, A.V., Konovalenko, L.A., 1988. Parch 3 – a new Mesolithic site on the Vychegda. In: Sites of the Stone and Metal Ages in the Northern Urals Region. Materials on the Archaeology of the European Northeast, Iss. 11. Komi Research Centre, the Urals Division, RAS, Syktyvkar, pp. 19–32 (in Russian).
Volokitin, A.V., Kosinskaya, L.L., 1999. Mesolithic dwellings in the northeast of Europe. In: Komi Research Centre, the Urals Division, RAS. Nauchnye Doklady (Scientific Reports), Iss. 414. Syktyvkar 36 pp. (in Russian).
Volokitin, A.V., Tkachev, Y.A., 2004. Reconstructions of environmental conditions of the Mesolithic people on the Vychegda River. Archaeology, Ethnography and Anthropology of Eurasia 2, 2–10.
Vorobey, I.E., 1996. Druchak-Vetrenyi site. In: Archaeological Survey in the North of the Far East. Northeastern Interdisciplinary Research Institute, Far Eastern Division, RAS, Magadan, pp. 24–51 (in Russian).
Vorobey, I.E., 1999. On the Paleolithic finds on the Omolon River. In: Archaeological Studies of the Northern Far East. Northeastern Interdisciplinary Research Institute, Far Eastern Division, RAS, Magadan, pp. 4–15 (in Russian).
Vorobey, I.E., 2003. Druchak microblade industries of northeast Asia. Current Research in the Pleistocene 11, 81–83.
Vorobey, I.E., 2007. The ancient hunters' access to sea: a problem or a figure of speech? (Late Neolithic – Middle Ages). In: University in the Region: Science, Education, Culture. Materials of International Conference, December 5–6, 2006. Northern International Univ. Publ., Magadan, pp. 148–150 (in Russian).
Vorobey, I.E., 2008. Magdavena Mouth 1 – Late Paleolithic locality in the upper reaches of the Arman' River. In: Materials of Dikov's Conference V. Magadan, March 18–20, 2008. Kordis Publ., Magadan, pp. 66–68 (in Russian).
Vorren, K.-D., Elverland, E., Blaauw, M., Ravna, E.K., Jensen, C.A.H., 2009. Vegetation and climate c. 12,300–9000 cal. yr BP at Andøya, NW Norway. Boreas 38, 401–420.
Waters, M., Forman, S., Pierson, G., 1997. Diring-Yuriakh: a lower Paleolithic site in Central Siberia. Science 275, 1281–1284.
Waters, M.R., Forman, S.L., Pierson, J.M., 1999. Late Quaternary geology and geochronology of Diring Yuriakh, an Early Paleolithic site in Central Yakutia. Quaternary Research 51 (2), 195–211.
Weaver, T.D., 2009. The meaning of Neanderthal skeletal morphology. Proceedings of National Academy of Sciences (PNAS) 106 (38), 16028–16033.
Welinder, S., 1981. Den Kontinentaleuropeiska bakgrunden till Norges äldsta stenålder. Årbok Universitetets oldsaksamling, Oslo, pp. 21–34.
West, F.H., 1981. The Archaeology of Beringia. Columbia Univ. Press, New York.
West, F.H., 1996. The archaeological evidence. In: West, F.H. (Ed.), American Beginnings: The Prehistory and Palaeoecology of Beringia. Univ. of Chicago Press, Chicago, pp. 537–559.
West, F.H., Robinson, B.S., Curran, M.L., 1996a. Phipps site. In: West, F.H. (Ed.), American Beginnings: The Prehistory and Palaeoecology of Beringia. Univ. of Chicago Press, Chicago, pp. 381–386.

West, F.H., Robinson, B.S., West, C.F., 1996b. Whitmore Ridge. In: West, F.H. (Ed.), American Beginnings: The Prehistory and Palaeoecology of Beringia. Univ. of Chicago Press, Chicago, pp. 386–394.

Whitridge, P., 1999. The Construction of Social Difference in a Prehistoric Inuit Whaling Community (Ph.D. dissertation). Dept. Anthropology, Arizona State University, Tempe.

Williams, K.M., 1990. Paleolimnology of three Jackman Sound Lakes, Southern Baffin Island, based on down-core diatom analyses. Journal of Paleolimnology 4, 203–217.

Williams, L.D., Bradley, R.S., 1985. Paleoclimatology of the Baffin Bay region. In: Andrews, J.T. (Ed.), Quaternary Environments: The Eastern Canadian Arctic, Baffin Bay and West Greenland. Allen & Unwin, Boston, pp. 741–772.

Wohlfarth, B., Tarasov, P., Bennike, O., Lacourse, T., Subetto, D., Torssander, F., Romanenko, F., 2006. Late Glacial and Holocene palaeoenvironmental changes in the Rostov-Yaroslavl' area, West Central Russia. Journal of Palaeolimnology 35, 543–569.

Wood, R., 2006. Chronometric and Paleodietary Studies at the Mesolithic and Neolithic Burial Ground of Minino, Northwestern Russia (M.Sc. dissertation). Oxford University, Oxford.

Woodman, P., 1993. The Komsa Culture: a re-examination of its position in the Stone Age of Finnmark. Acta Archaeologica 63, 57–76.

Wright, J.V., 1995. A History of the Native People of Canada. Vol. 1: 10,000–1000 B.C. Mercury Series. Archaeology. No. 152. Canadian Museum of Civilization, Hull, Quebec. 506 pp.

Wygal, B.T., 2009. Prehistoric Colonization of Southcentral Alaska: Human Adaptations in a Post-glacial World (Unpublished Ph.D. dissertation). Department of Anthropology, Univ. of Nevada, Reno.

Yakhimovich, V.Ya., Nemkova, V.K., Sidnev, A.V., Suleimanova, F.I., Khabibullina, G.A., Shcherbakova, T.I., Yakovlev, A.G., 1987. Pleistocene of the Cis-Urals Region. Nauka, Moscow. 113 pp. (in Russian).

Yakhimovich, V.Ya., Nemkova, V.K., Yakovlev, A.G., 1988. Regional Units in the New Stratigraphic Scheme of the Cis-Urals Pleistocene and Some Key Sections. Bashkirian Research Centre, the Urals Division, USSR Academy of Sciences, Ufa. 65 pp. (in Russian).

Yakimov, V.P., 1950. A human skull of the Bronze Age from Yakutia. In: Okladnikov, A.P. (Ed.), Lena's Antiquities, Iss. 3, pp. 189–198 (in Russian).

Yakimov, V.P., 1953. Anthropological Characteristics of Skeletons from the Bolshoy Oleni Burials (The Barents Sea), Transaction Museum of Anthropology and Ethnography, vol. XV. Leningrad, pp. 449–455 (in Russian).

Yakimov, V.P., 1956. Initial stages of peopling of the Eastern Baltic region. In: Cheboksarov, N.N., Terentyeva, L.N. (Eds.), Baltic ethnographic Papers. (Trans. Institute of Etnography. V. 32). Academy of Sciences, USSR. Publ., Moscow, pp. 258–260 (in Russian).

Yakimov, V.P., 1960. Anthropological Materials from the Neolithic Burial Site on the Yuzhny Oleni Island (Onega Lake), Trans. Museum of Anthropology and Ethnography, vol. XIX. Leningrad, pp. 231–360 (in Russian).

Yakimov, V.P., 1961. Horizontal profiling of the facial skeleton of modern man. Voprosy Antropologii 4, 62–70 (in Russian).

Yakovlev, S.A., 1956. Fundamentals of Quaternary geology of the Russian Plain. In: Proc. VSEGEI, New Series, vol. 17. Gosgeoltekhizdat Publ., Moscow. 311 pp. (in Russian).

Yesner, D.R., 1994. Subsistence diversity and hunter-gatherer strategies in Late Pleistocene/ Early Holocene Beringia: evidence from the Broken Mammoth site, Big Delta, Alaska. Current Research in the Pleistocene 11, 154–156.

Yesner, D.R., 1996. Human adaptation at the Pleistocene/Holocene boundary in Eastern Beringia. In: Straus, L.G., Eriksen, B.V., Erlandson, J.M., Yesner, D.R. (Eds.), Humans at the End of the Ice Age: Archaeology of the Pleistocene/Holocene Transition. Plenum Press, New York, pp. 255–276.

Yesner, D.R., 2000. Human colonization of Eastern Beringia and the question of mammoth hunting. In: West, D.L. (Ed.), Mammoth Site Studies. Univ. of Kansas Publications in Anthropology No. 22, Lawrence, pp. 69–84.

Yesner, D.R., 2001. Human dispersal into interior Alaska: antecedent conditions, mode of colonization, and adaptations. Quaternary Science Reviews 20, 315–327.

Yesner, D.R., 2007. Faunal Extinction, Hunter-Gatherer Foraging Strategies, and Subsistence Diversity Among Eastern Beringian Paleoindians. In: Walker, R.B., Driskell, B.N. (Eds.), Foragers of the Terminal Pleistocene in North America. Univ. of Nebraska Press, Lincoln, pp. 15–31.

Yesner, D.R., Holmes, C.E., Crossen, K.J., 1992. Archaeology and paleoecology of the Broken Mammoth site, Central Tanana Valley, Interior Alaska, USA. Current Research in the Pleistocene 9, 53–57.

Yesner, D.R., Pearson, G.A., Stone, D.E., 2000. Additional organic artifacts from the Broken Mammoth site, Big Delta, Alaska. Current Research in the Pleistocene 17, 87–90.

Yesner, D.R., Crossen, K.J., Easton, N.A., 2011. Geoarchaeological and zooarchaeological correlates of Early Beringian artifact assemblages: insights from the Little John site, Yukon. In: Goebel, T., Buvit, I. (Eds.), From the Yenisei to the Yukon: Interpreting Lithic Assemblage Variability in Late Pleistocene/Early Holocene Beringia. Peopling of the Americas Publications Series. Texas A&M Univ. Press, College Station, pp. 308–322.

Yi, S., Clark, G.A., 1985. The "Dyuktai culture" and the new world origins. Current Anthropology 26 (1), 1–20.

Young, C., Gilbert-Young, S., 2007. A fluted projectile-point base from Bering Land Bridge national preserve, Northwest Alaska. Current Research in the Pleistocene 24, 154–156.

Yurgenson, E.A., 1958. On Siliceous Formations in Ordovician and Silurian Carbonaceous Rocks in Estonian SSR, No. 2. Transaction of Institute of Geology, Estonian Academy of Sciences, Tallinn, pp. 87–92 (in Russian).

Yurtsev, B.V., 1972. Steppe communities in the Chukotka tundra and the problem of thePleistocene "tundra-steppe". In: Problems of the Quaternary Studies. Nauka, Moscow, pp. 471–479 (in Russian).

Zabenskie, S., Gajewski, K., 2007. Post-glacial climatic change on Boothia Peninsula, Nunavut, Canada. Quaternary Research 68, 261–270.

Zagorska, I., 2006. Radiocarbon chronology of the Zvejnieki burials. In: Larsson, L., Zagorska, I. (Eds.), Back to the Origin. New Research in the Mesolithic–Neolithic Zvejnieki Cemetery and Environment, Northern Latvia. Acta Archaeologica Lundensia Series, vol. 8(52). Almqvist & Wiksell, Stockholm, pp. 91–114.

Zagorska, I.A., Zagorskis, F.A., 1977. Mesolithic in Latvia. Short Communications of the Institute of Archeology, vol. 149, pp. 69–75 (in Russian).

Zaitseva, G.I., Timofeyev, V.I., Zagorska, I.A., Kovalyukh, N.N., 1997. Radiocarbon dates of the Mesolithic sites in Eastern Europe. In: Kolchin, B.A. (Ed.), Radiocarbon and Archaeology. Yearbook of the Radiocarbon Laboratory, Iss. 2. Institute for Material Culture History, St. Petersburg, pp. 117–120 (in Russian).

Zaliznyak, L.L., 1989. Reindeer Hunters in the Ukrainian Polessye Region at the Final Paleolithic. Naukova Dumka Publ., Kiev. 173 pp. (in Russian).

Zaliznyak, L.L., 1991. Polessye Population in the Mesolithic Time. Naukova Dumka Publ., Kiev. 172 pp. (in Russian).

Zander, A., Frechen, M., Zykina, V., Boenigk, W., 2003. Luminescence chronology of the Upper Pleistocene loess record at Kurtak in Middle Siberia. Quaternary Science Reviews 22, 999–1010.

Zaretskaya, N.E., 2005. Radiocarbon and calendar chronology of multilayered settlements in peatlands of the Volga-Oka interfluve. In: Zhilin, M.G. (Ed.), Stone Age of the Forest Zone in Eastern Europe and Transuralian Regions. Academia Publ., Moscow, pp. 113–129 (in Russian).

Zaretskaya, N.E., Volokitin, A.V., Karmanov, V.N., Uspenskaya, O.N., 2007. New data on radiocarbon chronology and paleogeography of the Sindor geoarchaeological micro-region. In: Volokitin, A.V., Karmanov, V.N., Pavlov, P.Yu. (Eds.), Stone Age of the European North. Institute of Linguistic, Literature, and History, Komi Research Centre, the Urals Division, RAS, Syktyvkar, pp. 148–160 (in Russian).

Zarrina, E.P., 1991. Quaternary Deposits of the Northwestern and Central Regions of the European Part of the USSR. Nedra Publ., Leningrad. 187 pp. (in Russian).

Zavernyaev, F.M., 1972. Stone Age sites in the Khotylevo settlement vicinities on the Desna River. Bulletin of Commission for Study of the Quaternary 39, 90–97 (in Russian).

Zemlyakov, B.F., 1937a. Arctic Paleolithic in the North of the USSR. Transaction of Commission for Study of Quaternary 1, 69–87 (in Russian).

Zemlyakov, B.F., 1937b. Archaeological survey on the coasts of the Arctic Ocean. In: Proceedings of the INQUA Soviet Section, Iss. 3, pp. 62–74 (in Russian).

Zemlyakov, B.F., 1940. Arctic Paleolithic in the North of the USSR. Sovetskaya Arkheologiya 5, 107–143 (in Russian).

Zenin, V.N., 2002. Principal stages in the West Siberian Plain colonization by the Paleolithic man. Archaeology, Ethnography and Anthropology of Eurasia 4 (12), 22–44.

Zenin, V.N., Leshchinskiy, S.V., Zolotarev, K.V., Grutes, P.M., Nado, M.-H., 2006. Geoarchaeology and material culture characteristics of the Paleolithic site Lugovskoye. Archaeology. Ethnography and Anthropology of Eurasia 1, 41–53.

Zhilin, M.G., 2001. Relations between population of the Baltic region and that of the Upper Volga area in the Early Mesolithic. In: Tverskoy Arkheologicheskiy Sbornik, Iss. 4. Triada Publ., Tver', pp. 72–79 (in Russian).

Zhilin, M.G., 2002. On the Problem of the Pioneering Peopling of Southern Karelia and Finland in Early Holocene. Bulletin of Karelian Regional Museum (Petrozavodsk) (4), 3–15 (in Russian).

Zhilin, M.G., 2004. Mesolithic Population of the Central and Northeastern Forest Zone of Eastern Europe: Environments and Economy. Akademiya Publ., Moscow. 144 pp. (in Russian).

Zhilin, M.G., 2006. Mesolithic Peatland Sites in the Tver' Region on the Volga: Cultural Originality and the People Adaptation. Institute of Archaeology RAS, Moscow. 140 pp. (in Russian).

Zhilin, M.G., Kostyleva, E.L., Utkin, A.V., Engovatova, A.V., 2002. Mesolithic and Neolithic Cultures of the Upper Volga Region (Based of Materials from the Ivanovskoye VII Site). Nauka, Moscow. 246 pp. (in Russian).

Zhirov, E.V., 1940. Notes on the Skeletons from the Neolithic Burials on the Yuzhny Oleni Island. Short Communications of the Institute for the Material Culture History. Russian Academy of Sciences, Leningrad, pp. 51–54. Iss. 6 (in Russian).

Zhuravlev, A.P., 1982. Source of raw material for flint tools in the Onega Lake basin. Sovetskaya Arkheologiya 1, 204–207 (in Russian).

Zolotareva, I.M., 1974. Anthropological differentiation of the eastern Samoyedic peoples. In: Alekseyev, V.P. (Ed.), Anthropology and Gene Geography. Nauka, Moscow, pp. 215–231 (in Russian).

Zubov, A.A., 2004. Paleoanthropological Lineage of Man. Institute of Ethnography and Anthropology RAS, Moscow. 551 pp. (in Russian).

Zvelebil, M., 2008. Innovating hunter-gatherers: the Mesolithic of the Baltic. In: Bailey, G., Spikins, P. (Eds.), Mesolithic Europe. Cambridge Univ. Press, Cambridge, pp. 18–59.

Zykin, V.S., 2012. Stratigraphy and Evolution of the Environment and Climate in the Late Cenozoic in the South of West Siberia. "Geo" Publ., Siberian Branch RAS, Novosibirsk. 487 pp. (in Russian).

Zykin, V.S., Zykina, V.S., Orlova, L.A., 2003. Reconstructions of environmental and climatic changes in the south of West Siberia in the Late Pleistocene based on studies of the Lake Aksor deposits. Archaeology, Ethnography and Anthropology of Eurasia 4, 2–16.

Zykin, V.S., Zykina, V.S., Chirkin, K.A., Smolyaninova, L.G., 2005. Geological structure and the Upper Cenozoic stratigraphy in the vicinities of the Early Paleolithic site of Karama in the upper reaches of the Anuy River. (Northwestern Altai). Archaeology, Ethnography and Anthropology of Eurasia 3, 2–20.

Zykina, V.S., 1990. Environments at the Time of the Late Pleistocene Warming as Indicated by Paleopedological Data in West Siberia. Materials of the 7th All-Union Conference on the Quaternary 2, 32–33 Abstracts. Inst. of Geol., Estonian Ac. Sc (in Russian).

Zykina, V.S., 2006. Loess-Soil Sequence and Evolution of Pedogenesis in the Pleistocene in West Siberia (Dr.Sc. dissertation) Novosibirsk. 646 pp. (in Russian).

Zykina, V.S., Volkov, I.A., Dergacheva, M.I., 1981. The Upper Quaternary Deposits and Fossil Soils in the Novosibirsk Region of the Ob' Drainage Basin. Nauka, Moscow. 204 pp. (in Russian).

Index

'*Note*: Page numbers followed by "f" indicate figures, "t" indicate tables.'

D

J

K

L

M

Y

Z

FSC
www.fsc.org
MIX
Papier | Fördert gute Waldnutzung
FSC® C083411

Kolibri 360 GmbH
Ferdinand-Jühlke-Straße 7
99095 Erfurt, Deutschland
produktsicherheit@kolibri360.de